Genomic Designing of Climate-Smart Pulse Crops

Chittaranjan Kole
Editor

Genomic Designing of Climate-Smart Pulse Crops

Springer

Editor
Chittaranjan Kole
Raja Ramanna Fellow
Department of Atomic Energy, Government of India
ICAR-National Institute for Plant Biotechnology
New Delhi, India

ISBN 978-3-319-96931-2 ISBN 978-3-319-96932-9 (eBook)
https://doi.org/10.1007/978-3-319-96932-9

Library of Congress Control Number: 2019934793

This Springer imprint is published by the registered company Springer Nature Switzerland AG
The registered company address is: Gewerbestrasse 11, 6330 Cham, Switzerland

Dedicated to
(Late) Prof. Sukumar Dana
Formerly Professor in the Department of Genetics and Plant Breeding of the Bidhan Chandra Krishi Viswavidyalaya (Agricultural University) who supervised my Post-Graduate Thesis work on a leading pulse crop, mungbean (green gram), and during that period inculcated in my mind the urge and habit of exploration, collection, characterization and utilization of indigenous varieties, local landraces and allied wild species, and taught me the art and science of 'talking' to the plants.

Preface

The past 120 years have witnessed a remarkable evolution in the science and art of plant breeding culminating in quite a revolution in the second decade of the twenty-first century! A number of novel concepts, strategies, techniques and tools have emerged from time to time over this period and some of them deserve to be termed as milestones. Traditional plant breeding, immediately following the rediscovery of laws of inheritance, has been playing a spectacular role in the development of innumerable varieties in almost all crops during this entire period. Mention must be made on the corn hybrids, rust-resistant wheat, and obviously the high-yielding varieties in wheat and rice that ushered the so-called Green Revolution. However, the methods of selection, hybridization, mutation and polyploidy employed in traditional breeding during this period relied solely on the perceivable phenotypic characters. But most, if not all, of the economic characters in crops are governed by polygenes which are highly influenced by environmental fluctuations and hence phenotype-based breeding for these traits has hardly been effective.

Historical discovery of DNA structure and replication in 1953 was followed by a series of discoveries in the 1960s and 1970s that paved the way for recombinant DNA technology in 1973 facilitating the detection of a number of DNA markers in 1980 onwards and their utilization in construction of genetic linkage maps and mapping of genes governing the simply inherited traits and quantitative trait loci controlling the polygenic characters in a series of crop plants starting with tomato, maize and rice. Thus, a new crop improvement technique called as molecular breeding started in the later part of the twentieth century. On the other hand, genetic engineering led to the modification of crops for target traits by transferring alien genes, for example the *Bt* gene from the bacteria *Bacillus thuringiensis*. A large number of genetically modified crop varieties have thus been developed starting with the commercialization of 'flavr Savr' tomato in 1994.

Meantime, the manual DNA sequencing methodology of 1977 was being improved with regard to speed, cost-effectiveness and automation. The first-generation sequencing technology led to the whole genome sequencing of Arabidopsis in 2000 and followed by rice in 2002. The next-generation sequencing technologies were available over time and used for sequencing of genomes of many other models and crop plants. Genomes, both nuclear and organellar, of more than 100 plants have already been sequenced by now and the information thus generated are available in public database for most of them. It must be mentioned here that bioinformatics played a remarkable role in handling the enormous data being produced in each and every minute. It can be safely told that the 'genomics' era started at the beginning of the twenty-first century itself accompanying also proteomics, metabolomics, transcriptomics, and several other 'omics' technologies.

Structural genomics have thus facilitated annotation of genes, enumeration of gene families and repetitive elements and comparative genomics studies across taxa. On the other hand, functional genomics paved the way for deciphering the precise biochemistry of gene function through transcription and translation pathways. Today, genotyping-by-sequencing of primary, secondary and even tertiary gene pools; genome-wide association studies; and genomics-aided breeding are almost routine techniques for crop improvement. Genomic selection in crops is another reality today. Elucidation of the chemical nature of crop chromosomes has now opened up a new frontier for genome editing that is expected to lead the crop improvement approaches in near future.

At the same time, we will look forward to the replacement of transgenic crops by cisgenic crops through transfer of useful plant genes and atomically modified crops by employing nanotechnology that will hopefully be universally accepted for commercialization owing to their human-friendly and environment-friendly nature.

I wish to emphatically mention here that none of the technologies and tools of plant breeding is too obsolete or too independent. They will always remain pertinent individually or as complimentary to each other, and will be employed depending on the evolutionary status of the crop genomes, the genetic resources and genomics resources available, and above all the cost–benefit ratios for adopting one or more technologies or tools. In brief, utilization of these crop improvement techniques would vary over time, space and economy scales! However, as we stand today, we have all the concepts, strategies, techniques and tools in our arsenal to practice genome designing, as I would prefer to term it, of crop plants not just genetic improvement to address simultaneously food, nutrition, energy and environment security, briefly the FNEE security, I have been talking about for the past 5 years at different platforms.

Addressing FNEE security has become more relevant today in the changing scenario of climate change and global warming. Climate change will lead to greenhouse gas emissions and extreme temperatures leading to different abiotic stresses including drought or waterlogging on one hand and severe winter and freezing on the other. It will also severely affect uptake and bioavailability of water and plant nutrients and will adversely cause damage to physical, chemical and biological properties of soil and water in cropping fields and around. It is also highly likely that there will be emergence of new insects and their biotypes and of new plant pathogens and their pathotypes. The most serious concerns are, however, the unpredictable crop growth conditions and the unexpected complex interactions among all the above stress factors leading to drastic reduction in crop yield and quality in an adverse ecosystem and environment. Climate change is predicted to significantly reduce productivity in almost all crops. For example, in cereal crops, the decline of yield is projected at 12–15%. On the other hand, crop production has to be increased at least by 70% to feed the alarmingly growing world population, projected at about 9.0 billion by 2050 by even a moderate estimate.

Hence, the unpredictability of crop growing conditions and thereby the complexity of biotic and abiotic stresses warrant completely different strategies of crop improvement from those practiced over a century aiming mostly at one of the few breeding objectives at a time such as yield, quality, resistance to biotic tresses due to disease-pests, tolerance to abiotic stresses due to drought, heat, cold, flood, salinity, acidity or improved water and nutrient-use efficiency, etc. In the changing scenario of climate change, for sustainable crop production, precise prediction of the above limiting factors by long-term survey and timely sensing through biotic agents and engineering devices and regular soil and water remediation will play a big role in agriculture. We have been discussing on 'mitigation' and 'adaptation' strategies for the past few years to reduce the chances of reduction of crop productivity and improve the genome plasticity of crop plants that could thrive and perform considerably well in a wide range of growing conditions over time and space. This is the precise reason for adopting genomic designing of crop plants to improve their adaptability by developing climate-smart or climate-resilient genotypes.

Keeping all these in mind, I planned to present deliberations on the problems, priorities, potentials and prospects of genome designing for development of climate-smart crops in about 50 chapters, each devoted to a major crop or a crop group, allocated under five volumes on cereal, oilseed, pulse, fruit and vegetable crops. These chapters have been authored by more than 250 of eminent scientists from over 30 countries including Argentina, Australia, Bangladesh, Belgium, Brazil, Canada, China, Egypt, Ethiopia, France, Germany, Greece, India, Ireland, Japan, Malaysia, Mexico, New Zealand, Kenya, Pakistan, Philippines, Portugal, Puerto Rico, Serbia, Spain, Sri Lanka, Sweden, Taiwan, Tanzania, Tunisia, Uganda, UK, USA and Zimbabwe.

There are a huge number of books and reviews on traditional breeding, molecular breeding, genetic engineering, nanotechnology, genomics-aided breeding and gene editing with crop-wise and trait-wise deliberations on crop genetic improvement including over 100 books edited by me since 2006. However, I believe the present five book volumes will hopefully provide a comprehensive enumeration on the requirement, achievements and future prospects of genome designing for climate-smart crops and will be useful to students, teaching faculties and scientists in the academia and also to the related industries. Besides, public and private funding agencies, policymaking bodies and the social activists will also get a clear idea on the road travelled so far and the future roadmap of crop improvement.

I must confess that it has been quite a difficult task for me to study critically the different concepts, strategies, techniques and tools of plant breeding practiced over the past 12 decades that also on diverse crop plants to gain confidence to edit the chapters authored by the scientists with expertise on the particular crops or crop groups and present them in a lucid manner with more or less uniform outline of contents and formats. However, my experience gained over the past 7 years in the capacity of the Founding Principal Coordinator of the International Climate-Resilient Crop Genomics Consortium (ICRCGC) was highly useful while editing these books. I have the opportunity to interact with a number of leading scientists from all over the world almost on regular basis. Organizing and chairing the annual workshops of ICRCGC since 2012 and representing ICRCGC in many other scientific meetings on climate change agriculture offered me a scope to learn from a large number of people from different backgrounds including academia, industries, policymaking, and funding agencies and social workers. I must acknowledge here the assistance I received from all of them to keep me as a sincere student of agriculture specifically plant breeding.

This volume entitled Genomic Designing of Climate-Smart Pulse Crops includes 9 major crops including Common Bean, Pigeonpea, Chickpea, Lentil, Mungbean, Pea, Fava Bean, Bambara Groundnut and Grass Pea. These chapters have been authored by 80 scientists from 12 countries including Australia, Argentina, Brazil, China, Egypt, India, Malaysia, Pakistan, Puerto Rico, Spain, UK and USA. I place on record my thanks for these scientists for their contributions and cooperation.

My own working experience on pulse crops dates back to late 70s in the laboratory of (Late) Prof. Sukumar Dana in the Department of Genetics and Plant Breeding in the Bidhan Chandra Krishi Viswavidyalaya (Agricultural University), West Bengal, India. While working as a postgraduate student with him on genetics of mungbean also known as green gram, I learnt for the first time the importance of collection, characterization and utilization of indigenous varieties, local landraces and wild allied species in crop improvement. It is him who inculcated in me the 'love' for the plants and the art to 'care' them and 'talk' to them and guided me to become a plant breeder one day. Hence, I have dedicated this book to (Late) Prof. Dana as a token of my respect, thanks and gratitude.

New Delhi, India Chittaranjan Kole

Contents

Contributors

S. Álvarez-García Grupo Universitario de Investigación en Ingeniería y Agricultura Sostenible (GUIIAS), Instituto de Medio Ambiente, Recursos Naturales y Biodiversidad, Universidad de León, León, Spain

Ido Bar School of Natural Sciences, Environmental Futures Research Institute, Griffith University, Nathan, QLD, Australia

P. S. Basu ICAR-Indian Institute of Pulses Research, Kalyanpur, Kanpur, India

J. S. Beaver Department of Agro-Environmental Sciences, University of Puerto Rico, Mayaguez, Puerto Rico

A. Bedmar Villanueva National Plant Genetic Resources Centre (CRF), National Institute for Agricultural and Food Research and Technology, Alcalá de Henares, Spain

Navya beera Institute of Biotechnology, Telangana State Agricultural University, Hyderabad, India

Sajitha Biju Faculty of Veterinary and Agricultural Sciences, University of Melbourne, Dookie, VIC, Australia

Abhishek Bohra ICAR-Indian Institute of Pulses Research (IIPR), Kanpur, Uttar Pradesh, India

G. Carro-Huelga Grupo Universitario de Investigación en Ingeniería y Agricultura Sostenible (GUIIAS), Instituto de Medio Ambiente, Recursos Naturales y Biodiversidad, Universidad de León, León, Spain

P. A. Casquero Grupo Universitario de Investigación en Ingeniería y Agricultura Sostenible (GUIIAS), Instituto de Medio Ambiente, Recursos Naturales y Biodiversidad, Universidad de León, León, Spain

Hui Hui Chai Biosciences, University of Nottingham Malaysia, Crops for the Future, Semenyih, Selangor, Malaysia;
Crops for the Future Research Centre, Semenyih, Selangor, Malaysia

Yue Chang BGI-Shenzhen, Beishan Industrial Zone, Yantian District, Shenzhen, China;
China National GeneBank, BGI-Shenzhen, Shenzhen, China

Rama Harinath Reddy Dadu Faculty of Veterinary and Agricultural Sciences, University of Melbourne, Dookie, VIC, Australia

Poonam Devi Department of Botany, Panjab University, Chandigarh, India

A. M. De Ron Biology of Agrosystems, Misión Biológica de Galicia (MBG), National Spanish Research Council (CSIC), Pontevedra, Spain

L. De la Rosa National Plant Genetic Resources Centre (CRF), National Institute for Agricultural and Food Research and Technology, Alcalá de Henares, Spain

Sonali Dubey ICAR-Indian Institute of Pulses Research, Kalyanpur, Kanpur, India

M. Dworkin Molecular Genetics & EpiGenomics Laboratory, Center for Integrated Biological & Environmental Research (CIBER), Delaware State University, Dover, USA

Peter M. F. Emmrich John Innes Centre, Norwich Research Park, Norwich, UK; Biosciences Eastern and Central Africa - International Livestock Research Institute Hub, Nairobi, Kenya

M. Z. Galván Laboratorio de Biotecnología, CONICET, Estación Experimental Salta, INTA Cerrillos, Salta, Argentina

M. C. Gonçalves Vidigal Departamento de Agronomia, Universidade Estadual de Maringá, Maringá, Brazil

Ruijun Guo Institute of Crop Sciences, Chinese Academy of Agricultural Sciences, Beijing, China

Dorin Gupta Faculty of Veterinary and Agricultural Sciences, University of Melbourne, Dookie, VIC, Australia

Sanjeev Gupta ICAR-Indian Institute of Pulses Research, Kalyanpur, Kanpur, India

S. Gutiérrez Grupo Universitario de Investigación en Ingeniería y Agricultura Sostenible (GUIIAS), Instituto de Medio Ambiente, Recursos Naturales y Biodiversidad, Universidad de León, León, Spain

Yuhua He Institute of Grain Crops, Yunnan Academy of Agricultural Sciences, Kunming, China

Wai Kuan Ho Biosciences, University of Nottingham Malaysia, Crops for the Future, Semenyih, Selangor, Malaysia;
Crops for the Future Research Centre, Semenyih, Selangor, Malaysia

Xiangdong Hu Institute of Agricultural Economics and Development, Chinese Academy of Agricultural Sciences, Beijing, China

Uday C. Jha ICAR-Indian Institute of Pulses Research (IIPR), Kanpur, Uttar Pradesh, India

Mitchell Jones RMIT University, Melbourne, VIC, Australia

Mayank Kaashyap Cornell University, Ithaca, NY, USA

V. (K.) Kalavacharla Molecular Genetics & EpiGenomics Laboratory, Center for Integrated Biological & Environmental Research (CIBER), College of Agriculture, Science and Technology, Delaware State University, Dover, USA

Gita Kumari ICAR-Indian Institute of Pulses Research, Kalyanpur, Kanpur, India

Guan Li Institute of Crop Sciences, Chinese Academy of Agricultural Sciences, Beijing, China

Ling Li Liaoning Research Institute of Cash Crops, Liaoyang, Liaoning Province, China

Quanlan Liu Department of Bioengineering and Biotechnology, College of Marine Science and Biological Engineering, Qingdao University of Science & Technology, Qingdao, China

Rong Liu Institute of Crop Sciences, Chinese Academy of Agricultural Sciences, Beijing, China

A. Lorenzana Grupo Universitario de Investigación en Ingeniería y Agricultura Sostenible (GUIIAS), Instituto de Medio Ambiente, Recursos Naturales y Biodiversidad, Universidad de León, León, Spain

Jiliang Ma Institute of Agricultural Economics and Development, Chinese Academy of Agricultural Sciences, Beijing, China

Cathie Martin John Innes Centre, Norwich Research Park, Norwich, UK

Festo Massawe Biosciences, University of Nottingham Malaysia, Crops for the Future, Semenyih, Selangor, Malaysia;
Crops for the Future Research Centre, Semenyih, Selangor, Malaysia

Alok Kumar Maurya ICAR-Indian Institute of Pulses Research (IIPR), Kanpur, India

Sean Mayes Plant and Crop Sciences, Biosciences University of Nottingham, Sutton Bonington Campus, Loughborough, Leicestershire, UK;
Crops for the Future Research Centre, Semenyih, Selangor, Malaysia

S. Mayo-Prieto Grupo Universitario de Investigación en Ingeniería y Agricultura Sostenible (GUIIAS), Instituto de Medio Ambiente, Recursos Naturales y Biodiversidad, Universidad de León, León, Spain

Harsh Nayyar Department of Botany, Panjab University, Chandigarh, India

Shalini Pareek ICAR-Indian Institute of Pulses Research (IIPR), Kanpur, India

Swarup Kumar Parida National Institute of Plant Genome Research (NIPGR), Delhi, India

Prakash G. Patil ICAR-Indian Institute of Pulses Research (IIPR), Kanpur, India

T. Porch Department of Agro-Environmental Sciences, University of Puerto Rico, Mayaguez, Puerto Rico

Uma Shankar Prajapati ICAR-Indian Institute of Pulses Research, Kalyanpur, Kanpur, India

Aditya Pratap ICAR-Indian Institute of Pulses Research, Kalyanpur, Kanpur, India

Anju Rani Department of Botany, Panjab University, Chandigarh, India

Meenal Rathore ICAR-Indian Institute of Pulses Research, Kalyanpur, Kanpur, India

A. P. Rodiño Biology of Agrosystems, Misión Biológica de Galicia (MBG), National Spanish Research Council (CSIC), Pontevedra, Spain

A. Rodríguez-González Grupo Universitario de Investigación en Ingeniería y Agricultura Sostenible (GUIIAS), Instituto de Medio Ambiente, Recursos Naturales y Biodiversidad, Universidad de León, León, Spain

Parasappa Rajappa Saabale ICAR-Indian Institute of Pulses Research (IIPR), Kanpur, Uttar Pradesh, India;
Regional Research Centre, Indian Institute of Pulses Research, Dharwad, Karnataka, India

Ahmed Sallam Faculty of Agriculture, Department of Genetics, Assiut University, Assiut, Egypt

Prabhakaran Sambasivam School of Natural Sciences, Environmental Futures Research Institute, Griffith University, Nathan, QLD, Australia

Abhimanyu Sarkar John Innes Centre, Norwich Research Park, Norwich, UK

Ashutosh Sarker South Asia & China Program & Principal Food Legume Breeder, International Center for Agricultural Research in the Dry Areas (ICARDA), New Delhi, India

SJ Satheesh Naik ICAR-Indian Institute of Pulses Research (IIPR), Kanpur, India

Rachit Saxena International Crops Research Institute for the Semi-Arid Tropics (ICRISAT), Hyderabad, India

Baoyu Shen Liaoning Research Institute of Cash Crops, Liaoyang, Liaoning Province, China

Mohar Singh ICAR-National Bureau of Plant Genetic Resources Regional Station, Shimla, India

Parikshit Singh ICAR-Indian Institute of Pulses Research, Kalyanpur, Kanpur, India

Bo Song BGI-Shenzhen, Beishan Industrial Zone, Yantian District, Shenzhen, China;
China National GeneBank, BGI-Shenzhen, Shenzhen, China

V. Suárez-Villanueva Grupo Universitario de Investigación en Ingeniería y Agricultura Sostenible (GUIIAS), Instituto de Medio Ambiente, Recursos Naturales y Biodiversidad, Universidad de León, León, Spain

Suli Sun Institute of Crop Sciences, Chinese Academy of Agricultural Sciences, Beijing, China

Rakhi Tomar ICAR-Indian Institute of Pulses Research, Kalyanpur, Kanpur, India

Sami Ul-Allah Department of Plant Breeding and Genetics, College of Agriculture, BZU Bahadur Sub-Campus, Layyah, Pakistan

Rajeev K. Varshney International Crops Research Institute for the Semi-Arid Tropics (ICRISAT), Hyderabad, India;
The University of Western Australia, Crawley, WA, Australia

Trevor L. Wang John Innes Centre, Norwich Research Park, Norwich, UK

Tao Yang Institute of Crop Sciences, Chinese Academy of Agricultural Sciences, Beijing, China

Huijie Zhang Department of International Cooperation, Chinese Academy of Agricultural Sciences, Beijing, China;
Institute of Agricultural Economics and Development, Chinese Academy of Agricultural Sciences, Beijing, China

Tianjing Zhang Liaoning Research Institute of Cash Crops, Liaoyang, Liaoning Province, China

Xiaoyan Zhang Qingdao Academy of Agricultural Sciences, Qingdao, China

Zhendong Zhu Institute of Crop Sciences, Chinese Academy of Agricultural Sciences, Beijing, China

Xuxiao Zong Institute of Crop Sciences, Chinese Academy of Agricultural Sciences, Beijing, China

Abbreviations

β-ODAP	β-N-oxalyl-L-α,β-diaminopropionic acid
6PP	6-Pentyl-α-pyrone
AAC	1-Aminocyclopropane-1-caboylic acid
ADH	Alcohol dehydrogenase
AFLP	Amplified fragment length polymorphism
AOCC	African Orphan Crops Consortium
APX	Ascorbate peroxidase
ASAP	Allele specific amplified primer
ATP	Adenosin triphosphate
AUSPC	Area under symptom progress curve
AVRDC	Asian Vegetable Research and Development Centre
BA	6-Benzylaminopurine
BA	Benzyladenine
BAC	Bacterial artificial chromosome
BAP	Benzyl amino purine
BARI	Bangladesh Agricultural Research Institute
BC	Backcross
BCL	Blocked cotyledon lethal
BCMV	Bean common mosaic virus
BES	BAC-end sequence
BIBAC	Binary bacterial artificial chromosome
BLAST	Basic local alignment search tool
BLRV	Bean leaf roll virus
BM	Biomass
Bt	*Bacillus thuringiensis*
BUSCO	Benchmarking universal single-copy orthologs
BYMV	Bean yellow mosaic virus
CAPS	Cleaved amplified polymorphic sequence
CBD	Convention on Biological Diversity
CBN PMP	Conservatoire Botanique National des Pyrénées et de Midi Pyrénées

CcCDR	*Cajanus cajan* cold and drought regulatory
CcCYP	*Cajanus cajan* cyclophilin
CcHyPRP	*Cajanus cajan* hybrid proline-rich protein
cDNA	Complimentary DNA
CDS	Coding sequence
CFF	Crops for the Future
CGIAR	Consultative Group for International Agricultural Research
CID	Carbon isotope discrimination
CIM	Composite interval method
CLD	Crinkle leaf dwarf
ClYVV	Clover yellow vein virus
CMS	Cell membrane stability
CMS	Cytoplasmic male sterility
CNV	Copy number variation
COS	Conserved orthologous set
CP	Coat protein
CRISPR	Clustered regularly interspaced short palindromic repeats
CS	Climate-smart
CST	Climate-smart trait
CTDB	Chickpea Transcriptome Database
CTD	Canopy temperature depression
CWR	Crop wild relative
CWSI	Crop water stress index
DArT	Diversity arrays technology
DArT Seq	Diversity arrays technology Sequencing
dCAPS	Derived CAPS
DDS	Direct disease resistance
DEG	Differentially expressed gene
DH	Doubled haploid
DL	Dwarf lethal
DREB	Dehydration responsive element binding
DS	Dormant seeding
DTI	Drought tolerance index
DUS	Distinctiveness, Uniformity and Stability
EC	Electrical conductivity
EIAR	Ethiopian Institute of Agricultural Research
eIF(iso)4E	Isoform of eukaryotic translation initiation factor 4E
eIF4E	Eukaryotic translation initiation factor 4E
EMBL	European Molecular Biology Laboratory
EMS	Ethyl methane sulphonate
eQTL	Expression QTL
EST	Expressed sequence tag
EST-SSR	EST-derived SSR
ET	Ethylene
FAO	Food and Agriculture Organization

FBCM	Faba bean consensus map
FISH	Fluorescence in situ hybridization
FNEE	Food, Nutrition, Energy & Environment (Security)
Fpod	Number of filled pods per plot
FW	Fusarium wilt
GA3	Gibberelic acid
GABA	γ-Aminobutyric acid
GBS	Genotyping-by-sequencing
GC-MS	Gas chromatography–mass-spectrometry
GEBV	Genomic estimated breeding value
GEMs	Gene expression markers
GEO	Gene expression omnibus
GHMM	Generalized hidden Markov model
GLM	General linear model
GMO	Genetically modified organism
GMP	Geometric mean productivity
GP	Gene pool
GS	Genomic selection
GSI	Germination stress index
GSS	Genome survey sequences
GUS	β-Glucuronidase
GWAS	Genomewide association study
GY	Grain yield per plot
HI	Harvest index
HM	Harmonic mean
HMM	Hidden Markov model
HPL	Hydroperoxide lyase
HRM	High resolution melting
HS	Heat stress
HSP	Heat shock protein
HT	High temperature
IAA	Indolacetic acid
ICAR	Indian Council of Agricultural Research
ICARDA	International Center for Agriculture Research in the Dry Areas
ICMR	Indian Council of Medical Research
IIPR	Indian Institute of Pulses Research
IITA	International Institute of Tropical Agriculture
IL	Introgression line
InDel	Insertion/deletion
IPCC	Intergovernmental Panel on Climate Change
IRLC	Inverted repeat lacking clade
ISR	Induced systemic resistance
ISSR	Inter-simple sequence repeat
ITAP	Intron targeted amplified polymorphism

ITPGRFA	International Treaty on Plant Genetic Resources for Food and Agriculture
ITS	Internal transcribed spacer
JA	Jasmonic acid
KASP	Kompetitive allele specific PCR
KEGG	Kyoto Encyclopedia of Genes and Genomes
LC-MS	Liquid chromatography–mass-spectrometry
LD	Linkage disequilibrium
LG	Linkage group
LINE	Long interspersed element
LIS	Legume Information System
LMICs	Low and Medium Income Countries
LMP	Long mate pair
LOD	Logarithm of odds
LRR	Leucine-rich repeat
LRS	Likelihood ratio statistics
LTR	Long terminal repeat
MABC	Marker-assisted backcrossing
MABCB	Marker-assisted backcross breeding
MAB	Marker-assisted breeding
MAGIC	Multiparent advanced generation intercross
MAMP	Microbe-associated molecular pattern
MARS	Marker-assisted recurrent selection
MAS	Marker-assisted selection
MAT	Mutually agreed term
miRNA	MicroRNA
MLO	Mildew resistance locus *O*
MP	Mean productivity
MS	Murashige and Skoog (medium)
MTA	Marker-trait association
NAA	1-Naphthalenenacetic acid
NADH	Nicotinamide adenine dinucleotide
NAM	Nested association mapping
NARS	National Agricultural Research Systems (India)
NBPGR	National Bureau of Plant Genetic Resources (India)
NBS-LRR	Nucleotide binding site–leucine-rich repeat
NCBI	National Center for Biotechnology Information (USA)
NDRE	Normalized difference red edge
NDVI	Normalized difference vegetation index
NGS	Next-generation sequencing
NILs	Near-isogenic lines
NMR	Nodule mass ratio
NMR	Nuclear magnetic resonance
NUE	Nutrient-use efficiency
ODAP	β-N-Oxalyl-L-α,β-diaminopropionic acid

PAL	Phenylalanine ammonia-lyase
PAMP	Pathogen-associated molecular pattern
PAR	Photosynthetically active radiation
PAV	Presence and absence variation
PBA	Pulse Breeding Australia
PCR	Polymerase chain reaction
PDC	Pyruvate decarboxylase
PEG	Polyethylene glycol
PEMV	Pea enation mosaic virus
PGRFA	Plant Genetic Resources for Food and Agriculture
PIC	Prior informed consent
PLANEX	Plant co-expression database
PMV	Pea mosaic virus
PSbMV	Pea seedborne mosaic virus
QTL	Quantitative trait locus
QTLs	Quantitative trait loci
R/FR	Red/far red
RADP	Random amplified polymorphic DNA
RDp	Root depth
RFLP	Restriction fragment length polymorphism
RFS	Root frost susceptibility
RGA	Resistance gene analog
RGR	Relative growth rate
RIL	Recombinant inbred line
RNAi	RNA-interference
RNA-seq	RNA sequencing
ROS	Reactive oxygen species
RS ratio	Root–shoot ratio
RUE	Radiation use efficiency
RWC	Relative water content
SA	Salicylic acid
SAGE	Serial analysis of gene expression
SAR	Systemic acquired resistance
SCAR	Sequence characterized amplified region
SDG	Sustainable development goals
SDS-PAGE	Sodium dodecyl sulfate-polyacrylamide gel electrophoresis
SINE	Short interspersed nuclear element
SMD	Sterility mosaic disease
SMTA	Standard Material Transfer Agreement
SNF	Symbiotic nitrogen fixation
SNP	Single nucleotide polymorphism
SOD	Superoxide dismutase
SPLAT	Specific polymorphic locus amplification test
SRAPS	Sequence-related amplified polymorphism
SRR	Seed replacement rate

SSD	Single seed descent
SSH	Suppression subtractive hybridization
SSI	Stress susceptibility index
SSR	Simple sequence repeat
STI	Stress tolerance index
STS	Sequence-tagged site
SuSy	Sucrose synthase
TAG	Triacylglycerol
TALEN	Transcription activator-like effector nucleases
Tc	Canopy temperature
TDZ	Thidiazuron
TE	Transposable element
TF	Transcription factor
TI	Heat tolerance index
TILLING	Targeting induced local lesions in genome
TRIPS	Trade-Related Aspects of Intellectual Property Rights
TS	Total number of seeds per plot
TUE	Transpiration-use efficiency
TUS	Tentative unique sequence
UN	United Nations
USDA	United Sates Department of Agriculture
UV	Ultraviolet
VAM	Vesicular-arbuscular mycorrhizae
VEP	Variant effect predictor
VIGS	Virus-induced gene silencing
VNTR	Variable number tandem repeats
VPD	Water pressure deficit
VS	Visual scoring
WGRS	Whole genome re-sequencing
WGS	Whole genome sequencing
WHO	World Health Organization
WUE	Water-use efficiency
YAC	Yeast active chromosome
ZFN	Zinc finger nuclease

Chapter 1
Common Bean Genetics, Breeding, and Genomics for Adaptation to Changing to New Agri-environmental Conditions

A. M. De Ron, V. (K.) Kalavacharla, S. Álvarez-García, P. A. Casquero, G. Carro-Huelga, S. Gutiérrez, A. Lorenzana, S. Mayo-Prieto, A. Rodríguez-González, V. Suárez-Villanueva, A. P. Rodiño, J. S. Beaver, T. Porch, M. Z. Galván, M. C. Gonçalves Vidigal, M. Dworkin, A. Bedmar Villanueva and L. De la Rosa

Abstract Common bean (*Phaseolus vulgaris* L.) has become, over the last 20 years, a competitive crop in national, regional, and international markets. This situation presents a dynamic environment for producers and researchers of this crop and requires a rethinking of current strategies against research and production needs, the opportunities and challenges of the future, and adaptation to changing agri-

A. M. De Ron (✉) · A. P. Rodiño
Biology of Agrosystems, Misión Biológica de Galicia (MBG), National Spanish Research Council (CSIC), Pontevedra, Spain
e-mail: amderon@mbg.csic.es; antonio.de.ron@gmail.com

A. P. Rodiño
e-mail: aprodino@mbg.csic.es

V. (K.) Kalavacharla (✉)
Molecular Genetics & EpiGenomics Laboratory, Center for Integrated Biological & Environmental Research (CIBER), College of Agriculture, Science and Technology, Delaware State University, Dover, USA
e-mail: vkalavacharla@desu.edu

S. Álvarez-García · P. A. Casquero · G. Carro-Huelga · S. Gutiérrez · A. Lorenzana · S. Mayo-Prieto · A. Rodríguez-González · V. Suárez-Villanueva
Grupo Universitario de Investigación en Ingeniería y Agricultura Sostenible (GUIIAS), Instituto de Medio Ambiente, Recursos Naturales y Biodiversidad, Universidad de León, León, Spain
e-mail: pedro-casquero@unileon.es

S. Mayo-Prieto
e-mail: smayp@unileon.es

A. Rodríguez-González
e-mail: alrog@unileon.es

J. S. Beaver · T. Porch
Department of Agro-Environmental Sciences, University of Puerto Rico, Mayaguez, Puerto Rico
e-mail: james.beaver@upr.edu; j_beaver@hotmail.com

C. Kole (ed.), *Genomic Designing of Climate-Smart Pulse Crops*,
https://doi.org/10.1007/978-3-319-96932-9_1

environmental conditions. Improvement of the common bean means possessing in-depth knowledge of its genetic diversity, the genome and gene functions, to enable the analysis of pathways and networks in response to fluctuating environmental conditions. An important long-term challenge is the discovery of the gene(s) that control important production traits such as pest and disease resistance, abiotic stress tolerance, and biological fixation of nitrogen. This will need to be a cooperative worldwide effort that involves breeders, geneticists, and genomic and bioinformatics experts. Currently, new technologies built around the recently released common bean genome sequence are now being developed, and various genomic resources for common bean are available and include physical maps, bacterial artificial chromosome libraries, anchored physical and genetic maps, and expressed sequence tags. However, these approaches require precise phenotypic data. Complex interactions between the common bean crop genotype, environmental factors in combination with plant population dynamics and crop management greatly affect plant phenotypes in field experiments and are the key for the expansion of the productivity of this crop in traditional and nontraditional growing areas.

Keywords Abiotic stress tolerance · Agronomy · Diseases and pest resistance · Food legumes · Genetic resources · Genetic mapping · Molecular breeding · *Phaseolus vulgaris* L.

T. Porch
e-mail: Timothy.porch@ars.usda.gov

M. Z. Galván
Laboratorio de Biotecnología, CONICET, Estación Experimental Salta, INTA Cerrillos, Salta, Argentina
e-mail: martazgalvan@gmail.com

M. C. Gonçalves Vidigal
Departamento de Agronomia, Universidade Estadual de Maringá, Maringá, Brazil
e-mail: mcgvidigal@uem.br; mcgvidigal@gmail.com

M. Dworkin
Molecular Genetics & EpiGenomics Laboratory, Center for Integrated Biological & Environmental Research (CIBER), Delaware State University, Dover, USA

A. Bedmar Villanueva · L. De la Rosa
National Plant Genetic Resources Centre (CRF), National Institute for Agricultural and Food Research and Technology, Alcalá de Henares, Spain
e-mail: bedmar.ana@inia.es

L. De la Rosa
e-mail: rosa@inia.es

1.1 Challenges, Priorities, and Prospects of Recent Plant Breeding

1.1.1 Background

Understanding the effects of domestication on genetic diversity of common bean (*Phaseolus vulgaris* L.) is of great importance, not only for crop evolution but also for possible applications, such as the implementation of appropriate biodiversity conservation strategies, and the use of genetic variability in breeding programs. One of the most important and generalized features of plant domestication is the reduction in genetic diversity, not only during the initial domestication process but also during dispersion and adaptive radiation from the centers of domestication to other areas. The reduction of genetic diversity is usually more drastic in autogamous species such as common bean, which have restricted genetic recombination and presents a higher population structure as compared with allogamous species (Jarvis and Hodgkin 1999). This reduction is caused by both stochastic events (i.e., a bottleneck and genetic drift due to a reduction in the population size) and selection (i.e., adaptation to a novel agrosystem) (Vigouroux et al. 2002).

A recent hypothesis for the origin of the common bean defended a Mesoamerican origin (Bitocchi et al. 2012, 2013), based on the extensive diversity and population structure within the Mesoamerican gene pool, and the signature of pre-domestication bottlenecks in the south of the Andes detected in five gene fragments across 102 wild bean accessions. This novel structure of population not only evidences a Mesoamerican origin but also excludes an Andean origin of common bean. Additionally, these authors suggested that the wild common bean from northern Peru and Ecuador represents an old relict germplasm including a part of the genetic diversity of the ancestral common bean populations, displaying a type I phaseolin that probably was extinct in Mesoamerica. The resequencing of the genome of the common bean by Schmutz et al. (2014) recently confirmed this hypothesis.

Domestication took place after the formation of the Mesoamerican and Andean gene pools, and thus their structure is evident in both the wild and the domesticated forms (Papa and Gepts 2003; Papa et al. 2005, 2007, Rossi et al. 2009). This clear subdivision of the common bean germplasm is well documented, and it has been defined through several studies (Papa et al. 2007; Angioi et al. 2009; Bitocchi et al. 2012, 2013). However, the number of domestication events within each pool is still debated. Bitocchi et al. (2013) hypothesized a single domestication event within each gene pool and indicated the Oaxaca valley in Mesoamerica and southern Bolivia and northern Argentina as geographical areas of common bean domestication.

The exploration of The Americas by the Europeans, from the 15th century, marked the arrival into the Old World of many plant species such as common bean (*Phaseolus vulgaris* L.), peanuts (*Arachis hypogaea* L.), cocoa (*Theobroma cacao* L.), corn (*Zea mays* L.), potato (*Solanum tuberosum* L.), tomato (*Solanum lycopersicum* L.), etc. The introduction of these exotic species in a new agricultural area under different environmental conditions raises relevant questions about adaptation, taking into account the requirements of tolerance to several stresses, as well as competitiveness with other indigenous crops in production and economic value (De Ron et al. 2016).

No records of common bean earlier than 1543 have been found in European herbariums; however, as reported by Zeven (1997), in 1669 it was widely grown in many areas of Europe. The dispersion of the common bean to Europe probably started from the Iberian Peninsula (Spain and Portugal), where the species was introduced mainly from Central America around 1506 and from the southern Andes after 1532, through sailors and traders who brought with them the nicely colored and easily transportable seeds as a curiosity (Brücher and Brücher 1976; Debouck and Smartt 1995). The pathways of dissemination of the crop across Europe were very complex, with several introductions from America combined with direct exchanges between European and other Mediterranean countries (Papa et al. 2007). Over time, the dissemination across Europe surely occurred through seed exchanges among farmers being facilitated by territorial contiguity and similarity of environments. The protein marker phaseolin was used as a marker in describing the worldwide dissemination of common bean (Gepts 1988). A higher frequency of Andean types (T, C, H, and A) was recorded with respect to Mesoamerican ones (S, B, M) (Lioi 1989; Santalla et al. 2002).

As mentioned before, the common bean originated and was domesticated in tropical highlands. This means that abiotic and biotic conditions had an influence on the development of European varieties (Rodiño et al. 2006, 2007). In some cases, bean breeders have had to incorporate tolerances to abiotic stresses from sources outside the primary gene pool of common bean. For example, tepary bean could also provide tolerance to heat or drought, and runner bean, tolerance to low soil fertility (Miklas et al. 2006a, b). In the case of rhizobia symbiotic system, it is possible that migration of the species had not been parallel, so additional efforts are underway to achieve efficient symbiotic genotypes of common bean and rhizobia (Rodiño et al. 2011). As a result of plant-rhizobia coevolution, a spectrum of compatible specific rhizobia is recognized for one or more legume species.

1.1.2 The Common Bean as a Food Resource

Grain legumes (pulses) are considered an essential source of nutrients and are also recognized as poor man's meat, showing their importance for people of developing countries, where the consumption of animal protein is limited by nonavailability or is self-imposed because of religious or cultural habits. Furthermore, legume seeds contain many bioactive and/or antinutritional compounds, such as phytate, oligosaccharides, phenolic compounds, nonprotein amino acids, lectins, enzyme inhibitors that play metabolic roles in humans or animals that frequently assume these seeds. These effects may be regarded as positive, negative, or both (Champ 2002).

From a nutritional point of view, the amino acid profile of legume storage proteins reveals low amounts of the essential sulfur-containing amino acids (i.e., methionine and cysteine) and tryptophan, while lysine, another essential amino acid, is quite abundant. Legume proteins complement very well those of cereals, which are normally rich in sulfur amino acids and poor in lysine and threonine. Besides the composition in essential amino acids, the nutritional quality of seed proteins is also largely determined by their digestibility. In fact, amino acids composition only represents the potential nutritional quality of a protein, being their bioavailability critical for the supply of amino acids in the diet (Sparvoli et al. 2015).

The common bean is the third most important food legume crop worldwide, surpassed only by soybean (*Glycine max* (L.) Merr.) and peanut (*Arachis hypogea* L.), and it is the first one for direct human consumption. Beans are produced and consumed mainly as a dry food legume, due to the high protein content of the grain, but the use of the fresh pod (snap bean) is common in many countries. Common bean is highly consumed in many areas of Africa and Latin America (as the most important source of plant protein), as well as in traditional diets of the Middle East and Europe (Broughton et al. 2003; Casquero et al. 2006). This legume is part of the healthy diet of the European Mediterranean basin and gaining importance in the USA where consumption has been increasing due to public interest in ethnic and healthy foods (Blair and Izquierdo 2012).

Recently the role of bean in human diet is being focused not only in its protein content but in the functional properties also and some authors have reported that its consumption could contribute to reduce the risk of obesity, diabetes, cardiovascular diseases and colon, prostate, and breast cancer (Hangen and Bennink 2003; Thompson et al. 2009). These health benefits could be due to the fiber content in the grain but also to antioxidant compounds as the phenolic ones. All the molecules present in legumes having anticancer properties are soluble in aqueous-alcohol extracts, while resistant starches, present in high amount in legumes, together with non-starch polysaccharides, are primarily insoluble residues from aqueous-alcohol extracts (Sparvoli et al. 2015). Colon carcinogenesis was induced by azoxymethane treatment in obese ob/ob mice fed with a diet containing cooked navy beans (whole beans), the insoluble or soluble fraction of aqueous-alcohol extracts, or a standard diet (Bobe et al. 2008).

1.2 Prioritizing Climate Smart (CS) Traits

1.2.1 Disease Resistance

1.2.1.1 Introduction

The abnormal functioning of diseased plants generally leads to a reduction in quantity and quality of yield. Disease is the result of an interaction among the plant and its environment and it is often affected by biotic and abiotic factors (e.g., microorganisms, humidity, temperature, etc.) that are detected as signals for the activation of plant response mechanisms (American Phytopathological Society 2005).

When a plant is present in a stress situation (biotic or abiotic), it shows a minimum resistance to this situation, which will slow down their vital functions, reducing their development. This alarm phase is the one that will trigger all the mechanisms to get over it. If this situation persists, the plant will die. However, if it triggers some defense mechanisms, it will enter a resistance phase reaching a maximum level. If the stress continues, the plant will enter a phase of exhaustion. This phase may cause plant death if the stress does not disappear. Nevertheless, if the stress situation ends, plant recovers its physiological functions, being able to regenerate and to reach a new physiological state optimal for the present conditions, which corresponded to the regeneration phase (Tadeo and Gómez-Cadenas 2008).

Crops are affected by a wide diversity of fungal pathogens, for example, *Sclerotinia* spp., *Fusarium* spp., *Botrytis* spp., *Rhizoctonia* spp., etc., causing important economic losses (Mayo et al. 2017). A form of control to diseases is the application of synthetic fungicides. Its application on the seed or directly to the soil can be effective against fungi that affect the crops during or shortly after germination (Beebe and Corrales 1991) because they reduce its incidence and improve the emergence of plants (Valenciano et al. 2004). However, applications with fungicides aimed at avoiding damage caused by fungi that cause root rot or yellowing and wilting are often ineffective and usually impracticable due to the large volume of soil to which they should be directed. Actually, the number of authorized plant protection products has been reduced in order to ensure food safety and its sustainable in the long term. It is therefore proposed to prioritize nonchemical methods in integrated production, organic farming, and others (Mayo et al. 2017).

As a strategy to control plant infectious diseases, mainly those caused by fungi, the use of biocontrol agents can reduce the negative effects of plant pathogens and they also can promote positive responses in the plant (Shoresh et al. 2010). Biocontrol agents are perceived to have specific advantages over synthetic fungicides, including fewer nontarget and environmental effects, efficacy against fungicide-resistant pathogens, reduced probability of resistance development and use in organic farming situations where synthetic fungicides are restricted (Brimner and Boland 2003).

Bacterial species belonging to genera such as *Agrobacterium*, *Pseudomonas*, *Streptomyces*, and *Bacillus*, and fungal genera such as *Gliocladium*, *Trichoderma*, *Ampelomyces*, *Candida*, and *Coniothyrium*, are beneficial organisms that have shown good efficiency as biocontrol agents against pathogenic microorganisms (Vinale et al. 2008a).

1.2.1.2 *Trichoderma* spp.

Trichoderma spp. (Teleomorph: Hypocrea) is a fungal genus that is found in the soil, and it is a secondary fast growing opportunistic invasive (Mayo et al. 2016a, b) producer of chitinases, glucanases and proteases, and metabolites with antimicrobial activity (Lorito et al. 2010). Many *Trichoderma* species are also well known as biocontrol agents of important phytopathogenic fungi. The primary mechanisms of biocontrol used by *Trichoderma* in direct confrontation with pathogenic fungi are mycoparasitism, antibiosis, and competition for nutrients with the pathogen (Harman et al. 2004). *Trichoderma* species colonize the root surface and cause substantial changes in plant metabolism (Shoresh et al. 2010). The physical interaction between *Trichoderma* and plants is limited to the first cell layer of the epidermis and the root bark. In addition, *Trichoderma* biocontrol strains are able to induce the expression of genes involved in defense response and also to promote plant growth, root development, and nutrient uptake (Hermosa et al. 2012).

Trichoderma spp. is recognized for their important benefits to agriculture such as its ability to protect crops against diseases (Benítez et al. 2004) and increase crop yield under field conditions (Harman et al. 2004). Most species of *Trichoderma* have been linked to biocontrol and biotechnological applications (Monte 2001), and the versatility of *Trichoderma* strains to suppress diseases caused by pathogens (Howell 2003). Since *Trichoderma* strains grow and proliferate best when there are abundant healthy roots, they have evolved numerous mechanisms of action both to attack other fungi and to enhance plant and root growth (Benítez et al. 2004).

In a symbiotic relationship with *Trichoderma*, the transport of sucrose from plants with subsequent intracellular hydrolysis by *T. virens* has been shown (Fig. 1.1). This source–sink communication may be central to the mutualistic interaction, influencing the development of *Trichoderma* in the rhizosphere and root plant (Vargas et al. 2012).

Competition and Mycoparasitism

Competition between *Trichoderma* and pathogens (Fig. 1.1) would be established with the purpose to get more nutrients, oxygen, light, etc. (Paulitz 1990). *Trichoderma* is an excellent competitor for space and nutritional resources. It appears in almost all soils and in habitats that contain high amounts of organic matter. In those niches, it would be an excellent decomposer of plant and fungal material. Moreover, some species of the genus *Trichoderma* show great metabolomic versatility that allows them to grow using a wide range of nitrogen and carbon sources. Furthermore, *Trichoderma* has the ability to colonize the rhizosphere, and this skill might be essential for being used as an excellent biological control agent (Howell 2003).

Mycoparasitism (Fig. 1.1) consists in the recognition of the fungus, attacking it, and penetrating it with the purpose to cause its death. This process involves some different phases. Firstly, *Trichoderma* locates the pathogen without previous contact, beginning to enlarge toward the pathogen by tropism (Chet et al. 1981; Lu et al. 2004). During this process, *Trichoderma* secretes some enzymes that hydrolyze the cell wall of the pathogen (Howell 2003; Woo et al. 2006). It has been studied that *Trichoderma* releases an extracellular exochitinase (Brunner et al. 2003) that might cause the liberation of some oligomers from the fungus, which could induce the expression of toxic endochitinases that would diffuse and would start to attack to the pathogen, even before the physical contact had happened. Some enzymes belonging to these fungi have been purified and used for biocontrol. When they have been

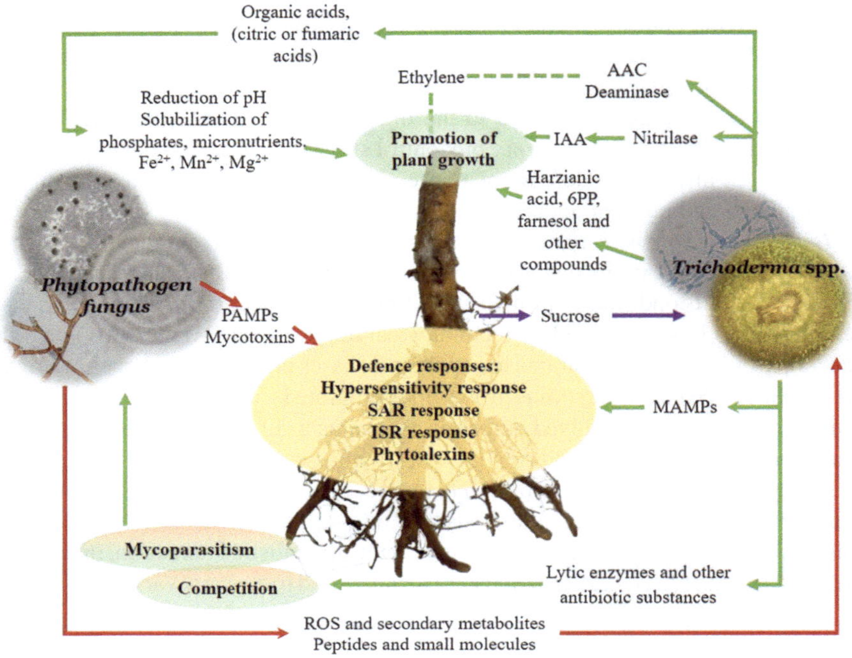

Fig. 1.1 Interactions between phytopathogen fungus, plant, and biocontrol agent *Trichoderma*. The green lines and circles are compounds and actions produced/induced by *Trichoderma*. The red lines are compounds and responses produced/caused by the phytopathogen fungus. The purple lines and circles are the compounds and plant responses produced/induced by the fungi (Altomar et al. 1999; Druzhinina et al. 2011; Howell 2003; Rubio et al. 2009; Vargas et al. 2011; Vinale et al. 2009; Vinale et al. 2008a, b) (*6PP* 6-pentyl-α-pyrone; *AAC* 1-aminocyclopropane-1-carboxylic acid; *IAA* indoleacetic acid; *ISR* induced systemic resistance; *MAMPs* microorganism-associated molecular patterns; *PAMPs* pathogen-associated molecular patterns; *ROS* reactive oxygen species; *SAR* systemic acquired resistance)

assessed, they have shown antifungal activity and have controlled a large number of pathogens, such as *Fusarium*, *Rhizoctonia*, *Alternaria*, *Ustilago*, *Venturia*, and *Colletotrichum* (Lorito et al. 1993; Lorito et al. 1994).

A major part of the *Trichoderma* antifungal system consists of a number of genes encoding an astonishing variety of secreted lytic enzymes (Sanz et al. 2004) including endochitinases, N-acetyl-β-glucosaminidases, chitin 1,4-β-chitobiosidases, proteases, glucan β-1,3-glucosidases, glucan β-1,6-glucosidases, glucan α-1,3-glucosidases, lipases, xylanases, mannanases, pectinases, pectin lyases, amylases, phospholipases, RNAses, DNAses, etc. Some of these proteins have been purified and their corresponding genes have been cloned and characterized by our group: protease PRA1 (Suarez et al. 2004), chitinases CHIT36 and CHIT37 (Viterbo et al. 2002), α-glucanases AGN13.1 (Ait-Lahsen et al. 2001) and AGN13.2 (Sanz et al. 2005), and β-1,6-glucanases BGN16.2 and BGN16.3 (Montero et al. 2005, 2007).

The direct confrontation assays were used to verify the ability of *Trichoderma* spp. to overgrow the pathogen and its capacity of mycoparasitism. In a study, the percentage of overgrow oscillated between 72.77 and 14.63%, according to the species of *Trichoderma* (Mayo et al. 2015).

Some *Trichoderma* spp. are selected because of their mycoparasitic mechanism but the most efficient biocontrol strains display, simultaneous or sequentially, more than one biocontrol strategy (Howell 2003). *Trichoderma* spp. can also exert marked antimicrobial activity (Vizcaino et al. 2005) due to the production of blends of secondary metabolites (Cardoza et al. 2005; Reino et al. 2007). Currently, better knowledge about *Trichoderma* has facilitated its use in biocontrol as whole microorganisms, able to be monitored in natural environments (Hermosa et al. 2001; Rubio et al. 2005), as enzyme formulations (Benítez et al. 2004) or as sources of genes for transgenic plant development. Since the early description of the capacity of *Trichoderma* to increase plant biomass production (Chang et al. 1986), several new general mechanisms for both biocontrol and plant growth increase have been demonstrated and it is now clear that there must be hundreds of separate genes and gene products involved in these processes.

There are compounds produced by *Trichoderma* that cause inhibitory effects on plants. For example, trichosetin, a secondary metabolite isolated from dual cultures of *T. harzianum*-*Catharanthus roseus* callus that is an antimicrobial compound with activity against *Staphylococcus aureus* and *Bacillus subtilis* (Marfori et al. 2002), but also inhibited root and shoot growth in some plant species (*Oryza sativa*, *Vigna radiata*, *Medicago sativa*, *Capsicum frutescens*, and *Lycopersicon esculentum*) (Marfori et al. 2003). Additional compounds with negative effects on plant growth (as necrosis in bean, tobacco, and corn) include trichocaranes (A, B, and C) (Macías et al. 2000), koninginins (B, C, E, and G) (Cutler et al. 1989; Parker et al. 1995), cyclonerodiol, and a laevorotatory form of harzianopyridone (Cutler and Jacyno 1991). *T. virens* also synthesizes negative plant growth promoters such as viridiol, a potent herbicidal compound, which is effective for weed control (Héraux et al. 2005).

Recently, they were identified other compounds with antimicrobial, antioxidant, and cytotoxicity activity. However, they inhibited germination of cabbage seeds as alternariol 1′-hydroxy-9-methyl ether, alternariol 9-methyl ether, alternariol, altechromone A, altenuene, 4′-epialtenuene, α-acetylorcinol, and cerebroside C (Zhang et al. 2017).

Promotion of Plant Growth

Trichoderma spp. has developed opportunistic mechanisms for their adaptation to abiotic stresses as well as for nutrient uptake and solute transport. In the plant, these processes are facilitated by the induction of cell wall extension and expansion, secondary root development, lateral root hair production and a higher photosynthetic rate (Shoresh et al. 2010; Hermosa et al. 2013).

Trichoderma produces some organic acids such as citric or fumaric acids that reduce soil pH and allow the solubilization of phosphates and other micronutrients such as iron, manganese, and magnesium (Fig. 1.1) (Benítez et al. 2004; Harman et al. 2004). On the other hand, there are some *in vitro* studies indicating that *T. harzianum* and other *Trichoderma* isolates could solubilize iron (III) oxide, manganese (IV) oxide, zinc, and phosphates, which are highly insoluble compounds or with low solubility, owing to chelation processes and oxidation-reduction activity (Altomare et al. 1999). The increment of all those nutrients, in particular, phosphorus, could favor the plant growth. It has been shown that *T. atroviride* produces and degrades indoleacetic acid (IAA), which in combination with ethylene by the microorganisms present in the rhizosphere causes a promotion of plant growth (Fig. 1.1) (Gravel et al. 2007).

The volatile pyrone 6-pentyl-α-pyrone (6PP) is a common *Trichoderma* compound that inhibits the growth of the pathogen such as *Fusarium oxysporum*. However, at low concentrations, 6PP significantly promotes the plant growth and it was able to induce the expression of plant defense genes (Viterbo et al. 2007; Vinale et al. 2008a).

Cremenolide is another compound that inhibits the development of plant pathogens. This compound significantly inhibited the growth of *F. oxysporum*, *Botrytis cinerea*, and *Rhizoctonia solani*. Furthermore, in tomato seedlings assays it promoted plant growth in terms of root length and fresh weight (Vinale et al. 2016).

Farnesol is produced by *Trichoderma* and is a signaling molecule that by accumulating in the extracellular space generates a response across the local fungal population. In another study, its effect on the development of bean plant was evaluated. This compound, which farnesol at concentrations of 10 and 100 μM farnesol showed a negative effect on growth of bean plants, which could be related to abscisic acid synthesis. However, 2 mM of farnesol has the opposite effect. Thus, at this concentration bean plants increased the development of aerial parts and root systems (Mayo et al. 2016a, b).

Defense Response

The relationships established between plants and microorganisms are very diverse. Plant's defense against pathogens is regulated through a complex network of signaling pathways involving several molecules such as reactive oxygen species (ROS), salicylic acid (SA), jasmonic acid (JA), and ethylene (ET) (Kunkel and Brooks 2002; Vitti et al. 2015) and some secondary metabolites with antimicrobial activity that can act also as signaling molecules (i.e., phytoanticipins and phytoalexins) (Mhlongo et al. 2016) (Fig. 1.1). When a plant is exposed to a pathogenic microorganism, the production of molecules associated with SA is increased, which is related to a systemic acquired resistance (SAR) response. The response of plants against non-pathogenic microorganisms is different, resulting in activation of signaling cascades that are dependent on JA and ET, such as hydroperoxide lyase, peroxidase, and phenylalanine ammonia-lyase, all related to an induced systemic resistance (ISR) response (Druzhinina et al. 2011). Other responses result in rapid cell death in infected tissues. Then, plants activate the hypersensitive response that involves the accumulation of salicylic acid, ROS, and an increased the influx of Ca^{2+} (Guerrero-González et al. 2011).

Hypersensitive Defense and Phytoalexins

Another response exhibited by plants is the necrotic defense or hypersensitive defense (Fig. 1.1) that induces the selective death of some cells to block the progress of phytopathogens through the plant tissues (Tadeo and Gómez-Cadenas 2008). These changes in hypersensitive reactions include loss of cell membranes permeability and increase in respiration and production of phytoalexins. Phytoalexins are not present in healthy plants but are synthesized in response to biotic stress as part of the plant defense response and are restricted to the tissue colonized by the fungus and the cells surrounding the infection site (Morrissey and Osbourn 1999). The result is death and collapse of the infected cells. The necrotic tissues isolate the phytopathogen causing its death because the pathogen depends entirely on the plant to survive. It is likely that the faster the host cells die after they have been infected, the more resistant they become to infection (Agrios 2002). For example, alfalfa (*Medicago sativa*) or barrel medic (*Medicago truncatula*) produce medicarpin, which is an isoflavone, in response to the pathogens *Colletotrichum trifolii* or *Phoma medicaginis*, respectively (Saunders and O'neill 2004; Jasiński et al. 2009). Other example is peanut (*Arachis hypogaea*) that produces resveratrol in response to *Aspergillus* spp., *Botryodiplodia theobromae*, *Ganoderma lucidum*, or *Rhizopus oligosporus* (Sobolev et al. 2009; Condori et al. 2010; Yang et al. 2010; Wu et al. 2011). Soybean (*Glycine max*) produces glyceollin, another phytoalexin, in response to the attack of *Macrophomina phaseolina*, *Sclerotinia sclerotiorum*, *Phytophthora sojae*, *Fusarium solani*, or *Aspergillus* spp. (Lozovaya et al. 2004; Feng et al. 2007; Simons et al. 2011a; Simons et al. 2011b; Eromosele et al. 2013).

In the tripartite interaction of plants with a pathogen and a biocontrol *Trichoderma* species, several changes are produced in the plant, such as the increase in phenolic acid and lignin, accumulation of phytoalexins, and down- or upregulation of defense-related genes expression (Guerrero-González et al. 2011; Mayo et al. 2015).

Different categories of defense-related genes whose expression is modulated by biotic stresses have been described in bean plants interacting with pathogenic and nonpathogenic microorganisms (Mayo et al. 2015).

SAR and ISR Responses

The perception of an external stimulus can active the response genes. There are some components that regulate many processes in response to stimuli (Fig. 1.1).

A component involved in the regulation of plant defense gene expression is WRKY transcription factors (TFs) (Rushton and Somssich 1998; Singh et al. 2002). They can function up- and downstream of hormones that are involved in the antagonist functions of SA and JA/ET. They also contribute to the development control processes via auxins, cytokinins, and brassinosteroids (Chen et al. 2010; Agarwal et al. 2011; Rushton et al. 2012; Bakshi and Oelmüller 2014). Several ROS-dependent responses are controlled by WRKY TFs, and they also regulate major changes in the plant transcriptome during early phases of root colonization with arbuscular mycorrhizal fungi (Gallou et al. 2012).

Thus, *WRKY33* has a role in biotic stress defense, where it regulates the balance between necrotrophic and biotrophic pathogen responses (Lippok et al. 2007; Pandey and Somssich 2009; Birkenbihl et al. 2012). A rapid pathogen-induced *WRKY33* expression did not require salicylic acid signaling but downregulation of this gene involved a direct activation of jasmonic acid (Bakshi and Oelmüller 2014). Other reports established that *WRKY33* is essential for defense against the necrotrophic fungus *B. cinerea* (Birkenbihl et al. 2012). Loss of *WRKY33* function results in inappropriate activation of the SA-related host response and elevated SA levels post-infection, and in the downregulation of JA-associated responses at later stages. This downregulation appears to involve direct activation of several JA *ZIM*-domain genes, encoding repressors of the JA response pathway, by loss of *WRKY33* function and by additional SA-dependent WRKY factors. Moreover, genes involved in redox homeostasis, SA signaling, ET-JA-mediated cross-communication, and camalexin biosynthesis were identified as direct targets of *WRKY33*. Although SA-mediated repression of the JA pathway may contribute to the susceptibility of *wrky33* plants to *B. cinerea*, it is insufficient for *WRKY33*-mediated resistance. Thus, *WRKY33* apparently directly targets other still unidentified components that are also critical for establishing full resistance toward this necrotroph (Birkenbihl et al. 2012).

In the work of Mayo et al. (2016a), when bean plants (*Phaseolus vulgaris*) were in contact with *T. velutinum* T028 without pathogen, the *WRKY33* gene expression was significantly upregulated while the *PR* genes expression (*PR2*, *PR3*, and *PR4*) was significantly downregulated, compared to expression levels in plants without *Trichoderma* treatment. However, in the same work, when the pathogen *R. solani* was added to the substrate, expression of *WRKY33* was significantly downregulated in plants with *Trichoderma* inoculation, while *PR2*, *PR3*, and *PR4* were downregulated. In the study by Mayo et al. (2015), the expression of *PR1*, *PR2*, *PR3*, and *PR4* was downregulated when beans (*P. vulgaris*) were inoculated with *R. solani*. An overexpression of *PR2* and *PR5* has also been observed in *Arabidopsis thaliana* inoculated by the necrotrophic bacteria *Erwinia carotovora* subsp. *carotovora* (Li et al. 2004). WRKY family members have been shown to be responsible for the regulation of expression of *PR2* and *PR5* in grapevine (Marchive et al. 2013) and *A. thaliana* (Li et al. 2004). *PR1*, together with *PR2*, *PR3*, *PR4*, and *PR5*, is considered marker for SAR.

WRKY33 is also involved in the regulation of expression of genes modulated by components of the ethylene signaling pathway. Expression of *ERF1* and *ERF5* reached similar significant values either with or without *Trichoderma* and/or *R. solani* in the substrate. *WRKY33* would act as a repressor of *ERF1* and *ERF5* expression. Thus, when the expression of *WRKY33* is increased, the expression of *ERF1* and

ERF5 is downregulated (Mayo et al. 2016b). In *Arabidopsis*, *ERF5* may contribute to plant innate immunity against biotrophic pathogens, by regulating SA signaling, while also affected plant resistance to necrotrophic pathogens by regulating JA signaling (Son et al. 2012).

The *CH5b* gene encodes an endochitinase precursor and it is related with the ethylene signaling pathway. In previous works, it has been shown that when this gene was over-expressed the *R. solani* symptoms were reduced in crops like *N. tabacum* and *Brassica napus* (Broglie et al. 1991). However, when *P. vulgaris* plants were in contact with *R. solani*, the expression of this gene was downregulated but not significantly, while treatment of these infected plants with *T. velutinum* resulted in its significant upregulation. These results are in agreement with previous data, showing that the pathogen represses its expression, and the presence of *Trichoderma* induced it (Mayo et al. 2015). Furthermore, expression of a chitinase encoding gene from *T. harzianum* in transgenic tobacco and potato plants and observed an increase in the resistance to *Alternaria alternata*, *R. solani*, and *B. cinerea*, a much wider protection spectrum than the one obtained when using plant chitinases (Lorito et al. 1998).

Osmotins have also plant protective effects against pathogen infection (Narasimhan et al. 2009). When *T. velutinum* or *R. solani* were present in the soil, the expression of *OSM34* was not significantly upregulated with respect to control plants, but when both fungi were in the soil at the same time, *OSM34* was slightly but significantly downregulated (Mayo et al. 2016b).

PAL (phenylalanine ammonia-lyase) plays an important role in plant defense; it is involved in the biosynthesis of salicylic acid, which is related to plant systemic resistance (Mauch-Mani and Slusarenko 1996; Nugroho et al. 2002; Chaman et al. 2003). *PAL* gene expression is also regulated in response to pathogen infection. The presence of *T. velutinum* and *R. solani* in the soil resulted in a significant downregulation of this gene compared with control plants (Mayo et al. 2016b). Similarly, potatoes inoculated with *T. harzianum* and/or *R. solani*, showed an upregulation *PR1* at 168 h post inoculation (hpi) and a slight upregulation of *PAL* at 96 hpi, in plants inoculated with *T. harzianum* alone (Gallou et al. 2009). This was in apparent contradiction with other studies in which a marked induction after a short time (24 hpi or 48 hpi) of *PAL*, hydroxyperoxide lyase (*HPL*), and *Lox*, *PAL*, ethylene receptor 1 (*ERF1*), ethylene-inducible *CTR1*-like protein kinase-encoding genes was observed (Yedidia et al. 2003; Shoresh et al. 2005). Such differences might be attributed to the absence of root cell penetration and colonization by the *Trichoderma* strain.

HPL (hydroperoxide lyase) is involved in the production of antimicrobial and defense signaling oxylipins (Noordermeer et al. 2001; Huang et al. 2010). The presence of *T. velutinum* and *R. solani* resulted in a downregulation of this gene expression when compared versus control plants. Thus, after 45 days of growth in contact with *T. velutinum* and/or *R. solani*, its expression was downregulated, indicating that the plant identifies *Trichoderma* and *Rhizoctonia* as two invader organisms. Some of the mechanisms activated against the presence of both are similar, independently of the final response that will be specifically activated in the plant by each one (Mayo et al. 2016b).

The expression of dependent genes of JA was studied in common bean plants inoculated by *T. harzianum* ALL-42. They also presented differential expression pattern for defense response such as *BCH1* (chitinase), *Glu1* (β-1-3-glucanase), *Lox* (lipoxygenase encoding gene), and *POD3* (peroxidase) in comparison to control plants, and with plants infected with *F. solani* or *R. solani.* This response is in agreement with previous works which showed that this is a typical host plant response to its colonization by a symbiotic or pathogenic microorganism (Harman et al. 2004; Shoresh et al. 2005; Shoresh et al. 2010). Plants challenged by *T. harzianum* ALL-42 showed upregulation of *Glu1*, *Lox*, and *POD3* compared with plants challenged by phytopathogenic fungi. *T. harzianum* ALL-42 also seems to potentiate common bean (*P. vulgaris*) response to the presence of the phytopathogenic fungus *R. solani*, as shown by the increase in the levels of *Glu1* and *POD3* for the double treatment (*Trichoderma* + pathogen) in comparison to that obtained for plants in the presence of *R. solan*i alone (Pereira et al. 2014).

The *CNGC* genes can be related to early plant defense responses due to changes in ion flux, including H^+ and Ca^{2+} influx and K^+ and Cl^- efflux (Atkinson et al. 1996). The upregulation of *CNGC2* confirms the importance of ion channels for the plant resistance response (Borges et al. 2012). *CNGC2* was downregulated in plants treated with *T. velutinum* (Mayo et al. 2016b).

GSTa (2,4-D inducible glutathione S-transferase) expression also responds to pathogen attack (Mauch and Dudler 1993) and can be induced by molecules such as salicylic acid, methyl jasmonate, abscisic acid, and H_2O_2 (Dixon et al. 2002; Moons 2005). In *Gossypium arboretum*, *GST* provides resistance to fungal pathogens and oxidative stress (Barthelson et al. 2010). *GST* expression was upregulated during fungal infection in barley, *Arabidopsis*, and cotton (Dowd et al. 2004; Durrant and Dong 2004; Lu et al. 2005). However, in banana *GST* was downregulated following *F. oxysporum* f specialis (f. sp.) *cubense* infection (Wang et al. 2013), which is in agreement with the downregulation of *GSTa* when *T. velutinum* and/or *R. solani* were present in the soil (Mayo et al. 2016b).

hGS encodes a homoglutathione synthetase that is involved in response to oxidative stress. There is not much information about the behavior of this gene in the plant. In the study of Mayo et al. (2016b), when bean plants (*P. vulgaris*) were in contact with *T. velutinum* and/or *R. solan*i, expression of this gene was significantly upregulated compared to control plants. In other studies, treatment of *Medicago truncatula* plants with compounds that release nitric oxide, a key signaling molecule, induced expression of *GST* but not *hGS* in roots (Innocenti et al. 2007). Similarly, common bean plants treated with H_2O_2 showed upregulation of *hGS* in nodules, whereas treatments with cadmium, sodium chloride, or jasmonic acid had no effect (Loscos et al. 2008).

Production of Secondary Metabolites: Changes in Plant Metabolism as Defense Response

When a plant is induced by exposure to a microorganism, it starts to produce diverse metabolites and enzymes. The physiological changes activated in the plant lead to the activation of various metabolic pathways, which will be different depending on the type and origin of these signaling natural products. Different secondary metabolites are synthesized after perception and recognition of the signals originating from plant or pathogenic microorganism elicitors produced during the first steps of plant defense reactions (Grotewold 2005; Boller and Felix 2009; Veitch 2009). Plant responds after the invasion of a phytopathogen or a biocontrol agent by activating disease-resistance responses (i.e., upregulation of defense-related genes) against the invasion (Mayo et al. 2016b). Also, plant produces some antimicrobial secondary metabolites such as phytoalexins (phenols, isoflavones, terpenes), and some substances that can block pathogen invasion and spread, such as lignin and callose (Chen et al. 2015). Some plants do not produce phytoalexins when are in contact with pathogens but release toxins that are normally stored as less toxic glycosides (Grayer and Kokubun 2001).

Trichoderma spp. are also considered as efficient producers of extracellular enzymes, and some of these enzymes have been involved in the biological control of plant diseases (Monte 2001; Harman et al. 2004). *Trichoderma* species also produce plant hormones and solubilize minerals in the soil, which help to promote plant growth and suppress the disease (Kim et al. 2006).

During the *Trichoderma*–plant interaction, various classes of metabolites could induce resistance such as proteins with enzymatic activity, low molecular weight compounds related to the fungal or the plant cell wall, which can be originated by the enzymatic activity of *Trichoderma* (Woo et al. 2006; Woo and Lorito 2007), and other secondary metabolites. These elements trigger plant defense responses against the pathogen (Hermosa et al. 2012; Malmierca et al. 2014), by inducing the expression of genes encoding for pathogenesis-related (PR) proteins, which further contribute to reduce the disease symptoms.

During the plant–*Trichoderma* interactions, the fungus participates actively in protecting and improving its ecological niche. Leucine-rich repeat (LRR)-containing proteins are signal receptors regulating plant development and defense (Afzal et al. 2008). Marra et al. (2006) observed that LRR proteins increased in bean leaves (*P. vulgaris*) interacting with *T. atroviride*, and that hydrophobins and ABC transporters were accumulated in the proteome of the fungus. Hydrophobins (Rosado et al. 2007) and ABC transporters (Ruocco et al. 2007) support the biocontrol activity of *Trichoderma* and its ability to colonize the roots. In a similar way, a *Trichoderma*-secreted swollenin (an expansin-like 5 protein) remarkably increased fungus plant root colonization efficiency. Due to a cellulose-binding domain was able to trigger defense responses in the plant and afforded pathogen protection, indicating that this domain might, therefore, be recognized by the plant as a microbe-associated molecular pattern (MAMP) in the *Trichoderma*–plant interaction (Brotman et al. 2008). At least four classes of substances that elicit plant defense responses have been identified in *Trichoderma*: polysaccharide oligomers, enzymes, low molecular weight proteins, and peptaibols. Some cell wall oligomers may act as elicitor molecules released by plants following pathogen attack (Woo et al. 2006). The overexpression of *Trichoderma* chitinase genes in tobacco plants generates innate defense responses and enhanced stress tolerance (Dana et al. 2006). Also, it was detected hydrophobin-like cysteine-rich low molecular weight secreted proteins Sm1 from *T. virens* and Epl1 from *T. atroviride* (Djonović et al. 2006; Seidl et al. 2006) that can trigger ISR but, with the exception of peptaibols as elicitors of plant defense responses (Viterbo et al. 2007), the role of secondary metabolites in this task remains unexplored. In fact, the peptaibol alamethicin produced by *T. viride* sprayed on *Phaseolus lunatus* plants activates ISR, resulting in the production of defense compounds against herbivores (Engelberth et al. 2000). A plausible explanation is that the peptaibols produced by *Trichoderma* spp. can affect its own plasma membrane functions, and that the lack of production of these metabolites by the mutant potentiates growth, leading to the production of more aerial mycelium (Velázquez-Robledo et al. 2011).

Trichothecenes are important mycotoxins, which in general have potent phytotoxicity, but they are also toxic for animals and humans. Some *Trichoderma* species can produce trichothecenes (Nielsen et al. 2005). Thus, *T. brevicompactum* produces trichodermin, a phytotoxic compound that enables this species to be used as a biocontrol agent (Tijerino et al. 2011). *T. arundinaceum* produces harzianum A, a trichothecene lacking phytotoxic activity when assayed *in vivo*, but with antifungal activity against *B. cinerea* and *R. solani* (Malmierca et al. 2013). Harzianum A also elicits systemic defense and priming responses in tomato plants (Malmierca et al. 2012). In the antagonistic interaction of *T. arundinaceum* and *B. cinerea*, the former produces harzianum A while the latter inhibits the expression of genes in the trichothecene biosynthetic cluster. *B. cinerea* on tomato activates a typical JA response in the plant; *T. arundinaceum* on tomato activates the expression of SA and JA signaling genes by the plant. In the interaction between *T. arundinaceum*, *B. cinerea*, and tomato, there is a dramatic increase in the expression of tomato plant defense-related genes belonging to the SA and JA pathways, compared to a background of *B. cinerea*–tomato and *T. arundinaceum*–tomato conditions (Malmierca et al. 2012).

In the work of Velázquez-Robledo and et al. (2011) suggest that hydrolytic enzymes and mycoparasitism are more relevant than antibiotics in the control of *R. solani* during seed protection. A similar observation was made in the case of a *T. virens* mutant that did not produce gliotoxin but remained efficient in the protection of plants against infection by *R. solani* (Howell and Stipanovic 1995).

1.2.1.3 Conclusions

Crops are affected by a wide diversity of fungal pathogens and a method of control is the application of synthetic fungicides. However, it is a priority to develop non-chemical methods in integrated production, organic farming, and others such as the use of biocontrol agents. *Trichoderma* is a fungal genus including a huge number of species and strains. A high percentage of these species have the abilities to protect crops against diseases and to increase crop yield under field conditions. Plant can response to attack of pathogen as a hypersensitive defense that induces the selective death of some cells, including loss of permeability of cell membranes, an increase in respiration, and production of phytoalexins. *Trichoderma* and/or a phytopathogen can cause an upregulation or a downregulated response that will depend on the function gene, plant age, tissue, etc.

1.2.2 Cold Tolerance

Low temperature is a collective term, incorporating two distinct but related stresses, chilling, and freezing. Chilling temperatures fall in the range of 0–15 °C, while freezing temperatures are below 0 °C. While there is some commonality between the metabolic impact of chilling and freezing, their physiological impacts differ. However, both chilling and freezing can have extremely harmful effects on plant functions (Thomashow 1999). The sensitivities of plants to low temperatures are broadly correlated with their agro-environmental distribution. Several visual symptoms of chilling injury are exhibited by sensitive plant species. The most noticeable of these is the wilting of aerial organs, resulting from reduced water retention capacity. Moreover, prolonged chilling exposure can cause accelerated aging that is characterized by a loss of leaf coloration (Lukatkin et al. 2012). However, the processes underpinning the initiation and regulation of programmed cell death are not yet fully understood (Van Durme and Nowack 2016).

Despite the proven benefits of legume utilization, yield increases have not kept pace with those of cereal crops. Global increases in legume production are a result of increased land usage, rather than a direct increase in crop productivity (Foyer et al. 2016). Pulse crops are members of a diverse family of plants, the ecological and nutritional characteristics of which are well matched to the varied challenges of climate change, calorific provision, and nutritional demand. However, in order to sufficiently address these challenges a greater level of research must be conducted into legume biology, with a specific focus on the enhancement of legume survival and productivity under stress conditions (Foyer et al. 2016). Low temperatures in particular place a significant constraint on global legume yields and those legumes of significant dietary importance must be studied further.

While the general mechanisms of low-temperature tolerance have been characterized in the plant kingdom, extensive research has not been conducted on the factors underpinning low-temperature tolerance in legumes. Recent evidence has emerged showing that cold tolerance may be enhanced through favorable interactions between plants and the soil microbiome (Subramanian et al. 2016). This finding is particularly interesting when considered in the context of legumes, which are characterized by their intimate links with the soil microbiome.

Low temperature is a phenomenon that impacts agricultural productivity on every continent. In the United States, an estimated 25% of the reduction in crop productivity was attributed to low temperatures (Boyer 1982). Exposure to cold is also a limiting factor in the agricultural distribution of legume crops in Australia (Maqbool et al. 2010) and Africa. Moreover, in Europe severe cold weather events limit overwintering legumes such as faba bean (*Vicia faba*) and chickpea (*Cicer arietinum*) (Link et al. 2010). As such, the development of low-temperature tolerant legume crops is of critical importance for the protection of food security (Link et al. 2010). Yield reduction is the dominant consequence of stress exposure. Plants are vulnerable to cold stress at all stages of development, with susceptibility being particularly high during seedling establishment and seed formation. However, plants employ numerous strategies for the survival of low-temperature stress. While the genetic and biochemical factors underpinning low-temperature tolerance have been extensively characterized in cereals (Winfield et al. 2010), limited research has been conducted on the mechanisms of low-temperature tolerance in legumes.

Biotechnology has provided some insight into the genetic factors contributing to stress tolerance; however, the focus has been placed on abiotic stresses. Moreover, the resolution of causative genetic factors tends only to extend to the level of genomic loci. As such, progress needs to be made in the elucidation of single gene location and function (Dita et al. 2006). However, some understanding of the mechanisms through which plants protect against abiotic stress exposure has been gained through transgenic studies. In legumes, the most susceptible stages are flowering, early pod formation, and seed-filling stages (Siddique et al. 1999). The cold stress can also lead to other problems, including increased vulnerability to pathogen entry, such as to bacterial blight, which requires a wound to infect the field pea plant. Genetic assessment for frost tolerance in pulses either under natural or controlled frost conditions is a relatively new area of research, with the majority of studies carried out on reproductive frost tolerance in barley (Reinheimer et al. 2004) and cold tolerance in chickpea (Clarke et al. 2004). The timing of the exposure to low temperature or frost is a key factor that determines the disruption of fertilization of flowers in legumes (Stoddard et al. 2006). However, international efforts to breed for frost tolerance, cold tolerance, freezing tolerance, and winter hardiness vary depending on the specific local climatic conditions, whereas the most severe damage may be caused at the seedling stage, the vegetative stage or the reproductive stage.

The genetic improvement strategies could include developing new screening and selection methodologies, including methods for marker-assisted backcrossing and genetic engineering (Stoddard et al. 2006). Only a limited number of studies have been carried out on tolerance in pulse crops (Margesin et al. 2007).

1.2.3 Drought Tolerance

Legumes rank among humanity's most important agricultural food crops. They are grown in almost every climatic region and on a wide range of soil types. Drought is one of the most common abiotic stresses reducing the yield of many crops including legumes. The yield of food legumes grown in arid to semiarid environments or drylands such as the Mediterranean (e.g., faba beans, chickpea, and lentil) is usually variable or low due to terminal droughts that characterize these areas (Mafakheri et al. 2010; Karou and Oweis 2012). Improving the tolerance of crops under water-limited environments is prerequisite if agricultural production is to keep pace with the expected demographic increase. Beyond productivity, the resilience of crops to water-limited environments, i.e., the capacity to yield even under very harsh conditions, will be increasingly important. The economically viable approaches to support crop production under drought are still limited. More importantly, it remains unclear how the impact of drought on legume production varies with legume species, regions, agroecosystems, soil texture, and drought timing.

Besides soil degradation and heat stress (Abate et al. 2012), drought is the abiotic factor that most adversely affects legume production. It turns out, however, that the largest producers of pulses (70% of global production) (Gowda et al. 2009) are located in regions that experience water shortage (Rockstrom et al. 2009) and their production are highly vulnerable to drought.

1.2.3.1 Differences in Species Response to Drought

There are significant differences among legume species with regard to their adaptability to drought as measured by their ability to maintain high yield following a period of water stress. Lentil and groundnut were the legumes that exhibited the lowest yield reduction (21.7% and 28.6% respectively) while faba bean had the highest yield reduction (40%) under the highest observed water reduction (>65%). Under slightly lower water reduction (60–65%), pigeon pea exhibited the lowest yield reduction (21.8%) followed by soybean (28.0%), chickpeas (40.4%), cowpeas (44.3%), and common beans (60.8%). There are some legume crops (soybeans and common beans) that have migrated successfully from their center of origin while others remain largely confined to their areas of origin. During the evolutionary history of domesticated species, the wild types generally adapt themselves to their environment of origin, ensuring their own survival and that of their progeny. At the same time, genetic variability may exist within a legume species, from extremely drought-sensitive to drought-resistant types. This origin, however, does not always correspond to the adaptability of a legume species to drought. This indicates that most legumes may have the potential to be modified into more drought-resistant species.

1.2.3.2 Differences in Drought Responses Under Different Plant Phenological Stages

Plant phenological stage affected the percentage of yield reduction observed in legume crops, with drought during the vegetative phase resulting in lowest yield reduction (15.5%) compared to drought that occurred during the early and late reproductive stages under the same amount of water reduction. Although drought during the very early vegetative stage may impair germination, most studies that examined the effect of drought usually allowed sufficient water to support good and uniform plant establishment. Therefore, drought that happens during the later vegetative periods was relatively more tolerable to plants even though they might experience retarded cell elongation, division, and differentiation (Farooq et al. 2009). They are still able to maintain their growth functions under stress because early drought may lead to immediate survival or acclimation where the plants modify their metabolic and structural capabilities mediated by altered gene expression (Chaves et al. 2002).

A number of drought-resistant cultivars/lines of different crops have been developed solely using conventional breeding approach. These drought-tolerant lines of different crops provide a sound testament that conventional plant breeding played a considerable role during the last century not only for improving the quality and yield of crops but also for improving abiotic stress tolerance including drought tolerance. While transferring desired genes from one plant to other through the conventional plant breeding, a number of undesired genes are also transferred. Furthermore, to achieve the desired gain through traditional breeding, a number of selection and breeding cycles may be required. The limited success in improving crop drought tolerance could be due to the reason that the drought tolerance trait is controlled by multiple genes having an additive effect (Thi Lang and Chi Buu 2008) and a strong interaction exists between the genes for drought tolerance and those involved in yield potential. Thus, there is a need to seek more efficient approaches for genetically tailoring crops for enhanced drought tolerance.

The role of polygenes in controlling a trait has been widely assessed by traditional means, but the use of DNA markers and quantitative trait locus (QTL) mapping has made it convenient to dissect the complex traits (Humphreys and Humphreys, 2005). Due to the intricacy of abiotic stress tolerance and the problems encountered in phenotypic-based selection, the QTL mapping has been considered as imperative to the use of DNA markers for improving stress tolerance (Ashraf et al. 2008). QTL mapping for the drought tolerance trait has been done in different crops, the most notable being maize, wheat, barley, cotton, sorghum, and rice (Bernier et al. 2008). Molecular mapping and a number of QTL associated with drought tolerance identified in different crops can be effectively used in appropriate breeding programs meant for improving crop drought tolerance. Marker-assisted breeding approach is a prospective alternative to traditional breeding, because of being less time-consuming and labor- and cost-effective. Molecular mapping and analysis of QTL have been carried out for a number of qualitative and quantitative traits including stress tolerance, which has undoubtedly resulted in a great magnitude of knowledge and better understanding of the causal genetic phenomena that regulate these traits.

1.2.4 Insect Resistance

1.2.4.1 Biological Control Agents Against Insect Pests

Nowadays, the priority in pest management is to select compounds with different modes of action, with greater selectivity and less persistence. Thus, to minimize side effects on auxiliary fauna, the environment, and public health, there is an increasing interest on the use of entomopathogenic fungi to control invertebrate pests, weeds, and plant diseases, as shown by the increasing number of commercial products available or under development (Rodríguez-González et al. 2017a). Entomopathogenic fungi have great potential as control agents, constituting a group with more than 750 species, disseminated in the environment and causing fungal infections to arthropods populations (Pucheta-Díaz et al. 2006). López-Llorca and Hans-Börje (2001) cite the following genera as the most important for arthropod control: *Metarhizium*, *Beauveria*, *Paecilomyces*, *Verticillium*, and *Trichoderma*. The field of biological control is an industry focused on the development of less harmful pest management strategies (Abdul-Wahid and Elbanna 2012). In recent years, this industry has started to use fungi to control populations of insect pests, specifically agricultural pests (Hajek 2004). The ability of entomopathogenic fungi to actively invade live insects through their cuticle and proliferate inside them, make these fungi unique and highly effective tools for the management of insect pests (Rodríguez-González et al. 2016).

Meyling and Eilenberg (2007) pointed out that in order to use entomopathogenic fungi as BCAs it is essential to use agricultural practices which enhance their establishment and development. For this reason, knowledge about the ecology of these fungi is of utmost importance. Different parameters which influence the ecology of these fungi are humidity, temperature, pathogenicity, virulence, and hosts range, among others. These pathogenic fungi have been searched and isolated in plants and crops affected by pests and/or diseases. Different *Trichoderma* species have been isolated and identified in bean seeds (*Phaseolus vulgaris*) (Campelo 2010). Rodriguez-Gonzalez and Carro-Huerga (unpublished data) have also been able to isolate and identify different *Trichoderma* species on vineyard soils and vine wood (*Vitis vinifera*) affected by *Xylotrechus arvicola* Olivier (Coleoptera: Cerambycidae).

Rumbos and Athanassiou (2017) described that most studies using entomopathogenic fungi to control post-harvest insects have been conducted with isolates of *Beauveria bassiana* and, to a lesser extent, *Metarhizium anisopliae* (Metschnikoff) Sorokin (Ascomycota: Hypocreales). These fungal pathogens have a wide host range and have been tested against most of the major storage pests under various conditions and crops.

Some studies have been conducted against *Sitophilus zeamais* (Motschulsky) (Barra et al. 2013), *A. obtectus* (Dal Bello et al. 2006), and *Callosobruchus maculatus* (F.) (Cherry et al. 2005) with *B. bassiana*, while *M. anisopliae* has been tested to control *Rhyzopertha dominica* (F.) (Athanassiou et al. 2008), and *Sitophilus oryzae* (L.) (Batta 2004). *B. bassiana* has shown high effectiveness on the control of other Coleoptera families, as for example *Enaphalodes rufulus* (Coleoptera: Cerambycidae) (Meyers et al. 2013), *Monochamus alternus* (Coleoptera: Cerambycidae) (Maehara and Kanzaki 2013) and *X. arvicola* (Coleoptera: Cerambycidae) under laboratory conditions (Rodríguez-González et al. 2016) or simulating field conditions in laboratory (Rodríguez-González et al. 2017b).

As described for *B. bassiana*, *Trichoderma* has shown good results in the control of different development stages of several insect pests within the orders Lepidoptera and Coleoptera. Examples of these results are shown in Alahmadi et al. (2012) *with Lucanus cervus* (Coleoptera: Lucanidae), Ghosh and Pal (2016) with *Leucinodes orbonalis* (Lepidoptera: Crambidae) and Rodríguez-González et al. (2017a, b) with *X. arvicola* (Coleoptera: Cerambycidae). For all this, the use of *Trichoderma* spp. as BCA against *A. obtectus* (Coleoptera: Chrisomelidae: Bruchinae) may be an economical, simple, and ecologically sustainable alternative.

1.2.4.2 *Trichoderma* spp.

Harman and Kubicek (2002) described *Trichoderma* spp. (Teleomorph: *Hypocrea*) as a genus of filamentous ascomycetes that is among the most commonly found saprophytic fungi in nature. These fungi frequently appear on the ground and grow on wood, bark, other fungi, and many other substrates, having high opportunistic potential and great adaptability to diverse ecological conditions. *Trichoderma* spp. produces chitinases, glucanases, and proteases, as well as other metabolites with antimicrobial activity (Lorito et al. 2010). Many *Trichoderma* species are also well known as biocontrol agents of important phytopathogenic fungi. Its two main mechanisms of biocontrol against these pathogens are mycoparasitism antibiosis (Papavizas 1985) and competition for nutrients with the pathogen (Harman and Kubicek 1998). *Trichoderma* species colonize plant root surface and cause substantial changes in plant metabolism (Harman et al. 2004).

There are several authors (Benítez et al. 2004) who have recognized *Trichoderma*'s important benefits to agriculture, such as its ability to protect crops against diseases and increase crop yield under field conditions (Harman et al. 2004). Benitez et al. (2004) described that once *Trichoderma* strains have grown and proliferated around abundant healthy roots, the fungus develops numerous mechanisms of action both to attack other fungi and to enhance plant and root growth.

1.2.4.3 The Bean Weevil, *Acanthoscelides Obtectus*

The bean weevil is an insect pest of neotropical origin (Fig. 1.2a) that feeds on wild and cultivated common bean (Paul et al. 2009; Thakur 2012; Vilca-Mallqui et al. 2013). Their larvae feed exclusively on the seeds and, cause considerable damage to them (Fig. 1.2b). The galleries they produce in the seed destroy the cotyledons, causing a significant reduction in its weight and germination rate (Gallo et al. 2002; Quintela 2002). Moreover, the commercial depreciation of the damaged beans is also due to the presence of insect excrements and death individuals. These remains favor the development of fungi and other pathogens inside the beans making them unsuitable for human consumption (Ramírez and Suris 2015).

The bean weevil is both a field and a storage pest, although major losses are caused when beans are in storage (Baier and Webster 1992). The bean weevil is a polyphagous species that affects around 35 species of legumes (Romero-Nápoles and Johnson 2004). Adults are straw colored, have an ovoid shape and, measure 2–4 mm in length. They are good flyers and can easily infect new beans, both in the field and in storage (Gallo et al. 2002). Gołebiowski et al. (2008) described that these insect populations grow exponentially when left untreated and can destroy stored crops within a few months. The management of the bean weevil in storage facilities is either nonexistent (by small farmers) or relies on the application of synthetic insecticides (in big storage facilities), such as phosphine, pyrethroids, and organophosphates (Daglish et al. 1993; Oliveira et al. 2013). The application of these compounds causes the development of pest resistance, environmental contamination and also

(a) (b)

Fig. 1.2 *A. obtectus* adults (left: female; right: male) (**a**). Damage caused in beans by *A. obtectus* larvae (**b**) (images from da Silva 2017)

threats human health (Subramanyam and Hagstrum 1995; Daglish 2008). Therefore, the use of synthetic insecticides has been recently questioned by a society that seeks sustainable alternatives for pest control (Regnault-Roger et al. 2012).

1.2.4.4 *Trichoderma* spp. Against the Bean Weevil

Four *Trichoderma* species were evaluated against bean weevil eggs showing high biocontrol activity. *T. harzianum* had an almost total ovicidal control (96.7% of eggs infected) (Fig. 1.3a). *T. atroviride* and *T. citrinoviride* also inhibited most of the tested eggs (Fig. 1.3b, c), whereas *T. longibrachiatum* was only able to infect half of the eggs (Fig. 1.3d).

T. harzianum has been described in previous reports as a control agent against insect immature stages (Alahmadi et al. 2012) using Trichodex® (Makhteshim Ltd., Makhteshim-Agan of North America, Inc., New York). Trichodex® is a commercial compound made from *T. harzinanum* that controlled *Lucanus cervus* (Coleoptera: Lucanidae) larvae. *T. citrinoviride* also showed a biocontrol effect on *A. obtectus* eggs. There are no previous studies where *T. citrinoviride* was applied to control immature stages of insect pests. Until now, this *Trichoderma* species has been used exclusively against plant diseases, so it may be interesting to test their insecticidal activity on other insect species. Even the lower effect shown by *T. atroviride* could be used to significantly reduce egg density and subsequently diminish the emergence of neonatal larvae in storage conditions. This inhibitory activity shown by *T. atroviride* has also been described by Razinger et al. (2014) treating *Delia radicum* L. (Diptera: Insecta) larvae. To date, no use of *T. citrinoviride* has been described to control insect pests, being its use limited to species of the Plantae kingdom (Mayo et al.

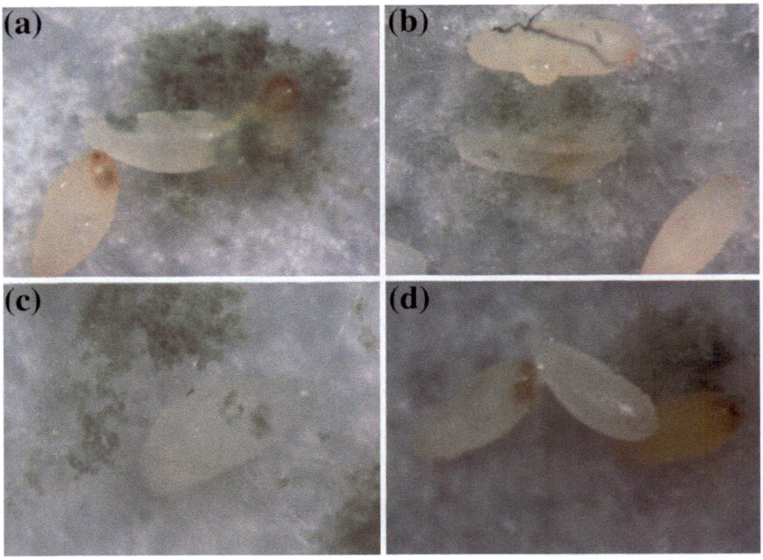

Fig. 1.3 Infection and sporulation of *T. harzianum* (**a**), *T. atroviride* (**b**), *T. citrinoviride* (**c**), *and T. longibrachiatum* (**d**) on bean weevil eggs (images from Rodríguez-González et al. 2017a)

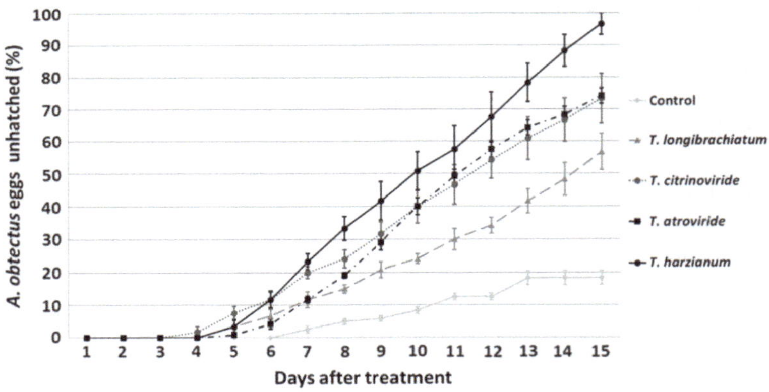

Fig. 1.4 Bean weevil eggs infected by *Trichoderma* species during 15 days after treatment. Upper and lower error bars are represented (image from Rodríguez-González et al. 2017a)

2015). Likewise, *T. longibrachiatum*, *T. citrinoviride*, and *T. longibrachiatum* have been used mainly on plants and have not been described as a biological control agents against insect pests, but for the control of *Leucinodes orbonalis* (Lepidoptera: Crambidae) larvae (Ghosh and Pal 2016).

The Fig. 1.4 shows the percentage of bean weevil eggs hatching during 15 days after they have been treated by the different *Trichoderma* species evaluated.

When these Trichoderma species were applied on bean weevil adults, the results showed that *T. citrinoviride* was able to control all adults evaluated. Furthermore, *T. longibrachiatum*, *T. harzianum*, and *T. atroviride* also showed a high performance, being able to control 98.3, 95.0, and 93.3% of adults evaluated, respectively (Da Silva 2017). *T. longibrachiatum* has also been used to control adult stages of insect pests such as *Bemisia tabaci* (Homoptera: Aleyrodidae) (Anwar et al. 2017) and *Leucinodes orbonalis* (Lepidoptera: Crambidae) (Ghosh and Pal 2016). As for *T. citrinoviride*, it has biological activity against the aphid *Rhopalosiphum padi* (Homoptera: Aphididae), an important pest of cereal crops (Ganassi et al. 2016). *T. atroviride*, on the other hand, proved to be useful against the cabbage root fly, *Delia radicum* (L.) (Diptera: Anthomyiidae) (Razinger et al. 2017). The cited species obtained high control rates in all cases.

Great attention is focused on developing these entomopathogenic fungal species as inundative biopesticides against insect and other arthropod pests (de Faria and Wraight 2007). Many authors highlight the multiple roles played by fungal entomopathogens as a promising potential for their indirect, multifaceted, and cost-effective use in sustainable agriculture (Jaber and Ownley 2018). For instance, they can be used as biofertilizers (Kabaluk and Ericsson 2007; Sasan and Bidochka 2012; Jaber and Enkerli 2016, 2017), as a vertically transmitted fungal endophytes (Quesada-Moraga et al. 2014; Lefort et al. 2016), and dual microbial control agents of plant diseases and arthropod pests (Vega et al. 2009; Ownley et al. 2010; Lacey et al. 2015). Several studies have shown that by inoculating *Trichoderma* on plants, insect behavioral changes occur due to plant emitted volatiles and, plant defensive responses are activated. Previous studies have shown changes on insect development and behavior by treating their plant hosts seeds with fungi. Akello and Sikora (2012) reported that inoculation of fungal isolates in bean seeds reduced the population of *Acyrthosiphon pisum* Harris (Homoptera: Aphididae) 33 fold compared to population growth observed in untreated samples. Menjivar-Barahona (2010) described the reduction of whitefly population in tomatoes inoculated with *T. atroviride*. More recently, Rodríguez-González et al. (2018), demonstrated that the application of different *Trichoderma* species (volatile producers and nonproducers) on beans changed the behavior of *A. obtectus* adults. Accordingly, a new line of research is opened for the control of insects by treating beans with *Trichoderma*. To date, the treatment of bean seeds with different *Trichoderma* spp. has been focused on the control of phytopathogenic fungi. This technique is easy, fast, and saves time and resources (Martínez et al. 2013).

In conclusion, these results show that the *Trichoderma* species evaluated against the bean weevil may be suitable for the control of this insect pest. *T. harzianum* shows good control activity against different *A. obtectus* stages. Meanwhile, *T. atroviride*, *T. citrinoviride*, and *T. longibrachiatum* exhibit high biological control activity only on adults. These fungi can be considered a highly effective tool for the control of this insect species.

1.2.5 *Other Crop-Specific Traits: Biological (Symbiotic) N Fixation*

Approximately 80% of Earth's atmosphere is nitrogen gas (N_2). Unfortunately, N_2 is unusable by most living organisms. All organisms use the ammonia (NH_3) form of nitrogen to manufacture amino acids, proteins, nucleic acids, and other nitrogen-containing components necessary for life. Biological nitrogen fixation is the process that changes inert N_2 into biologically useful NH_3. This process is mediated in nature only by N-fixing rhizobia bacteria (Sørensen and Sessitsch 2007). Other plants benefit from N-fixing bacteria when the bacteria die and release nitrogen to the environment. In legumes and a few other plants, the bacteria live in small growths on the roots called nodules. Within these nodules, nitrogen fixation is done by the bacteria, and the NH_3 they produce is absorbed by the plant.

The legumes provide a range of nutritional and agroecosystems services to the societies: as important sources of protein-rich food and feed, oil, fiber, minerals, and vitamins, improve soil fertility by contributing nitrogen through atmospheric N_2 fixation in symbiosis with rhizobia; improve soil structure and increase soil organic carbon status; reduce the incidence of pest and diseases in cropping systems; and increase the overall productivity and economic benefits of the production systems (Lupwayi et al. 2011). Legumes also contribute to mitigating the climate change effects by reducing fossil fuel use, ammonia fertilizer production or by providing feedstock for the emerging bio-based economies where fossil fuel sources of energy and industrial raw materials are replaced in part by sustainable and renewable biomass resources.

Thus, the legumes are key components of sustainable agriculture and their use in crop rotation leads to a reduction in agricultural CO_2 emissions and a decrease in nitrogen fertilizer application (Barton et al. 2014) for their capacity for nitrogen fixation. Interestingly, the nitrogen-fixing capacity of legumes is not a ubiquitous trait, with approximately 88% of described legumes showing this ability (Graham and Vance, 2003). The N_2 fixed by the legume crops represents a renewable source of nitrogen for agricultural soils. Globally, legumes in symbiosis with soil rhizobia are reported to fix 20–22 million tons of nitrogen each year in agricultural production systems (Herridge et al. 2008). Nitrogen fixation is achieved through symbiotic interactions with organisms in the soil microbiome, consisting of bacterial species, rhizobia, and arbuscular mycorrhizal fungi. The legumes are able to access atmospheric nitrogen fixed in the forms of ammonia (NH_3), nitrate (NO^{3-}) or ureides (Atkins 1987). This process of symbiotic nitrogen fixation is resulted from the complex interaction between the host plant and rhizobia. This mutualistic relationship is beneficial for both symbiotic partners; the host plant provides the rhizobia with carbon and source of energy for growth and functions while the rhizobia fix atmospheric N_2 and provide the plant with a source of reduced nitrogen in the form of ammonium. Thus, the process offers an economically attractive and ecologically sound mean of reducing external inputs and improving internal resources (Van Hameren et al. 2013).

Legume nodules are very complex organs, containing several interacting processes that operate at distinct levels, including, at least, nodule formation, carbon metabolism, oxygen supply, cellular redox, and transmembrane transport (Udvardi and Poole 2013). Nodule metabolism and regulation have been a topic of intensive research for quite a long time. Despite the enormous progress in this field, more research will still be required to provide a greater understanding of this fantastic process (Oldroyd and Dixon 2014). Pink or red nodules should predominate on a legume in the middle of the growing season. If white, gray, or green nodules predominate, little nitrogen fixation is occurring as a result of an inefficient rhizobia strain, poor plant nutrition, pod filling, or other plant stress. Factors like temperature and water availability may not be under the farmer's control, but nutrition stress (especially phosphorus, potassium, zinc, iron, molybdenum, and cobalt) can be corrected with fertilizers. An increase in soil concentration of nitrate can inhibit N_2 fixation quite severely. When nutritional stress is corrected, the legume responds directly to the nutrient and indirectly to the increased nitrogen nutrition resulting from enhanced nitrogen fixation. Poor nitrogen fixation in the field can be easily corrected by inoculation, fertilization, irrigation, or other management practices.

Plant breeding research in the 1980s and 1990s focused at combining high symbiotic nitrogen efficiency into improved genetic backgrounds in legumes, with some germplasm and breeding lines with high N_2 fixation being released. This becomes more important in the view of advances made in genomics of rhizobium and several model symbiotic nitrogen-fixing (SNF) legumes. Research on SNF suggests that several plant traits are associated with nitrogen fixation in grain legume crops, including nodule number and nodule weight, root and shoot weight, total biomass, and percent and total atmospheric N_2 fixed. An accurate estimation of the total atmospheric N_2 fixed and phenotyping of traits associated with nitrogen fixation is a prerequisite to detect genetic variation associated with nitrogen fixation in crop germplasm. Digital image analysis allows rapid and nondestructive phenotyping of various parameters after segmentation of an image and extraction of quantitative features from the segmented objects of interest (Hatem and Tan 2003). Gray et al. (2013) developed a minirhizotron imaging system as a novel in situ method for assessing the number, size, and distribution of nodules in field-grown soybean exposed to elevated atmospheric CO_2 and reduced precipitation. The performance of the symbiosis depends on the rhizobial attributes of competitiveness, infectiveness, and effectiveness. In the future, the success of SNF will depend on improving host plant, rhizobia, and environment system of the crop. Therefore, plant breeders should consider nitrogen fixation in the breeding programs as mandatory and a prerequisite for the future success of symbiosis.

The discovery of PCR-based DNA markers led to the construction of genetic linkage maps of varying intensity that has revolutionized the use of genomic-led approaches in applied crop breeding. Genetic research in the preceding paragraph clearly indicates that SNF is a complex trait and is possibly governed by various genes with varying effects, and dissecting its genetic basis may provide crop breeders more opportunities to harness marker (QTL)-trait association in crop improvement (Collard and Mackill 2008). A large number of specific genes influencing the legume–rhizobia interactions have been cloned or analyzed with forward and reverse genetics. Likewise, the sequence variations among rhizobium genomes may provide insights into the genetic basis of SNF. Several data are now available, through classical genetic experiments (screening of mutants, etc.) and whole genome sequences. However, no ultimate markers for the identification of the "best" strains can be defined, since the overall picture of gene interactions during the symbiotic processes is not fully understood, especially for those genes present in the dispensable genome fraction of rhizobial species. Consequently, more effort is needed toward the molecular characterization of gene functions and the modeling of genome–phenotype relationships. A large number of mutants with altered nodulation pattern (nod−, no nodulation; nod+/−, few nodules; fix−, ineffective nodulation; nod++, hypernodulation; nod++nts, hypernodulation even in the presence of otherwise inhibitory nitrate levels) have been reported in several grain legume crops (Bhatia et al. 2001).

Research showed that use of nodulation mutants has indeed contributed to the understanding of the genetic regulation of host–symbiotic interactions, and nodule development and nitrogen fixation (Sidorova et al. 2011). The use of DNA markers may, therefore, facilitate the identification of QTL associated with high SNF and their introgression into improved germplasm (Collard and Mackill 2008). Candidate genes associated with high nitrogen fixation have been identified in the genomes of common bean (Ramaekers et al. 2013), soybean (Schmutz et al. 2010), and model legume *M. truncatula* (Stanton-Geddes et al. 2013). Sequence variation of plant genes that determine the stability and effectiveness of symbiosis may be used for developing DNA markers that will facilitate breeding of legume cultivars with high symbiotic efficiency (Zhukov et al. 2010). The future of rhizobial biology is then directed toward the screening and collection of strains with interesting phenotypes and to link, under a systems biology view, such new or already known phenotypes with genomic information, providing genetic tools to screen and improve plant growth promoting performances of rhizobial strains.

1.3 Genetic Resources of CS Genes

1.3.1 Primary Gene Pool

Daryanto et al. (2015) reported that, among different grain legumes, common beans have among the greatest seed yield reductions in response to drought, with an estimated 70% of bean production areas affected by drought worldwide (Beebe et al. 2012). Middle American races, specifically Durango race bean lines, originating from higher altitude semiarid climatic zones and have the highest levels of drought tolerance (Singh 2001). Hybridization of Mesoamerican and Durango races has resulted in improvements in drought tolerance (Terán and Singh 2002; Frahm et al. 2004).

Limited sources of heat tolerance have been found in common bean, while most current production areas in Africa and Latin America are predicted to be unsuitable for bean production by 2100 (Ramirez-Cabral et al. 2016). The small-seeded Middle American race Mesoamerica has the highest levels of heat tolerance. Beebe et al. (2013) note that larger seeded Andean beans with determinate growth habits have little heat tolerance. There are a few exceptions such as G122 (collected in India), Sacramento (developed in California), and CELRK (developed in California) that have been selected under high-temperature production environments resulting in higher levels of heat tolerance. "Indeterminate Jamaica Red" (Román-Aviles and Beaver 2003), also originating from the same region as G122 in India, has among the highest levels of heat tolerance yet identified in Andean germplasm and has been used for introgression of this trait into different Andean seed classes, including indeterminate types, e.g., PR9920-171, and determinate types, TARS-HT1 and HT2. The indeterminate growth habit is a common type among Andean bean landraces collected in the Caribbean (Durán et al. 2005), while indeterminacy has been shown to be a source of yield stability under abiotic stress. Beebe et al. (2013) also noted that mid-season bean lines with indeterminate, prostrate habits tend to have better adaptation to intermittent drought. In addition, improved germplasm for drought often combines deep rooting and improved seed fill under stress. Although precise ideotypes for heat or drought have not been suggested, certain characteristics of the shoot and root architecture have been identified and associated with stress tolerance.

Seed size may be associated with abiotic stress tolerance in common bean with smaller seeded types associated with greater heat and drought tolerance. This association could be due to a number of causes including the Middle American geographic origin with inherent abiotic stress selection, reduced diversity in the domestication process (Beebe et al. 2001), shorter seed-filling period less exposed to intermittent stress, or indeterminate plant habit, among others. Beebe et al. (2013) noted that small-seeded beans in the tropics are often produced at lower altitudes where tolerance to both heat and drought are needed. More progress has been made in the development and release of small-seeded (small red, black, and white beans) with enhanced levels of heat and drought tolerance, while less effort has been dedicated to larger seeded Andean beans. Larger seeded beans generally have a lower relative growth rate (RGR), as compared to smaller seeded beans, which has been associated with lower biomass and yield. In Lima bean (*Phaseolus lunatus* L.) a similar relationship has been found in California production environments with Middle American sieva seed types having higher heat tolerance as compared to large-seeded Andean types (Long et al. 2014). As global temperatures rise, producers and consumers of Andean beans at higher altitudes may switch to smaller seeded beans to maintain productivity.

There may be limits for the genetic improvement of common bean for tolerance to drought and high temperature. It may be necessary to introgress genes for tolerance to abiotic stress from related species such as the tepary (*Phaseolus acutifolius* L.) or to consider shorter-season common beans, or altering planting dates to avoid peak periods of heat or drought. As the physiological and genetic basis of drought and heat tolerance is better understood, genome editing techniques may provide opportunities to enhance abiotic stress tolerance. For example, Baltes et al. (2017) inserted a promoter into maize (*Zea mays* L.) to increase the expression ARGOS genes (negative regulators of the ethylene response to drought and heat stress) that resulted in increased drought tolerance.

There may be greater use of irrigation to meet global demand for grain legumes; however, freshwater reserves are critically low in certain production zones and rainfall patterns are changing in others. Under these conditions, water use efficiency may gain importance as a criterion for selection by bean breeding programs. Beebe et al. (2013) noted that drought tolerance would be beneficial for irrigated production by reducing the amount of water required to produce the crop. In the tropics, bean production may move to higher altitudes where the risks of erosion and soil degradation are greater. Breeding for infertile soils or Al toxicity may need to be added to the list of breeding objectives since these conditions are more prevalent at higher altitudes.

1.3.2 Secondary Gene Pool

The scarlet runner bean (*Phaseolus coccineus* L.) is from the secondary gene pool and originates from high altitudes of Middle America. There are no reports of introgression of drought or heat tolerance from scarlet runner bean although it has been used extensively as a source of disease resistance (reviewed in Porch et al. 2013b) and recently to introgress tolerance to aluminum toxicity into common bean (Butare et al. 2011).

1.3.3 Tertiary Gene Pool

The tepary bean is recognized as having greater heat and drought tolerance than common beans (Federici et al. 1990; Teran and Singh 2002; Acosta-Gallegos et al. 2007). Rao et al. (2013) suggested that the tepary bean could be used as a model to improve drought tolerance of common beans. Beebe et al. (2009) reported that the tepary bean invests in early root growth, limited vegetative growth and efficient partitioning of photosynthates to the seed. Traub et al. (2017) noted that tepary beans have a slower increase in stomatal conductance in response to rainfall after a drought. They suggested that this would be advantageous to conserve water during periods of terminal drought. Souter et al. (2017) identified interspecific (*P. vulgaris* × *P. acutifolius*) lines that had superior performance in trials for drought and tolerance to low temperature.

Beebe et al. (2013) noted that Lima beans are very tolerant to heat and soil constraints. At present, it is not possible to introgress genes for traits such as heat tolerance from lima to common bean. Beebe et al. (2009) reported that crosses between common (*P. vulgaris*) and lima bean (*P. lunatus*) genotypes do not produce fertile hybrids. A better understanding of the physiological and genetic basis of abiotic stress tolerance in lima bean may lead to the identification of traits or breeding strategies that could be used to improve the abiotic stress tolerance of common bean.

In regions where high temperatures or drought stress are expected to become too extreme for common bean production, the commercial production of tepary or lima beans may become a viable alternative. Systematic plant breeding efforts to improve the tepary bean in the lowland tropics (Porch et al. 2013a) and heat tolerance of lima bean in temperate zones (Ernest et al. 2017) have been limited. However, lima bean and tepary bean may have appeal and potential for broader expansion in the Americas, the Caribbean, and Africa (Porch et al. 2013b) as production environments become increasingly marginal.

1.4 Glimpses on Classical Genetics and Traditional Breeding for CS Traits

1.4.1 Breeding Objectives

Rao et al. (2013) and Beebe et al. (2009) reported that globally almost 2/3 of the production areas planted in beans are vulnerable to drought. Singh (1995) noted that the degree and length of intermittent and terminal drought stress are associated with the reduction in common bean yield. Singh (2001) reported that daytime temperatures >30 °C and or nighttime temperatures >20 °C can limit bean production. In temperate bean production regions, a temporary heat wave during a critical period of reproductive development can reduce pod set and yield, especially for Andean beans such as snap beans with a determinate growth habit. Significantly greater yield reduction or complete crop failure would be expected with the occurrence of both heat and drought. Future climatic conditions in most bean production regions are expected to be warmer, drier, and more variable (Williams et al. 2007; McClean et al. 2011). Daryanto et al. (2015) concluded that the common bean could be the grain legume in greatest need of improved drought tolerance given its importance in world production and human nutrition.

As bean production expands in Central America into the tropical lowlands, heat tolerance has gained importance as a trait. Some bean diseases also become more important in higher-temperature environments. For example, bean cultivars lacking the dominant *I* and either the *bc-3* or *bc-1*^{2} recessive genes are susceptible to *Bean common mosaic necrosis virus* (BCMNV) at higher ambient temperatures (>30 °C) (Singh and Schwartz 2010). Resistance to BCMNV must be added to breeding objectives to lowland tropical bean breeding programs where this seed-borne virus is endemic or has the potential to emerge.

Plant breeders need to identify the most appropriate combination of traits needed for adaptation to specific geographic regions and/or cropping systems. It may be necessary to evaluate the performance of bean lines in dryer and or hotter environments, or a combination of both, than current production zones in order to successfully select for future environmental conditions. The USDA/ARS bean research program in Puerto Rico screens beans for drought on the southern coast of the island where conditions are much dryer than the regions where most beans are currently produced. In Honduras, the bean breeding program at Zamorano University screens beans for heat tolerance at a low altitude site near the Pacific coast that is considered too warm for bean production, but has resulted in the successful selection of improved cultivars for lowland production in Central America (Rosas et al. 2000). The USDA/ARS and University of Nebraska bean breeding programs have screened beans for drought in alternate generations in Puerto Rico and Nebraska. By conducting two field screenings each year, the development of improved bean breeding lines is accelerated. The screening of bean lines in contrasting environments may lead to more robust drought tolerance in breeding lines.

In some regions, beans are exposed to intermittent periods of drought whereas, in other regions, beans are more likely to suffer from terminal drought toward the end of the growing season (Omae et al. 2012). Breeding for these different types of drought will require the selection of distinct sets of traits. The sensitivity of bean to abiotic stress during reproductive development makes intermittent heat or drought during this period, or terminal drought, critically important. Beebe et al. (2009) noted that breeding for abiotic traits such as tolerance to drought and higher temperatures is challenging due to low genetic variability for these traits and the importance of genotype × environment interaction.

Plant breeders recognize that there may be trade-offs when focusing on the selection of specific traits related to greater drought or heat tolerance. Purcell (2009) noted that biomass accumulation in plants is directly linked with water loss due to transpiration. Traits such as deep rooting that increase the amount of water available for transpiration generally benefit crop growth and yield under drought conditions. Although deep and healthy root systems are considered desirable for all types of drought, it has been reported that shallow roots are more efficient in the absorption of nutrients such as P from the soil (Ho et al. 2005). Beebe et al. (2009) reported that deeper and more dense roots do not insure higher yields under drought conditions. In fact, too much investment in root growth may lead to lower seed yield potential. Blum (2009) argued that effective water use, involving improved harvesting of soil water by the plant and efficient use of that water in transpiration and biomass production would ensure continued yield gain, while selection for WUE and TUE (transpiration use efficiency) tends to result in drought resistance, but yield loss. Beebe et al. (2009) suggested that more efficient root systems that require less biomass accumulation may contribute to greater seed yield potential under drought stress.

Traub et al. (2017) suggested the use of lower stomatal conductance as a criterion for selection for drought tolerance. Greater stomatal conductance and lower leaf temperatures were associated with deeper rooting in beans under drought although no yield advantage was reported (Beebe et al. 2009). Taub et al. (2017) reported that the common bean line SER 16 and the tepary bean line TB1 had lower stomatal conductance under both drought and well-watered conditions. A trade-off of slightly lower net photosynthesis under nonstress conditions allowed for better performance under drought conditions and led to greater water use efficiency. Beebe (2012) noted that a more rapid stomatal recovery may be advantageous under conditions of intermittent drought, while Ramirez Builes et al. (2011) found stomatal response associated with yield under drought in the greenhouse and field. Pimentel et al. (1999) proposed the integration of a calculated photosynthetic rate based on the stomatal conductance through the use of intrinsic water use efficiency (WUE) that would allow for selection of efficient genotypes during key developmental stages. Instantaneous measurements of leaf temperatures using high-throughput phenotyping platforms have been used to estimate stomatal response to drought (Andrade-Sanchez et al. 2014), and instantaneous normalized difference vegetation index (NDVI) measurements have been correlated with yield in bean (Sankaran et al. 2018), and could potentially be used for selection of stress-responsive lines. Carbon isotope discrimination (CID), on the other hand, provides a cumulative assessment of WUE over the whole season and can be readily evaluated on the harvested seed. Due to its inverse relationship with WUE, selection for low CID has been recommended (Easlon et al. 2014).

Omae et al. (2012) reported that higher leaf water content was associated with greater drought and heat tolerance. Traub et al. (2017), however, noted that plants must balance the influx of CO_2 and the loss of water. Hoyos-Villegas et al. (2017) indicated that water conservation may limit photosynthesis and the growth and development of plants. In response to severe drought stress, bean plants may produce smaller or fewer leaves that can result in suboptimal leaf area and reduced net photosynthesis. Schneider et al. (1997) suggested selection for increased biomass under drought stress to avoid the reduction in photosynthetic capacity. This practice, however, may result in indirect selection for later maturity.

Daryanto et al. (2015) noted that osmotic regulation through increased solute concentration is less energy demanding than stomatal conductance and allows the roots to extract water at lower soil water potentials. A plant breeding challenge is the selection of an appropriate balance between water conservation and net photosynthesis for a specific cropping system? True tolerance is the ability of plants to withstand drought conditions by having low internal water potential. This trait, however, has limited utility since drought tolerance is more important for survival and is often associated with slow rates of growth and low productivity (Passioura 2012).

Beebe et al. (2013) noted that mechanisms to escape drought include early maturity, phenotypic plasticity, and rapid partitioning of photosynthate to seed. Selection for earlier maturity may help to avoid terminal drought but earliness may reduce seed yield potential during more favorable growing seasons. An early, defined, and un-reversible shift to reproductive development and a shorter period of pod filling could reduce the exposure during the sensitive reproductive period of development and shorten the growing season, thus increasing the chances of escape. In the highlands of Mexico, some bean genotypes use phenotypic plasticity adapt to intermittent periods of drought by delaying flowering until more humid conditions return (Acosta-Gallegos et al. 1989; Acosta-Gallegos and White 1995). Indeterminacy is often key for plasticity, providing for reproductive organ abscission as a result of abiotic stress and reflowering at new flower nodes, thus allowing for a degree of avoidance of short-term dry or hot conditions. However, split-sets in snap beans and late maturity in dry beans can result.

Selection for a greater harvest index has been a successful strategy to increase the yield potential of many crops (Unkovich et al. 2010). Foster et al. (1995) reported that greater partitioning or higher harvest index contributed to terminal drought tolerance. Cuellar-Ortíz et al. (2008) reported that carbohydrate partitioning toward seed fill is a useful drought tolerance strategy. Beebe et al. (2013) noted that accelerated partitioning of photosynthates toward reproductive development contributed to better adaptation and seed yield under both terminal and intermittent drought. Beebe et al. (2009) reported that pod harvest index (grain as percent of total pod biomass) to be consistently associated with seed yield under drought stress. Blum (2005) noted that selection for greater yield potential can contribute to better performance in environments with moderate levels of drought stress. Recent research has shown variability for pod harvest index, or the extent of dry matter translocation from the pod wall to the seed, and its effectiveness as a trait for breeding (Polania et al. 2016). Some tepary bean germplasm has shown efficiency for pod harvest index, short reproductive period and high harvest index, likely key abiotic stress tolerance mechanisms in the tepary bean ideotype that include a thick taproot, prostrate habit, small phototropic leaves, high pod number, and small seed size.

Screening for heat tolerance in the tropics is more predictable than selection for drought tolerance. Evaluations can be conducted at lower altitudes to ensure higher temperatures. On the other hand, screening may need to be conducted under controlled conditions for response to heat waves during critical stages of development that can occur in temperate regions. Porch (2006) noted that temperatures of >30 °C during the day or >20° at night result in the reduction of seed yields of most common beans. In the evaluation of response of common bean to high temperatures, Porch (2006) found that geometric mean (GM) and the stress tolerance index (STI), as described by (Fernández 1992), to be effective in the identification of lines with superior yields in stress and nonstress trials.

1.4.2 Classical Breeding Achievements

Terán and Singh (2002) noted that considerable progress has been made in breeding beans with greater adaptation to both intermittent and terminal drought. Most progress has been made in the selection of drought tolerance of bean races Durango and Mesoamerica. Crosses between Durango and Mesoamerican races have produced progenies with superior performance under drought, for example, SEA 5 (Terán and Singh 2002) and L88-63 (Frahm et al. 2004). Much less research and genetic progress have been made improving the drought tolerance of other bean races, especially beans of Andean origin. Limited gains in breeding for Andean abiotic stress tolerance may be a result of reduced efforts and to more limited genetic diversity in this gene pool (McClean et al. 2011). Bean germplasm or cultivars reported to have drought or heat tolerance are listed in Table 1.1.

Polania et al. (2016) measured seed yield, canopy biomass, stomatal conductance, and carbon isotope discrimination to evaluate the response of common bean lines to drought. The authors identified lines such as SER 16, ALB 60, ALB 6, BFS 10, BFS 29 that conserve water through lower rates of transpiration, moderate rates of growth,

Table 1.1 Release of bean germplasm and cultivars reported to have heat or drought tolerance

Identity	Seed type	Location (year of release)	Type of tolerance	Citations
Bella	White	Puerto Rico (2017)	Heat	Beaver et al. (2018). J. Plant Reg. 12:190–193
Verdín	Black	Mexico (2016)	Terminal drought	Tosquy Valle (2016). Rev. Mex. Cien. Agríc. 7:1775–1780.
DAB-53	Large-seeded red	CIAT germplasm	Andean bean with terminal drought tolerance	Mayor-Duran et al. (2016). Acta Agron. 65:431–437
CENTA EAC	Small red	El Salvador (2015)	Heat	Parada Cardona et al. (2015). CENTA 7 p.
SER 16	Small red	CIAT germplasm	Terminal drought tolerance	
TARS-LFR1	Small red	Puerto Rico (2014)	Heat	Porch et al. (2013a). J. Plant Reg. 8:177–182
INTA Sequía Precoz	Small red	Nicaragua (2013)	Terminal drought	
TARS-MST1	Black	Puerto Rico (2012)	Heat and drought	Porch et al. (2012). J. Plant Reg. 6:75–80
CENTA Pipil	Small red	El Salvador (2013)	Heat	
PR0401-259	Pink	Puerto Rico (2012)	Heat	Beaver et al. (2012). J. Plant Reg. 6:81–84.
TARS-HT1 and TARS-HT2	Dark and light red kidney	Puerto Rico (2010)	Andean beans with heat tolerance	Porch et al. (2010). HortSci. 45:1278–1280
Verano	White	Puerto Rico (2008)	Heat	Beaver et al. (2008). J. Plant Reg. 2:187–189.

(continued)

Table 1.1 (continued)

Identity	Seed type	Location (year of release)	Type of tolerance	Citations
Cornell 503	Snap bean	New York (2005)	Heat	Rainey and Griffiths (2005). J. Am. Soc. Hort. Sci. 130:700–706.
Amadeus 77	Small red	Honduras (2004)	Heat	Rosas et al. (2004). Crop Sci. 44:1867–1868
Indeterminate Jamaica Red	Striped pink kidney	Germplasm Landrace	Andean bean with heat tolerance	Román-Aviles and J. Beaver (2003). J. Agric. Univ. P. R. 87, 113–121.
UI-239	Small red	Idaho (1997)	Terminal drought	Singh (2007). Agron. J. 99:1219–1225
Pinto Villa	Pinto	Mexico (1995)	Phenotypic plasticity to intermittent drought	Acosta-Gallegos et al. (1995). Crop Sci. 35:1211

and more efficient partitioning of photosynthates. Other groups of bean lines such as NCB 280, NCB 226, SEN 56, SCR 2, SCR 16, SMC 141, RCB 593, and BFS 67 were able to avoid drought by having deep roots and more efficient use of available water by combining early maturity and greater harvest indices. Polania et al. (2016) noted that the former group would be most useful in environments that are prone to severe drought. The latter group would be more suited for intermittent drought and soils that have a greater water-holding capacity.

1.4.3 Limitations of Traditional Breeding and Rationale for Molecular Breeding

Heat and drought tolerance are quantitative traits that require the evaluation of large numbers of later-generation breeding lines. These evaluations should be conducted in several environments using replicated trials to obtain reliable estimates of the performance of the lines. Singh (2001) reported that seed yield remains the most reliable trait to evaluate the performance of common bean under drought stress. Beebe et al. (2013) recommended that sites for screening for drought tolerance should have uniform soil and management practices that are representative of the target production zone. In general, more replications are needed if the lines are to be screened under high levels of drought stress. Instead of screening bean lines at sites with multiple biotic and abiotic constraints, Beebe et al. (2013) recommended the sequential screening of bean lines for individual traits such as drought. The evaluation of advanced generation lines allows the simultaneous evaluation of several traits at different locations. Regional performance trials can be used to evaluate the performance over a wide range of environments.

Screening for drought tolerance is often conducted by comparing the performance of bean lines in drought and nonstress trials. Selection criteria include geometric mean of the seed yields from drought and nonstress trials, percent reduction in seed yield in relation to the nonstress environment, and drought susceptibility indices (Terán and Singh 2002). Schneider et al. (1997) noted that geometric means allows the identification of lines that perform well under drought and nonstress conditions. This should be followed in a breeding program by evaluating the seed yield under drought to confirm the performance of the selections under stress. These conventional plant breeding practices are costly and time-consuming.

Beebe et al. (2013) noted that yield loss depends on the timing, duration, and severity of the drought. The authors also noted that bean root growth and development is sensitive to soil compaction and low soil fertility. A better understanding of interactions among edaphic conditions, soil management practices, and the physiology of traits associated with drought tolerance should lead to the development of robust molecular markers.

Briñez et al. (2017) noted that the response of beans to drought tolerance is a complex quantitative trait controlled by many minor QTLs. Due to the importance of genotype × environmental interaction, Briñez et al. (2017) noted that the stability of QTL for drought tolerance needs to be confirmed across populations and a wide range of environments in which the type and severity of drought may occur. Due to the variable nature of rainfall patterns across years and locations and the importance of genotype × environment interaction, Beebe et al. (2013) pointed out the need to validate in the field the drought tolerance of bean lines selected using marker-assisted selection.

Purcell (2009) noted that a major limitation for the improvement of quantitative traits such as drought is the difficulty in phenotyping plants. At present, rapid and simple methods of evaluating phenotypes for quantitative traits such as drought and BNF are not available. Meta-analyses using data from different trials have been used to compare the response of beans to drought (Daryanto et al. 2005). The use of drones (Sankaran et al. 2015, 2018) or proximal sensing carts (White and Conley 2012) allow for the collection of field data from a large number of lines in a short period of time for traits such as leaf canopy temperature, NDVI, and normalized difference red edge (NDRE) index. Rapid, high-throughput phenotyping allows for a more representative comparison of bean lines for traits at a particular time and thus stress condition or at different times of the day (Andrade-Sanchez et al. 2014).

Purcell (2009) noted that for most legumes BNF is more sensitive to drought stress than photosynthesis, although both decrease with stress. Castellanos et al. (1996) reported that drought stress significantly decreased biological nitrogen fixation of common bean. High temperature also inhibits BNF (Hungria and Kaschuk 2014) in common bean. This represents a significant challenge to breeding for environments with multiple climatic and edaphic constraints. However, high-temperature tolerant BNF capacity is another trait that could be introduced from tepary bean, where a range of nitrogen fixation capacity has been identified (Vargas 2016).

In the tropics, charcoal rot caused by *Macrophomina phaseoli* tends to be more severe in drought conditions, but occurs frequently under high-temperature humid conditions. In more temperate climates, root rots caused by *Fusarium* spp. are more common during periods of low rainfall. Beebe et al. (2013) note that resistance to root rots is an important trait for beans produced in areas where drought stress is common. It is also important in high-temperature environments where there are higher rates of transpiration.

Dry and warm climatic conditions favor some pests such as leafhoppers (*Empoasca* spp), aphids (*Aphis* spp.) and whiteflies (*Bemisia* spp.). Resistance to leafhoppers is an especially important trait for beans cultivars growing under these conditions. Likewise, resistance to viral diseases vectored by one these pests, such as BGYMV, BCMNV, BCMV, and CTMV, may need to be included as breeding objectives.

Screening for local adaptation, seed size, and commercial seed type and other highly heritable traits can be conducted in earlier generations. There are numerous molecular markers available for major genes for resistance to specific diseases, and some pests, that could be used for marker-assisted selection in earlier generations (Miklas et al. 2006a, b).

In recognition of the difficulty to improve drought and heat tolerance of Andean beans, CIAT bean breeders have developed Durango race bean breeding lines that have seed types that mirror Andean seed types. This approach would allow breeders to take advantage of superior levels of drought tolerance found in the Durango race, while introducing biotic stress tolerance to regions where mostly Andean races of pathogens currently exist, e.g., sub-Saharan Africa.

Several studies have been conducted to identify QTLs associated with drought tolerance. Briñez et al. (2017) evaluated a RIL population from the cross "SEA 5 × AND 277" and reported that the drought-tolerant line SEA 5 had lower leaf temperature under drought conditions than AND 277. These results suggested that SEA 5 had a greater rate of transpiration than AND 277 in the presence of drought stress. All of the QTLs associated with drought were from SEA 5 including a QTL for seed weight under normal and drought conditions. The authors noted that greater seed weight may suggest better seed fill under drought. Hoyos-Villegas et al. (2017) identified significant QTL associated with drought tolerance that may be useful for MAS for this trait.

Marker-assisted selection of major QTLs associated with heat and drought tolerance in earlier generations would help reduce the number of breeding lines that would need to be screened in later generations. Gamete selection suggested by Singh (1994) may be useful to accumulate alleles for drought tolerance when robust molecular markers for this trait have been identified. Lines harboring key QTL for abiotic stress tolerance could be selected in the F_1 from double crosses, thus accelerating the pyramiding of key regions of interest. Large F_1 populations would be required necessitating many crosses, but critical QTL often affected by G × E could be combined. Beebe et al. (2013) and Hinkossa et al. (2013) noted that recurrent selection is an appropriate breeding approach for quantitative traits such as tolerance to drought and heat. Recurrent selection also provides for a gradual accumulation or pyramiding of key regions for quantitative traits through successive recombination of superior breeding lines.

A gene-based crop model has been developed that predicts vegetative and reproductive development based on genotype and weather data (Hwang et al. 2017). The development of more sophisticated models may facilitate the study of the interaction of traits related to drought tolerance with varying weather patterns and crop management practices.

1.5 Diversity Analysis

Occasional outcrossing, adaptation to particular environments (in terms of temperature, moisture, photoperiod, soil fertility, diseases, and insects), different cropping systems and strong selection for consumer preferences addressed to particular seed types, might have played a significant role in the evolution of new genetic variation in common bean. As a consequence, each country selected its own set of landraces able to respond to the needs and preferences of local populations. The common bean populations were involved in new evolutionary pathways that were not possible in the American center of origin, due to the spatial isolation between these two gene pools. Thus, new germplasm could have arisen from recombination events between Mesoamerican and Andean gene pools, better adapted to the conditions of the new agrosystems out of The Americas. Evidence of this phenomenon has been detected using phaseolins, allozymes, and morphological data (Santalla et al. 2002; Rodiño et al. 2006), and inter-simple sequence repeats (ISSRs) and simple sequence repeats (SSRs) from both the chloroplast and nuclear genomes (Sicard et al. 2005; Angioi et al. 2009). Gene flow between both gene pools appears to be relatively common in the Andean (Debouck et al. 1989; Beebe et al. 1997; Chacón et al. 2005) and European zones (Santalla et al. 2002; Sicard et al. 2005; Piergiovanni et al. 2006; Rodiño et al. 2006; Sánchez et al. 2008).

To date, considerable efforts have been made toward DNA polymorphisms discovery in common bean. Several thousand single-nucleotide polymorphisms (SNPs) and insertions–deletions (InDels) have been discovered through expressed sequence tags data mining and partial resequencing of several genotypes (Hyten et al. 2010; Souza et al. 2012; Felicetti et al. 2012; Blair et al. 2013; Goretti et al. 2014; Zou et al. 2014). At the transcriptional level, expressed sequence tags (ESTs), sequencing has been used to discover and identify genes differentially expressed under different conditions. Whole genome transcriptome analysis is also an effective way to exploit key factors for common bean responses to biotic and abiotic stress that are involved in transcriptional and metabolic activities. The data obtained from these technologies will serve as an invaluable genomic reference to further our knowledge about the common bean at the molecular level and can be applied to molecular breeding for plants with enhanced biotic and abiotic tolerance.

The genome of an Andean common bean genotype (G19833) was sequenced and recently released (Schmutz et al. 2014). A combination of Sanger, 454, and Illumina HiSeq 2000 reads and a genetic map based on 7015 SNP markers were used to assemble the common bean reference genome sequence (Schmutz et al. 2014), with a total genome size of 521 Mb that represents 89% of the 587 Mb bean genome. Also, a first draft of the entire common bean genome sequence of a Mesoamerican genotype (BAT93) was also developed by Vlasoba et al. (2016).

1.6 Molecular Mapping of CS Genes and QTLs

1.6.1 A Brief History of Mapping Efforts

Linkage maps are important genetic tools for common bean improvement and other biological approaches. These maps have been used in several types of studies, including cloning of agronomically important genes, marker-assisted selection (MAS), comparative mapping, and analysis of germplasm diversity (Gepts 1999). Accordingly, several linkage maps have been developed in common bean (Table 1.2), and they differ in several characteristics, such as the types of parents and segregating population used, the type of markers and traits segregating in each population, the total map length and the degree of genome saturation. However, a common feature among the first maps is that they were generated based on low-throughput markers, resulting in low-density maps. Therefore, to increase the precision of bean maps, researchers have exerted much effort in generating new genomic-based tools that are supported by bioinformatics. Different projects, such as the Phaseomics international consortium and the BeanCAP project (USDA Common Bean Coordinated Agricultural Project), were developed to establish the necessary framework of knowledge and materials for the advancement of bean genomics, transcriptomics, and proteomics (reviewed by Gepts et al. 2008; Hyten et al. 2010). As a result, genome sequencing and high-throughput genotyping approaches are enabling the development of high-density functional maps that assist in accelerating bean genetic improvement through MAS.

1.6.2 Evolution on Marker Types

Common bean genetic maps have evolved in parallel with the development of molecular marker technologies. Linkage maps were once based on phenotypic markers (Lamprecht 1961), though molecular markers greatly increase the number of poly-

Table 1.2 Molecular linkage maps in common bean

Parents	Map size (cM)	Markers/traits mapped[a]	References
XR235-1-1/Calima (BC_1)	960	224 RFLPs, 9 seed proteins, 9 isozymes, *P*	Vallejos et al. (1992)
BAT 93/Jalo EEP558 (F_2)	1226	194 RFLP, 24 RAPDs, 15 SSR/ ALS, ANT, CBB, *V*, *C*, rhizobium	Nodari et al. (1993), Gepts (1999), Yu et al. (2000a, b)
Corel/Ms8EO2 (BC_1)	567.5	51 RFLP, 100 RAPD, 2 SCAR/ANT	Adam-Blondon et al. (1994)
Midas/G 12873 (RIL)	1,111	77 RFLPs, 5 isozymes/domestication traits	Koinange et al. (1996)
DOR364/XAN176 (RIL)	930	147 RAPDs, 2 SCARs, 1 ISSR/ ASB, BGYMV, CBB, *R*, *V*, *Asp*, rust	Miklas et al. (1996, 1998, 2000)
BAC6/HT7719 (RIL)	545	75 RAPDs/CBB, WB, rust	Jung et al. (1996)
PC50/XAN159 (RIL)	426	168 RAPDs/ CBB, *C*, *V*, rust, WM	Jung et al. (1997), Park et al. (2001)
BAT 93/Jalo EEP558 (RIL)	1226	120 RFLP, 430 RAPD, 5 isozymes/ BCMV	Freyre et al. (1998)
BelNeb-RR-1/A55 (RIL)	755	172 RAPDs, 2 SCARs/BBS, HB, BCMV	Ariyarathne et al. (1999), Fourie et al. (2004)
Eagle/Puebla152 (RIL)	825	361 RAPDs/ RR	Vallejos et al. (2001)
Jamapa/Calima (RIL)	950	155 RAPDs, 88 RFLPs/RGA	Vallejos et al. (2001)
OACSeaforth/OAC 95-4 (RIL)	1,717	49 AFLPs, 43 RFLPs, 11 SSRs, 9 RAPDs, 1 SCAR/ CBB, agronomic traits	Tar'an et al. (2001, 2002)
CDRK/Yolano (RIL)	862	196 AFLPs, 8 RFLP/SY, *C*	Johnson and Gepts (2002)
DOR364/G19833 (RIL)	1,720	78 SSR, 48 RFLPs, 102 RAPDs, 18 AFLPs	Blair et al. (2003)
ICA Cerinza/G24404 (RIL)	869,5	80 SSRs, 1 SCAR/ *C*, *fin*, *st*, agronomic traits	Blair et al. (2006a, b)
G14519/G4825 (RIL)	915.4	46 RAPDs, 68 SSRs/seed Fe and Zn concentrations and contents	Blair et al. (2010)
BAT 93/Jalo EEP558 (RIL)	1,545	199 gene-based, 59 core and 17 other markers	Hanai et al. (2010), McConnell et al. (2010)

(continued)

Table 1.2 (continued)

Parents	Map size (cM)	Markers/traits mapped[a]	References
DOR364/BAT477 (RIL)	2,041	1,060 (SSR, EST-SSR, BES-SSR, gene-based markers)/SW, Y, DF, DM	Blair et al. (2012), Galeano et al. (2011, 2012)
IAC-UNA/CAL143 (RIL)	1,865.9	198 SSRs, 8 STS-DArT, 3 SCAR/ALS	Oblessuc et al. (2012, 2013)
SEA5/CAL96 (RIL)	1,351	2,122 SNPs/SW, Y	Mukeshimana et al. (2014)
Stampede/Red Hawk (RIL)		7,276 SSRs and SNPs	Schmutz et al. (2014)

[a]*ALS* angular leaf spot, *BCMV* bean common mosaic virus, *CBB* common bacterial blight, *HB* halo blight, *RR* root rot, *WM* white mold, *SW* seed weight, *SY* seed yield, *DF* days to flowering, *DM* days to maturity, *Y* yield, *fin* determinacy, *Ppd* gene for photoperiod sensitivity, *V* flower color, *C* seed color

morphic loci in mapping populations. Thus, the first maps were developed based on restriction fragment length polymorphism (RFLP) markers, a technique that involves DNA hybridization. Later, new markers based on polymerase chain reaction (PCR) were used for genetic mapping, including random amplified polymorphic DNA (RAPD) (Williams et al. 1990), simple sequence repeats (SSRs) (Tautz 1989), amplified fragment length polymorphisms (AFLPs) (Vos et al. 1995) and inter-simple sequence repeats (ISSRs) (Zietkiewicz et al. 1994).

Due to their great robustness and repeatability, RFLP markers have allowed the development of the first DNA-based genetic maps in common bean (Vallejos et al. 1992; Nodari et al. 1993); these markers have also been used to compare and integrate different genetic maps (Adam-Blondon et al. 1994; Koinange et al. 1996; Freyre et al. 1998; Gepts 1999; Yu et al. 2000a, b). In addition, PCR-based molecular markers have been employed for saturating RFLP maps and for generating new ones using additional mapping populations (Freyre et al. 1998; Ariyarathne et al. 1999; Yu et al. 2000a, b; Blair et al. 2003, 2010; Fourie et al. 2004). For example, the first RFLP-based genetic map was constructed with 224 RFLP marker loci; the seed and flower color marker *P*, nine seed proteins, and nine isozyme markers were also included (Vallejos et al. 1992). These markers were distributed into 11 linkage groups (LGs) spanning 960 cM of the common bean genome. A second RFLP-based genetic map was developed by Nodari et al. (1993). This map included 108 RFLPs, seven RAPDs, seven isozymes and 18 loci corresponding to 15 known genes, the *I* gene for bean common mosaic virus (BCMV) resistance, a flower color gene, and a seed color pattern gene; these loci are spread among 15 LGs covering 827 cM of the bean genome, with an average interval of 6.5 cM between markers. A third map constructed by Adam-Blondon et al. (1994) included 157 markers: 51 RFLPs, 100 RAPDs, 2 SCARs (sequence characterized amplified regions), and four morphological markers that covered 567.5 cM of the bean genome. Moreover, Adam-Blondon et al. (1994) established a preliminary correspondence with the map developed by Vallejos et al. (1992) because 19 RFLP markers were shared between these maps.

The first core linkage map of common bean was constructed by Freyre et al. (1998) on the basis of the shared RFLP markers among these previous maps (Vallejos et al. 1992; Nodari et al. 1993; Adam-Blondon et al. 1994). The Freyre et al. map involved 563 markers, including 120 RFLPs and 430 RAPDs, in addition to a few isozymes and phenotypic marker loci; the markers were grouped into 11 LGs spanning 1226 cM. In successive years, RFLP markers were replaced by SSR markers, which are highly polymorphic PCR-based markers, for anchoring of different genetic maps. Yu et al. (2000a, b) published the first successful assignment of 15 SSRs to a framework map based on RAPD and RFLP markers. Moreover, with the availability of common bean EST (expressed sequence tag) sequencing programs, several functional markers, which are specifically developed from coding genomic regions, were identified and incorporated into bean linkage maps. The linkage map produced by Blair et al. (2003) was the first to incorporate SSR markers developed from EST databases, integrating these markers into a base map comprising 246 loci (78 SSR, 48 RFLP, 102 RAPD, and 18 AFLP markers) spanning 1720 cM. Indeed, EST libraries have become an important source of gene-based markers, such as EST-SSRs, single-nucleotide polymorphisms (SNPs) and insertions/deletions (InDels), which are valuable markers because they represent transcribed sequences that can be associated with phenotypic characteristics (Hanai et al. 2010; Galeano et al. 2012; Oblessuc et al. 2012). Furthermore, because EST-based markers are highly conserved between species, they allow for synteny comparisons between the common bean genome and those of other species (McConnell et al. 2010).

Additionally, with the advent of next-generation sequencing (NGS) technology, the sequencing of complete plant genomes has become increasingly more accessible and routine. The whole genome of common bean has recently been sequenced, and the complete genomes of the Mesoamerican and Andean beans BAT93 and G19833 are also available (Schmutz et al. 2014; Vlasova et al. 2016). In general, whole genome sequence availability accelerates the development of markers for high-throughput genotyping in plant breeding and genetic studies promoting the identification of markers tightly linked to agronomically important traits (Moghaddam et al. 2014; Mukeshimana et al. 2014; Meziadi et al. 2016; Valentini et al. 2017).

1.6.3 Mapping Populations Used

As shown in Table 1.2, several segregating populations are employed for mapping in common bean. Considering that many different economic traits of interest have been considered in bean breeding programs, divergent parents were chosen in each case to maximize phenotypic variation and genetic polymorphism. Moreover, in most cases, the parents chosen belonged to different gene pools, as experiments have shown that polymorphism among genotypes markedly increases in that situation (Nodari et al.

1993; Haley et al. 1994). For example, the mapping population used by Vallejos et al. (1992) to develop the first linkage map consisted of backcross progeny (BC_1) between the Mesoamerican line XR-235-1-1 and the Andean cultivar Calima (XC). Adam-Blondon et al. (1994) also utilized a BC_1 population derived from an inter-gene pool cross between two European bean genotypes: Ms8EO2 and Corel (MsCo). In contrast, Nodari et al. (1993) applied an F_2 population derived from a cross between the Mesoamerican line BAT 93 and the Andean cultivar Jalo EEP558 (BJ).

In addition, recombinant inbred line (RIL) populations, which are derived from single-seed descent from F_2 individuals, have been widely used in bean mapping because of their advantages (Table 1.2). For example, the BJ F2 mapping population was advanced to an RIL for the generation of the first core linkage map of common bean (Freyre et al. 1998), which was later improved (McConnell et al. 2010; Hanai et al. 2010). Furthermore, the base map developed by Blair et al. (2003) using SSR markers was produced using an RIL from the cross between the Mesoamerican variety DOR364 and the Andean landrace G19833 (DG). Similarly, numerous RIL populations were developed during the following years and used for bean genetic mapping studies and QTL identification (Blair et al. 2006b, 2010; Hanai et al. 2010; Oblessuc et al. 2012; Mukeshimana et al. 2014). Overall, the RIL populations derived from BJ and DG inter-gene pool crosses have been widely employed for genetic mapping studies because they are considered core mapping populations (Freyre et al. 1998; Blair et al. 2003, 2006a; Galeano et al. 2009, 2011, 2012; McConnell et al. 2010; Hanai et al. 2010).

1.6.4 Enumeration of Simply Inherited CS Trait and CS QTL Mapping

1.6.4.1 Disease Resistance

Fungal Diseases

Resistance to anthracnose (ANT), caused by *Colletotrichum lindemuthianum* (Sacc. and Magnus) Briosi and Cavara, is conferred by single, independent genes named and mapped to date (Table 1.3). Most of these genes are identified with the *Co* symbol: *Co-1* with four alleles; *Co-2* and *Co-3* with four alleles; *Co-4* with two alleles; *Co-5* with one allele; *Co-6*, *Co-11*, *Co-12*, *Co-13*, *Co-14*, *Co-15*, *Co-16*, and *Co-17*; and a new genes provisionally named *Co-Pa* and *Co-AC* (Kelly and Vallejo 2004; Gonçalves-Vidigal et al. 2006, 2007, 2009, 2011, 2012, 2013, 2016; Alzate-Marin 2007; Rodrigues-Suarez et al. 2008; Campa et al. 2011; Sousa et al. 2014; Lacanallo and Gonçalves-Vidigal 2015; Trabanco et al. 2015; Lima Castro et al. 2017; Gilio et al. 2017). An additional allele of *Co-1*, provisionally named *Co-1^{HY}*, was published in 2017 (Chen et al. 2017). Other genes with the *Co* symbol include *Co-x*, *Co-w*, *Co-y*, *Co-z*, *Co-u*, *CoPv02*, *Co-v* (*Co-6*), and *CoPv09c* as well as a QTL named

PMBO225 (Geffroy 1997; Geffroy et al. 2008; Richards et al. 2014; Campa et al. 2014). However, some previously known single, independent genes were renamed based on new allelism tests: *Co-7* as $Co\text{-}3^5$, *Co-9* as $Co\text{-}3^3$, *Co-10* as $Co\text{-}3^4$, and *Co-6* as *Co-v* (Geffroy 1997; Geffroy et al. 2008; Sousa et al. 2014; Gonçalves-Vidigal et al. 2006; 2013; Richards et al. 2014). Eleven genes (*Co-1*, *Co-12*, *Co-13*, *Co-14*, *Co-15*, *Co-Pa*, *Co-AC*, *Co-x*, *Co-w*, *Co-y*, and *Co-z*) belong to the Andean gene pool; the other 15 genes belong to the Mesoamerican gene pool. Chromosomes containing clusters of ANT resistance genes (shown in parenthesis) include Pv-01, (*Co-14*, *Co-Pa*, *Co-x*, *Co-AC*, and *Co-w*), Pv-02 (*Co*-u and *CoPv02*), Pv-03 (*Co-13* and *Co-17*), Pv-04 (*Co-3*, *Co-15*, *Co-16*, *Co-y*, and *Co-z*), and Pv-07 (*Co-5*, *Co-6*, and *Co-v*). All ANT resistance genes on chromosome Pv-01 (*Co-1* and five alleles including $Co\text{-}1^{hy}$, *Co-14*, *Co-x*, and *Co-w*) and other genes for resistance to rust and angular leaf spot are present in cultivars belonging to the Andean gene pool (Gonçalves-Vidigal et al. 2011; 2013; Richards et al. 2014; Chen et al. 2017). Additionally, recent studies conducted by Azevedo et al. (2018) have revealed that COK-4, a putative kinase encoded in the ANT resistance locus *Co-4* that is transcriptionally regulated during the immune response, is highly similar to the kinase domain of FERONIA (FER) in *Arabidopsis thaliana*, a factor that has a role in balancing distinct signals to regulate growth and defense.

Several sources of resistance to angular leaf spot (ALS), which is caused by the fungus *Pseudocercospora griseola*, (Sacc.) Crous and Braun, have been identified in common bean. Furthermore, single, dominant resistance loci as well as QTLs conferring resistance to ALS have been reported (Miklas et al. 2006a, b; Mahuku et al. 2009, 2011; Gonçalves-Vidigal et al. 2011, 2013; Oblessuc et al. 2013; Keller et al. 2015). The genes conferring resistance to ALS formally accepted by the Bean Improvement Cooperative (BIC) Genetic Committee are presented in Table 1.3. *Phg-1* on chromosome Pv01 is tightly linked (0.0 cM) to the ANT locus $Co\text{-}1^4$ in cultivar AND 277, which led to the designation of the locus as $Phg\text{-}1/Co1\ ^4$(Gonçalves-Vidigal et al. 2011). The *Phg-1* locus was discovered using F_2 plants from crosses of AND 277 × Rudá and AND 277 × Ouro Negro inoculated with *P. griseola* race 63-23. A previous study conducted by Carvalho et al. (1998) used the name *Phg-1* before describing a resistance locus in AND 277 crossed with Rudá. The molecular markers $CV542014^{450}$ and $TGA1.1^{570}$ have been found to be linked with the $Co\text{-}1^4/Phg\text{-}1$ loci at 0.7 and 1.3 cM, respectively (Gonçalves-Vidigal et al. 2011).

The ALS resistance gene *Phg-2* in Mesoamerican cultivar Mexico 54 was discovered by Sartorato et al. (1999) using a cross between Mexico 54 × Rudá and *P. griseola* race 63-19. The authors identified RAPD markers $OPN02^{890}$, $OPAC14^{2400}$, and $OPE04^{650}$ as being linked to *Phg-2* at 5.9, 6.6 and 11.8 cM, respectively, on chromosome Pv08. Similarly, the RAPD marker OPE04 was found in all resistant individuals but was absent in those scored as susceptible based on virulence data (Namayanja et al. 2006). Additionally, an allelism test between Mexico 54 and BAT 332 inoculated with *P. griseola* race 63-39 showed that a single, dominant gene controls ALS resistance, suggesting that the resistance to ALS in Mexico 54 and

Table 1.3 Enumeration of mapping of simply inherited CS traits and CS QTLs associated with biotic stress resistance in common bean

Disease	Gene symbol	LG	Resistant parent	Reference
Angular Leaf spot (ALS)	*Phg-1*	1	AND277	Gonçalves-Vidigal et al. (2011)
	Phg-2	8	Mexico 54	Namayanja et al. (2006)
	Phg-2[2]		BAT332	Mahuku et al. (2011)
	Phg-3	4	Ouro Negro	Gonçalves-Vidigal et al. (2013)
	Phg-4	4	CAL143	Mahuku et al. (2009), Oblessuc et al. (2012)
	Phg-5	10	G5686	Keller et al. (2015)
Anthracnose (ANT)	*Co-1*	1	Michigan Dark Red Kidney	McRostie (1919)
	Co-1[2]		Kaboon	Melotto and Kelly (2000)
	Co-1[3]		Perry Marrow	Melotto and Kelly (2000)
	Co-1[4]		AND277	Vallejo and Kelly (2002)
	Co-1[5]		Widusa	Gonçalves-Vidigal and Kelly (2006)
	Co-AC		Amendoim Cavalo	Gonçalves-Vidigal et al. (2011)
	Co-14		Pitanga	Gonçalves-Vidigal et al. (2012); (2016), de Lima Castro et al. (2017) Gilio et al. (2017)
	Co-Pa		Paloma	

(continued)

Table 1.3 (continued)

Disease	Gene symbol	LG	Resistant parent	Reference
	Co-2	11	Cornell 49-242	Adam-Blondon et al. (1994)
	Co-3	4	Mexico 222	Geffroy et al. (1999); Mendéz-Vigo et al. 2005; Rodríguez-Suárez et al. (2008) Sousa et al. (2014) Coimbra-Gonçalves et al. (2016)
	Co-15		Corinthiano	
	Co-16		Crioulo 159	
	Co-4³/Co-3³	8, 4	PI207262	Alzate-Marin et al. (2007)
	Co-4	8	TO	Fouilloux (1979) Young et al. (1998) Awale and Kelly (2001)
	Co-4²		SEL1308	
	Co-5	7	TU	Gonçalves-Vidigal (1994), Young and Kelly (1996), Kelly and Young (1996), Young et al. (1998), Vallejo and Kelly (2009), Sousa et al. (2014)
	Co-5²		MSU 7-1	
	Co-6		AB136	
	Co-4²/Co-5²/Co-3⁵	8, 7, 4	G2333	Young et al. (1998)
	Co-12	–	Jalo Vermelho	Gonçalves-Vidigal et al. (2008)
	Co-11		Michelite	Gonçalves-Vidigal et al. (2007)
	Co-13	3	Jalo Listras Pretas SEL1308	Trabanco et al. (2014)
	Co-17			Lacanallo and Gonçalves-Vidigal (2015)

(continued)

Table 1.3 (continued)

Disease	Gene symbol	LG	Resistant parent	Reference
Rust	*Ur-3*, *Ur-6*, *Ur-7*, *Ur-11*, *Ur-Dorado53*, *Ur-BAC6*	11	P94207 P94232 Beltsville DOR 364 BAC6 BelNeb-RR-1	Stavely (1998), Miklas et al. (2002)
	Ur-5, *Ur-14*, *Ur-Dorado108*	4	DOR 364 Ouro Negro Mexico309	Miklas et al. (2000), Souza et al. (2011)
	Ur-4	6	BAT93	Miklas et al. (2002)
	Ur-9	1	PC50	Miklas et al. (2002)
	Ur-12	7	PC50	Jung et al. (1998)
	Ur-13	8	Kranskop	Mienie et al. (2005)
White mold (WM)	*WM1.1*, *WM7.1*	1, 7	G122	Miklas et al. (2001)
	WM2.1, *WM4.1*, *WM5.1*, *WM8.1*	2, 4, 5, 8	PC-50	Park et al. (2001)
	WM2.2, *WM7.2*	2, 7	Bunsi	Kolkman and Kelly (2003)
	WM2.3, *WM5.2*, *WM7.2*, *WM8.4*	2, 5, 7, 8	Bunsi	Ender and Kelly (2005)
	WM1.2, *WM2.4*, *WM8.2*, *WM8.3*, *WM9.1*	1, 2, 8, 9	G122	Maxwell et al. (2007)
	WM2.2, *WM5.4*, *WM6.1* *WM7.5*	2, 5, 6, 7	I9365-31 VA19	Soule et al. (2011) Vasconcellos et al. (2017)
	WM3.3, *WM7.5*, *WM9.2*, *WM11.1*	3, 7, 9, 11	Tacana PI 318695 PI 313850	Mkwaila et al. (2011)
	WM1.3, *WM3.1*, *WM6.2*, *WM7.1*, *WM7.4*	1, 3, 6, 7	Xana	Pérez-Vega et al. (2012), Vasconcellos et al. (2017)

(continued)

Table 1.3 (continued)

Disease	Gene symbol	LG	Resistant parent	Reference
Common bacterial blight (CBB)	*D2, D5, D7, D9*	2, 5, 7, 9	BAT93	Nodari et al. (1993)
	CBB-2LL, CBB-2S, CBB-2P, CBB-2FL, CBB-1LL,	1, 2, 3, 4, 5, 6	BAC 6	Jung et al. (1996)
	CBLEAF, CBPOD	1, 2, 9, 10	BelNeb-RR-1	Ariyarathne et al. (1999)
	Bng40, Bng139	7, 8	XR-235-1-1	Yu et al. (1998)
	CBB-GH-leaf, CBB-GH-pod, CBB-GH-field	7, 10	DOR 364	Miklas et al. (2000)
	SU91, SAP6, $Xa11.4^{OV1,OV3}$	8, 10, 11	Vax1, Vax3	Viteri et al. (2015)
	$Xa3.3^{SO}$	3	BOAC 09-3.	Xie et al. (2017)
Halo blight (HB)	*Rpsar-1, Rpsar-2*	8, 11	BAT93	Fourie et al. (2004)
	Pse-1, Pse-2, Pse-3, Pse-4, pse-5, Pse-6	2, 4, 10	UI-3 ZAA12 BelNeb-RR-1	Fourie et al. (2004), Miklas et al. (2009, 2011, 2014)
	HB4.1, HB6.1	4, 6	Cornell 49-242	Trabanco et al. (2014)
	HB83, HB16	2, 3, 4, 5, 9, 10	BelNeb-RR-1	Ariyarathne et al. (1999)
	SDC^{7}*-6,* $SAUDPC^{3}$*-2,* $PLAUDPC^{3}$*-2,* PDC^{3}*-2,* PDC^{4}*-2,* PDC^{5}*-2,* $PAUDPC^{3}$*-2,* $PAUDPC^{4}$*-2*	2, 6	P1037 PHA1037	González et al. (2016)
	HB4.2, HB5.1	4, 5	PI 150414 Rojo CAL 143	Tock et al. (2017)

(continued)

Table 1.3 (continued)

Disease	Gene symbol	LG	Resistant parent	Reference
BCMV/BCMNV	*I*	2	BelNeb-RR-1	Ariyarathne et al. (1999)
	bc-1^2, *bc-u*	3	Olathe Sierra	Strausbaugh et al. (1999)
	bc-3	6	BAT93	Johnson et al. (1997)
ClYVV	*cyv, desc*	3	Black Knight	Hart and Griffiths (2013)

BAT 332 is conditioned by the same resistance locus (Namayanja et al. 2006). The *Phg*-22 allele of BAT 332 is the only allele officially accepted by the BIC Genetics Committee.

Phg-3 was originally published as *Phg-ON*, as first described by Corrêa et al. (2001) in cultivar Ouro Negro. This cultivar is an important source of resistance for ALS and other diseases in common bean, such as ANT and rust. Inheritance studies in an F_2 population derived from the Ouro Negro × US Pinto 111 cross revealed one dominant resistance gene conferring resistance to race 63-39 (Corrêa et al. 2001). To investigate associations between *Co-3*^{4} (previously named *Co-10*) and the *Phg-3* genes, Gonçalves-Vidigal et al. (2013) conducted co-segregation analysis of resistance to *C. lindemuthianum* races 7 and 73 and *P. griseola* race 63-39 in Ouro Negro using an F_2 population from the Rudá × Ouro Negro cross and $F_{2:3}$ families from the AND 277 × Ouro Negro cross. This co-segregation analysis showed that *Co-3*^{4} and *Phg-3* are inherited together. Additionally, the genes *Phg-3* and *Co-3*^{4} were found to be tightly linked to marker g2303 at a distance of 0.0 cM (Gonçalves-Vidigal et al. 2013) on chromosome Pv04 (Fig. 1.5).

Furthermore, seven QTLs on five LGs have been reported by Oblessuc et al. (2012). Among these, the major QTL *ALS4.1*^{GS,UD} on Pv04 and *ALS10.1*^{DG,UC} and *ALS10.1*$^{DG,UC,\ GS}$ on Pv10, identified in G5686 and CAL143 (Mahuku et al. 2009; Oblessuc et al. 2012; Keller et al. 2015), have been recently named as *Phg-4* and *Phg-5* (Souza et al. 2016). The *Phg-4* locus was first discovered by evaluating the G5686 × Sprite F_2 population with race 31-0 and was published as Phg_{G5686A} (Mahuku et al. 2009). This QTL was later fine mapped to a 418-kb region on chromosome Pv04 and named *ALS4.1*^{GS,UC} (Keller et al. 2015). As this major locus had consistent and significant effects across different environments and populations (Mahuku et al. 2009; Oblessuc et al. 2012, 2013; Keller et al. 2015), the BIC genetics committee accepted the name QTL *ALS4.1*^{GS,UC} for *Phg-4* in G5686 (Souza et al. 2016). The resistance *Phg-5* locus on chromosome Pv10 was discovered using the CAL 143 × IAC-UNA RIL population. The RILs were evaluated under natural infection in the field and in the greenhouse inoculated with race 0-39, whereby QTL *ALS10.1* exhibited a strong effect in all environments (Oblessuc et al. 2012). Keller et al. (2015) confirmed the QTL *ALS10.1* in G5686. Because of its strong effect on resistance to

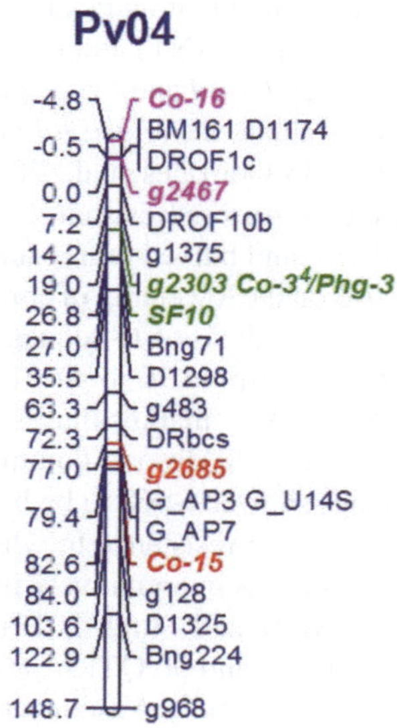

Fig. 1.5 Genetic distances and locations of the *Co-3*^{4} gene for resistance to common bean ANT, the *Phg-3* gene for resistance to ALS, and the molecular markers g2303 on linkage group Pv04 of *Phaseolus vulgaris* L. The map was drawn with MapChart (Voorrips 2002)

ALS in different environments, the BIC Genetics Committee proposed officially named *Phg-5 ALS10.1* in both G5686 and CAL143 (Souza et al. 2016).

Correspondingly, several genes conferring race-specific resistance to the rust pathogen *Uromyces appendiculatus* (Pers.) Unger have been identified, named, and mapped in different LGs in the common bean genome (Table 1.3). Indeed, three large clusters harboring a number of resistance genes located at the ends of chromosomes have been identified on Pv04, Pv10 and Pv11 of the *Phaseolus vulgaris* genome (Schmutz et al. 2014). Among these, one of the most complex disease-resistance clusters containing a large number of genes that confer resistance to various common bean pathogens has been identified at the end of the short arm of chromosome Pv04 (Geffroy et al. 2009; Richards et al. 2014). Moreover, 10 major rust resistance genes have been named and mapped in six different LGs of the common bean genome (Pv01, Pv04, Pv06, Pv07, Pv08, and Pv11) (Kelly et al. 1994, Miklas et al. 2002, Kelly and Vallejo 2004; Miklas et al. 2006a, b; Rodríguez-Suárez et al. 2008; Hurtado-Gonzales et al. 2017a, b). Mesoamerican rust resistance genes include *Ur-3*, *Ur-5*, *Ur-7*, *Ur-11* and *Ur-14* (Augustin et al. 1972; Ballantyne 1978; Stavely 1984; Stavely 1990; Souza et al. 2011). Andean rust resistance genes include *Ur-4*, *Ur-6*, *Ur-9*, *Ur-12* and *Ur-13* (Ballantyne 1978, Finke et al. 1986; Jung et al. 1998; Liebenberg and Pretorius 1997).

In addition, several genes conferring resistance to various common bean pathogens are arranged in clusters of tightly linked genes, often located at the end of the chromosomes. For example, *Ur-9* (Pv01), *Ur-5* (Pv04) and *Ur-3* (Pv11) co-localize with ANT resistance genes *Co-1* (Pv01), *Co3* (Pv04) and *Co-2* (Pv11), respectively (Geffroy et al. 1999, 2000; Kelly et al. 2003). Similarly, *Ur-13* maps close to the *Phg-2* gene for ALS resistance on Pv08 (Garzon and Blair 2014). Recently, co-segregation analysis inoculating $F_{2:3}$ families from the Rudá × Ouro Negro cross with of *C. lindemuthianum* (ANT) and *U. appendiculatus* (Rust) races reported the genetic linkage between *Ur-14* and *Co-3⁴* genes (Valentini et al. 2017). In this study, the authors did not evaluate the *P. griseola* in the $F_{2:3}$ families from the Rudá × Ouro Negro cross. Hurtado-Gonzales et al. (2017a, b) evaluated an F_2 population of Pinto 114 (susceptible) × Aurora (resistant *Ur-3*) for its reaction to four different races of *U. appendiculatus*, and bulked segregant analysis using the SNP chip BARCBEAN6K_3 placed *Ur-3* on the lower arm of chromosome Pv11. Specific SSR and SNP markers and haplotype analysis of 18 sequenced bean varieties positioned *Ur-3* in a 46.5-kb genomic region from 46.96 to 47.01 Mb on Pv11. The authors identified in this region the SS68 KASP marker that is tightly linked to *Ur-3*, and validation of SS68 using a panel of 130 diverse common bean cultivars containing all known rust resistance genes showed SS68 to be highly accurate.

Genetic resistance to white mold (WM), caused by the fungus *Sclerotinia sclerotiorum*, is quantitatively inherited, and several QTLs have been identified thus far (Schwartz and Singh 2013). A comparative map composed of 27 QTLs for WM resistance and 36 QTLs for disease-avoidance traits was developed by Miklas et al. (2013). Recently, Vasconcellos (2017) identified 37 QTLs condensed into 17 named loci (12 previously named and five new), nine of which were defined as meta-QTLs WM1.1, WM2.2, WM3.1, WM5.4, WM6.2, WM7.1, WM7.4, WM7.5, and WM8.3; these are robust consensus QTLs representing effects across different environments, genetic backgrounds, and related traits.

Bacterial Diseases

Xanthomonas axonopodis pv. phaseoli (*Xap*) and *X. fuscans subsp. fuscans* cause common bacterial blight (CBB), a damaging disease of common bean worldwide. CBB resistance has been reported to be quantitatively inherited, often involving QTLs with major and minor effects (Singh and Miklas 2015). More than 20 different QTLs responsible for CBB resistance have been reported across all 11 LGs of common bean (Singh and Miklas 2015; Viteri et al. 2015). Recently, Viteri et al. (2015) identified the major QTL $Xa11.4^{OV1,OV3}$, which explained up 51% of the phenotypic variance for CBB resistance in leaves. Moreover, a new isolate-specific QTL underlying CBB resistance and showing an additive effect with SU91 QTL was recently found on Pv03 (Xie et al. 2017).

Both qualitative and quantitative resistance genes have been reported for resistance to halo blight (HB), which is caused by *Pseudomonas syringae* pv. *phaseolicola* (Burkn.) Downs (Ariyarathne et al. 1999; Fourie et al. 2004; Miklas et al. 2014; Trabanco et al. 2014; González et al. 2016; Tock et al. 2017). Five dominant (*Pse-1*, *Pse-2*, *Pse-3*, *Pse-4* and *Pse-6*); one recessive (*pse-5*) gene has also been identified (Miklas et al. 2009, 2011, 2014). Furthermore, 76 main-effect QTLs were found to explain up to 41% of the phenotypic variation in HB resistance, and 101 epistatic

QTLs were identified by González et al. (2016). Moreover, Tock et al. (2017) recently found a major QTL of race-specific resistance (*HB5.1*) in cv. Rojo and a major QTL of race-nonspecific resistance (*HB4.2*) in PI 150414.

Viral Diseases

Recessive resistance to Bean common mosaic virus (BCMV) and Bean common mosaic necrosis virus (BCMNV) in common bean is controlled by four genes that include one strain-nonspecific helper gene, *bc-u*, and three strain-specific genes, *bc-1*, *bc-2*, and *bc*-3. (Drijfhout 1978). Moreover, there are two alleles each for *bc-1* (*bc-1* and $bc\text{-}1^2$) and *bc-2* (*bc-2* and $bc\text{-}2^2$). The *bc-u* and *bc-1* genes mapped at the end of Pv03; *bc-3* is located on Pv06 and belongs to the eIF4E gene family (Miklas et al. 2000; Naderpour et al. 2010; Meziadi et al. 2016). In addition, the dominant *I* gene mapping to Pv2 imparts resistance to all strains of BCMV (Drijfhout 1978). With regard to resistance to another potyvirus, Clover yellow vein virus (ClYVV), two recessive genes located on Pv06, *cyv* and *desc*, are reported to be allelic forms of *bc-3*, encoding eIF4E factors (Hart and Griffiths 2013; Meziadi et al. 2016).

Drought Resistance

Drought stress is the major limitation of common bean grown in subsistence farming systems worldwide. Several traits associated with drought tolerance have been identified, and different QTL studies have been conducted. Schneider et al. (Schneider et al. 1997) identified RAPD markers associated with yield under stress and non-stress conditions in Sierra × AC1028 and Sierra × Lef2RB populations across a broad range of environments. Additionally, Beebe et al. (2007) identified QTLs for yield under drought using an RIL population from the SEA 5 × MD 23-24 cross; this QTL also influenced yield in well-watered environments, suggesting that yield under both conditions may be influenced by the same factors. Later, Blair et al. (2012) utilized a Mesoamerican intra-gene pool RIL population derived from the cross of drought-tolerant BAT477 and drought-susceptible DOR364 to identify five QTLs associated with yield under irrigated conditions, with mapping to LGs Pv03 and Pv07 and explaining 11 and 19% of the phenotypic variance. When the same population was evaluated using mixed model analysis under eight environments differing in drought stress across Africa and South America, nine QTLs were detected for 10 drought stress tolerance mechanism traits and mapped to six of the 11 LGs (Asfaw et al. 2012a, b).

A total of 14 QTLs for performance under drought were consistently identified in different environments by Mukeshimana et al. (2014). In that study, an inter-gene pool RIL population from a cross of drought-tolerant lines SEA5 and CAL96 was evaluated for several years in Rwanda and Colombia under drought stress and nonstress. QTLs associated with phenology and seed weight traits were identified and mapped near previously reported QTL (Mukeshimana et al. 2014).

Two major QTLs, named *SY1.1*^{BR} and *SY2.1*^{BR}, that conditioned yield in an RIL population with consistent expression across multiple drought-stress environments were identified on Pv01 and Pv02 by Trapp et al. (2015). In this study, 140 RILs from the Buster × Roza cross were tested for yield under multiple stresses (intermittent drought, compaction, and low fertility) across numerous locations and years. The *SY1.1*^{BR} QTL explained up to 37% of the phenotypic variance for seed yield under multiple stresses and was defined by the marker SNP50809 (47.7 Mb). Moreover, when compared to QTLs identified for yield in previous studies, *SY1.1*^{BR} and *SY2.1*^{BR} displayed a larger effect (Asfaw et al. 2012a, b; Blair et al. 2012; Mukeshimana et al. 2014).

Recently, by analyzing 160 RILs derived from the cross between IAPAR 81 (drought-tolerant) and LP97-28 (susceptible to drought) under conditions of drought stress and nonstress for two years in Maringá PR, Brazil (Elias 2018), 16 QTLs were identified on five chromosomes (Pv01, Pv02, Pv07, Pv08 and Pv11) (Fig. 1.6). The author used 773 SNP markers to construct an LG covering 815.9 cM of the bean genome, with a distance of 1.34 cM between markers. The QTL *SY9*^{IL} associated with grain yield was identified on chromosome Pv09, three QTLs for grain yield per day were mapped to Pv07, Pv08, and Pv09, and QTLs linked to seed weight were found on chromosomes Pv07 and Pv08 (Elias 2018).

Another study of genotyping-by-sequencing analysis and 19 climatic characteristics obtained through the WorldClim site was carried out by Elias (2018), in which a set of 110 accessions of common bean previously genotyped using a sequencing genotyping methodology was evaluated, producing 28,823 SNPs. Through associative mapping, it was possible to detect loci of quantitative characteristics, for a total of 135 associations between characteristics vs. markers (Bonferroni test <0.5%). Of the 19 bioclimatic traits, eight exhibited significant associations, and associations for seasonality of temperature and precipitation in the driest quarter were found, both on Pv09, with $R^2 = 36.26$ and 36.46%, respectively. Associations between markers and climatic variables were distributed throughout common bean LGs, except for Pv08. The results show a correlation between markers and climatic characteristics on a national scale, helping to identify candidate genes for regional adaptation. These considerations are of great relevance for the conservation and exploration of genetic diversity between and within common bean accessions in Brazil (Elias 2018).

1.7 Marker-Assisted Breeding for CS Traits

1.7.1 Marker-Assisted Gene Introgression

Molecular mapping and tagging of important genes have contributed to significant advances in MAS of crop breeding. Since molecular markers are related to nucleotide sequence variations, many tags have been developed for different types of plant crops. They also have several advantages over the traditional phenotypic markers (Mohan

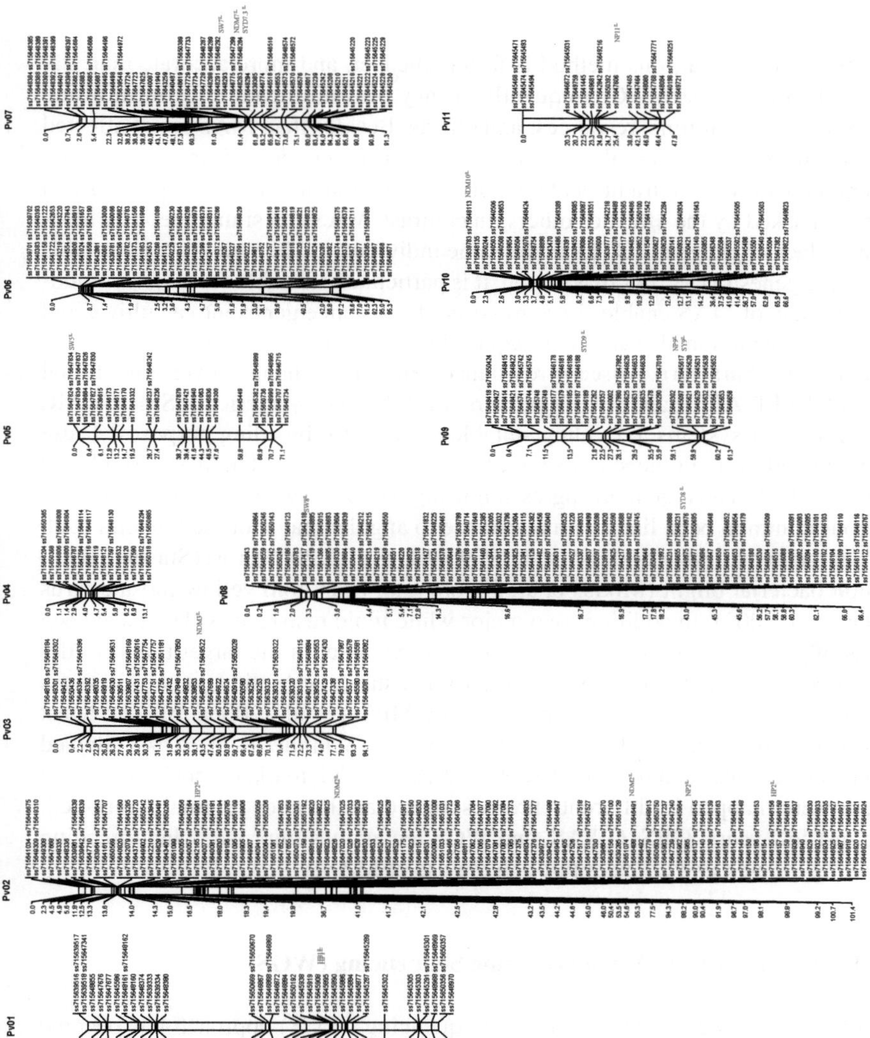

Pv01
Pv02
Pv03
Pv04
Pv05
Pv06
Pv07
Pv08
Pv09
Pv10
Pv11

◀**Fig. 1.6** Genetic mapping for the RIL population Iapar 81 × LP97-28 cross using 773 SNPs markers assigned to the 11 common bean linkage groups. QTL locations are mapped in the Iapar81/LP97-28 population, using the composite interval method (CIM) of the Win cartógrapher software and the LOD thresholds calculated based on 1000 permutations. A total of 16 QTLs were associated with the yield per day, weight of 100 grains, number of pods per plant, height of plant, number of days for flowering, and number of days for maturation under water stress condition

et al. 1997). In general, this method is faster, cheaper, and more accurate than traditional phenotypic assays. Consequently, it may provide higher effectiveness and efficiency in terms of time, resources, and efforts. Besides that, MAS is not affected by the environment, which allows the selection to be conducted under any environmental conditions. In traditional phenotypic selection, an individual gene or loci might be masked by the effect of others. In contrast, MAS can simultaneously identify and select single genes/QTLs in the same individuals, when traits are controlled by multiple genes/QTLs. For that reason, it is particularly feasible for gene pyramiding. The usage of MAS enables introgression of resistance genes into a cultivar and decrease of population size and time required to develop a new variety.

Methods to characterize disease-resistance genes have changed over time. Initial work with RFLP, AFLP, and RAPD markers was followed by a series of SSR, SCAR, and SNP marker systems, providing suitable markers for breeding purposes. These markers linked to single-gene traits have been successfully used in MAS (Singh and Schwartz 2010). Thus, gene introgression using MAS allowed the development of numerous common bean lines with resistance to angular leaf spot (de Oliveira et al. 2005), anthracnose (Alzate-Marin et al. 1999; Miklas et al. 2003), rust (Stavely 2000), common bacterial blight (Miklas et al. 2006b) and, bean gold yellow mosaic virus (Miklas et al. 2002). In addition, two major white mold resistance QTLs have been successfully introgressed using MAS with positive asset in the target traits (Ender et al. 2008). The use of MAS in breeding for resistance to biotic and abiotic stress in common bean has been widely reviewed by Miklas et al. (2006a, b). The latest publication about common bean reference genome (Schmutz et al. 2014) allowed mapping and comparison of several SSR, SCAR, and SNP markers' positions. Some of them were mapped in different chromosomes than the ones originally reported. In the last few years, GBS, GWAS, and WGS techniques improved plant breeding by making it quick and efficient through the use of MAS.

1.7.1.1 Common Bean Whole Genome Sequencing (WGS)

Over the past decade whole genome (re)sequencing (WGS) approach has become feasible due to its continuous cost reduction. Therefore, we currently gained a deep insight into the structure of nearly complete genomes across populations (Lobaton et al. 2018a, b). The history of common bean domestication at genomic level led to introgression of gene pools during the domestication of two independent lines (Andean and Mesoamerican) within a single species (Schmutz et al. 2014). More recently, a large number of inter-gene pool introgressions were identified, and inter-specific introgressions for disease resistance in breeding lines were also mapped (Lobaton et al. 2018a, b).

In 2014 the Joint Genome Institute, Department of Energy released the first chromosome scale version of *Phaseolus vulgaris* (v1.0) (http://www.phytozome.net/) (Schmutz et al. 2014). Interestingly, studies reported that databases like PhaseolusGenes (http://phaseolusgenes.bioinformatics.ucdavis.edu) are actually important tools to accelerate marker identification (Gonçalves-Vidigal et al. 2011; Lobaton et al. 2018a, b).

Another strategy to develop DNA markers is a combination of bulked segregant analysis (BSA) and high-throughput genotyping method. This mapping technique is able to screen many bulks with markers spread throughout the genome in a short period of time (Hyten et al. 2009). Many researchers have used the same procedures to determine the abundance of SSRs in the common bean genome and, developed candidate SSR database for common bean. The Infinium® assay is a newly developed high-throughput SNP genotyping method with higher level of capacity. Recently, the Illumina Infinium® beadchip was designed for soybean (Song et al. 2013) and, also for common bean. Illumina Infinium® beadchip (BARCBEAN 6k_3) was firstly designed for soybean and, was able to screen 5,399 SNPs (USDA-ARS, Maryland, USA). Hyten et al. (2010) developed the Illumina Golden Gate beadchip containing 1,536 SNPs. As a result, the use of Golden Gate assay successfully mapped a few SSRs linked with slow darkening trait onto bean linkage group 7 (Felicetti et al. 2012). Later, Song et al. (2015) generated a highly dense map of the common bean containing 7,040 SNP markers with BARCBean6K_1 and BARCBean6K_2 BeadChips. At the moment, common bean SNP cheap BARCBEAN6K_3 containing 5,398 SNPs (Song et al. 2015) is extensively used to develop specific molecular markers linked to resistance genes (Hurtado-Gonzales et al. 2017a, b).

The use of specific markers for population breeding through next-generation sequencing (NGS) became a common practice in plant breeding, since the development of reference genome sequences allows efficient identification of a large number of physically mapped new and/or different markers (Miller et al. 2018). Reference genomes of common bean have been recently released (Schmutz et al. 2014; Vlasova et al. 2016; http://www.beangenomics.ca/). The mentioned genomes were based on sequences of G19833 (Andean landrace), BAT93 (Mesoamerican breeding line), and OAC-Rex (Mesoamerican cultivar, introgressed with *P. acutifolius*).

The aforementioned databases provide the development of new markers for MAS use and map-based gene isolation. In addition, short genomic sequences for each breeding parent can be mapped on a reference genome and, new polymorphisms such as SSR, SNP and/or INDEL can be detected.

1.7.2 Gene Tagging and Marker-Assisted Selection for Bean Diseases

Conventional breeding methods used depend on visual to screening of genotypes to select for traits of economic importance. Nevertheless, successful using this method depends on its reproducibility and heritability of the characteristic. MAS is an excellent methodology for common bean breeders who also work to improve disease resistance. On behalf of MAS to be highly effective, a high association and tight linkage must exist between the genes for resistance to diseases and molecular markers and easy to evaluate (Yu et al. 2004). As mentioned in the previous section, associations between resistance genes and molecular markers are widely used for mapping genes to specific linkage groups. Since the last century, several studies have used molecular markers to select qualitative resistance to anthracnose (ANT), angular leaf spot (ALS), common bean mosaic virus (BCMV) and, rust diseases.

1.7.2.1 Anthracnose

Garzón et al. (2007) were the first to evaluate the efficiency of MAS to detect anthracnose resistance. For that purpose, a series of backcross plants, using PCR-based markers SAB3 and SAS13 linked to *Co-5* and *Co-4*^{2} genes were used. The authors concluded that *Co-5* is associated with SAB3, whereas *Co-4*^{2} is linked to SAS13.

Likewise, Vidigal Filho et al. (2008) evaluated backcross F_2BC_3 lines using $SAS13_{950}$ marker and observed that it was linked to *Co-4*^{2} allele. Two hundred and thirty-three BC_3F_2 near-isogenic lines containing *Co-4*^{2} resistance allele in various combinations were developed through marker-assisted selection (MAS) for the resistance genes and phenotypic selection for the anthracnose. The BC_3F_2 NILs having *Co-4*^{2} resistance allele showed a wider resistance spectrum and manifested increased levels of resistance to race 2047 of *C. lindemuthianum*. Out of the 233 BC_3F_2 lines analyzed by molecular markers, 80 of them revealed the presence of $SAS13_{950}$ linked to *Co-4*^{2} allele. Moreover, two Brazilian cultivars, both resistant to anthracnose, were obtained by five backcrossing with $SAS13_{950}$ marker through MAS (Gonçalves-Vidigal, unpublished data). These cultivars were released on the market in 2018.

1.7.2.2 Rust

On the subject of rust, the first resistance gene tagged in common bean was *Ur-4* gene (Miklas et al. 1993), using the molecular marker $OA14_{1100}$. This marker was used to perform assisted selection of plants containing *Ur-4* (Kelly et al. 1993). However, its usage is restricted to Mesoamerican cultivars, since progenies from a cross between Early Gallatin and Andean cultivar do not segregate for $OA14_{1100}$marker (Miklas et al. 1993). Previous studies reported limitations of molecular markers linked to *Ur-*

3 gene (Haley et al. 1994; Nemchinova and Stavely 1998; Stavely 2000). However, Valentini et al. (2017) developed several SSR markers linked to *Ur-3*, *Ur-4*, *Ur-5*, *Ur-11*, *Ur-14*, and *Ur-PI310762* genes. For that, accurate phenotyping for the inheritance of resistance studies, bulk segregant analysis (BSA) combined with high-throughput genotyping using the SNP chip BARCBEAN6K_3, were used. Following the same line of experiments, further SSR and SNP markers closely linked to *Ur-3* were developed based on BSA, SNP assay, and whole genome sequencing methodologies (Hurtado-Gonzales et al. 2017a, b). Interestingly, KASP SNP marker SS68 reliably distinguished cultivars containing *Ur-3* alone or in combination with other genes (Hurtado-Gonzales et al. 2017a, b). Recently, co-segregation analysis inoculating $F_{2:3}$ families from the Rudá × Ouro Negro cross with of *C. lindemuthianum* (ANT) and *U. appendiculatus* (Rust) races reported the genetic linkage between *Ur-14* and *Co-3*^{4} genes (Valentini et al. 2017). In this study, the authors did not evaluate the *P. griseola* in the $F_{2:3}$ families from the Rudá × Ouro Negro cross. A different approach was to investigate rust resistance in locus *Ur-14*, which is tightly linked to gene *Co-3*^{4} (Valentini et al. 2017b). The results allowed the construction of a genetic map linkage based on SNP, SSR and, KASP markers linked to *Ur-14*.

1.7.2.3 White Mold

QTLs for white mold on linkage groups Pv02 and Pv07 from an ICA Bunsi × Newport Middle American dry bean population were identified by Kolkman and Kelly (2003). In ICA Bunsi × Raven Middle American dry bean populations, QTLs were also detected and, mapped on linkage groups Pv02, Pv05, Pv07, and Pv08 (Ender and Kelly 2005). Later, Miklas et al. (2007) found two QTLs in a Pinto 3 navy bean (Aztec/ND88–106–04), which were mapped on linkage groups Pv02 and Pv03. Interestingly, the QTL described on Pv02 was identified previously in populations of ICA Bunsi 3 navy and ICA Bunsi 3 black bean RIL.

Further, a comparative study revealed the presence of QTLs in two separate populations, "Benton"/VA19 (BV) and "Raven"/I9365-31 (R31) crosses (Soule et al. 2011). For the first one, WM2.2 and WM8.3 were described for greenhouse straw test and field resistance. In contrast, WM2.2, WM4.2, WM5.3, WM5.4, WM6.1, WM7.3 were found in the Raven/I9365-31 (R31) for greenhouse straw test and field resistance.

In addition, two QTLs were characterized in "Tacana" × PI 318695 (linkage groups Pv04 and Pv07) and Tacana × PI 313850 (linkage groups Pv02 and Pv09) inbred backcross lines, using the greenhouse straw test (Mkwaila et al. 2011). Recently, an evaluation of RIL population from AN-37 × P02630 cross demonstrated the presence of 13 QTLs for agronomic and disease-related traits (Hoyos-Villegas et al. 2015).

1.7.2.4 Fusarium Root Rot

Resistance to FRR is quantitatively inherited and is strongly affected by environmental factors. QTLs associated with this disease varied between studies and populations. Due to limited genomic coverage of the available markers, a comparison of the physical positions of those QTLs was not suitable (Schneider et al. 2001; Chowdhury et al. 2002; Román-Avilés and Kelly 2005). In 2005, (Román-Avilés and Kelly 2005) identified nine QTLs in crosses "Negro San Luis" × "Red Hawk" and "Negro San Luis" × C97407. Later, five regions on linkage groups Pv03, Pv06, and Pv07 associated with QTL for FRR in an Eagle/Puebla 152 population were identified (Navarro et al. 2004). Most recently, two QTLs associated with FRR for greenhouse straw test and field resistance were mapped on Pv02 (Wang et al. 2018).

1.7.2.5 Common Bacterial Blight

In early 2000s, important historical research steps toward MAS were taken. PI 319443 resistance was introgressed into the common bean breeding line XAN 159. By doing that, two major QTLs for common bacterial blight resistance were defined: SCAR marker SU91 (Pedraza et al. 1997) found in Pv08, and BC420 marker detected in linkage group Pv06 (Yu et al. 2000a, b). Yu et al. (2000a, b) evaluated co-segregation of two polymorphic markers. Only $BC420_{900}$ revealed a significant association with a major QTL, which conferred resistance in HR67 to CBB. Following that, another major resistance QTL in OAC-Rex was mapped on Pv05 (Bai et al. 1997; Tar'an et al. 2001; Michaels et al. 2006). Recently was reported the full genome sequence of the common bean OAC-Rex with introgression from the tepary bean, *P. acutifolius* (Perry et al. 2013).

However, a negative association of seed yield with the SU9 marker linked with CBB resistance QTL derived from tepary bean was reported (O'Boyle et al. 2007). Furthermore, Miklas et al. (2009) addressed the presence of SH11.800, SR13.1150, and ST8.1350 markers linked to *Pse-1* and, mapped on Pv10.

1.7.2.6 Bean Common Mosaic Virus

Since BCMV resistance genes are independent in common bean, it contributes to the use of gene pyramiding as an approach for durable resistance (Tryphone et al. 2013). In 1994, Raven was released as the first common bean cultivar resistant to BCMV. The aforementioned cultivar carries two genes: one dominant hypersensitive *I* and one recessive *bc-3*, both confirmed by RAPD markers. This combination has been recognized for its durability over single-gene resistance to both BCMV and BCMNV (Kelly 1997). SCAR markers based on OC11350/420 (ROC11) and OC20460 RAPD markers linked to bc-3 gene were also developed (Johnson et al. 1997). However, the use of these markers in MAS have been limited in common bean because of a lack of polymorphism and, reproducibility across different genetic backgrounds and gene pools (Kelly et al. 2003).

Pedigree selection through the F7 generation based on superior agronic features (early maturity, erect plant architecture, and good pod set) and commercial seed type, Bella cultivar was created. Derived from cross "Verano"//PR0003-124/"Raven," Bella combines resistance to BCMV, BCMNV, BGYMV, and web blight (Beaver et al. 2018).

1.7.3 Gene Pyramiding

The conventional breeding methods involve complex selection of several genotypes harboring different resistance genes, which can affect the accuracy and efficiency of the process. However, pyramiding gene is a good strategy for durable resistance, and it can also facilitate MAS approach. This technique is a combination of multiple desirable genes from multiple parents into a single genotype for specific trait. Thus, this methodology enhances genetic resistance into bean cultivars.

Pyramiding of different genes was developed from a single cross between lines obtained in the introgression step, using either pedigree or backcross method. Currently, several resistant common bean cultivars were developed to improve resistance level to anthracnose, angular leaf spot, rust and, BCMV (Ragagnin et al. 2009).

A marker-assisted gene pyramiding approach was used to develop carioca bean elite lines harboring three different rust resistance genes (Souza et al. 2014). That was only possible because Rudá recurrent parent has a high-yield performance. On the subject of anthracnose and Pythium root rot resistance, genes were pyramided in four susceptible market class varieties using SCAR markers (Kiryowa et al. 2015). It was also shown that higher numbers of selected pyramided genes may indirectly affect yield by reducing the number of seeds per plant.

Through MAS, resistant pyramided lines to rust, anthracnose and, angular leaf spot were developed (Ragagnin et al. 2009). They showed resistance spectra equivalent to those of their respective donor parents. Besides that, yield tests showed that these lines were as productive as the best carioca-type common bean cultivar.

1.7.4 Limitations and Prospects of MAS and Marker-Assisted Backcrossing Breeding (MABCB)

MAS is an important tool to support plant breeders in crop improvement. It considerably increases the efficiency of breeding, when markers tightly linked to genes of interest are used. Despite its advantages, MAS might not be as successful as expected, when introgression of QTL is necessary (Fazio et al. 2003).

MAS is not always better or more cost-effective than direct disease resistance (DDS), especially for quantitatively inherited resistance to diseases. An efficiency comparison of these two techniques, regarding pyramiding and transfer of CBB resistance into dark red kidney bean, showed that DDS was significantly more effective than MAS (Duncan et al. 2012). Under greenhouse conditions of high disease pressure, DDS produced more resistant breeding lines with greater levels of resistance than MAS.

MAS is considered as smart breeding for different reasons. First of all, it is a non-transgenic biotechnological approach for plant improvement and is not subjected to rules/regulations that restrict its use. Second, disease-resistance selection without the use of pathogen is feasible, and off-season screening is possible. Finally, it is suitable to combine multiple sources of disease resistance for distinct pathogens.

1.8 Potential for the Role of Molecular Genetics, Transcriptomics, Epigenomics, and Bioinformatics as Tools to Address Climate Resiliency/CS Traits

1.8.1 Status of Common Bean Genomics

More than 100,000 years after the divergence of Mesoamerican and Andean gene pools a minimum of two separate domestications occurred ~8,200–8,500 years ago (Vlasova et al. 2016). The common bean (*Phaseolus vulgaris*) genome was originally released in 2014 (Schmutz et al. 2014). The Andean inbred landrace G19833 was used for this sequence. The second version of this genome is currently available (*Phaseolus vulgaris* v2.1, DOE-JGI, and USDA-NIFA, http://phytozome.jgi.doe.gov/). Approximately 537.2 Mb of the genome is arranged in 478 scaffolds. An estimated 99.1% of the genome is contained within 87 scaffolds of >50 kb in size. There are 27,433 coding sequence loci and 36,995 protein-coding transcripts; thus, there are 9,562 alternatively spliced variants. A second sequence was released two years after the G198333 genome was released for the Mesoamerican breeding line BAT93 (Vlasova et al. 2016). The Mesoamerican genome was found to be approximately 549.6 Mb, of which 81% is anchored within eleven linkage groups. The BAT93 genome was found to have 30,491 coding sequence loci, with 66,634 protein-coding transcripts that encode for 53,904 unique proteins (Vlasova et al. 2016).

The Mesoamerican genotype BAT93 has been identified to be more resistant to some diseases, including anthracnose, angular leaf spot, and bean common mosaic virus, and rust (Vlasova et al. 2016). Despite its decreased susceptibility, BAT93 was found to have fewer cytoplasmic NBS-LRR class resistance genes (234), than G19833 (376) (Vlasova et al. 2016). Functional enrichment analysis showed that BAT93 has undergone the largest gene expansion in genes related to cellular receptors with extracellular domains. Genes related to seed development and the ubiquitin pathway were also enriched in BAT93, compared to G19833 (Vlasova et al. 2016).

In the BAT93 genome, 35% was found to be composed of transposable elements (Vlasova et al. 2016). The G19833 genome is approximately 41% of transposable elements (*Phaseolus vulgaris* v2.1, DOE-JGI, and USDA-NIFA, http://phytozome.jgi.doe.gov/). Long noncoding RNAs (lncRNAs) were highly conserved between the two genotypes, with 94% of Mesoamerican lncRNAs also contained within the Andean genome (Vlasova et al. 2016). lncRNAs appear not to be highly conserved within legumes, as only a third were found to be conserved past soybean (*Glycine max*) (Vlasova et al. 2016).

Since the sequencing of the common bean genomes, numerous resequencing, transcriptomic, epigenomic, proteomic, and metabolomic projects have been conducted. A recent resequencing project identified introgression within the Mesoamerican and Andean common bean gene pools (Lobaton et al. 2018a, b). This project undertook sequencing of 35 common bean, 22 Mesoamerican and 13 Andean, accessions and one genotype each of the closely related species *P. acutifolius* and *P. coccineus* (Lobaton et al. 2018a, b). These lines were selected based on agriculturally significant traits, including resistance to a variety of biotic and abiotic stresses. The other *Phaseolus spp.* were selected as they have introgressed into some common bean cultivars (Lobaton et al. 2018a, b). A total of 203 possible introgression events were detected (Lobaton et al. 2018a, b). Surprisingly, it was determined that the Andean reference genome, G19833, contained a large Mesoamerican introgression on chromosome Pv08, which spans 24 Mbp. Additionally, there were three other Andean-derived genotypes that contained this same introgression. Other introgressions of over 1 Mbp were identified in other chromosomes (Lobaton et al. 2018a, b). Due to self-fertilization, heterozygosity rates were low, averaging 0.17% in Andean and 0.46% in Mesoamerican genotypes.

1.8.2 Gene, Genome, and Comparative Genome Databases (Phytozome, NCBI, LIS, EBI, CoGe, DAVID)

Phaseolus spp has been proposed to serve as a model for understanding crop evolution due to the multiple domestication events in Mesoamerica and South America and other characteristics (Bitocchi et al. 2017; Rendón-Anaya et al. 2017).

1.8.2.1 Databases

Vast information on common bean genes, genomes, and comparative genomics are widely publicly available. Phytozome is the "Plant Comparative Genomics portal of the Department of Energy's Joint Genome Institute" (Goodstein et al. 2012). Currently, the *Phaseolus vulgaris* version 2.1 genome is the most recent release. Previous versions of released genomes can be found at the "Download" section of Phytozome's dropdown menus. Genome v2.1 combined an 83.2x sequence coverage PacBio-based assembly that is annotated with their proprietary Gene Model Improvement (GMI) pipeline. Genes can be searched for via running a BLAST search or keyword search. The output yields genes and ontologies with a direct link to PANTHER. The gene section reveals the functional annotation, view in a genome browser (JBrowse), genomic, transcript, and coding sequences, protein homologs, gene ancestry, and gene expression and co-expression in various tissue types. PhytoMine allows users to search for information of genomics, transcripts, proteins, comparative species, and expression based on a variety of input identifiers; which includes gene IDs, GO terms, and panther terms.

The National Center for Biotechnology Information (NCBI) is a repository for several different data types. NCBI hosts categories of information classified as Literature (books, journal articles, and reports), Genes (ESTs, genes, homologs, phylogenetics, unigenes, functional genomics), Genetics (clinical, genotype/phenotype, human-related), Proteins (conserved domains, sequences, clusters, structure), Genomes (genome assembly, biosamples/projects, SRA, nucleotide sequences, proves, taxonomy, and Chemicals (molecular pathways, screening, deposited information). The current genome data was supplied by the Joint Genome Institute. NCBI's Sequence Read Archive hosts user-supplied next-generation sequencing data for public availability ("National Center for Biotechnology Information" n.d.).

The Legume Federation also serves as an information and tool repository to "facilitate collaborative development of software, methods, and standards…to help build a healthy research ecosystem." Tools that are offered or linked to include Legume Mines, Data Store at CyVerse, Transcript annotation, Genomic Context Viewer, Data Store at Legumeinfo, and upcoming CMap-js ("Legume Federation" n.d.). Legume Mines-BeanMine is a common bean database that provides gene expression, QTL, gene ontology (GO) terms, and QTL marker resources. Annotation data are available to download at the National Science Foundation-funded CyVerse ("CyVerse" n.d.) database and at the Legume Information System (LIS) ("Legume Information System" n.d.). CMap-js is a comparative genome software in alpha testing, which upon release will allow users to compare biological maps, which includes genetic, physical, cytogenetic, genomic, linkage groups, chromosomes, and scaffolds ("Legume Federation" n.d.).

The Legume Information Systems (LIS) is a legume-specific database with the intention of building on traits for crop improvement. LIS hosts unique tools for QTLs, germplasm resources, genetic maps, physical maps, and molecular markers. Some of these tools are accessible through the Legume Federation website. The Transcript Annotation tool allows the user to upload nucleotide or protein sequences and run the sequences. The Genomic Context Viewer is a comparative genome viewer that allows the user to input a variety of gene identifiers and the output includes "Macro-Synteny" and "Micro-Synteny" tracks to visualize chromosomal patterns or conserved gene function-specific functions (Cleary and Farmer 2018). Phylotree is a gene family search tool allowing users to search gene family IDs, gene descriptions, or by count, the results for each of the legumes are displayed. A "list" of genes can be built for users to save for future analysis; which serves as a convenient organizational tool for complicated data analysis.

The European Bioinformatics Institute (EBI) and the Wellcome Trust Sanger Institute jointly host plant-specific information and tools including pHMMER, BLAST, comparative genomics, variant effect predictor (VEP), assembly converter, and ID history converter. The user can search the database for genomes and metagenomes, nucleotide and protein sequences, macromolecular structures, bioactive molecules, gene and protein expression, molecular interactions, reactions and pathways, protein families, enzymes, literature, and samples and ontologies, which totals over 1.3 million results.

CoGe is a comparative genomics platform, containing over 47,000 genomes from over 18,000 organisms. Genomes can be viewed in a browser with GC content, coding sequence (CDS), gene annotations, rRNA, and tRNA. Features unique to CoGe, which are not included in JBrowse include: filter track list by name, data type, manage experiments, export track data, search features by name, search tracks, combine tracks by dragging and dropping, convert search results into marker tracks, and save search results as new experiments in CoGe. CoGeBlast allows the user to perform a BLAST search against selected genomes. Multiple common bean genomes are available to search in this database. SynMap is a tool that allows the user to find homologs among two or more species.

1.8.2.2 Diversity Panels and Seed Banks

Common bean diversity panels are assemblies of germplasm for breeding and crop improvement purposes (Cichy et al. 2015a, b). Domesticated Andean bean genotypes have less genetic diversity than domesticated Mesoamerican genotypes due to a bottleneck event that occurred before domestication events (Cichy et al. 2015a, b). Because of the lack of diversity in Andean genotypes, breeding among this gene pool is limited in comparison to progress made in Mesoamerican genotypes (Cichy et al. 2015a, b).

An Andean diversity panel (ADP) was developed in 2015, consisting of 396 accessions; 349 Andean, 21 Mesoamerican, and 26 admixed accessions collected globally. Information can be accessed about this diversity panel at http://arsftfbean.uprm.edu/bean/ (accessed 15 May 2015). Diversity panels have been used in many types of studies, including those screening for flooding tolerance (Soltani et al. 2018), drought tolerance (Asfaw et al. 2017), resistance to root rot (Binagwa et al. 2016), population structure in Uganda (Okii et al. 2014), cooking time (Cichy et al. 2015a, b), gene-based microsatellites (Blair et al. 2009), SNPs between common bean and tepary bean (Gujaria-Verma et al. 2016), and agronomic traits (Moghaddam et al. 2016). A Middle American diversity panel was developed to include 280 Middle American cultivars from the BeanCAP diversity panel (Moghaddam et al. 2016).

The Consultative Group on International Agricultural Research (CGIAR) hosts an international database, Genebank Platform, which allows researchers to request 750,000 accessions of various plant species ("Genebank Platform" n.d.). CGIAR partners with AfricaRice, Bioversity International, International Center for Tropical Agriculture (CIAT), International Maize and Wheat Improvement Center CIMMYT, International Potato Center, Crop Trust, International Center for Agricultural Research in the Dry Areas (ICARDA), International Crops Research Institute for the Semi-Arid Tropics (ICRISAT), International Institute of Tropical Agriculture (IITA), International Livestock Research Institute (ILRI), International Rice Research Institute (IRRI), and the World Agroforestry Center. CIAT's missions are to develop crops for food security and improved nutrition, profits for small-scale farmers and food accessibility, and developing resilient crops. CIAT currently contains nearly 38,000 *Phaseolus spp.* accessions, 32,375 of which are common bean genotypes. Accessions can be searched by species, location and collection features, characterization features (growth habit, seed color, shape, brightness, and weight, days to flowering, first and last harvest, and use), reactions to biotic and abiotic stresses, and/or nutritional and technological traits. CIAT scientists and collaborations have led to the development and release of more than 550 bean varieties, beans that are tolerant to $\geq$3 °C higher average temperatures, 3$\times$ higher yielding climbing beans, and beans that accumulate higher iron.

The IITA's research programs are based in four areas, mainly impacting natural resources in Africa: 1. biotechnology and genetic improvement, 2. natural resource management, 3. Social science and agribusiness, and 4. Plant production, plant health, nutrition, and food technology ("International Institute of Tropical Agriculture" n.d.). The Genesys Plant Genetic Resources (PGR) database was created by Bioversity International and is the largest plant genetic resource repository, containing more than 2.8 million accessions, more than 54,000 are common bean accessions. Bioversity International's goal is to establish community seed banks to benefit small-scale farmers ("Genesys Plant Genetic Resources" n.d.).

Crop Trust was also developed by Bioversity International on behalf of CGIAR and the UN Food and Agriculture Organization (FAO). The goal of this initiative is to conserve diverse crop genetic material for food security ("Crop Trust" n.d.). The Svalbard Global Seed Vault contains nearly 1 million seed samples, from almost 6,000 species ("Svalbard Global Seed Vault" n.d.). National Plant Germplasm System (NPGS) is a collaborative initiative of the United States Department of Agriculture Agricultural Research Service (USDA-ARS) to protect genetic diversity ("National Plant Germplasm System" n.d.).

The National Agricultural Research Organisation (NARO) organizes agricultural research in Uganda, which includes the National Agricultural Research System (NARS) ("National Agricultural Research Organisation" n.d.). The National Bureau of Plant Genetic Resources (NBPGR) is headquartered in New Delhi, India, where researchers work toward conserving germplasm and to provide tools for breeders ("National Bureau of Plant Genetic Resources" n.d.). Some web application tools that are provided by the NBPGR include the PGR portal for information on germplasm, PGR searchable map, an herbarium, intellectual property, Cryo database, crop wild relatives portal, genebank, climate smart management, and genetic resources.

The European Cooperative Programme for Plant Genetic Resources (ECPGR) which aims to conserve germplasm for breeding purposes and functions under

Bioversity International ("European Cooperative Programme for Plant Genetic Resources" n.d.). The germplasm repository, the European Search Catalogue for Plant Genetic Resources (EURISCO), contains more than 52,000 accessions in the *Phaseolus* genus, more than 46,000 accessions are common bean. ECPGR provides a platform to acquire accessions by allowing requests to be directed to institutions that are associated with the accession.

The Genetic Resources Center, National Agriculture and Food Research Organization (NARO) Genebank Project is a conservation effort coordinated in Japan ("NARO Genebank Project" n.d.). This database contains 915 germplasm accessions (accessed July 25, 2018) which can be searched/filtered by many physical characteristics. There are other international crop resources including the Australian Temperate and Field Crops Collection, the Chinese Crop Germplasm Information System (CGRIS), Leibniz Institute of Plant Genetics and Crop Plant Research, the National Institute of Agrobiological Sciences Genebank, and the Asian Vegetable Research and Development Center.

1.8.3 Gene Expression Databases

Many databases that were described previously, in Sect. 13.2.1 are used by researchers who perform high-throughput RNA sequencing methodologies. Some of these databases allow users to upload their generated data, among the most frequently used is NCBI's Short Read Archive (SRA) hosts many user-supplied gene expression data, which can be openly downloaded by other users. The SRA Toolkit includes many free programs that can be used for analyzing sequencing data. Similarly, to NCBI, EBI allows users to submit high-throughput sequencing data to the database as well as searching existing projects and downloading previously submitted data.

Phytozome released the common bean genome, which also contains gene expression data on many tissue types, reported as fragments per kilobase of transcript per million mapped reads (FPKM). Phytozome also contains gene ancestry and co-expression. This information is useful in determining whether selected genes are expressed in target tissues, which can serve as a confirmation for RNA sequencing data. The *Phaseolus vulgaris* Gene Expression Atlas (PvGEA) database hosts downloadable data for common bean tissues harvested at several developmental stages. Expression data for roots, nodules, leaves, stems, flowers, seeds, and pods are available. The user can download normalized and/or raw data or view gene expression data by performing a keyword or sequence search ("PvGEA" n.d.).

1.8.4 *Protein and Metabolome Databases (NCBI, EBI, UniProt, PvTFDB, KEGG)*

1.8.4.1 Protein

The integration of proteomic and genomic approaches, termed "proteogenomics," has been developing into a powerful tool to better understand the molecular mechanisms that are activated in plants during stress (Zargar et al. 2017). However, proteome studies in common bean are lacking and underrepresented among other legumes (Zargar et al. 2017). These types of studies are important for determining genes as related to stress tolerance, and growth and development of plants and seed (Zargar et al. 2017). To date, most studies on legume proteomics have involved gel-based approaches, which are considered to be low-throughput (Zargar et al. 2017).

Posttranslational modifications are yet another factor in proteomics, for example, phosphorylation of a dehydrin in responding to and recovering from osmotic stress (Zargar et al. 2015). Changes in phosphorylation of phaseolin proteins were found to be implicated in seed dormancy transition to germination (Zargar et al. 2015). Developing a "proteome atlas" to detect rare proteins may prove to be a powerful identification tool to target pathways involved in response to specific stresses (Zargar et al. 2015).

Biotic and abiotic stresses can cause changes in plant protein expression (Zargar et al. 2017)

Databases like NCBI and EBI contain information and tools to search protein sequences, but there are some databases that provide more insight into common bean-specific protein structure and function; including UniProt and PvTFDB. UniProt is a protein database, which contains more than 32,000 protein entries for common bean, 159 of which have been manually annotated and reviewed (accessed July 17, 2018). UniProt provides information on function (catalytic activity, cofactors, enzyme regulation, binding and active sites, gene ontology (GO) molecular functions, and links to other enzyme databases), taxonomy/aliases, subcellular location, pathology, post-translational modifications/processing, interactions, and structure ("UniProt" n.d.).

PvTFDB is a database that houses information on 2,370 putative transcription factors (TFs) in common bean ("Phaseolus Vulgaris Transcription Factor Database" n.d.). The authors of this database suggested that transcription factors are the most important target in terms of developing stress-tolerant crops (Bhawna et al. 2016). PvTFDB also provides other useful data on these TFs including tissue-specific gene expression, *cis*-regulatory elements, phylogeny, gene ontology, and functional annotations (Bhawna et al. 2016). This database has downloadable information for each transcription factor family, which includes the DNA sequence, coding sequence (CDS), primary transcript, amino acid sequence, and the 2 kb region upstream from the transcription start site (Bhawna et al. 2016).

1.8.4.2 Metabolome

An estimated 100,000 to 1 million metabolites are present in all plants, of which 5,000 or more are unique to each species (Alseekh et al. 2018). The most widely used tools to study metabolomics are nuclear magnetic resonance (NMR), gas chromatography mass spectrometry (GC-MS), and liquid chromatography mass spectrometry (LC-MS) (Alseekh et al. 2018). Each methodology comes with positives and pitfalls. NMR is limited by its ability to only detect abundant metabolites, or those extracted from copious amounts of tissue. LC-MS requires samples to be treated prior to testing. GC-MS analytes are largely unannotated (Alseekh et al. 2018).

The Kyoto Encyclopedia of Genes and Genomes (KEGG) database models signaling pathways in biological systems. For common bean, KEGG currently has 134 pathways available (accessed June 25, 2018), which are mostly related to metabolism, but other pathways are represented as well ("Kyoto Encyclopedia of Genes and Genomes" n.d.). It is established that symbiotic relationships with microbes can change the physiology of the host common bean plant (Figueiredo et al. 2008; Mwenda et al. 2018; Sánchez et al. 2014).

Nitrogen fixation is a metabolic process that is characteristic of legumes (Ramalingam et al. 2015). Common bean was crossed with *Phaseolus acutifolius* yielded a common bacterial blight resistant genotype; this consequently led to decreased ability to form symbiotic relationships with bacteria (Farid et al. 2017). Rhizobia are the microbes responsible for establishing the symbiosis of fixing nitrogen in the soil. Due to pleiotropic effects, tracking many phenotypes and physical characteristics will be an important component of breeding studies moving forward as crop improvement projects continue.

Another important group of compounds is phytochemicals, which have positive health benefits for humans (Thompson et al. 2017). A study conducted in rats with cancer showed that triacylglycerol (TAG) precursors were reduced in the mammary glands of the bean-fed rats compared to the control group (Mensack et al. 2012). The results of this study suggest that lipid metabolism is a target of bioactive chemicals in dry beans (Mensack et al. 2012). However, the lack of plant metabolome coverage continues to be a challenge in this area of study, as well as annotation of metabolites (Alseekh et al. 2018).

1.8.4.3 Role of Microbial Interactions

In addition to resources specifically involving common bean, genomes and resources of symbionts may prove to be useful as well. There are distinct differences in the interaction between allowing symbiotic relationships with some microbes versus defense mechanisms against potentially pathogenic microbes. A recent study showed that increased disease resistance in common bean showed a decreased ability to form symbiotic relationships (Farid et al. 2017).

Rhizobium spp. are gram-negative bacteria that form symbiotic relationships with legumes (Carrasco-Castilla et al. 2018). Common bean is a promiscuous host, meaning it can form nodules with multiple species of rhizobia. Currently, it is known that common bean can be nodulated by at least 27 rhizobia species (Mwenda et al. 2018). Plant growth-promoting rhizobacteria (PGPR) are thought to play a role on plant growth by different mechanisms, including alteration of hormones within the plant, increasing solubilization of nutrients and nitrogen uptake, iron chelation, and negative effects on plant pathogens (Figueiredo et al. 2008; Sánchez et al. 2014). The third mechanism may prove to be a promising disease management practice to increase yield (Figueiredo et al. 2008). Beans co-inoculated with nonpathogenic *Rhizobium tropici* (CIAT 899) and *Paenibacillus polymyxa* (DSM36) were found to form more efficient symbiotic associations (Figueiredo et al. 2008). This study examined co-inoculation of CIAT 899 with other PGPRs and the results showed significant differences in phytohormone activity and cytokinin content in the host.

Another recent study described the relationship between rhizobial infection, nodulation, and bean expression of annexin (Carrasco-Castilla et al. 2018). Rhizobia secrete lipo-chitooligosaccharides, or nod factors, which are detected by bean root hairs to induce the formation of the infection thread. This thread is a channel that allows the rhizobia to cross the root hair cell to ultimately lead to nodulation and nitrogen fixation (Carrasco-Castilla et al. 2018). Bean annexins have been shown to play wide-ranging roles, including abiotic stress, biotic stress, growth and development, immunity, and symbiotic microbial relationships (Carrasco-Castilla et al. 2018).

Complete genome sequences of eight *Rhizobium* symbionts associated with common bean (Santamaría et al. 2017). Interestingly, the *Rhizobium etli* and *Rhizobium phaseoli* isolates were found to be rather different in their genomic lineages, despite all being associated with common bean nodules and nitrogen fixation. Beneficial microbes are able to establish symbiotic relationships by secretion of effector molecules that interact with the host, which can lead to downregulation of plant immunity genes (Seidl and Thomma 2017).

Coevolution with pathogens has been noted in several studies including the fungal pathogens *Colletotrichum lindemuthianum* (Geffroy et al. 1999; Luana et al. 2017), *Colletotrichum lindemuthianum* (Padder et al. 2017) *Uromyces appendiculatus* (Cooper and Campbell 2017; Odogwu et al. 2016), *Pseudocercospora griseola* (Ddamulira et al. 2014; Chilagane et al. 2016), and bacterial pathogen *Pseudomonas syringae* (O'Leary et al. 2016), (Vlasova et al. 2016). Transposable elements (TEs) are one of the major contributing factors to coevolution of plants and pathogens (Seidl and Thomma 2017). TEs provide opportunities to substantially impact the structure of the host's genome and this is discussed in more detail in other section.

1.8.5 History of Epigenetics/Epigenomics

The idea of epigenetics is considered to have started in the 1930s, by Waddington, who was interested in embryology. He wanted to determine what happens during development to allow an adult to form from an embryo (Nicoglou and Merlin

2017). In the 1940s and 1950s, McClintock observed "coordinated transposition" in maize and chromatin organization effects on gene expression. "Cellular memory" was introduced by Nanney in the 1950s, which was described as mitotically stable phenomenon; meaning that the same genotype can display different phenotypes. In 1961, the operon model of gene expression was introduced by Jacob and Monod. This model describes the induction of enzymes when a substrate is present. Britten and Davidson introduced the gene-batter model in 1969, which stated that noncoding sequences regulate gene expression. In the 1970s, Riggs and Holliday independently hypothesized about DNA methylation affecting gene expression. DNA experiments in methylation and histone modifications and their effect on gene expression started to appear in the 1990s (Nicoglou and Merlin 2017). Presently, it is widely known that there are several epigenetic mechanisms that contribute to control of gene expression, which include DNA methylation, histone modifications, and noncoding RNAs. Plants, including common bean, have the relatively unique capability to have widespread, extensive DNA methylation in three different motifs, CG, CHG, and CHH (Crampton et al. 2016; Kim et al. 2015).

The link between evolution and the development of organisms is abbreviated as "evo-devo." This was essentially the 1990 s-2000 s version of the "epigenetics" concept, particularly in explaining differences in phenotypic variation and maps (Abouheif et al. 2014; Nicoglou and Merlin 2017). "Eco-evo-devo" incorporates ecological/environmental impact on an organism's genes and development (Abouheif et al. 2014).

Plants are unique as they comprise the highest number of polyploid/alloploid species found in nature. Polyploidy events can cause gene silencing, loss of redundant genes, chromosomal recombination, and TE bursts (Wendel et al. 2018). Genome fractionation and chromosomal restructuring can occur following a polyploidy/whole genome duplication event. Ancient genome duplications and fractionation have led to the current status of the common bean genome (Schmutz et al. 2014). Gene and genome duplications are a major driver of species evolution. Whole genome duplication events can cause other downstream functions to occur that further the evolution of genes and genomes (Wendel et al. 2018). Genome duplications can cause transposable element (TE) bursts, which is the unpredictable mobilization of TEs (Galindo-González et al. 2017; Wendel et al. 2018). This event can cause major mutagenesis leading to chromosome rearrangements (Wendel et al. 2018). Mechanisms that control chromosome conformation and gene expression are affected by genome duplication, specifically, these are small RNAs and DNA and histone modifications (Wendel et al. 2018).

Transposable elements used to be thought of as almost exclusively parasitic DNA in genomes (Galindo-González et al. 2017). TEs are present in significant proportions in plant genomes, from 14% in *Arabidopsis thaliana*, 41% in *Phaseolus vulgaris*, to 80% in *Zea mays* (DOE-JGI 2018; Galindo-González et al. 2017). TEs are classified as type/class 1 (retrotransposons), which spread via "copy-and-paste" and type/class 2 (DNA transposons), which move via "cut-and-paste" (Paszkowski 2015; Gao et al. 2016). Within type/class 1 TEs there are long terminal repeats (LTRs) and non-LTRs (Paszkowski 2015). LTRs are further categorized as Ty1-*copia* and Ty3-*gypsy* (Gao et al. 2014). Non-LTR retrotransposons are categorized as either short interspersed nuclear elements (SINEs) or long interspersed elements (LINES) (Gao et al. 2014). The most prevalent TEs in common bean genome are retrotransposons, which comprise 35% of the total genome (DOE-JGI 2018). DNA transposons comprise about 5.3% of the genome, with 0.7% as "unclassified transposons" (DOE-JGI 2018).

1.8.6 Integration of "Omic" Datasets

Because gene and protein expression are complicated processes, the integration of multiple "omic" analyses has proven to be a powerful tool. There are many recent studies that involve the integration of multiple "omic" datasets; such as histone modifications (Ayyappan et al. 2015), proteomics, metabolomics, genome resequencing (Vlasova et al. 2016), DNA methylation, and small RNA sequencing, which are combined with mRNA sequencing.

Before the release of the reference genome, a multi-omics study was conducted on navy bean and white kidney genotypes from both centers of domestication (Mensack et al. 2010). The combination of transcriptomics, proteomics, and metabolomics allowed the authors to classify the cultivars to the correct center of domestication, which also suggests inherent differences in gene expression, protein expression, and metabolism (Mensack et al. 2010).

Omics approaches have also been useful in biotic stress when looking at the host and pathogen. The microbial–host interaction is complex, as common bean plants must make a differentiation between friend and foe. Since there is coevolution between pathogens and common bean, integrated omics studies are even more appealing.

1.9 Social, Political, and Regulatory Issues

This section of the chapter addresses social, political, and regulatory issues related to common bean genetic resources and associated traditional knowledge.

The importance of plant genetic resources for food and agriculture (PGRFA) for achieving food security worldwide and for sustainable development of agriculture in the context of poverty alleviation and climate change is widely recognized. PGRFA are maintained in situ, on farm, and ex situ.

PGRFA have been used and exchanged since the beginnings of agriculture, some 10,000 years ago. Consequently, nowadays all countries depend to some extent on genetic diversity that originated elsewhere. There is a continued need for exchange of PGRFA for research, breeding and conservation for ensuring continued ability to adapt to climatic changes, pest and disease resistance, reduced soil fertility, and ultimately, food security. In fact, while studies suggest that the average degree of genetic interdependence among countries for their most important crops is around

70% (Palacios 1998), in the light of climate change, it is expected that this interdependency will increase considerably. Awareness about the importance of continuous access to PGRFA led to the creation during the last few decades of different international instruments, agreements, and institutions to ensure its management, especially in those aspects related to PGRFA shared use (Chiarolla et al. 2012; Esquinas-Alcázar et al. 2012; Halewood 2014). Some examples of these include the Convention on Biological Diversity (CBD), its Nagoya Protocol on Access to Genetic Resources and the Fair and Equitable Sharing of Benefits Arising from the Utilization (hereinafter referred to as the Nagoya Protocol), the International Treaty on Plant Genetic Resources for Food and Agriculture (ITPGRFA) and different forms of intellectual property rights.

The CBD, adopted in 1992, is the first legally binding international instrument that recognized the sovereignty of the States over their genetic resources regarding their conservation and sustainable use, the traditional knowledge of the indigenous and local communities and the distribution of benefits derived from their use with these communities. The Nagoya Protocol, adopted in 2010, established a legal framework for the implementation of the third objective of the CBD: the fair and equitable sharing of benefits arising out of the utilization of genetic resources and associated traditional knowledge, including by appropriate access to them. Implementing this third objective should contribute to the conservation of biological diversity and the sustainable use of its components, the other two objectives of the CBD. The ITPGRFA, adopted in 2001, established an international legal framework, in harmony with the CBD, for the conservation and sustainable use of plant genetic resources for food and agriculture and the fair and equitable sharing of the benefits arising from their use. Both international agreements are meant to be implemented in complementarity. That is, the Nagoya Protocol does not apply for the Parties to the ITPGRFA in respect of the PGRFA covered by and for the purpose of the Treaty. The Nagoya Protocol and the ITPGRFA are, however, based on two different models of access and benefit sharing systems. On the one hand, the Nagoya Protocol establishes that, in accordance with national legislation, access to genetic resources and to its associated traditional knowledge for their utilization is subject to obtaining the prior informed consent (PIC) from the provider and to the establishment of mutually agreed terms (MAT), which are to be agreed between the user and the provider. The ITPRFA, on the other hand, creates a "multilateral system of access and benefit sharing" whereby countries agree to virtually pool and grant facilitated access to "all PGRFA listed in Annex I of the Treaty that are under the management and control of the Contracting Parties and in the public domain." The Treaty's Annex I includes 64 crops and forages that were selected according to criteria of food security and interdependence. This facilitated access under the ITPGRFA is provided under the terms and conditions of the Standard Material Transfer Agreement (SMTA) when the intended use of the genetic resource is its conservation and sustainable use for research, breeding, and training for food and agriculture. Common bean is part of the crops listed in Annex I of the ITPGRFA. Therefore, access to common bean genetic resources by any legal or private person from any Contracting Party to the ITPGRFA should be facilitated under the conditions established in the SMTA when the intended uses are those cover by the ITPGRFA.

Indigenous and local communities, farmers, researchers, and breeders worldwide have all contributed throughout history to the range of crop diversity that is currently the base of the world's production systems. The development of new varieties is in general a costly and time-consuming process. As a result, intellectual property rights were created as a mean to promote investments in knowledge creation and business innovation by granting exclusive rights to right-holders to prevent others from using newly developed technologies, goods, and services without their permission.

The Agreement on Trade Related Aspects of Intellectual Property Rights (TRIPS), binding on all the members of the World Trade Organization, is a multilateral agreement on intellectual property. TRIPS establishes that plant varieties must be provided with some form of intellectual property system, either patents or an effective sui generis system (a system especially designed for its purposes). As a result, countries worldwide are progressively adopting a plant variety protection law in line with the regulations established by the International Convention for the Protection of New Varieties of Plants (hereinafter referred to as the UPOV Convention). The UPOV Convention is a sui generis system designed specifically to protect the rights of plant breeders over new plant varieties. Its first Act was drafted in 1961 and was later revised in 1972, 1978 and 1991. As of July 2018, 73 countries (plus the African Intellectual Property Organization and the European Union) were members of this organization (www.upov.int). The UPOV Convention offers protection to the breeder of a plant variety (according to the definition established by the UPOV Convention), in the form of a "breeder's right," if his plant variety satisfies the conditions set out in the UPOV Convention (i.e., novelty, distinctness, uniformity, and stability).

The breeder's right is granted for a period of not less than 20 years from the date of grant or, in the case of trees and vines, for not less than 25 years. An authorization of the breeder is therefore needed for the use of the reproduction or multiplication material. The breeder's right under the UPOV Convention, however, does not extend to acts done privately and for noncommercial purposes, acts done for experimental purposes and acts done for the purpose of breeding other varieties and, for the purpose of exploiting these new varieties provided the new variety is not a variety essentially derived from another protected variety (UPOV 1991).

Common bean is a self-pollinated crop or, in other words, it is easily copied. Therefore, there are no great incentives for farmers to buy seeds from the breeder (or the producer under contract). Nonetheless, common bean was among the 13 botanical genera included in the first list to which the UPOV procedures were to be applied. As of August 2018, 11.492 varieties of genus Phaseolus had been included in the UPOV PLUTO database, including information provided by 61 countries, the African Intellectual Property Organization and the European Union (last accessed in July 2018—available at http://www.upov.int/pluto/en/).

Under the formal seed sector, breeding programs are usually focused on producing plant varieties for high-input commercial agriculture that perform well in uniform environments. As a consequence, these varieties are usually not suitable for the nonuniform conditions typical of marginal areas or for those farmers who cannot afford to purchase additional inputs (Ceccarelli and Grando 2007; Assefa et al. 2005). In many developing countries, common bean constitutes the staple food and it is mainly produced by smallholder farmers who grow the crop in small areas. In fact, in many of these countries, both the production and the market for certified seed under the responsibility of the formal sector is still limited. Depending on the crop and country, between 60 and 90% of the seed sown comes from the informal system (Almekinders and Louwaars 2002). Studies show that technologies developed for smallholder farmers without their own participation or without taking into account their own knowledge are rarely adopted (Trutmann et al. 1996). As a result, there is an increasing number of initiatives aiming at creating linkages between the formal and informal seed systems through collective initiatives such as participatory plant breeding and participatory variety selection. These approaches join farmers and professional breeders, local and formal conditions, and the rural communities' experience and traditional knowledge to identify varieties that perform well in specific agroecological systems and that are attractive to farmers. Some examples of participatory breeding in common bean have been carried out in Rwanda (Waldman et al. 2014; Isaacs et al. 2016), Kenya (Ojwang et al. 2009), Central Africa (Trutmann et al. 1996) and in Ethiopia (Asfaw et al. 2012a, b; Balcha and Tigabu 2015). The involvement of farmers can take place during the definition of breeding objectives and priorities. These include hosting trials on their land, contributing during the selection of lines for further crossing or in the planning for the following year's activities, etc.

In the same lines, the potential of community seed banks for both contributing to link in situ and ex situ conservation and to the interaction and integration of the informal and formal seed systems is increasingly being recognized. Defined as "locally governed and managed, mostly informal, institutions whose core function is to maintain seeds for local use" (Sthapit 2013), community seed banks play different functions in the community. Examples of these are preserving seeds, providing seed access for members of the community, and generating a degree of food security and food sovereignty (Vernooy et al. 2015), contributing at the same time to the implementation of farmers' rights through the recognition of farmers' knowledge of local biodiversity, their participation in decision-making for its conservation and benefit sharing (Sthapit 2013).

The importance of involving farmers in conservation and breeding activities also relies on the internationally recognized contribution made by local and indigenous communities and farmers to the conservation, management, and development of plant genetic resources for food and agriculture. This international recognition has led to the relatively recent appearance of different tools aiming at supporting the implementation at the local level of the international agreements on access and benefit sharing of genetic resources and its associated traditional knowledge. An example of these instruments are the so-called biocultural community protocols, which are formally recognized by the Nagoya Protocol, whereby Parties committed to take into account community protocols and other community rules and procedures where traditional knowledge associated with genetic resources is concerned. When driven and designed by the communities, the development of these documents has the potential to strengthen community cohesion and the capacity to make visible their connection with the agrobiodiversity of their surroundings. Moreover, these protocols have the potential to simultaneously advance the communities' interests in both obtaining an equitable share of benefits when their genetic resources are accessed and used, and gaining access to, and being able to use genetic diversity from elsewhere (by taking, for example, advantage of the multilateral system of the ITPGRFA when PGRFA are concerned) for use in their own agricultural production systems.

In addition, there is an increasing number of efforts and initiatives worldwide aiming at compiling traditional knowledge related to genetic resources. In Spain, for example, where the traditional agricultural practices have almost completely disappeared, a national inventory of traditional knowledge related to biodiversity has been currently developed (Pardo de Santayana et al. 2014) focused on wild diversity of plant, animal, and fungus.

1.10 Future Perspectives

Common bean has become, over the last 20 years, in a competitive crop in national, regional, and international markets. This situation presents a dynamic environment for producers and researchers of this crop and requires a rethinking of current strategies against research and production needs, the opportunities, and challenges of the future.

The secondary diversification of the common bean and the existence of new recombinant types between the Andean and Mesoamerican genetic pools open the door for new opportunities for the genetic improvement of the species. Breeders can cross between Mesoamerican and Andean gene pools, as well among races, although it is well known that there are constraints to the crosses between Mesoamerican and Andean germplasm due to genetic barriers [blocked cotyledon lethal (BCL), crinkle leaf dwarf (CLD) and dwarf lethal (DL)] (Singh and Gutierrez 1984; Hannah et al. 2007). González et al. (2009) reported successful interracial and interpool crosses for the development of new common bean varieties in Europe. Since the Mesoamerican germplasm usually display resistance to pathogens and some Andean varieties have high seed quality, the use of the European recombinant germplasm as bridge parents in interpool crosses to overcome the interpool genetic barriers provides an interesting opportunity for introgression of relevant genes in the common bean varieties currently grown in Europe. Breeding can also involve gene introgression from additional genes pools, such as the secondary and tertiary gene pools, covering a range of environments from cool moist highlands to hot semiarid regions, and from drought periods to more wet conditions.

An important long-term challenge is the discovery of the gene(s) that control important production traits. This will need to be a cooperative worldwide effort that involves breeders, geneticists, and genomic and bioinformatics experts. Breeders provide the essential skills of phenotyping and the identification and development of genetic populations. Connecting phenotyping with the functional gene requires the skills of pathologists, physiologists, and those with a deep knowledge of plant anatomy. Those skilled with genomics and bioinformatics provide the expertise to link the phenotypic and genotypic data with candidate genes. Once a candidate gene is defined and the causative mutation is discovered, breeders will then have access to best possible marker, one that is in the gene controlling the important phenotype.

Currently, new technologies built around the recently released common bean genome sequence (Schmutz et al. 2014; Vlasova et al. 2016) are now being developed. Regarding the new breeding technologies, genetic transformation causes some public concern in many countries, but novel breeding material obtained by mutagens are more acceptable to consumers, breeders, and governments. In this context, Targeting Induced Local Lesions in Genome (TILLING) technology has been developed as a new powerful breeding methodology (De Ron et al. 2015). TILLING is a non-transgenic method that uses gene-specific primers for the identification of mutants of a gene of interest from a large mutagenesis population (McCallum et al. 2000). TILLING has gained popularity as a reverse genetic approach because it can produce a series of mutants, including knockouts, and it does not rely on the transformation method for gene discovery and verification. Significant advances have been made in the development of a TILLING platform in common bean, but the protocol for this crop has yet to be optimized. Induced mutation breeding is an effective method to increase the common bean genetic variability available to the plant breeders. Additionally, renewed interest is being generated in induced mutations since the sequence of the common bean genome is already available and it will bring new opportunities for functional genomics research. Therefore, induced mutagenesis will probably become a powerful tool for the isolation and functional characterization of interesting genes, which can be used in common bean genetic improvement.

Improvement of the common bean means possessing in-depth knowledge of its genetic diversity, the genome and gene functions, to enable the analysis of pathways and networks in response to fluctuating environmental conditions. Various genomic resources for common bean are available and include physical maps, bacterial artificial chromosome libraries, anchored physical and genetic maps, expressed sequence tags, and the recently published complete genome sequence (Schmutz et al. 2014; Vlasova et al. 2016). However, these approaches require precise phenotypic data. Complex interactions between the crop genotype, environmental factors in combination with plant population dynamics and crop management greatly affect plant phenotypes in field experiments. Hence, novel techniques should be kept cost-effective and robust under varying field conditions and should allow for the monitoring of various and complex traits.

References

Abate T, Alene AD, Bergvinson D, Shiferaw B, Silim S et al (2012) Tropical Grain Legumes in Africa and South Asia: Knowledge and Opportunities. International Crops Research Institute for the Semi-Arid Tropics, Nairobi, Kenya, 112p

Abdul-Wahid OA, Elbanna SM (2012) Evaluation of the insecticidal activity of *Fusarium solani* and *Trichoderma harzianum* against cockroaches. Periplaneta americana. Afr J Microbiol Res 6(5):1024–1032

Abouheif E, Favé M-J, Ibarrarán-Viniegra AS, Lesoway MP, Rafiqi AM et al (2014) Advances in experimental medicine and biology. In: Landry CR, Aubin-Horth N (eds) Ecology and the evolution of genes and genomes. Springer, Dordrecht, Netherlands, pp 107–125

Acosta-Gallegos JA, Kelly JD, Gepts P (2007) Prebreeding in common bean and use of genetic diversity from wild germplasm. Crop Sci 47(S3):S44–S59

Acosta-Gallegos JA, Ochoa-Marquez R, Arrieta-Montiel MP (1989) Registration of Pinto Villa common bean. Crop Sci 35:1211

Acosta-Gallegos JA, White JW (1995) Phenological plasticity as an adaptation by common bean to rainfed environments. Crop Sci 35:199–204

Adam-Blondon AF, Sevignac M, Dron M, Bannerot H (1994) A genetic map of common bean to localize specific resistance genes against anthracnose. Genome 37:915–924

Afzal AJ, Wood AJ, Lightfoot DA (2008) Plant receptor-like serine threonine kinases: Roles in signaling and plant defense. Mol Plant-Microbe Interact 21(5):507–517

Agarwal P, Reddy MP, Chikara J (2011) WRKY: its structure, evolutionary relationship, DNA-binding selectivity, role in stress tolerance and development of plants. Mol Biol Rep 38(6):3883–3896

Agrios GN (2002) Fitopatología, 2a edn. Limusa, Ciudad de México, México

Ait-Lahsen H, Soler A, Rey M, de La Cruz J, Monte E et al. (2001) An antifungal exo-alpha-1,3-glucanase (AGN13.1) from the biocontrol fungus *Trichoderma harzianum.* Appl Environ Microbiol 67(12):5833–5839

Akello J, Sikora R (2012) Systemic acropedal influence of endophyte seed treatment on *Acyrthosiphon pisum* and *Aphis fabae* offspring development and reproductive fitness. Biol Control 61:215–221

Alahmadi S, Ouf A, Ibrahim A, El-Shaikh A (2012) Possible control of data palm stag beetle, *Lucanus cervus* (L.) (Coleoptera: Lucanidae), using gut protease inhibitors of different biocontrol agents. Egyp Soc Biol Control Pests 22(2):93-101

Almekinders CJM, Louwaars NP (2002) The importance of the Farmers' Seed Systems in a Functional National Seed Sector. J New Seeds 4(1–2):15–33. https://doi.org/10.1300/J153v04n01_02

Alseekh S, Alisdair RF (2018) Metabolomics 20 years on: what have we learned and what hurdles remain? Plant J 94: 933–942. http://doi.wiley.com/10.1111/tpj.13950

Altomare C, Norvell WA, Björkman T, Harman GE (1999) Solubilization of phosphates and micronutrients by the plant-growth- promoting and biocontrol fungus *Trichoderma harzianum* Rifai 1295-22. Appl Environ Microbiol 65(7):2926–2933

Alzate-Marin AL, Menarin H, Cardoso de Arruda MC, Chagas JM, Gonçalves de Barros E et al (1999) Backcross assisted by RAPD markers for the introgression of *Co-4* and *Co-6* anthracnose resistant genes in common bean cultivars. Annu Rep Bean Improv Coop 42:15–16

Alzate-Marin AL, Souza KA, Silva MGM, Oliveira EJ, Moreira MA et al (2007) Genetic characterization of anthracnose resistance genes *Co-4*^{3} and *Co-9* in common bean cultivar Tlalnepantla 64 (PI207262). Euphytica 154:1–8

American Phytopathological Society (2005) Compendium of bean diseases. 2nd ed., eds. Schwartz HF, Steadman JR, Hall R, Forster RL. APS Press, American Phytopathological Society, 104 pp

Andrade-Sanchez P, Gore MA, Heun JT, Thorp KR, Carmo-Silva AE et al (2014) Development and evaluation of a field-based high-throughput phenotyping platform. Funct Plant Biol 41:68–79

Angioi SA, Rau D, Rodriguez M, Logozzo G, Desiderio F et al (2009) Nuclear and chloroplast microsatellite diversity in *Phaseolus vulgaris* L. from Sardinia (Italy). Mol Breed 23:413–429

Anwar W, Nawaz K, Haider MS, Shahid AA, Iftikhar S (2017) Biocontrol Potential of *Trichoderma longibrachiatum* as an entomopathogenic fungi against *Bemisia tabaci*. Can J Plant Pathol 39(4):559–559

Ariyarathne HM, Coyne DP, Jung G, Skroch PW, Vidaver AK et al (1999) Molecular mapping of disease resistance genes for halo blight, common bacterial blight, and bean common mosaic virus in a segregating population of common bean. J Amer Soc Hort Sci 124:654–662

Asfaw A, Almekinders CJM, Blair MW, Struik PC (2012a) Participatory approach in common bean (*Phaseolus vulgaris* L.) breeding for drought tolerance for southern Ethiopia. Plant Breed 131(1): 125-134

Asfaw A, Blair MW, Struik PC (2012b) Multienvironment quantitative trait loci analysis for photosynthate acquisition, accumulation, and remobilization traits in common bean under drought stress. Genes Genomes Genet 2:579–595

Asfaw A, Ambachew D, Trushar S, Blair MW (2017) Trait associations in diversity panels of the two common bean (*Phaseolus vulgaris* L.) gene pools grown under well-watered and water stress conditions. Front Plant Sci 8: 1–15. http://journal.frontiersin.org/article/10.3389/fpls.2017.00733/full

Ashraf M, Athar HR, Kwon Harris PJC (2008) Some prospective strategies fo improving crop salt tolerance. Adv Agron 97:45–110

Assefa T, Abebe G, Fininsa C, Tesso B, Al-Tawaha ARM (2005) Participatory bean breeding with women and small holder farmers in eastern Ethiopia. World J Agri Sci 1(1):28–35

Athanassiou CG, Kavallieratos NG, Vayias BJ, Tsakiri JB, Mikeli NH et al (2008) Persistence and efficacy of *Metarhizium anisopliae* (Metschnikoff) Sorokin (Deuteromycotina: Hyphomycetes) and diatomaceous earth against *Sitophilus oryzae* (L.) (Coleoptera: Curculionidae) and *Rhyzopertha dominica* (F.) (Coleoptera: Bostrychidae) on wheat and maize. Crop Protec 27:1303–1311

Atkins C (1987) Metabolism and translocation of fixed nitrogen in the nodulated legume. Plant and Soil. 100:157–169

Atkinson MM, Midland SL, Sims JJ, Keen NT (1996) Syringolide 1 triggers Ca^2 influx, K efflux, and extracellular alkalization in soybean cells carrying the disease-resistance gene *Rpg4*. Plant Physiol 112(1):297–302

Augustin E, Coyne DP, Schuster ML (1972) Inheritance of resistance in *Phaseolus vulgaris* to *Uromyces phaseoli typica* Brazilian rust race B11 and of plant habit. J Amer Soc Hort Sci 97:526–529

Awale HE, Kelly JD (2001) Development of SCAR markers linked to *Co-4*^{2} gene in common bean. Annu Rep Bean Improv Coop 44:119–120

Ayyappan V, Kalavacharla V, Thimmapuram J, Bhide KP, Venkateswara SR et al (2015) Genome wide profiling of histone modifications (H3K9me2 and H4K12ac) and gene expression in rust (*Uromyces appendiculatus*) inoculated common bean (*Phaseolus vulgaris* L.). PLoS ONE 10(7):e0132176. http://dx.plos.org/10.1371/journal.pone.0132176

Azevedo RF, Gonçalves-Vidigal MC, Oblessuc PR, Melotto M (2018) The common bean COK-4 and the Arabidopsis FER kinase domain share similar functions in plant growth and defense. Mol Plant Pathol 9:1–14

Bai Y, Michaels TE, Pauls KP (1997) Identification of RAPD markers linked to common bacterial blight resistance genes in *Phaseolus vulgaris* L. Genome 40:544–551

Baier AH, Webster BD (1992) Control of *Acanthoscelides obtectus* Say (Coleoptera: Bruchidae) in *Phaseolus vulgaris* L. seed stored on small farms. Evaluation of damage. J Stored Prod Res 28:289–293

Bakshi M, Oelmüller R (2014) WRKY transcription factors. Plant Signal Behav 9(2):e27700

Balcha A, Tigabu R (2015) Participatory varietal selection of common bean (*Phaseolus vulgaris* L.) in Wolaita, Ethiopia'. Asian J Crop Sci 7(4):295–300

Ballantyne BJ (1978) The genetic bases of resistance to rust, caused by *Uromyces appendiculatus* in bean (*Phaseolus vulgaris* L.). PhD Thesis. Sydney. University of Sydney, Australia

Baltes NJ, Gil-Humanes J, Voytas DF (2017) Genome engineering and agriculture: Opportunities and challenges. In: Weeks DP, Yang B (eds). Increasing Resistance to Abiotic Stress Elsevier, Amsterdam, pp 1–26

Barra P, Rosso L, Nesci A, Etcheverry M (2013) Isolation and identification of entomopathogenic fungi and their evaluation against *Tribolium confusum, Sitophilus zeamais,* and *Rhyzopertha dominica* in stored maize. J Pest Sci 86:217–226

Barthelson RA, Qaisar U, Galbraith DW (2010) Functional analysis of the *Gossypium arboreum* genome. Plant Mol Biol Report 28(2):334–343

Barton L, Thamo T, Engelbrecht D, Biswas WK (2014) Does growing grain legumes or applying lime cost effectively lower greenhouse gas emissions from wheat production in a semi-arid climate? J Cleaner Prod 83:194–203

Batta YA (2004) Control of rice weevil (*Sitophilus oryzae* L., Coleoptera: Curculionidae) with various formulations of *Metarhizium anisopliae*. Crop Protec 23:103–108

Beaver JS, Estévez de Jensen C, Lorenzo-Vázquez G, González A, Martínez H et al (2018) Registration of 'Bella' white-seeded common bean cultivar. J Plant Reg 12:190–193

Beaver JS, Zapata M, Alameda M, Porch TG, Rosas JC (2012) Registration of PR0401-259 and PR0650-31 Dry Bean Germplasm Lines. J Plant Reg 6:81–84

Beaver JS, Porch TG, Zapata M (2008) Registration of 'Verano' white bean. J Plant Reg 2:187–189

Beebe S, Rao I, Terán H, Cajiao C (2007) Breeding concepts and approaches in food legumes: The example of common bean. In: Food and Forage Legumes of Ethiopia: Progress and Prospects. Proceedings of the Workshop on Food and Forage Legumes. Addis Abeba, Ethiopia, pp 23–29

Beebe S, Toro O, González AV, Chacón MI, Debouck D (1997) Wildweed-crop complex of common bean (*Phaseolus vulgaris* L., Fabaceae) in the Andes of Peru and Colombia, and their implications for conservation and breeding. Genet Resour Crop Evol 44:73–91

Beebe SE (2012) Common bean breeding in the tropics. Plant Breed Rev 36:357–426

Beebe SE, Corrales MP (1991) Breeding for disease resistance. In: Schoonhoven O, van Voysest A (eds) Common beans: research for crop improvement. CAB International, Wallingford, United Kingdom, pp 561–617

Beebe SE, Rao IM, Blair MW, Acosta-Gallegos JA (2013) Phenotyping common beans for adaptation to drought. Front Physiol 4:1–20

Beebe SE, Rao IM, Blair MW, Butare L (2009) Breeding for abiotic stress tolerance in common bean: Present and future challenges. SABRAO J Breed Genet 41:1–11

Beebe SE, Rengifo J, Gaitan E, Duque MC, Tohme J (2001) Diversity and origin of Andean landraces of common bean. Crop Sci 41:854–862

Beebe SE, Rao IM, Mukankusi C, Buruchara R (2012) Improving resource use efficiency and reducing risk of common bean production in Africa, Latin America, and the Caribbean. In: Hershey CH (ed) Eco-efficiency: From Vision to Reality. CIAT, Cali, Colombia, pp 117–134

Benítez T, Rincón AM, Limón MC, Codón AC (2004) Biocontrol mechanisms of *Trichoderma* strains. Int Microbiol 7(4):249–260

Bernier J, Atlin GN, Serraj R, Kumar A, Spaner D (2008) Breeding upland rice for drought resistance. J Sci Food Agri 88:927–939

Bhatia CR, Nichterlein K, Maluszynski M (2001) Mutants affecting nodulation in grain legumes and their potential in sustainable cropping systems. Euphytica 120:415–432

Bhawna VSB, Prasad Gajula MNV (2016) PvTFDB: a *Phaseolus vulgaris* transcription factors database for expediting functional genomics in legumes. Database: JBiolDatabasesCuration:1–6. https://doi.org/10.1093/database/baw114

Binagwa PH, Conrad BK, Msolla SN (2016) Evaluation of common bean (*Phaseolus vulgaris*) genotypes for resistance to root rot disease caused by *Pythium aphanidermatum* and *Pythium splendens* under screen house conditions. J Nat Sci Res 6(6):36–43

Birkenbihl RP, Diezel C, Somssich IE (2012) Arabidopsis WRKY33 is a key transcriptional regulator of hormonal and metabolic responses toward *Botrytis cinerea* infection. Plant Physiol 159(1):266–285

Bitocchi E, Bellucci E, Giardini G, Rau R, Rodriguez M et al (2013) Molecular analysis of the parallel domestication of the common bean in Mesoamerica and the Andes. New Phytol 197:300–313

Bitocchi E, Nanni L, Bellucci E, Rossi M, Giardini A et al (2012) Mesoamerican origin of the common bean (*Phaseolus vulgaris* L.) is revealed by sequence data. Proc Natl Acad Sci USA 109:E788–E796

Bitocchi E, Rau D, Bellucci E, Rodriguez M, Murgia et al (2017) Beans (*Phaseolus ssp.*) as a Model for Understanding Crop Evolution. Front Plant Sci 8:722. https://doi.org/10.3389/fpls.2017.00722

Blair MW, Cortes AJ, Penmetsa RV, Farmer A, Carrasquilla-Garcia N et al (2013) A high-throughput SNP marker system for parental polymorphism screening, and diversity analysis in common bean (*Phaseolus vulgaris* L.). Theor Appl Genet 126:535–548

Blair MW, Galeano CH, Tovar E, Torres M, Velasco A et al (2012) Development of a Mesoamerican intra-genepool genetic map for quantitative trait loci detection in a drought tolerant × susceptible common bean (*Phaseolus vulgaris* L.) cross. Mol Breed 29:71–88

Blair MW, Giraldo MC, Buendia HF, Tovar E, Duque MC et al (2006a) Microsatellite marker diversity in common bean (*Phaseolus vulgaris* L.). Theor Appl Genet 113:100–109

Blair MW, Iriarte G, Beebe S (2006b) QTL analysis of yield traits in an advanced backcross population derived from a cultivated Andean x wild common bean (*Phaseolus vulgaris* L.) cross. Theor Appl Genet 112:1149–1163

Blair MW, Izquierdo P (2012) Use of the advanced backcross-QTL method to transfer seed mineral accumulation nutrition traits from wild to Andean cultivated common beans. Theor Appl Genet 125:1015–1031

Blair MW, Medina JI, Astudillo C, Rengifo J, Beebe SE et al (2010) QTL for seed iron and zinc concentration and content in a Mesoamerican common bean (*Phaseolus vulgaris* L.) population. Theor Appl Genet 121:1059–1070

Blair MW, Pedraza F, Buendia HF, Gaitán-Solís E, Beebe SE et al (2003) Development of a genome-wide anchored microsatellite map for common bean (*Phaseolus vulgaris* L.) Theor Appl Genet 107:1362–74

Blair MW, Torres MM, Giraldo MC, Pedraza F (2009) Development and diversity of Andean derived, gene-based microsatellites for common bean (*Phaseolus vulgaris* L.). BMC Plant Biol 9:1–14

Blum A (2005) Drought resistance, water-use efficiency, and yield potential-are they compatible, dissonant, or mutually exclusive? Aust J Agri Res 56:1159–1168

Blum A (2009) Effective use of water (EUW) and not water-use efficiency (WUE) is the target of crop yield improvement under drought stress. Field Crops Res 112:119–123

Bobe G, Barrett KG, Mentor-Marcel RA, Saffiotti U, Young MR et al (2008) Dietary cooked navy beans and their fractions attenuate colon carcinogenesis in azoxymethane-induced ob/ob mice. Nutr Cancer 60:373–381

Boller T, Felix G (2009) A Renaissance of elicitors: perception of microbe-associated molecular patterns and danger signals by pattern-recognition receptors. Annu Rev Plant Biol 60(1):379–406

Borges A, Melotto M, Tsai SM, Caldas DGG (2012) Changes in spatial and temporal gene expression during incompatible interaction between common bean and *Anthracnose* pathogen. J Plant Physiol 169(12):1216–1220

Boyer JS (1982) Plant productivity and environment. Science 218:443–448

Brimner TA, Boland GJ (2003) A review of the non-target effects of fungi used to biologically control plant diseases. Agric Ecosys. Environ 100(1):3–16

Briñez B, Morini Küpper Cardoso Perseguini J, Santa Rosa J, Bassi D, Ribeiro Gonçalves JG et al (2017) Mapping QTLs for drought tolerance in a SEA 5 x AND 277 common bean cross with SSRs and SNP markers. Genet Mol Biol 40:813–823

Broglie K, Chet I, Holliday M, Cressman R, Biddle P et al (1991) Transgenic plants with enhanced resistance to the fungal pathogen *Rhizoctonia solani*. Science 254(5035):1194–1197

Brotman Y, Briff E, Viterbo A, Chet I (2008) Role of swollenin, an expansin-like protein from *Trichoderma*, in plant root colonization. Plant Physiol 147(1):779–789

Broughton WJ, Hernández G, Blair M, Beebe S, Gepts P et al (2003) Beans (*Phaseolus* spp.) - model food legumes. Plant Soil 252:55–128

Brücher OB, Brücher H (1976) The South American wild bean (*Phaseolus aborigineus* 'Burk'.), as ancestor of the common bean. Econ Bot 30:257–272

Brunner K, Peterbauer CK, Mach RL, Lorito M, Zeilinger S et al (2003) The Nag1 N-acetylglucosaminidase of *Trichoderma atroviride* is essential for chitinase induction by chitin and of major relevance to biocontrol. Curr Genet 43(4):289–295

Butare L, Rao IM, Lepoivre P, Polania J, Cajiao C et al (2011) New genetic sources of resistance in the genus *Phaseolus* to individual and combined aluminium toxicity and progressive soil drying stresses. Euphytica 181:385–404

Campa A, Giraldez R, Ferreira JJ (2011) Genetic analysis of the resistance to eight anthracnose races in the common bean differential cultivar Kaboon. Phytopathology 101:757–764

Campa A, Rodriguez-Suarez C, Giraldez R, Ferreira JJ (2014) Genetic analysis of the response to eleven *Colletotrichum lindemuthianum* races in a RIL population of common bean (*Phaseolus vulgaris* L.). BMC Plant Biol 14:115. PMID: 24779442. https://doi.org/10.1186/1471-2229-14-115

Campelo MP (2010) Estudio de la microbiotica patógena presente en semillas de "Alubia de León" (*Phaseolus vulgaris* L.) y los métodos de control. PhD Thesis, Univ de León, León, Spain

Cardoza RE, Vizcaíno JA, Hermosa MR, Sousa S, González FJ et al (2005) Cloning and characterization of the *erg1* gene of *Trichoderma harzianum*: Effect of the *erg1* silencing on ergosterol biosynthesis and resistance to terbinafine. Fungal Genet Biol 43(3):164–178

Carrasco-Castilla J, Ortega-Ortega Y, Jáuregui-Zúñiga D, Juárez-Verdayes MA, Arthikala M-K et al (2018) Down-regulation of a Phaseolus vulgaris annexin impairs rhizobial infection and nodulation. Environ Exp Bot 153:108–119. https://doi.org/10.1016/j.envexpbot.2018.05.016

Carvalho GA, Paula Junior TJ, Alzate-Marin AL, Nietsche S, Barros EG et al (1998) Herença da resistência da linhagem AND-277 de Feijoeiro-comun à raca 63-23 de *Phaeoisariopsis griseola* e identificaçao de marcador RAPD ligado ao gene de resistencia. Fitopatol Bras 23:482–485

Casquero PA, Lema M, Santalla M, De Ron AM (2006) Performance of common bean landraces from Spain in the Atlantic and Mediterranean environments. Genet Resour Crop Evol 53:1021–1032

Castellanos JZ, Pena-Cabriales JJ, Acosta-Gallegos JA (1996) 15N-determined dinitrogen fixation capacity of common bean (*Phaseolus vulgaris*) cultivars under water stress. J Agri Sci 126:327–333

Ceccarelli S, Grando S (2007) Decentralized-participatory plant breeding: an example of demand driven research. Euphytica 155:349–360

Chacón MI, Pickersgill S, Debouck D (2005) Domestication patterns in common bean (*Phaseolus vulgaris* L.) and the origin of Mesoamerican and Andean cultivated races. Theor Appl Genet 110:432–444

Chaman ME, Copaja SV, Argandoña VH (2003) Relationships between salicylic acid content, Phenylalanine Ammonia-Lyase (PAL) activity, and resistance of barley to aphid infestation. J Agric Food Chem 51(8):2227–2231

Champ MM (2002) Non-nutrient bioactive substances of pulses. Br J Nutr 88:S307–S319

Chang YC, Chang YC, Baker R, Kleifeld O, Chet I (1986) Increased growth of plants in the presence of the biological control agent *Trichoderma harzianum*. Plant Dis 70(2):145–148

Chaves MM, Pereira JS, Maroco J, Rodrigues ML, Ricardo CPP et al (2002) How plants cope with water stress in the field. Photosynthesis and growth. Ann Bot 89:907–916

Chen H, Lai Z, Shi J, Xiao Y, Chen Z et al (2010) Roles of arabidopsis WRKY18, WRKY40 and WRKY60 transcription factors in plant responses to abscisic acid and abiotic stress. BMC Plant Biol 10(1):281

Chen M, Wu J, Wang L, Mantri N, Zhang X et al (2017) Mapping and genetic structure analysis of the anthracnose resistance locus Co-1HY in the common bean (*Phaseolus vulgaris* L.). PLoS One 12:1–18

Chen X, Qi X, Duan LX (2015) Overview. In: Qi X, Chen X, Wang Y (eds) Plant Metabolomics. Springer, Dordrecht, Netherlands, pp 1–24

Cherry AJ, Abalo P, Hell K (2005) A laboratory assessment of the potential of different strains of the entomopathogenic fungi *Beauveria bassiana* (Balsamo) Vuillemin and *Metarhizium anisopliae* (Metschnikoff) to control *Callosobruchus maculatus* (F.) (Coleoptera: Bruchidae) in stored cowpea. J Stored Prod Res 41:295–309

Chet I, Harman GE, Baker R (1981) *Trichoderma hamatum*: Its hyphal interactions with *Rhizoctonia solani* and *Pythium* spp. Microb Ecol 7(1):29–38

Chiarolla C, Louafi S, Schloen M (2012) An analysis of the relationship between the Nagoya protocol and instruments related to genetic resources for food and agriculture and farmers' rights. In: Buck M, Morgera E, Tsoumani E (eds) The 2010 Nagoya protocol on access and benefit-sharing: Implications for international law and implementation challenges. Brill Academic Publisher, Leiden, Boston, USA

Chilagane L, Nchimbi-Msolla S, Kusolwa P, Porch T et al (2016) Characterization of the common bean host and *Pseudocercospora griseola*, the causative agent of angular leaf spot disease in Tanzania. Afric. J. Plant Sci. 10:238–245. https://doi.org/10.5897/AJPS2016.1427

Chowdhury MA, Yu K, Park SJ (2002) Molecular mapping of root rot resistance in common bean. Annu Rep Bean Improv Coop 45:96–97

Cichy KA, Porch TG, Beaver JS, Cregan P, Fourie D et al (2015a) A *Phaseolus vulgaris* diversity panel for Andean bean improvement. Crop Sci 55(5):2149–2160

Cichy KA, Wiesinger JA, Mendoza FA (2015b) Genetic diversity and genomewide association analysis of cooking time in dry bean (*Phaseolus vulgaris* L.). Theor Appl Genet 128:1555–1567. https://doi.org/10.1007/s00122-015-2531-z

Clarke HJ, Khan TN, Siddique KHM (2004) Pollen selection for chilling tolerance at hybridisation leads to improved chickpea cultivars. Euphytica 139:65–74

Cleary A, Farmer A (2018) Genome context viewer: visual exploration of multiple annotated genomes using microsynteny. Bioinformatics 34(9):1562–1564

Coimbra-Gonçalves GK, Gonçalves-Vidigal MC, Coelho RT, Valentini G, Vidigal Filho PS et al (2016) Characterization and mapping of anthracnose resistance genes in mesoamerican common bean cultivar Crioulo 159. Crop Sci 56:2904–2915

Colás Sánchez A, Torres Gutiérrez R, Cupull Santana R, Rodríguez Urrutia A, Fauvart M et al (2014) Effects of co-inoculation of native *Rhizobium* and *Pseudomonas* strains on growth parameters and yield of two contrasting *Phaseolus vulgaris* L. genotypes under Cuban soil conditions. Eur J Soil Biol 62:105–112. https://doi.org/10.1016/j.ejsobi.2014.03.004

Collard BCY, Mackill DJ (2008) Marker-assisted selection: an approach for precision plant breeding in 21st century. Phil Trans Roy Soc Lond Biol Sci 363:557–572

Condori J, Sivakumar G, Hubstenberger J, Dolan MC, Sobolev VS et al (2010) Induced biosynthesis of resveratrol and the prenylated stilbenoids arachidin-1 and arachidin-3 in hairy root cultures of peanut: Effects of culture medium and growth stage. Plant Physiol Biochem 48(5):310–318

Cooper B, Campbell KB (2017) Protection against common bean rust conferred by a gene silencing method. Phytopathology 107(8):920–927. http://apsjournals.apsnet.org/doi/10.1094/ PHYTO-03-17-0095-R

Corrêa RX, Good-God PIV, Oliveira MLP, Nietsche S, Moreira MA et al (2001) Herança da resistência à mancha-angular do feijoeiro e identificação de marcadores moleculares flanqueando o loco de resistência. Fitopatol Bras 26:27–32

Crampton M, Sripathi VR, HossainK, KalavacharlaV (2016) Analyses ofmethylomes derived from Meso-American common bean (*Phaseolus vulgaris* L.) using MeDIP-Seq and whole genome sodium bisulfitesequencing. Front Plant Sci 7: 447

Cuellar-Ortiz SM, de la Paz Arrieta-Montiel M, Acosta-Gallego JA, Covarrubias AA (2008) Relationship between carbohydrate partitioning and drought resistance in common bean. Plant Cell Environ 31:1399–1409

Cutler HG, Himmelsbach DS, Arrendale RF, Cole PD, Cox RH (1989) Koninginin A: A novel plant growth regulator from *Trichoderma koningii*. Agric Biol Chem 53(10):2605–2611

Cutler HG, Jacyno JM (1991) Biological activity of (-)-Harzianopyridone isolated from *Trichoderma harzianum*. Agric Biol Chem 55(10):2629–2631

Da Silva PH (2017) Control biológico del gorgojo de la judía *Acanthoscelides obtectus* Say (Coleoptera: Chrysomelidae: Bruchinae) en la región de Castilla y Léon-España. PhD Thesis, University of León, León, Spain

Daglish GJ (2008) Impact of resistance on the efficacy of binary combinations of spinosad, chlorpyrifos-methyl and s-methoprene against five stored-grain beetles. J Stored Prod Res 44:71–76

Daglish GJ, Hall EA, Zorzetto MJ, Lambkin TM, Erbacher JM (1993) Evaluation of protectants for control of *Acanthoscelides obtectus* (Say) (Coleoptera: Bruchidae) in navy beans (*Phaseolus vulgaris* (L.)). J Stored Prod Res 29:215–219

Dal Bello G, Padín S, Juárez P, Pedrini N, De Giusto M (2006) Biocontrol of *Acanthoscelides obtectus* and *Sitophilus oryzae* with diatomaceous earth and Beauveria bassiana on stored grains. Biocontrol Sci Technol 16:215–220

Dana M de las M, Pintor-Toro JA, Cubero B (2006) Transgenic tobacco plants overexpressing chitinases of fungal origin show enhanced resistance to biotic and abiotic stress agents. Plant Physiol 142:722–730

Daryanto S, Wang L, Jacinthe PA (2015) Global synthesis of drought effects on food legume production. PLoS ONE 10(6):e0127401. https://doi.org/10.1371/journal.pone.0127401

Ddamulira G, Mukankusi C, Ochwo-Ssemakula M, Edema R, Sseruwagi P et al (2014) Distribution and variability of *Pseudocercospora griseola* in Uganda. J Agric Sci 6(6):16–29. http://www.ccsenet.org/journal/index.php/jas/article/view/33535

de Faria MR, Wraight SP (2007) Mycoinsecticides and mycoacaricides: a comprehensive list with worldwide coverage and international classification of formulation types. Biol Control 43:237–256

De Ron AM, González AM, Rodiño AP, Santalla M, Godoy L et al (2016) History of the common bean crop: its evolution beyond its areas of origin and domestication. Arbor 192(779):a317

De Ron AM, Papa R, Bitocchi E, González AM, Debouck et al (2015) Common bean. In: De Ron AM (ed) Grain Legumes, Series: Handbook of Plant Breeding. Springer Science+Business Media, New York, USA, pp 1-36

Debouck DG, Araya Villalobos R, Ocampo Sánchez RA, González UWG (1989) Collecting *Phaseolus* in Costa Rica. Plant Genet Resour Newsl 78:44–46

Debouck DG, Smartt J (1995) Beans. In: Smartt J, Simmonds NW (eds) Evolution of Crop Plants. Longman Scientific & Technical, London, United Kingdom, pp 287–296

Dita MA, Rispail N, Prats E, Rubiales D, Singh KB (2006) Biotechnology approaches to overcome biotic and abiotic stress constraints in legumes. Euphytica 147:1–24

Dixon DP, Lapthorn A, Edwards R (2002) Plant glutathione transferases. Genome Biol 3(3):3004.1-3004.10

Djonović S, Pozo MJ, Dangott LJ, Howell CR, Kenerley CM (2006) Sm1, a proteinaceous elicitor secreted by the biocontrol fungus *Trichoderma virens* induces plant defense responses and systemic resistance. Mol Plant-Microbe Interact 838(8):838–853

DOE-JGI (2018) Panicum Virgatum v2.1. http://phytozome.jgi.doe.gov/

Dowd C, Wilson IW, McFadden H (2004) Gene expression profile changes in cotton root and hypocotyl tissues in response to infection with *Fusarium oxysporum* f. sp. *vasinfectum*. Mol Plant-Microbe Interact 17(6):654–667

Drijfhout E (1978) Genetic interaction between *Phaseolus vulgaris* and bean common mosaic virus with implications for strain identification and breeding for resistance. Centre for Agricultural Publishing and Documentation, Wageningen, The Netherlands

Druzhinina IS, Seidl-Seiboth V, Herrera-Estrella A, Horwitz BA, Kenerley CM et al (2011) *Trichoderma*: The genomics of opportunistic success. Nat Rev Microbiol 9(10):749–759

Duncan RW, Gilbertson RL, Singh SP (2012) Direct and marker-assisted selection for resistance to common bacterial blight in common bean. Crop Sci 52:1511

Durán L, Blair MW, Giraldo MC, Macchiavelli R, Prophete E et al (2005) Morphological and molecular characterization of common bean landraces and cultivars from the Caribbean. Crop Sci 45:1320–1328

Durrant WE, Dong X (2004) Systemic acquired resistance. Annu Rev Phytopathol 42:185–209
Easlon HM, Nemali KS, Richards JH, Hanson DT, Juenger TE et al (2014) The physiological basis for genetic variation in water use efficiency and carbon isotope composition in *Arabidopsis thaliana*. Photosynth Res 119:119–129
Elias JCF (2018) Association Analysis for Characteristics Related to Drought Tolerance in Common Bean (*Phaseolus vulgaris* L.). Doctoral Thesis. Universidade Estadual de Maringá: Programa de Pós-Graduação em Agronomia. Brazil. 105 pp
Ender M, Kelly JD (2005) Identification of QTL associated with white mold resistance in common bean. Crop Sci 45:2482–2490
Engelberth J, Koch T, Kühnemann F, Boland W (2000) Channel-forming peptaibols are potent elicitors of plant secondary metabolism and tendril coiling. Angew Chemie Int Ed 39(10):1860–1862
Ernest EG, Wisser RJ, Johnson GC (2017) Physiological effects of heat stress on lima bean (*Phaseolus lunatus*) and development of heat tolerant screening techniques. Annu Rep Bean Improv Coop 60:101–102
Eromosele O, Bo S, Ping L (2013) Induction of phytochemical glyceollins accumulation in soybean following treatment with biotic elicitor (*Aspergillus oryzae*). J Funct Foods 5(3):1039–1048
Esquinas-Alcázar JT, Hilm A, López Noriega I (2012) A brief history of the negotiations on the International Treaty on Plant Genetic Resources for Food and Agriculture. Crop genetic resources as a global commons. Routledge, London, United Kingdom, pp 147–161
Farid M, Earl HJ, Pauls K, Navabi A (2017) Response to selection for improved nitrogen fixation in common bean (*Phaseolus vulgaris* L.). Euphytica 213(99):1–13. http://dx.doi.org/10.1007/s10681-017-1885-5
Farooq M, Wahid A, Kobayashi N, Fujita D, Basra SMA (2009) Plant drought stress: effects, mechanisms and management. Agron Sustain Dev 29:185–212
Fazio G, Staub JE, Stevens MR (2003) Genetic mapping and QTL analysis of horticultural traits in cucumber (*Cucumis sativus* L.) using recombinant inbred lines. Theor Appl Genet 107:864–874
Federici CT, Ehdaie B, Waines JC (1990) Domesticated and wild tepary bean: field performance with and without drought-stress. Agron J 82:896–900
Felicetti E, Song Q, Jia G, Cregan P, Bett KE et al (2012) Simple sequence repeats linked with slow darkening trait in pinto bean discovered by single nucleotide polymorphism assay and whole genome sequencing. Crop Sci 52:1600–1608
Feng S, Saw CL, Lee YK, Huang D (2007) Fungal-stressed germination of black soybeans leads to generation of oxooctadecadienoic acids in addition to glyceollins. J Agric Food Chem 55(21):8589–8595
Fernández GCJ (1992) Effective selection criteria for assessing stress tolerance. In: Kuo CG (ed) Proceedings of the international symposium on adaptation of vegetables and other food crops in temperature and water stress, Tainan, Taiwan
Figueiredo MVB, Martinez CR, Burity HA, Chanway CP (2008) Plant growth-promoting rhizobacteria for improving nodulation and nitrogen fixation in the common bean (*Phaseolus vulgaris* L.). World J Microbiol Biotechnol 24(7):1187–93
Finke ML, Coyne DP, Steadman JR (1986) The inheritance and association of resistance to rust, common bacterial blight, plant habit and foliar abnormalities in *Phaseolus vulgaris* L. Euphytica 35:969–982
Foster EF, Pajarito A, Acosta-Gallegos JA (1995) Moisture stress impact on N partitioning, N remobilization and N-use efficiency in beans (*Phaseolus vulgaris* L.). J Agri Sci 124:27–37
Fouilloux G (1979) New races of bean anthracnose and consequences in our breeding programs. In: Maraite H, Meyer JA (eds) International symposiumof diseases of tropical food crops. Universite Catolique de Louvain-La Neuve, Belgium, pp 221–235
Fourie D, Miklas P, Ariyarathne HM (2004) Genes conditioning halo blight resistance to races 1, 7 and 9 occur in a tight cluster. Annu Rep Bean Improv Coop 47:103–104
Foyer CH, Lam H-M, Nguyen HT, Siddique KHM, Varshnevy RK et al (2016) Neglecting legumes has compromised human health and sustainable food production. Nat Plants 2:16112

Frahm MA, Rosas JC, Mayek-Perez N, Lopez-Salinas E, Acosta-Gallegos JA et al (2004) Breeding beans for resistance to terminal drought in the lowland tropics. Euphytica 136:223–232

Freyre R, Skroch PW, Geffroy V, Adam-Blondon AF, Shirmohamadali A et al (1998) Towards an integrated linkage map of common bean. 4. Development of a core linkage map and alignment of RFLP maps. Theor Appl Genet 97:847–856

Galeano CH, Cortés AJ, Fernández AC, Soler A, Franco-Herrera N et al (2012) Gene-based single nucleotide polymorphism markers for genetic and association mapping in common bean. BMC Genetics 13:48. https://doi.org/10.1186/1471-2156-13-48

Galeano CH, Fernandez AC, Franco-Herrera N, Cichy KA, McClean PE et al (2011) Saturation of an intra-gene pool linkage map: Towards a unified consensus linkage map for fine mapping and synteny analysis in common bean. PLoS One 6:e28135

Galeano CH, Fernández AC, Gómez M, Blair MW (2009) Single strand conformation polymorphism based SNP and Indel markers for genetic mapping and synteny analysis of common bean (*Phaseolus vulgaris* L.). BMC Genomics 10:629

Galindo-González L, Mhiri C, Deyholos MC, Grandbastien MA (2017) LTR-Retrotransposons in plants: engines of evolution. Gene 626:14–25

Gallo D, Neto SS, Carvalho RPL, Baptista GC, Filho EB et al (2002) Entomologia Agrícola. Fundação de Estudos Agrários Luiz de Queiróz, Piracicaba, Brazil

Gallou A, Cranenbrouck S, Declerck S (2009) *Trichoderma harzianum* elicits defence response genes in roots of potato plantlets challenged by *Rhizoctonia solani*. Eur J. Plant Pathol 124(2):219–230

Gallou A, Declerck S, Cranenbrouck S (2012) Transcriptional regulation of defence genes and involvement of the WRKY transcription factor in arbuscular mycorrhizal potato root colonization. Funct Integr Genomics 12(1):183–198

Ganassi S, Grazioso P, De Cristofaro A, Fiorentini F, Sabatini MA et al (2016) Long chain alcohols produced by *Trichoderma citrinoviride* have phagodeterrent activity against the bird cherry-oat aphid *Rhopalosiphum padi*. Front Microbiol 7:29

Gao D, Abernathy B, Rohksar D, Schmutz J, Jackson SA (2014) Annotation and sequence diversity of transposable elements in common bean (*Phaseolus vulgaris*). Front Plant Sci 5: 1–9. http://journal.frontiersin.org/article/10.3389/fpls.2014.00339/abstract

Gao D, Zhao D, Abernathy B, Iwata-Otsubo A, Herrera-Estrella A et al (2016) Dynamics of a novel highly repetitive CACTA family in common bean (*Phaseolus vulgaris*). Genes Genomes Genet 6: 2091–2101. http://g3journal.org/lookup/doi/10.1534/g3.116.028761

Garzon LN, Blair MW (2014) Development and mapping of SSR markers linked to resistance-gene homologue clusters in common bean. Crop J. 2:183–194

Garzón LN, Ligarretoa GA, Blair MW (2007) Molecular marker-assisted backcrossing of anthracnose resistance into andean climbing beans (*Phaseolus vulgaris* L.). Crop Sci 48:562–570

Geffroy V (1997) Dissection Génétique de La Résistance à Colletotrichum Lindemuthianum, Agente de L' Anthracnose, Chez Deux Génotypes Représentatifs des Pools Géniques de *Phaseolus vulgaris*. Thèse de Doctorat. Paris-Grignon: Institut National Agronomique Paris Grignon, France, 263 pp

Geffroy V, Macadre C, David P, Pedrosa-Harand A, Sévignac M et al (2009) Molecular analysis of a large subtelomeric nucleotide-binding-site-leucine-rich-repeat family in two representative genotypes of the major gene pools of *Phaseolus vulgaris*. Genetics 181:405–419

Geffroy V, Sévignac M, Billant P, Dron M, Langin T (2008) Resistance to *Colletotrichum lindemuthianum* in *Phaseolus vulgaris*: A case study for mapping two independent genes. Theor Appl Genet 116:407–415

Geffroy V, Sevignac M, De Oliveira JCF, Fouilloux G, Skroch P et al (2000) Inheritance of partial resistance against *Colletotrichum lindemuthianum* in *Phaseolus vulgaris* and co-localization of quantitative trait loci with genes involved in specific resistance. Mol Plant Microbe Interact 13:287–296

Geffroy V, Sicard D, de Oliveira JCF, Sévignac M, Séverine C et al (1999) Identification of an ancestral resistance gene cluster involved in the coevolution process between *Phaseolus vulgaris* and its fungal pathogen *Colletotrichum lindemuthianum*. Mol Plant Microbe Interact 12(9): 774–784. http://apsjournals.apsnet.org/doi/10.1094/MPMI.1999.12.9.774

Gepts P (1988) Phaseolin as an evolutionary marker. In: Gepts P (ed) Genetic resources of *Phaseolus* beans. Kluwer, Dordrecht, The Netherlands, pp 215–241

Gepts P (1999) Development of an integrated linkage map. In: Singh SP (ed) Common Bean Improvement in the Twenty-First Century. Developments in Plant Breeding, Springer, Dordrecht, The Hetherlands, pp 53–91

Gepts P, Aragão FJ, De Barros E, Blair MW, Brondani R et al (2008) Genomics of *Phaseolus* beans, a major source of dietary protein and micronutrients in the tropics. In: Moore PH, Ming R (eds) Genomics of tropical crop plants. Springer, New York, pp 113–143

Ghosh SK, Pal S (2016) Entomopathogenic potential of *Trichoderma longibrachiatum* and its comparative evaluation with malation against the insect pest *Leucinodes orbonalis*. Environ Monitor Assess 188:37–44

Gilio TAS, Hurtado-Gonzales OP, Valentini G, Castro SAL, Elias HT et al (2017) Fine mapping the broad spectrum anthracnose resistance gene in Amendoim Cavalo. Annu Rep Bean Improv Coop 60:3–4

Gołebiowski M, Malinski E, Nawrot J, Stepnowski P (2008) Identification and characterization of surface lipid components of the dried-bean beetle *Acanthoscelides obtectus* (Say) (Coleoptera: Bruchidae). J Stored Prod Res 44:386–388

Gonçalves-Vidigal MC (1994) Herança da Resistência às Raças Alfa, Delta e Capa de *Colletotrichum lindemuthianum* (Sacc. et Magn.) Scrib. no Feijoeiro (*Phaseolus vulgaris* L.), Ph.D. Thesis, Universidade Federal de Viçosa, Viçosa, Brazil

Gonçalves-Vidigal MC, Cruz AS, Garcia A, Kami J, Vidigal Filho PS et al (2011) Linkage mapping of the *Phg-1* and *Co-1*^{4} genes for resistance to angular leaf spot and anthracnose in the common bean cultivar AND 277. Theor Appl Genet 122:893–903

Gonçalves-Vidigal MC, Cruz AS, Lacanallo GF, Vidigal Filho PS, Sousa LL et al (2013) Cosegregation analysis and mapping of the anthracnose *Co-10* and angular leaf spot *Phg-ON* disease-resistance genes in the common bean cultivar Ouro Negro. Theor Appl Genet 126:2245–2255

Gonçalves-Vidigal MC, Kelly JD (2006) Inheritance of anthracnose resistance in the common bean cultivar Widusa. Euphytica 151:411–419

Gonçalves-Vidigal MC, Lacanallo GF, Vidigal Filho PS (2008) A new Andean gene conferring resistance to anthracnose in common bean (*Phaseolus vulgaris* L.) cultivar Jalo Vermelho. Plant Breed 127:592–596

Gonçalves-Vidigal MC, Meirelles AC, Poletine JP, Sousa LL, Cruz AS et al (2012) Genetic analysis of anthracnose resistance in Pitanga dry bean cultivar. Plant Breed 131:423–429

Gonçalves-Vidigal MC, Pacheco CMNA, Vidigal Filho PS, Lacanallo GF, Sousa LL et al (2016) Genetic mapping of the anthracnose resistance gene *Co-14* in the common bean cultivar Pitanga. Annu Rept Bean Improv Coop 59:85–86

Gonçalves-Vidigal MC, Silva C, Vidigal Filho PS, Gonela A, Kvitschal MV (2007) Allelic relationships of anthracnose (*Colletotrichum lindemuthianum*) resistance in the common bean (*Phaseolus vulgaris* L.) cultivar Michelite and the proposal of a new anthracnose resistance gene, *Co-11*. Genet Mol Biol 30:589–593

Gonçalves-Vidigal MC, Vidigal Filho PS, Medeiros AF, Pastor-Corrales MA (2009) Common bean landrace Jalo Listras Pretas is the source of a new andean anthracnose resistance gene. Crop Sci 49:133–138

González AM, Godoy L, Santalla M (2017) Dissection of resistance genes to *Pseudomonas syringae* pv. *phaseolicola* in UI3 common bean cultivar. Int J Mol Sci 18:2503

González AM, Rodiño AP, Santalla M, De Ron AM (2009) Genetics of intra-gene pool and inter-gene pool hybridization for seed traits in common bean (*Phaseolus vulgaris* L.) germplasm from Europe. Field Crops Res 112:66–76

González AM, Yuste-Lisbona FJ, Godoy L, Fernández-Lozano A, Rodiño P et al (2016) Exploring the quantitative resistance to *Pseudomonas syringae* pv. *phaseolicola* in common bean (*Phaseolus vulgaris* L.). Mol Breed 36: 166 https://doi.org/10.1007/s11032-016-0589-1

Goodstein DM, Shu S, Howson R, Neupane R, Hayes RD et al (2012) Phytozome: a comparative platform for green plant genomics. Nucl Acids Res 40(D1):1178–1186

Goretti D, Bitocchi E, Bellucci E, Rodríguez M, Rau D et al (2014) Development of single nucleotide polymorphisms in *Phaseolus vulgaris* and related *Phaseolus* spp. Mol Breed 33:531–544

Gowda CLL, Parthasarathy Rao P, Bhagavatula S (2009) Global trends in production and trade of major grain legumes. International Conference on Grain Legumes: Quality Improvement, Value Addition and Trade; Indian Society of Pulses Research and Development and Indian Institute of Pulses Research, Kanpur, India, pp 282–301

Graham PH, Vance CP (2003) Legumes: Importance and constraints to greater use. Plant Physiol 131(3):872–877

Gravel V, Antoun H, Tweddell RJ (2007) Growth stimulation and fruit yield improvement of greenhouse tomato plants by inoculation with Pseudomonas putida or T*richoderma atroviride:* Possible role of indole acetic acid (IAA). Soil Biol Biochem 39(8):1968–1977

Gray SB, Strellner RS, Puthuval KK, Christopher N, Shulman RE et al (2013) Minirhizotron imaging reveals that nodulation of field-grown soybean is enhanced by free-air CO_2 enrichment only when combined with drought stress. Funct Plant Biol 40:137–147

Grayer RJ, Kokubun T (2001) Plant–fungal interactions: the search for phytoalexins and other antifungal compounds from higher plants. Phytochemistry 56(3):253–263

Grotewold E (2005) Plant metabolic diversity: a regulatory perspective. Trends Plant Sci 10(2):57–62

Guerrero-González ML, Rodríguez-Kessler M, Rodríguez-Guerra R, González-Chavira M, Simpson J et al (2011) Differential expression of *Phaseolus vulgaris* genes induced during the interaction with *Rhizoctonia solani*. Plant Cell Rep. 30(8):1465–1473

Gujaria-Verma N, Ramsay L, Sharpe AG, Sanderson L-A, Debouck DG et al (2016) Gene-based SNP discovery in tepary bean (*Phaseolus acutifolius*) and common bean (*P. vulgaris*) for diversity analysis and comparative mapping. BMC Genomics 17:1–16. https://doi.org/10.1186/s12864-016-2499-3

Hajek A (2004) Natural Enemies: An Introduction to Biological Control. Cambridge University Press, Cambridge, United Kingdom

Halewood M (2014) International efforts to pool and conserve crop genetic resources in times of radical legal change. Intellectual property rights: legal and economic challenges for development, Oxford University Press, United Kingdom, pp 288-322

Haley SD, Afanador LK, Miklas PM, Stavely JR, Kelly JD (1994) Heterogeneous inbred populations are useful as sources of near-isogenic lines for RAPD marker localization. Theor Appl Genet 88:337–342

Hanai LR, Santini L, Camargo LEA, Fungaro MHP, Gepts P et al (2010) Extension of the core map of common bean with EST-SSR, RGA, AFLP, and putative functional markers. Mol Breed 25:25–45

Hangen L, Bennik MR (2003) Consumption of black beans and navy beans (*Phaseolus vulgaris*) reduced azoxymethane-induced colon cancer in rats. Nutr Cancer 44:60–65

Hannah MA, Krämer KM, Geffroy V, Kopka J, Blair MW et al (2007) Hybrid weakness controlled by the dosage-dependent lethal (DL) gene system in common bean (*Phaseolus vulgaris*) is caused by a shoot-derived inhibitory signal leading to salicylic acid-associated root death. New Phytol 176:537–549

Harman GE, Howell CR, Viterbo A, Chet I, Lorito M (2004) *Trichoderma* species-opportunistic, avirulent plant symbionts. Nat Rev 2(1):43–56

Harman GE, Kubicek CP (1998) *Trichoderma* and *Gliocladium*. Volume 1: Enzymes, biological control and commercial applications. CRC Press, London, United Kingdom, p 393

Harman GE, Kubicek CP (2002) *Trichoderma* and *Gliocladium*. Volume 2: Basic biology, taxonomy and genetics. CRC Press London, United Kingdom, p 300

Hart JP, Griffiths PD (2013) A series of eIF4E alleles at the *Bc-3* locus are associated with recessive resistance to Clover yellow vein virus in common bean. Theor Appl Genet 126:2849–2863

Hatem I, Tan J (2003) Image analysis. In: Heldman DR (ed) Encyclopedia of agriculture, food, and biological engineering. Marcel Dekker, NewYork, USA, pp 517–523
Héraux FM, Hallett SG, Ragothama KG, Weller SC (2005) Composted chicken manure as a medium for the production and delivery of *Trichoderma virens* for weed control. HortScience 40(5):1394–1397
Hermosa MR, Grondona I, Díaz-Mínguez JM, Iturriaga EA, Monte E (2001) Development of a strain-specific SCAR marker for the detection of *Trichoderma atroviride* 11, a biological control agent against soilborne fungal plant pathogens. Curr Genet 38(6):343–350
Hermosa MR, Viterbo A, Chet I, Monte E (2012) Plant-beneficial effects of *Trichoderma* and of its genes. Microbiology 158(1):17–25
Hermosa R, Belén Rubio M, Cardoza RE, Nicolás C, Monte E et al (2013) The contribution of *Trichoderma* to balancing the costs of plant growth and defense. Int Microbiol 16(2):69–80
Herridge DF, Peoples MB, Boddey RM (2008) Global inputs of biological nitrogen fixation in agricultural systems. Plant Soil 311:1–18
Hinkossa A, Gebeyehu S, Zeleke H (2013) Generation mean analysis and heritability of drought resistance in common bean (*Phaseolus vulgaris* L.). Afric J Agri Res 8:1319–1329
Ho MD, Rosas JD, Brown KM, Lynch JP (2005) Root architectural tradeoffs for water and phosphorus acquisition. Funct Plant Biol 32:737–748
Howell CR (2003) Mechanisms employed by *Trichoderma* species in the biological control of plant diseases: the history and evolution of current concepts. Plant Dis 87(1):4–10
Howell CR, Stipanovic RD (1995) Mechanisms in the biocontrol of *Rhizoctonia solani*-induced cotton seedling disease by *Gliocladium virens*: antibiosis. Phytopathology 85:469–472
Hoyos-Villegas V, Song Q, Kelly JD (2017) Genome-wide association analysis for drought tolerance and associated traits in common bean. Plant Genome 10:1–17
Hoyos-Villegas VW, Mkwaila PB, Cregan P, Kelly JD (2015) QTL Analysis of white mold avoidance in pinto bean (*Phaseolus vulgaris*). Crop Sci 55:2116–2129
Huang FC, Studart Witkowski C, Schwab W (2010) Overexpression of hydroperoxide lyase gene in *Nicotiana benthamiana* using a viral vector system. Plant Biotechnol J 8(7):783–795
Humphreys MO, Humphreys MW (2005) Breeding for stress resistance: general principles. In: Ashraf M, Harris PJC (eds) Abiotic stresses: plant resistance through breeding and molecular approaches. Haworth, New York, USA, pp 19–46
Hungria M, Kaschuk G (2014) Regulation of N2 fixation and NO3−/NH4+ assimilation in nodulated and N-fertilized *Phaseolus vulgaris* L. exposed to high temperature stress. Environ Exp Bot 98:32–39
Hurtado-Gonzales OP, Valentini G, Gilio TAS, Martins AM, Song Q, Pastor-Corrales MA (2017a) Fine Mapping of *Ur-3*, a historically important rust resistance locus in common bean. Genes Genomes Genet 7:557–569
Hurtado-Gonzales OP, Valentini G, Gilio TAS, Quigley C, Song Q, Gonçalves-Vidigal MC, Pastor-Corrales MA (2017b) Fine mapping of genes conferring resistance to rust and anthracnose of common bean. Annu Rep Bean Improv Coop 61:27–28
Hwang C, Correll MJ, Gezan SA, Zhang L, Bhakta MS et al (2017) Next generation crop models: A modular approach to model early vegetative and reproductive development of the common bean (*Phaseolus vulgaris* L). Agri Syst 155:225–239
Hyten DL, Smithb JR, Frederickc RD, Tuckera ML, Song Q et al (2009) Bulked segregant analysis using the goldengate assay to locate the *Rpp3* locus that confers resistance to soybean rust in soybean. Crop Sci 49:265–271
Hyten DL, Song Q, Fickus EW, Quigley CV, Lim JS et al (2010) High through-put SNP discovery and assay development in common bean. BMC Genom 11:475–483
Innocenti G, Pucciariello C, Le Gleuher M, Hopkins J, de Stefano M et al (2007) Glutathione synthesis is regulated by nitric oxide in *Medicago truncatula* roots. Planta 225(6):1597–1602
Isaacs KB, Snapp SS, Chung K, Waldman KB (2016) Assessing the value of diverse cropping systems under a new agricultural policy environment in Rwanda. Food Security: The Science, Sociology and Economics of Food Production and Access to Food 8(3):491–506

Jaber LR, Enkerli J (2016) Effect of seed treatment duration on growth and colonization of *Vicia faba* by endophytic *Beauveria bassiana* and *Metarhizium brunneum*. Biol Control 103:187–195

Jaber LR, Enkerli J (2017) Fungal entomopathogens as endophytes: can they promote plant growth? Biocontrol Sci Technol 27:28–41

Jaber LR, Ownley BH (2018) Canwe use entomopathogenic fungi as endophytes for dual biological control of insect pests and plant pathogens? Biol Control 116:36–45

Jarvis DI, Hodgkin T (1999) Wild relatives and crop cultivars: detecting natural introgression and farmer selection of new genetic combinations in agro-ecosystem. Mol Ecol 8:S159–S173

Jasiński M, Kachlick P, Rodziewicz P, Figlerowicz M, Stobiecki M (2009) Changes in the profile of flavonoid accumulation in *Medicago truncatula* leaves during infection with fungal pathogen *Phoma medicaginis*. Plant Physiol Biochem 47(9):847–853

Johnson W, Gepts P (2002) The role of epistasis in controlling seed yield and other agronomic traits in an (*Phaseolus vulgaris* L.). Euphytica 125:69–79

Johnson W, Guzmán P, Mandala D, Mkandawire ABC, Temple S et al (1997) Molecular tagging of the *bc-3* gene for introgression into Andean common bean. Crop Sci 37:248–254

Jung G, Coyne D, Scroch P, Nienhuis J, Bokosi J et al (1996) Molecular markers associated with plant architecture and resistance to common blight, web blight, and rust in common beans. J Am Soc Hort Sci 121:794–803

Jung G, Coyne DP, Bokosi JM, Steadman JR, Nienhuis J (1998) Mapping genes for specific and adult plant resistance to rust and abaxial leaf pubescence and their genetic relationship using random amplified polymorphic DNA (RAPD) markers in common bean. J. Am. Soc. Hort. Sci. 123:859–863

Jung G, Skroch P, Coyne D, Nienhuis J, Arnaud-Santana E et al (1997) Molecular-markers-based genetic analysis of tepary bean derived common bacterial blight resistance in different developmental stage of common bean. J Am Soc Hort Sci 122:329–337

Kabaluk JT, Ericsson JD (2007) Seed treatment increases yield of field corn when applied for wireworm control. Agron J 99:1377–1381

Karou M, Oweis T (2012) Water and land productivities of wheat and food legumes with deficit supplemental irrigation in a Mediterranean environment. Agri Water Manag 107:94–103

Keller B, Manzanares C, Jara C, Lobaton JD, Studer B et al (2015) Fine–mapping of a major QTL controlling angular leaf spot resistance in common bean (*Phaseolus vulgaris* L.). Theor Appl Genet 128:813–826

Kelly JD (1997) A review of varietal response to bean common mosaic potyvirus in *Phaseolus vulgaris*. Plant Var Seeds 10:1–6

Kelly JD, Gepts P, Miklas PN, Coyne DP (2003) Tagging and mapping of genes and QTL and molecular marker-assisted selection for traits of economic importance in bean and cowpea. Field Crops Res 82:135–154

Kelly JD, Hosfield GL, Varner GV, Uebersax MA, Haley SD et al (1994) Registration of "Raven" black bean. Crop Sci 34:1406–1407

Kelly JD, Stavely R, Mikla P, Afanador L, Haley SD (1993) Pyramiding rust resistance genes using RAPD markers. Annu Rept Bean Improv Coop 36:166–167

Kelly JD, Vallejo V (2004) A comprehensive review of the major genes conditioning resistance to anthracnose in common bean. HortScience 39:1196–1207

Kelly JD, Young RA (1996) Proposed symbols for anthracnose resistance genes. Annu Rept Bean Improv Coop 39:20–24

Kim KC, Fan B, Chen Z (2006) Pathogen-induced *Arabidopsis* WRKY7 is a transcriptional repressor and enhances plant susceptibility to *Pseudomonas syringae*. Plant Physiol 142(3):1180–1192

Kim KD, Baidouri ME, Abernathy B, Iwata-Otsubo A, Chavarro C et al (2015) A comparative epigenomic analysis of polyploidy-derived genes in soybean and common bean. Plant Physiol 168(4):1433–1447. https://doi.org/10.1104/pp.15.00408

Kiryowa M, Nkalubo ST, Mukankusi C, Talwana H, Gibson P, Tukamuhabwa P (2015) Effect of marker aided pyramiding of anthracnose and *Pythium* root rot resistance genes on plant agronomic characters among advanced common bean genotypes. J Agric Sci 7:98–104

Koinange EMK, Singh SP, Gepts P (1996) Genetic control of the domestication syndrome in common bean. Crop Sci 36:1037–1045

Kolkman JM, Kelly JD (2003) QTL conferring resistance and avoidance to white mold in common bean. Crop Sci 43:539–548

Kumar J, Choudhary AK, Solanki RK, Pratap A (2011) Towards marker-assisted selection in pulses: a review. Plant Breed 130:297–313

Kunkel BN, Brooks DM (2002) Cross talk between signaling pathways in pathogen defense. Curr Opin Plant Biol 5(4):325–331

Lacanallo GF, Gonçalves-Vidigal MC (2015) Mapping of an Andean gene for anthracnose resistance (*Co-13*) in common bean (*Phaseolus vulgaris* L.) JaloListras Pretas landrace. Aust J Crop Sci 9:394–400

Lacey LA, Grzywacz D, Shapiro-Ilan DI, Frutos R, Brownbridge M et al (2015) Insect pathogens as biological control agents: back to the future. J Invertebr Pathol 132:1–41

Lamprecht H (1961) Weitere Kopplungsstudien an *Phaseolus vulgaris* mit einer Ubersicht ber die Koppelungsgruppen. Agri Hort Genet 19:319–332

Lefort MC, McKinnon AC, Nelson TL, Glare TR (2016) Natural occurrence of the entomopathogenic fungi *Beauveria bassiana* as a vertically transmitted endophyte of *Pinus radiate* and its effect on above- and below-ground insect pests. NZ Plant Protec 69:68–77

Li J, Brader G, Palva ET (2004) The WRKY70 transcription factor: a node of convergence for jasmonate-mediated and salicylate-mediated signals in plant defense. Plant Cell 16(2):319–331

Liebenberg MM, Pretorius ZA (1997) A review of angular leaf spot of common bean (*Phaseolus vulgaris* L.). Afr Plant Prot 3:81–106

Lima Castro SA, Gonçalves-Vidigal MC, Gilio TAS, Lacanallo GF, Valentini G et al (2017) Genetics and mapping of a new anthracnose resistance locus in andean common bean Paloma. BMC Genomics 18:306

Link W, Balko C, Stoddard FL (2010) Winter hardiness in faba bean: physiology and breeding. Field Crops Res 115:287–296

Lioi L (1989) Varition in the storage protein phaseolin in common bean (*Phaseolus vulgaris* L.) from the Mediterranean area. Euphytica 44:151–155

Lippok B, Birkenbihl RP, Rivory G, Brümmer J, Schmelzer E et al (2007) Expression of *AtWRKY33* encoding a pathogen-or PAMP-responsive WRKY transcription factor is regulated by a composite DNA motif containing W box elements. Mol Plant-Microbe Interact 20(4):420–429

Lobaton JD, Miller T, Gil J, Ariza D, de la Hoz JF et al (2018a) Resequencing of common bean identifies regions of inter–gene pool introgression and provides comprehensive resources for molecular breeding. Plant Genome 11(2):1–21

Lobaton JD, Miller T, Gil J, Ariza D, de la Hoz JF et al (2018b) Resequencing of common bean identifies regions of inter–gene pool introgression and provides comprehensive resources for molecular breeding. Plant Genome 11:170068

Long R, Temple S, Meyer R, Schwankl L, Godfrey L et al. (2014) Lima bean production in California. UC ANR Publication 8505. http://beans.ucanr.edu/files/204221.pdf

López-Llorca LV, Hans-Börje J (2001) Biodiversidad del suelo: control biológico de nematodos fitopatógenos por hongos nematófagos. Cuaderno Biodiv 3(6):12–15

Lorito M, Harman GE, Hayes CK, Broadway RM, Tronsmo A et al (1993) Chitinolytic enzymes produced by Trichoderma harzianum : antifungal activity of purified endochitinase and chitobiosidase. Phytopathology 83(3):302–307

Lorito M, Hayes CK, Di Pietro A, Woo SL, Harman GE (1994) Purification, characterization, and synergistic activity of a glucan 1,3-β-glucosidase and an N-acetyl-β-glucosaminidase from *Trichoderma harzianum*. Phytopathology 84(4):398–405

Lorito M, Woo SL, Garcia I, Colucci G, Harman GE et al (1998) Genes from mycoparasitic fungi as a source for improving plant resistance to fungal pathogens. Proc Natl Acad Sci USA 95(14):7860–7865

Lorito M, Woo SL, Harman GE, Monte E (2010) Translational research on *Trichoderma*: From 'omics to the field. Annu Rev Phytopathol 48(1):395–417

Loscos J, Matamoros MA, Becana M (2008) Ascorbate and homoglutathione metabolism in common bean nodules under stress conditions and during natural senescence. Plant Physiol 146(3):1282–1292

Lozovaya VV, Lygin AV, Zernova OV, Li S, Hartman GL, Widholm JM (2004) Isoflavonoid accumulation in soybean hairy roots upon treatment with *Fusarium solani*. Plant Physiol Biochem 42(7–8):671–679

Lu Z, Tombolini R, Woo S, Zeilinger S, Lorito M et al (2004) *In vivo* study of *Trichoderma*-pathogen-plant interactions, using constitutive and inducible green fluorescent protein reporter systems. Appl Environ Microbiol 70(5):3073–3081

Lu ZX, Gaudet DA, Frick M, Puchalski B, Genswein B et al (2005) Identification and characterization of genes differentially expressed in the resistance reaction in wheat infected with *Tilletia tritici*, the common bunt pathogen. BMB Rep 38(4):420–431

Luana MD, Gonela A, Elias HT, da Silva CR, Pastre HHet al (2017)Common bean germplasm resistant to races 73 and 2047 of *Colletotrichum lindemuthianum*. Afr J Biotechnol 16(19):1142–49. http://academicjournals.org/journal/AJB/article-abstract/79EE2BB64180

Lukatkin A, Brazaityte A Bobinas C. Duchovskis P (2012). Chilling injury in chilling-sensitive plants: a review. Zemdirbyste Agri 99:111-124

Lupwayi NZ, Kennedy AC, Chirwa RM (2011) Grain legume impacts soil biological processes in sub-Saharan Africa. Afr J Plant Sci 5:1–7

Macías FA, Varela RM, Simonet AM, Cutler HG, Cutler SJ et al (2000) Bioactive carotanes from *Trichoderma virens*. J Nat Prod 63(9):1197–1200

Maehara N, Kanzaki N (2013) Effect of aging in adult *Monochamus alternatus* (Coleoptera: Cerambycidae) on the susceptibility of the beetle to *Beauveria bassiana* (Ascomycota:Hypocreales). J Forest Res 19(3):357–360

Mafakheri A, Siosemardeh A, Bahramnejad B, Struik PC, Sohrabi E (2010) Effect of drought stress on yield, proline and chlorophyll contents in three chickpea cultivars. Aust J Crop Sci 4(8):580–585

Mahuku GS, Henríquez MA, Montoya C, Jara C, Teran H et al (2011) Inheritance and development of molecular markers linked to angular leaf spot resistance genes in the common bean accession G10909. Mol Breed 28:57–71

Mahuku GS, Iglesias ÁM, Jara C (2009) Genetics of angular leaf spot resistance in the Andean common bean accession G5686 and identification of markers linked to the resistance genes. Euphytica 167:381–396

Malmierca MG, Barua J, Mccormick SP, Izquierdo-Bueno I, Cardoza RE et al (2014) Novel aspinolide production by *Trichoderma arundinaceum* with a potential role in *Botrytis cinerea* antagonistic activity and plant defence priming. Environ Microbiol 17(4):1103–1118

Malmierca MG, Cardoza RE, Alexander NJ, McCormick SP, Collado IG et al (2013) Relevance of trichothecenes in fungal physiology: Disruption of *tri5* in *Trichoderma arundinaceum*. Fungal Genet Biol 53:22–33

Malmierca MG, Cardoza RE, Alexander NJ, McCormick SP, Hermosa R et al (2012) Involvement of *Trichoderma* trichothecenes in the biocontrol activity and induction of plant defense-related genes. Appl Environ Microbiol 78(14):4856–4868

Maqbool A, Shafiq S, Lake L (2010) Radiant frost tolerance in pulse crops - a review. Euphytica 172:1–12

Marchive C, Léon C, Kappel C, Coutos-Thévenot P, Corio-Costet MF et al (2013) Over-expression of *VvWRKY1* in grapevines induces expression of jasmonic acid pathway-related genes and confers higher tolerance to the downy mildew. PLoS One 8(1):e54185

Marfori EC, Kajiyama S, Fukusaki E, Kobayashi A (2002) Trichosetin, a novel tetramic acid antibiotic produced in dual culture of *Trichoderma harzianum* and *Catharanthus roseus* callus. Z Naturforsch C 57(5–6):465–470

Marfori EC, Kajiyama S, Fukusaki E, Kobayashi A (2003) Phytotoxicity of the tetramic acid metabolite trichosetin. Phytochemistry 62(5):715–721

Margesin R, Neuner G, Storey KB (2007) Cold-loving microbes, plants, and animals fundamental and applied aspects. Naturwissenschaften 94:77–99

Marra R, Ambrosino P, Carbone V, Vinale F, Woo SL et al (2006) Study of the three-way interaction between *Trichoderma atroviride*, plant and fungal pathogens by using a proteomic approach. Curr Genet 50(5):307–321

Martínez B, Infante I, Ii D, Reyes Y (2013) *Trichoderma* spp. y su función en el control de plagas en los cultivos. Rev Protec Vegetal 28:1–11

Mauch F, Dudler R (1993) Differential induction of distinct glutathione-S-transferases of wheat by xenobiotics and by pathogen attack. Plant Physiol 102(4):1193–1201

Mauch-Mani B, Slusarenko AJ (1996) Production of salicylic acid precursors is a major function of Phenylalanine Ammonia-Lyase in the resistance of *Arabidopsis* to *Peronospora parasitica*. Plant Cell 8(2):203–212

Maxwell JJ, Brick MA, Byrne PF, Schwartz H, Shan X et al (2007) Quantitative trait loci linked to white mold resistance in common bean. Crop Sci 47:2285–2294

Mayo S, Cominelli E, Sparvoli F, González-López O, Rodríguez-González A et al (2016a) Development of a qPCR strategy to select bean genes involved in plant defense response and regulated by the *Trichoderma velutinum* – *Rhizoctonia solani* interaction. Front Plant Sci 7:1109

Mayo S, Gutiérrez S, Cardoza RE, Hermosa R, Monte E et al (2017) *Trichoderma* species as biocontrol agents in legumes. In: Clemente A, Jiménez-López JC (eds) Legumes for global food security. Nova Science Publishers, New York, USA, pp 73–100

Mayo S, Gutierrez S, Malmierca MG, Lorenzana A, Campelo MP et al (2015) Influence of *Rhizoctonia solani* and *Trichoderma* spp. in growth of bean (*Phaseolus vulgaris* L.) and in the induction of plant defense-related genes. Front Plant Sci 6:685–696

Mayo S, Izquierdo H, González-López Ó, Rodríguez-González Á, Lorenzana A et al (2016b) Effect of farnesol, a compound produced by *Trichoderma* when growing on bean (*Phaseolus vulgaris* L.). Planta Med 82(S01):S1–S381

Mayor-Duran VM, Raatz B, Blair MW (2016) Desarrollo de líneas de frijol (*Phaseolus vulgaris* L.) tolerante a sequía a partir de cruces inter-acervo con genotipos procedentes de diferentes orígenes (Mesoamericano y Andino). Acta Agron 65:431–437

McCallum CM, Comai L, Greene EA, Henikoff S (2000) Targeting induced local lesions in genomes (TILLING) for plant functional genomics. Plant Physiol 123:439–442

McClean PE, Burridge J, Beebe S, Rao IM, Porch TG (2011) Crop improvement in the era of climate change: An integrated, multi-disciplinary approach for common bean (*Phaseolus vulgaris* L.). Funct Plant Biol 38:927–933

McConnell M, Mamidi S, Lee R, Chikara S, Rossi M et al (2010) Syntenic relationships among legumes revealed using a gene-based genetic linkage map of common bean (*Phaseolus vulgaris* L.).Theor Appl Genet 121(6):1103–1116

McRostie GP (1919) Inheritance of anthracnose resistance as indicated by a cross between a resistant and a susceptible bean. Phytopathology 9:141–148

Melotto M, Kelly JD (2000) An allelic series at the Co-1 locus for anthracnose in common bean of Andean origin. Euphytica 116:143–149

Mendéz-Vigo B, Rodríguez C, Pañeda A, Ferreira JJ, Giraldez R (2005) Molecular markers and allelic relationships of anthracnose resistance gene cluster B4 in common bean. Euphytica 141:237–245

Menjivar-Barahona RD (2010) The systemic activity of mutualistic endophytic fungi in Solanaceae and Cucurbitaceae plants on the behaviour of the phloem-feeding insects Trialeurodes vaporariorum, *Aphis gossypii* and *Myzus persicae*. Inaugural dissertation PhD. Rhenish FriedrichWilhelm University, Bonn, Germany, 120 pp

Mensack MM, Fitzgerald VK, Ryan E, Lewis MR, Thompson HJ et al. (2010) Evaluation of diversity among common beans (*Phaseolus vulgaris* L.) from two centers of domestication using 'omics' technologies. BMC Genomics 11(1):686. http://www.biomedcentral.com/1471-2164/11/686/

Mensack MM, McGinley JN, Thompson HJ (2012) Metabolomic analysis of the effects of edible dry beans (*Phaseolus vulgaris* L.) on tissue lipid metabolism and carcinogenesis in rats. Brit J Nutr 108 Suppl 1S155-165

Meyers JM, Stephen FM, Haavik LJ, Steinkraus DC (2013) Laboratory and field bioassays on the effects of *Beauveria bassiana* Vuillemin (Hypocreales: Cordycipitaceae) on red oak borer, *Enaphalodes rufulus* (Haldeman) (Coleoptera: Cerambycidae). Biol Control 65:258–264

Meyling NV, Eilenberg J (2007) Ecology of the entomopathogenic fungi *Beauveria bassiana* and *Metarhizium anisopliae* in temperate agroecosystems: potential for conservation biological control. Biol Control 43(2):145–155

Meziadi C, Richard MMS, Derquennes A, Thareau V, Blanchet S, Gratias A et al (2016) Development of molecular markers linked to disease resistance genes in common bean based on whole genome sequence. Plant Sci 242:351–357

Mhlongo MI, Steenkamp PA, Piater LA, Madala NE, Dubery IA (2016) Profiling of altered metabolomic states in *Nicotiana tabacum* cells induced by priming agents. Front Plant Sci 7:1527

Michaels TE, Smith TH, Larsen J, Beattie AD, Pauls KP (2006) OAC Rex common bean. Can. J. Plant Sci. 86:733736

Mienie CM, Liebenberg MM, Pretorius ZA, Miklas PN (2005) SCAR markers linked to the common bean rust resistance gene *Ur-13*. Theor Appl Genet 111:972–979

Miklas PN, Stavely JR, Kelly JD (1993) Identification and potential use of a molecular marker for rust resistance in common bean. Theor Appl Genet 85:745–749

Miklas P, Fourie D, Trapp J, Larsen RC, Chavarro C, Blair MW, Gepts P (2011) Genetic characterization and molecular mapping *Pse-2* gene for resistance to halo blight in common bean. Crop Sci 51:2439–2448

Miklas P, Fourie D, Wagner J, Larsen RC, Mienie CMC (2009) Tagging and mapping *Pse-1* gene for resistance to halo blight in common bean host differential cultivar UI-3. Crop Sci 49:2009

Miklas PN, Delorme R, Stone V, Stavely J, Steadman J et al (2000) Bacterial, fungal, virus disease loci mapped in a recombinant inbred common bean population ('Dorado/XAN176'). J Am Soc Hort Sci 125:476–481

Miklas PN, Fourie D, Trapp J, Davis J, Myers JR (2014) New loci including *Pse-6* conferring resistance to halo bacterial blight on chromosome Pv04 in common bean. Crop Sci 54:2099–2108

Miklas PN, Johnson E, Stone V, Beaver JS, Montoya C, Zapata M (1996) Selective mapping of QTL conditioning disease resistance in common bean. Crop Sci 36:1344–1351

Miklas PN, Johnson WC, Delorme R, Gepts P (2001) QTL conditioning physiological resistance and avoidance to white mold in dry bean. Crop Sci. 41:309–315

Miklas PN, Kelly JD, Beebe SE, Blair MW (2006a) Common bean breeding for resistance against biotic and abiotic stresses: From classical to MAS breeding. Euphytica 147:105–131

Miklas PN, Kelly JD, Singh SP (2003) Registration of anthracnose resistant Pinto Bean germplasm line USPT-ANT-1. Crop Sci 43:1889–1890

Miklas PN, Larsen KM, Terpstra KA, Hauf DC, Grafton KF, Kelly JD (2007) QTL analysis of ICA Bunsi-derived resistance to white mold in a pinto x navy bean cross. Crop Sci 47:174–179

Miklas PN, Pastor-Corrales MA, Jung G, Coyne DP, Kelly JD, Mcclean PE, Gepts P (2002) Comprehensive linkage map of bean rust resistance genes. Annu Rep Bean Improv Coop 45:125–129

Miklas PN, Porter LD, Kelly JD, Myers JR (2013) Characterization of white mold disease avoidance in common bean. Eur J Plant Pathol 135:525–543

Miklas PN, Smith JR, Singh SP (2006b) Registration of common bacterial blight resistant dark red kidney bean germplasm line USDK-CBB-15. Crop Sci 46:1005

Miklas PN, Stone V, Urrea CA, Johnson E, Beaver JS (1998) Inheritance and QTL analysis offield resistance to ashy stem blight. Crop Sci 38:916–921

Miller T, Gepts P, Kimmo S, Arunga E, Chilagane LA, Nchimbi-Msolla S, Namusoke A, Namayanja A, Tedla YR (2018) Alternative markers linked to the *Phg-2* angular leaf spot resistance locus in common bean using the phaseolus genes marker database. African Journal Biotechnology 17:818–828

Mkwaila W, Terpstra KA, Ender M, Kelly JD (2011) Identification of QTL for resistance to white mold in wild and landrace germplasm of common bean. Plant Breed 130:665–672

Moghaddam SM, Mamidi S, Osorno JM, Lee R, Brick M et al. (2016) Genome-wide association study identifies candidate loci underlying agronomic traits in a middle american diversity panel of common bean. Plant Genome 9(3): 1–21. https://dl.sciencesocieties.org/publications/tpg/abstracts/9/3/plantgenome2016.02.0012

Moghaddam SM, Song Q, Mamidi S, SchmutzJ Lee R et al (2014) Developing market class specific InDel markers from next generation sequence data in *Phaseolus vulgaris* L. Front Plant Sci 5:1–14

Mohan M, Nair S, Bhagwat A, Krishna TG, Yano M et al (1997) Genome mapping, molecular markers and marker-assisted selection in crop plants. Mol Breed 3:87–103

Monte E (2001) Understanding Trichoderma : between biotechnology and microbial ecology. Int Microbiol 4:1–4

Montero M, Sanz L, Rey M, Llobell A, Monte E (2007) Cloning and characterization of bgn16·3, coding for a β-1,6-glucanase expressed during *Trichoderma harzianum* mycoparasitism. J Appl Microbiol 103(4):1291–1300

Montero M, Sanz L, Rey M, Monte E, Llobell A (2005) BGN16.3, a novel acidic β-1,6-glucanase from mycoparasitic fungus *Trichoderma harzianum* CECT 2413. FEBS J. 272(13):3441–3448

Moons A (2005) Regulatory and functional interactions of plant growth regulators and plant glutathione S-transferases (GSTs). Vitam Horm 72:155–202

Morrissey JP, Osbourn AE (1999) Fungal resistance to plant antibiotics as a mechanism of pathogenesis. Microbiol Mol Biol Rev 63(3):708–724

Mukeshimana G, Butare L, Cregan PB, Blair MW, Kelly JD (2014) Quantitative trait loci associated with drought tolerance in common bean. Crop Sci 54:923–938

Mwenda GN, O'Hara GW, De Meyer SE, Hwieson JG, Terpolilli JJ (2018) Genetic diversity and symbiotic effectiveness of Phaseolus Vulgaris-nodulating rhizobia in Kenya". Syst Appl Microbiol 41(4):1–9. https://doi.org/10.1016/j.syapm.2018.02.001

Naderpour M, Lund OS, Larsen R, Johansen E (2010) Potyviral resistance derived from cultivars of *Phaseolus vulgaris* carrying *bc-3* is associated with the homozygotic presence of a mutated eIF4E allele. Mol Plant Pathol 11:255–263

Namayanja A, Buruchara R, Mahuku G, Rubaihayo P, Kimani P, Mayanja S, Eyedu H (2006) Inheritance of resistance to angular leaf spot in common bean and validation of the utility of resistance linked markers for marker assisted selection out side the mapping population. Euphytica 151:361–369

Narasimhan ML, Bressan RA, D'Urzo MP, Jenks MA, Mengiste T (2009) Unexpected turns and twists in structure/function of PR-proteins that connect energy metabolism and immunity. Adv Bot Re. 51:439–489

Nemchinova YP, Stavely JR (1998) Development of SCAR primers for the *Ur-3* rust resistance gene in common bean. Phytopathology 88:S67

Nicoglou A, Merlin F (2017) Epigenetics: a way to bridge the gap between biological fields. Studies in history and philosophy of science part C: Stud Hist Philos Biol Biomedl Sci 66:73–82

Nielsen KF, Gräfenhan T, Zafari D, Thrane U (2005) Trichothecene production by T*richoderma brevicompactum*. J Agric Food Chem 53(21):8190–8196

Nodari RO, Tsai SM, Gilbertson RL, Gepts P (1993) Towards an integrated linkage map of common bean. 1. Development of an RFLP-based linkage map. Theor Appl Genet 85:513–520

Noordermeer MA, Veldink GA, Vliegenthart JFG (2001) Fatty acid hydroperoxide lyase: a plant cytochrome P450 enzyme involved in wound healing and pest resistance. Chembiochem 2(7–8):494–504

Nugroho LH, Verberne MC, Verpoorte R (2002) Activities of enzymes involved in the phenylpropanoid pathway in constitutively salicylic acid-producing tobacco plants. Plant Physiol Biochem 40(9):755–760

Oblessuc P, Baroni R, Garcia AA, Chioratto AF, Carbonell SA, Camargo LE, Benchimol L (2012) Mapping of angular leaf spot resistance QTL in common bean (*Phaseolus vulgaris* L.) under different environments. BMC Genetics 13:50

Oblessuc PR, Perseguini JMKC, Baroni RM, Chioratto AF, Carbonell SA, Mondero JM, Vidal RO, Camargo LE, Benchimol L (2013) Increasing the density of markers around a major QTL controlling resistance to angular leaf spot in common bean. Theor Appl Genet 126:2451–2465

O'Boyle PD, Kelly JD, Kirk WW (2007) Use of marker-assisted selection to breed for resistance to common bacterial blight in common bean. J. Amer Soc Hort Sci 132:381–386

Odogwu BA, Nkalubo ST, Mukanski C, Paparu P, Payrick R et al. (2016) Prevalence and variability of the common bean rust in Uganda." Afri J Agri Res 11(49):4990–99. http://academicjournals.org/journal/AJAR/article-abstract/E7B3D8E62040

Ojwang PPO, Melis R, Songa JM, Githiri M, Bett C (2009) Participatory plant breeding approach for host plant resistance to bean fly in common bean under semi-arid Kenya conditions. Euphytica 170(3):383–393

Okii D, Tukamuhabwa P, Kami J, Namayanja A, Paparu P et al (2014) The genetic diversity and population structure of common bean (*Phaseolus vulgaris* L) germplasm in Uganda. Afr J Biotechnol 13(29): 2935–49. http://academicjournals.org/journal/AJB/article-abstract/59E971146091

Oldroyd GE, Dixon R (2014) Biotechnological solutions to the nitrogen problem. Curr Opin Biotechnol 26:19–24

O'Leary BM, Neale HC, Geilfus CM, Jackson RW, Arnold DL et al (2016) Early changes in apoplast composition associated with defence and disease in interactions between *Phaseolus vulgaris* and the halo blight pathogen *Pseudomonas Syringae* Pv. *phaseolicola*. Plant Cell Environ 39(10):2172–84

Oliveira EJ, Alzate-Marin AL, Borém A, Azeredo Fagundes S, Barros EG, Moreira MA (2005) Molecular marker-assisted selection for development of common bean lines resistant to angular leaf spot. Plant Breed 124:572–575

Oliveira MRC, Corrêa AS, de Souza GA, Guedes RNC, Oliveira LO (2013) Mesoamerican origin and pre- and post-columbian expansions of the ranges of *Acanthoscelides obtectus* Say, a cosmopolitan insect pest of the common bean. PLoS One 8:e70039

Omae H, Kumar A, Shono M (2012) Adaptation to high temperature and water deficit in the common bean (*Phaseolus vulgaris* L.) during the reproductive period. J Bot (2012) ID 803413. https://doi.org/10.1155/2012/803413

Osorno JM, McClean PE (2014) Common bean genomics and its applications in breeding programs. In: Gupta S, Nadarajan N, Gupta D (eds) Legumes in the Omic Era. Springer, New York, pp 185–206

Ownley B, Gwinn KD, Vega FE (2010) Endophytic fungal entomopathogens with activity against plant pathogens: ecology and evolution. Biocontrol 55:113–128

Padder BA, Sharma PN, Awale HE, Kelly JD (2017) *Colletotrichum lindemuthianum*, the causal agent of bean anthracnose. J Plant Pathol 99(2):317–330. https://doi.org/10.4454/jpp.v99i2.3867

Palacios XF (1998) Contribution to the estimation of countries' interdependence in the area of plant genetic resources'. Commission on Genetic Resources for Food and Agriculture, Background Study Paper No. 7, Rev. 1 (Food and Agriculture Organization of the United Nations), Rome, Italy

Pandey SP, Somssich IE (2009) The role of WRKY transcription factors in plant immunity. Plant Physiol 150(4):1648–1655

Papa R, Acosta J, Delgado-Salinas A, Gepts PA (2005) A genome-wide analysis of differentiation between wild and domesticated *Phaseolus vulgaris* from Mesoamerica. Theor Appl Genet 111:1147–1158

Papa R, Belluci E, Rossi M, Leonardi S, Rau D et al (2007) Tagging the signatures of domestication in common bean (*Phaseolus vulgaris*) by means of pooled DNA samples. Ann Bot 100:1039–1051

Papa R, Gepts P (2003) Asymmetry of gene flow and differential geographical structure of molecular diversity in wild and domesticated common bean (*Phaseolus vulgaris* L.) from Mesoamerica. Theor Appl Genet 106:239–250

Papavizas GC (1985) Trichoderma and Gliocladium: biology, ecology, and potential for biocontrol. Annu Rev Phytopathol 23(1):23–54

Pardo de Santayana M, Morales R, Aceituno-Mata L, Molina M (eds.) (2014) Inventario español de los conocimientos tradicionales relativos a la biodiversidad. Ministerio de Agricultura, Alimentación y Medio Ambiente. Madrid, España, 411 p

Park SO, Coyne DP, Steadman JR, Skroch PW (2001) Mapping of QTL for resistance to white mold diseases in common bean. Crop Sci 41:1253–1262

Parker SR, Cutler HG, Schrelner PR (1995) Koninginin C: A biologically active natural product from *Trichoderma koningii.* Biosci Biotechnol Biochem 59(6):1126–1127

Passioura JB (2012) Phenotyping for drought tolerance in grain crops: When is it useful to breeders? Funct Plant Biol 39:851–859

Paszkowski J (2015) Controlled activation of retrotransposition for plant breeding. Curr Opin Biotechnol 32:200–206. https://doi.org/10.1016/j.copbio.2015.01.003

Paul UV, Lossini JS, Edwards PJ, Hilbeck A (2009) Effectiveness of products from four locally grown plants for the management of *Acanthoscelides obtectus* (Say) and *Zabrotes subfasciatus* (Boheman) (both Coleoptera: Bruchidae) in stored beans under laboratory and farm conditions in Northern Tanzania. J Stored Prod Res 45:97–107

Paulitz TC (1990) Biochemical and ecological aspect of competition in biological control. In New directions in biological control. Alternatives for suppressing agricultural, Pests and diseases, Baker RR, Dunn PE (eds) Wiley-Liss Inc, New York, USA, p 837

Pedraza F, Gallego G, Beebe S, Tohme J (1997) Marcadores SCAR y RAPD para la resistencia a la bacteriosis comun (CBB). In: Singh SP, Voysest O (eds) Taller de mejoramiento de frijol para el siglo XXI, bases para una estrategia para America Latina. CIAT, Cali, Colombia, pp 130–134

Pereira JL, Queiroz RML, Charneau SO, Felix CR, Ricart CAO et al (2014) Analysis of *Phaseolus vulgaris* response to its association with *Trichoderma harzianum* (ALL-42) in the presence or absence of the phytopathogenic fungi *Rhizoctonia solani* and *Fusarium solani.* PLoS One 9(5):e98234

Pérez-Vega E, Pascual A, Campa A, Giraldez R, Miklas PN, Ferreira JJ (2012) Mapping quantitative trait loci conferring partial physiological resistance to white mold in the common bean RIL populationXana x Cornell 49242. Mol Breed 29:31–41

Perry G, DiNatale C, Xie W, Navabi A, Reinprecht Y et al (2013) A comparison of the molecular organization of genomic regions associated with resistance to common bacterial blight in two *Phaseolus vulgaris* genotypes. Front Plant Sci 4:318. https://doi.org/10.3389/fpls.2013.00318

Piergiovanni A, Taranto G, Losavio FP, Pignone D (2006) Common bean (*Phaseolus vulgaris* L.) landraces from Abruzzo and Lazio regions (Central Italy). Genet Resour Crop Evol 53:313–322

Pimentel C, Laffray D, Louguet P (1999) Intrinsic water use efficiency at the pollination stage as a parameter for drought tolerance in *Phaseolus vulgaris.* Physiol Plant 106:184–189

Polania JA, Poschenrieder C, Beebe SE, Rao IM (2016) Effective use of water and increased dry matter partitioned to grain contribute to yield of common bean improved for drought resistance. Front Plant Sci 7:660. https://doi.org/10.3389/fpls.2016.00660

Porch TG (2006) Application of stress indices for heat tolerance screening of common bean. J Agron Crop Sci 192:390–394

Porch TG, Beaver JS, Brick MA (2013a) Registration of tepary germplasm with multiple-stress tolerance, TARS-Tep 22 and TARS-Tep 32. J Plant Reg 7:358–364

Porch TG, Beaver JS, Debouck DG, Jackson SA, Kelly JD, Dempewolf H (2013b) Use of wild relatives and closely related species to adapt common bean to climate change. Agronomy 3:433–461

Porch TG, Smith JR, Beaver JS, Griffiths PD Canaday CH (2010) TARS-HT1 and TARS-HT2 heat-tolerant dry bean germplasm. HortScience 45:1278-1280

Porch TG, Urrea CA, Beaver JS, Valentin S, Peña PA, Smith JR (2012) Registration of TARS-MST1 and SB-DT1 multiple-stress-tolerant black bean germplasm. J Plant Reg 6:75–80

Pucheta-Díaz M, Flores-Macías A, Rodríguez-Navarro S, De La Torre M (2006) Mecanismo de acción de los hongos entomopatógenos. Interciencia 31:856–860

Purcell LC (2009) Physiological responses of N2 fixation to drought and selecting genotypes for improved N2 fixation. In: Emerich DW, Krishnan HB (eds) Nitrogen fixation in crop production. American Society of Agronomy, Madison, USA, pp 211–238

Quesada-Moraga E, López-Díaz C, Landa BB (2014) The hidden habit of the entomopathogenic fungus *Beauveria bassiana*: first demonstration of vertical plant transmission. PLoS One 9(2):e89278
Quintela ED (2002) Manual de identificação dos insetos e outros invertebrados pragas do feijoeiro. Santo Antônio de Goiás: Embrapa Arroz e Feijão 142:51
Ragagnin VA, De Souza TLPO, Sanglard DA, Arruda KMA, Costa MR, Alzate-Marin AL, Carneiro JE, De Barros EG (2009) Development and agronomic performance of common bean lines simultaneously resistant to anthracnose, angular leaf spot and rust. Plant Breed 128:156–163
Rainey KM, Griffiths PD (2005) Inheritance of heat tolerance during reproductive development in snap bean (*Phaseolus vulgaris* L.). J Amer Soc Hort Sci 130:700–706
Ramaekers L, Galeano CH, Garzón N, Vanderleyden J, Blair MW (2013) Identification of quantitative trait loci for symbiotic nitrogen fixation capacity and related traits in common bean. Mol Breed 31:163–180
Ramalingam A, Kudapa H, Pazhamala LT, Weckwerth W, Varshney RV (2015) Proteomics and metabolomics: two emerging areas for legume improvement. Front Plant Sci 1:21. http://journal.frontiersin.org/Article/10.3389/fpls.2015.01116/abstract
Ramirez Builes VH, Porch TG, Harmsen EW (2011) Genotypic differences in water use efficiency of common bean under drought stress. Agron J 103:1206–1215
Ramírez S, Suris M (2015) Ciclo de vida de *Acanthoscelides obtectus* (Say.) sobre frijol negro (*Phaseolus vulgaris* L.) en condiciones de laboratorio. Rev Protec Veg 30(2): 158-160
Ramirez-Cabral NYZ, Kumar L, Taylor S (2016) Crop niche modeling projects major shifts in common bean growing areas. Agri For Meteorol 218:102–113
Rao IM, Beebe SE, Polania J, Ricaurte J, Cajiao C et al (2013) Can tepary bean be a model for improvement of drought resistance in common bean? Afr Crop Sci J 21:265–281
Razinger J, Lutz M, Schroers HJ, Urek G, Grunder J (2014) Evaluation of insect associated and plant growth promoting fungi in the control of cabbage root flies. J Econ Entomol 107(4):1348–1354
Razinger J, Zerjav M, Zemljic-Urbancic M, Modic S, Lutz M et al (2017) Comparison of Sauliflower-insect-fungus interactions and pesticides for cabbage root fly control. Insect Sci 24(6):1057–1064
Regnault-Roger C, Vincent C, Arnason JT (2012) Essential oils in insect control: low-risk products in a high-stakes world. Annu Rev Entomol 57:405–424
Reinheimer JL, Barr AR, Eglinton JK (2004) QTL mapping of chromosomal regions conferring reproductive frost tolerance in barley (*Hordeum vulgare* L.). Theor Appl Genet 109:265–273
Reino JL, Guerrero RF, Hernández-Galán R, Collado IG (2007) Secondary metabolites from species of the biocontrol agent *Trichoderma*. Phytochem Rev 7(1):89–123
Rendón-Anaya M, Montero-Vargas JM, Saburido-Álvarez S, Vlasova A, Capella-Gutierrez, et al (2017) Genomic history of the origin and domestication of common bean unveils its closest sister species. Genome Biol 18(60):1–17
Richards MS, Pflieger S, Sévignac M, Thareau V, Blanchet S et al (2014) Fine mapping of *Co-x*, an anthracnose resistance gene to a highly virulent strain of *Colletotrichum lindemuthianum* in common bean. Theor Appl Genet 127:1653–1666
Rockström J, Falkenmark M, Karlberg L, Hoff H, Rost S et al (2009) Future water availability for global food production: the potential of green water for increasing resilience to global change. Water Resour Res 45:W00A12, https://doi.org/10.1029/2007wr006767
Rodiño AP, De La Fuente M, De Ron AM, Lema MJ, Drevon JJ et al (2011) Variation for nodulation and plant yield of common bean genotypes and environmental effects on the genotype expression. Plant Soil 346:349–361
Rodiño AP, Riveiro M, Santalla M, De Ron AM (2007) Sources of variation of common bean for drought tolerance. Annu Rept Bean Improv Coop 50:163–164
Rodiño P, Santalla M, González AM, De Ron AM, Singh SP (2006) Novel genetic variation in common bean from the Iberian Peninsula. Crop Sci 46:2540–2546
Rodríguez-González A, Casquero PA, Suárez-Villanueva V, Carro-Huerga G, Mayo-Prieto S et al (2018) Effect of trichodiene production by *Trichoderma harzianum* on *Acanthoscelides obtectus*. J Stored Prod Res 77:231–239. https://doi.org/10.1016/j.jspr.2018.05.001

Rodríguez-González A, Mayo S, González-López O, Reinoso B, Gutiérrez S et al (2017a) Inhibitory activity of *Beauveria bassiana* and *Trichoderma* spp. on the insect pests *Xylotrechus arvicola* (Coleoptera: Cerambycidae) and *Acanthoscelides obtectus* (Coleoptera: Chrisomelidae: Bruchinae). Environ Monitor Assess 189:12

Rodríguez-González A, Peláez HJ, GonzálezNúñez M, Casquero PA (2017b) Control of egg and neonate larvae of *Xylotrechus arvicola* (Coleoptera: Cerambycidae), a new vineyard pest, under laboratory conditions. Aust J Grape Wine Res 23:112–119

Rodríguez-González A, Peláez HJ, Mayo S, González-López O, Casquero PA (2016) Egg development and toxicity of insecticides to eggs, neonate larvae and adults of *Xylotrechus arvicola*, a pest in Iberian grapevines. Vitis 5:83–93

Rodríguez-Suárez C, Ferreira JJ, Campa A, Pañeda A, Giradles R (2008) Molecular mapping and intra-cluster recombination between anthracnose race-specific resistance genes in the common bean differential cultivars Mexico 222 and Widusa. Theor. Appl. Genet. 116:807–814

Román-Aviles B, Beaver JS (2003) Inheritance of heat tolerance in common bean of Andean origin. J Agri Univ Puerto Rico 87:113–121

Román-Avilés B, Kelly JD (2005) Identification of quantitative trait loci conditioning resistance to fusarium root rot in common bean. Crop Sci. 45:1881–1890

Romero-Napoles J, Johnson CD (2004) Database BRUCOL. Programa de Entomología, Instituto de Fitosanidad, Colegio de Postgraduados, México

Rosado IV, Rey M, Codón AC, Govantes J, Moreno-Mateos MA, Benítez T (2007) QID74 Cell wall protein of *Trichoderma harzianum* is involved in cell protection and adherence to hydrophobic surfaces. Fungal Genet Biol 44(10):950–964

Rosas JC, Beaver JS, Escoto D, Perez CA, Llano A et al (2004) Registration of 'Amadeus 77' small red common bean. Crop Sci 44:1867–1868

Rosas JC, Castro A, Beaver JS, Perez CA, Morales A et al (2000) Mejoramiento genético para tolerancia a altas temperaturas y resistencia al mosaico dorado en frijol común. Agron Mesoam 11:1–10

Rossi M, Bitocchi E, Belluci E, Nanni L, Rau D et al (2009) Linkage disequilibrium and population structure in wild and domesticated populations of *Phaseolus vulgaris* L. Evol Appl 2:504–522

Rubio MB, Hermosa MR, Keck E, Monte E (2005) Specific PCR assays for the detection and quantification of DNA from the biocontrol strain *Trichoderma harzianum* 2413 in soil. Microb Eco 49(1):25–33

Rubio MB, Hermosa R, Reino J, Collado I, Monte E (2009) Thctf1 transcription factor of *Trichoderma harzianum* is involved in 6-pentyl-2H-pyran-2-one production and antifungal activity. Fungal Genet Biol 46(1):17–27

Rumbos CI, Athanassiou CG (2017) Use of entomopathogenic fungi for the control of stored-product insects: can fungi protect durable commodities? J Pest Sci 90:839–854

Ruocco M, Lanzuise S, Vinale F, Marra R, Turrà D et al (2007) Identification of a new biocontrol gene in *Trichoderma atroviride*: the role of an ABC transporter membrane pump in the interaction with different plant-pathogenic fungi. Mol Plant-Microbe Interact 44(3):950–964

Rushton DL, Tripathi P, Rabara RC, Lin J, Ringler P et al (2012) WRKY transcription factors: key components in abscisic acid signalling. Plant Biotechnol J 10(1):2–11

Rushton PJ, Somssich IE (1998) Transcriptional control of plant genes responsive to pathogens. Curr Opin Plant Biol 1(4):311–315

Sánchez E, Sifres A, Casañas F, Nuez F (2008) The endangered future of organoleptically prestigious European landraces: Ganxet bean (*Phaseolus vulgaris* L.) as an example of a crop originating in the Americas. Genet Resour Crop Evol 55:45–52

Sankaran S, Khot LR, Zúñiga Espinoza C, Jarolmasjed S, Sathuvalli VR et al (2015) Low-altitude, high-resolution aerial imaging systems for row and field crop phenotyping: a review. Eur J Agron 70:112–123

Sankaran S, Zhou J, Khot LR, Trapp JJ, Mndolwa E et al (2018) High-throughput field phenotyping in dry bean using small unmanned aerial vehicle based multispectral imagery. Comput Electron Agric 151:84–92

Santalla M, Rodiño AP, De Ron AM (2002) Allozyme evidence supporting southwestern Europe as a secondary center of genetic diversity for common bean. Theor Appl Genet 104:934–944
Santamaría RI et al. 2017. "Complete Genome Sequences of Eight Rhizobium Symbionts Associated with Common Bean (*Phaseolus Vulgaris*)." Genom Announ 5(30):1–2. e00645-17. https://doi.org/10.1128/genomea.00645-17
Sanz L, Montero M, Grondona I, Vizcaíno JA, Llobell A et al (2004) Cell wall-degrading isoenzyme profiles of *Trichoderma* biocontrol strains show correlation with rDNA taxonomic species. Curr Genet 46(5):277–286
Sanz L, Montero M, Redondo J, Llobell A, Monte E (2005) Expression of an α-1,3-glucanase during mycoparasitic interaction of *Trichoderma asperellum*. FEBS J. 272:493–499
Sartorato A, Nietsche S, Barros EG, Moreira MA (1999) SCAR marker linked to angular leaf spot resistance gene in common bean. Annu Rept Bean Improv Coop 42:23–24
Sasan RK, Bidochka MJ (2012) The insect-pathogenic fungus *Metarhizium robertsii* (Clavicipitaceae) is also an endophyte that stimulates plant root development. Amer J Bot 99:101–107
Saunders J, O'Neill N (2004) The characterization of defense responses to fungal infection in alfalfa. BioControl 49(6):715–728
Schmutz J, Cannon SB, Schlueter J, Ma J, Mitros T et al (2010) Genome sequence of the palaeopolyploid soybean. Nature 463:178–183
Schmutz J, McClean PE, Mamidi S, Wu GA, Cannon SB et al (2014) A reference genome for common bean and genome-wide analysis of dual domestications. Nat Genet 46:707–713
Schneider KA, Grafton KF, Kelly JD (2001) QTL analyses of resistance to *Fusarium* root rot in bean. Crop Sci 41:535–542
Schneider KA, Rosales-Serna R, Ibarra-Perez F, Cazares-Enriquez B, Acosta-Gallegos JA et al (1997) Improving common bean performance under drought stress. Crop Sci 37:43–50
Schwartz HF, Singh SP (2013) Breeding common bean for resistance to white mold: a review. Crop Sci 53:1832–1844
Seidl MF, Thomma BPHJ (2017) Transposable elements direct the coevolution between plants and microbes. Trends Genet 33(11):842–851. https://doi.org/10.1016/j.tig.2017.07.003
Seidl V, Marchetti M, Schandl R, Allmaier G, Kubicek CP (2006) Epl1, the major secreted protein of *Hypocrea atroviridis* on glucose, is a member of a strongly conserved protein family comprising plant defense response elicitors. FEBS J 273(18):4346–4359
Shoresh M, Harman GE, Mastouri F (2010) Induced systemic resistance and plant responses to fungal biocontrol agents. Annu Rev Phytopathol 48:21–43
Shoresh M, Yedidia I, Chet I (2005) Involvement of jasmonic acid/ethylene signaling pathway in the systemic resistance induced in cucumber by *Trichoderma asperellum* T203. Phytopathology 95(1):76–84
Sicard D, Nanni L, Porfiri O, Bulfon D, Papa R (2005) Genetic diversity of *Phaseolus vulgaris* L. and *Phaseolus coccineus* L. landraces in central Italy. Plant Breed 124:464–472
Siddique KHM, Loss SP, Regan KL, Jettner RL (1999) Adaptation and seed yield of cool season grain legumes in Mediterranean environments of south-western Australia. Aust J Agri Res 50:375–387
Sidorova KK, Shumny VK, Vlasova EY, Glyanenko MN, Mishchenko TM (2011) Symbiogenetics and breeding of microsymbionts for increased nitrogen fixation capacity with special reference to the pea (*Pisum sativum* **L.**). Russ J Genet Appl Res 1:73–87
Simons R, Vincken JP, Bohin MC, Kuijpers TFM, Verbruggen MA et al (2011a) Identification of prenylated pterocarpans and other isoflavonoids in *Rhizopus* spp. elicited soya bean seedlings by electrospray ionisation mass spectrometry. Rapid Commun Mass Spectrom 25(1):55–65
Simons R, Vincken JP, Roidos N, Bovee TFH, van Iersel M et al (2011b) Increasing soy isoflavonoid content and diversity by simultaneous malting and challenging by a fungus to modulate estrogenicity. J Agric Food Chem 59(12):6748–6758
Singh KB, Foley RC, Oñate-Sánchez L (2002) Transcription factors in plant defense and stress responses. Curr Opin Plant Biol 5(5):430–436

Singh SP (1994) Gamete selection for simultaneous improvement of multiple traits in common bean. Crop Sci 34:352–355

Singh SP (1995) Selection for water-stress tolerance in interracial populations of common bean. Crop Sci 35:118–124

Singh SP (2001) Broadening the genetic base of common bean cultivars: a review. Crop Sci 41:1659–1671

Singh SP (2007) Drought resistance in the race Durango dry bean landraces and cultivars. Agron J 99:1219–1225

Singh SP, Gutierrez JA (1984) Geographical distribution of the Dl1, and Dl2 genes causing hybrid dwarfism in *P. vulgaris* L., their association with seed size, and their significance to breeding. Euphytica 33:337–345

Singh SP, Miklas PN (2015) Breeding common bean for resistance to common blight: a review. Crop Sci 55:971–984

Singh SP, Schwartz HF (2010) Breeding common bean for resistance to diseases: A review. Crop Sci 50:2199–2223

Sobolev VS, Neff SA, Gloer JB (2009) New stilbenoids from peanut (*Arachis hypogaea*) seeds challenged by an *Aspergillus caelatus* strain. J Agric Food Chem 57(1):62–68

Soltani A, MafiMoghaddam S, Olazad-Abbasabadi A, Walter K, Kearns PJ et al (2018) Genetic analysis of flooding tolerance in an andean diversity panel of dry bean (*Phaseolus vulgaris* L.). Front Plant Sci 9(767):1–15

Son GH, Wan J, Kim HJ, Nguyen XC, Chung WS et al (2012) Ethylene-responsive element-binding factor 5, *ERF5*, is involved in chitin-induced innate immunity response. Mol Plant-Microbe Interact 25(1):48–60

Song Q, Hyten DL, Jia G, Quigley CV, Fickus EW et al (2013) Development and evaluation of SoySNP50K, a high-density genotyping array for soybean. Plos One 8:1–12

Song Q, Jia G, Hyten DL, Jenkins J, Hwang EY et al (2015) SNP assay development for linkage map construction, anchoring whole genome sequence and other genetic and genomic applications in common bean. Gene Genome Genet 5:2285–2290

Sørensen J, Sessitsch A (2007) Plant-associated bacteria-lifestyle and molecular interactions. In: van Elsas JD, Jansson JK, Trevors JT (eds) Modern soil microbiology, 2nd edn. CRC Press, Taylor and Francis Group, Boca Raton, USA, pp 211–236

Soule M, Porter L, Medina J, Santana GP, Blair MW et al (2011) Comparative QTL map for white mold resistance in common bean, and characterization of partial resistance in dry bean lines VA19 and I9365-31. Crop Sci 51:123–139

Sousa LL, Cruz AS, Vidigal Filho PS, Vallejo VA, Kelly JD et al (2014) Genetic mapping of the resistance allele *Co-5*^{2} to *Colletotrichum lindemuthianum* in the common bean MSU 7-1 line. J Crop Sci 8:317–323

Souter JR, Gurusamy V, Porch TG, Bett KE (2017) Successful introgression of abiotic stress tolerance from wild tepary bean to common bean. Crop Sci 57:1160–1171

Souza TLPO, Barros EG, Bellato CM, Hwang E-Y, Cregan PC et al (2012) Single nucleotide polymorphism discovery in common bean. Mol Breed 30:419–428

Souza TLPO, Dessaune SN, Sanglard DA, Moreira MA, Barros EG (2011) Characterization of the rust resistance gene present in the common bean cultivar Ouro Negro the main rust resistance source used in Brazil. Plant Pathol 60:839–845

Souza TLPO, Gonçalves-Vidigal MC, Raatz B et al (2016) Major loci controlling resistance to the angular leaf spot of common bean. Annu Rept Bean Improv Coop 59:49–50

Souza TLPO, Ragagnin VA, Dessaune SN, Sanglard DA, Carneiro JES et al (2014) DNA marker-assisted selection to pyramid rust resistance genes in "Carioca" seeded common bean lines. Euphytica 199:303–316

Sparvoli F, Bollini R, Cominelli E (2015) Nutritional value. In: De Ron AM (ed) Grain legumes, Series: Handbook of Plant Breeding. Springer Science+Business Media, New York, USA, pp 291–325

Stanton-Geddes J, Paape T, Epstein B, Briskine R, Yoder J et al (2013) Candidate genes and genetic architecture of symbiotic and agronomic traits revealed by whole-genome, sequence-based association genetics in *Medicago truncatula*. PLoS One 8:e65688

Stavely JR (1990) Genetics of rust resistance in *Phaseolus vulgaris* plant introduction PI181996. Phytopathology 80:1056

Stavely JR (1984) Pathogenic specialization in *Uromyces phaseoli* in the United States and rust resistance in beans. Plant Dis 68:95–99

Stavely JR (1998) Recombination of two major dominant rust resistance genes that are tightly linked in repulsion. Annu Rept Bean Improv Coop 41:17–18

Stavely JR (2000) Pyramiding rust and viral resistance genes using traditional and marker techniques in common bean. Annu Rept Bean Improv Coop 43:1–4

Sthapit B (2013) Emerging theory and practice: community seed banks, seed system resilience and food security. In: Shrestha P, Vernooy R, Chaudhary P (eds) Community seedbanks in Nepal: Past, present, future. Proceedings of a National Workshop, 14–15 June 2012, Pokhara, Nepal. Local Initiatives for Biodiversity, Research and Development, Pokhara, Nepal, and Bioversity International, Rome, Italy, pp 16–40

Stoddard FL, Balko C, Erskine W, Khan HR, Link W et al (2006) Screening techniques and sources of resistance to abiotic stresses in cool-season food legumes. Euphytica 147:167–186

Strausbaugh C, Myers J, Forster R, McClean P (1999) *bc-1* and *bc-u*—two loci controlling bean common mosaic virus resistance in common bean are linked. J Am Soc Hort Sci 124:644–648

Suarez B, Rey M, Castillo P, Monte E, Llobell A (2004) Isolation and characterization of PRA1, a trypsin-like protease from the biocontrol agent *Trichoderma harzianum* CECT 2413 displaying nematicidal activity. Appl Microbiol Biotechnol 65(1):46–55

Subramanian VV, MacQueen AJ, Vader G, Shinohara M, Sanchez A et al (2016) Chromosome synapsis alleviates Mek1-dependent suppression of meiotic DNA repair. PLoS Biol 14(2):e1002369

Subramanyam B, Hagstrum DW (1995) Resistance measurement and management. In: Subramanyam B, Hagstrum DW (eds) Integrated management of insects in stored products. Marcel Dekker, New YorkUSA, pp 331–397

Tadeo FR, Gómez-Cadenas A (2008) Fisiología de las plantas y el estrés. In: Azcón-Bieto J, Talón M (eds) Fundamentos de fisología vegetal. McGraw-Hill, Barcelona, Spain, pp 577–597

Tar'an B, Michaels TE, Pauls KP (2001) Mapping genetic factors affecting the reaction to Xanthomonas axono- podis pv. phaseoli in *Phaseolus vulgaris* L. under field conditions. Genome 44:1046–1056

Tar'an B, Michaels TE, Pauls KP (2002) Genetic mapping of agronomic traits in common bean. Crop Sci 42:544–556

Tautz D (1989) Hypervariability of simple sequences as a general source for polymorphic DNA markers. Nucleic Acids Res 17:6463–6471

Terán H, Singh SP (2002) Comparison of sources and lines selected for drought resistance in common bean. Crop Sci 42:64–70

Thakur DR (2012) Taxonomy, distribution and pest status of Indian biotypes of *Acanthoscelides obtectus* (Coleoptera: Chrysomelidae: Bruchinae). A new record. Pak J Zool 44:189–195

Thi Lang N, Chi Bou B (2008) Fine mapping for drought tolerance in rice (*Oryza sativa* L.). Omonrice 16:9–15

Thomashow MF (1999) Plant cold acclimation: freezing tolerance genes and regulatory mechanisms. Annu Rev Plant Physiol Plant Mol Biol 50:571–599

Thompson HJ, McGinley JN, Neil ES, Brick M (2017) Beneficial effects of common bean on adiposity and lipid metabolism. Nutrients 9(9):1–12

Thompson MD, Brick MA, McGinley JN, Thompson HJ (2009) Chemical composition and mammary cancer inhibitory activity of dry bean. Crop Sci 49:179–186

Tijerino A, Hermosa R, Cardoza RE, Moraga J, Malmierca MG et al (2011) Overexpression of the *Trichoderma brevicompactum tri5* gene: Effect on the expression of the trichodermin biosynthetic genes and on tomato seedlings. Toxins 3(9):1220–1232

Tock AJ, Fourie D, Walley PG, Holub EB, Soler A et al (2017) Genome-wide linkage and association mapping of halo blight resistance in common bean to Race 6 of the globally important bacterial pathogen. Front Plant Sci 8:1–17

Tosquy Valle OH, López Salinas E, Villar Sánchez B, Acosta Gallegos JA, Rodríguez-Rodríguez JR (2016) Verdín: variedad de frijol negro tolerante a sequía terminal para Veracruz y Chiapas, México. Rev Mex de Cienc Agrí 7:1775–1780

Trabanco N, Asensio-Manzanera MC, Pérez-Vega E, Ibeas A, Campa A et al (2014) Identification of quantitative trait loci involved in the response of common bean to *Pseudomonas syringae* pv. *phaseolicola*. Mol Breed 33:577–588

Trabanco N, Campa A, Ferreira JJ (2015). Identification of a new chromosomal region involved in the genetic control of resistance to anthracnose in common bean. Plant Genome: 8(2). https://doi.org/10.3835/plantgenome2014.10.0079

Trapp JJ, Urrea CA, Cregan PB, Miklas PN (2015) Quantitative trait loci for yield under multiple stress and drought conditions in a dry bean population. Crop Sci 55:1596–1607

Traub J, Kelly JD, Loescher W (2017) Early metabolic and photosynthetic responses to drought stress in common and tepary bean. Crop Sci 57:1670–1686

Trutmann P, Voss J, Fairhead J (1996) Local knowledge and farmer perceptions in bean diseases in the Central African highlands. Agri Hum Values 13:64–67

Tryphone GM, Chilagane L, Protas D, Kusolwa P, Nchimbi-Msolla S (2013) Marker assisted selection for common bean diseases improvements in Tanzania: prospects and future needs. In Plant breeding from laboratories to fields. Chapter 5. Andersen SB (ed). InTech. 10.5772/3362

Udvardi M, Poole S (2013) Transport and metabolism in legume-rhizobia symbioses. Annu Rev Plant Biol 64:781–805

Unkovich M, Baldock J, Forbes M (2010) Variability in harvest index of grain crops and potential significance for carbon accounting: examples from Australian agriculture. Adv Agron 105:173–219

UPOV (1991) International Convention for the Protection of New Varieties of Plants. Geneva

Valenciano JB, Casquero PA, Boto JA (2004) Influence of sowing techniques and pesticide application on the emergence and the establishment of bean plants (*Phaseolus vulgaris* L.). Agronomie 24(2):113–118

Valentini G, Gonçalves-Vidigal MC, Hurtado-Gonzales OP, de Lima Castro SA, Cregan PB et al (2017) High-resolution mapping reveals linkage between genes in common bean cultivar Ouro Negro conferring resistance to the rust, anthracnose, and angular leaf spot diseases. Theor Appl Genet 130:1705–1722

Vallejo V, Kelly JD (2002) The use of AFLP analysis to tag the *Co-1*^{2} gene conditioning resistance to bean anthracnose. In: Proceedings of the X conference on plant and animal genome. http://www.intl-pag.org/pag/10/abstracts/PAGX_P233.html

Vallejo V, Kelly JD (2009) New insights into the anthracnose resistance of common bean landrace G 2333. Open Hortic J 2:29–33

Vallejos C, Skroch P, Nienhuis J (2001) *Phaseolus vulgaris*-the common bean. Integration of RFLP- and RAPD-based linkage maps. In: Phillips R, Vasil I (eds) DNA-based markers in plants. Kluwer Academic Publishers, Dordrecht, The Netherlands, pp 301–317

Vallejos CE, Sakiyama NS, Chase CD (1992) A molecular marker-based linkage map of *Phaseolus vulgaris* L. Genetics 131:733–740

Van Durme MK, Nowack M (2016) Mechanism of developmentally controlled cell death in plants. Curr Opin Plant Biol 29:29–37

Van Hameren B, Hayashi S, Gresshoff PM, Ferguson BJ (2013) Advances in the identification of novel factors required in soybean nodulation, a process critical to sustainable agriculture and food security. J Plant Biol Soil Health 1(1):6

Vargas A (2016) Estudio de la reacción al virus del mosaico necrótico común del frijol (BCNMV) y la habilidad de fijación biológica del nitrógeno (FBN) en frijol tépari (*Phaseolus acutifolius* A. Gray) e introgresión de la FBN al frijol común (*Phaseolus vulgaris* L.). Masters Thesis, Universidad de Puerto Rico, Mayaguez, Puerto Rico

Vargas WA, Crutcher FK, Kenerley CM (2011) Functional characterization of a plant-like sucrose transporter from the beneficial fungus *Trichoderma virens*. Regulation of the symbiotic association with plants by sucrose metabolism inside the fungal cells. New Phytol 189(3):777–789
Vargas WA, Sanz Martin JM, Rech GE, Rivera LP, Benito EP et al (2012) Plant defense mechanisms are activated during biotrophic and necrotrophic development of *Colletotricum graminicola* in maize. Plant Physiol 158(3):1342–1358
Vasconcellos RCC, Oraguzie OB, Soler A, Arkwazee H, Myers JR et al (2017) Meta-QTL for resistance to white mold in common bean. PLoS One 12:1–22
Vega FE, Goettel MS, Blackwell M, Chandler D, Jackson MA et al (2009) Fungal entomopathogens: new insights on their ecology. Fung Ecol 2:149–159
Veitch NC (2009) Isoflavonoids of the Leguminosae. Nat Prod Rep 26(6):776
Velázquez-Robledo R, Contreras-Cornejo HA, Macías-Rodríguez L, Hernández-Morales A, Aguirre J et al (2011) Role of the 4-phosphopantetheinyl transferase of *Trichoderma virens* in secondary metabolism and induction of plant defense responses. Mol Plant-Microbe Interact 24(12):1459–1471
Vernooy R, Shrestha P, Sthapit B (eds) (2015) Community seed banks: origins, evolution and prospects. Taylor & Francis, Routledge, United KIngdom
Voss J (1989) Integrating social science research into the development and testing of new agricultural technology: the case of CIAT's Great Lakes Bean Project. In: Groenfeldt D, Moock JL (eds) Social science perspectives in managing agricultural technology. IIMI, Colombo, Sri Lanka, pp 57–62
Vidigal Filho PS, Gonçalves-Vidigal MC, Silva CR, Gonela A, Lacanallo GF (2008) Identification of anthracnose resistance genes in common bean cultivars from Paraná State, Brazil. Annu Rept Bean Improv Coop 51:64–65
Vigouroux Y, MacMullen M, Hittinger CT, Houchins K, Schulz L et al (2002) Identifying genes of agronomic importance in maize by screening microsatellites for evidence of selection during domestication. Proc Natl Acad Sci USA 99:9650–9655
Vilca-Mallqui KS, Oliveira EE, Guedes RNC (2013) Competition between the bean weevils *Acanthoscelides obtectus* and *Zabrotes subfasciatus* in common beans. J Stored Prod Res 55:32–35
Vinale F, Flematti G, Sivasithamparam K, Lorito M, Marra R et al (2009) Harzianic acid, an antifungal and plant growth promoting metabolite from T*richoderma harzianum*. J Nat Prod 72(11):2032–2035
Vinale F, Sivasithamparam K, Ghisalberti EL, Marra R, Barbetti MJ et al (2008a) A novel role for *Trichoderma* secondary metabolites in the interactions with plants. Physiol Mol Plant Pathol 72(1–3):80–86
Vinale F, Sivasithamparam K, Ghisalberti EL, Marra R, Woo SL et al (2008b) *Trichoderma*–plant–pathogen interactions. Soil Biol. Biochem 40(1):1–10
Vinale F, Strakowska J, Mazzei P, Piccolo A, Marra R et al (2016) Cremenolide, a new antifungal, 10-member lactone from *Trichoderma cremeum* with plant growth promotion activity. Nat Prod Res 30(22):2575–2581
Viterbo A, Montero M, Ramot O, Friesem D, Monte E et al (2002) Expression regulation of the endochitinase chit36 from *Trichoderma asperellum* (*T. harzianum* T-203). Curr Genet 42(2):114–122
Viterbo A, Wiest A, Brotman Y, Chet I, Kenerley C (2007) The 18mer peptaibols from *Trichoderma virens* elicit plant defence responses. Mol Plant Pathol 8(6):737–746
Viteri DM, Cregan PB, Trapp JJ, Miklas P, Singh SP (2015) A new common bacterial blight resistance QTL in VAX 1 common bean and interaction of the new QTL, SAP6, and SU91 with bacterial strains. Crop Sci 54:1598
Vitti A, La Monaca E, Sofo A, Scopa A, Cuypers A et al (2015) Beneficial effects of *Trichoderma harzianum* T-22 in tomato seedlings infected by Cucumber mosaic virus (CMV). BioControl 60(1):135–147
Vizcaino JA, Sanz L, Basilio A, Vicente F, Gutiérrez S et al (2005) Screening of antimicrobial activities in *Trichoderma* isolates representing three *Trichoderma* sections. Mycol Res 109(12):1397–1406

Vlasova A, Capella-Gutiérrez S, Rendón-Anaya M, Hernandez-Oñate M, Minoche AE et al (2016) Genome and transcriptome analysis of the Mesoamerican common bean and the role of gene duplications in establishing tissue and temporal specialization of genes. Genome Biol 17:32

Voorrips RE (2002) Mapchart: software for the graphical presentation of linkage maps and QTLs. J Hered 93:77–78

Vos P, Hogers R, Bleeker M, Reijans M, Lee TVD et al (1995) AFLP: a new technique for DNA fingerprinting. Nucleic Acids Res 23:4407–4414

Waldman KB, Kerr JM, Isaacs KB (2014) Combining participatory crop trials and experimental auctions to estimate farmer preferences for improved common bean in Rwanda. Food Policy 46:183–192

Wang W, Jacobs JL, Chilvers MI, Mukankusi CM, Kelly JD et al (2018) QTL analysis of Fusarium root rot resistance in an Andean × Middle American common bean RIL population. Crop Sci 58:1–15

Wang Z, Huang S, Jia C, Liu J, Zhang J et al (2013) Molecular cloning and expression of five glutathione S-transferase (*GST*) genes from Banana (*Musa acuminata* L. AAA group, cv. *cavendish*). Plant Cell Rep 32(9): 1373–1380

Wendel JF, Lisch D, Hu G, Mason AS (2018) The long and short of doubling down: polyploidy, epigenetics, and the temporal dynamics of genome fractionation. Curr Opin Genet Dev 49:1–7. https://doi.org/10.1016/j.gde.2018.01.004

White JW, Conley MM (2012) A flexible, low-cost cart for proximal sensing. Crop Sci 53:1646–1649

Williams JGK, Kubelik AR, Livak KJ, Rafalski JA, Tingey SV (1990) DNA polymorphisms amplified by arbitrary primers are useful as genetic markers. Nucleic Acids Res 18:6531–6535

Williams JW, Jackson ST, Kutzbach JE (2007) Projected distributions of novel and disappearing climates by 2100 AD. Proc Natl Acad Sci USA 104:5738–5742

Winfield MO, Lu C, Wilson ID, Coghill JA, Edwards KJ (2010) Plant responses to cold: Transcriptome analysis of wheat. Plant Biotechnol J 8:749–771

Woo SL, Lorito M (2007) Exploiting the interactions between fungal antagonists, pathogens and the plant for biocontrol. In: Vurro M, Gressel J (eds) Novel biotechnologies for biocontrol agent enhancement and management. Springer, The Ntherlands, pp 107–130

Woo SL, Scala F, Ruocco M, Lorito M (2006) The molecular biology of the interactions between *Trichoderma* spp., phytopathogenic fungi, and plants. Phytopathology 96(2): 181–185

Wu Z, Song L, Huang D (2011) Food grade fungal stress on germinating peanut seeds induced phytoalexins and enhanced polyphenolic antioxidants. J Agric Food Chem 59(11):5993–6003

Xie W, Khanal R, McClymont S, Stonehouse R, Kirstin B et al (2017) Interaction of quantitative trait loci for resistance to common bacterial blight and pathogen isolates in *Phaseolus vulgaris* L. Mol Breed 37:55

Yang MH, Lin YJ, Kuo CH, Ku KL (2010) Medicinal mushroom *Ganoderma lucidum* as a potent elicitor in production of t -Resveratrol and t -Piceatannol in peanut calluses. J Agric Food Chem 58(17):9518–9522

Yedidia I, Shoresh M, Kerem Z, Benhamou N, Kapulnik Y et al (2003) Concomitant induction of systemic resistance to *Pseudomonas syringae* pv. *lachrymans* in cucumber by *Trichoderma asperellum* (T-203) and accumulation of phytoalexins. Appl Environ Microbiol 69(12):7343–7353

Young R, Kelly JD (1996) RAPD markers linked to three major anthracnose resistance genes in common bean. Crop Sci 37:940–946

Young RA, Melotto M, Nodari RO, Kelly JD (1998) Marker-assisted dissection of the oligogenic anthracnose resistance in the common bean cultivar, 'G2333'. Theor Appl Genet 96:87–94

Yu K, Park SJ, Poysa V (2000a) Marker-assisted selection of common beans for resistance to common bacterial blight: Efficacy and economics. Plant Breed 199:300–304

Yu K, Park SJ, Poysa V, Gepts P (2000b) Integration of simple sequence repeat (SSR) markers into a molecular linkage map of common bean (*Phaseolus vulgaris* L.). J Hered 91:429–434

Yu K, Park SJ, Zhang B, Haffner M, Poysa V (2004) An SSR marker in the nitrate reductase gene of common bean is tightly linked to a major gene conferring resistance to common bacterial blight. Euphytica 138:89–95

Yu Z, Stall R, Vallejos C (1998) Detection of genes for resistance to common bacterial blight of beans. Crop Sci 38:1290–1296

Zargar SM, Mahajan R, Nazir M, Nagar P, Kim ST et al (2017) Common bean proteomics: present status and future strategies. J Proteom 169:239–248. https://doi.org/10.1016/j.jprot.2017.03.019

Zargar SM, Nazir M, Rai V, Hajduch M, Agrawal GK et al (2015) Towards a common bean proteome atlas: looking at the current state of research and the need for a comprehensive proteome. Front Plant Sci 6:201. https://doi.org/10.3389/fpls.2015.00201

Zeven AC (1997) The introduction of the common bean (*Phaseolus vulgaris* L.) into Western Europe and the phenotypic variation of dry beans collected in The Netherlands in 1946. Euphytica 94:319–328

Zhang J, Chen GY, Li XZ, Hu M, Wang BY et al (2017) Phytotoxic, antibacterial, and antioxidant activities of mycotoxins and other metabolites from *Trichoderma* sp. Nat Prod Res 31(23):2745–2752

Zhukov VA, Nemankin TA, Ovchinnikova ES, Kuznetsova EV, Zhernakov AI et al (2010) Creating a series of gene-specific molecular markers for comparative mapping of the genome of pea (*Pisum sativum* L.) and diploid alfalfa (*Medicago truncatula* Gaertn.). In: Kunakh VA (ed) Factors of experimental evolution of organisms. Ukraine, Logos, Kiev, pp 30–34

Zietkiewicz E, Rafalski A, Labuda D (1994) Genome fingerprinting by simple sequence repeat (SSR)-anchored polymerase chain reaction amplification. Genomics 20:176–183

Zou XL, Shi C, Austin RS, Merico D, Munholland S et al (2014) Genome-wide single nucleotide polymorphism and insertion-deletion discovery through next-generation sequencing of reduced representation libraries in common bean. Mol Breed 33:769–778

Chapter 2
Genomic Interventions to Improve Resilience of Pigeonpea in Changing Climate

Abhishek Bohra, Shalini Pareek, Mitchell Jones, Uday C. Jha, SJ Satheesh Naik, Mayank Kaashyap, Prakash G. Patil, Alok Kumar Maurya, Rachit Saxena and Rajeev K. Varshney

Abstract Pigeonpea is an important food legume crop for rainfed agriculture in developing countries, particularly in India. Productivity gains in pigeonpea have remained static, and the challenge of improving pigeonpea yield is further aggravated by increasingly uncertain climatic conditions. Improved pigeonpea cultivars

---

A. Bohra (✉) · S. Pareek · U. C. Jha · S. Satheesh Naik · P. G. Patil · A. K. Maurya
ICAR-Indian Institute of Pulses Research (IIPR), Kanpur, India
e-mail: abhi.omics@gmail.com

S. Pareek
e-mail: shalinijnu@gmail.com

U. C. Jha
e-mail: u9811981@gmail.com

S. Satheesh Naik
e-mail: satheeshnaikagri@gmail.com

P. G. Patil
e-mail: patilbt@gmail.com

A. K. Maurya
e-mail: alokmaurya994@gmail.com

M. Jones
RMIT University, Melbourne, VIC 3000, Australia
e-mail: mitchell.jones@student.rmit.edu.au

M. Kaashyap
Cornell University, Ithaca, NY 14853, USA
e-mail: s3462450@student.rmit.edu.au; mayankkaashyap@gmail.com

R. Saxena · R. K. Varshney
International Crops Research Institute for the Semi-Arid Tropics (ICRISAT), Hyderabad, India
e-mail: R.Saxena@cgiar.org

R. K. Varshney
e-mail: r.k.varshney@cgiar.org

R. K. Varshney
The University of Western Australia, 35 Stirling Highway, Crawley, WA 6009, Australia

C. Kole (ed.), *Genomic Designing of Climate-Smart Pulse Crops*,
https://doi.org/10.1007/978-3-319-96932-9_2

with favourable traits, allowing them to cope with climatic adversities, are urgently required. Modern genomic technologies have the potential to rapidly improve breeding traits that confer resistance to biotic and abiotic stresses. Recent advances in pigeonpea genomics have led to the development of large-scale genomic tools to accelerate breeding programs. Availability of high-density genotyping assays and high-throughput phenotyping platforms motivate researchers to adopt new breeding techniques like genomic selection (GS) for improving complex traits. Accurate GS predictions inferred from multilocation and multiyear data sets also open new avenues for 'remote breeding' which is very much required to achieve genotype selection for future climates. Speed breeding pigeonpea with deployment of rapid generation advancement (RGA) technologies will improve our capacity to breed cultivars endowed with resilient traits. Once such climate-resilient cultivars are in place, their rapid dissemination to farmer's fields will be required to witness the real impact. Equally important will be the acceleration of varietal turnover to keep pace with the unpredictably changing climatic conditions so that cultivars are constantly optimized for the climatic conditions at any given time.

Keywords Pigeonpea · Resilience · Gene · Hybrid · Sequencing · QTL

2.1 Background

Current trends in global crop production display a mere 28% increase from 1985 to 2005 (Ray et al. 2012) and highlight the magnitude of the challenge of feeding the world by 2050. Ensuring food security for nine billion people will require agricultural crop production to be doubled by 2050. Efforts to increase global agricultural production have been hindered by growing limitations imposed by the changing climate. Climate change has resulted in increased sea levels, increasing CO_2 concentrations, drought, floods, storms, extreme temperatures, melting of glaciers, reduced availability of drinking water, etc. These factors limit agricultural production through shifts in growing seasons, reduced crop yields, crop damage, a decrease in arable land area and increase in soil salinity and basicity which result in price instability. A large-scale study by Ray et al. (2015) attributed up to 38% of crop yield variability (maize, rice, wheat, soybean) to variation in climatic conditions, most notably temperature and precipitation and the interaction of the two. Some regions experienced more than 60% variation in crop yield due to climate variability.

Smallholding farmers in developing countries are likely to be hit the hardest by climate change (Haussmann et al. 2012; Atlin et al. 2017). Food security and poverty are interconnected, especially in countries such as India where agriculture is the backbone of the economy (Kumar and Sharma 2013). From 1969 to 2005, temperature rise following the rainy season exceeded the average monthly temperature rise in India. This adversely affected cereal and pulse yields, especially those of pigeonpea, chickpea, wheat and rice (Birthal et al. 2014). A steady reduction was reported in *kharif* (July–October) crops in 2009 following a decline in rainfall. Climate change poses a high risk to the cultivation of soy, a medium risk to sesame and a low risk to pigeonpea production in various geographical regions of Mozambique (USAID 2017).

Changes in micro- and global climate have dramatically affected plant–pathogen interactions from the genetic to the environmental level (Garrett et al. 2006; Pande and Sharma 2011). Elevation in atmospheric CO_2 concentration has resulted in physiological and anatomical changes in plants including increased leaf area and number, and dense plant growth, providing an optimal environment for pathogen growth and survival (Pritchard et al. 1999). Geographical distributions of insect pests have changed in response to changing climate conditions, and increased attacks have hampered crop production by up to 40% (Sharma 2016).

Although climate change is adversely affecting agricultural production, pigeonpea has the potential to play it smart and can largely mitigate the adverse effects of the changing climate. Cultivation of pigeonpea in semiarid or arid regions and marginalized environment enables it to sustain itself under harsh environmental conditions. Like other grain legumes, the nitrogen-fixing ability of pigeonpea allows improvement in soil health, particularly for non-leguminous crops during crop rotation (Yadav 2017). Pigeonpea has also been demonstrated to respond positively to elevated CO_2 levels (Vanaja et al. 2010; Sreeharsha et al. 2015).

Improved knowledge of the resilience or adaptation mechanisms of pigeonpea in changing climates is key to meeting the challenge of improving yield gains. Modern genomic technologies and improved breeding systems could contribute greatly to the development of climate-smart pigeonpea. Further, the deployment of new cultivars and their timely replacement by a steady stream of climate-resilient cultivars will be critical to the delivery of a climate change adapted crop production system based on pigeonpea. In this chapter, we highlight the increasing relevance of crops such as pigeonpea to growing climatic adversities. We focus on the role of genomic technologies in improving resilience traits in pigeonpea crops. Finally, we highlight the challenges that lie ahead, with a brief discussion of varietal turnover of pigeonpea as a key driver for delivering climate change adaptation.

2.2 Changing Climates and Food Security

Globally 85% of agriculture is rainfed, and India ranks highest among the rainfed agricultural countries most prone to climate change (Asha latha et al. 2012). The detrimental effects of climate change pose a serious threat to agricultural production and in turn global food security. Climate change also adversely affects the livelihood of farmers and revenue generated through agriculture. However, efficient crop, soil and water management; development and cultivation of stress-tolerant crops including legumes (especially in rainfed areas); and the use modern breeding techniques can help to overcome the deleterious effects of climate change (Dar and Gowda 2013). Pigeonpea is an excellent source of dietary proteins and can help to strengthen food security, especially is Asia and Africa. Pigeonpea also serves as protein-rich fodder for ruminants, resulting from an increase of 30% crude protein in fresh forage to 6-month-old hay (Onim et al. 1985). Rao et al. (2002) evaluated the nutritional value of three ecotypes of pigeonpea, ICP8151, ICPX910007 and PBNA, as forage and found that the ecotype ICPX910007 accumulated highest dry matter (16 Mg ha^{-1}) while mean N concentration (28.6 g kg^{-1}) and digestible dry matter (614 g kg^{-1}) were found to be highest in PBNA. The yield of dry leaf matter for all three ecotypes ranged from 2360 to 2600 kg ha^{-1}. Therefore, pigeonpea could be a primary or supplementary forage stock when the productivity of other forages is low.

Pigeonpea has proven to be a food boon in the Malawi regions where farmers are experiencing the most significant detrimental effects of drought. In 2014, FAOSTAT reported that the production of cereals declined in Malawi due to drought followed by floods which resulted in heavy losses to the food supply and economic security of the people in the region. Conversely, pigeonpea and other legumes remained less affected by the changes in climate.

In recent years, pigeonpea hybrids have been developed to meet growing food demand and to provide enhanced food and nutritional security (Bohra et al. 2017a). The early maturing varieties of pigeonpea such as ICPL 88039 have been reported to support the agricultural industry by ensuring higher yields despite drought-like conditions. Early maturing pigeonpea varieties have been reported to be very useful for farmers and were found to have high yields with low input (Saxena et al. 2019). The photo- and thermo-insensitivity of the super early genotypes of pigeonpea such as ICPL 11255, ICPL 20340 and ICPL 20338 renders them highly suitable for cultivation in various agroclimatic conditions. Availability of short-duration pigeonpea provides sufficient time for land preparation, and their shorter life cycle helps them to escape terminal drought, insect and pest attacks and waterlogging.

2.3 Pigeonpea as a Climate-Resilient Crop

Pigeonpea is a climate-resilient crop. It is an important drought-tolerant semiarid grain legume (Valenzuela and Smith 2002) that can survive without additional soil moisture after it has passed the seedling stage. It can extract moisture and nutrients with its deep root system which helps it to survive during severe droughts (Flower and Ludlow 1986). It has some grain yield even during drought conditions while other legumes experience wilting and drying (Okiror 1986). Its leaves show osmotic adjustments (Subbarao et al. 2000) and maintain better photosynthesis than cowpea (Lopez et al. 1987). It recycles nutrients efficiently, stores moisture and fixes more nitrogen per unit than other legumes. The crop, therefore, requires very little inorganic fertilization (Emefiene et al. 2013). Some varieties of pigeonpea were found to be Fusarium wilt (FW) resistant resulting from their ability to secrete four fungitoxic isoflavonoid phytoalexins namely, hydroxygenistein, genistein, cajanin and cajanol, (cajanol being the main antifungal phytoalexin) (Marley and Hillocks 1993). Piscidic acid present in pigeonpea roots has also been found to increase solubilisation and uptake of phosphorus. It is harvested two or more times in a season and is, therefore, a good option for food and environmental balance (Gwata 2012).

Some pigeonpea varieties have also shown waterlogging tolerance. Waterlogging-tolerant genotypes of pigeonpea form lenticels, aerenchyma cells and adventitious roots, which are not present in sensitive genotypes (Hingane et al. 2015). CO_2 is the most common greenhouse gas and has a positive effect on pigeonpea. Pigeonpea plants exposed to elevated CO_2 levels exhibited higher growth, 52.3% higher radiation use efficiency due to higher carbon fertilization and 12% increased seed yield (Saha et al. 2012). Elevated CO_2 levels in pigeonpea plants resulted in a 58% increase in root nodule and biomass but delayed flowering due to lack of photosynthetic acclimatization and increased carbohydrate-nitrogen reserves (Sreeharsha et al. 2015). Despite low infection rates by vesicular-arbuscular mycorrhizae (VAM) the effect of VAM was highest in pigeonpea when compared to cowpea and groundnut (Ahiabor and Hirata 1994) and was found to increase P and Zn nutrition in pigeonpea (Wellings et al. 1991). Pigeonpea is also capable of utilizing Fe-bound Phosphorus. It has been reported that only a few crops have this ability. Pigeonpea root exudates contain piscidic acid, and its p-O-methyl derivative, which by chelating Fe^{+3} helps in the release of P from Fe-bound P. As such, pigeonpea could be a better option for use in cropping systems with low P content in the soil (Ae et al. 1990; Subbarao et al. 1997). Pigeonpea also responds well to acidic soil conditions (Ogata et al. 1988). Pigeonpea varieties ICP 24 and ICP 99 were found to be moderately resistant to root knot nematode *Meloidogyne javanica* (Sharma et al. 1994). Based on a multilocation study conducted in Tanzania, the pigeonpea genotypes with medium duration (ICP 7035, ICPL 90094, Kat 50 and QP37) were found to be the best adapted to diverse climatic conditions with rainfall and stress index varying from 322 to 1297 mm and 57–89, respectively (Mligo and Craufurd 2005).

Intercropping of pigeonpea with maize helped to recirculate the N and P in soil and contributed to the maintenance of soil fertility, which facilitated better maize yields in Africa, where it is a staple crop (Myaka et al. 2006). Pigeonpea has been found to be the best-suited crop for alley cropping, crop rotation, no-till cropping systems, as a cover crop, e.g. in maize and coffee plantations and for intercropping (Lal et al. 1978; Sakala et al. 2000; Sogbedji et al. 2006; Venzon et al. 2006). Pigeonpea is also used in rotation with soil containing *Stringa*, a major problem in African agriculture, as it helps to control and decrease *Stringa* infestation in the soil (Oswald and Ransom 2001).

2.4 Effects of Climate Change on the Pigeonpea Crop

Results of climate change viz., rise and fall in temperature, drought, flood, heat, salinity and UV-radiation have been reported to affect growth and yield of pigeonpea in several ways. Although pigeonpea is more resilient than other crops, production is still affected to some extent. Increases in pests and insects are an indirect result of climate change, exerting considerable impact on pigeonpea yield. The studies that have experimentally proved the effect of climate change on this crop are detailed below.

2.4.1 Temperature Extremities

Temperature extremities, as low as 15 °C and as high as 40 °C, are reported to negatively affect the germination of pigeonpea genotypes (Shibairo et al. 1995). Decreases in temperature from 20 to 12.5 °C significantly increased the time required for seed germination and seedling emergence while seedling vigour and the number of plants produced from mature green pods reduced, except in the PR 2 variety which showed early flowering at 40 °N latitude (Velez-Colon and Garrison 1989). Research reported by Hetherington et al. (1989) demonstrated pigeonpea to be one of the most intolerant crops to chilling treatment (0 °C, dark and high relative humidity for several hours). Pigeonpea also showed the lowest ratio of chilling tolerance to photo-inhibition at 7 °C. Marsh et al. (2006) reported that for the lowest temperature regimes (20 °C day/10 °C night temperature) the height of pigeonpea plants was stunted and biological N_2 fixation stopped completely. At 30 °C day/20 °C night temperature the symbiotic association of the pigeonpea plant and *Bradyrhizobium* worked well. Temperature and rainfall were found to significantly influence the fluxes of CO_2 and N_2O generated by crops produced in rainfed semi-arid regions like pigeonpea and sorghum (Prasad et al. 2015).

2.4.2 Drought and Heat Stress

Daryanto et al. (2015) reported a direct correlation between drought intensity and yield reduction based on the meta-analysis of the data on drought stress in food legumes reported from 1980 to 2014. However, the authors observed that the extent of damage depended on the legume species and the growth state with the reproductive stage witnessing the highest yield reduction during drought. Importantly, yield reduction in pigeonpea along with other legumes like lentils, and groundnut was found to be lower than that of cowpea and green gram. In extra-short-duration pigeonpea, the impact of drought stress is evident at the vegetative, flowering and pod-filling stages (Nam et al. 2001). Also, the germination and seedling growth, dry and fresh mass of seedlings, seedling vigour are also adversely affected by drought stress in pigeonpea. Hypocotyl length is affected more than the radical length (Kumar et al. 2011). Tolerance to drought in pigeonpea, particularly of short-duration pigeonpea, can be ascribed to the crop's ability to maintain total dry matter, a small pod size, few seeds in the pod, high seed mass and low flowering synchronization (Lopez et al. 1996).

2.4.3 Salinity and Metal Toxicity

Wild relatives of pigeonpea (*Atylosia*, *Rynchosia* and *Dunbaria*) show greater tolerance to salt stress than cultivated ones (ICPL 227 and HY 3C) (Subbarao et al. 1991). Among *Cajanus* species, *Cajanus platycarpus, C. scaraboides* and *C. sericea* were found to be tolerant to salinity stress whereas *C. acutifolius, C. cajanifolius* and *C. lineata* showed susceptible reaction to salinity stress (Srivastava et al. 2006). Lesser accumulation of Na^+ ions in the stem could be attributed to salt tolerance in pigeonpea. A more recent study examining the impact of pre-sowing gamma irradiation on pigeonpea associated a range of factors including favourable carbon partitioning between source to sink, lower partitioning of Na^+ and abundance of K^+ with enhanced salt tolerance of irradiated plants (Kumar et al. 2017). Irradiated pigeonpea showed greater tolerance than unirradiated plants. Heavy metal (Cadmium and Chromium) toxicity and water stress induced the formation of antioxidant enzymes in roots and shoots of pigeonpea (Battana and Gopal 2014). Aluminium toxicity induces lipid peroxidation and therefore inhibits pigeonpea plant growth and seed yield. Application of 24-epibrassinolide removes this toxicity and restores normal plant growth and yield by inducing the activity of antioxidant enzymes (Sri et al. 2016).

2.4.4 Growing Prevalence of Pests and Diseases

Global warming and climate change have caused a change in the geographical distribution of insect pests across the world. An increase of 0.74 °C in average global temperature in the last 100 years has posed a serious threat to agricultural production worldwide. According to Gautam et al. (2013), rising temperature has enhanced the risk of attacks by pathogens, pests and insects on food and staple crops, thus leading to a noticeable reduction in crop yields. Mathukumalli et al. (2016) suggested a possible rise in the incidence of *Helicoverpa armigera* on pigeonpea resulting from increased temperatures in future climates viz. 2020, 2050 and 2080. Outbreaks of *Helicoverpa armigera*, *Maruca vitrata* and *Ceroplastodes cajani* are increasingly recorded in India and have caused serious losses in grain legume crops (Sharma et al. 2015). Besides influencing pest incidence, temperature rise has a dramatic impact on the natural populations of predators and parasitoids feeding on pests and insects that attack plants. Examples include a decline in the population of *Hymenopteran* parasitoids (Prasad and Bambawale 2010). This reduced abundance of natural enemies may eventually lead to an outbreak of pests and insects which will threaten agricultural production. The population of insects infesting vegetative and flowering stages of pigeonpea such as leaf webber and flower thrips has been affected both positively and negatively by a range of minimum and maximum temperatures and relative humidity. Temperature, relative humidity, wind velocity and rainfall were found to have a negative effect on the population of jassid and bud weevil, whereas sunshine and water evaporation showed a positive effect, although the effects were nonsignificant in nature. In contrast, positive effects of relative humidity, wind velocity and rainfall were recorded on blister beetle populations (Kumar and Nath 2005). A recent study demonstrated the impact of climatic factors on insect pest incidence and subsequent crop damage (Pathania et al. 2014). Concerning planting dates, the pigeonpea crop sown during August suffered maximum loss due to infestation by insects (particularly *Helicoverpa armigera*) followed by that sown in July. In contrast, a greater incidence of plume moth (*Exelastis atomosa*), pod sucking bugs (*Clavigralla* spp.) and thrips was recorded in the late (September) sown crop. Late-planted pigeonpea plants showed significantly reduced damage from *Maruca vitrata*.

Although pigeonpea cultivation is severely affected by the attack of pod fly (*Melanagromyza obtusa* Malloch) and pod wasp (*Tanaostigmodes cajaninae* La Salle) in Asia and Africa, Sharma et al. (2003) reported that the wild accessions of pigeonpea belonging to *C. scarabaeoides*, *C. sericeus*, *Rhynchosia bracteata*, *C. acutifolius, C. lineatus* and *C. albicans* showed resistance to pod fly, while *C. scarabaeoides*, *C. albicans*, *Flemingia stricta* and *R. bracteata* tolerated damage by pod wasp. Wild accessions that showed resistance to both pod fly and pod wasp damage were ICPW 14, ICPW 27, ICPW 141, ICPW 202, ICPW 214 and ICPW 280.

In addition to the growing incidence of insect- pests, the rise in the disease incidence of Phytophthora blight, Alternaria blight and Macrophomina blight in pigeonpea has also been witnessed in response to the changing climate. High rainfall for short durations and elevated CO_2 level predispose pigeonpea to Phytophthora blight and sterility mosaic disease (Sharma 2016).

2.4.5 Elevated Atmospheric CO_2 Levels

The atmospheric concentration of CO_2 is rising and is projected to reach 1000 ppm by 2100 (IPCC 2007). Higher CO_2 levels support plants' growth by increasing photosynthetic rates and reducing water loss per unit leaf area through reduced stomata conductance, especially in the case of C3 plants. C4 plants tend to remain unresponsive to enhanced CO_2 levels owing to higher CO_2 levels within the bundle sheath cells (Taub 2010). These observations are supported by the recently developed large-scale test technique 'free air CO_2 enrichment (FACE)' that allows plant's responses to elevated CO_2 levels to be recorded in a more natural setting, thus overcoming the potential limitations associated with studies conducted in enclosures (Ainsworth and Long 2005). Legumes are more responsive to elevated CO_2 because of their ability to establish mutual relationships with N_2-fixing bacteria. This ability enables legumes to maximize the benefits associated with elevated CO_2 concentrations by 'matching increased carbon supply with additional N_2 fixation' (see Rogers et al. 2009 and references therein). Rogers et al. (2009) highlighted that growth potential is greatly constrained by nutrient deficiency, low temperature and drought. Elevated CO_2 concentration (700 μmol/mol) was reported to exert a significantly enhanced positive impact on the growth of pigeonpea, with increased grain yield and harvest index

(Vanaja et al. 2010). Under N-limiting conditions, the absence of photosynthetic acclimation to elevated CO_2 levels (550 μmol mol^{-1}) was demonstrated in pigeonpea through greater photosynthetic carboxylation capacity (V_{cmax}) and ribulose-1, 5 bis P regeneration rates (Jmax) (Sreeharsha et al. 2015). In response to elevated CO_2, pigeonpea displayed an increase of 58% in the nodule mass ratio (NMR) over the ambient plants. Authors reported that higher growth of pigeonpea in the presence of greater CO_2 levels led to delayed flowering. This is similar to the response of other nodulating leguminous species to elevated CO_2 under N-deficiency, the excess carbon was shunted to the roots and subsequently to root nodules in pigeonpea, thus reflecting a greater sink capacity of the crop (Sreeharsha and Reddy 2015). This data highlights the capacity of pigeonpea, like other nodulating legumes, to capitalize on the benefits of increasing CO_2 environments through increased growth and yield. In contrast, Saha et al. (2012) suggested the possibility of limited yield gains under higher CO_2 concentrations due to a decline in harvest index. Nevertheless, the authors reported higher biomass production and radiation use efficiency (RUE) in pigeonpea exposed to elevated CO_2 levels (580 ppm) over ambient CO_2 (380 ppm).

2.4.6 Ultraviolet (UV) Irradiation

Environmental stress is reported to generate reactive oxygen species (ROS) in plants. 'Oxidative stress' in plants arises from a failure to establish a balance between ROS generation and scavenging. Antioxidant defense systems in plants involve diverse secondary metabolites such as flavonoids and enzymes such as superoxide dismutase (SOD). Wei et al. (2013) while evaluating the effect of irradiation on post-harvest pigeonpea leaves found that phenolic and antioxidant activities are more significantly increased in leaves treated with UV-B and -C than leaves treated with UV-A. This had a long-lasting effect on phenolic levels and antioxidant activity when irradiated with UV-B. The UV-absorbing tendency of phenolics suggests their potential as protectants against UV irradiation in changing climates, and UV irradiation of pigeonpea leaves could be harnessed to enhance antioxidant activities.

2.4.7 Waterlogging

Transient or permanent waterlogging is frequently encountered by plants under natural conditions (Parent et al. 2008). Intense and irregular rainfall in an increasingly variable climate is leading to the more frequent occurrence of waterlogging which adversely affects agricultural production. Cultivation of pigeonpea in deep vertisols and in areas encountering annual rainfall from 600 to 1500 mm coupled with sowing in monsoon season (June–September) renders the crop vulnerable to waterlogging. Following waterlogging, rapid changes occur in soil environment including soil physio-chemical properties such as a decline in soil redox potential (Eh), an increase in soil pH towards neutrality, etc. (Parent et al. 2008) and a concurrent increase in the concentrations of other gases and toxic substances like CO_2, ethylene, methane and hydrogen sulphide (Setter and Belford 1990).

As with other pulse crops like green gram and black gram, plants are most significantly damaged when waterlogging coincides with their early stages of development including seed germination and seedling establishment (Singh et al. 1986). Prolonged waterlogging has been reported to cause considerable damage in pigeonpea roots, thus increasing resistance to water flow and stomatal conductance and reducing transpiration rates and net photosynthesis. It has also been reported to cause chlorosis, senescence and abscission of lower leaves (Takele and McDavid 1995).

Attempts have been made to illustrate the morphological, physiological and biochemical basis of waterlogging tolerance in pigeonpea. Upon exposure to flooded conditions, the tolerant pigeonpea genotypes are reported to develop aerenchyma cells, hypertrophied lenticels and adventitious roots (Hingane et al. 2015). As highlighted by Parent et al. (2005), hypertrophied lenticels might play a key role in shoot water homeostasis in addition to being facilitators of O_2 in plant roots. Chlorosis of younger terminal leaves has been reported as the first visible symptom of water logging in susceptible genotypes. In addition, Setter and Belford (1990) reported reduced growth, premature senescence and leaf drop as other visible symptoms of damage in waterlogged plants. Leaf nitrogen balance index (NBI) and root capacitance declined more significantly in the sensitive genotype (ICPL 7035) than in the tolerant genotype (IPAC 79). The authors suggest leaf pigments and root capacitance could be promising indicators of waterlogging tolerance in pigeonpea (Datta et al. 2017).

Concerning the metabolic responses of pigeonpea in flooded conditions, greater enzymatic activity and gene expression were recorded for alcohol dehydrogenase (*ADH*) and sucrose synthase (*SuSy*) in waterlogging-tolerant genotypes (ICPL 84023 and ICP 301) than susceptible genotypes (ICP 7035 and Pusa 207) (Kumutha et al. 2008). Susceptible genotypes showed a mutation in the CAAT box region of the ADH promoter. This suggested that waterlogging tolerance of pigeonpea is attributable to SuSy activity which provides reducing sugars during glycolysis and ADH enabling reduction of nicotinamide adenine dinucleotide (NADH), thus ascertaining more efficient glycolysis to maintain sufficient energy and sugar reserves under oxygen-limiting conditions.

Also, activation of antioxidant enzymes (SOD, ascorbate peroxidase, glutathione reductase and catalase) in tolerant ICP 301 compared to Pusa 207 under waterlogged conditions could help establish an appropriate response to waterlogging in pigeonpea (Kumutha et al. 2009). Waterlogging triggered antioxidant enzymes, NADH oxidase and expression of *Cu/Zn-SOD* and *APX* in tolerant plants more than susceptible plants (Sairam et al. 2009). Krishnamurthy et al. (2012) reported that the varieties ranged from waterlogging-tolerant to moderate and susceptible. Out of the seven elemental deficiencies studied (N, Mg, Cu, Ca, Fe, Mn and Al), pigeonpea was found to be deficient for N, Mg and Mn under waterlogged conditions. No Al toxicity was observed due to waterlogging (Srivastava et al. 2010). A study conducted by Bansal and Srivastava (2015) on the effect of waterlogging on resistant

(ICPL 84023) and susceptible (MAL 18) genotypes of pigeonpea revealed that both genotypes demonstrated a decline in CO_2 exchange rates, stomatal conductance, transpiration rates and intracellular CO_2 concentrations. Whereas high efficiency of carboxylation, increased chlorophyll content, starch availability, ADH activity and membrane stability helped ICPL 84023 to survive under waterlogged conditions. Duhan et al. (2017) reported that the combined effects of waterlogging and salinity are more harmful to pigeonpea than any of the individual stresses alone and that the roots were more sensitive to waterlogging stress than the leaves. Authors also observed an increase in the activities of SuSy and ADH and formation of aerenchyma in roots under waterlogged conditions and combined treatments.

As described above, the formation of aerenchyma cells, hypertrophied lenticels and adventitious roots are important anatomical and morphological changes that occur in plants exposed to waterlogged conditions. The enzymes such as ADH and pyruvate decarboxylase (PDC) are among the important anaerobic proteins (AINs) that are induced during waterlogging stress. The abundance and subsequent utilization of carbohydrate reserves also constitute an important factor influencing plant tolerance to waterlogged conditions (Setter and Belford 1990; Setter et al. 1997).

2.5 Genetic and Genomic Resources to Improve Adaptive Plasticity

A range of genetic and genomic tools have been developed in pigeonpea to support breeding programs that target improvement of climate-resilient traits and crop yield to meet the growing demand for pigeonpea in the face of the changing climate (Varshney et al. 2013, 2018b; Bohra et al. 2014; Bohra and Singh 2015; Bohra et al. 2017b). Concerning genetic resources, ICRISAT, one of the largest repositories of pigeonpea germplasm, is reported to hold more than 13,000 active pigeonpea collections, and core and mini-core collections comprising 1,290 and 146 accessions respectively (Upadhyaya et al. 2008; Bohra et al. 2010). A comprehensive collection of pigeonpea comprising of nearly 1,000 accessions has been developed. As a manageable resource of global genetic diversity, the collection offers access to sources of resistance and tolerance to various biotic and abiotic stress and important breeding traits such as photoperiod response, early flowering and maturity, growth habit etc. (Upadhyaya et al. 2011). A genome-wide catalogue of genetic variants (56 K SNP array) developed recently in pigeonpea will greatly help to harness the gains associated with vast genetic diversity available in these diverse germplasm resources (Saxena et al. 2018).

Advances in genomics have led to the establishment of cost-efficient and large-scale marker technologies in pigeonpea to elucidate the genetic makeup of traits that hold relevance to climate change adaptation. For genotyping applications, the first set of large-scale DNA markers in pigeonpea was developed by Bohra et al. (2011). The authors developed more than 3,000 SSR markers from BAC end sequences

(BESs), which were used for genetic linkage mapping and QTL analysis in pigeonpea. Later, high-throughput genotyping assays were developed following identification of genome-wide SNP markers by applying next-generation sequencing (NGS). For instance, 752 and 1616 SNPs were used for design GoldenGate and KASP genotyping assay, respectively (Kassa et al. 2012; Saxena et al. 2012). Identification of such genome-wide DNA markers led to the development of high-density linkage map in pigeonpea with 875 SNPs spanning a length of 996.21 cM (Saxena et al. 2012).

Analysis of expressed sequence tags (ESTs) from FW- and sterility mosaic disease (SMD)—responsive pigeonpea revealed differentially expressed genes (DEGs) for these two important diseases, i.e. FW (19) and SMD (20) (Raju et al. 2010). Various transcriptome assemblies have been reported in pigeonpea in recent years (Dubey et al. 2011; Dutta et al. 2011; Kudapa et al. 2012). Priyanka et al. (2010a, b) reported a set of 75 high-quality ESTs in pigeonpea, 20 of which were found to be stress inducible. Further, functional validation of abiotic stress responsive genes namely, *Cajanus cajan* hybrid-proline-rich protein (*CcHyPRP*), *C. cajan* cyclophilin (*CcCYP*) and *C. cajan* cold and drought regulatory (*CcCDR*) were demonstrated by expressing them in *Arabidopsis thaliana*. A total of 105 high-quality ESTs were generated from root tissues of pigeonpea by Kumar et al. (2014), out of which four genes namely, *S-adenosylmethionine synthetase, phosphoglycerate kinase, serine carboxypeptidase* and *methionine aminopeptidase* were further validated.

Application of NGS allowed authors to assemble 72.7% (605.78 Mb) of the pigeonpea genome (833.07 Mb). The pigeonpea genome shows the presence of a total of 48,680 genes, of which 111 genes were suggested to be drought-responsive (Varshney et al. 2012). In a similar attempt, another group identified 1,213 and 152 genes responding to disease and abiotic stress, respectively in the genome through analyzing a 510-MB genome assembly (Singh et al. 2012). Resequencing of 20 pigeonpea genotypes (18 cultivated and two wild accessions) was performed following establishment of the reference genome sequence. The accessions represent parents of recombinant inbred line (RIL), introgression line (IL), multiparent advanced generation intercross (MAGIC) and nested association mapping (NAM) populations. Alignment of the resequencing data to the reference genome facilitated development of the first generation HapMap of *Cajanus* species, offering information on 5.5 million genome-wide polymorphic sites including large structural variations, i.e. copy number variations (CNVs) and presence and absence variations (PAVs). Importantly, a set of accession-specific variants (SNPs and InDels) was also identified in the study (Kumar et al. 2016). More recently, whole genome resequencing (WGRS) of 292 pigeonpea accessions including wild relatives, landraces and breeding lines followed by GWAS analysis revealed presence of specific genomic regions on pseudomolecule CcLG09 that were affected during the process of domestication and modern breeding. The study also suggested a less intense genetic bottlenecking from landraces to breeding lines, which might be an outcome of 'limited intensive breeding history' (Varshney et al. 2017).

In recent years, application of NGS protocols in mapping populations have led to the development of high-density genetic maps in pigeonpea for both intra-specific (Arora et al. 2017; Saxena et al. 2017a, b) and interspecific crosses (Saxena et al.

2012). Availability of such genome maps coupled with extensive phenotyping data has been instrumental to elucidate the genetic architecture of several important traits. For example, four non-synonymous (ns) single nucleotide polymorphisms (SNPs) each for FW (four candidate genes) and SMD (three candidate genes) were identified using NGS-based quantitative trailt locus (QTL) Seq of susceptible and resistant extremes of a recombinant inbred population (ICPL 20096 × ICPL 332) coupled with the whole genome resequencing data. Furthermore, the candidate genes imparting resistance against these two important diseases were further validated using qRT-PCR assay, i.e. '*C. cajan* _03203' for FW and '*C. cajan* _01839' for SMD (Singh et al. 2016b). More recently, Saxena et al. (2017a) identified a set of ten QTLs controlling SMD resistance through analyzing three mapping populations, and the phenotypic variations accounted to these QTL varied between 3.6 and 34.3%. Earlier, QTLs associated with SMD resistance were reported in pigeonpea by Gnanesh et al. (2012). Similarly, eight QTLs with the PV in the range of 6.55 (*qFW1.1*) to 14.67% (*qFW3.1*) were detected for FW resistance from three mapping populations (Saxena et al. 2017b). A combination of association mapping and bi-parental linkage analysis showed an association of three simple sequence repeat (SSR) markers with resistance to FW (Patil et al. 2017a, b). More recently, the authors demonstrated the utility of DNA markers in pigeonpea improvement through validating a set of SSRs having association with important traits like plant ideotype, earliness and growth habit (Patil et al. 2018). Table 2.1 details DNA markers/QTL that have been identified in pigeonpea for trait improvement.

Ten housekeeping genes (*EF1α, UBQ10, GAPDH, 18SrRNA, 25SrRNA, TUB6, ACT1, IF4α, UBC* and *HSP90*) were selected and validated as reference genes for expression studies in response to drought, heat and salt stress in pigeonpea (Sinha et al. 2015b, c, 2016). These reports provided sets of highly stable reference genes in pigeonpea for analyzing three important abiotic constraints [IF4α and HSP90 for drought, *UBC, HSP90, GAPDH* for heat and *GAPDH UBC, HSP90* for salt stress].

Genome-wide characterization of Hsp100 family genes—ClpB was achieved during a study of drought and thermal tolerance in pigeonpea (Danekar et al. 2014). Maibam et al. (2015) reported heat shock factor (Hsf) genes viz., *CcHsfA-1d* and *CcHsfA-2* as highly upregulated in pigeonpea during heat stress. Enhanced expression of genes for heat shock protein 90 (HSP 90) and dehydration responsive element binding (DREB) was also exhibited by pigeonpea in response to FW and SMD (Agarwal et al. 2016).

2.6 Transgenic Approaches for Trait Improvement

A transgene approach has been used in pigeonpea for improving tolerance to stresses, most notably insect tolerance (Table 2.2). Similarly, various genes for abiotic stress resistance from pigeonpea have been successfully transferred and expressed in other systems. Using a subtracted cDNA library of drought-stressed pigeonpea plants Priyanka et al. (2010a, b), Sekhar et al. (2010) and Tamirisa et al. (2014) isolated and

Table 2.1 DNA markers associated with important traits in pigeonpea

Traits	Mapping populations/Diverse germplasm lines	Number of QTL/genes/markers	[a]Phenotypic variation explained (PVE) by QTL	Reference
Sterility mosaic disease (SMD)	ICPL 20096 × ICPL 332 (RIL) ICPL 20097 × ICP 8863 (RIL) ICP 8863 × ICPL 87119 (F_2)	10 QTL	34.3	Saxena et al. (2017a)
	ICPL 20096 × ICPL 332 (RIL)	Candidate gene (*C. cajan_01839*)	–	Singh et al. (2016b)
	ICPL 20096 × ICPL 332 (RIL)	2 InDels		Singh et al. (2017)
	ICP 8863 × ICPL 20097 (F_2) TTB 7 × ICP 7035 (F_2)	6 QTL	24.72	Gnanesh et al. (2012)
Fusarium wilt (FW)	ICPB 2049 × ICPL 99050 (RIL) ICPL 20096 × ICPL 332 (RIL) ICPL 85063 × ICPL 87119 (F_2)	14 QTL	56.45	Saxena et al. (2017b)
	ICPL 20096 × ICPL 332 (RIL)	Candidate gene (*C.cajan_03203*)	–	Singh et al. (2016b)
	ICPL 20096 × ICPL 332 (RIL)	3 InDels		Singh et al. (2017)
	Set of 89 germplasm lines	Six SSRs	6%	Patil et al. (2017a)
	Bahar × KPL 43 (F_2)/89 germplasm lines	Three SSRs	15.7	Patil et al. (2017b)

(continued)

Table 2.1 (continued)

Traits	Mapping populations/Diverse germplasm lines	Number of QTL/genes/markers	[a]Phenotypic variation explained (PVE) by QTL	Reference
Determinacy, flowering time and plant height	142 pigeonpea germplasm lines	Candidate gene (*CcTFL1*)	77	Mir et al. (2012)
Plant type and earliness	Pusa Dwarf × HDM04-1 ($F_{2:3}$)	13 QTL	51.4	Kumawat et al. (2012)
	Pusa Dwarf × H2001-4 (RIL)	4 QTL	18.1	Geddam et al. (2014)
Growth habit	ICP 5529 × ICP 11605 (F_2)	5 QTL	61.6	Saxena et al. (2017c)
Fertility restoration	ICPA 2039 × ICPR 2447 (F_2) ICPA 2043 × ICPR 2671 (F_2) ICPA 2043 × ICPR 3467 (F_2)	4 QTL	24.17	Bohra et al. (2012)
	ICPA 2039 × ICPL 87119 (F_2)	3 QTL	28.5	Saxena et al. (2018)
Cytoplasmic male sterility (A_4-CMS)	–	10-bp deletion in nad7a gene	–	Sinha et al. (2015a)

[a]Only highest PVE has been shown

Table 2.2 Transgenic system to improve trait imparting stress tolerance in pigeonpea

Trait	Source of gene	Gene	Explant used	Method	Reference
Helicoverpa armigera	*Bacillus thuringiensis*	cry1Ac and cry2Aa	Embryonic axis	Agrobacterium tumefaciens-mediated	Ghosh et al. (2017)
Insect pests	*Bacillus thuringiensis*	*cry1Ab* gene	Adventitious shoot buds	Agrobacterium tumefaciens-mediated	Sharma et al. (2006)
Helicoverpa armigera		*cry*1AcF	–	–	Ramu et al. (2012)
Salinity stress	*Vigna aconitifolia*	*P5CSF129A*	Embryonic structures	Agrobacterium tumefaciens-mediated	Surekha et al. (2014)
Helicoverpa armigera	*Bacillus thuringiensis*	*cry1Ac*	Embryo axes	*Agrobacterium*-mediated	Kaur et al. (2016)
Insect pests and disease	Pigeon pea	–	Shoot apices and cotyledonary node	Agrobacterium tumefaciens-mediated	Geetha et al. (1999)
Helicoverpa armigera	*Bacillus thuringiensis*	*cry1Ac*	Embryonic axes	*Agrobacterium*-mediated	Krishna et al. (2011)
Insect pest	Cowpea	*pCPI*	Embryonic axes	*Agrobacterium*-mediated	Lawrence and Koundal, (2001)

characterized a hybrid-proline-rich protein-encoding gene (*CcHyPRP*), cyclophilin (*CcCYP*) and cold and drought regulatory gene (*CcCDR*) and expressed them individually in Arabidopsis where they conferred multiple abiotic stress tolerances. Transgenic rice produced by Mellacheruvu et al. (2016) conferred resistance to the fungus *Magnaporthe grisea* and tolerance to drought, salinity and heat by expressing gene *CcHyPRP* from pigeonpea.

Transgenic approaches have been particularly useful in incorporating insect and pest resistance to susceptible pigeonpea. Transgenic pigeonpea lines were developed which expressed the synthetic *BT cry1Ab* gene, synthetic *cry*1AcF and chimeric *Bt Cry1Aabc* genes and conferred resistance to pod borer (*Helicoverpa armigera.* Hubner) (Sharma et al. 2006; Ramu et al. 2012; Das et al. 2016). Successful transformation of embryonal segments of pigeonpea was completed to express the synthetic *cry* I E-C in transgenic plants, which conferred protection against the larvae of the insect *Spodoptera litura* (Surekha et al. 2005). To protect against fungal attack, Kumar et al. (2004) genetically transformed pigeonpea using the rice chitinase gene.

2.7 Genomic Tool for Accelerating Productivity Gains in Pigeonpea Hybrid Breeding

With regard to adaptability, the heterozygosity of genotypes offers the opportunity to capitalize on 'individual buffering' capacity (Haussmann et al. 2012). Cytoplasmic male sterility (CMS) technology in pigeonpea is a cost-efficient means to harness hybrid vigour. Heterosis levels extending up to 40% have been reported in pigeonpea using CMS technology (Saxena et al. 2013, 2015). In addition to offering a considerable yield advantage over the traditional varieties, hybrids have been reported to bear traits that confer tolerance or resistance to abiotic and biotic stresses. For instance, hybrids (ICPH 2431, ICPH 2740, ICPH 2671, ICPH 4187) have shown waterlogging tolerance and resistance to FW and SMD (Sultana et al. 2012; Saxena et al. 2013; Hingane et al. 2015). Increasing the cultivation area of these high-yielding pigeonpea hybrids will be crucial for achieving yield stability in unpredictably variable environments. Hybrids also offer an advantage with respect to rapid seed replacement in farmers' fields, a key factor for delivering climate change adaptation to agricultural systems.

The CMS technology is based on three lines viz male sterile (A)-, maintainer (B)- and restorer (R)-lines. Robust restoration system (R line) is key to successful CMS hybrid breeding (Bohra et al. 2016; Saxena et al. 2015). Development or identification of restorers is a cumbersome procedure relying on extensive field testing (Bohra et al. 2017a). Identification of molecular markers tightly associated with CMS restoration trait can facilitate fast track and precise introgression of the genomic segments that restore fertility to the F_1 hybrids. In pigeonpea, Bohra et al. (2012) identified four QTL for A4-CMS restoration through QTL analysis of the pollen fertility data and genotyping data recorded in three F_2 mapping populations (ICPA 2039 × ICPR 2447, ICPA 2043 × ICPR 2671, ICPA 2043 × ICPR 3467). The phenotypic variation accounted to these four QTL, QTL-RF-1, 2, 3 and 4 were 14.85, 15.84, 20.89 and 24.17%, respectively. More recently, genotyping-by-sequencing (GBS) assay was performed in an F_2 population (ICPA 2039 × ICPL 87119) that segregated for pollen fertility (Saxena et al. 2018). A total of 306 SNPs were mapped following linkage analysis, and the genetic linkage map spanned a total length of 981.9 cM. Further, QTL analysis led authors to discover one major QTL (flanked by S8_7664779 and S8_6474381) on CcLG08 controlling up to 28.5% phenotypic variance (PV) for restoration trait. Based on the comparison of the genomic positions of QTL detected in the two studies, the authors found one common region on CcLG8 that harboured major QTL for A_4-CMS restoration. In addition, DNA markers have also been identified in pigeonpea to carry out genetic purity tests in order to assist conventional grow out test (Bohra et al. 2011, 2015, 2017c).

2.8 Need for Accelerated Breeding Cycles and Rapid Varietal Replacement Systems

A recent commentary by Atlin and colleagues (2017) emphasized the need for rapid varietal replacement systems to deliver climate change adaptation to crops. The authors opined climate change adaptation as an 'unintended benefit' of an accelerated variety replacement system. Increasing development and deployment of new resilient crop cultivars accompanied by their dissemination and further replacement constitutes a key step towards attaining cultivars that achieve higher yields in changing climates. Breeding systems that enable efficient synthesis of new cultivars by reducing the length of crop breeding cycles and optimizing allelic combinations according to the current climate facilitate higher crop yields. Modern genomic technologies and rapid generation advancement tools remain central to this goal. However, delivery of true climate adaptation would require rapid dissemination of these newly developed resilient cultivars to farmers' fields in conjunction with the withdrawal of the old and obsolete varieties.

Concerning the national seed system of pigeonpea in India, 69.8% of indented breeder seed is shared by the seven popular cultivars BSMR 736, Maruti (ICP 8863), Bahar, Narendra Arhar 1 (NA 1), Asha (ICPL 87119), TJT 501, Malviya Chamatkar. Most of these varieties are 20-30 years old (Singh et al. 2016a). Importantly, the three leading cultivars BSMR 736 (released in 1996), TJT 501 (released in 2008) and Maruti (released in 1985) contribute 12%, 10% and 11%, respectively towards the indented breeder seed (Chauhan et al. 2016). The seed replacement rate (SRR) of pigeonpea in 2014-15 was reported to be nearly 41% that is less than the recommended level of SRR (50%). The SRR recommends the use of certified and quality crop seed by the stakeholders.

The scenario presented above details a preference for older varieties, and therefore, efforts are needed to speed up the varietal turnover in pigeonpea to sustain crop yields in future climates. To improve the accessibility of quality seed for farmers, the Department of Agriculture, Cooperation and Farmer Welfare, Ministry of Agriculture, Government of India has taken initiatives to create nearly 150 seed hubs targeting production of quality seed and pulse varieties released during the last 10 years. In addition, seed subsidies in India are only provided for varieties released during the last 10 years (Chauhan et al. 2016), this remains in line with suggestions proposed by Atlins et al. for accelerating varietal turnover worldwide.

2.9 Future Trends and Challenges Ahead

Agriculture has a great impact on the social and economic life of people, especially in developing countries like India. A projected rise in temperature of 3.5 °C by 2050 will adversely affect crop production systems. Temperature and water stress will affect the vegetative and reproductive growth of crops. Also, warmer temperatures resulting from global warming are likely to promote the growth of fertilizer-resistant crops. Quality of crops including food crops will also be compromised. Soil fertility will be challenged due to reduced organic matter in the soil and the risk of pest and insect attacks will also increase (Devendra 2012). A shift in agriculture from sorghum, millets and legumes to vegetables, which have greater profitability, has also reduced fodder production and has led to the depletion of soil nitrogen content and mineral nutrition (Zade et al. 2013). Substantial variability witnessed in crop yields worldwide resulting from climate change demonstrates the need for climate-smart cultivars that can support food demands and provide food security to the growing population. Given the increasing risks associated with sole cropping, cultivation of climate-smart crops like pigeonpea should be promoted in addition to mixed cropping, crop rotation, alley cropping, etc. to increase food production and contribute to the restoration of soil fertility and increased economic output. Improved arrangements for rainwater harvesting should also be implemented so that this water can be utilized in times of reduced rainfall.

Cultivation of climate-smart crop genotypes along with the implementation of effective measures to reduce global warming is necessary to support the food security and economic health of agriculture-based countries. Modern omics technologies could contribute significantly towards mitigating the negative impacts of climate change on agriculture and food security. Applications of next-generation sequencing techniques in conjunction with precise phenotyping assays have rendered a wide range of functional DNA markers accessible to pigeonpea breeders that can significantly improve our ability to develop new cultivars endowed with climate change adaptation traits (Varshney et al. 2018a). The availability of rapid generation advancement (RGA) and genomic selection (GS) also broaden the applications of breeding programs. The potential of RGA techniques has been recently demonstrated in pigeonpea using immature seeds and the single seed descent (SSD) method (Saxena et al. 2017a, b, c). As illustrated in Fig. 2.1, breeding programs should allow assimilation of growing 'omics' information together with harnessing benefits of RGA technologies and accurate phenotyping systems to fast-track development of resilient cultivars. Accurate genome-wide predictions stemming from multilocation and multiyear data sets open new avenues for 'remote breeding' which is essentially required to accomplish selections across countries facing climate change (Manickavelu et al. 2017). Once such climate-resilient pigeonpea cultivars are in place, their dissemination to farmer's fields will be required to witness the real impact. Equally important will be the acceleration of varietal turnover to keep pace with the unpredictably changing climatic conditions.

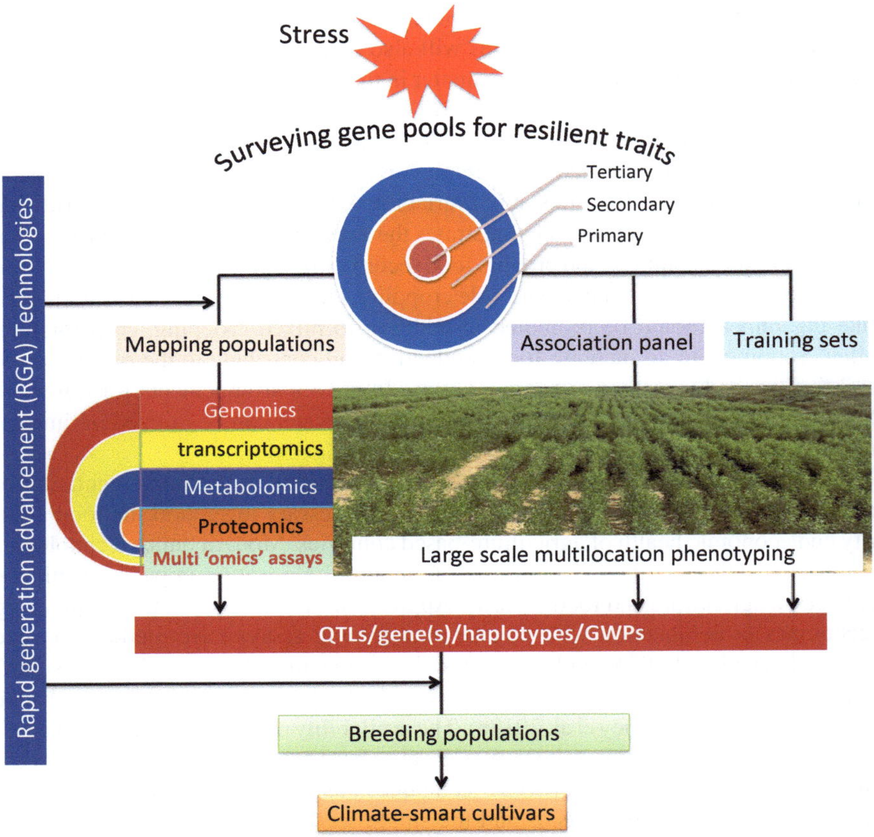

Fig. 2.1 Development of climate-resilient pigeonpea through integrating omics and phenotyping technologies with speed breeding tools

References

Ae N, Arihara J, Okada K, Yoshihara T, Johansen C (1990) Phosphorus uptake by pigeon pea and its role in cropping systems of the Indian subcontinent. Science 27:477–480

Agarwal G, Garg V, Kudapa H, Doddamani D, Pazhamala LT et al (2016) Genome-wide dissection of AP2/ERF and HSP90 gene families in five legumes and expression profiles in chickpea and pigeonpea. Plant Biotechnol J 14:1563–1577

Ahiabor BD, Hirata H (1994) Characteristic responses of three tropical legumes to the inoculation of two species of VAM fungi in Andosol soils with different fertilities. Mycorrhiza 5(1):63–70

Ainsworth EA, Long SP (2005) What have we learned from 15 years of free-air CO_2 enrichment (FACE)? A meta-analytic review of the responses of photosynthesis, canopy properties and plant production to rising CO_2. New Phytol 165:351–3702

Arora S, Mahato AK, Singh S, Mandal P, Bhutani S, Dutta S, Kumawat G, Singh BP, Chaudhary AK, Yadav R, Gaikwad K, Sevanthi AM, Datta S, Raje RS, Sharma TR, Singh NK (2017) A high-density intraspecific SNP linkage map of pigeonpea (*Cajanas cajan* L. Millsp.). PLoS One 12(6):e0179747

Atlin G, Cairns GE, Das B (2017) Rapid breeding and varietal replacement are critical to adaptation of cropping systems in the developing world to climate change. Glob Food Sec 12:31–37

Asha latha KV, Gopinath M, Bhat ARS (2012) Impact of climate change on rainfed agriculture in India: a case study of Dharwad. Int J Environ Sci Dev 3(4):368–371

Bansal R, Srivastava JP (2015) Effect of waterlogging on photosynthetic and biochemical parameters in pigeonpea. Russ J Plant Physiol 62(3):322–327

Battana S, Gopal GR (2014) Antioxidative enzyme responses under single and combined effect of water and heavy metal stress in two pigeon pea cultivars. J Sci Innov Res 3(1):72–80

Birthal P, Khan MT, Negi DS, Agarwal S (2014) Impact of climate change on yields of major food crops in India: Implications for food security. Agri Econ Res Rev 27(2):145–155

Bohra A, Mallikarjuna N, Saxena KB, Upadhyaya H, Vales I, Varshney RK (2010) Harnessing the potential of crop wild relatives through genomics tools for pigeonpea improvement. J Plant Biol 37:83–98

Bohra A, Dubey A, Saxena RK, Penmetsa RV, Poornima KN, Kumar N et al (2011) Analysis of BAC-end sequences (BESs) and development of BES-SSR markers for genetic mapping and hybrid purity assessment in pigeonpea (*Cajanus* spp.). BMC Plant Biol 11:56

Bohra A, Saxena RK, Gnanesh BN, Saxena KB, Byregowda M, Rathore A, KaviKishor PB, Cook DR, Varshney RK (2012) An intra-specific consensus genetic map of pigeonpea [*Cajanus cajan* (L.) Millspaugh] derived from six mapping populations. Theor Appl Genet 125:1325–1338

Bohra A, Saxena RK, Saxena KB, Sameerkumar CV, Varshney RK (2014) Advances in pigeonpea genomics. In: Gupta S, Nadarajan N, Sen Gupta D (eds) Legumes in the omics era. Springer, New York, Heidelberg, London, pp 95–110

Bohra A, Singh NP (2015) Whole genome sequences in pulse crops: a global community resource to expedite translational genomics and knowledge-based crop improvement. Biotechnol Lett 37:1529–1539

Bohra A, Singh IP, Yadav AK, Pathak A, Soren KR, Chaturvedi SK et al (2015) Utility of informative SSR markers in the molecular characterization of cytoplasmic genetic male sterility-based hybrid and its parents in pigeonpea. Natl Acad Sci Lett 38:13–19

Bohra A, Jha UC, Adhimoolam P, Bisht D, Singh NP (2016) Cytoplasmic male sterility (CMS) in hybrid breeding in field crops. Plant Cell Rep 35:967–993

Bohra A, Jha A, Singh IP, Pandey G, Pareek S, Basu PS, Chaturvedi SK, Singh NP (2017a) Novel CMS lines in pigeonpea [*Cajanus cajan* (L.) Millspaugh] derived from cytoplasmic substitutions, their effective restoration and deployment in hybrid breeding. Crop J 5:89–94

Bohra A, Pareek S, Jha R, Saxena RK, Singh IP, Pandey G et al (2017b) Modern genomic tools for pigeonpea improvement: status and prospects. In: Varshney RK, Saxena RK, Scott J (eds) The pigeonpea genome. Springer, Cham, pp 41–54

Bohra A, Jha R, Pandey G, Patil PG, Saxena RK, Singh IP, Singh D, Mishra RK, Mishra A, Singh F, Varshney RK, Singh NP (2017c) New hypervariable SSR markers for diversity analysis, hybrid purity testing and trait mapping in Pigeonpea [*Cajanus cajan* (L.) Millspaugh]. Front Plant Sci 8:1–15

Chauhan JS, Singh BB, Gupta S (2016) Enhancing pulses production in India through improving seed and variety replacement rates. Indian J Genet 76:1–10

Danekar P, Tyagi A, Mahto A, Krishna KG, Singh A, Raje RS, Gaikwad K, Singh NK (2014) Genome wide characterization of Hsp 100 family genes from pigeonpea. Indian J Genet 74(3):325–334

Dar WD, Gowda CLL (2013) Declining agricultural productivity and global food security. J Crop Improv 27(2):242–254

Daryanto S, Wang L, Jacinthe P-A (2015) Global synthesis of drought effects on food legume production. PLoS ONE 10(6):e0127401

Das A, Datta S, Sujayan GK, Kumar M, Kumar A, Arpan S et al (2016) Expression of chimeric Bt gene, Cry1Aabc in transgenic pigeonpea (cv. Asha) confers resistance to gram pod borer (*Helicoverpa armigera* Hubner.) Plant Cell Tiss Org Cult 127(3):705–715

Datta D, Bohra A, Satheesh Naik SJ et al (2017) Change in leaf pigment and root capacitance are effective indicators for waterlogging tolerance in pigeonpea. In: National symposium on pulses for nutritional security and agricultural sustainability, December 2–4, 2017. ICAR-Indian Institute of Pulses Research, Kanpur, Uttar Pradesh, India

Devendra C (2012) Climate change threats and effects: challenges for agriculture and food security. ASM series on climate change, pp 1–66

Dubey A, Farmer A, Schlueter J, Cannon SB, Abernathy B, Tuteja R et al (2011) Defining the transcriptome assembly and its use for genome dynamics and transcriptome profiling studies in pigeonpea (*Cajanus cajan* L.) DNA Res 18:153–164

Duhan S, Sheokand S, Kumari A, Sharma N (2017) Independent and interactive effects of water-logging and salinity on carbohydrate metabolism and root anatomy in pigeonpea genotypes at different growth stages. Indian J Agri Res 51(3):197–205

Dutta S, Kumawat G, Singh BP, Gupta DK, Singh S, Dogra V et al (2011) Development of genic-SSR markers by deep transcriptome sequencing in pigeonpea [*Cajanus cajan* (L.) Millspaugh]. BMC Plant Biol 11:17

Emefiene ME, Salaudeen AB, Yaroson AY (2013) The use of pigeon pea (*Cajanus cajan*) for drought mitigation in Nigeria. Intl Lett Nat Sci 1:6–16

Flower DJ, Ludlow MM (1986) Contribution of osmotic adjustment to the dehydration tolerance of water-stressed pigeonpea (*Cajanus cajan* (L.) Millsp.) leaves. Plant Cell Environ 9:33–40

Garrett KA, Dendy SP, Frank EE, Rouse MN, Travers SE (2006) Climate change effects on plant disease: genomes to ecosystems. Annu Rev Phytopathol 44:489–509

Gautam HR, Bhardwaj ML, Kumar R (2013) Climate change and its impact on plant diseases. Curr Sci 105(12)

Geddam SB, Raje RS, Prabhu KV et al (2014) Validation of QTLs for earliness and plant type traits in pigeonpea (*Cajanus cajan* (L.) Millsp.). Indian J Genet 74:471–477

Geetha N, Venkatachalam P, Lakshmi Sita G (1999) *Agrobacterium* mediated genetic transformation of pigeon pea (*Cajanus cajan* L.) and development of transgenic plants via direct organogenesis. Plant Biotechnol 16:213–218

Ghosh G, Ganguly S, Purohit A, Chaudhuri RK, Das S, Chakraborti D (2017) Transgenic pigeonpea events expressing Cry1Ac and Cry2Aa exhibit resistance to *Helicoverpa armigera*. Plant Cell Rep 36:1037–1051

Gnanesh BN, Bohra A, Sharma M et al (2011) Genetic mapping and quantitative trait locus analysis of resistance to sterility mosaic disease in pigeonpea [*Cajanus cajan* (L.) Millsp.]. Field Crops Res 123:53–61

Gwata E (2012) The potential of pigeonpea (*Cajanus cajan*) for producing important components of renewable energy and agricultural products. Geophys Res Abstr 14, EGU2012-1857

Haussmann BIG, Rattunde HF, Weltzien-Rattunde E, Traore PSC, vom Brocke K, Parzies HK (2012) Breeding strategies for adaptation of pearl millet and sorghum to climate variability and change in West Africa. J Agron Crop Sci 198:327–339

Hetherington SE, He J, Smillie RM (1989) Photoinhibition at low temperature in chilling-sensitive and -resistant plants. Plant Physiol 90:1609–1615

Hingane AJ, Saxena KB, Patil SB, Sultana R, Srikanth S, Mallikarjuna N, Vijaykumar R, Kumar CVS (2015) Mechanism of water-logging tolerance in pigeonpea. Indian J Genet 75(2):208–214

Kassa MT, Penmetsa RV, Carrasquilla-Garcia N et al (2012) Genetic patterns of domestication in pigeonpea (*Cajanus cajan* (L.) Millsp.) and wild Cajanus relatives. PLoS One 7:e39563

Kumawat G, Raje RS, Bhutani S et al (2012) Molecular mapping of QTLs for plant type and earliness traits in pigeonpea (*Cajanus cajan* L. Millsp.). BMC Genet 13:84

Kaur A, Sharma M, Sharma C, Kaur H, Kaur N, Sharma S, Arora R, Singh I, Sandhu JS (2016) Pod borer resistant transgenic pigeon pea (*Cajanus cajan* L.) expressing *cry1Ac* transgene generated through simplified *Agrobacterium* transformation of pricked embryo axes. Plant Cell Tiss Org Cult 127:717–727

Krishna GG, Reddy PS, Ramteke PW, Rambabu P, Tawar KB, Bhattacharya P (2011) *Agrobacterium*-mediated genetic transformation of pigeon pea [*Cajanus cajan* (L.) Millsp.] for resistance to legume pod borer *Helicoverpa armigera*. J Crop Sci Biotechnol 14:197–204

Krishnamurthy L, Upadhyaya HD, Saxena KB, Vadez V (2012) Variation for temporary waterlogging response within the mini core pigeonpea germplasm. J Agri Sci 150(3):357–364

Kudapa H, Bharti AK, Cannon SB, Farmerb AD, Mulaosmanovic B, Kramerb R et al (2012) A comprehensive transcriptome assembly of pigeonpea (*Cajanus cajan* l.) using Sanger and second-generation sequencing platforms Mol Plant 5(5):1020–1028

Kumar A, Sharma P (2013) Impact of climate variation on agricultural productivity and food security in rural India. Economics discussion papers no. 2013-43

Kumar RR, Karjol K, Naik GR (2011) Variation of sensitivity to drought stress in pigeon pea (*Cajanus cajan* [L.] Millsp) cultivars during seed germination and early seedling growth. World J Sci Technol 1(1):11–18

Kumar RR, Yadav S, Joshi S, Bhandare PP, Patil VK, Kulkarni PB, Sonkawade S, Naik GR (2014) Identification and validation of expressed sequence tags from pigeonpea (*Cajanus cajan* L.) root. Intl J Plant Genom 2014:651912

Kumar SM, Kumar BK, Sharma KK, Devi P (2004) Genetic transformation of pigeonpea with rice chitinase gene. Plant Breed 123:485–489

Kumar A, Nath P (2005) Insect pests of early pigeon-pea in relation to weather parameters. Ann Plant Protec Sci 13(1):23–26

Kumar V, Khan AW, Saxena RK, Garg V, Varshney RK (2016) First-generation HapMap in *Cajanus* spp. reveals untapped variations in parental lines of mapping populations. Plant Biotechnol J 14(8):1673–1681

Kumar P, Sharma V, Atmaram CK, Singh B (2017) Regulated partitioning of fixed carbon (^{14}C), sodium (Na^+), potassium (K^+) and glycine betaine determined salinity stress tolerance of gamma irradiated pigeonpea [*Cajanus cajan* (L.) Millsp]. Environ Sci Pollut Res 24:7285–7297

Kumutha D, Ezhilmathi K, Sairam RK, Srivastava GC, Deshmukh PS, Meena RC (2009) Waterlogging induced oxidative stress and antioxidant activity in pigeonpea genotypes. Biol Plant 53(1):75–84

Kumutha D, Sairam RK, Ezhilmathi K, Chinnusamy V, Meena RC (2008) Effect of waterlogging on carbohydrate metabolism in pigeon pea (*Cajanus cajan* L.): Upregulation of sucrose synthase and alcohol dehydrogenase. Plant Sci 175(5):706–716

Lal R, Wilson GF, Okigbo BN (1978) No-till farming after various grasses and leguminous cover crops in tropical alfisol. I. Crop performance. Field Crops Res 1:71–84

Lawrence PK, Koundal KR (2001) *Agrobacterium tumefaciens* mediated transformation of pigeon pea (*Cajanus cajan* L. Millsp.) and molecular analysis of regenerated plants. Curr Sci 80:1428–1432

Lopez FB, Johansen C, Chauhan YS (1996) Effects of timing of drought stress on phenology, yield and yield components of short-duration pigeonpea. J Agron Crop Sci 177(5):311–320

Lopez FB, Setter TM, McDavid CR (1987) Carbon dioxide and light responses of photosynthesis in cowpea and pigeonpea during water deficit and recovery. Plant Physiol 85(4):990–995

Maibam A, Tyagi A, Satheesh V, Mahato AK, Jain N, Raje RS, Rao AR, Gaikwad K, Singh NK (2015) Genome-wide identification and characterization of heat shock factor genes from pigeonpea (*Cajanus cajan*). Mol Plant Breed 6(7):1–11

Manickavelu A, Hattori T, Yamaoka S, Yoshimura K, Kondou Y, Onogi A et al (2017) Genetic nature of elemental contents in wheat grains and its genomic prediction: toward the effective use of wheat landraces from Afghanistan. PLoS ONE 12:e0169416

Marle PS, Hillocks RJ (1993) The role of phytoalexins in resistance to fusarium wilt in pigeon pea (*Cajanus cajan*). Plant Pathol 42:212–218

Marsh LE, Baptiste R, Marsh DB, Trinklein D, Kremer RJ (2006) Temperature effects on bradyrhizobium spp. growth and symbiotic effectiveness with pigeonpea and cowpea. J Plant Nutr 29:331–346

Mathukumalli SR, Dammua M, Sengottaiyan V, Ongolu S, Birada AK, Kondrua VR, Karlapudia S, Bellapukonda MKR, Chitiprolu RRA, Cherukumalli SR (2016) Prediction of *Helicoverpa armigera* Hubner on pigeonpea during future climate change periods using MarkSim multimodel data. Agri For Meteorol 228:130–138

Mellacheruvu S, Tamirisa S, Vudem DR, Khareedu VR (2016) Pigeonpea hybrid-proline rich protein (CcHyPRP) confers biotic and abiotic stress tolerance in transgenic rice. Front Plant Sci 6:1167

Mir RR, Saxena RK, Saxena K, Upadhyaya HD, Kilian A, Cook DR, Varshney RK (2012) Whole-genome scanning for mapping determinacy in pigeonpea (*Cajanus* spp.) Plant Breed 132(5):472–478

Mligo JK, Craufurd PQ (2005) Adaptation and yield of pigeonpea in different environments in Tanzania. Field Crops Res 94(1):43–53

Myaka FM, Sakala WD, Adu-Gyamfi JJ, Kamalongo D, Ngwira A, Odgaard R, Nielsen RE, Høgh-Jensen H (2006) Yields and accumulations of N and P in farmer-managed intercrops of maize–pigeonpea in semi-arid Africa. Plant Soil 285(1–2):207–220

Nam NH, Chauhan YS, Johansen C (2001) Effect of timing of drought stress on growth and grain yield of extra-short-duration pigeonpea lines. J Agri Sci 136(2):179–189

Ogata S, Adu-Gyamfi S, Fujita K (1988) Effect of phosphorus and pH on dry matter production, dinitrogen fixation and critical phosphorus concentration in pigeon pea (*Cajanus cajan* (L) millsp.). Soil Sci Plant Nutr 34(1):55–64

Okiror MA (1986) Breeding for resistance to Fusarium wilt of pigeonpea (*Cajanm cajan* (L.) Millsp.) in Kenya. PhD thesis, University of Nairobi, 202 p

Onim JFM, Semenye PP, Fitzhugh HA, Mathuva M (1985) Research on feed resources for small ruminants on smallholder farms in Western Kenya. In: Kategile JA, Said AN, Dzowela BH (eds) Animal feed resources for small-scale livestock producers. Proceedings of the 2nd PANESA workshop, held in Nairobi, Kenya

Oswald A, Ransom JK (2001) Striga control and improved farm productivity using crop rotation. Crop Protec 20(2):113–120

Pande S, Sharma M (2011) Climate change and changing scenario of plant diseases in semi arid tropics, plant pathology in India: Vision 2030. Indian Phytopathological Society, New Delhi, pp 128–131

Parent C, Capelli N, Berger A, Crèvecoeur M, Dat JF (2008) An overview of plant responses to soil waterlogging. Plant Stress 2:20–27

Patil PG, Dubey J, Bohra A, Mishra RK, Saabale PR, Das A, Rathore M, Singh NP (2017a) Association mapping to discover significant marker-trait associations for resistance against Fusarium wilt variant 2 in pigeonpea [*Cajanus cajan* (L.) Millspaugh] using SSR markers. J Appl Genet 58:307–319

Patil PG, Bohra A, Dubey J, Saabale PR, Mishra RK, Pandey G, Das A, Rathore M, Singh F, Singh NP (2017b) Genetic analysis and molecular resistance to race 2 of Fusarium wilt in pigeonpea [*Cajanus cajan* (L.) Millsp.]. Crop Prot 100:117–123

Patil PG, Bohra A, Satheesh Naik SJ, et al (2018) Validation of QTLs for plant ideotype, earliness and growth habit traits in pigeonpea (*Cajanus cajan* Millsp.). Physiol Mol Biol Plants 24:1245–1259

Pathania M, War AR, Munghate RS, Nagraja T, Sharma HC (2014) Effects of climatic factors on pest incidence in pigeonpea. In: National conference on pulses: challenges and opportunities under changing climate scenario, 29th Sept–1st Oct 2014

Prasad JVNS, Srinivasa Rao CH, Ravichandra K, Naga Jyothi CH, Prasad Babu MBB, Babu VR et al (2015) Green house gas fluxes from rainfed sorghum (*Sorghum bicolour*) and pigeonpea (*Cajanus cajan*)—Interactive effects of rainfall and temperature. J Agrometeorol 17(1):17–22

Prasad YG, Bambawale OM (2010) Effects of climate change on natural control of insect pests. Indian J Dryland Agric Res Dev 25(2):1–12

Pritchard SG, Rogers HH, Prior SA, Peterson CM (1999) Elevated CO_2 and plant structure: a review. Glob Change Biol 5(7):807–837

Priyanka B, Sekhar K, Reddy VD, Rao KV (2010a) Expression of pigeonpea hybrid-proline-rich protein encoding gene (CcHyPRP) in yeast and *Arabidopsis* affords multiple abiotic stress tolerance. Plant Biotechnol J 8:76–87

Priyanka B, Sekhar K, Sunita T, Reddy VD, Rao KV (2010b) Characterization of expressed sequence tags (ESTs) of pigeonpea (*Cajanus cajan* L.) and functional validation of selected genes for abiotic stress tolerance in *Arabidopsis thaliana*. Mol Genet Genom 283:273–287

Raju NL, Gnanesh BN, Lekha P, Jayashree B, Pande S, Hiremath PJ, Byregowda M, Singh NK, Varshney RK (2010) The first set of EST resource for gene discovery and marker development in pigeonpea (*Cajanus cajan* L.). BMC Plant Biol 10:45

Ramu SV, Rohini S, Keshavareddy G, Gowri Neelima M, Shanmugam NB, Kumar ARV, Sarangi SK, Kumar PA, Udayakumar M (2012) Expression of a synthetic *cry*1AcF gene in transgenic pigeon pea confers resistance to *Helicoverpa armigera*. J Appl Entomol 36:675–687

Rao SC, Coleman SW, Mayeux HS (2002) Forage production and nutritive value of selected pigeonpea ecotypes in the southern Great Plains. Crop Sci. 42(4):1259–1263

Ray DK, Ramankutty N, Mueller ND, West PC, Foley JA (2012) Recent patterns of crop yield growth and stagnation. Nat Commun 3:1293

Ray DK, Gerber JS, MacDonald GK, West PC (2015) Climate variation explains a third of global crop yield variability. Nat Commun 6:5989

Saha S, Sehgal VK, Nagarajan S, Pal M (2012) Impact of elevated atmospheric CO_2 on radiation utilization and related plant biophysical properties in pigeon pea (*Cajanus cajan* L.). Agri For Meteorol 158–159:63–70

Sairam RK, Kumutha D, Ezhilmathi K, Chinnusamy V, Meena RC (2009) Waterlogging induced oxidative stress and antioxidant enzyme activities in pigeon pea. Biol Plant 53(3):493–504

Sakala WD, Cadisch G, Giller KE (2000) Interactions between residues of maize and pigeonpea and mineral N fertilizers during decomposition and N mineralization. Soil Biol Biochem 32(5):679–688

Saxena KB, Choudhary AK, Srivastava RK, Bohra A, Saxena RK, Varshney RK (2019) Origin of early maturing pigeonpea germplasm and its impact on adaptation and cropping systems. Plant Breed. https://doi.org/10.1111/pbr.12696

Saxena KB, Kumar RV, Tikle AN et al (2013) ICPH2671—the world's first commercial food legume hybrid. Plant Breed 132:479–485

Saxena KB, Singh IP, Bohra A et al (2015) Strategies for breeding, production, and promotion of pigeonpea hybrids in India. J Food Legume 28:190–198

Saxena KB, Saxena RK, Varshney RK (2017a) Use of immature seed germination and single seed descent for rapid genetic gains in pigeonpea. Plant Breed 136:954–957

Saxena RK, Kale SM, Kumar V, Parupalli S, Joshi S, Singh VK, Garg V, Das RR, Sharma M, Yamini KN, Ghanta A, Rathore A, Sameer Kumar CV, Saxena KB, Varshney RK (2017b) Genotyping-by-sequencing of three mapping populations for identification of candidate genomic regions for resistance to sterility mosaic disease in pigeonpea. Sci Rep 7:1813

Saxena RK, Singh VK, Kale SM, Tathineni R, Parupalli S, Kumar V, Garg V, Das RR, Sharma M, Yamini KN, Muniswamy S, Ghanta A, Rathore A, Sameer Kumar CV, Saxena KB, Kavi Kishor PB, Varshney RK (2017c) Construction of genotyping-by-sequencing based high-density genetic maps and QTL mapping for fusarium wilt resistance in pigeonpea. Sci Rep 7:1911

Saxena RK, Penmetsa RV, Upadhyaya HD, Kumar A, Carrasquilla-Garcia N, Schlueter JA, Farmer A, Whaley AM, Sarma BK, May GD, Cook DR, Varshney RK (2012) Large-scale development of cost-effective single-nucleotide polymorphism marker assays for genetic mapping in pigeonpea and comparative mapping in legumes. DNA Res 19:449–461

Saxena RK, Patel K, Sameer Kumar CV, Tyagi K, Saxena KB, Varshney RK (2018) Molecular mapping and inheritance of restoration of fertility (Rf) in A_4 hybrid system in pigeonpea (*Cajanus cajan* (L.) Millsp.). Theor Appl Genet. https://doi.org/10.1007/s00122-018-3101-y

Setter TL, Belford R (1990) Waterlogging—how it reduces plant growth and how plants can overcome its effects. J Agri 31:51–55

Setter TL, Ellis M, Laureles EV, Ella ES, Senadhira D, Mishra SB, Sarkarung S, Datta S (1997) Physiology and genetics of submergence tolerance in rice. Ann Bot 79:67–77

Sekhar K, Priyanka B, Reddy VD, Rao KV (2010) Isolation and characterization of a pigeonpea cyclophilin (*CcCYP*) gene, and its over-expression in *Arabidopsis* confers multiple abiotic stress tolerance. Plant Cell Environ 33:1324–1338

Sharma HC (2016) Climate change *vis-a-vis* pest management. In: Conference on national priorities in plant health management, 4–5 Feb 2016, Tirupati, India

Sharma HC, Pampapathy G, Reddy LJ (2003) Wild relatives of pigeonpea as a source of resistance to the pod fly (*Melanagromyza obtusa* Malloch) and pod wasp (*Tanaostigmodes cajaninae* La Salle). Genet Resour Crop Evol 50(8):817–824

Sharma KK, Lavanya M, Anjaiah V (2006) *Agrobacterium*-mediated production of transgenic pigeonpea (*Cajanus cajan* L. Millsp.) expressing the synthetic BT *cry1Ab* gene. In Vitro Cell Dev Biol Plant 42(2):165–173

Sharma SB, Mohiuddin M, Jain KC, Remanandan P (1994) Reaction of pigeonpea cultivars and germplasm accessions to the root-knot nematode, *Meloidogyne javanica*. J Nematol 26(4S):644–652

Shibairo SI, Nyabundi JO, Otieno W (1995) Effects of temperature on germination of seeds of three pigeonpea (*Cajanus cajan*) genotypes. Discov Innov 7(3):283–287

Singh IP, Bohra A, Singh F (2016a) An overview of varietal development programme of pigeonpea in India. Legume Perspect 11:37–40

Singh VK, Khan AW, Saxena RK et al (2016b) Next-generation sequencing for identification of candidate genes for Fusarium wilt and sterility mosaic disease in pigeonpea (*Cajanus cajan*). Plant Biotechnol J 14:1183–1194

Singh VK, Khan AW, Saxena RK et al (2017) Indel-seq: a fast-forward genetics approach for identification of trait-associated putative candidate genomic regions and its application in pigeonpea (*Cajanus cajan*). Plant Biotechnol J 15:906–914

Singh K, Sharma SP, Singh TK, Singh Y (1986) Effect of waterlogging on growth, yield and nutrient concentration of black gram and green gram under subtropical condition of Varanasi. Ann Agri Res 7:169–177

Sinha P, Pazhamala LT, Singh VK, Saxena RK, Krishnamurthy L, Azam S, Khan AW and Varshney RK (2016) Identification and validation of selected universal stress protein domain containing drought-responsive genes in pigeonpea (*Cajanus cajan* L.). Front Plant Sci 6:1065

Sinha P, Saxena KB, Saxena RK et al (2015a) Association of nad7a gene with cytoplasmic male sterility in pigeonpea. Plant Genome 8:1–12

Sinha P, Saxena RK, Singh VK, Krishnamurthy L, Varshney RK (2015b) Selection and validation of housekeeping genes as reference for gene expression studies in pigeonpea (*Cajanus cajan*) under heat and salt stress conditions. Front Plant Sci 6:1071

Sinha P, Singh VK, Suryanarayana V, Krishnamurthy L, Saxena RK, Varshney RK (2015c) Evaluation and validation of housekeeping genes as reference for gene expression studies in pigeonpea (*Cajanus cajan*) under drought stress conditions. PLoS ONE 10:e0122847

Singh NK, Gupta DK, Jayaswal PK, Mahato AK et al (2012) The first draft of the pigeon pea genome sequence. J Plant Biochem Biotechnol 21:98–112

Sogbedji JM, van Es HM, Agbeko KL (2006) Cover cropping and nutrient management strategies for maize production in western Africa. Agron J 98:883–889

Sreeharsha RV, Reddy AR (2015) Dynamics of vegetative and reproductive growth patterns in Pigeonpea (*Cajanus cajan* L.) grown under elevated CO_2. Proced Environ Sci 29:147–148

Sreeharsha RV, Sekhar KM, Reddy AR (2015) Delayed flowering is associated with lack of photosynthetic acclimation in pigeon pea (*Cajanus cajan* L.) grown under elevated CO_2. Plant Sci 231:82–93

Sri ND, Mohan MM, Mahesh K, Raghu K, Rao SSR (2016) Amelioration of aluminium toxicity in pigeon pea [*Cajanus cajan* (l.) millsp.] plant by 24-epibrassinolide. Amer J Plant Sci 7:1618–1628

Srivastava JP, Singh P, Singh VP, Bansal R (2010) Effect of waterlogging on carbon exchange rate, stomatal conductance and mineral nutrients status in maize and pigeonpea. Plant Stress 4(1):94–99

Srivastava N, Vadez V, Upadhyaya HD, Saxena KB (2006) Screening for intra and inter specific variability for salinity tolerance in pigeonpea (*Cajanus cajan*) and its related wild species. SAT J 2(1)

Subbarao GV, Chauhan YS, Johansen C (2000) Patterns of osmotic adjustment in pigeonpea—its importance as a mechanism of drought resistance. Eur J Agron 12(3–4):239–249

Subbarao GV, Noriharu A, Otani T (1997) Genetic variation in acquisition, and utilization of phosphorus from iron-bound phosphorus in pigeonpea. Soil Sci Plant Nutr 43(3):511–519

Subbarao GY, Johansen C, Jana MK, Kumar Rao JVDK (1991) Comparative salinity responses among pigeonpea genotypes and their wild relatives. Crop Sci 3(l):415–418

Sultana R, Vales MI, Saxena KB et al (2012) Water-logging tolerances in pigeonpea (*Cajanus cajan* L. Millsp.) genotypic variability and identification of tolerant genotypes. J Agric Sci:1–13

Surekha CH, Beena MR, Arundhatia A, Singh PK, Tuli R, Dutta-Gupta A, Kirti PB (2005) Agrobacterium-mediated genetic transformation of pigeon pea (*Cajanus cajan* (L.) Millsp.) using embryonal segments and development of transgenic plants for resistance against Spodoptera. Plant Sci 169:1074–1080

Surekha C, Kumari KN, Aruna LV, Suneetha G, Arundhati A Kishor PBK (2014) Expression of the *Vigna aconitifolia P5CSF129A* gene in transgenic pigeonpea enhances proline accumulation and salt tolerance. Plant Cell Org Cult 116: 27–36

Rogers A, Ainsworth EA, Leakey ADB (2009) Will elevated carbon dioxide concentration amplify benefits of nitrogen fixation in legumes? Plant Physiol 151:1009–1016

Takele A, McDavid CR (1995) The response of pigeon pea cultivars to short durations of waterlogging. Afr Crop Sci J 3:51–58

Tamirisa S, Vudem DR, Khareedu VR (2014) Overexpression of pigeonpea stress-induced cold and drought regulatory gene (*CcCDR*) confers drought, salt, and cold tolerance in *Arabidopsis*. J Exp Bot 65(17):4769–4781

Taub D (2010) Effects of rising atmospheric concentrations of carbon dioxide on plants. Nat Edn Knowl 3:21

Upadhyaya HD, Gowda CLL, Sastry DVSSR (2008) Plant genetic resources management: collection, characterization, conservation and utilization. SAT J 6:1–15

Upadhyaya HD, Reddy KN, Shivali S (2011) Pigeonpea composite collection for enhanced utilization of germplasm in crop improvement programs. Plant Genet Resour Charact Util 9:97–108

Varshney RK, Chen W, Li Y, Bharti AK et al (2012) Draft genome sequence of pigeon pea (*Cajanus cajan*), an orphan legume crop of resource-poor farmers. Nat Biotechnol 30:83–89

Varshney RK, Murali Mohan S, Gaur PM et al (2013) Achievements and prospects of genomics-assisted breeding in three legume crops of the semi-arid tropics. Biotechnol Adv S0734–9750

Varshney RK, Pandey MK, Bohra A et al (2018b) Towards sequence-based breeding in legumes in the post-genome sequencing era. Theor Appl Genet. https://doi.org/10.1007/s00122-018-3252-x

Varshney RK, Saxena RK, Upadhyaya HD, Khan AW, Yu Y, Kim C et al (2017) Whole-genome resequencing of 292 pigeon pea accessions identifies genomic regions associated with domestication and agronomic traits. Nat Genet 49(7):1082–1088

Varshney RK, Thudi M, Pandey MK, Tardieu F, Ojiewo C, Vadez V, Whitbread AM, Siddique KHM, Nguyen HT, Carberry PS, Bergvinson D (2018a) Accelerating genetic gains in legumes for the development of prosperous smallholder agriculture: integrating genomics, phenotyping, systems modelling and agronomy. J Exp Bot. https://doi.org/10.1093/jxb/ery088

Valenzuela HR, Smith J (2002) CTAHR sustainable agriculture green manure crops series: pigeonpea. Univ Hawaii Coop Ext Serv SA-GM-8

Vanaja M, Reddy PR R, Lakshmi NJ, Razak SKA, Vagheera P, Archana G, Yadav SK, Maheswari M, Venkateswarlu B (2010) Response of seed yield and its components of red gram (*Cajanus cajan* L. Millsp.) to elevated CO_2. Plant Soil Environ 56(10):458–462

Velez-Colon R, Garrison SA (1989) Growth, maturity and flowering of pigeon peas, *Cajanus cajan* L Millsp., at high latitudes. J Agri Univ PR 73(3)

Venzon M, Rosado MC, Euzebio DE, Souza B, Schoerder JH (2006) Suitability of leguminous cover crop pollens as food source for the green lacewing *Chrysoperla externa* (Hagen) (Neuroptera: Chrysopidae). Neotrop Entomol 35(3):371–376

Wei Z, Luo M, Zhao C, Li C, Gu C, Wang W, Zu Y, Efferth T, Fu Y (2013) UV-induced changes of active components and antioxidant activity in postharvest pigeon pea [*Cajanus cajan* (L.) Millsp.] leaves. J Agri Food Chem 61(6):1165–1171

Wellings NP, Wearing AH, Thompson JP (1991) Vesicular-arbuscular mycorrhizae (VAM) improve phosphorus and zinc nutrition and growth of pigeonpea in a Vertisol. Aust J Agri Res 42:835–845

Yadav RC (2017) Cropping practice and makeup shortfall of pulse production with reduced emission of green house gas-nitrous oxide. Arch Chem Res 1:2

Zade D, Rao KB, Bendapudi R, D'souza M (2013) Towards resilient agriculture in a changing climate scenario. Watershed Organisation Trust 1–28

Chapter 3
Breeding and Genomics Approaches for Improving Productivity Gains in Chickpea Under Changing Climate

Uday C. Jha, Abhishek Bohra, Harsh Nayyar, Anju Rani, Poonam Devi, Parasappa Rajappa Saabale and Swarup Kumar Parida

Abstract Chickpea is a well-recognized global grain legume that plays an important role for providing plant-based protein security to global human population. Given the rising uncertainties in global climate coupled with growing occurrence of various pests and diseases and a range of abiotic stresses, global chickpea production is seriously challenged. Therefore, conventional breeding approaches are not adequate to meet the rising demand for chickpea. Evolving genomic technologies have yielded considerable success in accelerating molecular breeding program in various crops. To this end, unprecedented advances in genome sequencing technologies facilitated largely by next-generation sequencing (NGS) technologies have allowed decoding of whole genome sequences of both cultivated and wild species of chickpea. These developments have opened up great opportunity to improve the efficiency of chickpea breeding programs through deployment of large-scale genomic tools. Efforts are

U. C. Jha (✉) · A. Bohra · P. R. Saabale
ICAR-Indian Institute of Pulses Research (IIPR), Kanpur 208 024, Uttar Pradesh, India
e-mail: uday_gene@yahoo.co.in

A. Bohra
e-mail: abhi.omics@gmail.com

P. R. Saabale
e-mail: sparashu@gmail.com

H. Nayyar · A. Rani · P. Devi
Department of Botany, Panjab University, Chandigarh, India
e-mail: harshnayyar@hotmail.com

A. Rani
e-mail: thakur.aanjali@gmail.com

P. Devi
e-mail: heerpoonam03@gmail.com

S. K. Parida
National Institute of Plant Genome Research (NIPGR), Delhi, India
e-mail: swarup@nipgr.ac.in

P. R. Saabale
Regional Research Centre, Indian Institute of Pulses Research, Dharwad 580005, Karnataka, India

C. Kole (ed.), *Genomic Designing of Climate-Smart Pulse Crops*,
https://doi.org/10.1007/978-3-319-96932-9_3

underway to re-sequence multiple genomes for identifying new haplotypes of traits of breeding importance in the crop from wider germplasm resources such as the core collection and reference sets. Taken together, these massive genomic resources including the high-density genotyping assays have allowed chickpea breeders to embrace modern breeding techniques like genomic selection (GS) for enhancing genetic gain. This chapter focuses on the genomics-assisted improvement of chickpea, with an emphasis on the traits that impart resilience to changing climate. In addition to genomics, we highlight progress and possibilities of transgenic research for improving tolerance against biotic and abiotic stress resistance in chickpea. Moreover, the introduction of novel breeding schemes such as "speed breeding", CRISPR/Cas9-based genome editing holds great promise for accelerating the genetic gains projected to meet the ever-increasing demand for plant-based proteins.

Keywords Chickpea · Climate resilience · Genomics · QTL · NGS · Functional genomics

3.1 Introduction

Chickpea, the second most important grain legume in terms of global production next to dry beans grown worldwide, provides protein security to the vegetarian people especially in developing countries across the globe (Bohra et al. 2015; FAO 2016). As an important member of the legume family it is capable of fixing atmospheric nitrogen in association with rhizobia, thus enriching soil nitrogen content (Graham and Vance 2003; Bohra et al. 2014). In parallel, it also serves as an integral component of cereal-legume cropping ecosystem. Grown predominantly in the arid and semi-arid regions of the world, the crop sustains largely under residual soil moisture condition (Gaur et al. 2012). Among all chickpea-producing counties, India occupies the top position in terms of both area and production reported annually (FAO 2016). However, the current productivity of chickpea remains insufficient to meet the demand of plant-based dietary protein to the increasing human population across the globe. In the past, rigorous efforts using conventional breeding were devoted to developing high-yielding chickpea cultivars. All these efforts have met with limited success. Given the current uncertainties of global climate change, chickpea yield is increasingly challenged by growing incidences of various pests and diseases as well as various abiotic stresses (Bohra et al. 2013; Jha et al. 2014a; Gaur et al. 2014). Importantly, a significant change has been seen over four decades in terms of chickpea growing area shifting from cold-season environment to warm-season environment (Gaur et al. 2012). The shift in growing conditions coupled with the global climate change seeks an urgent response from researchers working on chickpea improvement (Gaur et al. 2014). To capture greater genetic diversity, germplasm resources in global gene bank should be explored properly and needs to be incorporated in the main breeding program through pre-breeding activities. In parallel, attention should be placed for incorporation of crop wild relatives and landraces for broadening the genetic base of cultivated chickpea. Unprecedented advances witnessed over the last decade in genome sequencing technologies have allowed the completion of draft genome sequences of chickpea (Jain et al. 2013; Varshney et al. 2013a). The availability of reference genome sequences has provided great opportunity to understand crop evolution and domestication, together with offering high-density and cost-efficient marker genotyping platforms that enable genomics-assisted selection. Recent efforts of genome re-sequencing have also given a new dimension to the discovery of the structural variations that could uncover genomic regions of breeding relevance. Afterward completion of reference chickpea genome sequence, important functional genomics milestones have been achieved in the context of various biotic, abiotic and several plant architecture and development related traits of breeding importance. Thus, increasing facilities of the genotyping platform have tremendously improved the genomic repertoire of chickpea that allows mapping of complex trait quantitative trait loci (QTL) viz., drought with high resolution (Jaganathan et al. 2015; Kale et al. 2015). Transgenics approaches have also been employed for the development for improved chickpea that shows a high level of resistance/tolerance to biotic as well as abiotic stresses. In parallel, the latest addition to the breeder's toolbox includes speed breeding (Li et al. 2018a), and genome editing like CRISPR/Cas9 system (Doudna and Charpentier 2014).

3.2 Biotic Stress

Among the major biotic stresses, Fusarium wilt (FW) caused by *Fusarium oxysporum* f. sp. *ciceris* and Ascochyta blight (AB) caused by *Ascochyta rabiei*, are the two most important diseases that cause yield loss up to 90% in chickpea (Sabbavarapu et al. 2013). Both dry and humid conditions are congenial for FW infection and the disease is estimated to cause 10–90% annual yield loss worldwide (Jimenez-Diaz et al. 1989). In India, FW causes 10–15% annual yield losses (Singh and Dahiya 1973), however complete crop failures have also been witnessed under favorable conditions (Halila and Strange 1996). Likewise, humid condition promotes AB infection (Pande et al. 2005). AB mainly prevails in northwestern parts of India including Punjab, Jammu, northern Rajasthan, and western Uttar Pradesh, causing huge loss to chickpea production. Additionally, dry root rot (caused by *Rhizoctonia bataticola*), collar rot (caused by *Sclerotium rolfsi*) and botrytis gray mold caused by *Botrytis cineria* Pres. are the other important emerging diseases that raise economic concerns among chickpea growers.

Efforts to elucidate the genetic makeup of the disease resistance traits have shed light on the genetic control of various races (eight races 0, 1A, 1B/C, 2, 3, 4, 5, and 6) of FW and AB diseases (Upadhyaya et al. 1983a, b; Singh et al. 1987; del Mar Jiménez-Gasco and Jiménez-Diaz 2003; Brinda and Ravikumar 2005). This understanding of FW (for various races) resistance gene(s)/QTLs was furthered with increasing use of molecular markers and their subsequent placement on to the genetic linkage maps (Sharma et al. 2004, 2005; Cobos et al. 2005, 2009; Gowda et al. 2009;

Sabbavarapu et al. 2013; Patil et al. 2014; Jingade and Ravikumar 2015). Examples include recent detection of QTLs/genes controlling AB resistance using various mapping populations (Flandez-Galvez et al. 2003; Millan et al. 2003; Udupa and Baum 2003; Iruela et al. 2006, 2007; Tar'an et al. 2007; Anbessa et al. 2009; Kottapalli et al. 2009; Aryamanesh et al. 2010; Bhardwaj et al. 2010; Madrid et al. 2013; Sabbavarapu et al. 2013). Following the development of high-density linkage maps and genome sequencing, FW and AB resistance gene/QTLs have been precisely delineated and cloned to allow gene pyramiding into the background of elite chickpea. Thus, the increasing repertoire of chickpea genomic resources is of immense importance when viewed from the point of accelerated breeding cycles.

3.3 Abiotic Stresses

Global climate change has also enhanced the impacts of abiotic stresses including drought, heat stress (HS), salinity, and cold stress that in turn pose a serious challenge to chickpea production worldwide.

3.3.1 Drought Stress

Chickpea is predominantly grown in arid and semi-arid regions prevailing in Asia, Africa, and Australia under rain-fed condition making it face water stress during its lifecycle. Drought remains the most important stress among the various abiotic stresses faced by chickpea and is reported to cause up to 50% yield loss annually in chickpea (Sabaghpour et al. 2006).

3.3.1.1 Breeding Efforts for Combating Drought Stress

Conventional breeding methods have been applied for identifying drought stress tolerant chickpea based on various morpho-physiological and traits of breeding importance. Promising chickpea genotypes implicating drought escape mechanism were screened out, which include ICC 96029 (Kumar and Rao 1996), ICCV 2 (Kumar and Abbo 2001), ACC 316, and ACC 317 (Canci and Toker 2009). Likewise, genotypes ICC 4958 and ICC 8261 (Gaur et al. 2008; Kashiwagi et al. 2008) have been incorporated as donor parents for introducing root related traits in chickpea breeding programs for increasing drought tolerance. However, the progress of developing drought-tolerant chickpea is limited owing to the complex genetics of drought and lack of accurate and automated phenotyping methods.

3.3.1.2 Genomics Resources for Drought Stress Tolerance

In the past, several efforts of conventional breeding approaches were intensely devoted to work out the genetic basis of drought tolerance in chickpea. However, success for breeding drought tolerance has not met with anticipated success as it is controlled by several gene(s)/QTLs and highly influenced by genotype × environment interaction (Fleury et al. 2010). Recent advancements in molecular marker technology have revealed various minor and major QTLs underlying drought tolerance in chickpea. A range of QTLs related to various traits that are directly or indirectly associated to drought stress have been elucidated in legume crops including chickpea (Singh et al. 2015). Considering the importance of phenological traits, several QTLs have been reported for days to flowering and days to maturity under drought stress (Rehman et al. 2011; Hamwieh et al. 2013; Varshney et al. 2014a; Jaganathan et al. 2015; Kale et al. 2015). In view of the significant role of root related traits such as higher root length density, root volume, maximum root depth (RDp), and higher root biomass in drought tolerance, several QTLs controlling these traits have been analyzed (Gaur et al. 2008; Varshney et al. 2013b, 2014a, b; Kale et al. 2015). Likewise, QTLs associated with relevant physiological traits viz., stomatal conductance (Rehman et al. 2011), canopy conductance (Sivasakthi et al. 2018), carbon isotope ratio (Kale et al. 2015), and plant vigor trait (Sivasakthi et al. 2018) have been detected that influence crop yield under drought stress. Most importantly, QTLs related to harvest index (Rehman et al. 2011; Varshney et al. 2014a; Jaganathan et al. 2015; Kale et al. 2015; Srivastava et al. 2016) has been reported under water stress condition.

Given the unprecedented advancements in linkage analysis in chickpea over the past decade, several genomic regions that influence traits of agro-economic importance such as drought tolerance have been uncovered in chickpea (Varshney et al. 2014a; Jaganathan et al. 2015; Kale et al. 2015). In this context, Varshney et al. (2014a) have reported a genomic segment in chickpea that harbors 13 main-effect QTLs controlling a wide range of drought-related traits, and explaining up to 58% phenotypic variance (PV). Interestingly, the same genomic region on CaLG4, referred to as "QTL-hotspot", was detected in two mapping populations. Application of high-density genotyping assays in chickpea narrowed down this QTL-containing segment to ~14 cM (Jaganathan et al. 2015) and more recently to ~300 Kb (Kale et al. 2015), finally leading to the partitioning of this genomic region into "*QTL-hotspot_a*" and "*QTL-hotspot_b*". Recently, Sivasakthi et al. (2018) uncovered QTLs related to canopy conductance and plant vigor within this *"QTL-hotspot*" region.

To circumvent the limitations of QTL mapping and to investigate novel alleles related to traits of importance available across the various crop germplasms, association genetics or genome-wide association study (GWAS) is a promising genomic scale approach (Bohra 2013). Following association analysis of 300 global chickpea collections, several important marker-trait associations (MTAs) related to physiological and agronomic traits were detected under drought stress (Thudi et al. 2014). Likewise, Li et al. (2018b) elucidated 38 significant MTAs related to various yield-related traits covering seed number to grain yield by conducting GWAS on 132 chickpea genotypes under drought stress. Authors reported a set of important candidate genes encoding "*auxin efflux carrier protein* (*PIN3*), *p-glycoprotein,* and *nodulin MtN21/EamA-like transporter*" that underlie the given genomic regions.

3.3.2 Temperature Extremities

Current uncertainties of global climate change render plants to experience abnormal temperature or temperature extremities that are either beyond or below the ambient temperature, causing serious impacts on proper plant growth and development (Jha et al. 2017). With the increasing evidence, HS is appearing as one of the important abiotic stresses which causes harmful effects and limits yield in crops including chickpea (Hasanuzzaman et al. 2013; Kaushal et al. 2013; Devasirvatham et al. 2012; Jha et al. 2014b). Various researchers have documented the reproductive stage as most sensitive to HS that leads to decreased fertilization resulting in flower abortion and reduced grain set (Wahid et al. 2007; Devasirvatham et al. 2012; Bhandari et al. 2016). The development of female (stigma-style, ovary) and male (pollen, anthers) parts are the most sensitive organs to abiotic stress in reproductive biology (Nayyar et al. 2005). Traditional approaches of plant breeding have been directed for developing heat-tolerant chickpea in order to sustain its yield under HS. Significant progress has been achieved in this regard through surveying genetic resources, and promising chickpea genotypes were identified for HS tolerance, such as ICCV 92944, ICC 4958, ICC 1205 (Devasirvatham et al. 2010, 2013; Krishnamurthy et al. 2011; Upadhyaya et al. 2011; Jha and Shil 2015; Jha et al. 2015, 2018a; Paul et al. 2018a).

Responses of different floral parts are different towards low-temperature stress, all effects of cold stress are impairing, resulting in the reduction in productivity (Staggenborg and Vanderlip 1996; Verheul et al. 1996). In chickpea, low temperature (below 15 °C) induced flower abscission and reduction in pod set in northern India and Australia (Srinivasan et al. 1999; Berger et al. 2004, 2006; Clarke and Siddique 2004; Nayyar et al. 2005). Considering low-temperature stress, several sources of low-temperature-tolerant chickpea genotypes have been suggested. A range of low temperature-tolerant chickpea genotypes have been identified based on different parameters such as ICCV 96029, ICCV 96030 based on escape mechanism (Sandhu et al. 2002; Kumar and Rao 1996), Sonali and Rupali for higher pollen fertility (Clarke et al. 2004), and ICCV 88502, ICCV 88503 for higher efficiency of pod setting at low temperature (Srinivasan et al. 1998, 1999). Likewise, Arslan et al. (2018) reported İnci chickpea genotype to show freezing tolerance considering various physiological and biochemical traits. These tolerant chickpea genotypes could be of great importance for developing improved cultivars having greater resilience to extreme temperatures.

3.3.2.1 Genomic Resources for Temperature Extremities

Significant genomic tools have been developed that are relevant to temperature extremities, especially for heat tolerance in chickpea (Thudi et al. 2014; Jha et al. 2018b, c; Jha 2018). Different molecular markers have been identified for QTLs/genes associated with a large number of traits (Gaur et al. 2014; Thudi et al. 2014) which can facilitate marker-assisted breeding for heat tolerance in chickpea. Availability of novel single nucleotide polymorphisms (SNPs) (Hiremath et al. 2011), high-density diversity array technology (Thudi et al. 2011), high-density genetic maps (Thudi et al. 2011) and transcriptome assemblies (Kudapa et al. 2014) could further accelerate molecular breeding for desired traits including HS tolerance in chickpea (Varshney et al. 2009; Mallikarjuna et al. 2017). Recombinant inbred lines (RILs), (292, $F_{8:9}$) developed by crossing ICC 4567 (heat sensitive) and ICC 15614 (tolerant) and these were later assessed for thermo-tolerance and used to identify molecular markers linked to QTLs related to heat tolerance traits (Devasirvatham et al. 2013; Paul et al. 2018a, b).

QTLs associated with chickpea heat tolerance at the reproductive stage are good indicators of high grain yield under HS heat and can be used in indirect selection for developing heat-tolerant cultivars. Genotyping by sequencing (GBS) based SNP markers have been used to identify key genomic regions (Paul et al. 2018b). Two potential genomic regions have been identified, which have four major QTLs each on *CaLG05* and *CaLG06* for several heat-responsive traits that are directly responsible for heat tolerance in chickpea. Digenic epistatic QTLs (19 nos.) were found to be associated with the six traits: biomass (BM), visual scoring (VS), number of filled pods per plot (Fpod), total number of seeds per plot (TS), grain yield per plot (GY) and % pod set for heat tolerance (Paul et al. 2018a). The epistatic QTLs loci were observed maximum for TS (nine), followed by % pod set (four) and some other loci such as *eqpodset2_1/eqts2_1, eqts2_1/eqpodset2_1,neqfpod4_5/neqts9_5,neqts9_5/neqfpod4_5* were collectively controlling more than one trait that shows the pleiotropy nature of the traits. QTLs for traits such as Fpod, TS, and GY were not expressed under non-stress condition confirming the fact that these QTLs were only expressed under high-temperature condition and were therefore efficient to be used as potential candidates for heat tolerance through marker-assisted breeding.

Three major QTLs, *Qefl1-2, Qefl3-3,* and *Qefl4-1*, corresponding to flowering time genes *efl-1* from ICCV 96029, *efl-3* from BGD 132, and *efl-4* from ICC 16641 were mapped on *CaLG04, CaLG08,* and *CaLG06*, respectively (Mallikarjuna et al. 2017). In chickpea four early flowering (*efl*) nonallelic genes have been reported that are *efl-1* from ICCV 2 and ICCV 96029 (Hegde 2010; Gaur et al. 2015; Kumar and van Rheenen 2000), *ppd* or *efl-2* from ICC 5810 (Hegde 2010; Or et al. 1999), *efl-3* from BGD 132 and *efl-4* from ICC 16641 and ICC 16644 (Gaur et al. 2015). Recently four major QTLs, *Qefl2-1, Qefl2-2, Qefl2-3,* and *Qefl2-4*, for flowering time were identified in the cross ICC 5810 × "CDC Frontier" on CaLG01, CaLG03, CaLG04, and CaLG08, respectively, in chickpea (Gaur et al. 2018). Accuracy of phenotyping would play an important role in the identification of molecular markers linked to QTLs associated with (HS) tolerance traits. QTLs detected in unique positions in the non-stress environment are stronger evidence that there is no correspondence between QTLs found in non-stress with the QTLs found in the heat stress environment. Effective breeding strategies by choosing *efl* gene or a combination of such genes based on the desired background and linkage relationships of the flowering time genes with other traits will be helpful in the development of better heat-tolerant chickpea varieties (Gaur et al. 2015). Several heat-tolerant progenies have been developed from multi-parent advanced generation inter-cross (MAGIC) populations at ICRISAT by rearranging of alleles. The development of breeding approach like the use of MAGIC populations/lines, are promising to combine favorable genes for genetic recombination with enhanced tolerance to HS and to bring about greater genetic diversity (Bandillo et al. 2013; Huang et al. 2015; Gaur et al. 2018). Therefore, by genetic improvement, heat tolerance in chickpea can be introduced/improved.

3.3.3 Salinity Stress

Increasing indiscriminate practices of irrigation water at various farmlands has caused a serious problem of salinity across the world. Significant yield loss has been recorded due to salinity in chickpea grown across the subtropical and semi-arid regions of the world (Ali et al. 2002; Kaashyap et al. 2017). Soil salinity stress above $3dSm^{-1}$ is reported to limit chickpea growth, reproductive development, and yield (Rao et al. 2002; Katerji et al. 2005; Vadez et al. 2012a; Turner et al. 2013; Pushpavalli et al. 2016).

Identification of salinity tolerant chickpea genotypes has been carried out through screening of a set of germplasm under salt-stressed environments leading to the identification of CSG 88101, CSG 8927, CSG 8962, and ICCV 96836, JG 62 (Dua and Sharma 1995; Vadez et al. 2007; for details see Jha et al. 2014a; Jha et al. 2019). However, the progress of developing salinity tolerant chickpea is hampered due to the genetically complex inheritance of salinity tolerance trait.

To elucidate the underlying salinity tolerance gene(s), some important genomic resources have been developed in chickpea (Kaashyap et al. 2017). To date, a limited number of QTLs conferring salinity tolerance have been recorded in chickpea (Vadez et al. 2012b; Pushpavalli et al. 2015). One major QTL related to seed yield explaining 19% PV on linkage group (LG) 03 was uncovered under salinity stress (Vadez et al. 2012b). Likewise, two major QTLs conferring salinity tolerance have been reported on LGs 5 and 7 from ICCV 2 × JG 11 mapping population (Pushpavalli et al. 2015) see in Table 3.1.

Table 3.1 List of trait QTLs in chickpea developed after chickpea genome sequence

Trait	Mapping population	QTL	Marker	LG groups	PV%	References
Drought	ICC 4958 × ICC1882	"QTL-hotspot"	SSR markers	CaLG04	58.2	Varshney et al. (2014a)
	ICC 283 × ICC 8261, ICC 4958 × ICC 1882	"QTL-hotspot" 164 main-effect QTLs including 24 novel QTLs	SNP, CAPS, dCAPS, SSR	CaLG04	10.1–67.71	Jaganathan et al. (2015)
	ICC 4958 × ICC 1882	QTL-hotspot_a (15 genes) QTL-hotspot_b (11 genes)	SNP	CaLG04 CaLG08	10.36–59.83	Kale et al. (2015)
	ICC 4958 × ICC 17163	*CaqYP2.1,CaqYP3.1* *CaqYP4.1, CaqYP5.1* *CaqYP6.1, CaqYP7.1* *CaqHI2.1, CaqHI3.1* *CaqHI4.1, CaqHI5.1* *CaqHI6.1, CaqHI7.1*	SNP	CaLG2, CaLG3 CaLG4, CaLG5, CaLG6, CaLG7	11.2–23.7	Srivastava et al. (2016)
	ICC 4958 × ICC 1882	21 major QTLs	SNP	CaLG1, CaLG4 CaLG03, CaLG05 CaLG06, CaLG07, CaLG08	53	Sivasakthi et al. (2018)
Salinity	JG 62 × ICCV 2		SSR	3, 6	37	Vadez et al. (2012b)
	JG 62 × ICCV 2		SSR	4	8.8–37.7	Vadez et al. (2012b)
	JG 62 × ICCV 2		SSR	6	34.6	Vadez et al. (2012b)
	ICCV2 × JG11		SSR, SNP	5, 7	12–17	Puspavalli et al. (2015)

(continued)

Table 3.1 (continued)

Trait	Mapping population	QTL	Marker	LG groups	PV%	References
Heat stress	ICC15614 × ICC4567	*qfpod02_5* *qfpod03_6* *qgy02_5* *qgy03_6*	SNP	CaLG05 CaLG06	6.5–11.5	Paul et al. (2018b)
Ascochyta blight	C 214 × ILC 3279	*AB-Q-SR-4-1 and AB-Q-SR-4-2)*	SSR	LG4	10.3–31.9	Sabbavarapu et al. (2013)
Ascochyta blight	C 214 × ILC 3279	*AB-Q-APR-6-1 and AB-Q-APR-6-2*	SSR	LG6	2.2–11.5	Sabbavarapu et al. (2013)
Ascochyta blight	ICCV 96029 × CDC	Eight QTLs	SNP	3 and 8	10–19	Daba et al. (2016)
Ascochyta blight	CPR-01 and CPR-02	Twelve QTLs	SNP	Ca1, Ca2, Ca4 Ca6, and Ca7	–	Deokar et al. (2018)
Ascochyta blight	–	*qABR4.1* and *qABR4.2* *qABR4.3*		4	–	Kumar et al. (2018)
Ascochyta blight	JG 62 × ICCV 05530	Two QTLs	SSR, SNP	–	–	Garg et al. (2018)
Fusarium wilt	C 214 × WR 315′	*FW-Q-APR-6-1, FW-Q-APR-6-2*	SSR	LG6	10.4–18.8	Sabbavarapu et al. (2013)
Fusarium wilt	JG 62 × ICCV 05530	Five QTLs	SSR, SNP	CaLG02, CaLG04, and CaLG06	6.63–31.55	Garg et al. (2018)

(continued)

Table 3.1 (continued)

Trait	Mapping population	QTL	Marker	LG groups	PV%	References
Seed traits	*C. arietinum* SBD377 × BGD112	Seven QTLs for seed weight Four QTLs for seed number Five QTLs for seeds per pod Four QTLs for pods per plant	SNP	LG1–LG8 except LG3	9.97–29.71	Verma et al. (2015a)
Seed weight	SBD377 × BGD112	*qSW-1* and *qSW-2*	ESTs and Potential intron Polymorphisms	LG1 and LG2	11.54–19.24	Gupta et al. (2015)
Plant height	SBD377 × BGD112	*qPH-1* and *qPH-2*	ESTs and Potential intron Polymorphisms	LG1 and LG2	13.98–12.17	Gupta et al. (2015)
Flowering time	ICCV 96029 × CDC Frontier, ICC 5810 × CDC Frontier BGD 132 × CDC Frontier ICC 16641 × CDC Frontier	*efl-1,efl-3,efl-4*	SSR	CaLG04, CaLG08, and CaLG06	–	Mallikarjuna et al. (2017)
Flowering time	CDC Frontier × ICCV 96029	*Early flowering1 (Efl1)*		Ca5	59	Ridge et al. (2017)

3.4 Other Stresses

Elevated levels of atmospheric carbon dioxide impair seed germination, seed emergence, and seedling vigor (Saha et al. 2015a) apart from exerting a negative influence on grain quality of chickpea by reducing protein content (Saha et al. 2015b). Likewise, increasing concentration of ozone gas is appearing as another global threat to crop production, causing inhibition in photosynthesis, and affecting yield and quality parameters (Ainsworth et al. 2012; Ainsworth 2017). Evidence of reduction in plant height and dry weight have been recorded in chickpea under ozone stress (Welfare et al. 2002). Given this, breeding efforts aimed to deliver chickpea cultivars that suit modern agriculture should place greater attention to these newly escalating production constraints.

3.5 Genetic and Genomic Resources for Improving Quality Traits in Chickpea

Chickpea serves as an important source of "plant-based protein" and essential dietary minerals ranging from iron, zinc, to vitamin A (Jukanti et al. 2012). Several important genotypes possessing high iron containing lines viz., CA 0469C025C, and Dwelley lines (Vandemark et al. 2018), higher zinc containing lines such as CA0790B0053C lines, Arerti, FLIP 07-27C, and FLIP 08-60C, Wolayita local landrace (Hidoto et al. 2017; Vandemark et al. 2018) and possessing high carotene content genotypes CDC Cory, CDC Jade, and CDC Verano (Rezaei et al. 2016) have been recorded. Concomitantly, given the unprecedented advancements in chickpea genomics, efforts are underway to elucidate the micronutrient content controlling genomic segments through employing chickpea genomics resources. Several (MTAs) controlling zinc and iron content have been uncovered on LG 1, 4, 6, and 7 via GWAS in a set of 94 chickpea genotypes (Diapari et al. 2014). Subsequently, Upadhyaya et al. (2016a) deciphered several genomic regions controlling these traits via Illumina GoldenGate assay in 92 chickpea germplasm. Moreover, to increase seed protein content (SPC) in chickpea, Jadhav et al. (2015) reported a total of 9 significant MTAs via GWAS. Likewise, a total of seven potential candidate gene controlling SPC was recovered from a large panel of 336 chickpea germplasm by recruiting 16,376 SNP markers (Upadhyaya et al. 2016b). With the availability of chickpea genome carotenoid biosynthesis governing genomic region has been elucidated (Rezaei et al. 2016). A total of 19 genes involved in carotenoid biosynthesis were also examined at the expression level in five different chickpea cultivars. Thus, collectively these chickpea genetic and genomic resources could help in designing biofortified chickpea genotype to overcome the increasing challenges of micronutrient deficiency related problems prevalent in the human population residing in the developing countries especially sub-Saharan Africa and Southeast Asia.

3.6 Crop Wild Relatives (CWRs) as a Rich Reservoir of Climate-Resilient Traits

Genetic bottlenecks occurred during the process of domestication accompanied by human-led artificial selection have considerably reduced the genetic diversity of present-day chickpea (Abbo et al. 2003a, b). CWRs maintain sufficient genetic variability, which may be largely credited to the limited attention that these have received during human-driven selections. Consequently, CWRs could serve as an important natural reservoir of the novel and rare alleles for various traits of breeding importance. To broaden the genetic base of chickpea, CWRs should be efficiently incorporated into breeding programs via enhanced pre-breeding activities. Eight annual wild species have been reported in chickpea, however, only *C. reticulatum* belonging to the primary gene pool, is crossable with cultivated chickpea. The other species remains non-crossable and need special techniques like embryo rescue for avoiding crossing incompatibility problems. Two important lines, ILC 10765 and ILC 10766, possessing cyst nematode resistance exemplify judicious exploitation of *C. reticulatum* accession, ILWC119 in chickpea breeding program (Malhotra et al. 2002). Likewise, incorporation of *C. reticulatum* and *C. echinospermum* wild species into chickpea breeding program resulted in the development of chickpea lines resistant to multiple biotic stresses including Fusarium wilt, root rot, and Botrytis gray mold (Singh et al. 2005; Ramgopal et al. 2012). Toker et al. (2007) have reported the presence of high drought tolerance level among perennial wild *Cicer* species, viz. *C. anatolicum*, *C. microphyllum*, *C. montbretti*, *C. oxydon* and *C. songaricum*, and accessions of annual wild species viz., *C. reticulatum*, *C. bijugum*, and *C. pinnatifidum* (Singh et al. 1995; Toker 2005). Breeding strategies that allow effective entry of CWRs into chickpea breeding program hold great opportunity to improve resilience against biotic and abiotic stresses.

3.7 Whole Genome Assemblies for Tapping the Novel Genetic Variants

Current advancements in genome sequencing technologies have enabled construction of de novo genome assemblies and reference genome sequencing in various legume species (Bohra and Singh 2015).

Apart from uncovering a plethora of genetic variants, whole genome assemblies shed new light into genome dynamics, domestication and evolutionary processes in chickpea (Jha 2018). The first reference genome sequence of cultivated Kabuli chickpea, measuring 738.09 Mb with 25,365 genes was released in the year 2013 (Varshney et al. 2013a). In 2013, another research group reported a 740.52-Mb genome assembly of desi type harboring 27,571 protein-coding genes (Jain et al.

2013). Most recently, Gupta et al. (2017) assembled the draft genome sequence of *Cicer reticulatum* L. into 327.07 Mb. These latest additions to genomic resource repertoire could greatly assist to accelerate the progress of the breeding program in chickpea.

3.8 Whole Genome Re-sequencing (WGRS) for Capturing Unknown Genomics Variants

With the availability of reference genome sequence in chickpea, attempts to re-sequence multiple genomes have elucidated novel genomic regions/structural variations related to various traits of economic significance through establishing a comparison between WGRS with the available reference genome. Thus this could lead to unveiling the genetic basis of phenotypic traits of agronomic importance in various cultivars at the genomic level. Several structural variations and the underlying candidate genes related to yield traits (Thudi et al. 2016a, b), Ascochyta blight resistance (Deokar et al. 2018; Kumar et al. 2018), pod-related traits (Das et al. 2016), and seed-related traits (Das et al. 2015) have been captured by applying WGRS (see Table 3.2). Furthermore, establishment of genome sequence assembly of CWRs of chickpea could help in unveiling novel genomic regions controlling traits of breeding importance.

Table 3.2 List of candidate genes identified from re-sequencing of chickpea genotypes

Trait	QTL/candidate gene	References
Seed weight	*CaqSW1.1*	Das et al. (2015)
Hundred seed weight	*Ca_04364* (on CaLG01) and *Ca_04607*	
Total dry root weight to total plant dry weight ratio (RTR)	*Ca_04586* (on CaLG04) for RTR	Singh et al. (2016)
Number of pods per plant	*CaqaPN4.1* and *CaqaPN4.2*	Das et al. (2016)
Thirteen yield-related traits under drought stress	*Auxin efflux carrier* protein (PIN3), *p-glycoprotein* (PGP), *nodulin MtN21/EamA-like transporter*	Li et al. (2018b)
Disease resistance, flowering time, and drought tolerance		Thudi et al. (2016b)
Both biotic and abiotic stress		Thudi et al. (2016a)
Ascochyta blight	*qABR4.1* and *qABR4.2* *qABR4.3*	Kumar et al. (2018)
Ascochyta blight	Six candidate genes	Deokar et al. (2018)
Ascochyta blight	Twelve candidate genes	Li et al. (2017)

3.9 GS and Hope for Increasing Genetic Gain in Chickpea

The recent development of genomic tools has strengthened the foundation for molecular breeding in chickpea. Modern molecular breeding techniques like marker-assisted backcrossing (MABC) and marker-assisted recurrent selection (MARS) have been deployed for transferring single and multiple QTLs, respectively, to elite cultivars in chickpea (Gaur et al. 2014). Notable instances of marker-assisted breeding in chickpea include the transfer of a major QTL with 30% phenotypic variation to elite chickpea cultivars (JG 11, KAK 2, and Chefe) (Gaur et al. 2014). Likewise, combined resistance to wilt race 1 and Ascochyta blight was achieved in the background of C 214 via MABC, and similarly, MARS enabled transferring of root traits and other drought-related traits to elite chickpea varieties (Varshney et al. 2013b). However, these approaches are effective for transferring single QTL with large phenotypic effects or accumulating a limited number of superior alleles. Improving complex traits such as yield, drought, and other traits governed by a large number of small effect QTLs demands new approaches like GS. Availability of high-density genotyping platform and draft genome sequences has provided great accessibility for harnessing high throughput SNP markers for performing GS in chickpea. This approach helps in selecting individuals from "target population" by estimating genomic estimated breeding value genomic estimated breeding value (GEBV) relying on the prediction model developed from "training population" with known genotyping and phenotyping information (Meuwissen et al. 2001; Meuwissen and Goddard 2007; Jannink et al. 2010; Bohra 2013).

This approach has been currently employed for estimating the "prediction accuracy" of two important yield-related traits including seed yield and 100-seed weight, and two phenological traits including days to 50% flowering and days to maturity (Roorkiwal et al. 2016, 2018). Likewise, Li et al. (2018a) computed GEBVs for predicting yield and yield-related traits under drought stress. Roorkiwal et al. (2018) applied GS models that integrate G $\times$ E interactions, which facilitated the selection of superior chickpea lines across different environments.

3.10 Advances in Functional Genomics

Accessibility to chickpea genome sequences have provided great opportunity to annotate gene function of various traits relevant to biotic and abiotic stresses, plant architecture and developmental traits related to both basic and applied chickpea research for accelerating development of improved cultivars (Jha 2018).

Initial efforts to generate functional genomics resources in chickpea mostly focused on abiotic stresses viz., drought, cold, and salinity, which revealed sets of differentially expressed genes or stress-responsive candidate genes. The studies include generation of expressed sequence tags (ESTs), and analyses like suppression subtractive hybridization (SSH), serial analysis of gene expression (SAGE), and microarray (Jayashree et al. 2005; Mantri et al. 2007; Molina et al. 2008, 2011; Varshney et al. 2009; Jain and Chattopadhyay 2010; Deokar et al. 2011). Mantri et al. (2007) reported several differentially expressed genes (DEGs) related to drought stress by using the microarray technique. Likewise, Varshney et al. (2009) leveraged functional genomics resources by developing 5,982 and 5,922 drought responsive ESTs from ICC 4958 and ICC 1882, respectively. To impart a global view of gene expression in response to drought stress, the massive scale of transcriptome data were developed from ICC 4958 and ICC 1882 by employing Roche/454 and Illumina/Solexa NGS technologies under drought stress (Hiremath et al. 2011). Along with the huge number of tentative unique sequences (TUSs), a total of 44,639 differentially expressed TUSs were recovered from the above mentioned two genotypes under drought stress. Given the transcriptome analysis of contrasting Fusarium wilt tolerance parents, the association of several important SNP/insertion/deletion (InDel) markers with Fusarium wilt tolerance were elucidated (Jain et al. 2015). Additionally, expression profiling of Fusarium tolerant contrasting parents suggested the role of various genes associated with G-protein signaling, plant defense signaling, and R gene-mediated defense against Fusarium wilt (Upasani et al. 2017). While, considering biotic stress, probable role of nucleotide binding site–leucine-rich repeat (NBS-LRR) gene contributing in Ascochyta blight resistances was demonstrated through various differential expression profiling (Leo et al. 2016; Li et al. 2017; Sagi et al. 2017).

Subsequently, unprecedented advancements in transcriptome profiling due to arrival of NGS based RNA-seq technology have further strengthen the current understanding of various stress-responsive regulatory gene(s) and their complex networks associated to various stresses with greater precision (Jain et al. 2014, 2015; Kudapa et al. 2014; Garg et al. 2015; Srivastava et al. 2016). Considering drought stress, RNA-seq analysis illuminated on the understanding regarding the global gene expression under drought in chickpea (Garg et al. 2016; Badhan et al. 2018; Kudapa et al. 2018; Mahdavi Mashaki et al. 2018). Following the development of gene expression atlas in chickpea, important candidate genes such as *Ca_04561*, *Ca_04564*, *Ca_04566*, *Ca_04567* were identified within QTL hot spot region (Kudapa et al. 2018). Likewise, important heat shock protein candidate genes *Ca_25811, Ca_23016, Ca_09743, Ca_17680,* and *Ca_25602* were elucidated from HS-treated vegetative and reproductive tissues (Agarwal et al. 2016). More importantly, availability of transcriptome atlas "*Cicer arietinum* Gene Expression Atlas (CaGEA)" can improve the knowledge about the gene expression pattern across a range of developmental phases in chickpea (Kudapa et al. 2018).

Emerging evidence of various non-coding RNAs and their participatory role have received significant attention in chickpea for drought (Khandal et al. 2017), salinity (Kohli et al. 2014; Singh et al. 2017) and other developmental traits (Khemka et al. 2016). Thus, the next grand challenge of chickpea functional genomics is to determine the biological functions of every gene for further chickpea genetic improvement.

3.11 Developing Web-Based Community Resources to Support Chickpea Improvement

In recent years, several web-based resources have been developed in chickpea such as (http://www.icrisat.org/gt1/cpest/home.asp.) (Jayashree et al. 2005) for ESTs, CicArMiSatDB markers (http://cicarmisatdb.icrisat.org) (Doddamani et al. 2014) for simple sequence repeats (SSRs), CicArVarDB (http://cicarvardb.icrisat.org/) (Doddamani et al. 2015) for genome-wide SNPs and InDels, Chickpea Transcriptome Database (CTDB) (http://www.nipgr.res.in/ctdb.html) (Verma et al. 2015b) for transcriptome related information, and CicerTransDB 1.0, (http://www.cicertransdb.esy.es) for transcription factor (TF) related information (Gayali et al. 2016), for comparative genomics http://www.nipgr.res.in/CGWR/home.php (Misra et al. 2014). Thus these web-based resources could enable chickpea research community to identify the various key regulatory gene(s) controlling various traits of importance.

3.12 Transgenic Systems in Chickpea for Developing Climate-Resilient Chickpea

Given the bottlenecks of traditional breeding to cope with current global climate change, transgenic approach stands to be among the most potent strategies that overcome sexual reproduction barriers to improve traits/gene(s) into any genetic background of choice. Considering major insect pests, *Helicoverpa armigera* remains the most devastating insect pest causing serious yield loss in chickpea. Several management practices including integrated pest management practices have been recruited to control this insect pest attack (Acharjee and Sarmah 2013). Though surveying natural germplasm resources encompassing diverse gene pools for stress resistance remains the most sustainable approach, lack of pod borer resistant germplasm has greatly restricted conventional breeding for developing *Helicoverpa* resistant chickpea cultivar. Thus, intervention of transgenics approaches could play a crucial role in the development of *Helicoverpa* resistant lines in chickpea. To date, deployment of *Bacillus thuringiensis* (Bt) insecticidal crystal protein genes through genetic transformation remains the most preferred transgenic technology for designing pod borer resistant chickpea (Sanyal et al. 2005; Lawo et al. 2008; Acharjee et al. 2010). In the context, several efforts were devoted previously and are underway for the development of pod borer resistant chickpea to sustain its yield. Initially, Sanyal et al. (2005) established the toxicity of cry1Ac gene isolated from Bacillus thuringiensis (*BT*) against Helicoverpa armigera. Subsequently, Acharjee et al. (2010) showed the effectiveness of *BT cry2Aa* gene against pod borer infection in chickpea. Likewise, Ganguly et al. (2014) introduced the cry1Ab/Ac insecticidal gene into the DCP 92-3 genotype under the control of pod-specific soybean *msg* promoter and *rice actin1* promoter through *Agrobacterium*-mediated transformation (see Table 3.3). The transformed lines possessing the abovementioned gene showed high lethality to

Table 3.3 Updates on transgenics development in chickpea

Trait	Source of gene	Gene	Explant used	Method	Reference
Helicoverpa armigera	*Bacillus thuringiensis*	*cry1Ac*	Cotyledonary nodes	*Agrobacterium tumefaciens* mediated	Sanyal et al. (2005)
Helicoverpa armigera	*Bacillus thuringiensis*	cry1Ab/Ac	–	*Agrobacterium tumefaciens* mediated	Ganguly et al. (2014)
Helicoverpa armigera	*Bacillus thuringiensis*	CryIA(c) delta-	Epicotyl	Particle gun bombard-ment mediated	Indurker et al. (2007)
Helicoverpa armigera	*Bacillus thuringiensis*	*cry2Aa*	–	*Agrobacterium tumefaciens* mediated	Acharjee et al. (2010)
Helicoverpa armigera	*Bacillus thuringiensis*	*Cry2A*	–	–	Lawo et al. (2008)
Callosobruchus maculatus and C. chinensis	–	bean αAI1 gene	–	*Agrobacterium*-mediated	Sarmah et al. (2004)
Salinity		*P5CS* gene	–	*Agrobacterium tumefaciens* mediated	Kiran Kumar Ghanti et al. (2011)
Iron content	Soybean	*CaNAS2* and *GmFER*	–	*Agrobacterium tumefaciens* mediated	Tan et al. (2018)
Stress tolerance	Chickpea	*CaFer1*	–	*Agrobacterium tumefaciens* mediated	Parveen et al. (2016)
Drought	–	P5CSF129A	Axillary meristem	*Agrobacterium tumefaciens* mediated	Bhatnagar-Mathur et al. (2009)
Drought	*Arabidopsis*	*DREB1A gene*	–	*Agrobacterium*-mediated	Anbazhagan et al. (2015)

the pod borer larvae. Likewise, the intervention of transgenics approach necessitates for drought tolerance (Bhatnagar-Mathur et al. 2009; Anbazhagan et al. 2015) and salinity tolerance (Kiran Kumar Ghanti et al. 2011) in chickpea for avoiding the complexities raised in conventional breeding techniques for transferring complex drought tolerant QTLs.

3.13 Advanced Breeding Techniques to Accelerate Chickpea Improvement

Innovative breeding techniques such as MAGIC (Gaur et al. 2014), nested association mapping (NAM), TILLING (Gaur et al. 2014) have been introduced in chickpea for trait mapping, complex QTL(s) discovery, for creating novel genetic variation. Latest interventions like speed breeding (Li et al. 2018a) that enable rapid generation to turn over are crucial to speed up the genetic gain. These techniques could help in introducing desired complex trait QTL(s) like drought and yield and yield-related genomic region into elite cultivars precisely. Likewise, emerging genome editing tools especially CRISPR/Cas9 (Pennisi 2013) has shown great promise for manipulating specific genome sequence and creating "targeted allele diversity" for improvement of various traits in crop plant including grain legume (Shen et al. 2017; Shi et al. 2017; Wang et al. 2017; Lemmon et al. 2018). Thus, a holistic approach encompassing various "omics" approaches and advanced breeding techniques (see Fig. 3.1) could help us to develop climate-resilient chickpea for safeguarding global food security.

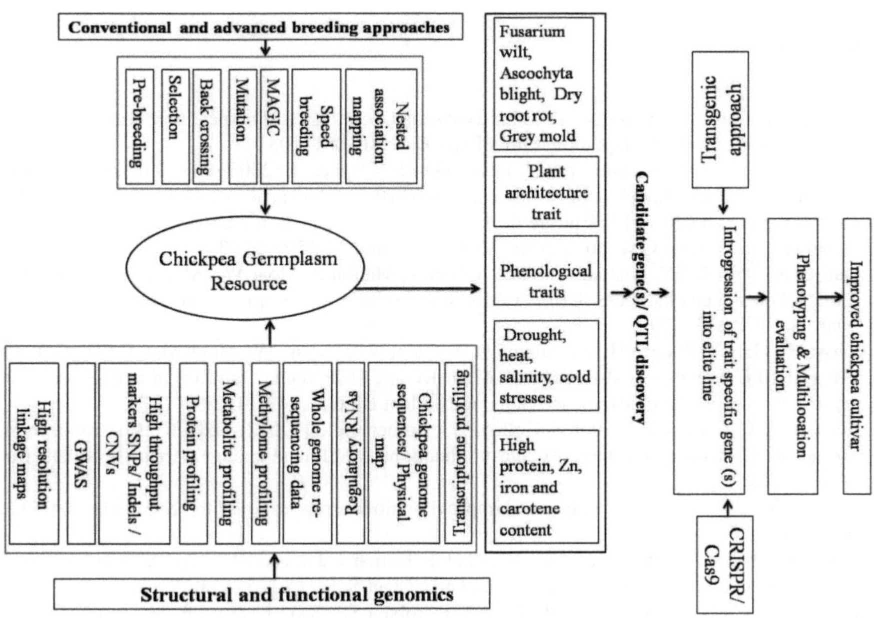

Fig. 3.1 Role of conventional breeding and 'omics' resources for development of climate resilient chickpea cultivar

3.14 Conclusion

Burgeoning human population worldwide along with global climate change has jeopardized achieving the goal of global food security. Conventional breeding approaches have enabled the development of several important chickpea cultivars over the decades. Latest developments in chickpea genomics facilitated increasing access to key genomic regions that influence various biotic and abiotic stress tolerance and grain yield. Toward this end, availability of chickpea reference genome sequence could be of great importance for addressing the domestication and evolution-related queries and functional dissection of traits of breeding relevance. Falling cost of genome sequence has provided great opportunity to perform re-sequencing of global chickpea core collection to capture the important haplotypes related to various biotic and abiotic stress tolerance along with yield contributing genomic regions. Translating modern "omics" knowledge into cultivar development with support from evolving technologies/methodologies such as genome editing and speed breeding will greatly reinforce breeding techniques for developing climate-resilient and nutrient-efficient chickpea.

References

Abbo S, Jens Berger J, Turner NC (2003a) Evolution of cultivated chickpea: four bottlenecks limit diversity and constrain adaptation. Funct Plant Biol 30:1081–1087

Abbo S, Shtienberg D, Lichtenzveig J, Lev-Yadun S, Gopher A (2003b) The chickpea, summer cropping, and a new model for pulse domestication in the ancient near east. Q Rev Biol 78:435–448

Acharjee S, Sarmah BK (2013) Transgenic *Bacillus thuringiensis* (Bt) chickpea: India's most wanted genetically modified (GM) pulse crop. Afr J Biotechnol 12:5709–5713

Acharjee S, Sarmah BK, Ananda Kumar P, Olsen K, Mahon R, Moar WJ, Moore A, Higgins TVJ (2010) Transgenic chickpeas (Cicer arietinum L.) expressing a sequence-modified cry2Aa gene. Plant Sci 178:333–339

Agarwal G, Garg V, Kudapa H, Doddamani D, Pazhamala LT, Khan AW, Thudi M, Lee SH, Varshney RK (2016) Genome-wide dissection of AP2/ERF and HSP90 gene families in five legumes and expression profiles in chickpea and pigeonpea. Plant Biotechnol J 14:1563–1577

Ainsworth EA, Yendrek CR, Sitch S, Collins WJ, Emberson LD (2012) The effects of tropospheric ozone on net primary productivity and implications for climate change. Annu Rev Plant Biol 63:637–661

Ainsworth EA (2017) Understanding and improving global crop response to ozone pollution. Plant J 90:886–897

Ali MY, Krishnamurthy L, Saxena NP, Rupela OP, Kumar J, Johansen C (2002) Scope for genetic manipulation of mineral acquisition in chickpea. Plant Soil 245:123–134

Anbazhagan K, Bhatnagar-Mathur P, Vadez V, Dumbala SR, Kavi Kishor PB, Sharma KK (2015) DREB1A overexpression in transgenic chickpea alters key traits influencing plant water budget across water regimes. Plant Cell Rep 34:199–210

Anbessa Y, Tara'n B, Warkentin TD, Tullu A, Vandenberg A (2009) Genetic analyses and conservation of QTL for Ascochyta blight resistance in chickpea (*Cicer arietinum* L.). Theor Appl Genet 4:757–765

Arslan O, Eyidogan F, Ekmekci Y (2018) Freezing tolerance of chickpea: biochemical and molecular changes at vegetative stage. Biol Plant 62:140–148

Aryamanesh N, Nelson MN, Yan G, Clarke HJ, Siddique KHM (2010) Mapping a major gene for growth habit and QTLs for Ascochyta blight resistance and flowering time in a population between chickpea and Cicer reticulatum. Euphytica 173:307–319

Badhan S, Kole P, Ball A, Mantri N (2018) RNA sequencing of leaf tissues from two contrasting chickpea genotypes reveals mechanisms for drought tolerance. Plant Physiol Biochem 129:295–304

Bandillo N, Raghavan C, Muyco PA, Sevilla MAL, Lobina IT, Dilla-Ermita J, Singh RK (2013) Multi-parent advanced generation inter-cross (MAGIC) populations in rice: progress and potential for genetics research and breeding. Rice 6:11

Berger JD, Turner NC, Siddique KHM, Knights EJ, Brinsmead RB, Mock I, Edmondson C, Khan TN (2004) Genotype by environment studies across Australia reveal the importance of phenology for chickpea (*Cicer arietinum* L.) improvement. Aust J Agri Res 55:1071–1084

Berger JD, Ali M, Basu PS, Chaudhary BD, Chaturvedi SK, Deshmukh PS, Kumar J (2006) Genotype by environment studies demonstrate the critical role of phenology in adaptation of chickpea (*Cicer arietinum* L.) to high and low yielding environments of India. Field Crops Res 98:230–244

Bhandari K, Siddique KH, Turner NC, Kaur J, Singh S, Agrawal SK, Nayyar H (2016) Heat stress at reproductive stage disrupts leaf carbohydrate metabolism, impairs reproductive function, and severely reduces seed yield in lentil. J Crop Improv 30:118–151

Bohra A, Jha UC, Singh B, Soren KR, Singh IP, Chaturvedi SK, Nadarajan N, Barh D (2013) Omics approaches in pulses. In: Barh D (ed) OMICS applications in crop science. Taylor & Francis, London, pp 102–138

Bohra A (2013) Emerging paradigms in genomics-based crop improvement. Sci World J 2013:585467

Bohra A, Pandey MK, Jha UC, Singh B, Singh IP, Datta D, Chaturvedi SK, Nadarajan N, Varshney RK (2014) Genomics assisted breeding in four major pulse crops of developing countries: present status and prospects. Theor Appl Genet 127:1263–1291

Bohra A, Singh NP (2015) Whole genome sequences in pulse crops: a global community resource to expedite translational genomics and knowledge-based crop improvement. Biotechnol Lett 37:1529–1539

Bohra A, Sahrawat KL, Kumar S, Joshi R, Parihar AK, Singh U, Singh D, Singh NP (2015) Genetics- and genomics-based interventions for nutritional enhancement of grain legume crops: status and outlook. J Appl Genet 56:151–161

Bhardwaj R, Sandhu JS, Kaur L, Gupta SK, Gaur PM, Varshney RK (2010) Genetics of Ascochyta blight resistance in chickpea. Euphytica 171:337–343

Bhatnagar-Mathur P, Vadez V, M Devi J, Lavanya M, Vani G, Sharma KK (2009) Genetic engineering of chickpea (*Cicer arietinum* L.) with the P5CSF129A gene for osmoregulation with implications on drought tolerance. Mol Breed 23:591–606

Brinda S, Ravikumar RL (2005) Inheritance of wilt resistance in chickpea: a molecular marker analysis. Curr Sci 88:12–19

Canci H, Toker C (2009) Evaluation of yield criteria for drought and heat resistance in chickpea (*Cicer arietinum* L.). J Agron Crop Sci 19:47–54

Clarke HJ, Siddique KHM (2004) Response of chickpea genotypes to low temperature stress during reproductive development. Field Crops Res 90:323–334

Clarke HJ, Khan TN, Siddique KHM (2004) Pollen selection for chilling tolerance at hybridisation leads to improved chickpea cultivars. Euphytica 139:65–74

Cobos MJ, Ferna'ndez MJ, Rubio J, Kharrat M, Moreno MT, Gil J, Milla'n T (2005) A linkage map of chickpea (*Cicer arietinum* L.) based on populations from Kabuli x Desi crosses: location of genes for resistance to Fusarium wilt race 0. Theor Appl Genet 110:1347–1353

Cobos MJ, Winter P, Kharrat M, Cubero JI, Gil J, Millan T, Rubio J (2009) Genetic analysis of agronomic traits in a wide cross of chickpea. Field Crops Res 111:130–136

Daba K, Deokar A, Banniza S, Warkentin TD, Tar'an B (2016) QTL mapping of early flowering and resistance to ascochyta blight in chickpea. Genome 59:413–425

Das S, Upadhyaya HD, Bajaj D, Kujur A, Badoni S, Kumar V, Tripathi S, Gowda CL, Sharma S, Singh S, Tyagi AK, Parida SK (2015) Deploying QTL-seq for rapid delineation of a potential candidate gene underlying major trait-associated QTL in chickpea. DNA Res 22:193–203

Das S, Singh M, Srivastava R, Bajaj D, Saxena MS, Rana JC, Bansal KC, Tyagi AK, Parida SK (2016) mQTL-seq delineates functionally relevant candidate gene harbouring a major QTL regulating pod number in chickpea. DNA Res 23:53–65

del Mar Jiménez-Gasco M, Jiménez-Díaz RM (2003) Development of a specific polymerase chain reaction-based assay for the identification of *Fusarium oxysporum* f. sp. *ciceris* and its pathogenic races 0, 1A, 5, and 6. Phytopathology 93(2):200–209

Diapari M, Sindhu A, Bett K, Deokar A, Wartkentin TD, Taran B (2014) Genetic diversity and association mapping of iron and zinc concentrations in chickpea (*Cicer arietinum* L.). Genome 57:459–468

Doddamani D, Mohan AVSK, Katta MAVSK, Khan AW, Agarwal G, Shah TM, Varshney RK (2014) CicArMiSatDB: the chickpea microsatellite database. BMC Bioinformatics 15:212

Doddamani D, Khan AW, Katta MAVSK, Agarwal G, Thudi M, Ruperao P, Edwards D, Varshney RK (2015) CicArVarDB: SNP and InDel database for advancing genetics research and breeding applications in chickpea. Database (Oxford) 19:2015

Deokar AA, Kondawar V, Jain PK, Karuppayil M, Raju NL, Vadez V, Varshney RK, Srinivasan R (2011) Comparative analysis of expressed sequence tags (ESTs) between drought-tolerant and -susceptible genotypes of chickpea under terminal drought stress. BMC Plant Biol 11:70

Deokar A, Sagi M, Daba K, Tar'an B (2018) QTL sequencing strategy to map genomic regions associated with resistance to ascochyta blight in chickpea. Plant Biotechnol J. https://doi.org/10.1111/pbi.12964

Devasirvatham V, Gaur PM, Mallikarjuna N, Raju TN, Trethowan RM, Tan DKY (2012) Effect of high temperature on the reproductive development of chickpea genotypes under controlled environments. Funct Plant Biol 39:1009–1018

Devasirvatham V, Gaur P, Mallikarjuna N, Raju TN, Trethowan RM, Tan DKY (2013) Reproductive biology of chickpea response to heat stress in the field is associated with the performance in controlled environments. Field Crops Res 142:9–19

Devasirvatham V, Tan DKY, Trethowan RM, Gaur PM, Mallikarjuna N (2010) Impact of high temperature on the reproductive stage of chickpea. In: Food Security from Sustainable Agriculture. Proceedings of the 15th Australian Society of Agronomy Conference. Lincoln, New Zealand, pp 15–18

Doudna JA, Charpentier E (2014) The new frontier of genome engineering with CRISPR-Cas9. Science 346:1258096

Dua RP, Sharma PC (1995) Salinity tolerance of kabuli and desi chickpea genotypes. Intl Chickpea Pigeonpea Newsl 2:19–22

Food and Agriculture Organization of the United Nations, FAOSTAT. Rome, Italy. FAO (2016). Available at: http://fao.org/faostat/en/#data/QC(Accessed13Jan2018)

Flandez-Galvez H, Ades PK, Ford R, Pang ECK, Tayler PWJ (2003) QTL analysis for Ascochyta blight resistance in an intraspecific population of chickpea (*Cicer arietinum* L.). Theor Appl Genet 107:1257–1265

Fleury D, Jefferies S, Kuchel H, Langridge P (2010) Genetic and genomic tools to improve drought tolerance in wheat. J Expt Bot 61:3211–3222

Ganguly M, Molla KA, Karmakar S, Datta K, Datta SK (2014) Development of pod borer-resistant transgenic chickpea using a pod-specific and a constitutive promoter-driven fused cry1Ab/Ac gene. Theor Appl Genet 127:2555–2565

Garg R, Bhattacharjee A, Jain M (2015) Genome-scale transcriptomic insights into molecular aspects of abiotic stress responses in chickpea. Plant Mol Biol Rep 33:388–400

Garg R, Shankar R, Thakkar B, Kudapa H, Krishnamurthy L, Mantri N, Varshney RK, Bhatia S, Jain M (2016) Transcriptome analyses reveal genotype- and developmental stage-specific molecular responses to drought and salinity stresses in chickpea. Sci Rep 6:19228

Garg T, Mallikarjuna BP, Thudi M, Samineni S, Singh S, Sandhu JS, Kaur L, Singh I, Sirari A, Basandrai AK, Basandrai D, Varshney RK, Gaur PM (2018) Identification of QTLs for resistance to Fusarium wilt and Ascochyta blight in a recombinant inbred population of chickpea (*Cicer arietinum* L.). Euphytica 214:1–11

Gaur PM, Krishnamurty L, Kashiwagi J (2008) Improvement drought-avoidance root traits in chickpea (Cicer arietinum L.)-current status of research at ICRISAT. Plant Prod Sci 3—11

Gaur PM, Jukanti AK, Varshney RK (2012) Impact of genomic technologies on chickpea breeding strategies. Agronomy 2:199–221

Gaur PM, Samineni S, Thudi M, Tripathi S, Sajja SB, Jayalakshmi V, Fikre A (2018) Integrated breeding approaches for improving drought and heat adaptation in chickpea (*Cicer arietinum* L.). Plant Breed. https://doi.org/10.1111/pbr.12641

Gayali S, Acharya S, Lande NV, Pandey A, Chakraborty S, Chakraborty N (2016) CicerTransDB 1.0: a resource for expression and functional study of chickpea transcription factors. BMC Plant Biol 16:169

Gaur PM, Thudi M, Srinivasan S, Varshney RK (2014) Advances in chickpea genomics. In: Gupta S, Nadarajan N, Gupta DS (eds) Legumes in the Omic Era. Springer, New York, pp 73–94

Gaur PM, Samineni S, Tripathi S, Varshney RK, Gowda CLL (2015) Allelic relationships of flowering time genes in chickpea. Euphytica 203:295–308

Graham PH, Vance CP (2003) Legumes: importance and constraints to greater use. Plant Physiol 131:872–877

Gowda SJM, Radhika P, Kadoo NY, Mhase LB, Gupta VS (2009) Molecular mapping of wilt resistance genes in chickpea. Mol Breed 24:177–183

Gupta S, Kumar T, Verma S, Bharadwaj C, Bhatia S (2015) Development of gene-based markers for use in construction of the chickpea (*Cicer arietinum* L.) genetic linkage map and identification of QTLs associated with seed weight and plant height. Mol Biol Rep 42:1571–1580

Gupta S, Nawaz K, Parween S, Roy R, Sahu K, Kumar Pole A, Khandal H, Srivastava R, Parida SK, Chattopadhyay D (2017) Draft genome sequence of *Cicer reticulatum* L., the wild progenitor of chickpea provides a resource for agronomic trait improvement. DNA Res 24:1–10

Halila MH, Strange RN (1996) Identification of the causal agent of wilt of chickpea in Tunisia as *Fusarium oxysporum* f. sp. *ciceri* race 0. Phytopathol Mediterr 35:67–74

Hamwieh A, Imtiaz M, Malhotra RS (2013) Multi-environment QTL analyses for drought-related traits in a recombinant inbred population of chickpea (*Cicer arietinum* L.) Theor Appl Genet 126:1025–1038

Hasanuzzaman M, Nahar K, Alam MM, Roychowdhury R, Fujita M (2013) Physiological, biochemical, and molecular mechanisms of heat stress tolerance in plants. Intl J Mol Sci 14:9643–9684

Hegde VS (2010) Genetics of flowering time in chickpea in a semi-arid environment. Plant Breed 129:683–687

Hidoto L, Tar'an B, Worku W, Mohammed H (2017) Towards Zinc Biofortification in chickpea: performance of chickpea cultivars in response to soil zinc application. Agronomy 7:11

Hiremath PJ, Farmer A, Cannon SB, Woodward J, Kudapa H, Tuteja R et al. (2011) Large-scale transcriptome analysis in chickpea (*Cicer arietinum* L.), an orphan legume crop of the semi-arid tropics of Asia and Africa. Plant Biotechnol J 9:922–931

Huang BE, Verbyla KL, Verbyla AP, Raghavan C, Singh VK, Gaur PM, Cavanagh CR (2015) MAGIC populations in crops: current status and future prospects. Theor Appl Genet 128:999–1017

Indurker S, Misra HS, Eapen S (2007) Genetic transformation of chickpea (*Cicer arietinum* L.) with insecticidal crystal protein gene using particle gun bombardment. Plant Cell Rep 26:755–763

Iruela M, Rubio J, Barro F, Cubero JI, Milla´n T, Gil J (2006) Detection of two quantitative trait loci for resistance to Ascochyta blight in an intra-specific cross of chickpea (*Cicer arietinum* L.): development of SCAR markers associated with resistance. Theor Appl Genet 112:278–287

Iruela M, Castro P, Rubio J, Cubero JI, Jacinto C, Milla´n T, Gil J (2007) Validation of a QTL for resistance to Ascochyta blight linked to resistance to Fusarium wilt race 5 in chickpea (*Cicer arietinum* L.). Eur J Plant Pathol 119:29–37

Jadhav AA, Rayate SJ, Mhase LB, Thudi M, Chitikineni A, Harer PN, Jadhav AS, Varshney RK, Kulwal PL (2015) Marker-trait association study for protein content in chickpea (*Cicer arietinum* L.). J Genet 94:279–286
Jaganathan D, Thudi M, Kale S, Azam S, Roorkiwal M, Gaur PM, Kishor PB, Nguyen H, Sutton T, Varshney RK (2015) Genotyping-by-sequencing based intra-specific genetic map refines a "QTL-hotspot" region for drought tolerance in chickpea. Mol Genet Genom 290:559–571
Jain D, Chattopadhyay D (2010) Analysis of gene expression in response to water deficit of chickpea (*Cicer arietinum* L.) varieties differing in drought tolerance. BMC Plant Biol 10:24
Jain M, Misra G, Patel RK, Priya P, Jhanwar S, Khan AW, Shah N, Singh VK, Garg R, Jeena G, Yadav M, Kant C, Sharma P, Yadav G, Bhatia S, Tyagi AK, Chattopadhyay D (2013) A draft genome sequence of the pulse crop chickpea (*Cicer arietinum* L.). Plant J 74:715–729
Jain M, Chevala VV, Garg R (2014) Genome-wide discovery and differential regulation of conserved and novel microRNAs in chickpea via deep sequencing. J Exp Bot 65:5945–5958
Jain M, Pole AK, Singh VK, Ravikumar RL, Garg R (2015) Discovery of molecular markers for Fusarium wilt via transcriptome sequencing of chickpea cultivars. Mol Breed 35:198
Jannink JL, Lorenz AJ, Iwata H (2010) Genomic selection in plant breeding: from theory to practice. Brief Funct Genom 9:166–177
Jayashree B, Buhariwalla HK, Shinde S, Crouch JH (2005) A legume genomics resource: the chickpea root expressed sequence tag database. Elec J Biotechnol 8:128–133
Jha UC, Chaturvedi SK, Bohra A, Basu PS, Khan MS, Barh D (2014a) Abiotic stresses, constraints and improvement strategies in chickpea. Plant Breed 133:163–178
Jha UC, Bohra A, Singh NP (2014b) Heat stress in crop plants: its nature, impacts and integrated breeding strategies to improve heat tolerance. Plant Breed 133:679–701
Jha UC, Shil S (2015) Association analysis of yield contributing traits of chickpea genotypes under high temperature condition. Trends Biosci 8:2335–2341
Jha UC, Basu PS, Singh DK (2015) Genetic variation and diversity analysis of chickpea genotypes based on quantitative traits under high temperature stress. Intl J Bio-Resour Stress Manage 6:700–706
Jha UC, Bohra A, Jha R, Parida S (2017) Integrated 'omics' approaches to sustain major global grain legume productivity under heat stress. Plant Breed 136:437–459
Jha UC (2018) Current advances in chickpea genomics: applications and future perspectives. Plant Cell Rep 37:947–965
Jha UC, Jha R, Singh NP, Shil S, Kole PC (2018a) Heat tolerance indices and their role in selection of heat stress tolerant chickpea (*Cicer arietinum* L) genotypes. Indian J Agri Sci 88(2):260–267
Jha UC, Jha R, Bohra A, Parida SK, Kole PC, Thakro V, Singh D, Singh NP (2018b) Population structure and association analysis of heat stress relevant traits in chickpea (*Cicer arietinum* L.). 3 Biotech 8:43
Jha UC, Kole PC, Singh NP (2018c) Genetic variability and marker trait association analysis of various phenological and yield Related traits for heat tolerance in chickpea (*Cicer arietinum* L.). Intl J Bioresour Stress Manage 9:345–352
Jha UC, Bohra A, Jha R, Parida SK (2019) Salinity stress response and 'omics' approaches for improving salinity stress tolerance in major grain legumes. Plant Cell Rep 38:255–277
Jimenez-Diaz RM, Trapero-Casas A, Cabrera de la, Colina J (1989) Races of *Fusarium oxysporum* f. sp. ciceris infecting chickpea in Southern Spain. In: Tjamos EC, Beckman CH (eds) Vascular Wilt Diseases of Plants. NATO ASI Series, vol H28. Springer, Berlin, pp 515–520
Jingade P, Ravikumar RL (2015) Development of molecular map and identification of QTLs linked to Fusarium wilt resistance in chickpea. J Genet 94:723–729
Jukanti AK, Gaur PM, Gowda CL, Chibbar RN (2012) Nutritional quality and health benefits of chickpea (Cicer arietinum L.): a review. Br J Nutr 1:S11–S26
Kale SM, Jaganathan D, Ruperao P, Chen C, Punna R, Kudapa H, Thudi M, Roorkiwal M, Katta MA, Doddamani D, Garg V, Kishor PB, Gaur PM, Nguyen HT, Batley J, Edwards D, Sutton T, Varshney RK (2015) Prioritization of candidate genes in "QTL-hotspot" region for drought tolerance in chickpea (*Cicer arietinum* L.). Sci Rep 5:15296

Kashiwagi J, Krishnamurty L, Gaur PM, Chandra S, Upadhyaya HD (2008) Estimation of gene effects of the drought avoidance root characteristics in chickpea (*C. arietinum* L.). Field Crops Res 105:64–69

Kaashyap M, Ford R, Bohra A, Kuvalekar, Mantri N (2017) Improving salt tolerance of chickpea using modern genomics tools and molecular breelding. Curr Genom 18:557–567

Katerji N, van Hoorn JW, Hamdy A, Mastrorilli M, Oweis T (2005) Salt tolerance analysis of chickpea, faba bean and durum wheat varieties. I. chickpea and faba bean. Agri Water Manag 72:177–194

Kaushal N, Awasthi R, Gupta K, Gaur PM, Siddique KHM, Nayyar H (2013) Heat-stress-induced reproductive failures in chickpea (*Cicer arietinum*) are associated with impaired sucrose metabolism in leaves and anthers. Funct Plant Biol 40:1334–1349

Khandal H, Parween S, Roy R, Meena MK, Chattopadhyay D (2017) MicroRNA profiling provides insights into post-transcriptional regulation of gene expression in chickpea root apex under salinity and water deficiency. Sci Rep 7:4632

Khemka N, Singh VK, Garg R, Jain M (2016) Genome-wide analysis of long intergenic non-coding RNAs in chickpea and their potential role in flower development. Sci Rep 6:33297

Kiran Kumar Ghanti S, Sujata KG, Vijay Kumar BM, Nataraja Karba N, Janardhan Reddy K, Srinath Rao M, Kavi Kishor PB (2011) Heterologous expression of P5CS gene in chickpea enhances salt tolerance without affecting yield. Biol Plant 55:634–640

Kohli D, Joshi G, Deokar AA, Bhardwaj AR, Agarwal M, Katiyar Agarwal S, Srinivasan R, Jain PK (2014) Identification and characterization of wilt and salt stress-responsive microRNAs in chickpea through high-throughput sequencing. PLoS ONE 9:e108851

Kottapalli P, Gaur PM, Katiyar SK, Crouch JH, Buhariwalla HK, Pande S, Gali KK (2009) Mapping and validation of QTLs for resistance to an Indian isolate of Ascochyta blight pathogen in chickpea. Euphytica 165:79–88

Krishnamurthy L, Gaur PM, Basu PS, Chaturvedi SK, Tripathi S, Vadez V, Rathore A, Varshney RK, Gowda CLL (2011) Large genetic variation for heat tolerance in the reference collection of chickpea (*Cicer arietinum* L.) germplasm. Plant Genet Res 9(01):59–69

Kudapa H, Azam S, Sharpe AG, Taran B, Li R, Deonovic B et al. (2014) Comprehensive transcriptome assembly of chickpea (*Cicer arietinum* L.) using sanger and next generation sequencing platforms: development and applications. PLoS One 9:e86039

Kudapa H, Garg V, Chitikineni A, Varshney RK (2018) The RNA-Seq-based high resolution gene expression atlas of chickpea (*Cicer arietinum* L.) reveals dynamic spatio-temporal changes associated with growth and development. Plant Cell Environ 2018 Apr 10. https://doi.org/10.1111/pce.13210

Kumar J, Abbo S (2001) Genetics of flowering time in chickpea and its bearing on productivity in semi arid environments. Adv Agron 72:107–138

Kumar J, Rao BV (1996) Super early chickpea developed at ICRISAT Asia center. Intl Chickpea Pigeonpea Newsl 3:17–18

Kumar J, van Rheenen HA (2000) A major gene for time of flowering in chickpea. J Hered 91:67–68

Kumar K, Purayannur S, Kaladhar VC, Parida SK, Verma PK (2018) mQTL-seq and classical mapping implicates the role of an AT-HOOK MOTIF CONTAINING NUCLEAR LOCALIZED (AHL) family gene in Ascochyta blight resistance of chickpea. Plant Cell Environ. https://doi.org/10.1111/pce.13177

Lawo NC, Mahon RJ, Milner RJ, Sarmah BK, Higgins TJ, Romeis J (2008) Effectiveness of *Bacillus thuringiensis*-Transgenic Chickpeas and the Entomopathogenic Fungus *Metarhizium anisopliae* in Controlling *Helicoverpa armigera* (Lepidoptera: Noctuidae). Appl Environ Microbiol 74:4381–4389

Lemmon ZH, Reem NT, Dalrymple J, Soyk S, Swartwood KE, Rodriguez-Leal D, Eck JV, Lippman ZB (2018) Rapid improvement of domestication traits in an orphan crop by genome editing. Nat Plants 766–770

Leo AE, Linde CC, Ford R (2016) Defence gene expression profiling to *Ascochyta rabiei* aggressiveness in chickpea. Theor Appl Genet 129:1333–1345

Li Y, Ruperao P, Batley J, Edwards D, Davidson J, Hobson K, Sutton T (2017) Genome analysis identified novel candidate genes for ascochyta blight resistance in chickpea using whole genome re-sequencing data. Front Plant Sci 8:1–13

Li H, Rasheed A, Hickey LT, He Z (2018a) Fast-forwarding genetic gain. Trends Plant Sci 23:183–186

Li Y, Ruperao P, Batley J, Edwards D, Khan T, Colmer TD, Pang J, Siddique KH, Sutton T (2018b) Investigating drought tolerance in chickpea using genome-wide association mapping and genomic selection based on whole-genome resequencing data. Front Plant Sci. https://doi.org/10.3389/fpls.2018.00190

Madrid E, Chen W, Rajesh PN, Castro P, Milla´n T, Gil J (2013) Allele-specific amplification for the detection of ascochyta blight resistance in chickpea. Euphytica 189:183–190

Mahdavi Mashaki K, Garg V, Nasrollahnezhad Ghomi AA, Kudapa H, Chitikineni A, Zaynali, Nezhad K et al. (2018) RNA-Seq analysis revealed genes associated with drought stress response in kabuli chickpea (*Cicer arietinum* L.). PLoS One 13:e0199774

Malhotra RS, Singh KB, Vito M, Greco N, Saxena, MC (2002) Registration of ILC 10765 and ILC 10766 chickpea germplasm lines resistant to cyst nematode. Crop Sci 42:1756. https://doi.org/10.2135/crop-sci2002.1756

Mallikarjuna BP, Samineni S, Thudi M, Sajja SB, Khan AW (2017) Molecular mapping of flowering time major genes and QTLs in chickpea (*Cicer arietinum* L.). Front Plant Sci 8:1140

Mantri NL, Ford R, Coram TE, Pang ECK (2007) Transcriptional profiling of chickpea genes differentially regulated in response to high-salinity, cold and drought. BMC Genom 8:303

Meuwissen THE, Goddard ME (2007) Multipoint identity-bydescent prediction using dense markers to map quantitative trait loci and estimate effective population size. Genetics 176:2551–2560

Meuwissen THE, Hayes BJ, Goddard ME (2001) Prediction of total genetic value using genome-wide dense marker maps. Genetics 157:1819–1829

Millan T, Rubio J, Iruela M, Daly K, Cubero JI, Gil J (2003) Markers associated with Ascochyta blight resistance in chickpea and their potential in marker-assisted selection. Field Crops Res 84:373–384

Misra G, Priya P, Bandhiwal N, Bareja N, Jain M, Bhatia S, Chattopadhyay D, Tyagi AK, Yadav G (2014) The chickpea genomic web resource: visualization and analysis of the desi-type *Cicer arietinum* nuclear genome for comparative exploration of legumes. BMC Plant Biol 14:315

Molina C, Rotter B, Horres R, Udupa SM, Besser B, Bellarmino L, Baum M, Matsumura H, Terauchi R, Kahl G, Winter P (2008) SuperSAGE: the drought stress-responsive transcriptome of chickpea roots. BMC Genom 9:553

Molina C, Zaman-Allah M, Khan F, Fatnassi N, Horres R, Rotter B, Steinhauer D, Amenc L, Drevon JJ, Winter P, Kahl G (2011) The salt-responsive transcriptome of chickpea roots and nodules via deepSuperSAGE. BMC Plant Biol 11:31

Nayyar H, Bains T, Kumar S (2005) Low temperature induced floral abortion in chickpea: relationship to abscisic acid and cryoprotectants in reproductive organs. Env Exp Bot 53:39–47

Or E, Hovav R, Abbo S (1999) A major gene for flowering time in chickpea. Crop Sci 39:315–322

Pennisi E (2013) The CRISPR craze. Science 341:833–836

Pande S, Siddique KHM, Kishore GK, Baya B, Gaur PM, Gowda CLL, Bretag T, Crouch JH (2005) Ascochyta blight of chickpea: biology, pathogenicity, and disease management. Aust J Agri Res 56:317–332

Parveen S, Gupta DB, Dass S, Kumar A, Pandey A, Chakraborty S, Chakraborty N (2016) Chickpea ferritin CaFer1 participates in oxidative stress response, and promotes growth and development. Sci Rep 6:31218

Patil BS, Ravikumar RL, Bhat JS, Soregaon D (2014) Molecular mapping of QTLs for resistance to early and late *Fusarium* wilt in chickpea. Czech J Genet Plant Breed 50:171–176

Paul PJ, Samineni S, Sajja SB, Rathore A, Das RR, Chaturvedi S K, Gaur PM (2018a) Capturing genetic variability and selection of traits for heat tolerance in a chickpea recombinant inbred line (RIL) population under field conditions. Euphytica 214:27

Paul PJ, Samineni S, Thudi M, Sajja SB, Rathore A, Das RR, Khan AW, Chaturvedi SK, Lavanya GR, Varshney RK, Gaur PM (2018b) Molecular mapping of QTLs for heat tolerance in chickpea. Intl J Mol Sci 19(8) pii:E2166

Pushpavalli R, Krishnamurthy L, Thudi M, Gaur PM, Rao MV, Siddique KH, Colmer TD, Turner NC, Varshney RK, Vadez V (2015) Two key genomic regions harbour QTLs for salinity tolerance in ICCV 2 × JG 11 derived chickpea (*Cicer arietinum* L.) recombinant inbred lines. BMC Plant Biol 15:124

Pushpavalli R, Quealy J, Colmer TD, Turner NC, Siddique KHM, Rao MV, Vadez V (2016) Salt stress delayed flowering and reduced reproductive success of chickpea (Cicer arietinum L.), a response associated with Na^+ accumulation in leaves. J Agron Crop Sci 202:125–138

Ramgopal D, Srivastava RK, Pande S, Rathore A, Jadhav DR, Sharma M et al. (2012) Introgression of Botrytis gray mold resistance gene from *Cicer reticulatum* (*bgmr1cr*) and *C. echinospermum* (*bgmr1ce*) to chickpea. Plant Genet Resour 1–6

Rao DLN, Giller KE, Yeo AR, Flowers TJ (2002) The effects of salinity and sodicity upon nodulation and nitrogen fixation in chickpea (Cicer arietinum). Ann Bot 89:563–570

Rehman AU, Malhotra RS, Bett K, Taran B, Bueckert R, Warkentin TD (2011) Mapping QTL associated with traits affecting grain yield in chickpea (*Cicer arietinum* L.) under terminal drought stress. Crop Sci 51:450–463

Rezaei MK, Deokar A, Tar'an B (2016) Identification and expression analysis of candidate genes involved in carotenoid biosynthesis in chickpea seeds. Front Plant Sci 7:1867

Ridge S, Deokar A, Lee R, Daba K, Macknight RC, Weller JL, Tar'an B (2017) The chickpea early flowering 1 (Efl1) locus is an ortholog of arabidopsis ELF3. Plant Physiol 175:802–815

Roorkiwal M, Rathore A, Das RR, Singh MK, Jain A, Srinivasan S, Gaur PM, Chellapilla B, Tripathi S, Li Y, Hickey JM, Lorenz A, Sutton T, Crossa J, Jannink JL, Varshney RK (2016) Genome enabled prediction models for yield related traits in chickpea. Front Plant Sci 7:1666

Roorkiwal M, Jarquin D, Singh MK, Gaur PM, Bharadwaj C, Rathore A, Howard R, Srinivasan S, Jain A, Garg V, Kale S, Chitikineni A, Tripathi S, Jones E, Robbins KR, Crossa J, Varshney RK (2018) Genomic-enabled prediction models using multi-environment trials to estimate the effect of genotype × environment interaction on prediction accuracy in chickpea. Sci Rep 8:11701

Sabbavarapu MM, Sharma M, Chamarthi SK, Swapna N, Rathore A, Thudi M et al (2013) Molecular mapping of QTLs for resistance to *Fusarium* wilt (race 1) and *Ascochyta* blight in chickpea (*Cicer arietinum* L.). Euphytica 193:121–138

Sabaghpour SH, Mahmodi AA, Kamel SM, Malhotra RS (2006) Study on chickpea drought tolerance lines under dryland condition of Iran. Indian J Crop Sci 1:70–73

Sagi MS, Deokar AA, Tar'an B (2017) Genetic analysis of NBS-LRR gene family in chickpea and their expression profiles in response to ascochyta blight infection. Front Plant Sci 8:838

Saha S, Chakraborty D, Sehgal VK, Pal M (2015a) Rising atmospheric CO_2: potential impacts on chickpea seed quality. Agri Ecosyst Environ 203:140–146

Saha S, Chakraborty D, Sehgal VK, Pal M (2015b) Potential impact of rising atmospheric CO_2 on quality of grains in chickpea (*Cicer arietinum* L.). Food Chem 187:431–436

Sandhu JS, Bains TS, Sidhu PS (2002) Evaluation of super early chickpea genotypes for vegetable purpose as a catch crop. Int Chickpea Pigeonpea Newsl 9:10–12

Sanyal I, Singh AK, Kaushik M, Amla DV (2005) *Agrobacterium*-mediated transformation of chickpea (*Cicer arietinum* L.) with *Bacillus thuringiensis cry1Ac* gene for resistance against pod borer insect *Helicoverpa armigera*. Plant Sci 168:1135–1146

Sarmah BK, Moore A, Tate W, Molvig L, Morton RL, Rees DP, Chiaiese P, Chrispeels MJ, Tabe LM, Higgins TJV (2004) Transgenic chickpea seeds expressing high levels of a bean alpha-amylase inhibitor. Mol Breed 14:73–82

Sharma KD, Winter P, Kahl G, Muehlbauer FJ (2004) Molecular mapping of *Fusarium oxysporum* f. sp. ciceris race 3 resistance gene in chickpea. Theor Appl Genet 108:1243–1256

Sharma KD, Chen W, Muehlbauer FJ (2005) Genetics of chickpea resistance to five races of Fusarium wilt and a concise set of race differentials for *Fusarium oxysporum* f. sp. *ciceris*. Plant Dis 89:385–390

Shen C, Que Z, Xia Y, Tang N, Li D, He R, Cao M (2017) Knock out of the annexin gene *OsAnn3* via CRISPR/Cas9-mediated genome editing decreased cold tolerance in rice. J Plant Biol 60:539–547

Shi J, Gao H, Wang H, Lafitte HR, Archibald RL, Yang M, Hakimi SM, Mo H, Habben JE (2017) ARGOS8 variants generated by CRISPR-Cas9 improve maize grain yield under field drought stress conditions. Plant Biotechnol J 15:207–216

Singh KB, Dahiya BS (1973) Breeding for wilt resistance in chickpea. Symposium on problem and breeding for wilt resistance in bengal gram. IARI, New Delhi, pp 13–14

Singh H, Kumar J, Smithson JB, Haware MP (1987) Complementation between genes for resistance to race 1 of *Fusarium oxysporum* f. sp. *ciceri* in chickpea. Plant Pathol 36:539–543

Singh KB, Malhotra RS, Saxena MC (1995) Additional sources of tolerance to cold in cultivated and wild Cicer species. Crop Sci 35:1491–1497

Singh S, Gumber RK, Joshi N, Singh K (2005) Introgression from wild *Cicer reticulatum* to cultivated chickpea for productivity and disease resistance. Plant Breed 124:477–480

Singh B, Bohra A, Mishra S, Joshi R, Pandey S (2015) Embracing new-generation 'omics' tools to improve drought tolerance in cereal and food-legume crops. Biol Plant 59:413–428

Singh VK, Khan AW, Jaganathan D, Thudi M, Roorkiwal M, Takagi H, Garg V, Kumar V, Chitikineni A, Gaur PM, Sutton T, Terauchi R, Varshney RK (2016) QTL-seq for rapid identification of candidate genes for 100-seed weight and root/total plant dry weight ratio under rainfed conditions in chickpea. Plant Biotechnol J 14 (11):2110–2119

Singh U, Khemka N, Rajkumar MS, Garg R, Jain M (2017) PLncPRO for prediction of long non-coding RNAs (lncRNAs) in plants and its application for discovery of abiotic stress-responsive lncRNAs in rice and chickpea. Nucl Acids Res 45:e183

Sivasakthi K, Thudi M, Tharanya M, Kale SM, Kholová J, Halime MH, Jaganathan D, Baddam R, Thirunalasundari T, Gaur PM, Varshney RK, Vadez V (2018) Plant vigour QTLs co-map with an earlier reported QTL hotspot for drought tolerance while water saving QTLs map in other regions of the chickpea genome. BMC Plant Biol 18:29

Srinivasan A, Johansen C, Saxena NP (1998) Cold tolerance during early reproductive growth of chickpea (*Cicer arietinum* L.): characterization of stress and genetic variation in pod set. Field Crops Res 57:181–193

Srinivasan A, Saxena NP, Johansen C (1999) Cold tolerance during early reproductive growth of chickpea (*Cicer arietinum* L.): genetic variation in gamete development and function. Field Crops Res 60:209–222

Srivastava R, Bajaj D, Malik A, Singh M, Parida SK (2016) Transcriptome landscape of perennial wild *Cicer microphyllum* uncovers functionally relevant molecular tags regulating agronomic traits in chickpea. Sci Rep 6:33616

Staggenborg SA, Vanderlip RL (1996) Sorghum grain yield reductions caused by duration and timing of freezing temperatures. Agron J 88:473–477

Tan GZH, Das Bhowmik SS, Hoang TML, Karbaschi MR, Long H, Cheng A, Bonneau JP, Beasley JT, Johnson AAT, Williams B, Mundree SG (2018) Investigation of baseline iron levels in Australian chickpea and evaluation of a transgenic biofortification approach. Front Plant Sci 9:788

Tar'an B, Warkentin TD, Tullu A, Vandenberg A (2007) Genetic mapping of ascochyta blight resistance in chickpea (*Cicer arietinum* L.) using a simple sequence repeat linkage map. Genome 50:26–34

Thudi M, Bohra A, Nayak SN, Varghese N, Shah TM, Penmetsa RV, Upadhyaya HD (2011) Novel SSR markers from BAC-end sequences, DArT arrays and a comprehensive genetic map with 1,291 marker loci for chickpea (*Cicer arietinum* L.). PLoS ONE 6:e27275

Thudi M, Upadhyaya HD, Rathore A, Gaur PM, Krishnamurthy L, Roorkiwal M, Nayak SN, Chaturvedi SK, Basu PS, Gangarao NV, Fikre A, Kimurto P, Sharma PC, Sheshashayee MS, Tobita S, Kashiwagi J, Ito O, Killian A, Varshney RK (2014) Genetic dissection of drought and heat tolerance in chickpea through genome-wide and candidate gene-based association mapping approaches. PLoS ONE 9:e96758

Thudi M, Khan AW, Kumar V, Gaur PM, Katta K, Garg V, Roorkiwal M, Samineni S, Varshney RK (2016a) Whole genome resequencing reveals genome-wide variations among parental lines of 16 mapping populations in chickpea (*Cicer arietinum)*. BMC Plant Biol 16:10

Thudi M, Chitikineni A, Liu X, He W, Roorkiwal M, Yang W, Jian J, Doddamani D, Gaur PM, Rathore A, Samineni S, Saxena RK, Xu D, Singh NP, Chaturvedi SK, Zhang G, Wang J, Datta SK, Xu X, Varshney RK (2016b) Recent breeding programs enhanced genetic diversity in both desi and kabuli varieties of chickpea (*Cicer arietinum* L.). Sci Rep 6:38636

Toker C (2005) Preliminary screening and selection for cold tolerance in annual wild *Cicer* species. Genet Resour Crop Evol 52:1–5

Toker C, Canci H, Yildirim T (2007) Evaluation of perennial wild *Cicer* species for drought resistance. Genet Resour Crop Evol 54:1781–1786

Turner NC, Colmer TD, Quealy J, Pushpavalli R, Krishnamurthy L, Kaur J, Singh G, Siddique KHM, Vadez V (2013) Salinity tolerance and ion accumulation in chickpea (*Cicer arietinum* L.) subjected to salt stress. Plant Soil 365:347–361

Udupa SM, Baum M (2003) Genetic dissection of pathotypespecific resistance to Ascochyta blight disease in chickpea (*Cicer arietinum* L.) using microsatellite markers. Theor Appl Genet 106:1196–1202

Upadhyaya H, Haware M, Kumar J, Smithson J (1983a) Resistance to wilt in chickpea. I. Inheritance of late-wilting in response to race 1. Euphytica 32:447–456

Upadhyaya H, Smithson J, Haware M, Kumar J (1983b) Resistance to wilt in chickpea. Further evidence for two genes for resistance to race 1. Euphytica 32:749–755

Upadhyaya HD, Dronavalli N, Gowda CLL, Singh S (2011) Identification and evaluation of chickpea germplasm for tolerance to heat stress. Crop Sci 51:2079–2094

Upadhyaya HD, Bajaj D, Das S, Kumar V, Gowda CL, Sharma S, Tyagi AK, Parida SK (2016a) Genetic dissection of seed-iron and zinc concentrations in chickpea. Sci Rep 6:24050

Upadhyaya HD, Bajaj D, Narnoliya L, Das S, Kumar V, Gowda CLL, Sharma S, Tyagi AK, Parida SK (2016b) Genome-wide scans for delineation of candidate genes regulating seed-protein content in chickpea. Front Plant Sci 7:302

Upasani ML, Limaye BM, Gurjar GS, Kasibhatla SM, Joshi RR, Kadoo NY, Gupta VS (2017) Chickpea-*Fusarium oxysporum* interaction transcriptome reveals differential modulation of plant defense strategies. Sci Rep 7:7746

Vadez V, Krishnamurthy L, Serraj R, Gaur PM, Upadhyaya HD, Hoisington DA, Varshney RK, Turner NC, Siddique KHM (2007) Large variation in salinity tolerance in chickpea is explained by differences in sensitivity at the reproductive stage. Field Crops Res 104:123–129

Vadez V, Rashmi M, Sindhu K, Muralidharan M, Pushpavalli R, Turner NC, Krishnamurthy L, Gaur PM, Colmer TD (2012a) Large number of flowers and tertiary branches, and higher reproductive success increase yields under salt stress in chickpea. Eur J Agron 41:42–51

Vadez V, Krishnamurthy L, Thudi M, Anuradha C, Colmer TD, Turner NC, Siddique KHM, Gaur PM, Varshney RK (2012b) Assessment of ICCV 2 9 JG 62 chickpea progenies shows sensitivity of reproduction to salt stress and reveals QTL for seed yield and yield components. Mol Breed 30:9–21

Vandemark GJ, Grusak MA, McGee PJ (2018) Mineral concentrations of chickpea and lentil cultivars and breeding lines grown in the U.S. Pacific Northwest. Crop J 6:253–262

Varshney RK, Hiremath PJ, Lekha P, Kashiwagi J, Balaji J, Deokar AA, Vadez V, Xiao Y, Srinivasan R, Gaur PM, Siddique KHM, Town CD, Hoisington DA (2009) A comprehensive resource of drought-and salinity responsive ESTs for gene discovery and marker development in chickpea (*Cicer arietinum* L.). BMC Genomics 10:523

Varshney RK, Song C, Saxena RK, Azam S, Yu S, Sharpe AG, Cannon S, Baek J, Rosen BD, Tar'an B, Millan T, Zhang X, Ramsay LD, Iwata A, Wang Y, Nelson W, Farmer AD, Gaur PM, Soderlund C, Penmetsa RV, Xu C, Bharti AK, He W, Winter P, Zhao S, Hane JK, Carrasquilla-Garcia N, Condie JA, Upadhyaya HD, Luo MC, Thudi M, Gowda CL, Singh NP, Lichtenzveig J, Gali KK, Rubio J, Nadarajan N, Dolezel J, Bansal KC, Xu X, Edwards D, Zhang G, Kahl G, Gil J, Singh

KB, Datta SK, Jackson SA, Wang J, Cook DR (2013a) Draft genome sequence of chickpea (*Cicer arietinum*) provides a resource for trait improvement. Nat Biotechnol 31:240–246

Varshney RK, Gaur PM, Chamarthi SK, Krishnamurthy L, Tripathi S, Kashiwagi J, Samineni S, Singh VK, Thudi M, Jaganathan D (2013b) Fast-track introgression of 'QTL-hotspot' for root traits and other drought tolerance traits in JG 11, an elite and leading variety of chickpea. Plant Genome. https://doi.org/10.3835/plantgenome2013.07.0022

Varshney RK, Thudi M, Nayak SN, Gaur PM, Kashiwagi J, Krishnamurthy L, Jaganathan D, Koppolu J, Bohra A, Tripathi S, Rathore A, Jukanti AK, Jayalakshmi V, Vemula A, Singh SJ, Yasin M, Sheshshayee MS, Viswanatha KP (2014a) Genetic dissection of drought tolerance in chickpea (*Cicer arietinum* L.). Theor Appl Genet 127:445–462

Varshney RK, Mohan SM, Gaur PM, Chamarthi SK, Singh VK, Srinivasan S, Swapna N, Sharma M, Singh S, Kaur L, Pande S (2014b) Marker-assisted backcrossing to introgress resistance to *Fusarium* wilt race 1 and *Ascochyta* blight in C214, an elite cultivar of chickpea. Plant Genome 7:1–11

Verheul MJ, Picatto C, Stamp P (1996) Growth and development of maize (*Zea mays* L.) seedlings under chilling conditions in the field. European J Agron 5:31–43

Verma S, Gupta S, Bandhiwal N, Kumar T, Bharadwaj C, Bhatia S (2015a) High-density linkage map construction and mapping of seed trait QTLs in chickpea (*Cicer arietinum* L.) using genotyping-by-sequencing (GBS). Sci Rep 5:17512

Verma M, Kumar V, Patel RK, Garg R, Jain M (2015b) CTDB: an integrated chickpea transcriptome database for functional and applied genomics. PLoS ONE 10:e0136880

Wahid A, Gelani S, Ashraf M, Foolad MR (2007) Heat tolerance in plants: an overview. Environ Exp Bot 61:199–223

Wang L, Chen L, Li R, Zhao R, Yang M, Sheng J, Shen L (2017) Reduced drought tolerance by CRISPR/Cas9-Mediated *SlMAPK3* mutagenesis in tomato plants. J Agri Food Chem 65:8674–8682

Welfare K, Yeo AR, Flowers TJ (2002) Effects of salinity and ozone, individually and in combination, on the growth and ion contents of two chickpea (*Cicer arietinum* L.) varieties. Environ Pollut 120:397–403

Chapter 4
Toward Climate-Resilient Lentils: Challenges and Opportunities

Dorin Gupta, Rama Harinath Reddy Dadu, Prabhakaran Sambasivam, Ido Bar, Mohar Singh, Navya beera and Sajitha Biju

Abstract Lentil among legumes has a significant place in crop production and rotation, and the nutritional security of growing human population. Current lentil cultivars have a narrow genetic base and are challenged with many biotic and abiotic stresses. The pressures from changing climate necessitate more efforts to find durable resistance sources for biotic and abiotic stresses. Distant landraces and wild lentil species which are less explored are known to possess such genes to develop resilient cultivars, one of the best adaptation strategies for climate change. The research efforts are currently focusing on enhancing lentil grain yield and resilience to climate change through introgression of desired genes from other gene pools. The current lentil-breeding efforts have concentrated upon conventional plant breeding techniques for the inclusion of the cultivated lentil gene pool only. Unlike other crops, genomics-assisted breeding remains one of the areas to be further explored to speed-up the climate-smart high-yielding cultivars development process, which is reliant on the extensive genomic resources. Several lentil linkage maps have been developed and quantitative trait loci for tolerance to biotic and abiotic stresses have been identified. However, advances in molecular markers, next-generation sequencing, genomewide sequencing, and bioinformatics will further help to precisely identify genes of interest that can be best utilized to breed climate-resilient cultivars for higher production and quality through genetic engineering and plant breeding.

Keywords Lentil · Wild · Gene pool · Climate-smart traits · Genomics

---

D. Gupta (✉) · R. H. R. Dadu · S. Biju
Faculty of Veterinary and Agricultural Sciences, University of Melbourne, Dookie, VIC, Australia
e-mail: doring@unimelb.edu.au; gupta.dorin@gmail.com

P. Sambasivam · I. Bar
School of Natural Sciences, Environmental Futures Research Institute, Griffith University, Nathan, QLD, Australia

M. Singh
ICAR-National Bureau of Plant Genetic Resources Regional Station, Shimla, India

N. beera
Institute of Biotechnology, Telangana State Agricultural University, Hyderabad, India

C. Kole (ed.), *Genomic Designing of Climate-Smart Pulse Crops*,
https://doi.org/10.1007/978-3-319-96932-9_4

4.1 Introduction

Lentil (*Lens culinaris* Medikus) like other food legumes offers a range of benefits from soil to human health and has become an integral part of current farming system as a valuable cash crop. However, lentil is still one of the neglected crops especially in developing countries, which has potential to be grown in more drier areas being fairly drought tolerant and highly nutritious. Legumes are accepted in farming system around the world but not to the extent as cereal crops. However, legumes including lentil which can fix atmospheric nitrogen and can minimize the nitrogen input requirements, provide pest breaks and weed control for subsequent cereal crop when used in crop rotations. Lentil offers many health benefits due to its low fat, high prebiotic carbohydrates, high fiber, and low glycemic index (Srivastava and Vasishtha 2012; Thavarajah 2017). Lentil grain provides 22–25% dietary protein (one-quarter of total caloric value), carbohydrates, vitamins, and a good balance of minerals (K, Mg, Fe, Zn), along with high contents of essential amino acids such as lysine and tryptophan (Erskine et al. 1990; Johnson et al. 2013; Faris et al. 2013; Ray et al. 2014). Staple cereals are high in sulfur-based amino acids which are lacking in lentils, therefore, when eaten together, cereal-lentil based diet can provide complete profile of the essential amino acids. Due to its high protein content, this grain is also regarded as cost-effective alternative to animal-based protein, especially in the Indian subcontinent where plant-based diet predominates due to religious believes, and less affordability of fresh meat. Lentil ranks sixth among important annual grain legumes with 5.4 million hectare worldwide area under cultivation leading to an annual production of 6.3 million tons (FAO 2016). It is grown in Canada, Australia, Southern Europe, Eastern, and Northern Africa, the drier regions of the Middle East and the Indian subcontinent during cooler season of the year. Lentil productivity has not seen tremendous rise over the past years and even has not crossed the mark of one ton per hectare globally (FAO 2016). Climate change will have a significant impact on global food production and food security of growing human population as clearly highlighted by Intergovernmental Panel on Climate Change (IPCC), if we look deeper, major implications will be through reduced soil fertility, reflection of reduced microbial activity and diversity, and carbon sequestration leading to less than optimal plant growth and yields (Dhankher and Foyer 2018). The future projections of drastic climatic events such as frequent droughts and floods, higher or lower temperatures, salt, and heavy metal stresses leading to higher incidences of pest infestations will significantly affect crop yields. Adaptation through diversified new crops and cultivars could be one of the strategies to combat climate change and sustain food production. Further, to be climate resilient, crop production system requires tailored solutions through inclusion of past and current knowledge about crops, beneficial crop rotations, their unique genetic make-up, and specific traits to be targeted for their inclusion in modern cultivars to cope and produce enough under various stresses. Lentil being one of the hardy crops can potentially yield higher, to meet the demands of quality food for growing human population. However, since past many years, crop is being grown on marginal lands especially in the developing countries due to preferential cereal-based cropping, which led to the loss of genes of higher productivity (Bejiga and Degago 2000). Along with cultivation on marginal lands which generally have low soil fertility, the crop is mostly grown as rainfed and is subjected to mainly terminal drought and heat stress (abiotic) and various fungal and bacterial diseases—Ascochyta blight, rust Stemphylium blight, collar rot, root rot, white mold, Fusarium wilt, and anthracnose (Kumar et al. 2013; Sharpe et al. 2013a). At parallel, current improved cultivars are bred to yield higher but not primarily to cope with various stresses, if resistant to one or the disease exists, it is not very durable and traits responsible for tolerance to various abiotic stresses are not introgressed as priory traits. These cultivars have narrow genetic base owing to handful desired cultivated germplasm parentage (Singh et al. 2014) whereas, not much has been explored in wild lentil relatives which are more diverse (Ford et al. 1997; Duran et al. 2004; Gupta and Sharma 2007; Singh et al. 2014). Therefore, these cultivars cannot yield higher due to their vulnerability to major biotic and abiotic stresses under climatic uncertainties and narrow genetic base. Farmers adopt new crop cultivars and change their practices to cope with changing environments. However, the pace of environmental change will be difficult to match along with the expected expansion of crops to new environments and lands suggesting strong need for research efforts to develop climate-resilient crops (Dhankher and Foyer 2018). In lentil, the identification and inclusion of climate resilient useful and diverse traits/genes to broaden the genetic base of the existing lentil cultivars from closely or distant relatives should be prioritized. Therefore, this chapter aims at understanding lentil's untapped sources of genetic variation, traits of importance, role of conventional, genomics, and modern molecular technologies for better use of such identified traits and their inclusion in breeding programs to breed and develop climate-resilient lentil cultivars.

4.2 Prioritizing Climate-Smart (CS) Traits

To sustain crop yields under uncertain environments, higher yielding climate-smart crop cultivars should possess multiple resistance and/or tolerance to stresses (biotic and abiotic). One of the major differences between two types of major stress categories is that the mechanisms controlling abiotic stresses are governed by multiple genes, therefore targeting germplasm in breeding programs which shows the potential for common defense mechanisms can address multiple stress tolerance in plants. The adaptation to climate change can be sought through the development of new cultivars with multiple tolerance to abiotic stresses such as heat, cold, frost, drought and salinity, and resistance to various diseases and pests. At parallel lentil, cultivars should possess adapted phenology (maturation times and responses) and different agro-morphological traits which will offset the new challenges of changes in growing season (shorten/longer than usual).

4.2.1 Flowering Time

The transition from vegetative to reproductive growth is an important trait and a major component of crop adaptation, particularly in rainfed environments (Subbarao et al. 1995; Gao et al. 2014). The timing of flowering is dependent upon the genotype, the seasonal temperature profile, photoperiod, light, nutrient levels, and vernalization responses of the plant. If flowering occurs prematurely under stressful environments, seed set and grain filling may be compromised. If flowering is delayed, the plant risks succumbing to terminal drought stress before producing any seed. A complex network of genetic pathways allows the plants to detect and integrate external or internal signals to initiate the floral transition (Bluemel et al. 2015). In indeterminate species like lentil, early flowering may enable the plants to prolong the reproductive phase, especially when the flowering duration is delimited by terminal drought stress that terminates seed set. The transition to flowering stage in lentil is proposed to be a function of both photoperiod and temperature, longer days and warmer temperatures accelerate flowering (Summerfield et al. 1985; Roberts et al. 1986; Barghi et al. 2013). Yuan et al. (2017) showed that the overall days to flowering of lentil genotypes were mainly influenced by the red/far-red (R/FR)-induced light quality change. While most of the wild lentil genotypes had reduced responses and flowering time, the cultivated lentil showed consistent, accelerated flowering in response to the low R/FR light environment together with three wild lentil genotypes (*L. orientalis* IG 72611, *L. tomentosus* IG 72830, and *L. ervoides* IG 72815). These genotypes would represent key genetic resources for developing lentil cultivars with better adaptation to variable light environments.

The role and importance of vernalization in floral induction for lentils, however, remains largely undefined. Summerfield et al. (1985) in his analysis of six lentil genotypes reported a variation in vernalization response with respect to flowering time, vernalized plants flowered earlier in all instances compared to nonvernalized plants. Roberts et al. (1986) in contrast proposed that the effect of vernalization on floral induction was negligible. It has also been suggested that for sensitive genotypes, vernalization exposure reduced the critical or nominal base photoperiod, required for floral induction (Summerfield et al. 1985; Roberts et al. 1986). Photoperiod-sensitive and insensitive phases can be identified through experiments in which individual plants can reciprocally be transferred in a time series from long to short days and vice versa in growth chambers. This will help to develop cultivars with shorter pre-inductive photoperiod-insensitive and sensitive phases to fit short growing seasonal regions. Exploitation of genetic variability for flowering time can assist in the development of high yielding early maturing cultivars that are able to adapt to changing environmental conditions. Exotic and indigenous lentil germplasm were screened to identify early flowering genotypes (Erskine et al. 1998; Asghar et al. 2010; Kumar and Solanki 2014; Kumar et al. 2014b; Singh et al. 2014). Sarker et al. (1999a, b) identified single recessive gene (*sn*) control for early flowering in lentil. The variants of early flowering at this locus could be useful for the development of early flowering cultivars for water-limited environments and can help to diversify the lentil genetic base.

4.2.2 *Root Characters*

Root characters are one of the important agronomic traits, which play vital roles in crop adaptation and productivity under stressed environments. Developing crops with better root systems is a promising strategy to ensure productivity in both optimum and stressed environments. A deep and proliferative root system extracts sufficient water and nutrients under stressed conditions. Well-developed root systems are linked to drought tolerance as an avoidance mechanism guaranteeing productivity of lentil under water-limited environments (Idrissi et al. 2015a, 2016; Sarker et al. 2005; Verslues et al. 2006; Gaur et al. 2008; Vadez et al. 2008). Drought-tolerant genotypes tend to elongate their rooting depth significantly more than sensitive ones under drought stress in lentil (Sarker et al. 2005). Specific rooting patterns can be associated with drought avoidance mechanisms that can be used in lentil breeding programs. Modifications in the root architecture allow the plants to increase their water extraction capacity and drought tolerance.

Gorim and Vandenberg (2017a) found significant differences for root traits and fine root distribution between and within selected wild lentil species and cultivated lentil. The authors also observed variability in nodule number and nodule shape within and between genotypes. Some genotypes used water more efficiently for either biomass or seed production. The allocation of resources to seed production also varied between genotypes. These findings could have an impact on the design of future lentil breeding CS traits in the context of strategies for managing changes in rainfall amount and distribution for lentil growing regions. The distribution pattern of root traits and nodulation at different soil depths in both wild and cultivated lentil genotypes were also analyzed by Gorim and Vandenberg (2017b). Their findings suggest that wild lentil genotypes from a particular gene pool might have similarity for root traits and nodule distribution in the soil. Furthermore, wild genotypes with deep root systems allocated their resources mostly toward biomass production implying that when interspecific hybridization and introgression become part of a long-term breeding strategy for lentil, it will be necessary to develop appropriate selection strategies for simultaneous selection of yield and root traits under stressed environments.

4.2.3 *Heat Tolerance*

Lentil is similar to other cool-season legumes in its susceptibility to rising temperatures (Summerfield et al. 1985; Ahmed et al. 1992; Porch and Jahn 2001; Croser et al. 2003; Choudhury et al. 2012; Bhandari et al. 2016; Sehgal et al. 2017). It requires cooler temperatures during the vegetative growth and warmer temperatures at maturity; the optimum temperature for lentil growth is 18–30 °C (Choudhury et al. 2012). Susceptibility of vegetative and reproductive stage in lentil crop to heat stress has been described by (Delahunty et al. 2015; Bhandari et al. 2016; Kumar et al. 2016a, b and Sita et al. 2017). Temperatures greater than 24.4 °C reduced the germination rate in lentil (Covell et al. 1986). Temperatures above 32/20 °C (max/min) during flowering and pod filling in lentil drastically reduced seed yield and resulted in 20–70% yield reductions, equating to $1000/ha loss, through flower drop and pod abortion (Delahunty et al. 2015; Kumar et al. 2016a, b). Heat stress in lentil causes a reduction in germination percentage, abnormal seedling growth, nodules degeneration, early flowering, reduction in plant biomass, loss in cell membrane stability and photosynthetic efficiency, and increase in lipid peroxidation, (Ellis and Barret 1994; Muehlbauer et al. 2006; Chakraborty and Pradhan 2010; Sehgal et al. 2017). Higher expression of ascorbate peroxidase (APX) has been linked with heat tolerance in lentil (Chakraborty and Pradhan 2010). Heat tolerance in lentil is attributed to superior pollen function and higher expression of leaf antioxidants (Sita et al. 2017). Heat stress especially when combined with drought stress, even for a few days during flowering and pod filling drastically reduces seed yield in lentil because of accelerated development, forced maturity, shortened reproductive period, and damage to reproductive organs leading to flower drop, pollen sterility, pod abortion, and reduced seed set (Siddique 1999; Boote et al. 2005; Choudhury et al. 2012; Gaur et al. 2015; Bhandari et al. 2016).

Even though only limited studies were conducted to screen lentil germplasm for heat tolerance in both laboratory and field conditions, genetic variations for heat tolerance have been identified in lentil and are listed in Table 4.1.

4.2.4 Cold/Frost Tolerance

Lentil is prone to radiant frost when compared with other legumes and are less prone to frost than peas but more susceptible than chickpeas (Murray et al. 1988). Frost tolerance for lentil at flowering is –2 to –3 °C. Lentil is least tolerant to frost injury at flowering due to the exposed nature of the flowers and the small size of pods. Frost injury symptoms in lentil include flower and pod abortion, damage to seed, and injuries to vegetative tissues. During the pod filling stage, frost can damage the seed coat and the developing seed. In severe frost events, leaves are damaged and stem wilts. Plant at the early vegetative stage can quickly recover from underground axillary buds, however, at the vegetative maturity stage or beyond, the plants will most likely die because axillary bud initiation will most likely not occur as the plant is moving into reproductive stage. Frost damage can also result in an increased vulnerability to entry of pathogen causing diseases like anthracnose and botrytis gray mold. Yield losses from frost damage can be severe for a high-value crop like lentil. Since 1980, considerable research efforts have been put into breeding and characterizing the genetics of frost tolerance of lentil (Erskine et al. 1981; Summerfield et al. 1985;

Table 4.1 Genetic variation for tolerance to heat, frost, and waterlogging in lentil

Type of stress	Accession	Selection criteria	References
Heat	IPL81, IPL406	Heat tolerance index (TI) and antioxidant activities	Chakraborty and Pradhan (2010)
	Ranjan, IC201710, IC208329, 14-4-1	Cell membrane thermostability	Choudhury et al. (2012)
	Qazvin	Cell membrane thermostability	Barghi et al. (2013)
	72578, 70548, 71457, 73838	Seed yield	Delahunty et al. (2015)
	ILL2181, ILL82, ILL5151, ILL5416 ILL4587, ILL956 ILL 598, FLIP2009-55L, ILL2507, LL4248	Pollen viability, grain yield	Gaur et al. (2015)
	FLIP2009-55L, IG2507, IG4258	Pollen viability	Kumar et al. (2016a, b)
	IG3745, IG4258, IG5146	Number of filled pods at higher temperature	Kumar et al. (2016a, b)
	LL931	Seed weight	Bhandari et al. (2016)
	GP2961, PL234, LKH2	Biological yield, grain yield, number of pods per plant, pod yield, and number of seeds per pod	Kumar et al. (2019)
	IG2507, IG3263, IG3297, IG3312, IG3327, IG3546, IG3330, IG3745, IG4258, FLIP2009	Pollen germination, pollen viability, ovular viability, pod number, nodulation, antioxidants, sucrose	Sita et al. (2017)
Frost/cold	LC9978057, LC9977006 LC9977116, LC9978013 ILL759, ILL1878, ILL4400 ILL7155, ILL8146, ILL8611, ILL9832, Kafcas, Cifei, Ubek	Winter survival rates, visual rate, damage percentage of survival	Hamdi et al. (1996)
	ILL5865, Balochistan local	Controlled freezing test	Ali et al. (1999)

(continued)

Table 4.1 (continued)

Type of stress	Accession	Selection criteria	References
	LL1878, ILL662, ILL857, ILL975, ILL1878	Winter hardiness	Sarker et al. (2002)
	Morton, WA8649041, WA8649090	Winter survival rates	Kahraman et al. (2004a)
	ILL662, ILL857, ILL975	Rapid ground cover Early vigor	Sarker et al. (2002)
Waterlogging/ flooding/ submergence	ILL6439, ILL6778, ILL6793	Stomatal conductance and biomass	Ashraf and Christi (1993)

Murray et al. 1988; Spaeth and Muehlbauer 1991; Kusmenoglu and Aydin 1995; Ali et al. 1999). More recently, several research studies have also been carried out in the aspects of winter hardiness and frost injury in lentil (Kahraman et al. 2004b; Barrios et al. 2007, 2010, 2016). Identified genetic variation for tolerance to frost is listed in Table 4.1.

4.2.5 Drought Tolerance

Lentil is considered as moderately tolerant to drought when compared to other legumes (Reda 2015). Even though lentil is a hardy crop requiring less water for its growth compared to other legumes, the plant productivity can decrease from 6 to 70% under drought conditions and can even lead to total crop failure (Saxena 1993; Johansen et al. 1994; Babayeva et al. 2014). Drought stress at reproductive stage led to 24% grain yield reduction and was 70% when drought occurred at pod development stage (Shrestha et al. 2006; Allahmoradi et al. 2013). Drought stress occurring at flowering or podding stage affects vegetative and reproductive growth leading to reduced leaf area (48–55%), total dry matter (32–50%), flower production (22–55%), and number of pods and seeds (27–66%), with significantly higher flower drop and aborted pods (Table 4.2) (Shrestha et al. 2006). Drought stress can also lead to fluctuation in concentration of photosynthetic pigments, osmoregulation, and antioxidant metabolism in lentil (Aksoy 2008; Öktem et al. 2008; Gokcay 2012; Muscolo et al. 2014; Mishra et al. 2016; Biju et al. 2017). The variable annual rainfall patterns threaten the sustainability of lentil production by increasing the frequency of drought periods during the cropping season (Dai 2011). Ninety percent of the world's lentil is produced in areas relying upon conserved, receding soil moisture and therefore, crop productivity is largely dependent on the efficient utilization of available soil moisture (Kumar and Van Rheenen 2000).

Lentil withstands drought stress through drought tolerance and drought avoidance mechanisms. Drought tolerance mechanisms in lentil include dense pubescence

Table 4.2 Identified sources of resistance to drought stress in lentil from literature

Accession	Selection criteria	References
ILL6439, ILL6451	Osmotic adjustment	Ashraf et al. (1992)
ILL1983, ILL2501, ILL2526	Seed yield	Hamdi et al. (1992)
MI30B, MI52, MI563	Leaf water traits	Salam and Islam (1994)
ILL1861, ILL784	Seed yield	Hamdi and Erskine (1996)
ILL590, ILL7200	Short duration, rapid biomass, leaf area development, high photosynthetically active radiation	Clements et al. (1997)
HUL35	Osmotic adjustment	Singh (2001)
ILL6002	Stem length, taproot length, number of lateral roots	Sarker et al. (2005)
TN1768	High yield	Salehi et al. (2008)
Naeen, Shiraz7	Stress tolerance index (STI), geometric mean productivity (GMP)	Rad et al. (2010)
TN1084, KC210034	GMP, harmonic mean (HM), STI, stress susceptibility index (SSI)	Siahsar et al. (2010)
Seyran	Antioxidant enzyme activities (APX, CAT, GR, and SOD), protein profiles	Gokcay (2012)
Cabralia inta	Shoot length, germination stress index (GSI)	Salehi (2012)
Land race	RWC, Fv/Fm, proline, stomatal resistance	Allahmoradi et al. (2013)
ILL10700, ILL10823, FLIP96-51	Seedling survivability, drought tolerance score, root and shoot length, fresh and dry weight of roots and shoots	Singh et al. (2013a)
Eston, Castelluccio	Seed germination, water content, root length	Muscolo et al. (2014)
ILL123613, ILL123466, ILL123613, ILL123466, ILL134466, ILL123684, ILL123679, ILL123648, ILL123629	Drought tolerance index (DTI)	Babayeva et al. (2014)
Eston, Castelluccio	Seed germination, RWC, root length, proline content, total soluble sugars	Muscolo et al. (2014)

(continued)

Table 4.2 (continued)

Accession	Selection criteria	References
PDL1, PDL2	Seed yield	Singh et al. (2016a, b)
Ranjan	Length of shoot and root, fresh weight of shoot and root and dry weight of shoot and root	Dash et al. (2017)
HUL57	Nodulation, yield traits, DSI, STI, mean productivity (MP)	Mishra et al. (2014, 2016, 2018)
Digger, Cumra, Indianhead, ILL5588, ILL6002, ILL5582	Crop water stress index (CWSI), canopy temperature depression (CTD), Root Shoot ratio, RWC, harvest index, and drought tolerance efficiency	Biju et al. (2018)
Binamasur10	Seed yield	www.icarda.org

of leaf, regulated stomatal closure, osmotic adjustment, increased antioxidant responses, and enhancement in yield components. Drought-avoidance strategy is shown by short duration genotypes of lentil such as BARI M4, BARI M5, BARI M6, Precoz, Idlib 3, and Bakaria at the reproductive stage as an adaptation to drought stress through early flowering, rapid root growth, and early growth vigor with high yield potential (Erskine and Saxena 1993; Silim et al. 1993a, b; Erskine et al. 1994; Shrestha et al. 2005). Shoot traits such as canopy structure, stem length, leaf surface, stomatal characteristics, and leaf movements also have significant roles in drought avoidance as reported in the lentil mutant line MI-30 (Salam and Islam 1994). Specific rooting patterns such as root-shoot ratio (RS ratio), can also be associated with drought avoidance mechanisms that can be used in lentil breeding programs (Idrissi et al. 2016; Biju et al. 2017). Drought escape was believed to be relatively insignificant in wild lentil genotypes when compared to cultivated ones (Hamdi and Erskine 1996). Contrary to this finding, recently, Gorim and Vandenberg 2017a) has identified the different drought mechanisms in wild lentil genotypes across species by assessing both above ground plant characteristics and their root systems. They found that wild lentil genotypes employed diverse strategies such as delayed flowering, reduced transpiration rates, reduced plant height, and deep root systems to either escape, evade, or tolerate drought conditions, based on the environmental conditions at their centers of origin. Interestingly, in some cases, more than one drought strategies were observed within the same genotype. The success of increasing lentil production in drier areas prone to terminal drought mainly depends upon the development of short-season cultivars that enable the crop to escape adverse soil–water scarcity (Siddique et al. 2001). Early sowing of lentil in Southwestern Australia and Northern Syria develops a large green canopy and rapid ground cover which absorbs a significant proportion of photosynthetically active radiation (PAR) early in season when vapour pressure deficits (VPD) or atmospheric demand for water are low and uses more water in post flowering phase thus producing good yield and biomass (Siddique et al. 1998; Chen et al. 2006). Hence, the selection for early flowering lines with pliability for the maturity that provides a massive yield under high moisture availability, is therefore required for severely drought-prone areas. The International Centre for Agricultural Research in Dry Areas (ICARDA) developed early maturing lentil genotypes with good yield and is deposited with 'International Drought Tolerant Nursery' to be shared with the national and international programs.

Changes in several morphological, physiological, and biochemical traits such as seedling survivability, seedling vigor, plant height, total root length, taproot length and number of lateral roots, total root weight, early flowering, maturity, pod number per plant, seed number per pod, grain yield, harvest index, relative water content, water use efficiency, stomatal conductance, and antioxidant activity have been used in screening genotypes for drought tolerance in lentil (Sarker et al. 2005; Shrestha et al. 2006; Chakherchaman et al. 2009; Kumar et al. 2012, 2013; Singh et al. 2017a; Biju et al. 2018). Well-developed vigorous shoot and root system at early seedling stage are important for drought tolerance (Mia et al. 1996; Aswaf and Blair 2012; Kumar et al. 2012; Idrissi et al. 2013, 2015a). Deep and well-developed roots will increase the uptake of water and nutrients in a low moisture soil under drought conditions (Wu and Cheng 2014). Thus, the selection of deep rooting is recommended to increase the yield of legumes including lentil under drought conditions (Buddenhagen and Richards 1988).

The extent of membrane damage and the enzymatic antioxidant activity appears to be a useful method for evaluating the level of plant drought stress. Simple screening tests like electrolyte leakage measurements after stress can be used for drought tolerance in cool-season food legumes. Cell leakage studies were performed in different lentil genotypes and found that drought-tolerant lentil genotypes exhibit lower cell membrane injury along with higher seedling growth, water use efficiency, and osmotic regulation (Stoddard et al. 2006). Similarly, germination stress index (GSI) and cell membrane stability (CMS) index can also be used as a good criterion prior to conducting a field screening for drought tolerance in lentil at a large scale (Salehi et al. 2008). Polyethylene glycol (PEG) based in vitro screening for drought tolerance at seedling stage has been proven to be another suitable method to effectively screen large sets of germplasm with good accuracy by analyzing the traits like germination percentage, germination rate, germination index, seedling length, root and shoot length of seedlings, seedling dry weight, relative water content, proline, and total soluble sugars (Salehi 2012; Muscolo et al. 2014; Keshtiban et al. 2015; Dash et al. 2017; Biju et al. 2017). A new phenotyping technique for drought tolerance assessment in lentil using hydroponics has been developed to screen many genotypes at seedling stage (Singh et al. 2013a). However, most of these methods are slow, laborious, time consuming, expensive, and influenced by environmental conditions. Most recently, it has been reported that canopy temperature (Tc) and crop water stress index (CWSI), both assessed using infrared thermal images, along with root-shoot (RS) ratio, relative water content (RWC), harvest index (HI), and other drought tolerance indices are useful in defining the drought stress tolerance variability within lentil genotypes (Biju et al. 2018). The water conservation traits, such as early partial stomatal closure under soil drying, and limited transpiration under high atmospheric vapor-pressure deficit have recently been proven to be useful in other legumes under drought stress (Devi et al. 2010; Zaman-Allah et al. 2011; Belko et al. 2012; Seversike et al. 2013; Ghanem et al. 2017) and these traits can be used in lentil for defining drought stress along with physiological screenings and mechanistic crop simulation modeling. Table 4.2 shows the identified sources of resistance to drought stress in lentil.

Early on partially closed stomata under moisture stress and high VPD will enable less transpiration loss and could be traits of importance for drought stress tolerance in lentil.

4.2.6 Flooding and Submergence Tolerance

Flooding and submergence are adverse environmental conditions, which severely constrain the growth and yield of legume crops growing in the fine-textured and duplex soils (Solaiman et al. 2007; El-Enany et al. 2013; Kang et al. 2017). Lentil is the most sensitive of all legumes to waterlogging (Solaiman et al. 2007; Singh et al. 2013a) and transient waterlogging is an important hindrance for lentil production, especially during the early developmental stages (Materne and Siddique 2009). Waterlogging in lentil affects yield at any growth stage during the growing season causing most damage (Materne and Siddique 2009). Waterlogging during germination can cause unsuccessful germination, late emergence, and suppression of root growth (Jayasundara et al. 1997). Flooding at vegetative stage can induce root system damage and led extensive leaf senescence and desiccation (Nessa et al. 2007). Lentil is most susceptible to waterlogging at flowering period causing flowers and pods to abort. The response of lentil to waterlogging is like its response to low light and low temperatures, all result in stunted growth and leaf senescence (turning yellowish to red), wither and eventually die. Lentil germplasm with waterlogging tolerance associated with their geographic origin was studied by Wiraguna et al. (2017) and reported that genotypes from Bangladesh are adapted to waterlogged soil at germination. Waterlogging-tolerant genotypes were characterized by its low biomass, higher stomatal conductance, early flowering and maturity, and high root porosity (Ashraf and Chishti 1993; Malik et al. 2015; Erskine et al. 2016). Formation of lysigenous cavities and aerenchyma are waterlogging responses found in lentil (Hamdi, 1987). Some management practices used to reduce the effects of waterlogging in lentil involve sowing time, paddock selection, seeding rate, and drainage (Toker and Mutlu 2011). Studies revealing the biochemical and physiological responses for waterlogging tolerance and possible measures to combat this abiotic stress in lentil still deserves more attention.

4.2.7 Salinity Tolerance

Salinity is a major abiotic stress for lentil production, especially under drought conditions in shallow subsoils of alkaline soils especially in the arid and semiarid regions of Australia, Canada, North Africa, and South Asia (Muehlbauer et al. 2006; Nuttall and Armstrong 2010). Lentil is considered as an extremely sensitive species to salinity than other legumes such as faba bean and soybean (Ashraf and Waheed 1990, 1993; Katerji et al. 2001, 2003; Sidari et al. 2008), whereas it has greater salinity tolerance than chickpea and field pea (Siddique 1999). Yield reduction due to salinity stress has been reported in lentil to be as high as 20% at an electrical conductivity (EC) of 2 dS/m and 90–100% at an EC of 3 dS/m by negatively affecting yield attributes (Ayoub 1977; Van Hoorn et al. 2001; Golezani and Yengabad 2012). In lentil, responses to salinity stress vary with both growth stage, salinity level, and environmental factors such as soil–water status, temperature relative humidity, and available nutrients (Lachaâl et al. 2002). Like all other legumes, lentil is more susceptible to salinity stress during seedling establishment and later growth stages (Ayoub 1977; Rahimi et al. 2009; Farooq et al. 2017). Lentil roots are highly sensitive to saline soils with limited root growth, root depth, and moisture extraction capabilities which, in turn, can badly affect the nodulation and nitrogen fixation probably by limiting the root hair growth and rhizobium infection (Rai and Singh 1999; Van Hoorn et al. 2001). Delays in seed germination, reduced seed germination percentage, reduced seed viability, and decreased seedling growth also occurs with increasing levels of salinity in lentil (AL-Quraan et al. 2014). Salinity intensifies anthocyanin pigmentation in leaves and stems in lentil resulting in necrosis of the outer margins and yellowing of the older leaves which ultimately leads to the death and withering of leaves due to excess accumulation of ions. Salinity also reduces flower production and pod setting in lentil (Van Hoorn et al. 2001). Increasing level of exchangeable sodium percentage (10–25%) under salinity stress decreased plant height, leaf area, leaf dry weight, total biomass production, chlorophyll a and b content, nitrate and nitrite reductase enzymes activities, DNA and RNA content and finally, the grain yield of lentil (Tewari and Singh 1991; Singh et al. 1993). Salinity stress also restricts lentil growth and morphology by adversely affecting various physiological and biochemical attributes such as photosynthesis (AL-Quraan et al. 2014), membrane damage (Hossain et al. 2017), ion homeostasis (Turan et al. 2007; Hossain et al. 2017), oxidative damage (Al-Quraan and Al-Omari 2017; Hossain et al. 2017), antioxidant responses (Bandeoglu et al. 2004), γ-aminobutyric acid (GABA) accumulation (Al-Quraan and Al-Omari 2017), osmolyte accumulation, and proline metabolism (Turan et al. 2007; Hossain et al. 2017) (Table 4.3). Recently, it has been reported that the excessive accumulation of betaine and choline in lentil plants might play a pivotal role in salt tolerance inducing osmotic adjustment or osmoregulation which causes a fall in water potential (Varshney and Singh 2017).

Table 4.3 Identified sources of resistance to salinity stress in lentil

Accession	Selection criteria	References
DL443, PantL406	Nitrogen fixation, grain yield	Rai et al. (1985)
ILL5845, ILL6451, ILL6788, ILL6793, ILL6796	Seed germination, biomass	Ashraf and Waheed (1990)
NEL2704	Seed germination, plant growth, grain yield	Mamo et al. (1996)
ILL6976	Biomass, soluble sugars, efficiency of potassium utilization	Asraf and Zafar (1997)
LC53, DLg103, Sehore74-3, LC50	Nodulation, seed germination, seed yield, plant height, root length, plant growth	Rai and Singh (1999)
ILL8006	Water use efficiency	Hamdi et al. (2000)
Masoor93, Mansehra89	Na/K ratio	Yasin et al. (2002)
LG128, ILL3534	Grain yield	Maher et al. (2003)
ILL5582	Proline, superoxide dismutase activity	Cicerali (2004)
DL443, Pant L406, ILL3534 LG 128, ILL6796	Grain yield and biomass	Materne and Reddy (2007)
Ustica, Pantelleria	Proline, sugar, amylase	Sidari et al. (2007)
Çağıl, Altın Toprak	Germination percentage, shoot and root length, shoot and root dry weight, and salt tolerance percentage	Kokten et al. (2010)
Nipper, PBA Flash, ILL2024	Biomass and grain yield	Siddique et al. (2013)
Siliana, Local oueslatia Nefza	Seed germination and seedling growth	Ouji et al. (2015)
Flash (CIPAL0411), Bounty CIPAL0415), Nipper (CIPAL0203)	Plant growth and yield traits	GRDC (2013)
Jordan 1	Seed germination, accumulation of reactive oxygen species, γ-aminobutyric acid (GABA) level	Al-Quraan and Al-Omari (2017)

(continued)

Table 4.3 (continued)

Accession	Selection criteria	References
Sapna, RLG258, RLG234	Dry matter yield, stress indices (TOL, SSI, STI, MP, GMP, YI, SSPI, and MSTI)	Kumawat et al. (2017)
Masoor2002, NL20-3-3, LN0188, M93, NL9775	Root and shoot length, root and shoot weight, total proteins contents, α-amylase, total soluble sugars, sodium ions (Na^+), potassium ions (K^+), sodium-to-potassium ratio (Na^+/K^+)	Aslam et al. (2017)
PDL1, PSL9, ILWL95	Seed germination, seedling growth, biomass accumulation, seedling survivability, salinity scores, root and shoot anatomy, sodium ion (Na^+), chloride ion (Cl^-), potassium ion (K^+) concentrations, proline, antioxidant activities	Singh et al. (2017a)

4.2.8 Disease Resistance

It is anticipated that climate change is likely to exert a substantial effect on various insect pest management programs including host-plant resistance, natural plant products, bio-pesticides, natural enemies, and efficacy of synthetic chemicals. Lentil crop is often affected by several diseases and economically important diseases include Ascochyta blight (*Ascochyta lentis*), botrytis gray mold (*Botrytis cinereal* and *Botrytis fabae*), rust (*Uromyces viciae-fabae*), anthracnose (*Colletotrichum lentis*), Stemphylium blight (*Stemphylium botryosum*), powdery mildew (*Erysiphe pisi* and *Erysiphe polygoni*), and Fusarium wilt (*Fusarium oxysporum*). In general, foliar diseases including ascochyta blight, rust, anthracnose, botrytis gray mold, Stemphylium blight, and powdery mildew cause premature leaf drop, stem girdling, and produce shriveled seeds that are unmarketable. While major losses by soil-borne disease Fusarium wilt are due to leaf curling, reduced root development, discoloration of vascular tissue and stunted growth. Moderate to heavy yield losses have been reported for major diseases while some diseases have less economic impact based on the conducive environment for disease infection and spread and its duration during the cropping season (Chen et al. 2009). Ascochyta blight in Australia alone has been reported to cause an estimated loss of $16.2 million AUD in the conducive years (Murray 2012). Hence, proper management of diseases is suggested to ensure the sustainable productivity of lentil. Climatic change will have huge implications on our food production system and impact will also be seen on aggressiveness of pathogen through its development and survival rates (optimal conditions for infection), simultaneously on host reaction to pathogen attack (host specificity and mechanisms of plant infection), which will significantly affect the impact of various diseases on crop growth and production (Elad and Pertot 2014). Among many, host-plant resistance is the most acceptable, environment–friendly, and economical control strategy to avert yield losses (Rubiales and Fondevilla 2012). In future also, to develop climate-resilient cultivars, reliance on durable, diverse, and novel host resistance will be the key to sustain crop production under various climatic pressures. Accordingly, partial to complete resistance sources have been identified within the cultivated species of lentil to various diseases and cultivars with improved resistance have been released.

Resistance sources to Ascochyta blight within the cultivated germplasm have been reported from several countries including India (Gurdip et al. 1982; Sugha et al. 1991), New Zealand (Cromey et al. 1987), Pakistan (Iqbal et al. 1990, 2010), Lebanon (Abi-Antoun et al. 1990), Syria (Erskine et al. 1996), Canada (Andrahennadi 1994), Australia (Nasir and Bretag 1998), and Ethiopia (Ahmed and Beniwal 1991). Several of these prominent sources are still being employed in the current breeding programs as a source of resistance to Ascochyta blight. Indianhead is still the major source of resistance in Australian and Canadian breeding programs (Tullu et al. 2010). Resistance for anthracnose disease was screened with 1771 accessions of which only 4 accessions from United States collection and 12 accessions from the German collection had resistance to race Ct1 (Buchwaldt et al. 2004). However, none of the accessions had resistant against most aggressive race Ct0 (Buchwaldt et al. 2004). Later, Shaikh et al. (2012) reported 23 genotypes were resistant to anthracnose in Canada. Of which, 15 genotypes were identified with Ct1 resistance, while 7 genotypes expressed Ct0 resistance and 1 genotype VIR 2633 from Georgia was found symptomless to both races. Significant yield losses associated with lentil rust disease led to evaluation of cultivated lentil germplasm for rust resistance and release of rust-resistant cultivars in countries where rust is prevalent including India (Singh et al. 1994), Bangladesh (Sarker et al. 1999a, b), Ethiopia (Negussie et al. 1998; Fikru et al. 2007), Morocco (Sakr et al. 2004), Chile (Peñaloza et al. 2007) and Pakistan (Sadiq et al. 2008). Likewise, several genotypes resistant to fusarium wilt have been identified across lentil growing countries such as India, Iraq, Ethiopia, Lebanon, Iran, Pakistan, Turkey, Syria, and Nepal as reviewed by Choudhary et al. (2013). Evaluation of lentil germplasm against botrytis gray mold resulted in moderate to high-resistant sources across botrytis gray mold predominant countries (Karki et al. 1993; Bretag and Materne 1999; Kuchuran et al. 2003; Lindbeck et al. 2008). Consequently, several cultivars were released with resistance to botrytis gray mold, such as Nipper, a selection from a cross between Indianhead (resistant) and Northfield (susceptible) was released in 2006 for cultivation in Australia by Pulse Breeding Australia (PBA) (Lindbeck et al. 2008). Efforts have been made by ICARDA in association with Bangladesh Agricultural Research Institute (BARI) to develop Stemphylium blight-resistant cultivars to boost the disease resistance and subsequent yields (Sarker et al. 1999a, b; Sarker et al. 2004). Recently, Kant et al. (2017) screened Australian lentil germplasm and found significant variation for Stemphylium blight resistance.

Nonetheless, several of the released lentil cultivars have been reported to have changed their respective reaction within a short period of commercial release. This may be explained by the possible selective adaptation of the pathogen population and hence selection of highly aggressive isolates that can overcome the resistance with changing climate. Loss of resistance in Australian cultivars Northfield and Nipper to *A. lentis* has been speculated as a case of selective adaptation of pathogen since several aggressive isolates of the pathogen have been recovered within these cultivars (Davidson et al. 2016). Additionally, the narrow pedigree of these cultivars with paralleled pathogen evolution, threaten the sustainability of several cultivars. Subsequently, accessions from exotic germplasm particularly wild species have been tested to various diseases for resistance. This revealed some great variations for resistance within the wild species that may be transferred to the cultigen as reviewed by Singh et al. (2018). A novel primary gene pool accession ILWL 180 has been found highly resistant to recently recovered highly aggressive *A. lentis* isolates from Australia (Dadu et al. 2017). Successful introgression of resistance to anthracnose from wild lentil to the cultivar has been reported from Canada (Fiala et al. 2009; Vail et al. 2012).

Viruses are known to affect lentils and at least 30 different species of the virus have been reported to naturally infect lentil. Among them, the most important viruses that can cause significant yield losses includes bean leafroll virus, bean yellow mosaic virus, beet western yellow virus, cucumber mosaic virus, faba bean necrotic yellow virus, pea enation mosaic virus-1, pea seed-borne mosaic virus, and pea streak virus (Kumari et al. 2009). They cause none or a minimum of 3% to a maximum of 61% yield losses in lentil depending on the conditions available during the cropping season (Kumari et al. 2009). Several sources of resistance and cultivars with resistance to different viruses have been identified and released (Makkouk and Kumari 1990; Kumari and Makkouk 1995; Makkouk et al. 2001; Latham and Jones 2001; Rana et al. 2016).

4.2.9 Insect Resistance

Effects of climate change on insect pests is of greater importance as the insects are involved in many biotic interactions such as plants, natural enemies, pollinators, and other organisms, which are the key players of the ecological functions (Boullis et al. 2015). Environmental effect will trigger diversified insect populations, changed geographical distribution, insect–plant interactions, activities and abundance of natural enemies, emergence of new biotypes, and crop losses associated with insect pests. Changes in geographical distribution, diversity, and abundance of insect pests will also be influenced by changes in the cropping pattern influenced by climate change. Major insect pests may move to temperate regions, leading to greater damage in crops. Geographical distribution of many tropical and subtropical insect pests will extend, along with shifts in production areas of their host plants (Gonzalez and Bell 2013; Sharma 2014). Among nearly 36 insect pests infecting lentil crop, aphids (*Aphis craccivora* and *Acyrthosiphon pisum*), leaf weevils (*Sitona* spp.), lygus bugs, (*Lyguss* spp.) and cutworm (*Agrotis ipsilon*) are of economic significance, some minor field pests such as thrips (*Thrips*, *Kakothrips*, and *Frankiniella*), bud weevils (*Apionarrogans*), pea moth, (*Cydia nigricana*), pod borers (*Helicoverpa armigera* and *Heliothis* spp.), lima bean pod borer (*Etiella zinckenella*), root aphids (*Smynthurodes betae*), and leaf miners (*Liriomyza* spp. and *Phytomyza* spp.) infest the crop (Stevenson et al. 2007). These minor pests may become a significance in future with changing climatic conditions. Stevenson et al. (2007) have summarized locations and regions around the world which specify the status of various insect pests, such as aphids and lima bean pod borer are major lentil pests in India, lima bean pod borer and leaf weevils in Turkey, whereas aphids, thrips, and leaf weevils are most prevalent in central Spain.

Aphids cause significant loss to the lentil as they feed directly on crop and act as a vector in transmitting plant viruses. Hossain et al. (2017) reported relative abundance of lentil aphids at different sowing dates during the winter season and its effect on lentil yield. Aphid population and infestation increased with the delayed sowing. The crop sown in November received less aphid infestation and consequently produced higher yield than the December-sown crop. Spotted aphid, and cowpea aphid population had negative impact of higher temperature. Sharma et al. (1995) also suggested that aphid population was sensitive to changing temperature and relative humidity. High humidity, moderate temperature, and low rainfall are conducive for growth and multiplication of aphids. In long run with changing patterns of weather and host–pest interactions, host resistance and effective biological control could be the best strategy instead of heavy reliance on chemical control. Few tolerant genotypes (2 and 23) have been reported based on 2 and 3 years of screening work, respectively, and were grouped as five distinctive groups for tolerance based on pedigree analysis (Kumari et al. 2007).

Leaf weevils could be another major threat with changing climate and can cause huge economic losses when abiotic stresses affect seedling growth along with weevil attack. As larvae feed on root nodules which leads to failure of atmospheric nitrogen fixation. However, climate change adaptation strategies like early sowing would be beneficial to escape terminal drought stress, studies have shown that nodule damage by larvae in early sown crops was higher than late sown lentil crop (Weigand et al. 1992; Stevenson et al. 2007). Future lentil cultivars with chemical defenses against adult weevil could be one of the important trait s to consider. So far, no genotypes have been found to be resistant to weevil infection in lentil germplasm (Erskine et al. 1994). Genetic engineering might help to transfer genes found in red clover which leads to expression of formononetin and related metabolites offering chemical defense against adult weevil. Pod borer is another serious pest on many crops, however, not a major threat to lentil. Pod borer incidence had significantly negative correlation with low temperature and rainfall. Though rising temperature might change the population dynamics, host resistance, and plant traits which act as physical barriers and transgenics for expression of defense chemicals are direct measures, and in general, resilient lentil cultivars for other biotic and abiotic stresses will indirectly equip lentil crop to sustain yields through adaption to changing insect pest infestations.

4.2.10 Nutrient and Water Use Efficiency (NUE and WUE)

World agricultural soils are deficient in one or more of the essential nutrients to support healthy and productive plant growth. Overall nutrient use efficiency in the plant is a function of capacity of the soil to supply adequate levels of nutrients, and the ability of the plant to acquire, transport in roots and shoot and to remobilize to other parts of the plant. Inter and intraspecific variation for plant growth and mineral nutrient use efficiency are known to be under genetic and physiological control and are modified by plant interactions with environmental variables. Identification of plant traits for nutrient absorption, transport, utilization, and mobilization in plant cultivars should greatly enhance nutrient use efficiency. Overall nutrient usage in the plants is governed by the flux of ions from the soil to the root surface and by the influx of ions into roots followed by their transport to the shoots and remobilization to plant organs. The root morphological traits such as length, thickness, surface area, and volume have profound effect on the plant's ability to acquire and absorb nutrients from the soil (Barber, 1995). Plant environment interaction (solar radiation, rainfall, and temperature) and their response to diseases, insects, and root microbes have a great influence on nutrient use efficiency of plants and their subsequent yields (Arkin and Taylor 1981; Fageria 1992; Barber 1995; Marschner 1995; Baligar 1997; Fageria and Baligar 1997).

Winter legumes require a neutral to alkaline soil pH for their optimum growth and yield. Root growth of legumes is particularly severely restricted in acid soils. Lentil is most sensitive to acidic pH followed by chickpea and field pea. Minor variations in soil pH drastically affect the availability of nutrients for crop growth and productivity. Sutaria et al. (2010) found that the extensive root system with balanced fertilization and organic matter in adequate amount assisted in the efficient absorption and utilization of other nutrients thereby optimizing nutrient use efficiency in lentil. Organic nutrients enhance macro and micronutrients availability in the root zone which improved nutrient use efficiency by creating microenvironment for root growth and number of nodules (Singh et al. 2001).

Water use efficiency (WUE) measures the water quality used by the crop during its growth period to produce a unit quantity of the crop yield. Therefore, the lower the water requirement per unit of crop yield, the higher the WUE. With climate change temperatures will rise and an increase in extremes of rainfall or drought will be evident in many areas where lentil is grown. Water availability and day length influence vital physiological processes and determine the input use efficiency of plants. Light and temperature affecting transpiration and dry matter production will further have implications on WUE according to weather changes. In short-season Mediterranean environments, species with early flowering, podding, and seed set have higher yields and WUE than those with later flowering, podding, and seed set (Siddique et al. 2001). When the yields and water use of chickpea and lentil were compared over 12 growing seasons at Tel Hadya in Syria, the WUE for grain yield varied from 1.9 to 5.5 (kg/ha/mm in chickpea and from 2.1 to 5.2 kg/ha/mm, respectively, depending on growing season.

Another important trait that increases WUE is partial stomatal closure, which generally reduces water loss more than it reduces CO_2 uptake, thus increasing dry matter accumulation per unit of water transpired. However, the factors that alter transpiration will have a direct impact on mass flow of water to the root surface, and with it, alter the mechanism of ion transport and possibly nutrient uptake also.

Farmers in Iran usually sow lentil in early spring (March) and harvest around July. Under these circumstances, the crop encounters low winter rainfall, low WUE, and often temperature stress and terminal drought during reproductive stages (Azimzadeh 2010). Recently, some farmers tried a dormant seeding management (DS) in lentil. In this management system, it is assumed that germination would take place once the initial soil moisture in the top-soil layer filled to the volumetric transpirable soil water. While the temperature of top-soil layer is above the base of 2 °C. Furthermore, the crop germination is stopped due to lower temperatures than base temperature of lentil. This method might increase grain yield, WUE, and duration of lentil growing season. The change in the management practice could be effective for saving water for the lentil and better exploiting from precipitations over growing season. The change in the sowing date management of lentil would be even more effective for higher grain yield and WUE when early maturing cultivar is selected.

4.3 Genetic Resources of CS Genes

Lentil is a self-pollinating true diploid ($2n = 2x = 14$) annual plant which belongs to tribe *Vicieae*, the genus *Lens* of family Fabaceae (Leguminosae) and has 4 Gbp genome size (Arumuganathan and Earle 1991). Based on seed size, lentil encompasses two groups—microsperma (small seeded of 2–6 mm size range) and macrosperma (large seeded of 6–9 mm size range). Cultivated lentil has been presumed to be originated from close wild species *L. orientalis* (Zohary 1972), cultivated and *L. orientalis* genotypes show high cross compatibility and fertile hybrids. Lentil is believed to be originated in the Near East around the Fertile Crescent which was further domesticated in Southern Turkey following Nile, Europe, Greece, and further Asia (Renfrew 1969; Ladizinsky 1979; Cubero 1984). Recent classification of genus *Lens* has classified it into four gene pools (primary, secondary, tertiary and quaternary) and have changed sub-species status of *orientalis*, *odomensis*, and *tomentosus* from earlier classification (Ferguson et al. 2000) to species level. Primary gene pool has one cultivated species (*L. culinaris*) and remaining six wild species belong to four genes pool such as *L. orientalis* and *L. tomentosus* (primary); secondary gene pool has *L. lamotte and L. odemensis; tertiary gene pool comprises L. ervoides* and quaternary gene pool has *L. nigricans* (Wong et al. 2015). Wild species from primary and secondary gene pools are easily crossable with cultivated lentil, unlike with the wild species from the remaining two gene pools (Gupta and Sharma 2007; Singh et al. 2013b).

For climate-resilient lentil cultivars combined resistance to major biotic stresses and/or abiotic stresses will help to sustain lentil yield in variable climate. Without any doubt, wild crop relatives offer an opportunity to be utilized for untapped rich source of desirable genes such as resistance to biotic and abiotic stresses Table 4.4. Improved root traits for better tolerance of stresses especially water and availability of nutrient for healthy crop growth will be key traits to target and wild lentils could be most appropriate ones to be explored for such traits. The research so far has shown that wild lentil species possess huge variation for various agro-morphological traits along with biotic and abiotic stresses, which is quite understandable as these untapped sources are preserved in nature and have not lost these genes during the process of domestication which emphasized more on selection for few important genes for high yields. Traditional and molecular approaches for gene pyramiding might be able to bring such traits in common genetic backgrounds to have climate-resilient lentil cultivars with a broad genetic base. Among few attempts to evaluate wild lentil accessions for useful climate-smart agro-morphological traits, *L. orientalis* accessions has some desirable traits such as early flowering and maturity (Hamdi et al. 1991; Gupta and Sharma 2007), higher leaves/plant, peduncles/plant, pods/plant, seeds/plant and leaf area (Ferguson and Robertson 1999) when cultivated and few more wild species were evaluated for various morphological, phenological and yield related characters. Another study revealed useful traits after evaluation of 405 wild lentil accessions from 4 gene pools which were collected from ICARDA gene bank (Singh et al. 2014). Among various abiotic stresses, cold stress could be detrimental and can limit lentil production due to injury to vegetative tissues with further damage to floral parts leading to flower and pod abortion (Eujayl et al. 1999 and Singh et al. 2018). *L. orientalis* accessions originating from high elevation areas *revealed* greater tolerance to cold stress than in the cultivated lentil (Hamdi et al. 1996). Finding diseases resistance sources is one of the key to develop disease-resistant cultivars which will be able to withstand new disease pressures, as these sources of resistance could be new and can provide long-term resistance to lentil cultivars. Among many diseases, fusarium wild is quite devastating and few researchers (Bayaa et al. 1995; Nasir 1998), found seeding and/or adult stage vascular wilt resistance from *L. orientalis*, *L. nigricans*, and *L. ervoides* accessions. ICARDA researchers found a good level of resistance from *L. orientalis* and *L. ervoides* (year 2000–2007) for Fusarium wilt and further evaluated them for agronomic traits at various locations to improve breeding strategies to develop better and well-adapted breeding lines. The first report of Ascochyta blight-resistant accessions from wild sources was from Bayaa et al. (1994) who found a fairly large number of accessions to be resistant from *L. orientalis*, *L. odemensis*, *L. nigricans* and *L. ervoides*. Ahmad et al. (1997a) identified sources of resistance to the major diseases of lentil, viz., rust, vascular wilt and Ascochyta blight in the wild relatives of lentil.

Hybridization efforts to transfer these useful CS traits from wilds to cultivated background to generate a wide spectrum of variability has not seen groundbreaking efforts. Among few attempts of crossing cultivated x wild lentil species from primary and secondary gene pools have most successful reports which include to greater extent *L. orientalis* accessions followed by *L. odomensis* to be readily crossable with cultivated lentil (Ladizinsky 1979; Ladizinsky et al. 1984; Muehlbauer et al. 1989; Vandenberg and Slinkard 1989; Ladizinsky and Abbo 1993; Hamdi and

Table 4.4 Useful wild germplasm for introgression of CS traits in cultivated lentil

Trait	Wild resource	References
Anthracnose resistance	*L. ervoides*, *L. lamottei*, *L. nigricans*	Tullu et al. (2006)
Ascochyta blight resistance	*L. ervoides*, *L. orientalis*, *L. odemensis* *L. nigricans*, *L. montbretii*	Bayaa et al. (1994) Tullu et al. (2006, 2010) Dadu et al. (2016, 2017)
Fusarium wilt resistance	*L. orientalis*, *L. ervoides*	Bayaa et al. (1995), Gupta and Sharma (2007)
Powdery mildew resistance	*L. orientalis*, *L. nigricans*	Gupta and Sharma (2007)
Rust resistance	*L. orientalis*, *L. ervoides*, *L. nigricans*, *L. odemensis*	Gupta and Sharma (2007)
Drought tolerance	*L. odemensis*, *L. ervoides*, *L. nigricans*	Hamdi and Erskine (1996), Gupta and Sharma (2007)
Cold tolerance	*L. orientalis*	Hamdi et al. (1996)
Yield attributes	*L. orientalis*	Gupta and Sharma (2007)
Resistance to orobanche	*L. ervoides*, *L. odemensis*, *L. orientalis*	Ferna' Ndez-Aparicio et al. (2009)
Resistance to sitona weevils	*L. odemensis*, *L. ervoides*, *L. nigricans*, *L. orientalis*	El-Bouhssini et al. (2008)

Erskine 1994; Fratini et al. 2004; Gupta and Sharma 2007; Kumari et al. 2018). Wide hybridization between cultivated and wild lentils does not always lead to successful crosses due to species, and genotypic level differences within species. Genetically distant remaining species from secondary, tertiary, and quaternary gene pools are not easily crossable with cultivated lentil and harbor genes for many climate-resilient traits. The fertilization barriers exist due to asynchronous flowering and mainly due to hybrid embryo abortion (Abbo and Ladizinsky 1991, 1994; Ahmad et al. 1995; Gulati et al. 2001; Gupta and Sharma 2005; Fratini and Ruiz 2006; Fiala 2006). Even some species of primary/secondary gene pool such as *L. tomentosis* (Ladizinsky 1999) has shown crossability barriers due to embryo abortion and hybrid fertility. To break these barriers, few remediations are researched and have had successful results for the inclusion of genotypes of these wild species into cultivated lentil gene base. Some examples include application of GA_3 growth hormone and embryo/ovule rescue techniques and understanding similarity of species for pollen and pistil morphology to overcome postfertilization barrier (Cohen et al. 1984; Ladizinsky et al. 1988; Ladizinsky 1993; Ahmad et al. 1995; Gupta and Sharma 2005; Fratini et al. 2006). Dadu et al. (2016) reported the success of approximately 100 crosses with 100 ppm GA_3 application immediately after pollination from a cross between AB resistant accession from *L. orientalis* and cultivated lentil.

The crossability potential and techniques to overcome some existing pre–postfertilization barriers suggest that these wild accessions with CS traits can be exploited for breeding climate-resilient cultivars to sustain lentil production under climatic variability.

Successful introgression through conventional or modified techniques does not reflect much for breeding programs unless filial generations are advanced and evaluated at field level. There are a handful of reports which evaluated fixed interspecific lines for various CS traits.

Among few attempts of interspecific hybridization, Gupta and Sharma (2007) developed interspecific hybrids and segregating generations (F_2, BC_1) from cultivated and *L. orientalis* and *L. odemensis* crosses and observed greater genetic variability with numerous transgressive segregants for various agro-morphological traits. Field evaluation of 76 advanced breeding lines (Gupta and Sharma 2007) and 20 intraspecific fixed lines for various agro-morphological traits revealed superiority of few lines for grain yield and related traits (Kumari et al. 2018). Anthracnose resistance genes identified from *L. ervoides* (Tullu et al. 2006) were introgressed into cultivated lentils using embryo rescue technique (Fiala et al. 2009). $F_{7:8}$ recombinant inbred lines exhibited resistance and validated successful introgression of anthracnose resistance genes from *L. ervoides* (Fiala et al. 2009).

Singh et al. (2013b) successfully crossed cultivated lentils with accessions from various gene pools (*L. orientalis*, *odemensis*, *lamottei*, and *ervoides*) and studied F_2 generations for yield and related traits indicating transgressive segregants with a potential for their inclusion in CS breeding program. Some progress has been made in introgression of alien genes for resistance to Ascochyta blight, anthracnose and cold in cultivated lentil (Hamdi et al. 1996; Ye et al. 2002; Fiala 2006; Dadu et al. 2017, 2018). In Canada, anthracnose resistance was transferred between different gene pools from *L. ervoides* to cultivated lentil and 150 recombinant inbred lines were developed. The same technique can be used to develop hybrids between cultivated lentil and *L. lamottei* (Fiala 2006). Gorim and Vandenberg (2017a) studies root and shoot traits of wild and cultivated lentils for drought tolerance and revealed their genetic diversity for drought tolerance. Segregation generations (F_3, F_4, and F_5) from two cultivated lentil and *L. orientalis* and *L. ervoides* crosses revealed substantial variation for most of the agronomic traits, whereas, F_5 recombinant inbred lines of one cross had resistance to wilt (Singh et al. 2017b).

4.4 Classical Mapping and Traditional Breeding for CS Traits

For the association of markers with different traits of interest, we need to develop biparental or multiparental mapping populations for classical mapping. The biparental mapping populations may be F_2, backcross, double haploid (DH), and recombinant inbred lines (RIL). In lentil, some efforts have been made by various labs around (USA, Australia, India, and Morocco) in the development of biparental mapping populations for desired traits and are used in marker trait association studies

Table 4.5 Recombinant inbred lines mapping populations developed for different traits

Trait	Cross	Population size	Organization
Drought	ILL7946 × ILL7979	174	ICARDA
Cold	ILL4605 × ILL10657	153	ICARDA
Earliness	ILL7115 × ILL8009	150	ICARDA
Rust	ILL5888 × ILL6002	152	ICARDA
Fusarium wilt	ILL213 × ILL5883, Precoz × Idlib 2	150	ICARDA
Zinc Content	ILL5722 × ILL9888	177	ICARDA
	ILL9888 × ILL5480	149	ICARDA
Iron content	ILL9932 × ILL9951	193	ICARDA
Early growth vigor	DPL15 × ILL7663	160	IIPR, India
Root traits	IPL98/193 × EC208362	160	IIPR, India
Earliness	Precoz × L4603	160	IIPR, India
Earliness	ILL10829 × ILWL30	180	NBPGR, India
Pod number and earliness	ILL8006 × ILWL62	185	NBPGR, India

Source Adapted from Kumar et al. (2015); *ICARDA* International Center for Agricultural Research in the Dry Areas, Morocco; *NBPGR* National Bureau of Plant Genetic Resources, New Delhi, India

(Table 4.5). RIL populations were developed from the crosses developed between contrasting parents through single seed descent (SSD) method. The Indian Institute of Pulses Research (IIPR) has recently developed a RIL population from a cross between ILL6002 and ILL7663 to identify and map early growth vigor genes. Further, the identification of markers linked to the genes or quantitative trait loci (QTLs) governing these traits will help in the development of genotype having high biomass at an early stage. Furthermore, the National Bureau of Plant Genetic Resources has also developed wide cross populations against pod number and earliness and validated these traits under multilocation testing under varied ecological conditions (Singh et al. 2017b). The first genetic map in lentil (linkage analysis) began during 1984 (Zamir and Ladizinsky 1984), the first map comprising DNA based markers was developed by Havey and Muehlbauer (1989). Subsequent maps were developed by several other workers in lentils (Table 4.6).

The classical manipulations refer to the transfer of genes through conventional hybridization. Most of the cultivars developed worldwide are only through intraspecific hybridization followed by pure line selection (Kumar et al. 2004b). The genetic manipulation of lentil is primarily based on the exploitation of two broad categories of cultivated lentils, i.e., macrosperma and microsperma through hybridization of desirable genes from one another (Chahota et al. 1996, 1997; Lal et al. 2000). The hybridization criteria are to introgress elite traits from macrosperma (erect growth habit and tolerance against prevailing biotic and abiotic stresses) and from microsperma (higher number of branches/plant, higher number of seeds/plant and higher seed yield/plant), which are considered important CS traits to address one or the

Table 4.6 List of various maps developed in lentil populations

Cross	Type and size (in parenthesis) of population	Type of marker mapped	Mapped length (cM) and no. of loci (in parenthesis)	References
L. culinaris × *L. orientalis*	RIL (14–180)	Isozyme and four morphological markers	– (20)	Tahir and Muehlbauer (1994)
L. culinaris × *L. orientalis*	F_2	Isozymes	– (10)	Zamir and Ladizinsky (1984)
L. culinaris × *L. ervoides*	F_3 (107) F_3 (22–56)	Isozymes 258(18)	258 (18)	Tadmor et al. (1987)
L. culinaris × *L. orientalis*	F_2 (113)	RAPD, ISSR, AFLP, SSR, CAPS, SRAPS, and morphological markers	2234 (200)	Duran et al. (2004), Fratini et al. (2004), de la Puente et al. (2013)
ILL5588 × L692-16-1 (s)	RIL (86)	SSR, AFLP	751 (283)	Hamwieh et al. (2005)
ILL5588 × ILL7537	F_2 (150)	RAPD, ISSR, and RGA	784 (114)	Rubeena et al. (2003a)
Eston × PI320937	RIL (94)	AFLP, RAPD, and SSR	1868 (207)	Tullu et al. (2008)
Precoz × WA8649041	RIL (94)	AFLP, ISSR, RAPD, and morphological markers	1396 (166)	Tanyolac et al. (2010)
ILL6002 × ILL5888	RIL (206)	SSR, RAPD, SRAP, and morphological markers	1565 (139)	Saha et al. (2013)
ILL5722 × ILL5588	RIL (94)	RAPD, ISSR, ITAP, and SSR	1392 (211)	Gupta et al. (2012a)
L830 × ILWL77	F_2 (114)	SSR, ISSR, and RAPD	3843 (199)	Gupta et al. (2012b)
CDC Robin × 964a-46	RIL (139)	SNP, SSR, and seed color genes	697 (561)	Sharpe et al. (2013a), Fedoruk et al. (2013)
Cassab × ILL2024	RIL (126)	SSR and SNP	1178 (318)	Kaur et al. (2014)
PI320937 × Eston	RIL (96)	AFLP, SSR, and SNP	840 (194)	Sever et al. (2014)
Precoz × WA8649041	RIL (101)	SNP	540 (519)	Temel et al. (2014)
ILL8006 × CDC Milestone	–	AFLP, SSR, and SNP	497 (149)	Aldemir et al. (2014)

(continued)

Table 4.6 (continued)

Cross	Type and size (in parenthesis) of population	Type of marker mapped	Mapped length (cM) and no. of loci (in parenthesis)	References
Precoz × L830	RIL (126)	SSR	1184 (219)	Verma et al. (2015)
Indianhead × Northfield; Indianhead × Digger; Northfield × Digger	RILs (117, 112, 114)	SNP	2429.6 (689)	Sudheesh et al. (2016)
L01-827A (*L. ervoides*) × IG 72815 (*L. ervoides*)	RIL (94)	SNP	740.9 (543)	Bhadauria et al. (2017)
ILL8006 × CDC Milestone	RIL (118)	SNP	497.1 (4177)	Aldemir et al. (2017)

Source Adapted from Kumar et al. (2015); Markers: *AFLP* Amplified fragment length polymorphism, *RAPD* Random amplified polymorphic DNA, *ISSR* Inter simple sequence repeat, *SSR* Simple sequence repeat, *CAPS* Cleaved amplified polymorphic sequences, *SRAPS* Sequence-related amplified polymorphism, *RGA* Resistance gene analog, *ITAP* Intron targeted amplified polymorphism, *SNP* Single nucleotide polymorphism

other stress. In lentil genetic improvement program, much has been reported about the creation of large amount of variation following hybridization of the microsperma and macrosperma lentils primarily for higher yields. Chahota et al. (2007) reported transgressive segregants for seed yield and other important agro-morphological traits from 77% of microsperma × macrosperma crosses. The prime advantage of such hybridization is that two classes are easily crossable, but this hybridization provides limited variability for further improvement (Muench et al. 1991; Ferguson 2000; Duran et al. 2004). In many crops, the wild relatives still possess useful variation and source of the desirable trait that no longer exist in these cultivated counterparts.

4.5 Diversity Analysis

Since the middle of twentieth century, breeders have been successful in improving the performance of the germplasm with the higher yield potential, adaptation to mechanization, and new agricultural practices (Perez-de-Castro et al. 2012). However, breeding cultivars for higher yield potential gradually prompted replacement of traits useful to future climates in the cultivated crop community (Grassini et al. 2013). Hence, continuous development of new CS cultivars that can withstand and perform against the environmental changes without compromising on the genetic gain is needed. However, the genetic gain within a progeny is always dependent on the amount of variation existing between the parents that are selected for hybridization (Roy et al. 2013). Therefore, an estimate of genetic diversity for a given trait is sought to allow selection of better parents from the existing plant genetic resources.

4.5.1 Phenotype-Based Diversity Analysis

Visually accessible morphological traits are used to estimate the phenotypic frequencies within and between the populations of lentil (Singh et al. 2014). Traits that were routinely phenotyped may be classified into three major categories such as qualitative, phenological, and yield related. Qualitative traits included growth habit, leaf pubescence, leaflet size, stem pigmentation, flower petal color, tendrils, pod indehiscence, cotyledon color, seed coat color, seed coat pattern, and seed shape. Traits such as time to emergence, days to flowering, days to 50% flowering, and days to maturity were recorded to understand the variation for phenology within the lentil germplasm. While yield-related traits such as plant height, number of branches/plant, number of flowers/peduncle, number of nodes/plant, number of pods/plant, number of seeds/plant, 100 seed weight, biomass/plant, and yield/plant were used to estimate the genetic divergence for yield potential of the lentil germplasm. The first noted assessment of genetic variability for lentil was made by Barulina (1930), who reported variations between accessions for various morphological characters. Since then, several authors made useful contributions to unravel the genetic diversity through agro-morphological traits (Hoffman et al. 1988; Lázaro et al. 2001; Roy et al. 2013; Choudhary et al. 2017). Variations are evident for almost all the morphological traits within the species and among different species of lentil and thus allow for an effective selection. Diversity assessments of 405 accessions collected from 7 lentil species revealed remarkable variations for traits such as leaf pubescence, leaflet size, tendril length, and seed coat pattern both within and between the species (Singh et al. 2014).

Lentil germplasm also exhibited quite a variation for various phenological traits. Considerable variation was demonstrated within a global collection of 1370 accessions for days to flower and maturity (Erskine et al. 1989). It was also observed that the accessions varied with the changes in temperature and photoperiod for the time taken to flower (Erskine et al. 1990; Erskine et al. 1994; Bicer and Sakar 2008). Understandably, maximum number of studies were undertaken to decipher the genetic divergence for yield and yield contributing traits (Erskine and Choudhary 1986; Tullu et al. 2001; Zaccardelli et al. 2012; GAAD et al. 2018). Significant variation has been reported for seed yield and traits such as number of pods/plant, number of seeds/plant, and biomass/plant that are said to have a positive relationship with yield. Alternatively, significant and positive correlations between seed yield and traits including biological yield/plant, plant height, number of pods/plant, and number of seeds/pod have been reported (Bicer and Sakar, 2008; Zaccardelli et al. 2012). This implies that a greater potential still exists within lentil germplasm to mine and select for yield and yield contributing traits.

Lentil is confounded with several production constraints including biotic and abiotic stresses. Diseases that cause substantial yield lose. Interestingly, several sources of resistance to each disease have been detected within the cultivated, landraces, and wild species of lentil as reviewed by Chen et al. (2009). Similarly, significant differences within the germplasm were reported for boron toxicity, a problem in arid areas of West Asia (Yau and Erskine 2000; Hobson et al. 2006). In addition, the evidence of ample genetic diversity within the germplasm for various minerals, mainly, iron (Fe) and Zinc (Zn) concentrations demonstrated a likely strategy to address the problem of micronutrient deficiencies usually associated with cultivars of lentil (Karaköy et al. 2012; Kumar et al. 2014a, 2018c; Shrestha et al. 2018).

4.5.2 Genotype-Based Diversity Analysis, Molecular Markers Applied

Although morphological characterization made useful contributions to the genetic diversity of lentil, these traits are often influenced by environment and display phenotypic plasticity (Bicer and Sakar 2008; Mondini et al. 2009; Govindaraj et al. 2015). Alternatively, biochemical and molecular markers offer numerous advantages over traditional morphological traits (Govindaraj et al. 2015). Biochemical markers involve analysis of seed storage proteins and isozymes (allele variants of an enzyme), and provide genotypic frequencies within and among the populations at functional gene level. Polymorphisms within number and molecular weight of polypeptides revealed through SDS-PAGE of seed storage proteins showed evidence for a greater genetic variation within the lentil germplasm (de la Rosa and Jouve 1992; Echeverrigaray et al. 1998; Piergiovanni and Taranto 2005; Zaccardelli et al. 2012). Additionally, proteomic technology using two-dimensional electrophoresis aided to analyze substantially higher number of proteins and demonstrated useful variations within lentil landraces of Italy (Scippa et al. 2008, 2010; Ialicicco et al. 2012). Isozyme and allozyme markers highlighted the differences within the functions of an enzyme between individuals and are routinely used to detect the differences within the lentil germplasm prior to the introduction of molecular markers (Zamir and Ladizinsky 1984; Hoffman et al. 1986; Erskine and Muehlbauer 1991; Ferguson et al. 1998b; Sultana and Ghafoor 2008).

The introduction and gradual evolution of molecular markers along with the shortcomings associated with morphological and biochemical markers observed the integration of various molecular markers to analyze and characterize the lentil germplasm. Molecular markers differentiate individuals by highlighting the differences within the genome caused due to either by an insertion/deletion/translocation/duplication/point mutation, etc. In addition, they are highly stable and detectable in all the plant tissues regardless of growth and development. Significant amount of variation has been reported within lentil germplasm by using various types of molecular markers such as restriction-hybridization-based restriction fragment length polymorphisms (RFLPs) (Havey and Muehlbauer 1989; Muench et al. 1991) PCR-based random amplified polymorphic DNAs (RAPDs) (Abo-Elwafa et al. 1995; Ford et al. 1997; Ferguson et al. 1998a; Sonnante and Pignone 2001; Sultana and Ghafoor 2008), and amplified fragment polymorphisms (AFLPs) (Sharma et al. 1996; Alghamdi et al. 2013; Idrissi et al. 2015b), microsatellite variable number tandem repeats (VNTRs) (Závodná et al. 2000) and inter-simple sequence repeats (ISSRs) (Sonnante and Pignone 2001; de la Vega and Durán 2004; Sonnante and Pignone 2007; Scippa et al. 2008; Fikiru et al. 2007; Toklu et al. 2009; El-Nahas et al. 2011; Seyedimoradi and Talebi 2014; Datta et al. 2016), genomic SSRs (Jin et al. 2008; Hamwieh et al. 2009; Babayeva et al. 2009; Zaccardelli et al. 2012; Kumar et al. 2014b; Verma et al. 2014; Idrissi et al. 2015a; Roy et al. 2015; Koul et al. 2017) and expressed sequence tag (EST)-derived simple sequence repeats (SSRs) (Dikshit et al. 2015a; Kumar et al. 2018a). Utilizing comparative genomics, cross-genera SSR markers derived from ESTs sequences of *Medicago truncatula*, *Pisum sativum* and *Triolium pratense* have been used to characterize lentil germplasm (Reddy et al. 2010; Alo et al. 2011). More recently, the highly abundant genome-wide and gene-based single-nucleotide polymorphisms (SNPs) have been used to assess the genetic diversity of lentil (Lombardi et al. 2014; Basheer-Salimia et al. 2015). Additionally, an exome capture array targeting the protein-coding genes was developed and applied in lentil to evaluate the variation within and among the lentil species (Ogutcen et al. 2018).

4.5.3 Relationship with Wild Relatives

An understanding of the intra- and interspecies relationships in the genus and multiplicity of the taxa is needed for the improvement and climate-resilient lentil cultivars. This may be because all taxa are morphologically similar and differ only for a few (Galasso 2003). Thereafter, several studies attempted to revise the classification and thereby relationships among the species by using biochemical and molecular methods. These included isozymes (Hoffman et al. 1986; de la Rosa and Jouve 1992; Ferguson and Robertson 1996), SDS-PAGE (Ahmad and McNeil 1996; Ahmad et al. 1997b; Zimniak-Przybylska et al. 2001), chloroplast DNA (Muench et al. 1991; Mayer and Soltis 1994), RFLP (Havey and Muehlbauer 1989), RAPD (Abo-Elwafa et al. 1995; Sharma et al. 1995; Ahmad and McNeil 1996; de la Vega and Durán 2004), AFLP (Sharma et al. 1996), FISH karyotype (Galasso, 2003), ISSR (de la Vega and Durán 2004) and ITS (Mayer and Bagga 2002; Sonnante et al. 2003) and genomic and EST-SSRs (Alo et al. 2011; Dikshit et al. 2015b) and genome-wide SNPs (Wong et al. 2015). While the outcomes of all the studies did not agree with each other, the most agreed facts of all these studies has been that (i) *L. orientalis* is the progenitor of the cultivated lentil; (ii) *L. nigricans* is the distant relative as supported by the crossability experiments (Ladizinsky et al. 1984; Fiala et al. 2009); (iii) the relationships among the remaining taxa need reassessment. Recently, classification and four gene pool categories (Wong et al. 2015) were validated through an exome capture array method, which represents the coding fraction of the genome (Ogutcen et al. 2018). The results also supported that *Lens nigricans* as a distant relative to the cultivated species as it showed only a 70% alignment similarity with the exome of the cultivated species.

4.5.4 Relationship with Geographical Distribution

Lentil is one of the oldest domesticated crops (Ladizinsky 1979). The oldest remains of lentil found in Greece and Syria dated back to 11,000 BC and 8500–7500 BC, respectively (Erskine 1997). Ferguson et al. (1998a) mapped the highest genetic diversity for wild progenitor *L orientalis* within southeast Turkey and northwest Syria using the PCR-based markers such as RAPDs. Similarly, southern Syria, coastal border region between Syria and Turkey and west Turkey are suggested to be the centers for maximum variation and unique diversity for taxa *Lens odemensis*, *L ervoides*, and *L nigricans*, respectively.

Interestingly, lentil adapted well to the conditions in South Asia region and subsequently emerged as a major contributor to world's lentil production (Erskine et al. 1998). While lentil cultivation in countries like Canada and Australia has been relatively new but took over Indian subcontinent as major producers of lentil with the help of high-yielding cultivars supplemented by mechanization and advanced agricultural management practices. Genetic distinctness between the South Asian landraces and other region landraces were made evident through morphological, phenological, biochemical, and molecular markers. Based on the morphological variation, lentil landrace collection was divided into three major regional groups such as levantine group (Egypt, Jordon, Lebanon, and Syria), northern group (Greece, Iran, Turkey, and USSR), and Indian group, which included Indian subcontinent and Ethiopian collections (Erskine et al. 1989). However, there was a clear differentiation between Indian and Ethiopian collections at gene level as diagnosed by RAPD marker analysis (Ferguson et al. 1998b). Additionally, accessions from Afghanistan were clustered along with South Asian group and thus conclude that lentil was introduced into Indian subcontinent from West Asia through Afghanistan. A similar observation of germplasm relatedness between Afghanistan and South Asian was also made by Khazaei et al. (2016) at gene level using SNP markers. Nevertheless, the diversity within the South Asian group was predicted as low and is affected by limited introductions (Erskine et al. 1998; Lombardi et al. 2014).

While the landraces collected from the Mediterranean region, especially from countries Turkey and Greece demonstrated higher diversity and suggest the presence of substantial level of genetic variation within the germplasm (Lombardi et al. 2014). Several other authors also reported higher genetic diversity nature of Mediterranean region compared to Asia and USA (Erskine et al. 1989; Piergiovanni and Taranto 2003; Toklu et al. 2009). Alternatively, similarities were found among the collections from Mediterranean, North Africa, and Chile (Ferguson et al. 1998b; Lombardi et al. 2014; Khazaei et al. 2016). Northern temperate group was recently proposed based on the differences in agro-ecological regions around the world where lentil is grown (Khazaei et al. 2016). Assessment of the variation within the northern temperate region, especially of Canada, currently top producer of lentil, showed a narrow genetic variability among the breeding lines (Khazaei et al. 2016). A similar trend was observed within the Australian lentil germplasm and is attributable to the limited introductions and also selection pressure for higher yield and specific adaptations such as disease resistance (Ford et al. 1997; Lombardi et al. 2014).

4.5.5 *Extent of Genetic Diversity*

In the process of domestication, lentil has been understood to have lost approximately 40% of genetic diversity (Alo et al. 2011). Evidently, breeding programs around the world possess a limited diversity within the cultivated lentil (Ferguson et al. 1998b; Lombardi et al. 2014; Khazaei et al. 2016). Especially, the diversity of South Asian, Canadian, and Australian germplasm is low as estimated by several authors using different methods. An assessment of a historic collection of Indian lentil accessions including cultivars released since 1975, advanced breeding lines, ready for release and a collection of germplasm lines using 260 SSR markers could reveal a mean polymorphic information content of 0.30 (Kumar et al. 2018b). This again resulted in 48–74% of genetic similarity between the genotypes and thus indicated a narrow genetic base. Contrary to this, the germplasm within the Mediterranean region demonstrated higher genetic diversity. The landraces from Turkey and Greece within the Mediterranean region showed greater divergence to that of other region landraces including America, Africa, Northern Europe, and Middle-East at gene level (Lombardi et al. 2014). Similarly, two ancient landraces (Capracotta and Conca casale) collected from South Central Italy showed greater variation between themselves and commercial cultivars at morphological, protein and DNA level (Scippa et al. 2008). While genetic variation within the wild species of lentil was found to be high compared to that of cultivated species at morphological, quantitative, protein, and DNA level (Havey and Muehlbauer 1989; de la Rosa and Jouve 1992; Singh et al. 2014). These evidence suggest the presence of substantial variation within the cultivated and wild species that could be mined for widening the genetic base, particularly of South Asia, Australia, and Canada regions (Dikshit et al. 2015a).

Estimation of the extent of genetic diversity also depends on the method used for analyzing the diversity as significant differences were claimed between different methods for their ability to detect the polymorphism. Assessment of lentil diversity observed an evolution of type of method used from morphological characters to SNP markers and with each upgradation, the polymorphism detectability power increased. Morphological evaluation of 405 wild accessions revealed only a variation of 18.97% but 98.26% of genetic dissimilarity was estimated using quantitative traits (Singh et al. 2014). A comparison of SDS-PAGE and ISSR marker techniques revealed greater differences between the two methods as seed proteins showed only a low level of genetic diversity as compared to that of ISSR markers (El-Nahas et al. 2011). Likewise, ISSR markers revealed a higher degree of variation within a collection of Italian landraces compared to RAPDs (Sonnante and Pignone, 2001). Interestingly, genome-derived SSRs revealed a higher average number of alleles and genetic diversity compared to EST derived SSRs within a collection of accessions from three species of lentil (Dikshit et al. 2015b). Sequence-based, and genome-wide

SNP markers have become preferred alternatives to the other markers because of their abundance throughout the genome, highly polymorphic status, and suitability for use in high-throughput genotyping and automated analysis (Rafalski 2002).

4.6 Molecular Mapping of CS Genes and QTLs

During the past several years, tremendous progress has been made for the development of molecular markers in lentil. These markers associated and tightly linked to gene/QTL controlling a trait of interest can be used to introgress that gene/QTL in the background of improved lines through marker-assisted selection (MAS) and breeding. Genetic linkage map construction has become a necessary tool for molecular genetics and plant breeding programs (Tanyolac et al. 2010). The availability of large numbers of molecular markers and large mapping populations is the first step for the construction of genetic linkage maps. These maps have served many purposes in basic and applied research. They have become a key tool for physical mapping of genomes and high-density linkage maps are directly used in breeding researches (Tanksley et al. 1989; Hamwieh et al. 2005). In lentil, most genome maps have been created with anonymous and dominant RAPD, AFLP, and ISSR markers. Eujayl et al. (1998b) first identified markers suitable for the selection of a simply inherited resistance trait loci for Fusarium wilt resistance (Fw). Subsequently, Ford et al. (1999) identified RAPD markers that were close and flanking the major dominant locus for Ascochyta blight resistance in the ILL5588 accession (Ral1/AbR1$_{)}$. Chowdhury et al. (2001) also developed RAPD markers that flanked the recessive Ascochyta blight resistance locus in the cultivar Indianhead (*ral2*). Rubeena et al. (2006) identified markers that flank the codominant Ascochyta blight resistance loci in ILL7537. Tullu et al. (2003) identified markers linked to the anthracnose resistance locus in accession PI320937 (*LCt-2*) (Eujayl et al. 1997, 1998a; Rubeena et al. 2003b). Nevertheless, these first-generation maps served as foundations upon which more detailed maps have been and will be generated. To maximize polymorphism for map construction in lentil, interspecific hybrid populations have been used (Paterson et al. 1991; Eujayl et al. 1997; Durán et al. 2004). Such populations have also been used to map quantitative traits related to plant structure, growth habit, and yield in lentil (Fratini et al. 2007). Though the use of F_2 populations in the identification of QTLs has been done widely in lentil, their use in marker-trait analysis has led to identification of only major QTLs. Thus, several minor QTLs were overlooked in such populations and identification of environmental responsive QTLs was difficult. Because quantitative traits are influenced by both genetic and environmental effects, RILs or near-isogenic lines (NILs) are more suitable populations to accurately dissect their components.

In lentil, although molecular markers linked to desirable genes/QTLs have been reported, only those with tight association (<1.0 cM) and positive effect can be used in MAS. Among CS traits, other than biotic and abiotic stresses, agro-morphological traits also play an important role being directly or indirectly related to complex trait like yield. Duran et al. (2004) detected five QTLs each for the height of the first ramification and flowering time, three for plant height, seven for pod dehiscence, and one each for shoot number and seed diameter. Other studies identified several QTLs using biparental mapping populations that control flowering time in lentil (Tahir et al. 1994; Fratini et al. 2007; Tullu et al. 2008; Kahraman et al. 2015). One QTL each for the seed weight (*qSW*) and seed size (*qSS*) traits explaining 48.4% and 27.5% of phenotypic variance, respectively, were identified. These QTLs were located on an average at 5.48 cM from markers indicating close marker-trait association and hence can be useful in marker-assisted breeding for improving the seed size and weight (Verma et al. 2015). Morphological markers, viz., cotyledon (Yc), anthocyanin in stem (Gs), pod indehiscence (Pi), seed coat pattern (Scp), flower color (W), radiation frost tolerance locus (Rf), early flowering (Sn), and ground color of the seed (Gc) were mapped as qualitative markers (Eujayl et al. 1998a; Duran et al. 2004; Hamwieh et al. 2005; Tullu et al. 2008).

QTLs for biotic and abiotic stress tolerances will play a key role for tagging genes of interest to develop CS cultivars which can harbor more than one key traits. For Ascochyta blight disease, three QTLs each were detected for resistance at seedling and maturity stages (Gupta et al. 2012a). These accounted for 34 and 61% of the total assessed phenotypic variation and demonstrated that resistance at different stages is potentially conditioned by different genomic regions. The flanking markers identified may be useful for MAS and pyramiding of potentially resistance genes into elite genetic backgrounds that are resistant throughout the cropping period. Tullu et al. (2003) mapped for anthracnose disease resistance (Lct-2). Whereas, Taran et al. (2003) identified lines with combined resistance to Ascochyta blight resistance (AbR1 and ral1) and Anthracnose (OPO61250) using gene pyramiding approach for developing cultivars resistance to both Ascochyta blight and anthracnose in lentil. Recently, QTLs conferring resistance to Stemphylium blight and rust using RIL populations were identified (Saha et al. 2010a, b).

Among abiotic stresses, Kahraman et al. (2004b) identified the QTLs for winter survival and winter injury, using a RIL population of 106 lines and showed that tolerance to low temperature is a multigenic trait. QTLs related to frost response were also related to yield under winter-sown conditions as reported by Barrios et al. (2007). In continuation with this finding, Barrios et al. (2017) also found that QTLs with a major effect for winter hardiness and yield seem to be closely located within a single linkage group, and they are tracked by using some molecular markers. Super-SAGE (serial analysis of gene expression) genomic analysis was used to analyze the allele-specific differential expression of transcripts potentially involved in frost tolerance by bulk segregant analysis among 90 F_9 RILs derived from the Precoz × WA8649041 lentil cross (Barrios et al. 2010). QTLs (qHt ss and qHt_ps, with 12.1 and 9.23% phenotypic variance) and its molecular mapping for heat tolerance in lentil based on seedling survival and pod set per plant under hydroponic assay were reported by Singh et al. (2017c). These QTLs would provide further opportunities to dissect the candidate genes and the development of molecular markers for improving lentil with heat tolerance. Kaur et al. (2014) identified QTLs for boron tolerance in Cassab × ILL2024 mapping population. The flanking markers identified may be useful for MAS and pyramiding of potentially different resistance genes into elite backgrounds that are resistant throughout the cropping season. Recently, some considerable progress has been made in identifying QTLs related to drought tolerance in lentil. Genetic control and linkage of SSR markers for drought tolerance in lentil were first reported by Singh et al. (2016a, b). They identified a molecular marker associated with *Sdt* locus controlling seedling survival drought tolerance in lentil. These linked markers could be used in molecular breeding programs for introgression of seedling survival drought tolerance gene in high-yielding genotypes. A linkage map, fortified with 291 SSR markers and 75 QTLs for drought tolerance and yield-related traits were established in lentil using intraspecific RIL mapping population (L830 × Precoz) (Rana et al. 2016).

Subsequently, 18 QTLs for root and shoot traits (dry root biomass, number of lateral roots, RS ratio, and specific root length) associated with drought tolerance in a lentil recombinant inbred line population (RIL), ILL 6002 × ILL 5888, was identified by Idrissi et al. (2016) as a promising step toward a MAS approach. The authors also confirmed the stability of detected QTLs by performing the analysis on two consecutive seasons. They also identified a QTL-hotspot genomic region related to a number of root and shoot characteristics associated with drought tolerance such as dry root biomass, root surface area, lateral root number, dry shoot biomass, and shoot length was identified. Results from various studies could be used for marker-assisted selection in lentil breeding programs targeting CS traits for further genetic enhancement of this crop species (Tables 4.5 and 4.6). Further, the application of the next-generation sequencing (NGS) and genotyping by sequencing (GBS) technologies have facilitated speeding up the lentil genome or transcriptome sequencing projects and large discovery of genome-wide SNP markers for genetic and association mapping.

4.7 Marker-Assisted Breeding for CS Traits

The use of cost-effective DNA markers derived from the fine mapped position of the genes for important agronomic traits, biotic and abiotic stress tolerance regions, and MAS strategies will provide opportunities for breeders to develop high-yielding, climate smart, and better-quality genotypes. Marker-assisted backcross breeding (MABCB) will be more effective to integrate major genes or QTLs with large effect into widely grown genotypes.

4.7.1 Germplasm Characterization and Distinctiveness, Uniformity, and Stability (DUS) Test

Characterization of germplasm plays a vital role in identifying desirable genotypes to enhance yield and crop improvement. A Distinctiveness, Uniformity, and Stability (DUD) test is a descriptive assessment that establishes the identity of the new cultivar, by using morphological traits, as well as its uniformity and stability. The new cultivar is compared with the existing cultivars to establish its distinctness (Kwon et al. 2005). Remarkable variations among the traits for use in breeding and selection programs have been reported (Ramgiry et al. 1989; Tullu et al. 2001). Barulina (1930) first reported the detailed morphological descriptions of lentil landraces and species from Asia. Morphological markers like color of stem, flower and foliage color, plant habit, cotyledon and testa color, and testa pattern are important for testing hybridity and keeping genetic purity to be used in MAS. Different lentil cultivars were found to be distinct, uniform and stable for different seed, seedling, and flowering traits (Dixit et al. 2009; ul Hussan et al. 2018). Conventionally, morphological descriptors are routinely used for establishing the identity of cultivars. But these morphological descriptors have many drawbacks, such as influence of environment on trait expression, epistatic interactions, pleiotropic effects, etc. Recently, molecular marker techniques are used for varietal identification, differentiation between species, and in resolving many breeding problems in lentil (Lombardi et al. 2014). The most commonly used methods for DNA profiling and genotype characterization by determining their distance and uniformity are the RFLP, PCR-based techniques (RAPD, AFLP, and SSR). They are used selectively depending on the crop species and genetic constitution of the genotype. Several types of molecular markers including RAPD, RFLP, STS, SCAR, SNP, CAPS, AFLP, ISSR, and resistance gene analogue (RGA) markers have been identified and effectively used in lentil genotyping (Eujayl et al. 1998a; Rubeena et al. 2003a; Hamwieh et al. 2005; Saha et al. 2010a; Sharpe et al. 2013a). The transcriptome sequencing approach has generated EST databases, delivering large numbers of EST-derived SSR and SNP markers (Kaur et al. 2011; Sharpe et al. 2013b). Diverse promising interspecific and intraspecific lentil genotypes have also been studied for useful genetic variability and genetic diversity using morphological and molecular markers (Kumari et al. 2018; Tsanakas et al. 2018). Genetic linkage maps are essential tools for genomic and genetic studies, especially in mapping phenotypic traits. Several genetic linkage maps of lentil have been constructed using a range of molecular marker systems and mapping populations (Eujayl et al. 1998a; Gupta et al. 2012b; Rubeena et al. 2003a), including SSR (Hamwieh et al. 2005; Phan et al. 2007) and SNP markers (Fedoruk et al. 2013; Kaur et al. 2014; Sharpe et al. 2013b; Rodda et al. 2017).

4.7.2 Scope of Marker-Assisted Breeding (MAB) and Marker-Assisted Backcrossing (MABC)

As conventional breeding system requires more number of breeding cycles to combine many target traits in a genotype. Molecular-assisted breeding programs have reported twice the rate of genetic gain over phenotypic selection for various traits such as yield, biotic and abiotic stress resistance and quality attributes (Oliveira et al. 2008). A high correlation must exist between the desirable gene and molecular markers for practicability and success of MAS and the markers must be stable, reproducible and easy to assay (Yu et al. 2004). MAS has been effectively used for detecting, tracking, retaining, combining, and pyramiding different desirable genes for biotic and abiotic stresses (O'Boyle et al. 2007). However, MAS has not been employed successfully in lentil breeding program due to the absence of tightly linked markers. Inspite of huge potential as described earlier in the chapter, various CS traits have been mapped and tagged on linkage map which potentially through fine mapping can be used in MAS for breeding climate-resilient cultivars. Expression QTL (eQTL) can be identified for desirable traits by using suitable genetic materials and global genome expression profiling. The markers linked to this eQTLs will have huge potential in MAS compared to the markers identified by traditional QTL analysis since eQTL affect the expression of the genes for the desirable traits (Ford et al. 2018).

Simultaneous expression of more than one genes in a cultivar to develop durable resistance against biotic and abiotic stresses in crops will require stacking of multiple genes from multiple parents also known as gene pyramiding (Shi et al. 2009). In this technique, genetic markers are employed to identify and select specific genes or combine multiple resistance genes (Brahm and Friedt 2000; Richardson et al. 2006). The concept of gene pyramiding was proposed by Nelson (1978) to develop crop cultivars with few to several different oligo genes for durable disease resistance. This technique has been named as multitrait introgression, since genes governing two or more traits are often introgressed into a single recurrent parent (Rana et al. 2019). Gene pyramiding involves different methods such as multiple parent crossing, backcrossing, and recurrent section (Ribaut et al. 2010). Gene pyramiding using molecular markers depends upon several factors such as the number of genes/QTLs, the number of parents containing the target genes/QTLs, the heritability of target genes/QTLs, marker-target gene associations, duration needed to complete the gene assembly, and relative cost. It is a realistic approach that can be exploited in lentil breeding programs for the development of genetic stocks and precise development of CS traits. The possible breeding schemes that can be used for gene pyramiding involving MAS and the required population size in each segregating population have been discussed in lentil (Gupta et al. 2010). Pyramiding genes for resistance to Ascochyta blight and anthracnose in lentil were done by Taran et al. (2003) and Sari et al. (2018). Marker-assisted gene pyramiding has been used in other cereals and legumes for combining multiple genes/QTLs controlling both qualitative and quantitative stress resistance (Concibido et al. 2004; Richardson et al. 2006; Shi et al.

2009; Li et al. 2010; Wang et al. 2007; Luo et al. 2016). To date, no information is available on pyramiding genes for resistance to abiotic stresses in lentil crops. There is a great opportunity to take advantage of gene pyramiding in lentil, to develop elite lines, combining traits from multiple parents, particularly for resistance to biotic and abiotic stresses. MABC using trait-linked markers may also be used to develop superior lines once a major gene or QTL is identified and validated in the donor, as it will facilitate retaining the whole genome of the recurrent parent. MABC is a good choice when phenotyping of a trait of interest is expensive or difficult, the heritability of desirable trait is low, the expression of trait is in late stages of plant development, or traits controlled by a recessive gene or multiple genes need to combine for one or more traits. In chickpea, root traits, drought tolerance score, canopy temperature differential, and seed size in chickpea are governed by many QTLs (Varshney et al. 2013). The same QTLs hold for yield and yield-contributing characters such as seed number and seed weight. These traits will get more attention in the final selection of genotypes for abiotic stress tolerance. Under such situations, Marker-assisted recurrent selection (MARS), which involves intercrossing among selected individuals in each cycle of selection, may be used to avoid the limitations of MABC. The initial cost of using markers in MABC would be more expensive compared to conventional breeding in the short term, however, time savings could lead to an accelerated cultivar release which could translate into much profits in the long term.

4.8 Map-Based Cloning of CS Genes

Ideally, the genes controlling a trait of interest are the perfect marker for MAS. However, this is often made difficult because cloning of a gene is labor intensive and time consuming. Alternatively, marker(s) that are tightly linked to and flanking a gene locus that conditions a sizable genetic variation for the trait may be selected for with the premise that the associated chromosomal region contains the functional gene(s). Often, genetically linked markers to traits of interest are identified by coarse mapping and these have limited use in MAS because of the distance and hence chance of recombination between the marker and actual gene locus. Therefore, genomic regions where the trait is mapped should be fine mapped at high resolution and be validated across genetic backgrounds to determine their utility in MAS. Also, physical characterization of genomic regions of interest will facilitate cloning of the gene to develop direct markers (candidate genes) and/or physically closer markers to the gene, increasing the reliability for MAS. The most useful marker system for MAS should be locus specific, highly reproducible and easy to discern. These include sequence tagged site (STS), sequence characterized amplified region (SCAR) or allele specific amplified primer (ASAP), specific polymorphic locus amplification test (SPLAT), and PCR-based RFLP markers. When locus-specific markers are not polymorphic among the parental lines used in the breeding programs, sequence discriminative methods are required. These include SNP, cleaved amplified polymorphic site (CAPS), and derived CAPS (dCAPS) markers. More recently, a cleaved amplified polymorphic sequences marker was developed to facilitate breeding and establishes a basis for map-based cloning of Ruv2 and breeding for rust resistance in cowpea and other legume crops (Wu et al. 2018).

In the last decade, few transcriptome sequencing works (Kaur et al. 2011; Verma et al. 2013) aid in the marker discovery and SNP-based linkage maps (Sharpe et al. 2013b, Temel et al. 2014). However, a comprehensive genome-wide physical map, and its integration with genetic maps possessing QTLs for important targeted traits and draft genome of lentil, is the need of the hour for facilitating cloning of candidate genes and enhancing molecular breeding programs. Most recently, a high-density consensus map was constructed using three different RIL populations based on DArT markers (Ates et al. 2018). The consensus map could provide insight into the lentil genome, also help to construct a physical map using a Bacterial Artificial Chromosome library and map-based cloning studies. To identify the genes responsible for the target QTL, fine mapping and map-based cloning strategies are necessary (Salvi and Tuberosa 2005).

4.9 Genome Libraries

Large-insert genomic DNA libraries are essential genomic resources for physical mapping, positional cloning, and genome sequencing of higher eukaryotes (Tanksley et al. 1995; Zhang et al. 1996). The BAC cloning system has become an invaluable tool in genomic studies because of its ability to stably maintain large DNA fragments and its ease of manipulation (Wang et al. 1995; Zhang et al. 1996). BAC libraries are an important resource for the development of molecular markers that can be used for MAS for desirable agronomic traits. The development of SSR markers from BAC-end sequences is very cost-effective (Temnykh et al. 2001) and offers genome-wide coverage as all repeat types are systematically sampled in the randomly selected BACs (Cho et al. 2004). Since the development of the BAC vector (O'Connor et al. 1989), many BAC libraries have been developed for the major crop species, such as wheat, rice, corn, and soybean. In recent years, however, BAC libraries have also been developed for several pulse crops including mungbean (*Vigna radiata* L.), cowpea (*V. unguiculata* L.), lupin (*Lupinus angustifolius* L.), chickpea (*Cicer arietinum* L.), pigeonpea (*Cajanus cajan* L.), field pea (*Pisum sativum* L.), lima bean (*Phaseolus lunatus* L.), and common bean (*P. vulgaris* L.).

Integrated physical, genetic and genome map should provide a foundation for cloning and isolation of QTLs/genes for molecular dissection of traits as well as markers for molecular breeding for lentil improvement. A physical map of chickpea was developed for the reference chickpea genotype (ICC 4958) using BAC libraries targeting 71,094 clones (~12 × coverage). Comprehensive analysis of markers in abiotic and biotic stress tolerance QTL regions led to identification of 654, 306, and 23 genes in drought tolerance 'QTL-hotspot' region, Ascochyta blight resistance QTL region and Fusarium wilt resistance QTL region, respectively (Varshney et al. 2017). In addition, several large-insert BAC and binary bacterial artificial chromosome

(BIBAC) based libraries were also constructed earlier for chickpea (Lichtenzveig et al. 2005; Zhang et al. 2010).

Most of the BAC applications in pulse crops to date are of structural genomics nature; however, the application of BACs in functional genomics analysis of pulses also has great potential. Since large-insert clones in BAC vectors are more likely to contain the necessary promoter, enhancer, and silencer combination, mimicking the natural expression of the gene of interest, the advantages of the BAC transgenic approach are significant compared to the conventional transgenic approach (Yang and Gong 2005). However, this has not been applied yet on lentil due to non-availability of BAC or YAC libraries. The need of the hour is to develop BAC/BIBAC or YAC libraries to facilitate map-based cloning of genes in lentil. Alternatively, the genome libraries developed in the closely related model legumes chickpeas and *Medicago*, will help lentil breeders to speedup the understanding of lentil genomes and assist map-based cloning of genes.

4.10 Genetic Transformations

Transgenic approach uses functional genes which are not available within the cross-able gene pool. Thus, cloned genes are important genomic resources for making genetic manipulation through transformation. Commonly, the particle bombardment and the *Agrobacterium tumefaciens* infection methods have been used to introduce genes with novel functions. With the explosion of sequence information available in the databases, transformation systems have also become useful tools to study gene function via RNA interference 'knockout,' T-DNA insertion or transforming a genotype lacking a particular gene. Thus, a robust, reproducible, and efficient transformation system combined with a protocol to regenerate complete fertile plants from transformed cells is essential to fully study the plant gene functions. To date, the transformation of lentil has been reported through *A. tumefaciens*-mediated gene transfer (Lurquin et al. 1998) and biolistic transformation including electroporation (Chowrira et al. 1996) and particle bombardment (Gulati et al. 2002; Mahmoudian et al. 2002). Warkentin and McHughen (1992) reported the susceptibility of lentil to *A. tumefaciens*. All explants showed transient b-glucuronidase (GUS) expression at the wound sites except cotyledonary nodes, which were subsequently transformed by Sarker et al. (2003). Oktem et al. (1999) reported the first transient and stable chimeric transgene expression on cotyledonary lentil nodes using particle bombardment. Gulati et al. (2002) reported regeneration of the first fertile transgenic lentil plants on MS medium with 4.4 μM benzyladenine (BA), 5.2 μM gibberellic acid (GA3), and chlorsulfuron (5 nM for 28 days and 2.5 nM for the rest of the culture period), followed by micrografting and transplantation in soil. The first successful work was reported by Barton et al. (1997), using pCGP1258 plasmid construct on four lentil genotypes. Khatib et al. (2007) have developed herbicide-resistant lentil through *A. tumefaciens* mediated transformation. This was achieved with the same plasmid construct pCGP1258, harboring the gene conferring resistance to the herbicide glufosinate ammonium that was transformed using *A. tumefaciens* strain AgL0. Akcay et al. (2009) reported the production of transgenic lentil plants via *Agrobacterium*-mediated transformation and the stable transmission of the *npt*II and *gus*A genes in the subsequent generations. However, these studies were mostly confined to establish transformation techniques rather than the introduction of genes into improved cultivars. Khatib et al. (2011) reported for the first time the introduction of the *DREB1A* gene into lentil for enhancing drought and salinity tolerance. The results showed that mRNA was accumulated and thus, the DREB1A gene was expressed in the transgenic plants.

Advanced molecular technology has enabled plant modifications at the genomic level. Several horizontal gene transfer approaches have addressed the issues related to challenges and limitations of genome boundary in transferring the alien gene of interest through vertical gene transfer methods. Techniques such as genetic transformation (*Agrobacterium*-mediated transformation and direct gene delivery system) have opened new pathways to transfer functional genes precisely from any organism into plant genome.

Trans-mitochondrial gene expression can be studied using reverse genetics when transformation strategy targets mitochondria instead of nucleus (Havey et al. 2002), which can target mitochondrial genes for transgenic crops. Kemble et al. (1988) put an effort to transform *Brassica napus* hybrid mitochondria through polyethylene glycol (PEG) or electroporation mediated protoplast fusion using recombinant vectors. Among other organelles, plastids with small genome size are used to construct suitable vectors by targeting their specific sequences for genetic transformation. Boynton et al. (1988) were the first to report the transformation of Chlamydomonas chloroplast. Since then there are many reports of transformation of new genes from chloroplast genomes via organogenesis in several plant species (Skarjinskaia, et al. 2003; Khan and Maliga 1999; Hou et al. 2003; Kumar et al. 2004a).

Plastid genetic engineering has seen success in crops of economic importance. Complete legume genome sequences will be essential for comparing intergenic spacer regions to develop transformation vectors for plastid genetic engineering as plastid genome information is not fully understood (Sabir et al. 2014). Fabaceae (legumes) in Papilionoids have certain level of variation for cell structural features and inverted repeat lacking clade (IRLC) offers opportunity to enhance understanding of genomic evolution mechanisms and its feasibility for genetic improvement (Sabir et al. 2014), which is mainly due to comprehensive knowledge of the genomes for vector construct followed by stable intergenic integration site selection in transplastomic crop legume species (Dufourmantel et al. 2004, 2006; Wei et al. 2011). Six new IRLC plastomes have complete sequences and lentil is among few which has most repetitive sequences, these findings highlight plastome evolution, transfer of functional genes over time, losses of introns indicative of new genomic rearrangements (Sabir et al. 2014).

To fast track gene discoveries plant metabolomics offers huge potential to identify novel genes relate to biosynthetic pathway mechanisms of plant-based natural products. Metabolomics aided with transcriptomics has paved the way to identify various genes functions and their characterization (Saito and Matsuda 2010). Among legumes, most of the studies have concentrated in model legumes only. The traits described below are important for climate-resilient crops and shows the potential of this technology to be implemented in lentil crop. A decrease in oxylipins in Medicago was due to the effect of rhizobial node factor (Nod) (Zhang et al. 2012). Survival of salt-tolerant Lotus species involved successive changes for metabolic adjustments of shoot components (Sanchez et al. 2011), whereas, large number of mitochondria associated metabolites were identified for flooding stress in soybean which suggests requirement of higher levels of metabolites (amino acids, NAD, and NADH along with depleted free ATPs) for respiration and glycolysis (Komatsu et al. 2011). Specific metabolite markers (threonate, asparagine/ornithine and alanine/homoserine) for stresses like drought and salinity were developed through metabolite phenotyping of four Mediterranean lentil genotypes under drought and salinity stress (Muscolo et al. 2015). Metabolomics has huge potential though various challenges including metabolite identification at a large scale, limits its application.

Gene silencing which limits the mRNA availability for translation and eventually reduces the protein amount is another powerful technology for desired trait development. Different RNA silencing strategies as tools are available for selectively knocking down of specific genes/functions. MicroRNAs (miRNAs) are involved in the plant development process as well as in various stress responses, affecting the gene expression at the posttranscriptional level (Zhang et al. 2006). Therefore, under stress, increased gene expression of tolerant genotypes can be correlated to changes in miRNAs, which makes them good candidates for enhancing crop stress tolerance through transgenic breeding. Drought tolerance related miRNAs are discovered for various crops, 11 of them are identified in cowpea (Barrera-Figueroa et al. 2011) and heat stress response related eight miRNAs are being identified in common bean (Jyothi et al. 2015). RNA silencing has evolved as a natural defense to protect plants against viruses. Virus-induced gene silencing (VIGS) is promising to suppress plant gene expression using virus vectors with host gene's target region (Baulcombe 2004; Britt and May, 2003), though not used extensively in legumes. Vertical and horizontal approaches including RNAi and VIGS can be explored to understand the molecular mechanisms of host resistance in lentil. Cisgenesis offers the opportunity to modify genetic constitution of host plant via gene present naturally in a crossable and sexually compatible donor plant. Many genes from crop wild relatives and distant landraces of various crops have been identified which code for abiotic and biotic stress tolerance and resistance, various agronomical and quality traits, and been introgressed into the desired genotypes of crops. Such genes are known as cisgenes to separate them from the transgenes (Sprink et al. 2016) and cisgenesis take care of undesirable issues of linkage drag (Podevin et al. 2012), and introgression of desired genes into the host genotypes without affecting their other desirable traits. Abiotic stress tolerance is controlled by many genes and is complex, therefore, one gene or QTL introgression will not be enough for the introduction of stress tolerant (Hartung and Schiemann, 2014). Cisgenesis still need to emerge and can off-set concerns of genetically modified crops and technology at least for those traits which are still present in distant relatives of the crops.

4.11 Role of Bioinformatics

4.11.1 Gene and Genome Databases

With the advent of molecular approaches for plant breeding, based on genetic markers and genes, a need emerged for comprehensive sequence databases that will enable the annotation of these genomic features into functional proteins or transcription regulators such as transcription factors, methylation sites, or ncRNAs. This need was particularly crucial for non-model crops such as lentil, which lack the genomic resources available for well-studied model organisms. One of the first publicly accessible sequence databases, emerged in the early 1990s with the development of the internet, is the American National Center for Biotechnology Information (NCBI) GenBank collection. Three decades after its development it is still considered the most comprehensive and updated database, thanks to the International Nucleotide Sequence Database Collaboration, along with the DNA DataBank of Japan and the European Nucleotide Archive of The European Bioinformatics Institute in the European Molecular Biology Laboratory (EMBL-EBI). The NCBI databases now hold hundreds of trillions of existing cDNA, RNA, DNA, and protein sequences from collections spanning all available phyla groups (Cochrane et al. 2016). Since its foundation, the GenBank collection offered web-based platform equipped with a suite of bioinformatics tools for querying of genes of interest and performing homology-based searches, most notably the BLAST suite of tools, to find and retrieve the closest available sequences and provide certain functional and taxonomic annotation of the results (Camacho et al. 2009). The era of next-generation-sequencing (NGS), which introduced massively parallel high-throughput sequencing in 2005 and led to an explosion of sequencing projects that were submitted to NCBI's databases, also introduced reduced accuracy in the annotation of the submitted sequences, which were mostly annotated using high-throughput computational methods (Bidartondo 2008; Schnoes et al. 2009). Despite its reduced annotation accuracy, NCBI's databases are still widely used for annotation of sequences from non-model species, thanks to their unmatched coverage of sequences and taxonomy groups.

In the early 2000s, as sequencing technologies evolved and became more accessible and affordable, a new type of databases was developed and deployed, ones that were dedicated to specific species or narrow taxonomic groups and covered the entire (or close to) gene repertoire. These databases, however, were initially developed for just a handful of model plant species, which benefitted from fully sequenced, annotated, and curated genomes, such as Arabidopsis, rice, poplar, corn and in the legume family, the wild *Lotus japonicus* and cultivated alfalfa and soy (Yon Rhee et al. 2003; Retzel et al. 2007; Yamazaki et al. 2008; Sjödin et al. 2009; Grant et al. 2010; Andorf et al. 2016; Mun et al. 2016). As it was for GenBank, utilizing these databases for nonmodel crop research was still useful, by means of comparative genomics, or using homology-based searches to annotate an unknown gene and infer its function based on its closest annotated relatives.

4.11.2 Comparative Genome Databases

The shortcoming of using species-specific databases for comparative genomics is that it relies on prior knowledge of the evolutionary relationship between the crop and model species to select the most suitable database. In addition, this approach requires multiple comparisons against different databases, each using a potentially different interface and producing results in a different format, making the entire procedure extremely complicated, cumbersome and labor intensive. To overcome this, 'themed' databases were developed, combining information from multiple genomes, often focusing on a taxonomic group of interest. These databases provide advanced bioinformatic tools for comparing gene sequences and functions between species, as well as genome browsers, genetic maps and known genetic variants, markers, and even QTLs. This allows for a more targeted approach for annotating and comparing unknown genes and markers across crop plants. Notable comparative genome databases include the Phytozome Plant Comparative Genomics portal (https://phytozome.jgi.doe.gov/pz/portal.html, USA Department of Energy's Joint Genome Institute), which currently encompasses genomes of 64 plant species (including 8 legume species) (Goodstein et al. 2012). Another example of plant-specific database is Plaza (https://bioinformatics.psb.ugent.be/plaza, Ghent University), which covers 55 species of dicots (including 7 legume species) and 29 monocots (Van Bel et al. 2017). The Gramene database (http://www.gramene.org/, Gramene project), a resource for plant and crop comparative genomics, is based on Ensembl technology with collaboration with EMBL-EBI and offers access to curated genomic data both via its web portal and through data mining and programmatic access tools (Tello-Ruiz et al. 2018). More relevant to lentil are the Cool-Season Food Legume Crop Database (https://www.coolseasonfoodlegume.org/, Washington State University), which provides comparative genomics and genetics tools for chickpea, pea, lentil, and faba bean, though it only includes the full genome of chickpea; and KnowPulse (http://knowpulse.usask.ca/portal/, University of Saskatchewan Pulse Crop Research Group) which currently hosts the only publicly available annotated draft genome of lentil (Sanderson et al. 2011).

4.11.3 Protein and Pathway Databases

Relying on nucleotide sequences alone for homology-based functional annotation of unknown genes is limited to well-conserved genes which were previously identified and characterized in closely related species. When these requirements are not met, a more general approach is needed, based on the conservation of the protein amino acid sequence, which generally diverges in a slower pace than the nucleotide sequence, due to selection pressure to preserve the protein's function.

In addition to its nucleotide collections, NCBI hosts a broad protein database, named RefSeq, with over 121 million annotated proteins from 84,276 species

(Release 90, September 17, 2018), which can be searched against a query sequence. The European-based Universal Protein Resource (https://www.uniprot.org/), a collaboration between EMBL-EBI, the Swiss Institute of Bioinformatics and the Protein Information Resource, offers a similar computationally-annotated protein database (TrEMBL), but in addition, a smaller manually curated and reviewed protein collection (Swiss-Prot), which can be used with high confidence for functional annotations (The UniProt Consortium 2008). A plant-specific protein annotation project in underway at UniProt, to identify protein families unique to plants, which so far includes 39,669 entries from 1,998 species of Viridiplantae.

When a whole-protein approach is still unable to identify a candidate homologous gene, it is possible to perform homology searches against databases of protein subdomains to identify at least some elements of the gene that can be annotated and associated with a known function. Such search is performed using a profile hidden Markov model (profile HMM) algorithm and the available databases include the Protein Family database (http://pfam.xfam.org/) and the all-inclusive InterPro (http://www.ebi.ac.uk/interpro/, EMBL-EBI) database, which integrates protein families, domains and functional sites from a diverse range of source databases.

Once a protein or its domains are annotated, its functional role in molecular pathways can be depicted from pathway databases such as the Gene Ontologies (http://www.geneontology.org/), EggNOG (http://eggnogdb.embl.de), the Kyoto Encyclopedia of Genes and Genomes (https://www.genome.jp/kegg/) and Reactome (https://reactome.org/) databases (GO Consortium 2013; Huerta-Cepas et al. 2016; Kanehisa et al. 2016; Fabregat et al. 2018). The Plant Reactome (http://plantreactome.gramene.org/, Gramene project) enables a focused pathway search within the plants kingdom (Naithani et al. 2017), however, given the generalized nature of the protein-based approach, and the relatively modest computational resources required compared to nucleotide-based homology searches, it might be useful not to restrict the search to a particular phyla.

4.11.4 Gene Expression Databases

The actual function of genes of interest cannot always be inferred based on their nucleotide and protein sequences and domains, especially if they share little similarity to known annotated genes. In these cases, it is helpful to observe the gene's expression profiles under different environmental and biotic conditions and relate it to well-described molecular pathways by clustering with other genes who share similar expression patterns and their role had been previously established. For this purpose, gene expression databases were developed to collate and combine expression information from multiple species, under multiple experimental design. As it is for genomic data, the NCBI's Gene Expression Omnibus (GEO) database (https://www.ncbi.nlm.nih.gov/gds) is leading in terms of sheer breadth of stored data, originating from high-throughput microarray and RNA-Sequencing experiments (Clough and Barrett 2016). Following closely behind is the Expression

Atlas (https://www.ebi.ac.uk/gxa/plant/experiments, EMBL-EBI), which allows focusing on plant species and offers expression sets of 'baseline' and 'differential' experiments (Papatheodorou et al. 2018). Additional plant-only gene expression databases include Plexdb (http://www.plexdb.org), which in addition to plant species includes expression profiles of common plant pathogens, but unfortunately it was last updated in 2011 and is now outdated (Dash et al. 2012); and PLANEX (http://planex.plantbioinformatics.org/), a server offering analysis of co-expressed genes across plant species, based on the GEO database (Yim et al. 2013). Other useful resources for species-specific gene expression are the aforementioned model species genomic portals such as the Arabidopsis Information Resource (https://www.arabidopsis.org/), the Maize Genetics and Genomics Database (https://www.maizegdb.org/), SoyBase (https://soybase.org/soyseq/), and others.

As is the case for genomic resources, the vast majority of expression datasets in all of these databases focus on several model species, while only a single experiment, containing just 10 lentil samples, was found in NCBI's GEO (accession GSE11374, Mustafa et al. 2009). The same challenges exist therefore, when attempting to use gene expression databases for annotation of lentil genes and they require reliance on less than ideal datasets of closely related model species such as alfalfa and soybean.

4.11.5 Integration of Different Data

The genomic databases detailed in the previous sections offer different data types and strategies to query it, but their overarching aim is similar: to annotate and characterize genomic features. The abundance of distributed databases which often compete, however, complicates the annotation efforts. Several web portals were developed to streamline this process, by bringing together multiple databases and using a common system to query them, identify genes and smoothly transition results from one analysis to another.

The Gramene project (http://www.gramene.org/) brings together genome sequences, gene expression data and pathway databases for a range of crop and model plant species. In addition to a suite of data accessing and querying tools, the portal provides a tool to predict the functional consequences of known and unknown variants uploaded by the user (Tello-Ruiz et al. 2018).

Another web portal, the Legume Information System (LIS; https://legumeinfo.org/, National Center for Genome Resources), integrates legume genomes, gene families, protein domains, gene expression data, QTL, and genetic maps; and phenotyping data as a one-stop shop for legume researchers. LIS advocates use of common data templates, formats, schemas, and interfaces to facilitate data acquisition and analysis across all users and data types (Dash et al. 2016). A continued collaboration effort toward building genomic resources and capacity for crop legumes, as being done by KnowPulse, LIS and to a lesser extent the Cool-Season Food Legume Crop Database, is vital to fill in the gap and equip legume and lentil researchers with tools for molecular-based breeding methods.

4.12 Conclusion

Lentil gene pools consist of many wild relatives offering resistance to abiotic and biotic stresses as well as other important agronomic traits. Further, continuous efforts have been made in the past in cultivated x wild lentil genotype hybridization and few successful examples are there in which promising efforts were made to transfer CS targeted traits into cultivated lentils. However, so far, conventional breeding approaches have helped to utilize the available genetic variability of target traits within cultivated genepool, resulting in the development of several cultivars of lentil with tolerance or resistance to biotic and abiotic stresses. Recently, the linkage maps have provided the basis for development and increase the availability of genetic markers for genome studies such as the construction of physical mapping and map-based gene cloning. Limited population size, low heritability, lack of lentil-specific candidate genes, and nonavailability of genome libraries (BAC/YAC) are the main limiting factors in lentil genomics and thus reducing the pace of the genome-aided cultivar development. The access to high-throughput phenotyping and genotyping, construction of high-density maps with desirable markers and sequencing technologies are expected to speedup cultivar development with improved CS traits.

References

Abbo S, Ladizinsky G (1991) Anatomical aspects of hybrid embryo abortion in the genus *Lens* L. Bot Gaz 152(3):316–320

Abbo S, Ladizinsky G (1994) Genetical aspects of hybrid embryo abortion in the genus *Lens* L. Heredity 72(2):193

Abi-Antoun M, Kiwan P, Erskine W (1990) Talia 2-a lentil cultivar for Lebanon. Lens 17:3–4

Abo-Elwafa A, Murai K, Shimada T (1995) Intra-and inter-specific variations in *Lens* revealed by RAPD markers. Theor Appl Genet 90:335–340

Ahmad M, Mcneil DL (1996) Comparison of crossability, RAPD, SDS-PAGE and morphological markers for revealing genetic relationships within and among *Lens* species. Theor Appl Genet 93:788–793

Ahmad MAGF, Fautrier AG, McNeil DL, Burritt DJ, Hill GD (1995) Attempts to overcome post-fertilization barrier in interspecific crosses of the genus *Lens*. Plant Breed 114(6):558–560

Ahmad M, Fautrier A, Burritt D, Mcneil D (1997a) Genetic diversity and relationships in *Lens* species and their F_1 interspecific hybrids as determined by SDS-PAGE. NZ J Crop Hort 25:99–108

Ahmad M, Mcneil DL, Sedcole JR (1997b) Phylogenetic relationships in *Lens* species and their interspecific hybrids as measured by morphological characters. Euphytica 94:101–111

Ahmed S, Beniwal S (1991) Ascochyta blight of lentil and its control in Ethiopia. Intl J Pest Manage 37:368–373

Ahmed FE, Hall AE, DeMason DA (1992) Heat injury during floral development in cowpea (*Vigna unguiculata*, Fabaceae). Am J Bot 9(7):784–791

Akcay UC, Mahmoudian M, Kamci H, Yucel M, Oktem H (2009) *Agrobacterium tumefaciens*-mediated genetic transformation of a recalcitrant grain legume, lentil (*Lens culinaris* Medik). Plant Cell Rep 28:407–417. https://doi.org/10.1007/s00299-008-0652-4

Aksoy EM (2008) Effect of drought and salt stresses on the gene expression levels of antioxidant enzymes in lentil (*Lens culinaris* M.) seedlings. Yüksek Lisans Tezi, Orta Doğu Teknik Üniversitesi Fen Bilimleri Enstitüsü, Ankara

Aldemir, S, Ateş D, Temel HY, Yağmur B, Alsaleh A, Kahriman A, Özkan H, Vandenberg A, Tanyolaç MB (2017) QTLs for iron concentration in seeds of the cultivated lentil (*Lens culinaris* Medic.) via genotyping by sequencing. Turk J Agric For 41(4):243–255

Aldemir SB, Sever T, Ates D, Yagmur B, Kaya HB, Temel HY (2014) QTL mapping of genes controlling Fe uptake in lentil (*Lens culinaris* L.) seed using recombinant inbred lines. In: Plant and animal geonme conference XXII, 3360 pp. https://pag.confex.com/pag/xxii/webprogram/Paper9689.html

Alghamdi SS, Khan AM, Ammar MH (2013) Phenological, nutritional and molecular diversity assessment among 35 introduced lentil (*Lens culinaris* Medik.) genotypes grown in Saudi Arabia. Intl J Mol Sci 15:277–295

Ali A, Johnson DL, Stushnoff C (1999) Screening lentil (*Lens culinaris*) for cold hardiness under controlled conditions. J Agric Sci 133(3):313–319

Allahmoradi P, Mansourifar C, Saiedi M, Jalali Honarmand S (2013) Effect of different water deficiency levels on some antioxidants at different growth stages of lentil (*Lens culinaris* L.). Adv Environ Biol 7(4):535–543

Alo F, Furman BJ, Akhunov E, Dvorak J, Gepts P (2011) Leveraging genomic resources of model species for the assessment of diversity and phylogeny in wild and domesticated lentil. J Hered 102:315–329

Al-Quraan NA, Al-Omari HA (2017) GABA accumulation and oxidative damage responses to salt, osmotic and H_2O_2 treatments in two lentil (*Lens culinaris* Medik) accessions. Plant Biosyst 151(1):148–157

Al-Quraan NA, Al-Sharbati M, Dababneh Y, Al-Olabi M (2014) Effect of temperature, salt and osmotic stresses on seed germination and chlorophyll contents in lentil (*Lens culinaris* Medik). Acta Hort 1054:47–54

Amin MR, Karim MA, Islam MR, Aktar S, Hossain MA (2016) Effect of flooding on growth and yield of mungbean genotypes. Bang J Agric Res 41(1):151–162

Andorf CM, Cannon EK, Portwood JL (2016) MaizeGDB update: new tools, data and interface for the maize model organism database. Nucl Acids Res 44:D1195–D1201. https://doi.org/10.1093/nar/gkv1007

Andrahennadi C (1994) Genetics and linkage of isozyme markers and resistance to seedborne Ascochyta infection in lentil. MSc thesis, Department of Crop Science and Plant Physiology, University of Saskatchewan, Canada

Arkin GF, Taylor HM (eds) (1981) Modifying the root environment to reduce crop stress. ASAE Monograph No. 4. American Society of Agricultural Engineers, St. Joseph, MI

Arumuganathan K, Earle ED (1991) Nuclear DNA content of some important plant species. Plant Mol Bio Report 9(3):208–218

Asghar MJ, Abbas G, Shah TM (2010) Study of genetic diversity in some local and exotic lentil (*Lens culinaris* Medik) genotypes. Pak J Bot 42:2681–2690

Ashraf M, Chishti SN (1993) Waterlogging tolerance of some accessions of lentil (*Lens culinaris* Medic.). Trop Agric 70(1):60–67

Ashraf M, Waheed A (1990) Screening of local/exotic accessions of lentil (*Lens culinaris* Medic.) for salt tolerance at two growth stages. Plant Soil 128(2):167–76

Ashraf M, Waheed A (1993) Responses of some local/exotic accessions of lentil (*Lens culinaris* Medic.) to salt stress. J Agron Crop Sci 170(2):103–112

Ashraf M, Bokhari H, Cristiti SN (1992) Variation in osmotic adjustment of lentil (*Lens culinaris* Medic) in response to drought. Acta Bot Neerlandica 41:51–62

Asraf M, Zafar ZU (1997) Effect of potassium deficiency on growth and some biochemical characteristics in two lines of lentil (*Lens culinaris* Medic.). Acta Physiol Plant 19(1):9–15

Aslam M, Maqbool MA, Zaman QU, Shahid M, Akhtar MA, Rana AS (2017) Comparison of different tolerance indices and PCA biplot analysis for assessment of salinity tolerance in lentil (*Lens culinaris*) genotypes. Intl J Agric Biol 19(3):1

Aswaf A, Blair M (2012) Quantitative trait loci for rooting pattern traits of common beans grown under drought stress versus non-stress conditions. Mol Breed 30:681–695. https://doi.org/10.1007/s11032-011-9654-y

Ates D, Aldemir S, Alsaleh A, Erdogmus S, Nemli S, Kahriman A, Ozkan H, Vandenberg A, Tanyolac B (2018) A consensus linkage map of lentil based on DArT markers from three RIL mapping populations. PLoS ONE 3(1):e0191375

Ayoub AT (1977) Salt tolerance of lentil (*Lens esculenta*). J Hort Sci 52(1):163–168

Azimzadeh S (2010) Effect of planting dates, seed rates and row spacing on grain yield and yield components of a lentil (*Lens culinaris*) genotype in northern khorasan dry land condition. J Res Crop Sci 3(9):121–135

Babayeva S, Akparov Z, Abbasov M, Mammadov A, Zaifizadeh M, Street K (2009) Diversity analysis of Central Asia and Caucasian lentil (*Lens culinaris* Medik.) germplasm using SSR fingerprinting. Genet Resour Crop Evol 56:293

Babayeva S, Akparov Z, Damania A, Izzatullayeva V, Aslanova G, Abbasov M (2014) Genetic diversity for drought tolerance in lentils from Central Asia and the Caucasus: CACLentil. Alb J Agric Sci 13(2):1–8

Baligar VC (1997) Nutrient use efficiency in acid soils: nutrient management and plant use efficiency. In: Monitz AC, Furlani AMC, Fageria NK, Rosolem CA, Cantarells H (eds) Plant-soil interactions at low ph: sustainable agriculture and forestry production. Brazilian Soil Science Society Compinas, Brazil, pp 75–93

Bandeoğlu E, Eyidoğan F, Yücel M, Öktem HA (2004) Antioxidant responses of shoots and roots of lentil to NaCl-salinity stress. J Plant Growth Regul 42(1):69–77

Barber SA (1995) Soil nutrient bioavailability: a mechanistic approach, 2nd edn. Wiley, New York, NY

Barghi SS, Mostafaii H, Peighami F, Zakaria RA, Nejhad RF (2013) Response of in vitro pollen germination and cell membrane thermostabilty of lentil genotypes to high temperature. Intl J Agric 3(1):13

Barrera-Figueroa BE, Gao L, Diop NN, Wu Z, Ehler JD, Roberts PA (2011) Identification and comparative analysis of drought-associated microRNAs i two cowpea genotypes. BMC Plant Biol 11:127. https://doi.org/10.1186/1471-2229-11-127

Barrios A, Kahraman A, Aparicio T, Rodríguez M, Mosquera P, García P, McPhee K, Pérez de la Vega M, Caminero C (2007) Preliminary identification of QTLS for winter hardiness, frost tolerance and other agronomic characters in lentil (*Lens culinaris* Medik.) for Castilla y León (SPAIN) region. In: Proceedings of the 6th European conference on grain legumes, Lisbon, pp 12–16

Barrios A, Martin-Sanz A, Ramos S, Rotter B, Horres R, Plotner A, Kretzdorn N, Constantino Saldaña C, Perez de la Vega M, Winter P (2010) Allele-specific differential expression of transcripts potentially involved in cold tolerance QTL in lentil revealed by bulked extremes SuperTag digital gene expression (BE-STDGE) profiling and qRT-PCR. In: Proceedings of the 5th international food legumes research conference and 7th European conference on grain legumes, Antalya, pp 26–30

Barrios A, Aparicio T, Rodríguez MJ, de la Vega MP, Caminero C (2016) Winter sowing of adapted lines as a potential yield increase strategy in lentil (*Lens culinaris* Medik.). Span J Agric Res 14(2):0702

Barrios A, Caminero C, García P, Krezdorn N, Hoffmeier K, Winter P, de la Vega MP (2017) Deep super-SAGE transcriptomic analysis of cold acclimation in lentil (*Lens culinaris* Medik.). BMC Plant Biol 17(1):111

Barton J, Klyne A, Tennakon D, Francis C, Hamblin J (1997) Development of a system for gene transfer to lentils. In: Proceedings of international food legume research conference III, Adelaide, SA

Barulina O (1930) Lentil of USSR and of other countries. Bull Appl Bot Genet Plant Breed 40:1–319

Basheer-Salimia R, Camilli B, Scacchi S, Noli E, Awad M (2015) Assessment of genetic diversity in lentils (*Lens culinaris* Medik.) based on SNPs. Genet Mol Res 14(2):5870–5878

Baulcombe (2004) DNA silencing in plants. Nature 431(7006):356–363

Bayaa B, Erskine W, Hamdi A (1994) Response of wild lentil to *Ascochyta fabae* f. sp. *lentis* from Syria. Genet Resour Crop Evol 41(2):61–65

Bayaa B, Erskine W, Hamdi A (1995) Evaluation of a wild lentil collection for resistance to vascular wilt. Genet Resour Crop Evol 42(3):231–235

Bejiga G, Degago Y (2000) Region 4: Sub-Sahara Africa. In: Knight R (ed) Linking research and marketing opportunities for pulses in the 21st century. Springer, Dordrecht, pp 99–105

Belko N, Zaman-Allah M, Cisse N, Diop ND, Zombre G. Ehler JD (2012). Low soil moisture threshold for transpiration decline under water deficit correlates with lower canopy conductance and higher transpiration efficiency in drought tolerant cowpea. Funct Plant Biol 39:306–322. https://doi.org/10.1071/fp11282

Bhadauria V, Ramsay L, Bett KE, Banniza S (2017) QTL mapping reveals genetic determinants of fungal disease resistance in the wild lentil species *Lens ervoides*. Sci Rep. https://doi.org/10.1038/s41598-017-03463-9

Bhandari K, Siddique KH, Turner NC, Kaur J, Singh S, Agrawal SK, Nayyar H (2016) Heat stress at reproductive stage disrupts leaf carbohydrate metabolism, impairs reproductive function, and severely reduces seed yield in lentil. J Crop Improv 30(2):118–151

Bicer BT, Sakar D (2008) Studies on variability of lentil genotypes in South eastern Anatolia of Turkey. Notulae Bot Hort Agrobot Cluj-Napoca 36:20–24

Bidartondo MI (2008) Preserving accuracy in GenBank. Science 319:1616. https://doi.org/10.1126/science.319.5870.1616a

Biju S, Fuentes S, Gupta D (2017) Silicon improves seed germination and alleviates drought stress in lentil crops by regulating osmolytes, hydrolytic enzymes and antioxidant defense system. Plant Physiol Biochem 119:250–264

Biju S, Fuentes S, Gupta D (2018) The use of infrared thermal imaging as a non-destructive screening tool for identifying drought-tolerant lentil genotypes. Plant Physiol Biochem 127:11–24

Bluemel M, Dally N, Jung C (2015) Flowering time regulation in crops—what did we learn from Arabidopsis? Curr Opin Biotechnol 32:121–129

Boote KJ, Allen LH, Prasad PVV, Baker JT, Gesch RW, Snyder AM (2005) Elevated temperature and CO_2 impacts on pollination, reproductive growth and yield of several globally important crops. J Agric Meteorol 60:469–474

Boullis A, Francis F, Verheggen FJJ (2015) Climate change and tritrophic interactions: will modifications to greenhouse gas emissions increase the vulnerability of herbivorous insects to natural enemies? Environ Entomol 44(2):277–286

Boynton JE, Gillham NW, Harris EH, Hosler JP, Johnson AM, Jones AR et al (1988) Chloroplast transformation in Chlamydomonas with high velocity microprojectiles. Science 240:1534–1538

Brahm L, Friedt W (2000) PCR-based markers facilitating marker assisted selection in sunflower for resistance to downy mildew. Crop Sci 40(3):676–682

Bretag T, Materne M (1999) Evaluation of 36 lentil lines for resistance to ascochyta blight and botrytis grey mould. In: Summary of field research in the Wimmera, pp 38–40

Britt AB, May G (2003) Re-engineering plant gene targeting. Trends Plant Sci (in press)

Buchwaldt L, Anderson K, Morrall R, Gossen B, Bernier C (2004) Identification of lentil germ plasm resistant to *Colletotrichum truncatum* and characterization of two pathogen races. Phytopathology 94:236–243

Buddenhagen IW, Richards RA (1988) Breeding cool season food legumes for improved performance in stress environments. In: Summerfield RJ (ed) World crops: cool season food legumes. Springer, Dordrecht, pp 81–95

Camacho C, Coulouris G, Avagyan V (2009) BLAST+: architecture and applications. BMC Bioinform 10:421. https://doi.org/10.1186/1471-2105-10-421

Chahota RK, Gupta VP, Sharma SK (1996) Combining ability and genetic architecture of macrosperma × microsperma derived crosses. J Hill Res 9:341–346

Chahota RK, Sharma SK, Gupta VP (1997) Heterotic and inbreeding effects in lentil crosses involving macrosperma and microsperma. Crop Improv 24:74–77

Chahota RK, Kishore N, Dhiman KC, Sharma TR, Sharma SK (2007) Predicting transgressive segregants in early generation using single seed descent method-derived micro-macrosperma gene pool of lentil. Euphytica 156:305–310

Chakherchaman SA, Mostafaei H, Imanparast L, Eivazian MR (2009) Evaluation of drought tolerance in lentil advanced genotypes in Ardabil region, Iran. J Food Agric Environ 7(3/4):283–288

Chakraborty U, Pradhan D (2010) Biochemical responses of lentil (*Lens culinaris* Medik.) to elevated temperature stress. J Pharm Biol Chem Sci 1(3):575–585

Chen C, Miller P, Muehlbauer F, Neill K, Wichman D, McPhee K (2006) Winter pea and lentil response to seeding date and micro-and macro-environments. Agron J 98(6):1655–1663

Chen W, Basandrai AK, Basandrai D, Banniza S, Bayaa B, Buchwaldt L, Davidson J, Larsen R, Rubiales D, Taylor PWJ (2009) Diseases and their management. In: Erskine W, Muehlbauer FJ, Sarker A, Sharma B (eds) The lentil: botany, production and uses. CABI, Wallingford, pp 269–270

Cho S, Chen W, Muehlbauer FJ (2004) Pathotype-specific genetic factors in chickpea (*Cicer arietinum* L.) for quantitative resistance to ascochyta blight. Theor Appl Genet 109(4):733–739

Choudhary A, Kumar S, Patil B (2013) Narrowing yield gaps through genetic improvement for fusarium wilt resistance in three pulse crops of the semi-arid tropics. SABRAO J Breed Genet 45:341–370

Choudhary R, Verma S, Panwar R, Chourasiya V, Pandey D (2017) Morphological characterization of lentil (*Lens culinaris* Medikus.) varieties based on six qualitative traits. J Pharmacogn Phytochem 6:1611–1615

Choudhury DR, Tarafdar S, Das M, Kundagrami S (2012) Screening lentil (*Lens culinaris* Medik.) germplasms for heat tolerance. Trends Biosci 5(2):143–146

Chowdhury MA, Andrahennadi CP, Slinkard AE, Vandenberg A (2001) RAPD and SCAR markers for resistance to acochyta blight in lentil. Euphytica 118(3):331–337

Chowrira GM, Akella V, Fuerst PE, Lurquin PF (1996) Transgenic grain legumes obtained by in planta electroporation-mediated gene transfer. Mol Biotechnol 5:85–96. https://doi.org/10.1007/bf02789058

Cicerali IN (2004) Effect of salt stress on antioxidant defense systems of sensitive and resistant cultivars of lentil (*Lens culinaris* M.). Doctoral dissertation, MSc thesis, Middle East Technical University, Ankara, Turkey

Clements JC, Haqqani AM, Siddique KH, French RJ (1997) Drought tolerance in lentil (*Lens culinaris*). Intl Food Legume Res Conf 3:22–26

Clough E, Barrett T (2016) The gene expression omnibus database. Methods Mol Biol 1418:93–110. https://doi.org/10.1007/978-1-4939-3578-9_5

Cochrane G, Karsch-Mizrachi I, Takagi T (2016) The international nucleotide sequence database collaboration. Nucl Acids Res 44:48–50. https://doi.org/10.1093/nar/gkv1323

Cohen D, Ladizinsky G, Ziv M, Muehlbauer FJ (1984) Rescue of interspecific Lens hybrids by means of embryo culture. Plant Cell Tissue Organ Cult 3(4): 343–347

Concibido VC, Diers BW, Arelli PR (2004) A decade of QTL mapping for cyst nematode resistance in soybean. Crop Sci 44(4):1121–1131

Covell S, Ellis RH, Roberts EH, Summerfield RJ (1986) The influence of temperature on seed germination rate in grain legumes: I. A comparison of chickpea, lentil, soyabean and cowpea at constant temperatures. J Exp Bot 37(5):705–715

Cromey MG, Mulholland RI, Russell AC, Jermyn WA (1987) *Ascochyta fabae* f. sp. *lentis* on lentil in New Zealand. NZ J Exp Agric 15:235–238

Croser JS, Clarke HJ, Siddique KH, Khan TN (2003) Low-temperature stress: implications for chickpea (*Cicer arietinum* L.) improvement. Crit Rev Plant Sci 22(2):185–219

Cubero JI (1984) Taxonomy, distribution and evolution of the lentil and its wild relatives. In: Genetic resources and their exploitation—chickpeas, faba beans and lentils. Springer, Dordrecht, pp 187–203

Dadu RHR, Ford R, Sambasivam P, Gupta D (2016) Screening of wild lentil germplasm to identify novel Ascochyta lentis resistance sources. Second international legume society conference, legumes for a sustainable world, pp 11–14 Toria, Portugal

Dadu RHR, Ford R, Sambasivam P, Gupta D (2017) A novel *Lens orientalis* resistance source to the recently evolved highly aggressive Australian *Ascochyta lentis* Isolates. Front Plant Sci 8:1038. https://doi.org/10.3389/fpls.2017.01038

Dai A (2011) Drought under global warming: a review. WIREs Clim Change 2:45–65. https://doi.org/10.1002/wcc.81

Dash S, Van Hemert J, Hong L, Wise RP, Dickerson JA (2012) PLEXdb: gene expression resources for plants and plant pathogens. Nucl Acids Res 40:D1194–D1201. https://doi.org/10.1093/nar/gkr938

Dash S, Campbell JD, Cannon EKS (2016) Legume information system (LegumeInfo.org): a key component of a set of federated data resources for the legume family. Nucl Acids Res 44:D1181–D1188. https://doi.org/10.1093/nar/gkv1159

Dash AP, De Kumar D, Mohanty S, Lenka D (2017) Screening of Lentil (*Lens culinaris* Medik.) genotypes and correlation analysis under PEG imposed water stress condition. Int J Bio-Res Stress Manage 8(4):539–547

Datta S, Prasoonpal G, Mayank K, Kumar S (2016) Genetic diversity analysis of lentil (*Lens culinaris* Medik) cultivars using inter simple sequence repeats markers. Mol Plant Breed 7(23):1–9

Davidson J, Smetham G, Russ MH (2016) Changes in aggressiveness of the Ascochyta lentis population in Southern Australia. Front Plant Sci 7:393

de la Puente R, Garcia P, Polanco C, Perez de la Vega M (2013) An improved intersubspecific genetic map in *Lens* including functional markers. Span J Agric Res 11:132–136

De La Rosa L, Jouve N (1992) Genetic variation for isozyme genes and proteins in Spanish primitive cultivars and wild subspecies of *Lens*. Euphytica 59:181–187

De La Vega MP, Durán Y (2004) Assessment of genetic variation and species relationships in a collection of "Lens" using RAPD and ISSR. Span J Agric Res 538–544

Delahunty A, Nuttall J, Nicolas M, Brand J (2015) Genotypic heat tolerance in lentil. In: Proceedings of the 17th ASA conference, 20–24 Sept 2015, Hobart, Australia, pp 20–24

Devi JM, Sinclair TR, Vadez V (2010) Genotypic variation in peanut for transpiration response to vapor pressure deficit. Crop Sci 50:191–196

Dhankher OP, Foyer CH (2018) Climate resilient crops for improving global food security and safety. Plant Cell Environ 41:877–884. https://doi.org/10.1111/pce

Dikshit HK, Singh A, Singh D (2015a) Genetic diversity in *Lens* species revealed by EST and genomic simple sequence repeat analysis. PLoS ONE 10:e0138101

Dikshit HK, Singh A, Singh D, Aski MS, Prakash P, Jain N, Meena S, Kumar S, Sarker A (2015a) Genetic diversity in *Lens* species revealed by EST and genomic simple sequence repeat analysis. PLoS ONE e0138101

Dixit G, Katiyar P, Singh B, Kumar S (2009) Lentil varieties in India. All India Coordinated Research Project on MULLaRP, IIPR, Kanpur, India

Dufourmantel N, Pelissier B, Garcon F, Peltier G, Ferullo JM, Tissot G (2004) Generation of fertile transplastomic soybean. Plant Mol Biol 55:479–489

Dufourmantel N, Tissot G, Garcon F, Pelissier B, Dubald M (2006) Stability of soybean recombinant plastome over six generations. Transgen Res 15:305–311

Duran Y, Fratini R, Garcia P, De la Vega MP (2004) An intersubspecific genetic map of Lens. Theor Appl Genet 108:1265–1273

Echeverrigaray S, Oliveira A, Carvalho M, Derbyshire E (1998) Evaluation of the relationship between lentil accessions using comparative electrophoresis of seed proteins [*Lens culinaris* Medik.]. J Genet Plant Breed 52(1):89–94

Elad Y, Pertot I (2014) Climate change impacts on plant pathogens and plant diseases. J Crop Improv 28(1):99–139

El-Bouhssini M, Sarker A, Erskine W, Joubi A (2008) First sources of resistance to Sitona weevil (*Sitona crinitus Herbst*) in wild Lens species. Genet Resour Crop Evol 55(1):1–4

El-Enany AE, AL-Anazi AD, Dief N, Wafa'a A (2013) Role of antioxidant enzymes in amelioration of water deficit and waterlogging stresses on *Vigna sinensis* plants. J Biol Earth Sci 3(1):144–153

Ellis RH, Barrett S (1994) Alternating temperatures and rate of seed germination in lentil. Ann Bot 74:519–524

El-Nahas A, El-Shazly H, Ahmed SM, Omran A (2011) Molecular and biochemical markers in some lentil (*Lens culinaris* Medik.) genotypes. Ann Agric Sci 56:105–112

Erskine W (1997) Lessons for breeders from land races of lentil. Euphytica 93:107–112

Erskine W, Choudhary M (1986) Variation between and within lentil landraces from Yemen Arab Republic. Euphytica 35:695–700

Erskine W, Muehlbauer F (1991) Allozyme and morphological variability, outcrossing rate and core collection formation in lentil germplasm. Theor Appl Genet 83:119–125

Erskine W, Saxena MC (1993) Problems and prospects of stress resistance breeding in lentil. In: Singh KB, Saxena MC (eds) Breeding for stress tolerance in cool-season food legumes. Wiley, New York, pp 51–62

Erskine W, Myveci K, Izgin N (1981) Screening a world lentil collection for cold tolerance. Lens Newsl 8:5–8

Erskine W, Adham Y, Holly L (1989) Geographic distribution of variation in quantitative traits in a world lentil collection. Euphytica 43:97–103

Erskine W, Ellis R, Summerfield R, Roberts E, Hussain A (1990) Characterization of responses to temperature and photoperiod for time to flowering in a world lentil collection. Theor Appl Genet 80:193–199

Erskine W, Hussain A, Tahir M, Bahksh A, Ellis RH, Summerfield RJ, Roberts EH (1994) Field evaluation of a model of photothermal flowering responses in a world lentil collection. Theor Appl Genet 88(3–4):423–428

Erskine W, Bayaa B, Saxena M (1996) Registration of ILL 5588 lentil germplasm resistant to vascular wilt and ascochyta blight. Crop Sci 36

Erskine W, Chandra S, Chaudhry M, Malik IA, Sarker A, Sharma B, Tufail M, Tyagi MC (1998) A bottleneck in lentil: widening its genetic base in South Asia. Euphytica 101(2):207–211

Erskine W, Sarker A, Kumari S (2016) Lentil breeding. In: Wrigley CW, Corke H, Seetharaman K, Faubion JM (eds) Encyclopedia of food grains, 2nd edn. Elsevier, Oxford, pp 317–324

Eujayl I, Baum M, Erskine W, Pehu E, Muehlbauer FJ (1997) The use of RAPD markers for lentil genetic mapping and the evaluation of distorted F2 segregation. Euphytica 96(3):405–412

Eujayl I, Baum M, Powell W, Erskine W, Pehu E (1998a) A genetic linkage map of lentil (*Lens* sp.) based on RAPD and AFLP markers using recombinant inbred lines. Theor Appl Genet 97(1/2):83–89

Eujayl I, Erskine W, Bayaa B, Baum M, Pehu E (1998b) Fusarium vascular wilt in lentil: inheritance and identification of DNA markers for resistance. Plant Breed 117:497–499. https://doi.org/10.1111/j.1439-0523.1998.tb01982.x

Eujayl I, Erskine W, Baum M, Pehu E (1999) Inheritance and linkage analysis of frost injury in lentil. Crop Sci 39(3):639–642

Fabregat A, Jupe S, Matthews L (2018) The reactome pathway knowledgebase. Nucl Acids Res 46:D649–D655. https://doi.org/10.1093/nar/gkx1132

Fageria NK (1992) Maximizing crop yields. Marcel Dekker, New York, NY

Fageria NK, Baligar VC (1997) Phosphorous—use efficiency by corn genotypes. J Plant Nutr 20:1267–1277

FAO (2016) The state of world fisheries and aquaculture. Contributing to food security and nutrition for all. Rome, Italy, 200 pp.

Faris ME, Takruri HR, Issa AY (2013) Role of lentils (*Lens culinaris* L.) in human health and nutrition: a review. Medit J Nutr Met 6(1):3–16

Farooq M, Gogoi N, Hussain M, Barthakur S, Paul S, Bharadwaj N, Migdadi HM, Alghamdi SS, Siddique KH (2017) Effects, tolerance mechanisms and management of salt stress in grain legumes. Plant Physiol Biochem 118:199–217

Fedoruk MJ, Vandenberg A, Bett KE (2013) Quantitative trait loci analysis of seed quality characteristics in lentil using single nucleotide polymorphism markers. Plant Genome 1(6):3

Ferguson M (2000) Conserved resources, priorities for collection and future prospects. In: Knight R (ed) Lining research and marketing opportunities for pulses in 21st century. Current plant science and biotechnology in agriculture, vol 34, pp 613–620

Ferguson ME, Robertson LD (1996) Genetic diversity and taxonomic relationships within the genus *Lens* as revealed by allozyme polymorphism. Euphytica 91:163–172

Ferguson M, Ford-Lloyd B, Robertson L, Maxted N, Newbury H (1998a) Mapping the geographical distribution of genetic variation in the genus *Lens* for the enhanced conservation of plant genetic diversity. Mol Ecol 7:1743–1755

Ferguson ME, Robertson LD, Ford-Lloyd BV, Newbury HJ, Maxted N (1998b) Contrasting genetic variation amongst lentil landraces from different geographical origins. Euphytica 102:265–273

Ferguson ME, Robertson LD (1999) Morphological and phenological variation in the wild relatives of lentil. Genet Resour Crop Evol 46(1):3–12

Ferguson ME, Maxted N, Van Slageren M, Robertson LD (2000) A re-assessment of the taxonomy of *Lens* Mill. (Leguminosae, Papilionoideae, Vicieae). Bot J Linn Soc 133:41–59

Fernández-Aparicio MF, Rispail N, Prats E, Morandi D, Garcia-Garrido JM, Dumas-Gaudot E, Rubiales D (2009) Parasitic plant infection is partially controlled through symbiotic pathways. Weed Res 1–7

Fiala JV (2006) Transferring resistance to Colletotrichum truncatum from wild lentil species to cultivated lentil species (*Lens culinaris subsp culinaris*). Doctoral dissertation, MSc thesis, University of Saskatchewan, Saskatoon, Canada

Fiala JV, Tullu A, Banniza S, Séguin-Swartz G, Vandenberg A (2009) Interspecies transfer of resistance to anthracnose in lentil. Crop Sci 49:825–830

Fikiru E, Tesfaye K, Bekele E (2007) Genetic diversity and population structure of Ethiopian lentil (*Lens culinaris* Medikus) landraces as revealed by ISSR marker. Afr J Biotechnol 6(12):1460–1468

Fikru E, Sarker A, Fikre A, Ahmed S, Ali K, Erskine W (2007) Registration of 'Assano' lentil. J Plant Registrat 1:41

Ford R, Pang E, Taylor P (1997) Diversity analysis and species identification in *Lens* using PCR generated markers. Euphytica 96:247–255. https://doi.org/10.1023/a:1003097600701

Ford R, Pang E, Taylor P (1999) Genetics of resistance to ascochyta blight (*Ascochyta lentis*) of lentil and the identification of closely linked RAPD markers. Theor Appl Genet 98(1):93–98

Ford R, Mustafa B, Sambasivam P, Baum M, Rajesh P (2018) Biotechnology and gene mapping in lentil. In: Grain legumes, vol 57. European Association for Grain Legume Research, pp 37–38

Fratini R, Ruiz ML (2006) Interspecific hybridization in the genus *Lens* applying in vitro embryo rescue. Euphytica 150:271–280

Fratini R, Ruiz ML, de la Vega MP (2004) Intra-specific and inter-sub-specific crossing in lentil. Can J Plant Sci 84:981–986

Fratini R, García P, Ruiz ML (2006) Pollen and pistil morphology, in vitro pollen grain germination and crossing success of *Lens* cultivars and species. Plant Breed 125(5):501–505

Fratini R, Durán Y, García P, De La Vega MP (2007) Identification of quantitative trait loci (QTL) for plant structure, growth habit and yield in lentil. Spanish J Agri Res 5(3):348–356

Gaad D, Laouar M, Abdelguerfi A, Gaboun F (2018) Collection and agro morphological characterization of Algerian accessions of lentil (*Lens culinaris*). Biodivers J Biol Divers 19:183–193

Galasso I (2003) Distribution of highly repeated DNA sequences in species of the genus *Lens* Miller. Genome 46:1118–1124

Gao J, Zhang Y, Zhang C, Qi F, Li X, Mu S, Peng ZO (2014) Characterization of the floral transcriptome of Moso bamboo (*Phyllostachys edulis*) at different flowering developmental stages by transcriptome sequencing and RNA-seq analysis. PLoS ONE 9(6):e98910

Gaur PM, Krishnamurthy L, Kashiwagi J (2008) Improving drought avoidance root traits in chickpea (*Cicer arietinum*)—current status of research at ICRISAT. Plant Prod Sci 11:3–11. https://doi.org/10.1626/pps.11.3

Gaur P, Saminen S, Krishnamurthy L (2015) High temperature tolerance in grain legumes. Legume Persp 7:23–24

Ghanem ME, Kibbou F, Guiguitant J, Sinclair TR (2017) Opportunities to improve the seasonal dynamics of water use in lentil (*Lens culinaris* Medik.) to enhance yield increase in water-limited environments. Chem Biol Technol Agric 4:22

GO Consortium (2013) Gene ontology annotations and resources. Nucl Acids Res 41:D530–D535. https://doi.org/10.1093/nar/gks1050

Gokcay D (2012) Physiological and biochemical screening of different Turkish lentil (*Lens culinaris* M.) cultivars under drought stress condition. Doctoral dissertation, Middle East Technical University, Ankara, Turkey

Golezani KG, Yengabad FM (2012) Physiological responses of lentil (*Lens culinaris* Medik) to salinity. Int J Agric Crop Sci 20:1531–1535

Gonzalez A, Bell G (2013) Evolutionary rescue and adaptation to abrupt environmental change depends upon the history of stress. Philos Trans R Soc B 368:20120079

Goodstein DM, Shu S, Howson R (2012) Phytozome: a comparative platform for green plant genomics. Nucl Acids Res 40:D1178–D1186. https://doi.org/10.1093/nar/gkr944

Gorim LY, Vandenberg A (2017a) Evaluation of wild lentil species as genetic resources to improve drought tolerance in cultivated lentil. Front Plant Sci 8:1129

Gorim LY, Vandenberg A (2017b) Root traits, nodulation and root distribution in soil for five wild lentil species and *Lens culin*aris (Medik.) grown under well-watered conditions. Front Plant Sci 8:1632

Govindaraj M, Vetriventhan M, Srinivasan M (2015) Importance of genetic diversity assessment in crop plants and its recent advances: an overview of its analytical perspectives. Genet Res Int 2015:431487

Grant D, Nelson RT, Cannon SB, Shoemaker RC (2010) SoyBase, the USDA-ARS soybean genetics and genomics database. Nucl Acids Res 38:D843–D846. https://doi.org/10.1093/nar/gkp798

Grassini P, Eskridge KM, Cassman KG (2013) Distinguishing between yield advances and yield plateaus in historical crop production trends. Nat Commun 4:2918

GRDC (2013) GRDC impact assessment report series: title: an economic analysis of GRDC investment in the lentil breeding program. http://www.grdc.com.au

Gulati A, Schryer P, McHughen A (2001) Regeneration and micrografting of lentil shoots. In Vitro Cell Dev Biol-Plant 37(6):798

Gulati A, Schryer P, McHughen A (2002) Production of fertile transgenic lentil (*Lens culinaris* Medik) plants using particle bombardment. Vitro Cell Dev Biol Plant 38:316–324. https://doi.org/10.1079/ivp2002303

Gupta D, Sharma SK (2005) Embryo-ovule rescue technique for overcoming post-fertilization barriers in interspecific crosses of Lens. J Lentil Res 2:27–30

Gupta D, Sharma SK (2007) Widening the gene pool of cultivated lentils through introgression of alien chromatin from wild *Lens* subspecies. Plant Breed 126:58–61

Gupta PK, Kumar J, Mir RR, Kumar A (2010) Marker-assisted selection as a component of conventional plant breeding. Plant Breed Rev 22:33–145

Gupta D, Taylor P, Inder P, Phan H, Ellwood S, Mathur P, Sarker A, Ford R (2012a) Integration of EST-SSR markers of *Medicago truncatula* into intraspecific linkage map of lentil and identification of QTL conferring resistance to ascochyta blight at seedling and pod stages. Mol Breed 1:429–439

Gupta D, Taylor PWJ, Inder P, Phan HTT, Ellwood SR, Mathur PN (2012b) Integration of EST-SSR markers of Medicago truncatula into intraspecific linkage map of lentil and identification of QTL conferring resistance to Ascochyta blight at seedling and pod stages. Mol Breed 30:429–439

Gupta M, Verma B, Kumar N, Chahota RK, Rathour R, Sharma SK, Bhatia S, Sharma TR (2012c) Construction of intersubspecific molecular genetic map of lentil based on ISSR, RAPD and SSR markers. J Genet 91(3):279–287

Gurdip S, Kuldip S, Gill A, Brar J (1982) Screening of lentil varieties/lines for blight resistance. Indian Phytopathol 35:678–679

Hamdi AH (1987) Variation in lentil (*Lens culinaris* Medik) in response to irrigation. Doctoral dissertation, Durham University, Durham, UK

Hamdi A, Erskine W (1994) Germplasm program legumes-annual report. ICARDA Aleppo, Syria
Hamdi A, Erskine W (1996) Reaction of wild species of the genus *Lens* to drought. Euphytica 91(2):173–179
Hamdi A, Erskine W, Gates P (1991) Relationships among economic characters in lentil. Euphytica 57:109–116
Hamdi A, Erskine W, Gates P (1992) Adaption of lentil seed yield to varying moisture supply. Crop Sci 32(4):987–990
Hamdi A, Küsmenoğlu I, Erskine W (1996) Sources of winter hardiness in wild lentil. Genet Resour Crop Evol 43(1):63–67
Hamdi A, Katerji N, Mastrorilli M, Ameen A (2000) Lentil (*Lens culinaris* Med.) sensitivity to salinity through the water use efficiency. International symposium on techniques to control salination for horticultural productivity 573:311–319
Hamwieh A, Udupa SM, Choumane W, Sarker A, Dreyer F, Jung C, Baum M (2005) A genetic linkage map of *Lens* sp. based on microsatellite and AFLP markers and the localization of fusarium vascular wilt resistance. Theor Appl Genet 110:669–677
Hamwieh A, Udupa SM, Sarker A, Jung C, Baum M (2009) Development of new microsatellite markers and their application in the analysis of genetic diversity in lentils. Breed Sci 59(1):77–86
Hartung F, Schiemann J (2014) Precise plant breeding using new genome editing techniques: opportunities, safety and regulation in the EU. Plant J 78(5):742–752. https://doi.org/10.1111/tpj.12413
Havey M, Muehlbauer F (1989) Variability for restriction fragment lengths and phylogenies in lentil. Theor Appl Genet 77:839–843
Havey MJ, Lilly JW, Bohanec B, Bartoszewski G, Malepszy S (2002) Cucumber: a model angiosperm for mitochondrial transformation? J Appl Genet 43:1–17
Hobson K, Armstrong R, Nicolas M, Connor D, Materne M (2006) Response of lentil (*Lens culinaris*) germplasm to high concentrations of soil boron. Euphytica 151:371–382
Hoffman DL, Soltis DE, Muehlbauer F, Ladizinsky G (1986) Isozyme polymorphism in *Lens* (Leguminosae). Syst Bot 392–402
Hoffman DL, Muehlbauer F, Ladizinsky G (1988) Morphological variation in *Lens* (Leguminosae). Syst Bot 87–96
Hossain MS, Alam MU, Rahman A, Hasanuzzaman M, Nahar K, Al Mahmud J, Fujita M (2017) Use of iso-osmotic solution to understand salt stress responses in lentil (*Lens culinaris* Medik.). South Afr J Bot 113:346–354
Hou BK, Zhou YH, Wan LH, Zhang ZL, Shen GF, Chen ZH, Hu ZM (2003) Chloroplast transformation in oilseed rape. Transgen Res 12:111–114
Huerta-Cepas J, Szklarczyk D, Forslund K, Cook H, Heller D, Walter MC, Rattei T, Mende DR, Sunagawa S, Kuhn M, Jensen LJ (2016) eggNOG 4.5: a hierarchical orthology framework with improved functional annotations for eukaryotic, prokaryotic and viral sequences. Nucl Acids Res 44:D286–D293. https://doi.org/10.1093/nar/gkv1248
Ialicicco M, Viscosi V, Arena S (2012) *Lens culinaris* Medik. seed proteome: analysis to identify landrace markers. Plant Sci 197:1–9
Idrissi O, Houasli C, Nsarellah N (2013) Comparaison de lignées avancées de lentille sous stress hydrique durant la phase de floraison et formation des gousses. Nat Technol 8:53–61
Idrissi O, Houasli C, Udupa SM, De Keyser E, Van Damme P, De Riek J (2015a) Genetic variability for root and shoot traits in a lentil (*Lens culinaris* Medik.) recombinant inbred line population and their association with drought tolerance. Euphytica 204(3):693–709
Idrissi O, Udupa SM, Houasli C, De Keyser E, Van Damme P, De Riek J (2015b) Genetic diversity analysis of Moroccan lentil (*Lens culinaris* Medik.) landraces using simple sequence repeat and amplified fragment length polymorphisms reveals functional adaptation towards agro-environmental origins. Plant Breed 134:322–332
Idrissi O, Udupa SM, De Keyser E, McGee RJ, Coyne CJ, Saha GC, De Riek J (2016) Identification of quantitative trait loci controlling root and shoot traits associated with drought tolerance in a lentil (*Lens culinaris* Medik.) Recombinant inbred line population. Front Plant Sci (7):1174

Iqbal S, Bakhsh A, Malik R (1990) Identification of resistant sources to ascochyta blight in lentil. Lens 17:26–27
Iqbal SM, Ghafoor A, Ali S, Ahmad I (2010) Identification of resistant source of ascochyta blight from local lentil germplasm. Pak J Phytopathol 22:29–33
Jayasundara HP, Thomson BD, Tang C (1997) Responses of cool season grain legumes to soil abiotic stresses. In: Advances in agronomy, vol 63. Academic Press, pp 77–151
Jin L, Jian-Ping G, Dong-Xu X, Zhang X-Y, Jing G, Xu-Xiao Z (2008) Genetic diversity and population structure in lentil (*Lens culinaris* Medik.) germplasm detected by SSR markers. Acta Agron Sin 34:1901–1909
Johansen C, Baldev B, Brouwer JB, Erskine W, Jermyn WA, Li-Juan L, Malik BA, Miah AA, Silim SN (1994) Biotic and abiotic stresses constraining productivity of cool season food legumes in Asia, Africa and Oceania. In: Muehlbauer FJ, Kaiser WJ (eds) Expanding the production and use of cool season food legumes. Kluwer Academic Publishers, Dordrecht, Netherlands, pp 175–194
Johnson CR, Combs Jr GF, Thavarajah P (2013) Lentil (*Lens culinaris* L.): a prebiotic-rich whole food legume. Food Res Intl 51(1):107–113
Jyothi MN, Rai DV, Nagesh Babu R (2015) Identification and characterization of high temperature stress responsive novel miRNAs in French bean (*Phaseolus vulgaris*). Appl Biochem Biotechnol 176:835. https://doi.org/10.1007/s12010-015-1614-2
Kahraman A, Kusmenoglu I, Aydin N, Aydogan A, Erskine W, Muehlbauer FJ (2004a) Genetics of winter hardiness in 10 lentil recombinant inbred line populations. Crop Sci 44(1):5–12
Kahraman A, Kusmenoglu I, Aydin N, Aydogan A, Erskine W, Muehlbauer FJ (2004b) QTL mapping of winter hardiness genes in lentil. Crop Sci 44(1):13–22
Kahraman A, Temel HY, Aydoğan A, Tanyolac MB (2015) Major quantitative trait loci for flowering time in lentil. Turk J Agric For 39(4):588–595
Kanehisa M, Sato Y, Kawashima M (2016) KEGG as a reference resource for gene and protein annotation. Nucl Acids Res 44:D457–D462. https://doi.org/10.1093/nar/gkv1070
Kang UG, Choi JS, Kim JJ, Cho JS (2017) Yield potentials of rice and soybean as affected by cropping systems in mid-mountainous paddy soils of Korea. 한국토양비료학회지 Kor J Soil Sci Fert 50(4):259–274
Kant P, Materne M, Rodda MS, Slater AT (2017) Screening lentil germplasm for stemphylium blight resistance. Australas Plant Pathol 46(2):1–8
Karaköy T, Erdem H, Baloch FS (2012) Diversity of macro-and micronutrients in the seeds of lentil landraces. Sci World J. http://dx.doi.org/10.1100/2012/710412
Karki PB, Joshi S, Chaudhary G, Chaudhary RN (1993) Studies on botrytis gray mold of chickpea in Nepal. In: Haware MP, Gowda CLL, McDonald D (eds) Recent advances in research on botrytis gray mold of chickpea. ICRISAT, Patancheru, AP, India, pp 11–13
Katerji N, Van Hoorn JW, Hamdy A, Mastrorilli M, Oweis T, Erskine W (2001) Response of two varieties of lentil to soil salinity. Agric Water Manage 47(3):179–190
Katerji N, Van Hoorn JW, Hamdy A, Mastrorilli M (2003) Salinity effect on crop development and yield, analysis of salt tolerance according to several classification methods. Agric Water Manage 62(1):37–66
Kaur S, Cogan NO, Pembleton LW, Shinozuka M, Savin KW, Materne M, Forster JW (2011) Transcriptome sequencing of lentil based on second-generation technology permits large-scale unigene assembly and SSR marker discovery. BMC Genomics 12(1):265
Kaur S, Cogan NI, Stephens A, Noy D, Butsch M, Forster J (2014) EST-SNP discovery and dense genetic mapping in lentil (*Lens culinaris* Medik.) enable candidate gene selection for boron tolerance. Theor Appl Genet 127:703–713
Kemble RJ, Barsby TL, Yarrow SA (1988) Ectopic expression of mitochondrial gamma carbonic anhydrase 2 causes male sterility by anther indehiscence. Plant Mol Biol 70(471):485
Keshtiban RK, Carvani V, Imandar M (2015) Effects of salinity stress and drought due to different concentrations of sodium chloride and polyethylene glycol 6000 on germination and seedling growth characteristics of lentil (*Lens culinaris* Medik). Adv Environ Biol 15:445–451

Khan MS, Maliga P (1999) Fluorescent antibiotic resistance marker for tracking plastid transformation in higher plants. Nat Biotechnol 17:910–915

Khatib F, Koudsieh S, Ghazal B, Barton J, Tsujimoto H, Baum M (2007) Developing herbicide resistant lentil (*Lens culinaris* Medikus subsp. *culinaris*) through *Agrobacterium*-mediated transformation. Arab J Plant Protec 25:185–192

Khatib F, Makris A, Yamaguchi-Shinozaki K, Kumar S, Sarker A, Erskine W (2011) Expression of the DREB1A gene in lentil (*Lens culinaris* Medik. subsp *culinaris*) transformed with the Agrobacterium system. Crop Pasture Sci 62:488–495. https://doi.org/10.1071/cp10351

Khazaei H, Caron CT, Fedoruk M, Diapari M, Vandenberg A, Coyne CJ, McGee R, Bett KE (2016) Genetic diversity of cultivated lentil (*Lens culinaris* Medik.) and its relation to the world's agro-ecological zones. Front Plant Sci 7:1093

Kokten K, Karakoy T, Bakoglu A, Akcura M (2010) Determination of salinity tolerance of some lentil (*Lens culinaris* M.) varieties. J Food Agric Environ 8(1):140–143

Komatsu S, Yamamoto A, Nakamura T, Nouri MZ, Nanjo Y, Nishizawa K (2011) Comprehensive analysis of mitochondria in roots and hypocotyls of soybean under flooding stress using proteomics and metabolomics techniques. J Proteome Res 10:3993–4004. https://doi.org/10.1021/pr2001918

Koul PM, Sharma V, Rana M, Chahota RK, Kumar S, Sharma TR, (2017) Analysis of genetic structure and interrelationships in lentil species using morphological and SSR markers. 3 Biotechnology 7(1):83

Kuchuran M, Banniza S, Vandenberg B (2003) Evaluation of lentil varieties for resistance to botrytis gray mould. In: Meeting the challenge: opportunities and pressures within the pulse industry. In: Proceedings of pulse field days, Saskatoon, Jan 6, pp 6–7

Kumar J, Solanki RK (2014) Evaluation of lentil germplasm for agro-morphological traits. J Food Legume 27(4):302–306

Kumar J, Van Rheenen HA (2000) Brief communication. A major gene for time of flowering in chickpea. J Hered 91(1):67–68

Kumar S, Dhingra A, Daniell H (2004a) Stable transformation of the cotton plastid genome and maternal inheritance of transgenes. Plant Mol Biol 56:203–216

Kumar S, Gupta S, Chandra S, Singh BB (2004b) How wide is the genetic base of pulse crop? Pulses in new perspective. In: Ali M, Singh BB, Kumar S, Dhar V (eds) Indian Society of pulses research and development. Indian Institute of Pulses Resserch, Kanpur, India, pp 211–222

Kumar J, Basu PS, Srivastava E, Chaturvedi SK, Nadarajan N, Kumar S (2012) Phenotyping of traits imparting drought tolerance in lentil. Crop Past Sci 63(6):547–554

Kumar SK, Barpete S, Kumar J, Gupta P, Sarker A (2013) Global lentil production: constraints and strategies. SATSA Mukhapatra-Annu Tech 17:1–3

Kumar H, Dikshit HK, Singh A (2014a) Characterization of grain iron and zinc in lentil (*Lens culinaris* Medikus *culinaris*) and analysis of their genetic diversity using SSR markers. Aust J Crop Sci 8:1005

Kumar J, Srivastva E, Singh M, Kumar S, Nadarajan N, Sarker A (2014b) Diversification of indigenous gene-pool by using exotic germplasm in lentil (*Lens culinaris* Medikus subsp. *culinaris*). Physiol Mol Biol Plants 20:125–132

Kumar S, Rajendra K, Kumar J, Hamwieh A, Baum M (2015) Current knowledge in lentil genomics and its application for crop improvement. Front Plant Sci 6:78

Kumar J, Gupta S, Gupta P, Dubey S, Tomar RSS, Kumar S (2016a) Breeding strategies to improve lentil for diverse agro-ecological environments. Indian J Genet 76:530–549

Kumar J, Kant R, Kumar S, Basu PS, Sarker A, Singh NP (2016b) Heat tolerance in lentil under field conditions. Legume Genom Genet 23:7

Kumar J, Gupta S, Biradar RS, Gupta P, Dubey S, Singh NP (2018a) Association of functional markers with flowering time in lentil. J Appl Genet 59(1):9–21

Kumar J, Gupta S, Dubey S, Gupta P, Gupta DS, Singh N (2018b) Genetic diversity changes in Indian lentils over the times. J Plant Biochem Biotechnol 27:415–422

Kumar S, Choudhary AK, Rana KS, Sarker A, Singh M (2018c) Bio-fortification potential of global wild annual lentil core collection. PLoS ONE 13:e0191122

Kumar J, Kumar S, Gupta DS, Dubey S, Gupta S, Gupta P (2019) Molecular marker assisted gene pyramiding in lentils. In: Singh M (ed) Lentils. Academic Press, Elsevier, pp 125–139

Kumari S, Makkouk K (1995) Variability among twenty lentil genotypes in seed transmission rates and yield loss induced by pea seed-borne mosaic potyvirus infection. Phytopathol Mediter 34:129–132

Kumari J, Ahmad R, Chandra S (2007) Identification and pedigree analysis of lentil (*Lens culinaris*) genotypes tolerant to black aphids (*Aphis craccivora*). Indian J Agric Sci 77(8):544–545

Kumari SG, Larsen R, Makkouk KM, Bashir M (2009) 19 Virus diseases and their control. In: Erskine W, Muehlbauer FJ, Sarker A, Sharma B (eds) The lentil: botany, production and uses. CABI, Wallingford, pp 269–270

Kumari M, Mittal RK, Chahota RK, Thakur K, Lata S, Gupta D (2018) Assessing genetic potential of elite interspecific and intraspecific advanced lentil lines for agronomic traits and their reaction to rust (*Uromyces viciae-fabae*). Crop Pasture Sci 69(10):999–1008

Kumawat KR, Gothwal DK, Singh D (2017) Salinity tolerance of lentil genotypes based on stress tolerance indices. J Pharmacogn Phytochem 4:1368–1372

Kusmenoglu I, Aydin N (1995) The current status of lentil germplasm exploitation for adaptation to winter sowing in the Anatolian highlands. Autumn-sowing of lentil in the Highlands of West Asia and North Africa. In: Keatinge JDH, Kusmenoglu I (eds) Autumn Sowing of lentil in the highlands of West Asia and North Africa. Central Research Institute for Field Crops (CRIFC), Ankara, Turkey, pp 63–71

Kwon YS, Lee JM, Yi GB, Yi SI, Kim KM, Soh EH, Bae KM, Park E, Song IH, Kim BD (2005) Use of SSR markers to complement tests of distinctiveness, uniformity, and stability (DUS) of pepper (*Capsicum annuum* L.) varieties. Mol Cells 19:428–435

Ladizinsky G (1993) Lentil domestication: on the quality of evidence and arguments. Econ Bot 47(1):60–64

Lachaâl M, Grignon C, Hajji M (2002) Growth rate affects salt sensitivity in two lentil populations. J Plant Nutr 25(12):2613–2625

Ladizinsky G (1979) The origin of lentil and its wild genepool. Euphytica 28:179–187

Ladizinsky G (1999) Identification of the lentil's wild genetic stock. Genet Resour Crop Evol 46:115–118

Ladizinsky G, Abbo S (1993) Cryptic speciation in *Lens culinaris*. Genet Resour Crop Evol 40(1):1–5

Ladizinsky G, Braun D, Goshen D, Muehlbauer F (1984) The biological species of the genus *Lens* L. Bot Gaz 145:253–261

Ladizinsky G, Pickersgill B, Yamamoto K (1988) Exploitation of wild relatives of the food legumes. In: Summerfield RJ (ed) World crops, cool season food legumes. Kluwer Academic Publishers, Dordrecht, pp 967–987

Lal C, Sharma SK, Chahota RK (2000) Combining ability and heterosis for growth parameters in microsperma × macrosperma lentil. Crop Res 19:115–121

Latham L, Jones R (2001) Incidence of virus infection in experimental plots, commercial crops, and seed stocks of cool season crop legumes. Aust J Agric Res 52:397–413

Lázaro A, Ruiz M, De La Rosa L, Martín I (2001) Relationships between agro/morphological characters and climatic parameters in Spanish landraces of lentil (*Lens culinaris* Medik.). Genet Resour Crop Evol 48:239–249

Li X, Han Y, Teng W, Zhang S, Yu K, Poysa V, Anderson T, Ding J, Li W (2010) Pyramided QTL underlying tolerance to Phytophthora root rot in mega-environments from soybean cultivars 'Conrad' and 'Hefeng 25'. Theor Appl Genet 121(4):651–658

Lichtenzveig J, Scheuring C, Dodge J, Abbo S, Zhang HB (2005) Construction of BAC and BIBAC libraries and their applications for generation of SSR markers for genome analysis of chickpea, *Cicer arietinum* L. Theor Appl Genet 110(3):492–510

Lindbeck K, Bretag T, Materne M (2008) Field screening in Australia of lentil germplasm for resistance to botrytis grey mould. Aust Plant Pathol 37:373–378

Lombardi M, Materne M, Cogan NO, Rodda M, Daetwyler HD, Slater AT, Forster JW, Kaur S (2014) Assessment of genetic variation within a global collection of lentil (*Lens culinaris* Medik.) cultivars and landraces using SNP markers. BMC Genet 15(1):150

Luo Y, Ma T, Zhang A, Ong KH, Li Z, Yang J, Yin Z (2016) Marker-assisted breeding of the rice restorer line Wanhui 6725 for disease resistance, submergence tolerance and aromatic fragrance. Rice 9(1):66

Lurquin PF, Cai Z, Stiff CM, Fuerst EP (1998) Half-embryo cocultivation technique for estimating the susceptibility of pea (*Pisum sativum* L.) and lentil (*Lens culinaris* Medik.) cultivars to *Agrobacterium tumefaciens*. Mol Biotechnol 9:175–179

Maher L, Armstrong R, Connor D (2003) Salt tolerant lentils—a possibility for the future. In: Proceedings of the 11th Australian agronomy conference, Feb, The Australian Society of Agronomy, Geelong, Victoria, pp 2–6

Mahmoudian M, Yucel M, Oktem HA (2002) Transformation of lentil (*Lens culinaris* M.) cotyledonary nodes by vacuum infiltration of *Agrobacterium tumefaciens*. Plant Mol Biol Rep 20:251–257 https://doi.org/10.1007/bf02782460

Makkouk K, Kumari S (1990) Variability among 19 lentil genotypes in seed transmission rates and yield loss induced by broad bean stain virus infection. Lens Newsl 17:31–33

Makkouk K, Kumari S, Sarker A, Erskine W (2001) Registration of six lentil germplasm lines with combined resistance to viruses. Crop Sci 41:931

Malik AI, Ailewe TI, Erskine W (2015) Tolerance of three grain legume species to transient waterlogging. AoB Plants 1:7. https://doi.org/10.1093/aobpla/plv040

Mamo T, Richter C, Heiligtag B (1996) Salinity effects on the growth and ion contents of some chickpea (*Cicer arietinum* L.) and lentil (*Lens culinaris* Medic.) varieties. J Agron Crop Sci 176(4):235–247

Marschner H (1995) Mineral nutrition of higher plants. Academic Press, San Diego, CA

Materne M, Reddy AA (2007) Commercial cultivation and profitability. In: Yadav SS, McNeil D, Stevenson PC (eds) The lentil—an ancient crop for modern times. Springer, Dordrecht, Rotterdam, The Netherlands, pp 173–186

Materne M, Siddique KH (2009) Agroecology and crop adaptation. In: Erskine W, Muehlbauer F, Sarker F, Sharma F (eds) The lentil: botany, production and uses. CABI, Wallingford, UK, pp 47–63

Mayer M, Bagga S (2002) The phylogeny of *Lens* (Leguminosae): new insight from ITS sequence analysis. Plant Syst Evol 232:145–154

Mayer M, Soltis P (1994) Chloroplast DNA phylogeny of *Lens* (Leguminosae): origin and diversity of the cultivated lentil. Theor Appl Genet 87:773–781

Mia MW, Yamauchi A, Kono Y (1996) Root system structure of six food legume species: inter- and intraspecific variations. Jpn J Crop Sci 65(1):131–140

Mishra BK, Srivastava JP, Lal JP (2014) Drought stress resistance in two diverse genotypes of lentil imposed at different phenophases. J Food Legume 27(4):307–314

Mishra BK, Srivastava JP, Lal JP, Sheshshayee MS (2016) Physiological and biochemical adaptations in lentil genotypes under drought stress. Russ J Plant Physiol 63(5):695–708

Mishra BK, Srivastava JP, Lal JP (2018) Drought resistance in lentil (*Lens culinaris* Medik.) in relation to morphological, physiological parameters and phenological developments. Intl J Curr Microbiol Appl Sci 7(1):2288–2304

Mondini L, Noorani A, Pagnotta MA (2009) Assessing plant genetic diversity by molecular tools. Diversity 1:19–35

Muehlbauer FJ, Weeden NF, Hoffman DL (1989) Inheritance and linkage relationships of morphological and isozyme loci in lentil (*Lens Miller*). J Heredity 80(4):298–303

Muehlbauer FJ, Cho S, Sarker A, McPhee KE, Coyne CJ, Rajesh PN, Ford R (2006) Application of biotechnology in breeding lentil for resistance to biotic and abiotic stress. Euphytica 147(1–2):149–165

Muench D, Slinkard A, Scoles G (1991) Determination of genetic variation and taxonomy in lentil (*Lens Miller*) species by chloroplast DNA polymorphism. Euphytica 56:213–218
Mun T, Bachmann A, Gupta V (2016) Lotus base: an integrated information portal for the model legume *Lotus japonicus*. Sci Rep 6:39447. https://doi.org/10.1038/srep39447
Murray GM (2012) diseases. Grains Research and Development Corporation, Victoria, Australia
Murray GA, Eser D, Gusta LV, Eteve G (1988) Winterhardiness in pea, lentil, faba bean and chickpea. In: Summerfield RJ (ed) World crops: cool season food legumes. Kluwer Academic Publishers, Dordrecht, Netherlands, pp 831–843
Muscolo A, Sidari M, Anastasi U, Santonoceto C, Maggio A (2014) Effect of PEG-induced drought stress on seed germination of four lentil genotypes. J Plant Interact 9(1):354–363
Muscolo A, Junker A, Klukas C, Weigelt-Fischer K, Riewe D, Altmann T (2015) Phenotypic and metabolic responses to drought and salinity of four contrasting lentil accessions. J Exp Bot 66:5467–5480. https://doi.org/10.1093/jxb/erv208
Mustafa BM, Coram TE, Pang ECK (2009) A cDNA microarray approach to decipher lentil (*Lens culinaris*) responses to Ascochyta lentis. Australas Plant Pathol 38:617–631. https://doi.org/10.1071/AP09048
Naithani S, Preece J, D'Eustachio P (2017) Plant reactome: a resource for plant pathways and comparative analysis. Nucl Acids Res 45:D1029–D1039. https://doi.org/10.1093/nar/gkw932
Nasir M (1998) Improvement of drought and disease resistance in lentil in Nepal, Pakistan and Australia. Mid-term Report 1995–1998. Victorian Institute for Dryland Agriculture, Agriculture Victoria, Horsham
Nasir M, Bretag T (1998) Reactions of lentil accessions from 25 different countries to Australian isolates of *Ascochyta lentis*. Genet Resour Crop Evol 45:297–299
Negussie T, Bejiga G, Million E (1998) Lentil rust outbreak in Gimbichu district. Ethiopia, AgriTopia
Nelson RR (1978) Genetics of horizontal resistance to plant diseases. Annu Rev Phytopathol 16(1):359–378
Nessa B, Islam MR, Haque MM, Ahmed JU (2007) Soil flooding tolerance in lentil at vegetative stage. Prog Agri 18(2):17–24
Nuttall JG, Armstrong RD (2010) Impact of subsoil physicochemical constraints on crops grown in the Wimmera and Mallee is reduced during dry seasonal conditions. Soil Res 48(2):125–139
O'Boyle PD, Kelly JD, Kirk WW (2007a) Use of marker-assisted selection to breed for resistance to common bacterial blight in common bean. J Am Soc Hort Sci 132(3):381–386
O'Boyle PD, Kelly JD, Kirk WW (2007b) Use of marker-assisted selection to breed for resistance to common bacterial blight in common bean. J Am Hort Sci 132(3):381–386
O'Connor M, Peifer M, Bender W (1989) Construction of large DNA segments in *Escherichia coli*. Science 44(491):1307–1312
Ogutcen E, Ramsay L, Von Wettberg EB, Bett KE (2018) Capturing variation in *Lens* (Fabaceae): development and utility of an exome capture array for lentil. Appl Plant Sci 6:e01165
Oktem H, Mahmoudian M, Yucel M (1999) GUS gene delivery and expression in lentil cotyledonary nodes using particle bombardment. Lens Newsl 26:2–6
Öktem HA, Eyidoðan F, Demirba D, Bayraç AT, Öz MT, Özgür E, Selçuk F, Yücel M (2008) Antioxidant responses of lentil to cold and drought stress. J Plant Biochem Biotechnol 17(1):15–21
Oliveira L, Melo L, Brondani C, Peloso M, Brondani R (2008) Backcross assisted by microsatellite markers in common bean. Genet Mol Res 7(4):000–010
Ouji A, El-Bok S, Mouelhi M, Younes MB, Kharrat M (2015) Effect of salinity stress on germination of five Tunisian lentil (*Lens culinaris* L.) genotypes. Eur Sci J 11(21):30
Papatheodorou I, Fonseca NA, Keays M (2018) Expression Atlas: gene and protein expression across multiple studies and organisms. Nucl Acids Res 46:D246–D251. https://doi.org/10.1093/nar/gkx1158
Paterson AH, Tanksley SD, Sorrells ME (1991) DNA markers in plant improvement. Adv Agron 46:39–90

Peñaloza E, Tay JU, France AI (2007) Calpun-INIA, a large seed size and rust-resistant lentil (*Lens culinaris* Medik.) cultivar. Agric Técnica 67:68–71

Perez-De-Castro A, Vilanova S, Cañizares J (2012) Application of genomic tools in plant breeding. Curr Genom 13:179–195

Phan HT, Ellwood SR, Hane JK, Ford R, Materne M, Olive RP (2007) Extensive macrosynteny between *Medicago truncatula* and *Lens culinaris* ssp. *culinaris*. Theor Appl Genet 114(3):549–558

Piergiovanni A, Taranto G (2003) Geographic distribution of genetic variation in a lentil collection as revealed by SDS-PAGE fractionation of seed storage proteins [*Lens culinaris* Medik.]. J Genet Breed (Italy) 57(1):39–46

Piergiovanni AR, Taranto G (2005) Assessment of the genetic variation in Italian lentil populations by electrophoresis (SDS-PAGE) of seed storage proteins. Plant Genet Resour Newsl (IPGRI/FAO) 41:33–38

Podevin N, Devos Y, Davies HV, Nielsen KM (2012) Transgenic or not? No simple answer! EMBO Rep 13(12):1057–1061

Porch TG, Jahn M (2001) Effects of high-temperature stress on microsporogenesis in heat-sensitive and heat-tolerant genotypes of Phaseolus vulgaris. Plant Cell Environ 24(7):723–731

Rad MR, Ghasemi A, Arjmandinejad A (2010) Study of limit irrigation on yield of lentil (*Lens culinaris*) genotypes of national plant gene bank of Iran by drought resistance indices. Am Eurasian J Agric Environ Sci 7(2):238–241

Rafalski A (2002) Applications of single nucleotide polymorphisms in crop genetics. Curr Opin Plant Biol 5:94–100

Rahimi A, Norton R, McNeil D, Hoseini SM (2009) Effects of salinity and temperature on germination, seedling growth and ion relations of two lentil (*Lens culinaris*) cultivars. Seed Technol 1:76–86

Rahman MM, Sarker A, Kumar S, Ali A, Yadav NK, Rahman L (2009) Breeding for short season environments. In: Erskine W, Muehlbauer F, Sarker A, Sharma B (eds) The lentil: botany, production and uses. CAB International, Wallingford, UK, pp 121–136

Rai R, Singh RP (1999) Effect of salt stress on interaction between lentil (*Lens culinaris*) genotypes and *Rhizobium* spp. strains: symbiotic N_2 fixation in normal and sodic soils. Biol Fertil Soils 29(2):187–195

Rai R, Nasar SK, Singh SJ, Prasad V (1985) Interactions between Rhizobium strains and lentil (*Lens culinaris* Linn.) genotypes under salt stress. J Agric Sci 104(1):199–205

Ramgiry S, Paliwal K Tomar S (1989) Variability and correlations of grain yield and other quantitative characters in lentil. Lens Newsl 16(1):129–210

Rana M (2016) Molecular mapping of quantitative trait loci for drought tolerance and yield traits in lentil. CSKHPKV, Palampur, India

Rana J, Gautam N, Gayacharan MS (2016) Genetic resources of pulse crops in India: an overview. Indian J Genet Pl Br 76:420–436

Rana M, Sood A, Hussain W, Kaldate R, Sharma TR, Gill R, Kumar S, Singh S (2019) Gene pyramiding and multiple character breeding. In: Singh M (ed) Lentils. Academic Press, pp 83–124

Ray H, Bett K, Tar'an B, Vandenberg A, Thavarajah D, Warkentin T (2014) Mineral micronutrient content of cultivars of field pea, chickpea, common bean, and lentil grown in Saskatchewan, Canada. Crop Sci 54(4):1698–1708

Reda A (2015) Lentil (*Lens culinaris* Medikus) current status and future prospect of production in Ethiopia. Adv Plants Agric Res https://doi.org/10.15406/apar.2015.02.00040

Reddy M, Rathour R, Kumar N, Katoch P, Sharma T (2010) Cross-genera legume SSR markers for analysis of genetic diversity in *Lens* species. Plant Breed 129:514–518

Renfrew JM (1969) The archaeological evidence for the domestication of plants: methods and problems. In: The domestication and exploitation of plants and animals, pp 149–172

Retzel EF, Johnson JE, Lamblin AF, Paule CE (2007) Legume resources: MtDB and Medicago.Org. In: Edwards D (ed) Plant bioinformatics: methods and protocols. Humana Press, Totowa, NJ, pp 261–274

Rhee SY, Beavis W, Berardini TZ, Chen G, Dixon D, Doyle A, Garcia-Hernandez M, Huala E, Lander G, Montoya M, Miller N (2003) The Arabidopsis Information Resource (TAIR): a model organism database providing a centralized, curated gateway to Arabidopsis biology, research materials and community. Nucl Acids Res 31(1):224–228

Ribaut JM, Hoisington D (1998) Marker-assisted selection: new tools and strategies. Trends Plant Sci 3(6):236–239

Ribaut JM, De Vicente MC, Delannay X (2010) Molecular breeding in developing countries: challenges and perspectives. Curr Opin Plant Biol 13(2):213–218

Richardson K, Vales M, Kling J, Mundt C, Hayes P (2006) Pyramiding and dissecting disease resistance QTL to barley stripe rust. Theor Appl Genet 113(3):485–495

Roberts EH, Summerfield RJ, Muehlbauer FJ, Short RW (1986) Flowering in lentil (*Lens culinaris* Medic.): the duration of the photoperiodic inductive phase as a function of accumulated daylength above the critical photoperiod. Ann Bot 58(2):235–248

Rodda MS, Davidson J, Javid M, Sudheesh S, Blake S, Forster JW, Kaur S (2017) Molecular breeding for ascochyta blight resistance in lentil: Current progress and future directions. Front Plant Sci 8:1136

Roy S, Islam M, Sarker A, Malek M, Rafii M, Ismail M (2013) Determination of genetic diversity in lentil germplasm based on quantitative traits. Aust J Crop Sci 7:14

Roy S, Ray B, Sarker A, Das S (2015) DNA fingerprinting and genetic diversity in lentil germplasm using SSR markers. Asian J Conserv Biol 12:2

Rubeena, TP, Taylor P, Ades P, Ford R (2006) QTL mapping of resistance in lentil (*Lens culinaris* ssp. *culinaris*) to ascochyta blight (*Ascochyta lentis*). Plant Breed 125(5):506–512

Rubeena TP, Ford R, Taylor PWJ (2003a) Construction of an intraspecific linkage map of lentil. Theor Appl Genet 107:910–916

Rubeena TP, Ford R, Taylor PWJ (2003b) Molecular mapping the lentil (*Lens culinaris* ssp. culinaris) genome. Theor Appl Genet 107:910–916

Rubeena A, Taylor PWJ, Ades PK, Ford R (2006) QTL mapping of resistance in lentil (*Lens culinaris* ssp. *culinaris*) to ascochyta blight (*Ascochyta lentis*). Plant Breed 125:506–512. https://doi.org/10.1111/j.1439-0523.2006.01259.x

Rubiales D, Fondevilla S (2012) Future prospects for ascochyta blight resistance breeding in cool season food legumes. Front Plant Sci 3:27

Sabir J, Schwarz E, Ellison N, Zhang J, Baeshen NA, Mutwakil M, Jansen R, Ruhlman T (2014) Evolutionary and biotechnology implications of plastid genome variation in the inverted-repeat-lacking clade of legumes. Plant Biotechnol 12:743–754

Sadiq MS, Haidar S, Haq MA, Abbas G (2008) A high yielding and disease resistant mutant of lentil developed through seed irradiation of an exotic germplasm. Can J Pure Appl Sci 2(2):411–416

Saha GC, Sarker A, Chen W, Vandemark GJ, Muehlbauer FJ (2010a) Inheritance and linkage map positions of genes conferring resistance to stemphylium blight in lentil. Crop Sci 50:1831–1839

Saha GC, Sarker A, Chen WD, Vandemark GJ, Muehlbauer FJ (2010b) Identification of markers associated with genes for rust resistance in *Lens culinaris* Medik. Euphytica 175:261–265. https://doi.org/10.1007/s10681-010-0187-y

Saha GC, Sarker A, Chen W, Vandemark GJ, Muehlbauer FJ (2013) Inheritance and linkage map positions of genes conferring agromorphological traits in *Lens culinaris*. Int J Agron. https://doi.org/10.1155/2013/618926

Saito K, Matsuda F (2010) Metabolomics for functional genomics, systems biology, and biotechnology. Annu Rev Plant Biol 61:463–489

Sakr B, Sarker A, El Hassan H, Kadah N (2004) Registration of 'Hamria' lentil. Crop Sci 44:686

Salam MA, Islam MT (1994) Growth, yield and leaf-water attributes of some advanced mutant lentil lines under different soil moisture regimes [*Lens culinaris*]. Lens Newsl 21:32–35

Salehi M (2012) The study of drought tolerance of lentil (*Lens culinaris* Medik) in seedling growth stage. Int J Agron Plant Prod 3(1):38–41

Salehi M, Haghnazari A, Shekari F, Faramarzi A (2008) The study of seed yield and seed yield components of lentil (*Lens culinaris* Medik) under normal and drought stress conditions. Pak J Biol Sci 11(5):758–762

Salvi S, Tuberosa R (2005) To clone or not to clone plant QTLs: present and future challenges. Trends Plant Sci 10(6):297–304

Sanchez DH, Pieckenstain FL, Escaray F, Erban A, Kraemer U, Udvardi MK (2011) Comparative ionomics and metabolomics in extremophile and glycophytic *Lotus* species under salt stress challenge the metabolic preadaptation hypothesis. Plant, Cell Environ 34:605–617. https://doi.org/10.1111/j.1365-3040.2010.02266.x

Sanderson L, Barlow E, Vijayan P (2011) Know pulse: a breeder-focused web portal that integrates genetics and genomics of pulse crops with model genomes. Can J Plant Sci 91:395–395. https://doi.org/10.4141/CJPS11501

Saraf CS, Patil RR, Prashad M (1985) Correlation and regression studies in lentil cultivars. LENS 12:11–12

Sari E, Bhadauria V, Ramsay L, Borhan MH, Lichtenzveig J, Bett KE, Vandenberg A, Banniza S (2018) Defense responses of lentil (*Lens culinaris*) genotypes carrying non-allelic ascochyta blight resistance genes to *Ascochyta lentis* infection. PLoS ONE 13(9):e0204124

Sarker A, Erskine W, Hassan M, Afzal M, Murshed A (1999a) Registration of 'Barimasur-4' Lentil. Crop Sci 39:876

Sarker A, Erskine W, Sharma B, Tyagi MC (1999) Inheritance and linkage relationship of days to flower and morphological loci in lentil (*Lens culinaris* Medikus subsp. *culinaris*). J Hered 90(2):270–275

Sarker A, Aydin A, Aydogan SH, Sabaghpour H, Ketata I, Kusmenoglu Erskine W (2002) Winter lentils promise improved nutrition and income in west Asian highlands. ICARDA Caravan 16:14–16

Sarker R, Mustafa BM, Biswas A, Mahbub S, Nahar M, Hashem R (2003) In vitro regeneration in lentil (*Lens culinaris* Medik.). Plant Tiss Cult 13:155–163

Sarker A, Erskine W, Abu Bakr M, Matiur Rahman M, Ali Afzal M, Saxena MC (2004) Lentil improvement in Bangladesh. A success story of fruitful partnership between the Bangladesh Agricultural Research Institute and the International Center for Agricultural Research in the Dry Areas. APAARI Publication 1, pp 1–38

Sarker A, Erskine W, Singh M (2005) Variation in shoot and root characteristics and their association with drought tolerance in lentil landraces. Genet Res Crop Evol 52(1):89–97

Saxena MC (1993) The challenge of developing biotic and abiotic stress resistance in cool-season food legumes. In: Singh KB, Saxena MC (eds) Breeding for stress tolerance in cool-season food legumes. Wiley, Chichester, Oxford, UK, pp 3–14

Saxena MC (2009) Plant morphology, anatomy and growth habit. In: Erskine W, Muehlbauer FJ, Sarker BS (eds) The lentil: botany, production and uses. CAB International, Wallingford, UK, pp 34–46

Scheben A, Wolter F, Batley J, Puchta H, Edwards D (2017) Towards CRISPR/Cas crops—bringing together genomics and genome editing. New Phytol 216(3):682–698

Schnoes AM, Brown SD, Dodevski I, Babbitt PC (2009) Annotation error in public databases: Misannotation of molecular function in enzyme superfamilies. PLoS Comput Biol 5. https://doi.org/10.1371/journal.pcbi.1000605

Scippa G, Trupiano D, Rocco M, Viscosi V, Di Michele M, D'andrea A, Chiatante D (2008) An integrated approach to the characterization of two autochthonous lentil (*Lens culinaris*) landraces of Molise (south-central Italy). Heredity 101:136

Scippa GS, Rocco M, Ialicicco M (2010) The proteome of lentil (*Lens culinaris* Medik.) seeds: discriminating between landraces. Electrophoresis 31:497–506

Sehgal A, Sita K, Kumar J, Kumar S, Singh S, Siddique KH, Nayyar H (2017) Effects of drought, heat and their interaction on the growth, yield and photosynthetic function of lentil (*Lens culinaris* Medikus) genotypes varying in heat and drought sensitivity. Front Plant Sci 17(8):1776

Sever T, Ates D, Aldemir SB, Yagmur B, Kaya HB, Temel HY (2014) Identification QTLs controlling genes to Se uptake in lentil seeds. In: Proceedings of the plant and animal genome XXII conference, San Diego, CA, p 359

Seversike TM, Sermons SM. Sinclair TR, Carter TE, Rufry TW (2013) Temperature interactions with transpiration response to vapor pressure deficit among cultivated and wild soybean genotypes. Physiol Plant 148:62–73. https://doi.org/10.1111/j.1399-3054.2012.01693.x

Seyedimoradi H, Talebi R (2014) Detecting DNA polymorphism and genetic diversity in Lentil (*Lens culina*ris Medik.) germplasm: comparison of ISSR and DAMD marker. Physiol Mol Biol Plants 20:495–500

Shaikh R, Diederichsen A, Harrington M, Adam J, Conner RL, Buchwaldt L (2012) New sources of resistance to Colletotrichum truncatum race Ct0 and Ct1 in *Lens culinaris* Medikus subsp. culinaris obtained by single plant selection in germplasm accessions. Genet Resour Crop Evol 60:193–201

Sharma HC (2014) Climate change effects on insects: Implications for crop protection and food security. J Crop Improv 28:229–259

Sharma SK, Dawson IK, Waugh R (1995) Relationships among cultivated and wild lentils revealed by rapd analysis. Theor Appl Genet 91:647–654. https://doi.org/10.1007/BF00223292

Sharma SK, Knox MR, Ellis THN (1996) AFLP analysis of the diversity and phylogeny of Lens and its comparison with RAPD analysis. Theor Appl Genet 93:751–758. https://doi.org/10.1007/bf00224072

Sharpe AG, Ramsay L, Sanderson LA, Fedoruk MJ, Clarke WE, Li R, Kagale S, Vijayan P, Vandenberg A, Bett KE (2013a) Ancient orphan crop joins modern era: gene-based SNP discovery and mapping in lentil. BMC Genomics 14(1):192

Sharpe AG, Ramsay L, Sanderson LA, Fedoruk MJ, Clarke WE, Rong L (2013b) Ancient orphan crop joins modern era: gene-based SNP discovery and mapping in lentil. BMC Genomics 14:1–13

Shi A, Chen P, Li D, Zheng C, Zhang B, Hou A (2009) Pyramiding multiple genes for resistance to soybean mosaic virus in soybean using molecular markers. Mol Breed 23(1):113

Shizuya H, Birren B, Kim UJ (1992) Cloning and stable maintenance of 300-kilobase-pair fragments of human DNA in Escherichia coli using an F-factor-based vector. Proc Natl Acad Sci USA 89(18):8794–8797

Shrestha R, Siddique KH, Turner NC, Turner DW, Berger JD (2005) Growth and seed yield of lentil (*Lens culina*ris Medikus) genotypes of West Asian and South Asian origin and crossbreds between the two under rainfed conditions in Nepal. Aust J Agric Res 56(9):971–981

Shrestha R, Turner NC, Siddique KH, Turner DW, Speijers J (2006) A water deficit during pod development in lentils reduces flower and pod numbers but not seed size. Aust J Agric Res 57(4):427–438

Shrestha R, Rizvi AH, Sarker A, Darai R, Paneru RB, Vandenberg A, Singh M (2018) Genotypic variability and genotype × environment interaction for iron and zinc content in lentil under Nepalese environments. Crop Sci 58:1–8. https://doi.org/10.2135/cropsci2018.05.0321

Siahsar BA, Ganjali S, Allahdoo M (2010) Evaluation of drought tolerance indices and their relationship with grain yield of lentil lines in drought stressed and irrigated environments. Aust J Basic Appl Sci 4(9):4336–4346

Sidari M, Muscolo A, Anastasi U, Preiti G, Santonoceto C (2007) Response of four genotypes of lentil to salt stress conditions. Seed Sci Technol 35(2):497–503

Sidari M, Santonoceto C, Anastasi U, Preiti G, Muscolo A (2008) Variations in four genotypes of lentil under NaCl-salinity stress. Am J Agric Biol Sci 3(1):410–416

Siddique KHM (1999) Abiotic stresses of cool season pulses. In: Trials OF (ed) Birchip cropping. Birchip, Vicxtoria, Australia

Siddique KH, Loss SP, Pritchard DL, Regan KL, Tennant D, Jettner RL, Wilkinson D (1998) Adaptation of lentil (*Lens culinaris* Medik.) to Mediterranean-type environments: effect of time of sowing on growth, yield, and water use. Aust J Agric Res 49(4):613–626

Siddique KH, Regan KL, Tennant D, Thomson BD (2001) Water use and water use efficiency of cool season grain legumes in low rainfall Mediterranean-type environments. Eur J Agron 15(4):267–280
Siddique KH, Erskine W, Hobson K, Knights EJ, Leonforte A, Khan TN, Paull JG, Redden R, Materne M (2013) Cool-season grain legume improvement in Australia—use of genetic resources. Crop Pasture Sci 64(4):347–360
Silim SN, Saxena MC, Erskine W (1993a) Adaptation of lentil to the Mediterranean environment. I. Factors affecting yield under drought conditions. Exp Agric 29(1):9–19
Silim SN, Saxena MC, Erskine W (1993b) Adaptation of lentil to the Mediterranean environment. II. Response to moisture supply. Exp Agric 29(1):21–28
Singh KP (2001) Evaluation of osmoitic drought tolerance in lentil genotypes during seed germination. Ann Plant Physiol 15(1):31–33
Singh DP, Singh BB (1991) Evaluation of exotic germplasm in lentil. Narendra Deva J Agric Res 6:304–306
Singh BB, Tewari TN, Singh AK (1993) Stress studies in lentil (*Lens esculenta* Moench) III. Leaf growth, nitrate reductase activity, nitrogenase activity and nodulation of two lentil genotypes exposed to sodicity. J Agron Crop Sci 171(3):196–205
Singh I, Singh J, Singh A, Chauhan M (1994) Pant Lentil 4: a high yielding, rust-, wilt- and blight-resistant variety for the North-Western Plains of India. Lens Newsl 21:8–9
Singh RA, Roy NK, Haque MS, Pusa SI (2001) Changes in growth and metabolic activity in seedlings of lentil (*Lens culinaris* Medic) genotypes during salt stress. Indian J Plant Physiol 6(4):406–410
Singh D, Dikshit HK, Singh R (2013a) A new phenotyping technique for screening for drought tolerance in lentil (*Lens culinaris* Medik.). Plant Breed 132(2):185–90
Singh M, Rana MK, Kumar K, Bisht IS, Dutta M, Gautam NK, Bansal KC (2013b) Broadening the genetic base of lentil cultivars through inter sub-specific and interspecific crosses of Lens taxa. Plant Breed 132:667–675
Singh M, Bisht IS, Dutta M, Kumar K, Kumar S, Bansal KC (2014) Genetic studies on morpho-phenological traits in lentil (*Lens culinaris* Medikus) wide crosses. J Genet 93(2):561–566
Singh D, Singh CK, Tomar RS, Chaturvedi AK, Shah D, Kumar A, Pal M (2016) Exploring genetic diversity for heat tolerance among lentil (*Lens culinaris* Medik.) genotypes of variant habitats by simple sequence repeat markers. Plant Breed 135(2):215–223
Singh D, Singh CK, Tomar RSS, Taunk J, Singh R, Maurya S, Chaturvedi AK, Pal M, Singh R, Dubey SK (2016) Molecular assortment of *Lens* species with different adaptations to drought conditions using SSR markers. PLoS ONE 11(1):e0147213
Singh D, Singh CK, Taunk J, Tomar RS, Chaturvedi AK, Gaikwad K, Pal M (2017a) Transcriptome analysis of lentil (*Lens culinaris* Medikus) in response to seedling drought stress. BMC Genomics 18(1):206
Singh M, Rana JC, Saxena DC, Saxena A, Sarker A (2017b) Comparative agronomic performance and reaction to Fusarium wilt of *Lens culinaris* × *L. orientalis* and *L. culinaris* × *L. ervoides* derivatives. Front Plant Sci. https://doi.org/10.3389/fpls.2017.01162
Singh D, Singh CK, Singh Tomar RS, Pal M (2017c) Genetics and molecular mapping of heat tolerance for seedling survival and pod set in lentil. Crop Sci 57(6):3059–3067
Singh M, Sharma SK, Singh B, Malhotra N, Chandora R, Sarker A, Singh K, Gupta D (2018) Widening the genetic base of cultivated gene pool following introgression from wild Lens taxa. Plant Breed 137(4):470–485
Sita K, Sehgal A, Kumar J, Kumar S, Singh S, Siddique KH, Nayyar H (2017) Identification of high-temperature tolerant lentil (*Lens culinaris* Medik.) genotypes through leaf and pollen traits. Front Plant Sci 19(8):744
Sjödin A, Street NR, Sandberg G, Gustafsson P, Jansson S (2009) The Populus Genome Integrative Explorer (PopGenIE): a new resource for exploring the Populus genome. New Phytol 182(4):1013–1025

Skarjinskaia M, Svab Z, Maliga P (2003) Plastid transformation in *Lesquerella fe*ndleri, an oilseed Brassicacea. Transgen Res 12(1):115–122

Solaiman Z, Colmer TD, Loss SP, Thomson BD, Siddique KH (2007) Growth responses of cool-season grain legumes to transient waterlogging. Aust J Agric Res 58(5):406–412

Solanki R, Singh S, Kumar J (2010) Molecular marker assisted testing of hybridity of F1 plants in lentil. Food Legumes 23:21–24

Sonnante G, Pignone D (2001) Assessment of genetic variation in a collection of lentil using molecular tools. Euphytica 120:301–307

Sonnante G, Pignone D (2007) The major Italian landraces of lentil (*Lens culinaris* Medik.): their molecular diversity and possible origin. Genet Resour Crop Evol 54:1023–1031

Sonnante G, Galasso I, Pignone D (2003) ITS sequence analysis and phylogenetic inference in the genus *Lens* Mill. Ann Bot 91:49–54

Spaeth SC, Muehlbauer F (1991) Registration of three germplasms of winter hardy lentil. Crop Sci 31(5):1395

Sprink T, Eriksson D, Schiemann J, Hartung F (2016) Regulatory hurdles for genome editing: process vs. product-based approaches in different regulatory contexts. Plant Cell Rep 35(7):1493–1506. https://doi.org/10.1007/s00299-016-1990-2

Srivastava RP, Vasishtha H (2012) Saponins and lectins of Indian chickpeas (*Cicer arietinum*) and lentils (*Lens culinaris*). Indian J Agric Biochem 25(1):44–47

Stevenson PC, Dhillon MK, Sharma HC, Bouhssini ME (2007) Insect pests of lentil and their management. In: Yadav SS, McNeil DL, Stevenson PC (eds) Lentil. Springer, Dordrecht, pp 331–348

Stoddard FL, Balko C, Erskine W, Khan HR, Link W, Sarker A (2006) Screening techniques and sources of resistance to abiotic stresses in cool-season food legumes. Euphytica 147(1–2):167–186

Storer NP, Thompson GD, Head GP (2012) Application of pyramided traits against Lepidoptera in insect resistance management for Bt crops. GM Crops Food 3(3):154–162

Subbarao GV, Johansen C, Slinkard AE, Nageswara Rao RC, Saxena NP, Chauhan YS, Lawn RJ (1995) Strategies for improving drought resistance in grain legumes. Crit Rev Plant Sci 14(6):469–523

Sudheesh S, Rodda MS, Davidson J, Javid M, Stephens A, Slater AT, Cogan NO, Forster JW, Kaur S (2016) SNP-based linkage mapping for validation of QTLs for resistance to ascochyta blight in lentil. Front Plant Sci 7:1604

Sugha S, Singh B, Sharma S (1991) Performance of lentil varieties/germplasm lines against blight. Lens 18:34–35

Sultana T, Ghafoor A (2008) Genetic diversity in ex-situ conserved *Lens culinaris* for botanical descriptors, biochemical and molecular markers and identification of landraces from indigenous genetic resources of Pakistan. J Intl Plant Biol 5:484–490

Summerfield RJ, Roberts EH, Erskine W, Ellis RH (1985) Effects of temperature and photoperiod on flowering in lentils (*Lens culinaris* Medic.). Ann Bot 56(5):659–671

Sutaria GS, Akbari KN, Vora VD, Hirpara DS, Padmani, DR (2010) Response of legume crops to enriched compost and vermicompost on vertic ustochrept under rain fed agriculture. Legume Res: An Int J 33(2)

Tadmor Y, Zamir D, Ladizinsky G (1987) Genetic mapping of an ancient translocation in the genus *Lens*. Theor Appl Genet 73:883–892

Tahir M, Muehlbauer FJ (1994) Gene-mapping in lentil with recombinant inbred lines. J Hered 85:306–310

Tahir M, Muehlbauer FJ, Spaeth SC (1994) Association of isozyme markers with quantitative trait loci in random single seed descent derived lines of lentil (*Lens culinaris* Medik.). Euphytica 75(1–2):111–119

Tanksley SD, Young ND, Paterson AH, Bonierbale MW (1989) RFLP mapping in plant breeding: new tools for an old science. Biotechnology 7(3):257

Tanksley SD, Ganal MW, Martin GB (1995) Chromosome landing: a paradigm for map-based gene cloning in plants with large genomes. Trends Genet 11(2):63–68

Tanyolac B, Ozatay S, Kahraman A, Muehlbauer FJ (2010) Linkage mapping of lentil genome using recombinant inbred lines revealed by AFLP, ISSR, RAPD and some morphologic markers. J Agric Biotechnol Sustain Develop 2:1–6

Taran B, Buchwaldt L, Tullu A, Banniza S, Warkentin TD, Vandenberg A (2003) Using molecular markers to pyramid genes for resistance to ascochyta blight and anthracnose in lentil (*Lens culinaris* Medik). Euphytica 134(2):223–230

Taran B, Warkentin TD, Vandenberg A (2013) Fast track genetic improvement of ascochyta blight resistance and double podding in chickpea by marker-assisted backcrossing. Theor Appl Genet 126(6):1639–1647

Tello-Ruiz MK, Naithani S, Stein JC et al (2018) Gramene 2018: unifying comparative genomics and pathway resources for plant research. Nucl Acids Res 46:D1181–D1189. https://doi.org/10.1093/nar/gkx1111

Temel HY, Gol D, Kahriman A, Tanyolac MB (2014) Construction of linkage map trough genotyping by sequencing in lentil. In: Proceedings of plant and animal genome conference XXII, San Diego, CA, p 358

Temnykh SG, DeClerck A, Lukashova L, Lipovich S, Cartinhour, McCouch S (2001) Computational and experimental analysis of microsatellites in rice (*Oryza sativa* L.): frequency, length variation, transposon associations, and genetic marker potential. Genome Res 11(8):1441–1452

Tewari TN, Singh BB (1991) Stress studies in lentil (*Lens esculenta* Moench). Plant Soil 136(2):225–230

Thavarajah D (2017) Lentil (*Lens culinaris* Medikus): a whole food rich in prebiotic carbohydrates to combat global obesity. InTechOpen

The UniProt Consortium (2008) The universal protein resource (UniProt). Nucl Acids Res 36:D190–D195. https://doi.org/10.1093/nar/gkm895

Thiel T, Michalek W, Varshney R, Graner A (2003) Exploiting EST databases for the development and characterization of gene-derived SSR-markers in barley (*Hordeum vulgare* L.). Theor Appl Genet 106:411–422

Toker C, Mutlu N (2011) Breeding for abiotic stresses. In: Pratap A, Kumar J (ed) Biology and breeding of food legumes. CAB International, Wallingford, UK, pp 1:241–1:260

Toklu F, Karaköy T, Haklı E, Bicer T, Brandolini A, Kilian B, Özkan H (2009) Genetic variation among lentil (*Lens culinaris* Medik) landraces from Southeast Turkey. Plant Breed 128(2):178–186

Tsanakas GF, Mylona PV, Koura K, Gleridou A, Polidoros AN (2018) Genetic diversity analysis of the Greek lentil (*Lens culinaris*) landrace 'Eglouvis' using morphological and molecular markers. Plant Genet Resour 16(5):1–9

Tullu A, Kusmenoglu I, McPhee K, Muehlbauer F (2001) Characterization of core collection of lentil germplasm for phenology, morphology, seed and straw yields. Genet Resour Crop Evol 48(2):143–152

Tullu A, Buchwaldt L, Warkenti T, Taran B, Vandenberg A (2003) Genetics of resistance to anthracnose and identification of AFLP and RAPD markers linked to the resistance gene in PI 320937 germplasm of lentil (*Lens culinaris* Medikus). Theor Appl Genet 106(3):428–434

Tullu A, Tar'an B, Breitkreutz C, Banniza S, Warkentin TD, Vandenberg A, Buchwaldt L (2006) A quantitative-trait locus for resistance to ascochyta blight [*Ascochyta lentis*] maps close to a gene for resistance to anthracnose [*Colletotrichum truncatum*] in lentil. Can J Plant Pathol 28(4):588–595

Tullu A, Taran B, Warkentin T, Vandenberg A (2008) Construction of an intraspecific linkage map and QTL analysis for earliness and plant height in lentil. Crop Sci 48:2254–2264

Tullu A, Banniza S, Tar'an B, Warkentin T, Vandenberg A (2010) Sources of resistance to ascochyta blight in wild species of lentil (*Lens culinaris* Medik.). Genet Resour Crop Evol 57:1053–1063

Turan MA, Turkmen N, Taban N (2007) Effect of NaCl on stomatal resistance and proline, chlorophyll, Na, Cl and K concentrations of lentil plants. J Agron 6:378–381

Udupa SM, Robertson LD, Weigand F, Baum M, Kahl G (1999) Allelic variation at (TAA)n microsatellite loci in a world collection of chickpea. Mol Gen Genet 261:354–363 ul Hussan S, Khuroo N, Lone A, Dar Z, Dar S, Dar M (2018) Study of variability and association analysis for various agromorphological traits in lentil (*Lens culinaris* M.). J Pharm Phytochem 7(2): 2172–2175

Vadez V, Rao S, Kholova J, Krishnamurthy L, Kashiwaji J, Ratnakumar P (2008) Root research for drought tolerance in legumes: quo-vadis? J Food Legumes 21:77–85

Vail S, Strelioff J, Tullu A, Vandenberg A (2012) Field evaluation of resistance to Colletotrichum truncatum in *Lens culinaris, Lens ervoides, and Lens ervoides* × *Lens culinaris* derivatives. Field Crops Res 126:145–151

Van Bel M, Diels T, Vancaester E, Kreft L, Botzki A, Van de Peer Y, Coppens F, Vandepoele K (2017) PLAZA 4.0: an integrative resource for functional, evolutionary and comparative plant genomics. Nucl Acids Res 46(D1):D1190–D1196

Van Hoorn JW, Katerji N, Hamdy A, Mastrorilli M (2001) Effect of salinity on yield and nitrogen uptake of four grain legumes and on biological nitrogen contribution from the soil. Agric Water Manage 51(2):87–98

Vandenberg A (2006) Sources of resistance to anthracnose *Colletotrichum truncatum* in wild *Lens* species. Genet Resour Crop Evol 53:111–119

Vandenberg A, Slinkard AE (1989) Inheritance of four new qualitative genes in lentil. J Heredity 80(4):320–322

Varshney NK, Singh NB (2017) Synthesis of choline and glycine betaine and alleviation of sodium chloride toxicity. Indian J Sci Res 7(2):173–176

Varshney RK, Song C, Saxena RK, Azam S, Yu S, Sharpe AG, Cannon S, Baek J, Rosen BD, Tar'an B, Millan T (2013) Draft genome sequence of chickpea (*Cicer arietinum*) provides a resource for trait improvement. Nat Biotechnol 31(3):240

Varshney RK, Saxena RK, Jackson SA (2017) Future prospects. In: Varshney RK, Saxena RK, Jackson SA (eds) The pigeonpea genome. Springer, Cham, Switzerland, pp 99–104

Verma P, Shah N, Bhatia S (2013) Development of an expressed gene catalogue and molecular markers from the de novo assembly of short sequence reads of the lentil (*Lens culinaris* Medik.) transcriptome. Plant Biotechnol J 11:894–905. https://doi.org/10.1111/pbi.12082

Verma P, Sharma TR, Srivastava PS, Abdin MZ, Bhatia S (2014) Exploring genetic variability within lentil (*Lens culinaris* Medik.) and across related legumes using a newly developed set of microsatellite markers. Mol Biol Rep 41:5607–5625. https://doi.org/10.1007/s11033-014-3431-z

Verma P, Goyal R, Chahota RK, Sharma TR, Abdin MZ, Bhatia S (2015) Construction of a genetic linkage map and identification of QTLs for seed weight and seed size traits in lentil. PLoS ONE e0139666

Verslues PE, Agarwal M, Katiyar-Agarwal S, Zhu J, Zhu JK (2006) Methods and concepts in quantifying resistance to drought, salt and freezing, abiotic stresses that affect plant water status. Plant J 45(4):523–539

Wang GL, Holsten TE, Song WY, Wang HP, Ronald PC (1995) Construction of a rice bacterial artificial chromosome library and identification of clones linked to the Xa21 disease resistance locus. Plant J 7(3):525–533

Wang J, Chapman SC, Bonnett DG, Rebetzke GJ, Crouch J (2007) Application of population genetic theory and simulation models to efficiently pyramid multiple genes via marker-assisted selection. Crop Sci 47(2):582–588

Warkentin TD, McHughen D (1992) *Agrobacterium tumefaciens*-mediated beta-glucuronidase (GUS) gene expression in lentil (*Lens culinaris* Medik.) tissues. Plant Cell Rep 11:274–278

Wei Z, Liu Y, Lin C, Wang Y, Cai QA, Dong Y, Xing (2011) Transformation of alfalfa chloroplasts and expression of green fluorescent protein in a forage crop. Biotechnol Lett 33(12):2487–2494

Weigand S, Pala M, Saxena MC (1992) Effect of sowing date, fertilizer and insecticide on nodule damage by *Sitona crinitus Herbst* (Coleoptera: Curculionidae) and yield of lentil (*Lens culinarie* Medik.) in northern Syria. J Plant Dis Prot 174–181

Williams DJ, McHughen A (1986) Plant regeneration of the legume *Lens culinaris* Medik. (lentil) in vitro. Plant Cell Tiss Organ Cult 7(2):149–153

Wiraguna E, Malik AI, Erskine W (2017) Waterlogging tolerance in lentil (*Lens culinaris* Medik. subsp. culinaris) germplasm associated with geographic origin. Genet Resour Crop Evol 64(3):579–586

Wong MM, Gujaria-Verma N, Ramsay L, Yuan HY, Caron C, Diapari M, Vandenberg A, Bett KE (2015) Classification and characterization of species within the genus *Lens* using genotyping-by-sequencing (GBS). PLoS ONE 10(3):e0122025

Wu W, Cheng S (2014) Root genetic research, an opportunity and challenge to rice improvement. Field Crop Res 165:111–124. https://doi.org/10.1016/j.fcr.2014.04.013

Wu X, Li G, Wang B, Hu Y, Wu X, Wang Y, Lu Z, Xu P (2018) Fine mapping Ruv2, a new rust resistance gene in cowpea (*Vigna unguiculata*), to a 193-kb region enriched with NBS-type genes. Theor Appl Genet 17:1

Yamazaki Y, Yoshimura A, Nagato Y, Kurata N (2008) Oryzabase: an integrated rice science database. In: Khush GS (ed) Advances in rice genetics. World Scientific Publishing Company, Toh Tuck Link, Singapore, pp 380–383

Yang XW, Gong S (2005) An overview on the generation of BAC transgenic mice for neuroscience research. Curr Protoc Neurosci 31(1):5–20

Yasin M, Zahid MA, Ghafoor A, Ahmad Z (2002) Genotypic behavior of lentil (*Lens culinaris* Medik) towards salinity. In: Ahmad R, Malik KA (eds) Prospects for saline agriculture. Springer, Dordrecht, Netherlands, pp 231–235

Yau SK, Erskine W (2000) Diversity of boron-toxicity tolerance in lentil growth and yield. Genet Resour Crop Evol 47:55–62

Ye G, McNeil DL, Hill GD (2002) Breeding for resistance to lentil ascochyta blight. Plant Breed 121(3):185–191

Yim WC, Yu Y, Song K, Jang CS, Lee BM (2013) PLANEX: the plant co-expression database. BMC Plant Biol 13(1):83

Yu JK, Dake TM, Singh S, Benscher D, Li W, Gill B, Sorrells ME (2004) Development and mapping of EST-derived simple sequence repeat markers for hexaploid wheat. Genome 47(5):805–818

Yuan HY, Saha S, Vandenberg A, Bett KE (2017) Flowering and growth responses of cultivated lentil and wild Lens germplasm toward the differences in red to far-red ratio and photosynthetically active radiation. Front Plant Sci 21(8):386

Zaccardelli M, Lupo F, Piergiovanni AR, Laghetti G, Sonnante G, Daminati MG, Sparvoli F, Lioi (2012) Characterization of Italian lentil (*Lens culinaris* Medik.) germplasm by agronomic traits, biochemical and molecular markers. Genet Resour Crop Evol 59(5):727–738

Zaman MW, Mian MAK, Rahman MM (1989) Variability and correlation studies in local germplasm of lentil in Bangladesh. Lens Newsl 16:17–19

Zaman-Allah M, Jenkinson DM, Vadez V (2011) A conservative pattern of water use, rather than deep or profuse rooting, is critical for the terminal drought tolerance of chickpeas. J Exp Bot 62:4239–4252. https://doi.org/10.1093/jxb/err139

Zamir D, Ladizinsky G (1984) Genetics of allozyme variants and linkage groups in lentil. Euphytica 33:329–336

Závodná M, Kraic J, Paglia G, Gregova E, Morgante M (2000) Differentiation between closely related lentil (*Lens culinaris* Medik.) cultivars using DNA markers. Seed Sci Technol 28:217–219

Zhang HB, Choi S, Woo SS, Li Z, Wing RA (1996) Construction and characterization of two rice bacterial artificial chromosome libraries from the parents of a permanent recombinant inbred mapping population. Mol Breed 2(1):11–24

Zhang X, Yuan YR, Pei Y, Lin SS, Tuschl T, Patel DJ, Chua NH (2006) Cucumber mosaic virus-encoded 2b suppressor inhibits Arabidopsis Argonaute1 cleavage activity to counter plant defense. Genes Dev 20(23):3255–3268

Zhang X, Scheuring CF, Zhang M, Dong JJ, Zhang Y, Huang JJ, Chen W (2010) A BAC/BIBAC-based physical map of chickpea, *Cicer arietinum* L. BMC Genomics 11(1):501

Zhang N, Venkateshwaran M, Boersma M, Harms A, Howes-Podoll M, den Os D, Ané JM, Sussman MR (2012) Metabolomic profiling reveals suppression of oxylipin biosynthesis during the early stages of legume–rhizobia symbiosis. FEBS Lett 586(19):3150–3158

Zimniak-Przybylska Z, Przybylska J, Krajewski P (2001) Electrophoretic seed globulin patterns and species relationships in the genus *Lens* Miller. J Appl Genet 42:435–447

Zohary D (1972) The wild progenitor and the place of origin of the cultivated lentil: *Lens culinaris*. Econ Bot 26:326–332

Chapter 5
Towards Development of Climate Smart Mungbean: Challenges and Opportunities

Aditya Pratap, Sanjeev Gupta, P. S. Basu, Rakhi Tomar, Sonali Dubey, Meenal Rathore, Uma Shankar Prajapati, Parikshit Singh and Gita Kumari

Abstract Mungbean is a quantitative short-day plant and grown across environments, locations, and seasons. It has inherent intrinsic tolerance mechanisms to many of the environmental stresses. However, being grown so widely, suffers from high temperatures, terminal moisture stress, soil salinity, and photo-thermo period sensitivity. Significant advancements have been made in the past 3–4 decades towards the development of input responsive, high yielding, disease-resistant, and short-duration varieties in mungbean. However, breeding for abiotic stress resistance has largely remained untouched and consequently, these pose serious constraints to mungbean production. Abiotic stresses such as heat, drought, salinity, etc. have deleterious effects on the morphology, physiology, and reproductive ability of the plants and ultimately reduce their plasticity and adaptation to changing climates, thereby affecting the quality and quantity of the produce significantly. Ample genetic and genomic resources are now available in mungbean and related *Vigna* crops, which can be exploited for the development of climate smart mungbean cultivars. Through various breeding approaches, climate smart traits can be incorporated in mungbean which will lead them to adapt to changing climate and perform well across environments. This chapter focuses on the development of climate smart mungbean and highlights gaps which need to be filled to this effect.

Keywords *Vigna radiata* · Climate resilience · Genomic resources · Genetic transformation · Biotic stresses · Abiotic stresses

5.1 Introduction

More than a dozen pulse crops are grown globally which form an integral part of cropping systems in many countries and offer great significance in sustainability of cereal-based agriculture. Besides, these ensure food and nutritional security in pre-

---

A. Pratap (✉) · S. Gupta · P. S. Basu · R. Tomar · S. Dubey · M. Rathore · U. S. Prajapati · P. Singh · G. Kumari
ICAR-Indian Institute of Pulses Research, Kalyanpur, Kanpur, India
e-mail: Aditya.Pratap@icar.gov.in; adityapratapgarg@gmail.com

C. Kole (ed.), *Genomic Designing of Climate-Smart Pulse Crops*,
https://doi.org/10.1007/978-3-319-96932-9_5

dominantly vegetarian countries like India. Among all pulses, mungbean or green gram (*Vigna radiata* (L.) Wilczek) has a unique position due to its shorter life cycle, high per day productivity and its use in numerous food preparations (Singh et al. 2017). It has a wider adaptability being cultivated in spring, summer and rainy seasons and low input requirements having an inherent ability to fix the atmospheric nitrogen in symbiotic association with *Rhizobium* bacteria. While mungbean is cultivated since ancient times in India, it is also widely grown in Southeast Asia and also in Africa, South America and Australia. In Australia, mungbean cultivation started in 1930s primarily for forage use and as a green manure crop to improve soil fertility in cereal-based farming systems (Chauhan and Williams 2018). In India alone, >4.0 million ha area is currently under cultivation of this crop and a phenomenal increase has been observed in area, production as well as productivity in the last five decades. The total area under this crop increased from 1.84 million ha in 1965–1966 to 4.50 million ha in 2016–2017 while the production increased from 0.55 to 2.83 million tons during the same period (Project Coordinator, MULLaRP Report 2018). With increased irrigation facilities through new irrigation projects, remunerative prices and availability of short-duration cultivars, this crop now occupies considerable area during spring and summer seasons in several parts of India (Gupta and Pratap 2016). Simultaneously, development of new cultivars with shorter maturity duration (60–65 days), high yield (1.0–1.5 t/ha), photo-thermo period insensitivity, synchronous maturity, and resistance to Mungbean Yellow Mosaic disease during the last two decades has promoted mungbean cultivation in summer season (Gupta and Pratap 2016). Summer mungbean can ideally be grown in irrigated tracts of northern and eastern India as catch-cum-cash crop diversifying rice-wheat system, canal command areas of Gujarat and Madhya Pradesh as a bonus crop and new delta areas of Cauvery basin of Tamil Nadu sandwiched between two rice crops. Success in summer cultivation has not only increased the mungbean production but also helped in defeating malnutrition, crop diversification, sustaining agricultural production and increasing household income of poor farmers of India.

Nevertheless, with an expansion in mungbean area in different climates, challenges have also increased manifolds towards the development of widely adaptable climate smart varieties which can perform well across seasons and environments. Despite an average yield potential of >1.2 t/ha for most of the released mungbean varieties, the average productivity is still <0.7 t/ha in India and <1.0 t/ha in several other mungbean growing countries. The yield gap is not likely due to subsistence farming but due to several biotic and abiotic factors including insect pests, diseases, and postharvest losses as well as heat, drought, salinity and photoperiod sensitivity. Mungbean Yellow Mosaic Virus and leaf crinkle during Kharif season in northern part of India and powdery mildew in southern coastal part of India during winter season cause considerable losses. Among the insect-pests, defoliators such as hairy caterpillars, semilooper and sucking pests such as thrips and aphids are common. Activity of thrips starts at the bud stage and poses serious problem when the crop is in flowering stage, resulting in considerable flower drop. Thrips also transmit Groundnut Bud Necrosis Virus (GBNV). To make mungbean a remunerative crop, such varieties are required to be developed which can perform reasonably well even in unfavorable climates. Designing new plant types which are resilient to changes in climate and can perform uniformly well across a series of environments will solve the problems of food and nutritional security.

5.2 Climate Change and Mungbean

The present trend of climate change indicates drought or drought-like situations occurring more frequently and rainfed agriculture is expected to suffer most as a result of water crisis due to delayed monsoon, uneven distribution and above all, complete failure of rains as a result of climate change (Singh et al. 2013). Reduction in yields is predicted to be more pronounced when drought and high temperature will interact together, and the damaging effects of both the stresses will be far more severe than their individual effects. Among pulses, mungbean being a warm season crop and grown under irrigated conditions is likely to be affected less by climate change due to its relatively higher tolerance towards high temperature and assured the availability of water during most of the cultivation period. However, the impact of climate change on mungbean may be serious when its reproductive phase coincides with terminal heat and drought stresses, especially during spring/summer seasons. In *Vigna* crops, the thermal regimes do not change drastically from vegetative to reproductive phase as the total crop duration is very short. In contrast, the cool season pulses (*rabi* crops) witness a clear-cut phase transition from one thermal regime to another when these crops shift from cool temperature vegetative phase (November–January) to reproductive stage at high temperature (February–March). Therefore, winter pulses such as chickpea, lentil, and field pea are more sensitive to abrupt changes in the temperature coinciding with podding stage as compared to crops like mungbean. Various abiotic stresses, such as temperature, drought, and salinity affect the growth of legumes at different developmental stages (Suzuki et al. 2014). Abiotic stresses result in a series of morphological, physiological, biochemical and molecular alterations, which negatively influence plant growth, productivity, and yield (Bita and Gerats 2013).

5.3 Sources of Climate Smart Traits in Mungbean

Germplasm resources are valuable repositories of useful genes which can be exploited for the development of improved cultivars in crop plants. While germplasm collection of pulse crops in India was initiated at the beginning of twentieth century by Botanical Section of the Imperial Agricultural Research Institute at Pusa (Bihar), the systematic efforts were made after the establishment of the All India Coordinated Pulses Improvement Project (AICPIP) in 1966–1967. Later, exploration and germplasm collection was continued by National Bureau of Plant Genetic Resources (NBPGR), New Delhi, India and the state agricultural universities of India. Simul-

Table 5.1 Current status of germplasm resources (wild and related species) at global and national level

Crop	Global holdings	National holdings at NBPGR		
		Indigenous	Exotic	Total
Mungbean	24,918	3567	537	4104
Black gram	3767	3127 + (13)	6	3146
Rice bean	–	1883	179	2062
Wild Vigna	–	490	–	490

Adopted from Singh et al. (2017)

taneously, germplasm collections were also made by United States Department of Agriculture (USDA) and several other international research institutes. Globally, the mungbean germplasm collections are maintained at different places including Indian Council of Agricultural Research (ICAR)-NBPGR; the University of the Philippines; The World Vegetable Center (erstwhile Asian Vegetable Research and Development Center, AVRDC), Taiwan; the Institute of Crop Germplasm Resources of the Chinese Academy of Agricultural Sciences; and the Plant Genetic Resources Conservation Unit of the University of Georgia, USA (Ebert 2013). The University of Philippines and the Rural Development Administration (RDA), Korea also hold duplicates of the mungbean germplasm collection of AVRDC-The World Vegetable Center. The current status of germplasm resources available at national and global level in *Vigna* species are given in Table 5.1. Many of these germplasm accessions are being utilized in national crossing programme for introgression of traits like resistance to diseases and insect pests, wider adaptability, earliness, high yield, large seed, long pod, heat tolerance, etc. (Table 5.2).

Information on gene pools helps the plant breeders in effective utilization of different species for introgression of desirable traits in cultivates species from even distant backgrounds (Pratap et al. 2015a, b). Wild relatives of *Vigna* can offer sources for imparting resistance to several biotic and abiotic stresses besides improving yield and quality traits (Pratap et al. 2012a) (Table 5.3). Realizing the importance of wild relatives, extensive exploration-cum-collection trips have been organized and collections of wild *Vigna* accessions have been maintained at ICAR-NBPGR, New Delhi and ICAR-Indian Institute of Pulses Research (IIPR), Kanpur. All these collections have also been evaluated for various plant traits. In a collection of 206 accessions of 14 wild *Vigna*, species-wide genetic diversity was observed for 45 morphological characters (Bisht et al. 2005). The sub-gene pool of wild types in accession PLN 5 of *V. radiata* var. *Sublobata* (Singh and Ahuja 1977) and IW 3390 of *V. mungo* var. *sylvestris* (Reddy and Singh 1993) has been identified as potential sources of MYMV resistance, and TC 1966 of *V. radiata* var. *sublobata* was identified to carry bruchid tolerance gene (Tomooka et al. 1992). In cowpea, resistance to post-flowering insect pests including legume pod borers and pod-sucking bugs was reported in *V. vexillata* (Fatokun 1991). Similarly, variation for yield components and Mungbean Yellow Mosaic Virus (MYMV) resistance was reported in *V. mungo* var. *sylvestris* and a few accessions of the wild progenitor *V. radiata* var. *sublobata* (Singh 1990). A wild

Table 5.2 Promising trait-specific germplasm of mungbean

Trait(s)	Accession(s)	Country of origin
Wide adaptability, earliness, and resistance to Tungro Mosaic Virus	EC 118889, EC 118894, EC 118895, EC 162584, EC 158782, EC 159734	Taiwan
Resistance to charcoal rot, Leaf Crinkle, tolerance to drought, flood, photoperiod insensitivity	EC 318985-319057	Taiwan
High yielding	EC 391170-75	Indonesia
Large seeded, long podded with shiny green seed coat	EC 393407-10	Bangladesh
Heat tolerant, short and long duration	EC 397138, EC 396394-396423	Thailand
High yielding	EC 390990-93	Taiwan
High yielding	EC 428862	Nepal
Resistance to MYMV	EC 564801-818, EC 565626-633	Taiwan
Early maturity	EC 512780-793	USA
Resistance to powdery mildew	EC 605445	Australia

Table 5.3 Potential sources of alien variation in *Vigna* spp

Character	Species	References
Low trypsin inhibitor activity	*V. tenuicaulis*	Konarev et al. (2002)
Chymotrypsin absence	*V. grandiflora*	Konarev et al. (2002)
High methionine content	*V. radiata* var. *sublobata*	AVRDC (1987), Babu et al. (1988)
High photosynthetic efficiency and drought tolerance	*V. radiata* var. *sublobata*	Ignacimuthu and Babu (1987)
Drought tolerance	*V. aconitifolia*	Jain and Mehra (1980)
Heat tolerance	*V. aconitifolia*	Tomooka et al. (2001)
	V. riukinensis	Egawa et al. (1999)
Insect resistance	*V. unguiculata* ssp. *dekindtiana* var. *pubescens*	Ehlers and Hall (1997)
YMV resistance	*V. radiata* var. *sublobata*	Singh and Ahuja (1977)
High tolerance to saline and alkaline soils	*V. radiata* var. *sublobata*	Lawn et al. (1988)
High no. of seeds/plant and pods/plant	*V. radiata* var. *sublobata*	Reddy and Singh (1990)
Resistance to Yellow Mosaic Virus	*V. radiata* var. *sublobata*	Reddy and Singh (1990), Pal et al. (2000)
	V. trilobata	Nagaraj et al. (1981)
	V. umbellata, V. trilobata, V. mungo	Pandiyan et al. (2008)
Photo-thermoinsensitivity	*V. umbellata, V. glabrescens*	Pratap et al. (2014)

accession of *V. radiata* var. *sublobata*, PLN 15, was found to be the potential donor for pods per plant and seeds per pod (Reddy and Singh 1990). Resistance to MYMV has also been reported in *V. umbellata, V. trilobata and V. mungo* (Nagaraj et al. 1981; Singh et al. 2003). *Vigna mungo* var. *silvestris* has been reported to be immune to bruchids (Fujii et al. 1989; Dongre et al. 1996). Rice bean (*V. umbellata*) was identified as highly useful being a cultivated species and also because many of its accessions show complete resistance or immunity to the bruchids; therefore, gene transfer from rice bean into mungbean and urdbean may be comparatively easy. IC251442 of rice bean and IC 251372 of *V. glabrescens* were reported to be photo-thermo period insensitive (Pratap et al. 2014) and may be utilized for the development of widely adaptable varieties. Hybridization between the cultivated Vignas and their wild relatives in secondary and tertiary gene pools is constrained by crossability barriers and therefore, their successful utilization in crop improvement programmes requires special efforts such as deploying embryo rescue, colchicine treatment, reciprocal crossing, hormonal manipulations, and use of bridge species (Pratap et al. 2015a, b).

5.4 Physiological Characteristics and Crop Phenology

Mungbean, despite being a warm season crop, is grown in diverse climates. Therefore, several physiological and phenological factors influence its yield and stability. Mungbean has epigeal germination and the cotyledons come out of the soil to support the growing plant. Therefore, soils with deficient moisture and hard texture may limit the initial growth of the plant restricting its overall growth and development. Likewise, high initial growth vigor is advocated to be one of the criteria for good summer crop as the crop may suffer from terminal moisture and high temperature stress, especially at the time of flowering and pod formation (Pratap et al. 2013a) and high initial vegetative growth may support the plant at such times. Nevertheless, this has been reported to have no direct relationship with final yield in mungbean (Tekrony and Egli 1991). Variable germination of seeds also has a direct relation with optimum plant stand as a plant stand of about 30–35 plants/m^2 is the optimum to obtain maximum yield in mungbean. Poor plant stand, due to poor germination as one of the factors, is expected to affect the final yield of mungbean negatively, especially in marginal environments (Harris et al. 2005). After successful establishment, the yielding ability of a plant depends upon the ability of the crop to produce and partition dry matter into grain yield which is directly dependent upon several developmental stages in a plant which are further dependent upon its response to photoperiod and temperature (Chauhan and Williams 2018). Plant canopy, leaf area index, biomass accumulation, light interception, conversion of absorbed radiation into assimilates, and partitioning of the assimilates into roots, leaves, pods, and seeds are the major physiological determinants of grain yield in mungbean. A manipulation in the efficiencies of these processes is ultimately dictated by the response of plants to varying photoperiods, available moisture, and changing temperature regimes.

(a) (b)

Fig. 5.1 Popular mungbean cultivars **a** Samrat and **b** Virat. Both these cultivars have narrower leaves and vertical arrangement

The time taken by a mungbean cultivar from sowing to maturity and also harvest is the prime determinant of the quantity and quality of the produce and it also determines the cropping system in which this crop can fit (Chauhan and Williams 2018). Temperature and photoperiod as well as growing conditions affect the crop duration and also its other phenological processes and therefore, the crop may behave variably in different seasons. Mungbean is a quantitative short-day plant and has broad, trifoliate leaves which generally have an overlapping and horizontal orientation. Therefore, narrower mungbean leaflets are expected to allow better light interception. The most popular Indian cultivars of mungbean, viz., "Samrat" and "Virat" (Fig. 5.1) have narrower leaves and vertical orientation allowing better light interception and this may be the reason why these despite being short-duration cultivars (55–60 days) yield high. Lee et al. (2004) reported a mungbean cultivar with narrower leaves and higher light interception to have a higher yield potential as compared to mungbean with broader leaves. Rachaputi et al. (2015) observed that sowing of mungbean at narrow rows of 0.5 m can accelerate closer canopy development and achieve better light interception as compared to wider rows of 1 m. Kuo et al. (1977) reported inadequate leaf area development as one of the limiting factors to yield increase in mungbean. Drought is also reported to affect leaf expansion and appearance rates (Lawn 1982).

5.4.1 Salinity

Legumes are highly salt-sensitive crops and high concentrations of Na and Cl ions around the root zone in water-scarce areas limit the geographical range of legumes in arid and semiarid climates where evapo-transpiration exceeds precipitation (HanumanthaRao et al. 2016). Salinity affects the crop growth and yield in three ways: (a) osmotic stress, (b) ion toxicity, and (c) reduced nodulation and therefore reduced nitrogen-fixing ability. Salinity stress has been reported to cause a significant reduction in mungbean yield (Abd-Alla et al. 1998; Saha et al. 2010). This can occur through reduction in seed germination, root and shoot growth and seedling vigor and the yield reduction level may vary in different genotypes (Promila and Kumar 2000; Misra and Dwivedi 2004). Other pronounced symptoms, viz., enhanced chlorosis, necrosis, and decreased content of chlorophyll and carotenoids are also noticed in mungbean (Gulati and Jaiwal 1993; Wahid and Ejaz 2004). NaCl stress has been reported to have more deleterious effect on roots than shoots, with a sudden dip in root growth associated traits (Friedman et al. 2006; Saha et al. 2010).

During germination under saline conditions, high osmotic pressure of saline water is created due to capillary rise leading to more salt density at seed depth than at lower soil profile, which reduces time and rate of germination (Mudgal et al. 2010). In mungbean seedlings, high salt concentration causes increased H_2O_2 content in both roots and leaves, hence salts should be removed to ensure proper growth and development (Saha et al. 2010). Chakrabarti and Mukherji (2002) reported a decrease in total leaf area and stomatal opening due to high salinity. Likewise, Arulbalachandran et al. (2009) reported a reduction in total chlorophyll content, sugar, starch, and peroxidise enzyme activity in roots and shoots of three different species of *Vigna* (*V. radiata, V. mungo* and *V. unguuiculata*). Germination percentage, seedling growth rate as well as photosynthesis were observed to decrease with increasing levels of NaCl in all the species while the growth decrease was higher in mungbean as compared to the other two species.

Increased salinity was also observed to have a profound effect on nodulation and nitrogen metabolism in mungbean. Salt stress significantly affected initiation, weight, and nitrogen-fixing ability of the root nodules and also lead to inhibition of root colonization by Rhizobium (Mudgal et al. 2010). The ill effect on nitrogen metabolism was more prominent on above ground parts as compared to roots (Munns and Tester 2008).

Maas (1986) observed that soil salinity also delays and reduces flowering and yield of crop plants. Mungbean shows decreased plant growth, photosynthesis as well as yield at higher levels of salinity which leads to delayed pod ripening during the spring season (Sehrawat et al. 2013a, b, c, d). Excessive salt may lead to injury to leaves by entering into the transpiration stream and subsequently may lead to reduced photosynthesis (Hossain and Fujita 2010). Misra and Dwivedi (1995) reported that a salt tolerant mungbean cultivar was characterized by higher levels of total soluble carbohydrates than a salt-sensitive cultivar irrespective of salinity level. While less intense salinity level in the field affects the mungbean crop a little, moderate and uniform salinity leads to restricted vegetative and reproductive growth. Nevertheless, intense form of salinity leads to drastic reduction in plant growth and vigor, flower initiation, and yield.

5.4.2 Temperature

Temperature is an important factor affecting seed yield and quality in legumes. For subtropical and tropical crops, heat stress occurs when the temperatures cross 32–35 °C (Bita and Gerats 2013). For cool season legumes, a daily maximum temperature above 25 °C is considered as the upper threshold for heat stress (Wahid et al. 2007). High temperatures can adversely affect growth, reduce yield, and truncate crop cycles (Araujo et al. 2015). Photosynthesis may be adversely affected by heat stress (Crafts-Brandner and Salvucci 2002) and carbon assimilation associated chloroplast functions coupled with impaired vigor, cellular respiration, N fixation, and metabolism may be affected (Buxton 1996). Mungbean grows well at a mean temperature of 28–30 °C (Carberry 2007; Chauhan et al. 2010). High temperature stress is reported to affect reproductive development in mungbean (Tzudir et al. 2014; HanumanthaRao et al. 2016) and it negatively affects flower initiation, pollen viability, stigma receptivity, ovule viability, ovule size, fertilization, fruit set, seed composition, grain filling as well as seed quality (Barnábas et al. 2008). The sudden rise in the temperature beyond 35 °C causes an increase in the respiration rates and unusually high degradation of stored starch as major chloroplasts carbon source was observed. As a result, failure to set pods, reduced or incomplete grain development at high temperature could be partly due to the inadequate supply of carbon and nitrogen from leaves or by a decrease in the activity of sucrose synthase, the key enzyme playing a crucial role in grain development. Poor partitioning of carbon and nitrogen at high temperature leads to low harvest index and low productivity in mungbean.

Abscission of reproductive organs was ascribed as the primary determinant of yield reduction in heat stress in many grain legumes by Rainey and Griffiths (2005). While terminal high temperature stress is a serious problem in spring/summer grown mungbean, early stage heat stress is observed to occur in *kharif* season crop. These lead to a drastic reduction in crop yield due to pollen sterility, lack of fertilization, flower drop, and embryo/endosperm degeneration leading to the development of undersized seeds. On contrary, low temperature stress, especially at the time of sowing and early growth stage of spring crop, may lead to delayed and reduced germination and slow plant growth. Low temperature stress at early growth stage and sudden rise in temperature during reproductive phase which usually occurs in spring grown crop in northern India has a deleterious effect on mungbean and seriously hampers its growth and development. During flower development, male and female organs are sensitive to high temperature, especially at temperatures beyond 30 °C (Sita et al. 2017). Heat stress severely affects flower bud initiation, and this sensitivity may prevail for 10–15 days (Hedhly et al. 2009; Bita and Gerats 2013). Heat stress influences the reproductive stage by decreasing the number and size of flowers, deforming floral organs, resulting in loss of flowers and young pods, and hence reduction in seed yield (Morrison and Stewart 2002) as reported in mungbean (Tickoo et al. 1996).

Reduced fertilization is a common problem associated with heat stress in many food legumes due to disruption of meiosis and fertilization (Kaushal et al. 2013; Jagadish et al. 2014). High temperature may arrest fertilization by inhibiting the development of male (Jain et al. 2007) and female gametophytes (Snider et al. 2009). Reduced fertilization efficiency due to heat stress has been attributed to increasing oxidative stress, reduced carbohydrates, ATP concentration in gynoecium, and decreased leaf photosynthesis in mungbean (Suzuki et al. 2001). Temperature fluctuations during seed filling have been reported to drastically reduce yield (Kaur et al.

2015a, b). High temperature stress causes yield loss in legumes (Canci and Toker 2009; Kumar et al. 2016) and other crops due to poor seed development (Hall 2004).

Male sterility has also been observed in many heat-stressed food legumes, including mungbean (Kaur et al. 2015a, b) and impaired pollen development has been a vital reason linked to yield losses due to heat stress (Wassmann et al. 2009). Anthers developing under high temperature showed cell proliferation arrest, distended vacuoles, altered chloroplast development, and mitochondrial abnormalities (Sakata et al. 2010). Heat stress decreases the accumulation of carbohydrates in pollen grains and stigmatic tissue by changing the partitioning of the assimilates and the proportion between symplastic and apoplastic loading of the phloem (Taiz and Zeiger 2006), which affects pollen viability (Kaushal et al. 2013). Heat stress decreases the activity of sucrose synthase and many cell wall and vacuolar invertases in developing pollen grains; as a result, the turnover of sucrose and starch turnover are impaired to reduce the accumulation of soluble carbohydrates in mungbean (Kaur et al. 2015a, b).

Temperature exceeding beyond 42 °C during summer, causes hardening of mungbean seeds due to incomplete sink development. Based on multilocation evaluation at Vamban (Tamilnadu) and Durgapura (Rajasthan) in India, 12 promising genotypes (IPM 02-16, IPM 9901-10, IPM 409-4, IPM 02-3, PDM 139, IPM 02-1, IPM 2-14, IPM 9-43-K, PDM 288, EC 470096, IPM 2K14-9, IPM 2K14-5) were identified which have been confirmed to be tolerant to heat and drought. Based upon sucrose synthase activity and protein profiling as biochemical markers, a few promising mungbean varieties were identified as heat tolerant which have been validated by repeated field trial across diverse agroclimatic zones prone to be affected by recurrent high temperature stress. These genotypes are PDM 139 (Samrat), IPM 02-1, PDM 288, IPM 05-3-21, ML-1257.

5.4.3 High Temperature and CO_2

High CO_2 induces closure of stomata and inhibits photosynthesis in mungbean. High temperature x CO_2 interaction studies revealed negative impact on mungbean plants. Results indicated formation of leaf starch at high CO_2 leading to poor assimilate export from source to sink and grain filling was adversely affected. The high level of carbon dioxide is however beneficial after setting of strong sinks, i.e., developing grains with high sucrose synthase activity.

The rate of light-saturated photosynthesis Pmax (PFD 1000 μmol photons m^{-2} s^{-1} at 20 °C) at elevated carbon dioxide (500 ppm) increased both at vegetative as well as grain filling stage (Table 5.4). At the podding stage, when sink demand is high, both photosynthesis and transpiration tremendously increased under elevated CO_2 and without any water-limiting conditions. However, high CO_2 level during vegetative stage contributed towards increased water use efficiency as compared to ambient CO_2 (300 ppm). High photosynthesis accompanied by increased transpiration and

Table 5.4 Relative changes in photosynthetic rates, stomal conductance and transpiration in mungbean at vegetative and podding stage

Crop stage	CO_2 condition (ppm)		Photosynthetic rate (mmol m^{-2} s^{-1})	Stomatal conductance	Transpiration rate	Ahs/Cs	Pn/Gs
Vegetative	Ambient	380	6.45	0.027	0.85	0.006	310.53
Vegetative	Elevated	500	8.96	0.030	1.33	0.009	738.71
Podding	Ambient	380	6.54	0.012	0.40	0.007	750.20
Podding	Elevated	500	13.84	0.235	3.86	0.023	227.20

stomatal conductance under elevated CO_2 supported high sink demand during grain filling.

If the duration of mungbean crop is reduced by 8–10 days without significant yield penalty, the losses caused by these stresses can be avoided in its major production base. Keeping this in view, two extra early maturing mungbean genotypes were developed by the ICAR-IIPR, Kanpur which matured in 50–55 days during Summer as well as rainy seasons (Pratap et al. 2013b). The variety IPM 205-7, popularly known as Virat was developed from the cross IPM 2-1 X EC 398889 and the genotype IPM 409-4 was developed from the cross PDM 288 X IPM 3-1. Both these genotypes also showed high resistance to yellow mosaic disease and were registered with ICAR-NBPGR as unique germplasm (Pratap et al. 2012b). Virat has been developed using identified heat-tolerant germplasm line EC 398889 and an early maturing and high yielding variety "Samrat". Samrat has synchronous podding and rapid grain filling. Physiological and molecular characterization of the heat-tolerant line EC 398889 differed significantly in respect to heat sensitive line LGG 460 when tested with the marker CEDG147 and pollen germination tested at 43 °C. One of the simplest approach to develop combined tolerance to drought and heat is to shorten the crop duration which may help escape terminal heat stress during summer season, induce synchronous podding, helping in single harvest and faster grain filling and integrating traits like osmotic adjustment or deep root system to avoid intermittent drought at early stages and terminal heat >40 °C during pod filling. A number of green gram accessions have been evaluated for heat tolerance and an exotic line EC 398889 has been identified having high levels of heat tolerance as compared to LGG 460. Molecular characterization of both of these accessions revealed significant differences for a specific marker that confirmed with high rate of in vivo pollen germination when pollen exposed to temperature above 44 °C for 2 h (Pratap et al. 2015a, b).

5.4.4 Drought

Mungbean can tolerate moderate temperature and soil moisture deficits and therefore has a definite role to play in drought-prone areas. A fairly regular supply of moisture is desirable for mungbean during growing period while complete dry conditions are required at harvest. Severe drought reduces vegetative growth, flower initiation, and pod set (Morton et al. 1982). It has the ability to extend its roots deeper in the soil in response to drought. The moderate soil moisture is needed for early growth till the onset of flowering and podding. Intermittent drought situations are very critical for this crop, particularly under dry conditions where air water deficit is higher and soil moisture loss is faster due to high evaporation demand. Soil water holding and the crop's water retention capacity both determine the ability of a mungbean cultivar to escape the drought stress before the reproductive stage. The mungbean has low water retention capacity by virtue of having low or inability for osmotically adjusted when subjected to drought, relatively higher lethal leaf water potential (less negative) in response to drought, as a result, the crop quickly loses turgor and stress symptoms may appear recurrently during early growth stages. Plant encounters recurrent transient drought stress when leaf water potential falls below −1.5 MPa but recovers thereafter when load of solar radiation and temperature are diminished. It may not be possible to revive the crop once it reaches lethal leaf water potential nearly to −2.5 MPa.

Some cultivars are more drought tolerant than others which could be due to enhanced ability to close the stomata in the leaves and reduce the rate of growth and leaf expansion during period of severe water stress. There is a variation in the root system in the cultivars of mungbean which can be exploited in breeding programme to develop varieties with delayed dehydration. Drought tolerance rating or sensitivity of pulses are as follows.

Lathyrus > Horsegram > Cowpea > Pigeonpea > Chickpea > Lentil > Mungbean> blackgram > Fieldpea > Rajmash

The lethal water potential is defined as the water status of leaf at the point where plant cannot survive any longer. Comparative studies showed that turgor loss in pulses occurs at much lower leaf water potential than wheat and potato indicating the high tolerance of pulses to drought (Table 5.5). However, as compared to pigeonpea, mungbean has four times less dehydration tolerance which needs to be improved further.

The degree of osmotic adjustment (OA) has also been shown to be correlated with yield under dryland conditions in pulses. From Table 5.6, it is clear that among pulses chickpea, pigeonpea and peanut are more tolerant to drought as compared to mungbean. Genetic diversity of OA can be exploited to inherit drought tolerance trait in mungbean as water demand is proportionately less if OA increases. Moreover, OA increases only when drought is intensified.

Table 5.5 Lethal leaf water potential for a range of grain legumes

Species	Crop	Lethal water potential (MPa)	Dehydration tolerance
Pigeonpea	Legume	−7.0 to −8.2	Very high
Groundnut	Legume	−3.4 to −8.2	Very high
Soybean	Legume	−5.0	High
Mungbean	Legume	−1.9	Moderate
Cowpea	Legume	−1.8 MPa	Moderate
Sorghum	Cereal	−3.0 MPa	High
Wheat	Cereal	0 to −2.0 MPa	Moderate

Table 5.6 Range of osmotic adjustment in grain legumes as compared to cereals and vegetables

Species (pulses)	Range of osmotic adjustment (MPa) in leaves	Degree of dehydration postponement	Species (cereals/vegetables)	Range in osmotic adjustment (MPa)	Degree of dehydration postponement
Groundnut	0.2 to 1.6	Very high	Sorghum	0.8 to 1.7	Very high
Pigeonpea	0.1 to 1.3	High	Wheat	0.2 to 1.5	High
Soybean	0.3 to 1.0	High	Barley	0.2 to 0.5	Moderate
Chickpea	0.0 to 1.3	High	Maize	0.1 to 0.4	Moderate
Lentil	0.0 to 0.6	Moderate	Potato	0.0 to 0.25	Low/sensitive
Mungbean	0.3 to 0.4	Moderate	Lupin	0.1 to 0.5	Moderate
Blackgram	to 0.5	Moderate	Fieldpea	0.0 to 0.4	Moderate
Cowpea	0.0 to 0.4	Moderate	Faba bean	0.0 to 0.2	Low/sensitive
			Lathyrus	0.0 to 0.1	Low/sensitive

5.5 Classical Genetic Studies

Several attempts have been made to understand the genetics of quantitative and qualitative traits related to climate resilience in mungbean besides studying the inheritance of resistance to biotic and abiotic stresses (Table 5.7). Since the first genetic studies on inheritance of color of ripe pods and seed coat surface by Bose (1939), numerous attempts were made to study the inheritance of morpho-physiological traits. Seed coat color is a highly variable trait in mungbean. Khattak et al. (1999) reported monogenic inheritance of this trait and black, black-spotted and dull-green seed coat colors were reported to be dominant over green, nonspotted and shiny green color, respectively. The inheritance of black and green seed colors was found controlled by a single gene (*B*), black being dominant over green (Chen and Liu 2001). For twining habit, a single dominant gene (*T*) was reported to be responsible (Khattak et al. 1999). However, Pathak and Singh (1963) reported a single recessive gene for this trait. Semi-spreading habit was reported to be dominant over erect habit and governed by

Table 5.7 Inheritance and gene action of economically important traits in mungbean

Trait	Inheritance	Reference
Plant type and growth habit	Single dominant/recessive gene, Semi-spreading is dominant over erect habit	Pathak and Singh (1963), Khattak et al. (1999)
Pubescence	Single dominant gene	Murty and Patel (1973)
Nodulation	Additive and non–additive gene action	Singh et al. (1985)
Leaf traits	Single dominant gene, large leaflet is dominant over small leaflet; lobbed is dominant over entire type	Singh and Singh (1995), Talukdar and Talukdar (2003)
Flower color	Single dominant gene	Bose (1939)
Pod color	Single dominant gene	Sen and Ghosh (1959), Murty and Patel (1973)
Pod shattering	Single dominant gene	Verma and Krishi (1969)
Seed coat color	One or few genes; mottling governed by single gene	Khattak et al. (1999), Chen and Liu (2001), Lambrides et al. (2004)
Seed coat surface	Two complementary genes	Sen and Ghosh (1959), Murty and Patel (1973)
Hard seededness	One or few dominant genes involved	Lambrides (1996), Humphry et al. (2005)
Preharvest sprouting	Additive and non–additive gene action; high G $\times$ E interaction	Durga and Kumar (1997)

a single dominant gene (Pathak and Singh 1963). For indeterminate growth habit, a single dominant gene which inherited independently from leaf shape was reported to be responsible (Talukdar and Talukdar 2003). Anthocyanin pigmentation is reported to be associated with drought resistance and heat tolerance in mungbean. On contrary, purple pigmentation on stem, petiole, and veins of the leaves was reported to be controlled by a single dominant gene "*Ppp1*" with pleiotropic effect. There are variable reports for inheritance of yield components in mungbean and it has been reported to be controlled by additive as well as nonadditive gene action in different studies (Dasgupta et al. 1998; Khattak et al. 2002). For seed weight, small seed has been reported dominant over large size (Sen and Murty 1960; Fatokun et al. 1992; Humphry et al. 2005) For leaf traits, narrow lanceolate leaf has been reported to be controlled by two recessive genes, "*nl1*" and "*nl2*". Several reports suggest that the trifoliate leaf is dominant over the entire leaf and this trait is governed by a single dominant gene (Chhabra 1990; Talukdar and Talukdar 2003). However, monogenic control was reported for pentafoliate leaf (Chhabra 1990). There are also a few reports of two dominant genes, "*Tlb1*" and "*Tlb2*" with duplicate gene action for trilobed leaves (Sareen 1985).

Pubescence has been ascribed to impart resistance to insect pests in many crop species. Pubescence of pods was reported to be dominant over nonpubescence and governed by independent duplicate genes (Khadilkar 1963). Seed hardiness is mostly observed in summer grown crops where the temperatures during pod formation and seed filling stage may go beyond 40 °C. Humphry et al. (2005) reported four loci to be responsible for hard seededness through quantitative trait locus (QTL) analysis among which two QTLs of hard seededness were found co-localized with the loci conditioning seed weight. For inflorescence type, the simple types were reported to be controlled by two dominant genes and compound types are double recessive and number of clusters controlled by single gene (Sen and Ghosh 1959; Singh and Singh 1970)

Among biotic stresses, resistance to MYMV in *Vigna* species is reported to be governed by two recessive genes, however, in few cases, resistance has also been reported due to a single dominant/recessive gene. The bacterial pustule in mungbean is due to a dominant gene. The discordance in the nature of inheritance could be ascribed to racial differences in these studies. The allelic relationships have been studied in the case of MYMV only. Resistant lines of mungbean, Tarai local, L-80, LM-214, and LM-294-1 had nonallelic genes for resistance to MYMV (Shukla and Pandya 1985). Resistance to bruchids in mungbean is dominant and is governed by few major genes (probably two) with some modifiers (Sarkar and Bhattacharya 2014).

5.6 Traditional Breeding Strategies

While breeding for developing climate smart mungbean, the inherent physiological attributes should be taken into consideration. The first step is to ensure why the crop is sensitive to a particular stress and what are the associated traits that are lacking or have a reduced expression. The second step is to explore available genetic variation for important traits associated with drought, heat and other climate variable traits and their intregression in the desired genetic background for improving the tolerance level. As suggested in Table 5.3, there is ample scope to improve OA, water use efficiency (WUE), biomass, and photosynthetic rates in mungbean as these features are inherently low in this crop. While numerous improved varieties have been bred in mungbean till date, most of these were developed for improvement of yield traits and resistance to major diseases. Limited attention was paid towards the development of climate smart mungbean. The early phase of systematic varietal development programme in mungbean targeted improving locally adapted but genetically variable populations, mainly by the methods of pure line and mass selections (Singh et al. 2017). This led to the development of several pure line varieties which became highly popular. The emphasis was gradually shifted towards hybridization and selection, later followed by distant hybridization and consequently, >150 mungbean varieties have been bred in mungbean till date in India. The first variety of mungbean, "Type 1" was developed in the year 1936 from a local selection of Muzaffarpur, Bihar and has been extensively utilized in hybridization programme to develop mungbean varieties T 2, K 851 and T 44 and Sunaina. Being a short-duration variety and possessing good seed quality, "T 44" became very popular in Spring/Summer season. Pusa Baisakhi was used later to develop PIMS 4 and Jyoti. Two varieties of mungbean ML 1 and ML 5 were developed from PAU, Ludhiana during the early 1970s and these were further used to develop ML 131, ML 267, ML 337 and ML 23. Large-seeded varieties of mungbean viz., Pusa Vishal, Pant moong 5 and SML 668 were developed from the selection of AVRDC material. Another landmark early duration variety, "Samrat" was also developed as a selection from local material and this variety became highly popular among the farmers.

The recent period has seen the development of largely the photo- and thermo period tolerant varieties in mungbean. Lately, the focus of breeders shifted towards development of short-duration, photo- and thermo period-insensitive varieties of mungbean coupled with resistance to major biotic stresses, viz, yellow mosaic disease and powdery mildew, which contributed significantly to the national mungbean production. For example, KM 2241, HUM 16, MH 2-15, and TMB 37 were other varieties developed through intraspecific hybridization and these became very popular among the farmers in short time (Singh et al. 2017). The variety PKV AKM 4 developed from a cross between BM4 X PS 16 has also been recommended for two zones, viz., Central Zone and South Zone of the country. IPM 02-3 was developed using IPM 99-125 and Pusa Bold 2 and recommended for both spring and *Kharif* seasons. IPM 410-3 (Shikha) has been recommended for Summer season in North West Plain Zone as well as Central zone while this has been performing very well in Kharif season also in North Hill Zone. Later other varieties for high yield, YMV resistance were released which were the products of intraspecific hybridization. IPM 2-14 is one such highly promising variety which has been released for spring cultivation in South Zone of the country and gained tremendous popularity. Another variety DGGV-2 developed from the cross Chinamung x TM-98-50 and Pusa 0672 developed from 11/395 × ML 267 were released for South Zone and North Hill Zone, respectively. Distant hybridization has also led to the development of three mungbean cultivars viz., HUM 1, Pant Moong 4 and IPM 99-125 in India. These cultivars have improved plant types in addition to high yellow mosaic resistance and synchronous maturity.

5.7 Genomic Resources

The availability of high-throughput and cost-effective next-generation sequencing (NGS) platforms as well as high-throughput genotyping technologies have facilitated the generation of massive genomic data for model as well as crop legumes. These platforms have been vital in producing the genome sequence assemblies for the mungbean (Kang et al. 2014). Whole genome-resequencing data are also becoming readily available for mining superior alleles. Genomic resources of mungbean, viz., whole genome/transcriptome sequences (Kang et al. 2014), chloroplast genome

(Tangphatsornruang et al. 2010) and mitochondrial genome sequence (Alverson et al. 2011) and available which are invaluable resource for mungbean research community. These resources may be tremendously useful in designing climate smart mungbean cultivars. A number of quantitative trait loci (QTL) have been reported in mungbean which can be of tremendous use in incorporating various yield and related traits for genetic improvement (Table 5.8).

5.7.1 Nuclear Genome

The development of molecular markers is critical for crop improvement programmes. Moreover, molecular markers are important for integrating useful alleles of wild genetic resources, such as MYMV and bruchid resistance, into domesticated mungbean (Chen et al. 2013). Although molecular marker resources are limited for mungbean, there have been several efforts to identify the genomic regions related to domestication-related traits, including seed size and seed germination (Isemura et al. 2012). Similarly, transcriptomics/gene expression studies, using a range of platforms, have been valuable for identifying candidate genes associated with tolerance/resistance to different stresses as well as several agronomic traits (Campbell et al. 2014; Brasileiro et al. 2015).

Sequencing of *Vigna radiata* genomic DNA was carried out using 454 Life Sciences technology on the Genome Sequencer (GS) FLX System (Tangphatsornruang et al. 2009). A total of 470,024 quality filtered sequence reads was generated with the average read length of 216 bases covering 100.5 Mb. Assembly of the obtained nucleotide sequence reads was performed using the Newbler, de novo sequence assembly software (Margulies et al. 2005; Kang et al. 2014) sequenced domesticated *V. radiata* var. *radiata*, its polyploid relative *V. reflexo-pilosa* var. *glabra* and its wild relative *V. radiata* var. *sublobata*. For *V. radiata* var. *radiata*, the pure line VC1973A was chosen for genome sequencing and a high-quality draft genome sequence of the diploid *V. radiata* var. *radiata* VC1973A with an estimated genome size of 579 Mb (1.2 pg per 2C) was constructed. A total of 22,427 genes were identified in the genome including 160 *Vigna* gene clusters and 1850 genes encoding transcription factors (TFs). Another 2310 noncoding genes were predicted, including 629 transfer RNAs, 280 ribosomal RNAs, 537 microRNAs, 717 small nucleolar RNAs, 110 small nuclear RNAs, and 37 regulatory RNAs. In contrast, the allotetraploid V1160 has a total of 41,844 genes, almost twice the number of mungbean genes. The estimated genome size of polyploid *Vigna* genome (968 Mb) is almost twice that of mungbean genome (579 Mb). The availability of this genome sequence can serve as a model to understand mungbean domestication.

Table 5.8 Major QTLs reported in mungbean for various traits

S.No.	QTL	Trait (s)	Location/linkage group	Amplifying marker	Reference
1	2 QTL (qPMR-1 and qPMR-2)	Powdery mildew/2 QTLs	LG-2 (20.10–57.1%)	RFLP	Kasettranan et al. (2010)
2	20 QTL collectively	Days to Ist flower, days to first pod maturity, days to harvest, 100 seed wt, no. of seed/pod, pod length	–	SSR	Kajonphol et al. (2012)
3	4 QTLs, 11 QTLs	Hard seededness, seed weight	–	RFLP	Humphry et al. (2005)
4	46 QTLs	Seed permeability (4 QTL)	LG-1, LG-2, LG-3, LG-4 (33.7%),	SSR	Isemura et al. (2012)
		Pod dehiscence (2 QTL)	LG-1, LG-7 (10.8–12.7%)		
		PDRW (3 QTL)	LG-1 (20%), LG-6, LG-7		
		Seed size related traits (5–7 QTL)	LG-8 (15.1–22.7%), LG-2 (11.4–16.6)		
		Pod length (5 QTL)	LG-2 (20.5%), LG-7, LG-8 (28.5%)		
		Pod width (4 QTL)			
		Primary leaf width (1 QTL)	–		
		Stem thickness (1 QTL)	LG-2 (10.2%)		
		Branch number (3 QTL)	LG-2 (22.3%), LG-4, LG-6		
		Flowering time (4 QTL)	LG-2 (32.9%), LG-4, LG-6, LG-11		
		Days to pod maturity (6 QTL)	LG-2 (20.3%), LG-4 (19.9%), LG-6, LG-7, LG-9, LG-11		
		Seed number/pod (2 QTL)	LG-1 (7%), LG-2 (9.1%)		
		Total number of pod (4 QTL)	LG-2, LG-4, LG-7 (5.8–12%)		
5	1 major QTL (QCLS)	Cercospora leaf spot resistance	LG-3 (65.5–80.53%)	SSR	Chankaew et al. (2011)
6	1 major QTL (qPMC72V18-1)	Powdery mildew	92.4%	ISSR, ISSR-RGA	Poolsawat et al. (2017)
7	1 QTL (qMYMV)	MYMV	LG-5 (47.43%), LG-10	SSR	Kitsanachandee et al. (2013)
8	1 QTL	Bruchid resistance	LG-8, 3.6 CM	RFLP	Young et al. (1992)

5.7.2 *Chloroplast Genome*

Chloroplast genome of Fabaceae family is known to have undergone more rearrangements than other angiosperms. Based on 454 pyro-sequencing technology, gene content and structural organization of mungbean chloroplast (cp) genome were reported to be similar to common bean, *Phaseolus vulgaris* (Tangphatsornruang et al. 2010). With an average AT content of 64.82%, they reported mungbean chloroplast genome to be 151.27 kb in length including pair of IRs of 26.474 kb separated by small single copy region of 17.427 kb and large single copy region of 80.896 kb. The genome contains 108 unique genes and 19 of these genes are duplicated in the IR. Of these, 75 are predicted protein-coding genes, 4 ribosomal RNA genes, and 29 tRNA genes.

5.7.3 *Mitochondrial Genome*

Mitochondria are rod-shaped organelles considered to be the power generator (ATP) of the cells and generally harbor circular double-stranded DNA molecules of variable sizes, structure, and sequence complexity. Based on shotgun Sanger sequencing, mungbean mitochondrial genome was reported to be protein-gene-poor 401.262 kb in length with that lacks large, recombinationally active repeats and *promiscuous* sequences from chloroplast and nuclear genomes (Liu et al. 2016). The contents of A, T, C, and G in the NM92 mitogenome were found to be 27.48%, 27.41%, 22.63%, and 22.48%, respectively. The NM92 mitogenome encoded 3 rRNAs, 16 tRNAs, and 33 proteins. Eight protein-coding genes (*nad1, nad2, nad4, nad5, nad7, rps3*, and *rps10*) certain introns. Among them, three (*nad1*, *nad2*, and *nad5*) are trans-spliced genes. A phylogenetic tree was reconstructed using the 21 protein-coding genes of 16 crops.

5.7.4 *Marker-Assisted Backcrossing (MABC)*

Success of a breeding programme greatly relies upon the selection efficiency. Selecting desirable plants from segregating and subsequent-segregating progenies that contain appropriate and desirable combinations of genes is a critical component of plant breeding (Ribaut and Betran 1999). Plant breeders typically work with hundreds or even thousands of populations which often contains large numbers (Ribaut and Betran 1999; Witcombe and Virk 2001). Marker-assisted selection may greatly increase the efficiency and effectiveness in plant breeding compared to conventional breeding methods. Once markers that are tightly linked to genes of interest have been identified, prior to field evaluation of large number of plants, breeders may use specific DNA marker alleles as a diagnostic tool to identify the plants carrying genes. Molecular marker-assisted backcross breeding (MABC) deploys foreground selection (FGS) and background selection (BGS) using genome-wide SSR markers for recovery of recurrent parent genome. This is an environment independent, precise, and quick approach for the development of cultivars of the trait of interest (Varshney et al. 2010). Therefore, a plant breeder would prefer to exercise marker-assisted backcrossing (MABC) for development of superior cultivar with desired traits. While considerable success has been reported in deploying MABC in legumes like chickpea (Pratap et al. 2017), this technology is yet to be successfully deployed in crops like mungbean. Nevertheless, success in chickpea has encouraged taking molecular marker-assisted breeding as a routine tool in crop improvement programme in mungbean also.

5.8 Genetic Transformation

Development of highly reproducible regeneration protocol is a prerequisite for widespread application of in vitro tissue culture techniques in legume improvement programmes (Pratap et al. 2018). Success of this technique also depends upon well characterized and cloned genes for target traits. Advancements in genetic engineering of crop plants have ensured recovery of improved plants with genes introgressed in them from across the species barrier (Pratap et al. 2018). Nevertheless, as legume species are largely recalcitrant to in vitro techniques, routine transformation protocols are limited in most of these species. Though efficient protocols for shoot regeneration have been worked on and established for mungbean since long (Gulati and Jaiwal 1992, 1994; Chandra and Pal 1995; Amutha et al. 2003; Khatun et al. 2008; Yadav et al. 2010a, b; Mookkan and Andy 2014), they vary based on genotype and age and type of explant(s). Different explants respond in a variable manner to phytohormones with change in genotype. Variables like explant type (hypocotyl, apical meristem, cotyledonary nodes, excised embryo, etc.), age of explant, basal media (MS with MS salts, B5 with B5 salts, MS with B5 salts), phytohormones (IAA, BA, zeatin, TDZ)-alone and in combination, presence of supplements ($AgNo_3$) decide the success and efficiency of standardized protocol. Literature reveals reports on both direct organogenesis as well as indirect organogenesis for regeneration. In *Vigna*, regeneration through callus has rarely been reported indicating that genetic factors affect regeneration ability. Literature on in vitro regeneration in mungbean is abundant, but its further utilization for genetic transformation and related studies is relatively less.

The first successful recovery of mungbean transgenic plants was reported from cotyledonary node explants in the mungbean cv. K-851 using *Agrobacterium tumefaciens* strain LBA4404 harboring pTOK233 vector carrying β-glucouronidase (*gusA*) and neomycin phosphotransferase II (*npt*II) marker gene at an overall efficiency of 0.9% (Jaiwal et al. 2001). However, transmission of transgenes (*GUS and npt*II) to the progeny was not confirmed. In another report, mungbean transgenic plants were regenerated via direct organogenesis from primary leaf explants of 10-day-old seedlings cv. K-851, cocultivated with disarmed *A. tumefaciens* strain C-58 harboring a pCAMBIA–1301 plasmid comprising β-glucouronidase (*GUS*) and hygromycin phosphotransferase (*hpt*) genes (Mahalakshmi et al. 2006). Hygromycin-selected shoots were rooted and transferred to glasshouse to produce seeds. Presence and stable inheritance of *gus* gene were confirmed by PCR and Southern hybridization and histochemical *GUS* assay confirmed the stable gene expression. However, lower regeneration efficiency of primary leaf explants and lot of escapes on hygromycin-containing medium may limit the use of these explants for routine introduction of desirable genes to mungbean. Tazeen and Mirza (2004) worked on varieties from Islamabad and regenerated shoots via callus. 2,4-D in B5 media was used to induce callus in explants. Sahoo et al. (2016) developed transgenic mungbean having AtNHX1 for salinity tolerance using cotyledonary node as explant with MSB5 media having BAP as sole phytohormone. TDZ was initially used in preculturing. An average transformation efficiency of 2.07% was documented. Reports of use of embryonic axis attached to cotyledon are also reported for transformation work. Mahalakshmi et al. (2006) reported an efficient genotype independent transformation protocol giving an efficiency of 65–75% based on GUS assays. They had used primary leaves cut at node as choice explant, from both 4- and 10-day-old seedlings, and regenerated them post-transformation on B5 media having only BAP as the phytohormone.

Sonia et al. (2007) reported an improved protocol of genetic transformation of mungbean (cv. Pusa 105) using phosphinothricin as selective agent and *Phaseolus vulgaris* α-amylase inhibitor-1 (αAI-1) gene for resistance to bruchids. Vijayan and Kirti (2012) generated transgenic mungbean (cv. ML-267) plants from cotyledonary node explants using kanamycin selection for enhancing resistance against seedling rot pathogen, *Rhizoctonia solani*. Yadav et al. (2012) reported standardization of different parameters for efficient *Agrobacterium*-mediated transformation in mungbean cv. ML267 using double cotyledonary node as explant of choice. A transformation efficiency of 4.2% was reported. They regenerated mungbean on MSB5 media having BAP as the lone phytohormone. Sahoo et al. (2016) employed *Arabidopsis thaliana* tonoplast Na^+/H^+ anti-porter (*AtNHX1*) gene in transgenic mungbean (cv. K-851) for incorporating enhanced salt tolerance based on kanamycin monosulphate selection. These transgenic lines exhibited enhanced tolerance to salt as confirmed by physiological and biochemical studies. Baloda et al. (2017) developed plants with salinity and drought tolerance plants by introducing a gene for an osmoprotectant glycine betaine.

5.9 Agronomic Manipulations

In most of the mungbean producing countries, consistent yields are obtained mainly by multiple harvesting of pods from multiple flushes. However, this kind of production is unsuitable for intensive and mechanized production systems where the row-to-row spacing is much higher than the manually harvested crops. Experimentation has proven that mungbean sown at narrow row spacing yield better. Even in mechanically harvested fields, the mungbeans planted at a row spacing of 50 cm produced better yields in 95% of the seasons in different locations in Australia (Rachaputi et al. 2015). Narrow row spacing reduces evaporative soil losses, especially during summer season and increases overall water use through transpiration and therefore water use efficiency (Chauhan and Williams 2018). Therefore, narrow row spacing may be especially beneficial during summer cultivation of mungbean. This has been well demonstrated by reducing row spacing in mungbean cultivar Virat which yielded better at a spacing of 15 × 7 cm as compared to 30 × 10 cm spacing. Increasing plant population may be anther avenues which may substantially increase its yield although the response to changes in plant density depends to a great extent on the specific characteristics of a variety (Muchow and Charles-Edwards 1982; Pookpakdi and Pataradilok 1993). In such a situation, while branching may be an important trait for maximizing grain yield and crop plasticity, limiting the number of branches may be necessary to maximize yield under closer planting (Chauhan and Williams 2018).

Growing mungbean in a season which has consistent day length and minimal day and night temperatures fluctuations is also one of the strategies to increased mungbean yield. Spring season provides such an opportunity, especially in northern and central parts of India where vast amounts of land are vacant after the harvest of wheat, potato, chickpea, and rapeseed mustard and temperatures during this season are not too high. Nevertheless, terminal temperature and moisture stress may adversely affect mungbean during its reproductive phase, leading to flower drop, less number of pods, shriveled, and hard seeds. Spring mungbean is becoming common in India and there has been tremendous increase in area and productivity in Spring/Summer mungbean in India during the last decade (Gupta and Pratap 2016) and in other parts of Asia (Ali and Kumar 2004).

5.10 Perspectives

Mungbean is a quantitative short-day and warm season plant. However, it is grown across several environments and climatic conditions and accordingly, the breeding efforts have been directed towards the development of varieties suitable to specific niches. In the past two decades, several achievements have been made towards the development of input responsive, high yielding, biotic and abiotic stress resistant, and short-duration varieties in mungbean. The biggest achievement has been made towards reducing the crop duration from 100–120 to 55–65 days which has made it possible to cultivate mungbean in several niches including rice fallows. The reduced duration has made mungbean a suitable candidate to grow as a catch crop during spring/summer season and also as a noncompeting intercrop in cash crops like sugarcane. Development of photo-thermo period-insensitive varieties like Shikha and IPM 2-3 ensured that a few promising varieties could be cultivated over large area without a need to change the varieties in different seasons. Likewise, synchronous maturity in modern-day varieties such as Samrat, Virat, IPM 2-3, HUM 1, HUM 12, MH 421, Pant Mung 5, Pusa Vishal, etc. made it possible to harvest the crop in a single go, thereby saving time and money involved in multiple pickings and also reduce the drudgery involved. The impact of such varieties has also been realized well in production as well as productivity of mungbean which showed a significant increase despite a reduction in its area. Simultaneously remarkable progress was also made in collection, evaluation, characterization, and documentation of germplasm Wild *Vigna* accessions were also collected and evaluated to great extent. There have also been remarkable success stories in transferring alleles from wild *Vigna* relatives to cultivated mungbean backgrounds and a few cultivars have been developed.

However, there are still a few gray areas which need attention, especially while developing a widely adaptable mungbean cultivar which may also suite to changing climates. Terminal heat stress is the major problem in spring/summer mungbean. While shortening crop duration is one of the strategies adopted to escape terminal heat stress, the crop length cannot be further shortened as it will lead to yield penalty. Therefore, developing heat-tolerant genotypes will help in mitigating the effects of high temperature. Breeding short-duration (52–55 days) varieties for spring/summer season with minimum yield penalty, longer duration genotypes (65–75 days) for Kharif season, and varieties with high initial growth vigor for rice fallow will promote this crop in new areas (Singh et al. 2017). Waterlogging at the early growth stage and preharvest sprouting at the time of maturity are the major limitations in Kharif grown mungbean. Soil salinity poses a significant threat in northern and western parts of India. The problem of storage pest, buchid, still remains largely untouched. A major thrust is required on incorporation of preharvest sprouting and bruchid resistance, pyramiding of genes for resistance to major insect pests (thrips, jassids and pod borer) and diseases (MYMV, powdery mildew and Cercospora leaf spot) for which resistance levels are not very high in cultivated germplasm. A number of cultivars have been developed which are resistant to yellow mosaic disease in recent past. However, its vector, whitefly (*Bemisia tabaci)* is considerably affected by prevailing environmental conditions. Natural transmission of YMD happens through whitefly, however, it can also be transmitted in the plants through *Agrobacterium*-mediated infectious clones. Temporal and spatial variations in reactions of mungbean cultivars to YMD are of common knowledge. This type of behavior of the cultivars may be attributed to one or more factors including mixed infection of the viruses, changing virus population, influence of weather, mixing of seeds of two or more cultivars, etc. Most of the resistant genotypes available so far have not been screened specifically against identified viruses and hence they may react differently against each or in combination of viruses. Meager information is available on gene/s expression pattern in YMD susceptible/resistant mungbean cultivars. Keeping in view the gap in the existing knowledge, there is a strong need to map yellow mosaic disease causing viruses in mungbean production hot spots, identify the host factors suppressing the virus multiplication, and ultimately silence the viruses causing yellow mosaic disease through genome editing.

The application of molecular marker technology for exploitation of favorable alleles in the wild *Vigna* relatives will provide an excellent opportunity for advances in mungbean improvement. Cost-effective, polymorphic, and reproducible markers such as SSRs, SNPs, etc., are available in plenty in mungbean now and can be deployed towards the development of improved cultivars employing marker-assisted breeding approaches. Establishing marker-trait association will enable the scientists to manipulate abiotic and biotic stresses constraining crop productivity. Simultaneously, high-throughput sequencing will accelerate the development of new genomic resources. These will together be useful in developing climate smart cultivars of mungbean.

References

Abd-Alla MH, Vuong TD, Harper JE (1998) Genotypic differences in nitrogen fixation response to NaCl stress in intact and grafted soybean. Crop Sci 38:72

Ali M, Kumar S (2004) Prospects of mungbean in rice-wheat cropping systems in Indo-Gangetic plains of India. In: Improving income and nutrition by incorporating mungbean in the cereal fallows of the Indo-Gangetic Plains of South Asia; Proceedings of the final workshop and planning meeting; AVRDC—The World Vegetable Center, Tainan, Taiwan, pp 246–254

Alverson AJ, Zhuo S, Rice DW, Sloan DB, Palmer JD (2011) The mitochondrial genome of the legume Vigna radiata and the analysis of recombination across short mitochondrial repeats. PLoS One 6(1)

Amutha S, Ganapathi A, Muruganantham M (2003) In vitro organogenesis and plant formation in Vigna radiata (L.) Wilczek. Plant Cell Tiss Org Cult 72:203

Araujo SS, Beebe S, Crespi M, Delbreli B, Gonzaliz EM, Gruber V et al (2015) Abiotic stress responses in legumes: strategies used to cope with environmental challenges. Crit Rev Plant Sci 34:237–280

Arulbalachandran D, Sankar GK, Subramani A (2009) Changes in metabolites and antioxidant enzyme activity of three *Vigna* species induced by NaCl stress. Amer Eur J Agron 2:109–116

AVRDC (1987) 1984 Progress report Asian vegetable research and development center, pp 196–199, Shanhua, Taiwan

Babu CR, Sharma SK, Chaterjee SR, Abrol YP (1988) Seed protein and amino acid compositions of wild *Vigna radiata* var. *sublobata* (Fabaceae) & two cultigens *V. mungo* and *V. radiata*. Econ Bot 42:54–61

Baloda A, Madanpotra S, Jaiwal P (2017) Transformation of mungbean plants for salt and drought tolerance by introducing a gene for an osmoprotectant glycine betaine. J Plant Stress Physiol 3:5–11

Barnabás B, Jäger K, Fehér A (2008) The effect of drought and heat stress on reproductive processes in cereals. Plant, Cell Environ 31:11–38

Bisht IS, Bhat KV, Lakhanpaul S, Latha M, Jayan PK, Biswas BK, Singh AK (2005) Diversity and genetic resources of wild species in India. Genet Resour Crop Evol 52:53–68

Bita CE, Gerats T (2013) Plant tolerance to high temperature in a changing environment: scientific fundamentals and production of heat stress tolerant crops. Front Plant Sci 4:273

Bose RD (1939) Studies in Indian pulses. IX. Contributions to the genetics of mung (*Phaseolus radiatus* Linn. Syn. *Ph. aureus* Roxb.). Indian J Agri Sci 9:575–594

Brasileiro AC, Morgante CV, Araujo AC, Leal-Bertioli SC, Silva AK, Martins AC, Vinson CC, Santos CMR, Bonfim O, Togawa RC, Saraiva MAP, Bertioli DJ, Guimaraes PM (2015) Transcriptome profiling of wild *Arachis* from water-limited environments uncovers drought tolerance candidate genes. Plant Mol Biol Rep 33:1876–1892

Buxton DR (1996) Quality related characteristics of for ages as influenced by plant environment and agronomic factors. Anim Feed Sci Technol 59:37–49

Campbell R, Pont SD, Morris JA, McKenzie G, Sharma SK, Hedley PE et al (2014) Genome-wide QTL and bulked transcriptomic analysis reveals new candidate genes for the control of tuber carotenoid content in potato (*Solanum tuberosum*). Theor Appl Genet 127:1917

Canci H, Toker C (2009) Evaluation of annual wild *Cicer* species for drought and heat resistance under field conditions. Genet Resour Crop Evol 56:1
Carberry P (2007) Crop development models. In: Encyclopedia of science encyclopedia of water science, 2nd edn. CRC Press, Boca Raton, FL, pp 121–124
Chandra M, Pal A (1995) Differential response of the two cotyledons of Vigna radiata in vitro. Plant cell Rep 15:248–253
Chakrabarti N, Mukherji S (2002) Growth regulator mediated changes in leaf area and metabolic activity in mungbean under salt stress condition. Indian J Plant Physiol 7:256–263
Chankaew S, Somta P, Sorajjapinun W, Srinives P (2011) Quantitative trait loci mapping of Cercospora leaf spot resistance in munbean, *Vigna radiata* (L.) Wilczek. Mol Breed 28(2):255–264
Chauhan YS, Williams R (2018) Physiological and agronomic strategies to increase mungbean yield in climatically variable environments of northern Australia. Agron 8(6):83
Chauhan YS, Douglas C, Rachaputi RCN, Agius P, Martin W, King K et al (2010) Physiology of mungbean and development of the mungbean crop model. In: Proceedings of the 1st Australian summer grains conference, Australia, Gold Coast, QL, 21–24 June
Chen H, Liu X (2001) Inheritance of seed color and lustre in mungbean (Vigna radiata). Agric Sci Technol Hunan 2:812
Chen HM, Schafleitner R, Bains TS, Kuo CG, Liu CA, Nair RM (2013) The major quantitative trait locus for mungbean yellow mosaic Indian virus resistance is tightly linked in repulsion phase to the major bruchid resistance locus in a cross between mungbean [*Vigna radiata* (L.) Wilczek] and its wild relative *Vigna radiata* ssp. *sublobata*. Euphytica 192:205–216
Chhabra AK (1990) Inheritance of lobed and pentafoliate leaf in mungbean. Indian J Pulses Res 3:69–72
Crafts-Brandner SJ, Salvucci ME (2002) Sensitivity of photosynthesis in a C4 plant, maize, to heat stress. Plant Physiol 129:1773–1780
Dasgupta T, Banik A, Das S (1998) Combining ability in mungbean. Indian J Pulses Res 11:28–32
Dongre TK, Pawar SE, Thakare RG, Harwalkar MR (1996) Identification of resistant source to cowpea weevil [*Callosobruchus maculatus* (F.)] in *Vigna* sp. and inheritance of their resistance in black gram (*Vigna mungo* var. *mungo*). J Stored Prod Res 32:201–204
Durga KK, Kumar SS (1997) Screening for pre-harvest sprouting in pulses. Legume Res 20:193–197
Dwivedi S, Singh DP (2013) Inheritance of purple pigmentation and seed colour in greengram. AGRIS Sci 13(1):63–66
Ebert AW (2013) Ex situ conservation of plant genetic resources of major vegetables In: Normah MN, Chin HF, Reed BM (eds) Conservation of tropical plant species, Springer Press, New York, NY, pp 373–417
Egawa Y, Takeda H, Suzuki K (1999) Research plan on crop heat tolerance at the crop introduction and cultivation laboratory. Japan International Re-search Center for Agricultural Sciences Working Report 14, pp 103–107
Ehlers JD, Hall AE (1997) Cowpea (*Vigna unguiculata* L. Walp.). Field Crops Res 53:187–204
Fatokun CA (1991) Wide hybridization in cowpea: problems and prospects. Euphytica 54:137–140
Fatokun CA, Menancio-Hautea D, Danesh D, Young ND (1992) Evidence for orthologous seed weight genes in cowpea and mungbean based on RFLP mapping. Genet 132:841–846
Friedman R, Altman A, Nitsa L (2006) The effect of salt stress on polyamine biosynthesis and content in mung bean plants and in halophytes. Physiol Plant 76:295–302
Fujii K, Ishimoto M, Kitamura K (1989) Pattern of resistance to bean weevil Bruchidae in *Vigna radiata-mungo-sublobata* complex inform the breeding of new resistant varieties. Appl Entomol Zool 24:126–132
Gulati A, Jaiwal PK (1992) In vitro induction of multiple shoots and plant regeneration from shoot tips of mung bean (Vigna radiata (L.) Wilczek). Plant Cell Tiss Org Cult 29:199
Gulati A, Jaiwal PK (1993) In vitro selection of salt resistant *Vigna radiata* (L.) Wilczek plants by adventitious shoot formation from cultured cotyledon explants. J Plant Physiol 142:99–102
Gulati A, Jaiwal PK (1994) Plant regeneration from cotyledonary node explants of mungbean (Vigna radiata (L.) Wilczek). Plant Cell Rep 13:523

Gupta S, Pratap A (2016) Mungbean-summer cultivation in India (Pocket Guide), AICRP on MULLaRP, Indian Institute of Pulses Research, Kanpur 208024, Extension Bulletin 42

Hall E (2004) Breeding for adaptation to drought and heat in cowpea. Eur J Agron 21:447–454

HanumanthaRao B, Nair RM, Nayyar H (2016) Salinity and high temperature tolerance in mungbean [*Vigna radiata* (L.) Wilczek] from a physiological perspective. Front Plant Sci 7:957

Harris D, Breese W, Rao JK (2005) The improvement of crop yield in marginal environments using 'on-farm' seed priming: nodulation, nitrogen fixation, and disease resistance. Aust J Agri Res 56:1211–1218

Hedhly A, Hormaza JI, Herrero M (2009) Global warming and sexual plant reproduction. Trends Plant Sci 14:30–36

Hossain MA, Fujita M (2010) Evidence for a role of exogenous glycine betaine and proline in antioxidant defense and methyl glyoxal detoxification systems in mungbean seedlings under salt stress. Physiol Mol Biol Plants 16:19–29

Humphry ME, Lambrides CJ, Chapman SC, Aitken EAB, Imrie BC, Lawn RJ, McIntyre CL, Liu CJ (2005) Relationships between hard-seededness and seed weight in mungbean (*Vigna radiata*) assessed by QTL analysis. Plant Breed 124:292–298

Ignacimuthu S, Babu CR (1987) Vigna radiata var. *sublobata* (Fabaceae): Economically useful wild relative of urd and mung beans. Econ Bot 41:418–422

Isemura T, Kaga A, Tabata S, Somta P, Srinives P et al (2012) Construction of a genetic linkage map and genetic analysis of domestication related traits in mungbean (*Vigna radiata*). PLoS ONE 7(8):e41304

Jagadish B, Craufurd PA, Shi W, Oane AR (2014) A phenotypic marker for quantifying heat stress impact during microsporogenesis in rice (Oryza sativa L.). Funct Plant Biol 41:48–55

Jain HK, Mehra KL (1980) Evaluation, adaptation, relationship and cases of the species of Vigna cultivation in Asia. In: Summerfield RJ, Butnting AH (eds) Advances in legume science, vol 273. Royal Botanical Garden, Kew, London, pp 459–468

Jain M, Prasad PVV, Boote KJ, Hartwell JAL, Chourey PS (2007) Effects of season-long high temperature growth conditions on sugar-to-starch metabolism in developing microspores of grain sorghum (Sorghum bicolor L. Moench). Planta 227:67

Jaiwal PK, Kumari R, Ignacimuthu S, Potrykus I, Sautter C (2001) Agrobacterium tumefaciens-mediated genetic transformation of mungbean (Vigna radiata L. Wilczek)—a recalcitrant grain legume. Plant Sci 161:239–247

Kajonphol T, Sangsiri C, Somta P, Toojinda T, Srinives P (2012) SSR map construction and quantitative trait loci (QTL) identification of major agronomic traits in mungbean (Vigna radiata (l.) Wilczek). SABRAO J Breed Genet 44(1):71–86

Kang YJ, Kim SK, Kim MY, Lestari P, Kim KH, Ha B, Jun TH, Hwang WJ, Lee T, Lee J, Shim S, Yoon MY, Jang YE, Han KS, Gwag JG, Moon JK, Lee YH, Park BS, Bombarely A, Doyle JJ, Jackson SA, Schafleitner R, Srinives P, Varshney RK, Lee SH (2014) Genome sequence of mungbean and insights into evolution within *Vigna* species. Nat Commun 5:5443

Kasettranan W, Somta P, Srinives P (2010) Mapping of quantitative trait loci controlling powdery mildew resistance in mungbean (Vigna radiata (L.) Wilczek). J Crop Sci Biotechnol 13(3):155–161

Kaur R, Bains TS, Bindumadhava H, Nayyar H (2015a) Responses of mungbean (*Vigna radiata* L.) genotypes to heat stress: effects on reproductive biology, leaf function and yield traits. Sci Hort 197:527–541

Kaur A, Shevkani K, Singh N, Sharma P, Kaur S (2015b) Effect of guar gum and xanthan gum on pasting and noodle-making properties of potato, corn and mungbean starches. J Food Sci Technol 52:8113

Kaushal N, Awasthi R, Gupta K, Gaur P, Siddique KH, Nayyar H (2013) Heat-stress-induced reproductive failures in chickpea (*Cicer arietinum*) are associated with impaired sucrose metabolism in leaves and anthers. Funct Plant Biol 40:1334–1349

Khadilkar BT (1963) Inheritance of two characters in green gram (*Phaseolus aureus* Roxb. Piper). Akola Agri Coll Mag 3:29–32

Khattak GSS, Haq MA, Rana SA, Srinives P, Ashraf M (1999) Inheritance of resistance to mungbean yellow mosaic virus (MYMV) in mungbean (*Vigna radiata* (L.) Wilczek). Thai J Agri Sci 32:49–54

Khattak GSS, Maq MA, Ashraf M, Tahir GR (2002) Triple test cross for some morphological traits in mungbean [*Vigna radiata* (L.) Welczek]. Euthopia 126:413–420

Khatun MK, Haque MS, Islam S, Nasiruddin KM (2008) In *vitro* regeneration of mungbean (*Vigna radiata* L.) from different explants. Progress Agri 19:2

Kitsanachandee R, Somta P, Chatchawankanphanich O, Akhtar KP, Shah TM, Nair RM, Bains TS, Sirari A, Kaur L, Srinives P (2013) Detection of quantitative trait loci for mungbean yellow mosaic India virus (MYMIV) resistance in mungbean (*Vigna radiata* (L.) Wilczek) in India and Pakisan. Breed Sci 63(4):367–373

Konarev AV, Tomooka N, Vaughan DA (2002) Proteinase inhibitor polymorphism in the genus *Vigna* Savi subgenus *Ceratotropis* and its biosystematic implications. Euphytica 123:165–177

Kumar J, Kant R, Kumar S, Basu PS, Sarker A, Singh NP (2016) Heat tolerance in lentil under field conditions. Legume Genom Genet 7:1–11

Kuo C, Wang L, Cheng A, Chou M (1977) Physiological basis for mungbean yield improvement. In: Proceedings of the International Mungbean Symposium, Los Banos Philippines, 16–19 Aug

Lambrides CJ (1996) Breeding for improved seed quality traits in mungbean (*Vigna radiata* L. Wilczek) using DNA markers. PhD Dissertation, University of Queensland, Brisbane, Australia

Lambrides CJ, Godwin ID, Lawn RJ, Imrie BC (2004) Segregation distortion for seed testa color in mungbean (*Vigna radiata* L. Wilczek). J Hered 95:532–535

Lawn R (1982) Response of four grain legumes to water stress in south-eastern Queensland. II. Plant growth and soil water extraction patterns. Aust J Agri Res 33:497–509

Lawn R, Williams W, Imrie BC (1988) Potential of wild germplasm as a source of tolerance to environmental stresses in mungbean. In: Proceedings of 2nd international symposium, AVRDC, Taiwan, pp 136–145

Lee YS, Lee JY, Kim DK, Yoon CY, Bak GC, Park IJ, Bang GP, Moon JK, Oh YJ, Min KS (2004) A new high-yielding mungbean cultivar, Samgang" with lobed leaflet. Kor Breed J 36:183–184

Liu C, Fan B, Cao Z, Su Q, Wang Y, Zhang Z, Wu J, Tian J (2016) A deep sequencing analysis of transcriptomes and the development of EST-SSR markers in mungbean (Vigna radiata). J Genet 95:527

Maas EV (1986) Salt tolerance of plants. Appl Agri Res 1:12–26

Mahalakshmi SL, Leela T, Kumar MS (2006) Enhanced genetic efficiency of mungbean by use of primary leaf explants. Curr Sci 91:93–98

Margulies M, Egholm M, Altman WE, Attiya S, Bader JS, Bemben LA, Berka J, Braverman MS, Chen YJ, Chen Z et al (2005) Genome sequencing in microfabricated high-density picolitre reactors. Nat 437(7057):376–380

Misra N, Dwivedi UN (1995) Carbohydrate metabolism during seed germination and seedling growth in green gram under saline stress. Plant Physiol Biochem 33:33–38

Misra N, Dwivedi UN (2004) Genotypic difference in salinity tolerance of green gram cultivars. Plant Sci 166:1135–1142

Mookkan M, Andy G (2014) $AgNO_3$ boosted high-frequency shoot regeneration in *Vigna mungo* (L.) Hepper. Plant Signal Behav 9(10)

Morrison MJ, Stewart DW (2002) Heat stress during flowering in summer *Brassica*. Crop Sci 42:797–803

Morton F, Smith RE, Poelman JM (1982) The mungbean. College of Agricultural Sciences, Department of Agronomy and Soils. Puerto Rico, Spec. Publ

Muchow R, Charles-Edwards D (1982) An analysis of the growth of mungbeans at a range of plant densities in tropical Australia. I. Dry matter production. Aust J Agri Res 33:41–51

Mudgal V, Madaan N, Mudgal A (2010) Biochemical mechanisms of salt tolerance in plants: a review. Intl J Bot 6:136–143

Munns R, Tester M (2008) Mechanisms of salinity tolerance. Annu Rev Plant Biol 59:651–681

Murty BK, Patel GJ (1973) Inheritance of some morphological characters in mungbean. Bansilal Amrital Coll Agri Mag 25:1–9

Nagaraj NC, Muniyappa V, Satyan BA, Shanmugam N, Jayarajan R, Vidhyasekaran P (1981) Resistance source for mungbean yellow mosaic virus. In: Proceedings of the national seminar on disease resistance in crop plants, pp 69–72

Pal SS, Singh JJ, Singh I (2000) Transfer of YMV resistance in cultivar SML32 of *Vigna radiata* from other related *Vigna* species. Plant Dis Res 15:67–69

Pandiyan M, Ramamoorthi N, Ganesh SK, Jebraj S, Pagarajan P, Balasubramanian P (2008) Broadening the genetic base and introgression of MYMY resistance and yield improvement through unexplored genes from wild relatives in mungbean. Plant Mutat Rep 2:33–38

Pathak GN, Singh B (1963) Inheritance studies in greengram. Indian J Genet 23:215–218

Pookpakdi A, Pataradilok H (1993) Response of genotypes of mungbean and blackgram to planting dates and plant population densities. Kasetsart J Nat Sci 27:395–400

Poolsawat O, Katiavat C, Arsakit K, Tantasawat PA (2017) Identification of quantitative trait loci associated with powdery mildew resistance in mungbean using ISSR and ISSR-RGA markers. Mol Breed 37(12):150

Pratap A, Gupta DS, Rajan N (2012a) Mungbean. In: Bharadwaj D (ed) Breeding Indian field crops. Agrobios Publishers, New Delhi, India, pp 208–227

Pratap A, Gupta SK, Kumar J, Solanki RK (2012b) Soybean. In: Gupta SK (ed) Technological innovations in major world oil crops, vol I. Breeding. Springer Science + Business Media, New York, pp 293–321

Pratap A, Gupta DS, Singh BB, Kumar S (2013a) Development of super early genotypes in greengram [*Vigna radiata* (L.) Wilczek]. Legume Res 36:105–110

Pratap A, Gupta DS, Singh BB, Kumar S (2013b) IPM 205-7 (IC0589309-IC0589310; INGR11043-INGR11044), a mungbean (*Vigna radiata* (L.) Wilczek) germplasm with super early maturity. Indian J Plant Genet Resour 26:89–90

Pratap A, Basu PS, Gupta S, Malviya N, Rajan N, Tomar R, Latha M, Nadarajan N, Singh NP (2014) Identification and characterization of sources for photo- and thermo-insensitivity in *Vigna* species. Plant Breed 133:756–764

Pratap A, Gupta S, Malviya N, Tomar R, Maurya R, Joseph JK, Singh NP (2015a) Genome scanning of Indian *Vigna* species through microsatellite variation for genetic diversity and population structure analysis. Mol Breed 35:1–3

Pratap A, Gupta S, Malviya N, Rajan N, Tomar R, Latha M, John JK, Singh NP (2015b) Genome scanning of Asiatic *Vigna* species for discerning population genetic structure based on microsatellite variation. Mol Breed 35:178

Pratap A, Chaturvedi SK, Tomar R, Rajan N, Malviya N, Thudi M, Saabale PR, Prajapati U, Varshney RK, Singh NP (2017) Marker-assisted introgression of resistance to fusarium wilt race 2 in Pusa 256, an elite cultivar of *desi* chickpea. Mol Genet Genom 137–1245. https://doi.org/10.1007/s00438-017-1343-z.292

Pratap A, Prajapati U, Singh CM, Gupta S, Rathore M, Malviya N, Tomar R, Gupta AK, Tripathi S, Singh NP (2018) Potential, constraints and applications of in vitro methods in improving grain legumes. Plant Breed 137:235–249

Project Coordinators Report (2018) All India coordinated research project on MULLaRP (Mungbean, Urdbean, Lentil, Lathyrus, Rajmash, Fieldpea). ICAR-Indian Institute of Pulses Research, Kalyanpur, Kanpur, p. 46

Promila K, Kumar S (2000) *Vigna radiata* seed germination under salinity. Biol Plant 43:423–426

Rachaputi RC, Chauhan Y, Douglas C, Martin W, Krosch S, Agius P, King K (2015) Physiological basis of yield variation in response to row spacing and plant density of mungbean grown in subtropical environments. Field Crops Res 183:14–22

Rainey KM, Griffiths PD (2005) Differential responses of common bean genotypes to high temperatures. J Amer Soc Hort Sci 130:18–23

Reddy KR, Singh DP (1990) The variation and transgressive segregation in the wide and varietal crosses of mungbean. Madras Agri J 77:12–14

Reddy MV, Singh KB (1993) Rate reducing resistance to *Ascochyta* blight in chickpeas. Plant Dis 77:231–233

Ribaut JM, Betran J (1999) Single large-scale marker-assisted selection (SLS-MAS). Mol Breed 5:531–541

Saha P, Chatterjee P, Biswas AK (2010) NaCl pre treatment alleviates salt stress by enhancement of antioxidant defense and osmolyte accumulation in mungbean (*Vigna radiata* L. Wilczek). Indian J Exp Biol 48:593–600

Sahoo DP, Kumar S, Mishra S, Kobayashi Y, Panda S, Sahoo L (2016) Enhanced salinity tolerance in transgenic mungbean overexpressing Arabidopsis antiporter (NHX1) gene. Mol Breed 36(10):144

Sakata T, Oshino T, Miura S, Tomabechi M, Tsunaga Y, Higashitani N (2010) Auxins reverse plant male sterility caused by high temperatures. Proc Natl Acad Sci USA 107:8569–8574

Sareen PK (1985) Further genetic analysis of the trilobate leaf mutants in mungbean (*Vigna radiata* var. *aureus* (L.) Wilczek). Curr Sci 54:930–931

Sarkar S, Bhattacharyya S (2014) Inheritance of bruchid resistance and morphological traits in greengram. Indian J Genet 74:98–102

Sehrawat N, Bhat KV, Sairam RK, Jaiwal PK (2013a) Screening of mungbean (*Vignaradiata* L. Wilczek) genotypes for salt tolerance. Intl J Plant Anim Environ Sci 4:36–43

Sehrawat N, Bhat KV, Sairam RK, Jaiwal PK (2013b) Identification of salt resistant wild relatives of mungbean (*Vigna radiata* L. Wilczek). Asian J Plant Sci Res 3:41–49

Sehrawat N, Bhat KV, Sairam RK, Tomooka N, Kaga A, Shu Y et al (2013c) Diversity analysis and confirmation of intra-specific hybrids for salt tolerance in mungbean (*Vigna radiata* L. Wilczek). Inlt J Integr Biol 14:65–73

Sehrawat N, Jaiwal PK, Yadav M, Bhat KV, Sairam RK (2013d) Salinity stress restraining mungbean (*Vigna radiata* L.Wilczek) production: gateway for genetic improvement. Intl J Agri Crop Sci 6:505–509

Sen NK, Ghosh AK (1959) Genetic studies in green gram. Indian J Genet 19:210–227

Sen NK, Murty ASN (1960) Inheritance of seed weight in green gram (*Phaseolus aurcus* Roxb.). Genet 45:1559–1562

Shukla GP, Pandya BP (1985) Resistance to yellow mosaic in greengram. SABRAO J Breed Genet 17:165–171

Singh DP (1990) Distant hybridization in hybrid genus *Vigna:* a review. Indian J Genet 50:268–276

Singh BV, Ahuja MR (1977) *Phaseolus sublobatus* Roxb. A source of resistance to yellow mosaic virus for cultivated mung. Indian J Genet 37:130–132

Singh TP, Singh KB (1970) Inheritance of clusters per node in mungbean (*Phaseolus aureus* Roxb.). Curr Sci 39:265

Singh BB, Singh DP (1995) Inheritance of a small leaf mutant in mungbean. Indian J Genet 55:69–70

Singh BB, Singh SR, Adjadi O (1985) Bruchid resistance in cowpea. Crop Sci 25:736–739

Singh KP, Monika, Sareen PK, Kumar A (2003) Interspecific hybridization studies in *Vigna radiata* L. Wilczek and *Vigna umbellata* L. Natl J Plant Improv 5:16–18

Singh BB, Pratap A, Basu PS (2013) Development of climate resilient pulse varieties. In: Proceedings of national conference of plant physiology on current trends in plant biology research. Directorate of Groundnut Research, Junagadh and Junagadh Agricultural University, Junagadh, India, pp 118–129, 13–16 Dec

Singh DP, Singh BB, Pratap A (2017) Genetic improvement of mungbean and urdbean and their role in enhancing pulse production in India. Indian J Genet 76:550–567

Sita K, Sehgal A, Kumar J, Kumar S, Singh S, Siddique KHM, Nayyar H (2017) Identification of high-temperature tolerant lentil (*Lens culinaris* Medik.) genotypes through leaf and pollen traits. Front Plant Sci 8:744

Snider JL, Oosterhuis DM, Skulman BW, Kawakami EM (2009) Heat stress-induced limitations to reproductive success in *Gossypium hirsutum*. Physiol Plant 137:125–138

Sonia SR, Singh RP, Jaiwal PK (2007) Agrobacterium tumefaciens mediated transfer of Phaseolus vulgaris α-amylase inhibitor-1 gene into mungbean Vigna radiata (L.) Wilczek using bar as selectable marker. Plant Cell Rep 26:187–198

Suzuki K, Takeda H, Tsukaguchi T, Egawa Y (2001) Ultrastructural study on degeneration of tapetum in anther of snap bean (*Phaseolus vulgaris* L.) under heat stress. Sex Plant Reprod 13:293–299

Suzuki N, Rivero RM, Shulaev V, Blumwald E, Mittler R (2014) Abiotic and biotic stress combinations. New Phytol 203:32–43

Taiz L, Zeiger E (2006) Stress physiology. In: Taiz L, Zeiger E (eds) Plant physiology. Sinauer Associates, Sunderland, pp 671–681

Talukdar T, Talukdar D (2003) Inheritance of growth habit and leaf-shape in mungbean [*Vigna radiata* (L.) Wilczek.]. Indian J Genet 63:165–166

Tangphatsornruang S, Somta P, Uthaipaisanwong P, Chanprasert J, Sangsrakru D, Seehalak W, Sommanas W, Tragoonrung S, Srinives P (2009) Characterization of microsatellites and gene contents from genome shotgun sequences of mungbean (*Vigna radiata* (L.) Wilczek). BMC Plant Biol 9:137

Tangphatsornruang S, Sangsrakru D, Chanprasert J, Uthaipaisanwong P, Yoocha T, Jomchai N, Tragoonrung S (2010) The chloroplast genome sequence of mungbean (*Vigna radiata*) determined by high-throughput pyrosequencing: structural organization and phylogenetic relationships. DNA Res 17:11–22

Tazeen S, Mirza B (2004) Factors affecting Agrobacterium tumefaciens mediated genetic transformation of Vigna radiata (L.) Wilczek. Pakistan J Bot 36:887–896

TeKrony DM, Egli DB (1991) Relationship of seed vigor to crop yield: a review. Crop Sci 31:816–822

Tickoo JL, Grajraj R, Matho M, Manji C (1996) Plant type in mungbean. In: Asthana AN, Kum H (eds) Proceedings of recent advances in mungbean. Indian Society of Pulses Research, Kanpur, India, pp 197–213

Tomooka N, Lairungruang C, Nakeeraks P, Egawa Y, Thavarasook C (1992) Development of bruchid resistant mungbean using wild mungbean germplasm in Thail. Plant Breed 109:60–66

Tomooka N, Vaughan DA, Xu RQ, Kashiwaba K, Kaga A (2001) Japanese native *Vigna* genetic resources. Jpn Agri Res Q 35:1–9

Tzudir L, Bera PS, Chakraborty PK (2014) Impact of temperature on the reproductive development in mungbean (Vigna radiata) varieties under different dates of sowing. Intl J Bioresour Stress Manage 5:194–199

Verma SNP, Krishi JN (1969) Inheritance of some qualitative characters in greengram (*Phaseolus aureus* Roxb.). Indian J Hered 1:105–106

Vijayan S, Kirti PB (2012) Mungbean plants expressing BjNPR1 exhibit enhanced resistance against the seedling rot pathogen, Rhizoctonia solani. Transgen Res 21:193

Wahid A, Ejaz MHR (2004) Salt injury symptom, changes in nutrient and pigment composition and yield characteristics of mungbean. Intl J Agri Biol 6:1143–1152

Wahid A, Gelani S, Ashraf MR, Foolad MR (2007) Heat tolerance in plants: an overview. Environ Exp Bot 61:199–223

Wassmann R, Jagadish S, Sumfleth K, Pathak H, Howell G, Ismail A (2009) Regional vulnerability of climate change impacts on Asian rice production and scope for adaptation. Adv Agron 102:91–133

Witcombe JR, Virk DS (2001) Number of crosses and population size for participatory and classical plant breeding. Euphytica 122:451–462

Yadav M, Chaudhary D, Sainger M, Jaiwal PK (2010a) Agrobacterium tumefaciens-mediated genetic transformation of sesame (Sesamum indicum L.). Plant Cell Tiss Org Cult 103:377

Yadav SP, Ibaraki Y, Dutta Gupta S (2010b) Estimation of the chlorophyll content of micropropagated potato plants using RGB based image analysis. Plant Cell Tiss Org Cult 100:183

Yadav SK, Katikala S, Yellisetty V, Kannepalle A, Narayana VM, Shanker AK, Bandi V, Bharadwaja KP (2012) Optimization of Agrobacterium mediated genetic transformation of cotyledonary node explants of Vigna radiata. Springer Plus 1:59

Young ND, Kumar L, Menancio-Hautea D, Danesh D, Talekar NS, Shanmugasundarum, Kim DH (1992) RFLP mapping of a major bruchid resistance gene in munbean (*Vigna radiata*, L. Wilczek). Theor and Appl Genet 84(7–8):839–844

Chapter 6
Genomic Designing for Climate-Smart Pea

Xuxiao Zong, Tao Yang, Rong Liu, Zhendong Zhu, Huijie Zhang, Ling Li, Xiaoyan Zhang, Yuhua He, Suli Sun, Quanlan Liu, Guan Li, Ruijun Guo, Xiangdong Hu, Baoyu Shen, Jiliang Ma and Tianjing Zhang

Abstract Pea (*Pisum sativum* L.), a diploid ($2n = 2x = 14$) annual cool-season legume crop adapted to a wide range of climates and altitudes, plays a very important role for sustainable agriculture as rotation and cash crops for food, vegetable, fodder, manure, etc. The genome size of a pea is estimated at 4.45 Gb comprising

X. Zong (✉) · T. Yang · R. Liu · Z. Zhu · S. Sun · G. Li · R. Guo
Institute of Crop Sciences, Chinese Academy of Agricultural Sciences, No. 12, Zhong Guan Cun South Avenue, Haidian District, Beijing, China
e-mail: zongxuxiao@caas.cn

T. Yang
e-mail: yangtao02@caas.cn

R. Liu
e-mail: liurong@caas.cn

Z. Zhu
e-mail: zhuzhendong@caas.cn

S. Sun
e-mail: sunsuli@caas.cn

G. Li
e-mail: liguan1221@163.com

R. Guo
e-mail: guo_2009_2013@nwafu.edu.cn

H. Zhang
Department of International Cooperation, Chinese Academy of Agricultural Sciences, No. 12, Zhong Guan Cun South Avenue, Haidian District, Beijing, China
e-mail: zhanghuijie@caas.cn

L. Li · B. Shen · T. Zhang
Liaoning Research Institute of Cash Crops, Liaoyang, Liaoning Province, China
e-mail: lilingsws@126.com

B. Shen
e-mail: shenby6336@126.com

T. Zhang
e-mail: 873613102@qq.com

C. Kole (ed.), *Genomic Designing of Climate-Smart Pulse Crops*,
https://doi.org/10.1007/978-3-319-96932-9_6

large amount of repetitive sequences with high complexity, so that the complete reference genome sequence of pea has not been published yet, which hindered the development of genome-assisted breeding in pea. This chapter discussed the challenges, priorities, and prospects of pea as climate-smart (CS) crop, in food, nutrition, energy, and environment security, effects on global warming and climate change to the industry and breeding of pea. For details, studies on CS agronomic traits of peas like flowering time, root characters, nutrient-use efficiency, water use efficiency, carbon and nitrogen sequestration, greenhouse gas emission, genome plasticity, as well as specific traits for vegetable purposes, were reviewed; CS stress tolerance/resistance traits studies of peas, like cold tolerance, drought tolerance, salinity tolerance, disease resistance, insect resistance were also reviewed. Pea-hosted biological nitrogen fixation (BNF) and soil resources, rhizobium for nodulation, characterization for rhizobium, interaction between pea and its anchored rhizobium, interaction between rhizobium and soil, optimized operation for rhizobium fertilization were illustrated. Utilizations of primary gene pool, secondary gene pool, tertiary gene pool, artificially induced/incorporated traits/genes in CS pea genetic development were reviewed. Of CS pea studies, classical mapping efforts, classical breeding achievements (yield, quality, stress resistance, etc.), limitations of traditional breeding and rationale for molecular breeding, genetic diversity analysis of *Pisum* genus using various means, such as association mapping studies between important traits and markers, molecular mapping of CS genes and QTLs, marker-assisted breeding for CS traits, genomic-aided breeding for CS traits, were all reviewed. Social, political, and regulatory issues concerning CS peas, for concerns and compliances, patent and IPR issues, disclosure of sources of GRs, access and benefit sharing, famers' rights, traditional knowledge, treaties and conventions, participatory breeding, in China and elsewhere were discussed. Peas, especially green pea production dramatically expanded and became increasingly important from the beginning of this century. Achievements on

X. Zhang
Qingdao Academy of Agricultural Sciences, No. 168, Wannianquan Road, Licang Distinct, Qingdao, China
e-mail: zxysea@126.com

Y. He
Institute of Grain Crops, Yunnan Academy of Agricultural Sciences, Kunming, China
e-mail: trbio@163.com

Q. Liu
Department of Bioengineering and Biotechnology, College of Marine Science and Biological Engineering, Qingdao University of Science & Technology, Zhengzhou Road 53, Qingdao, China
e-mail: liuquanlan@qust.edu.cn

H. Zhang · X. Hu · J. Ma
Institute of Agricultural Economics and Development, Chinese Academy of Agricultural Sciences, No. 12, Zhong Guan Cun South Avenue, Haidian District, Beijing, China
e-mail: huxiangdong@caas.cn

J. Ma
e-mail: majiliang@caas.cn

pea studies will lead to genomics, phenomics, and genome editing exploration to assist CS pea breeding purpose in the future.

Keywords Pea (*Pisum sativum* L.) · Genetic resources · Agronomic traits · Tolerance/resistance · Biological nitrogen fixation (BNF) · Genomics · Breeding

6.1 Introduction

Pea (*Pisum sativum* L.) is a diploid ($2n = 2x = 14$) annual cool-season food legume crop with an estimated genome size of about 4.45 Gb (Doležel et al. 1998; Doležel and Greilhuber 2010; Smýkal et al. 2012; Praça-Fontes et al. 2014). In addition, the genome of pea is comprised of a large amount of repetitive sequences ranging from 50 to 97% (Flavell et al. 1974; Murray et al. 1981; Macas et al. 2007; Novák et al. 2010; Smýkal et al. 2012). Due to the complexity of the pea genome mentioned above, the complete reference genome sequence of pea has not been published yet, which hindered the development of genome-assisted breeding in pea.

Adapted to a wide range of climates and altitudes, accessions of pea have been collected and maintained within several major well-characterized collections worldwide. These include but are not limited to collections held at National Genebank of CAAS (NGC), China; John Innes Centre (JIC), UK; Nordic Gene Bank (NGB), Sweden; United States Department of Agriculture (USDA), USA; International Center for Agricultural Research in the Dry Areas (ICARDA), Syria; and Vavilov Institute, Russia (Zong et al. 2009a, b). Archaeological evidence shows that pea appeared in the near east and Middle East about 10,000 years ago, and was domesticated by farmers in Neolithic era and is one of the earliest domesticated crops (Zohary and Hopf 1973; Smýkal et al. 2011, 2015). Pea is rich in various nutrients, and is as an important cash crop for food, vegetable and forage, which are cultivated worldwide (Smýkal et al. 2012; Tayeh et al. 2015a).

Due to the climate change and extreme climatic events as well as the evergrowing world population, crop adaptation to variable environments in production must be enhanced in combination with crop productivity increasing at a much greater rate (Godfray et al. 2010; Lobell et al. 2011; Bevan et al. 2017). To ensure and support the sustainability and food security, genetic improvements in crop performance and climate-smart (CS) traits are crucial for increasing crop productivity via marker-assisted breeding (Collard and Mackill 2008; Tilman et al. 2011; Lipper et al. 2014; Bevan et al. 2017).

According to FAO statistical database (FAOSTAT 2018), in 2016, the yield of both dry peas and fresh peas ranked second only to that of common beans. As we all know, pulses are important crops to support food and nutritional security worldwide, which play an important role not only in crop diversification but also in soil improvement by means of symbiotic nitrogen fixation (Cutforth et al. 2007; Pathak et al. 2017). Owing to its association with nitrogen-fixing bacteria, pea is a valuable component for low-input cropping systems. The biological nitrogen fixation capacity of pea ensures the sustainable development in ecological and cropping systems, and can effectively improve soil fertility and structure, at the same time, pea can also reduce diseases and pests through rotations with other nonlegume crops (Macwilliam et al. 2014; Tayeh et al. 2015a). In addition to the important economic value and ecological advantage, pea also has important theoretical research value. Since Mendel's discovery of the laws of heredity, pea has been widely used by geneticists as a model plant for hybridization experiments (Ellis et al. 2011; Reid and Ross 2011).

6.2 Challenges, Priorities and Prospects of Pea Industry and Breeding

As a cool-season legume crop, pea (*Pisum sativum* L.) has successfully colonized many areas of the planet due to its adaptability to grow in a wide range of conditions from spring growing areas to winter growing areas. The ancient humans have exploited in domestication of this species probably due to its successful symbiotic relationship with soil-borne *Rhizobium* bacteria that allow the plants to produce their own nitrogen nutrition, thereby extending the range of its growing habitats to soils of infertility. The dry and fresh seeds of pea contain high level of proteins with high food value, although this was not understood by the early cultivators of this species. The most significant historical hybridization efforts on pea was carried out by Mendel (1866), and his work was overlooked by researches for long period of time until it was rediscovered by several scientists (Bateson 1901; Druery and Bateson 1901), and this could be through the initiative of pea breeding in the world. Since then, the priorities of traditional pea breeding were earliness, standing ability at harvest, resistance to downy mildew, hundred-seed weight, protein content, which mainly aimed at high yield potential of varieties for white peas, large blue peas, maple peas, and marrowfat peas (Anthony 2017).

6.2.1 *Food, Nutrition, Energy, and Environment Security*

Pea has comprehensive and balanced nutrition (Table 6.1). Pea seeds are composed of seed coat, cotyledon, and embryo. Among them, protein, fat, carbohydrate, and mineral nutrients contained in cotyledon of dry pea accounted for 96%, 90%, 77% and 89% of the total amount of these nutrients in seeds, respectively (Zong 2002). Although the embryo is rich in protein and mineral elements, the seed coat contains most of the undigestible carbohydrates of the seeds, including high levels of calcium and phosphorus (Zong 2002).

Albumin, globulin, and gluten accounted for 21%, 66%, and 2%, respectively, in dry pea seeds (Zong 2002). In the amino acid composition of pea protein, sulfur amino

Table 6.1 Nutritional composition of pea (per 100 g)

Items	Dry pea	Fresh seeds	Snap and snow peas (green pods)
Water (g)	8.0–14.4	55.0–78.3	83.3
Protein (g)	20.0–24.0	4.4–11.6	3.4
Lipid (g)	1.6–2.7	0.1–0.7	0.2
Carbohydrate (g)	55.5–60.6	12.0–29.8	12.0
Crude fiber (g)	4.5–8.4	1.3–3.5	1.2
Ash (g)	2.0–3.2	0.8–1.3	1.1
Quantity of heat (kcal)	322–347	80–161	53.0

acid is the first limiting amino acid (Table 6.2), although it is more abundant than that in other food legumes. Pea protein contains more sulfur amino acids, including tryptophan, lysine, threonine, and methionine. However, its arginine, leucine, and phenylalanine were less abundant than globulin (Zong 2002).

It is reported that biological value (BV) of pea protein is 48–64%, and the efficacy ratio (PER) is 0.6–1.2, higher than that of soybean but lower than that of corn, rice, wheat, and peanut (Table 6.3). The reason why the nutritional value of pea protein cannot be fully exerted is its poor digestibility, the deficiency of sulfur-containing amino acids and the anti-nutritive substances (Zong 2002). The lack of sulfur-containing amino acids is the main reason for the low price of pea protein (Zong 2002). The addition of methionine to pea food can significantly increase the biological value of protein. Experiments have shown that albumin in pea protein is the most nutrient-rich relative to globulin and gluten because of the relatively high

Table 6.2 Essential amino acid content in pea protein (g/16 gN)

Amino acid composition	Results of analysis				WHO/FAO standard	Success rate (%)
	1	2	3	Mean		
Lysine	7.0	8.9	7.2	7.7	5.5	140.0
Threonine	3.8	4.2	3.7	3.9	4.0	97.0
Valine	4.8	6.5	4.8	5.4	5.0	108.0
Leucine	6.9	9.5	7.2	7.9	7.0	112.8
Isoleucine	4.5	7.4	4.0	5.3	4.0	132.5
Tryptophan	1.0	0.7	1.0	0.9	1.0	90.0
Methionine plus Cystine	0.8	1.3	2.0	1.4	3.5	40.0
Phenylalanine plus Tyrosine	4.5	4.6	7.6	5.6	6.0	93.3

Table 6.3 Nutritional quality of protein in pea and other common grains

Crops	Total protein/food caloric	Protein net utilization (%)	Available protein content (%)
Corn	0.110	43	4.7
Rice	0.090	54	4.9
Wheat	0.134	44	5.9
Peanut	0.163	39	6.4
Soybean	0.390	22	8.6
Pea	0.230	34	8.6

Table 6.4 Carbohydrates in pea seeds (%)

Composition	Round (smooth) peas	Wrinkled peas
Total sugar	5.3–8.7	10.2–15.1
Sucrose	2.3–2.4	2.3–4.2
Cottonseed sugar	0.3–0.9	10.2–1.6
Stachyose	2.2–2.9	2.9–5.5
Verbascose	1.7–3.2	2.2–4.2
Starch	36.9–48.6	24.0–36.6
Cellulose	0.9–4.9	1.2–4.2
Hemicellulose	1.0–5.1	0.9–6.6
Lignin	0.5–0.9	0.3–1.0
Total carbohydrate	44.6–68.2	36.6–63.5

level of sulfur-containing amino acids. Pea protein contains more lysine and is a good raw material for lysine extraction (Zong 2002).

Dry pea seeds contain about 60% carbohydrate, which includes starch, sugar, and coarse fiber. It also contains about 2% fat (Zong 2002). The sucrose content in round dry pea seeds was about 2.4%, accounting for 22–25% of the total sugar content (Zong 2002). The sugar content of crumpled peas was higher than that of round peas (with smooth seed coat) (Table 6.4). Dry pea seeds contain about 24–49% starch. Wrinkled peas have a relatively low starch content, while round peas have a relatively high starch content (Zong 2002). Generally speaking, amylose is the main starch in corrugated peas, while amylopectin is the main starch in rounded peas (Zong 2002). Pea starch diameter is generally about 40 microns (Zong 2002).

Crumpled peas had more hemicellulose and rounded peas had less (Zong 2002). The crude fiber of pea was mainly concentrated in seed coat, which accounted for 8.22% of seed weight, and contained 55.2% cellulose and 23.1% hemicellulose in the whole seed (Table 6.4). Crude fiber cannot be digested by human intestines and stomach, it is considered to be the least important component in dietary composition consequently, cellulose among them is digested hardest and the use that still can affect other nutrition especially protein (Zong 2002). However, in the diet of western

Table 6.5 Vitamins and minerals in dry pea seeds, green pea seeds and pod-eating pea pods (mg/100 g)

Nutrients	Dry pea	Fresh seeds	Snap and snow peas (green pods)
Vitamin B1	0.68–1.27	0.11–0.54	0.31
Vitamin B2	0.19–0.36	0.04–0.31	0.15
Nick acid	2.0–4.0	0.17–3.1	2.5
Folic acid	7.5		
Choline	235.0		
Vitamin C	4.0–9.0	9–38	25
Carotene	3.2–37.4	0.15–0.33	0.3
Vitamin PP	0.04–0.55		
Calcium	68–118	13–63	20
Phosphorus	307–471	71–127	80
Iron	4.4–8.3	0.8–1.9	1.5

developed countries, the importance of dietary fiber has been recognized in recent years because it can stimulate gastrointestinal motility (Zong 2002).

According to the analysis, the fat content in pea seeds is 1.1–2.8%, most of which exists in the state of oil. The fat content in seed coat was very little, and the fat content in cotyledons accounted for about 90% of the whole seed fat content (Zong 2002). The fat content in embryo is very high, but it is limited because of its small amount. Higher fat content is often associated with crumpled grain characteristics. Research shows that 60% of the fatty acids in pea seeds are unsaturated fatty acids (Zong 2002).

Dry pea seeds are rich in vitamins B_1, B_2, and niacin (Zong 2002). There is a big difference between different test results. In addition to genetic and growth environmental conditions and other factors, the operation of the test has a considerable impact on the final test results, such as the heating process, air oxidation, filtration loss and so on can destroy or lose part of the vitamin (Zong 2002). The total content of mineral elements in dry pea seeds is about 2.5%, which is a high-quality mineral nutrient source of potassium, iron, and phosphorus (Table 6.5). The edible part of green pea was up to 100%, and the edible part of the tender pod was 90–95%. Green pea and pod pea pods are not only rich in protein, carbohydrate, fat and other nutrients, but also rich in vitamins and mineral nutrients, and is a high-quality vegetable. Green peas and pod peas are rich in vitamin C, so they can effectively prevent gum bleeding, and can prevent colds (Zong 2002).

In addition to rich nutrients, pea also contains trypsin inhibitors, inositol hexaphosphate, plant thrombin, saponin, and other anti-nutrition solids, although the content is very small, the digestion and absorption of pea nutrition has a certain negative effect. Pea food processing can destroy most of the harmful ingredients and improve the nutritional value of pea (Zong 2002). In Chinese medicine, pea tasted a little bit sweet and has efficacy of antifebrile poison. Fresh pea juice drink can treat diabetes. Pea grinds besmeared the attacked skin surface of patient, the carbuncle swollen and haemorrhoids can be treated (Zong 2002).

The potential of biological nitrogen fixation (BNF) of peas can reduce the usage of fossil energy. The available manure N is taken into account in the future, combined with BNF, which can make sustainable use of arable land. Biological nitrogen fixation can be increased greatly in agriculture. Intensive use of peas in crop rotations could reduce the use of fertilizer N by 60% compared to current practices in China. It would reduce the usage of fossil energy, taking into account the energy use of machines. Nonlegumes can be replaced with legumes like peas in various ways, which would mean a 60–110 kg/ha decrease in the use of inorganic N compared to cultivation of cereals (Kankanen 2015). Furthermore, knowledge concerning legume crops which are able to replace fertilizer N for the following crops in different cropping situations. N fertilization replacement value of the legume crop is estimated 20–30 20 kg ha^{-1}, as the amount of chemical fertilizer that could be reduced after growth of legumes, without a reduction in the optimal growth of the subsequent nonlegume crop (Kankanen 2015). In Europe, the production of ammonia (NH_3) of which N fertilizers are almost entirely manufactured, consumes on an average 35.2 GJ t^{-1} of fossil energy per ton. However, the calculations were based on average 31.8 GJ t^{-1} of NH_3 which is the value for the most effective techniques in current factories (Kankanen 2015). Based on this, the total energy consumption of fertilizer N production and transportation is about 39 GJ t^{-1}. Big amount of energy will be saved when rotations involved pea and other legume crops for sustainable agriculture.

6.2.2 *Effects on Global Warming and Climate Change*

In recent years, traditional farming patterns have been hit by global climate anomalies. In order to better adapt to the changeable climate, we carried out the research on pea winter frost resistance. On the one hand, when low-temperature freezing damage weather occurs in southern winter (for example, in early 2008 and early 2012), traditional winter sowing peas will suffer from freezing damage, and there is a problem of "safe wintering". By selecting winter-hardy materials and locating winter-hardy genes for winter-hardy breeding of pea, farmers' loss of freezing damage can be reduced. On the other hand, warm winter weather appears in some winter years in northern China (such as early 2001 and early 2007), and the winter pea planting areas in China show a trend of gradual northward movement, that is, "northward movement of winter peas" will have a decisive impact on the development of pea production. "North winter peas" development is the core content of pea winter-hardy genes for winter-hardy breeding, winter planting area of pea expanded from latitude 33.5° to latitude 37.5°, which can cover the areas such as shandong, hebei, and tianjin, with full use of light, height, water, and soil resources under fruit trees or winter fallow field. So, winter-hardy pea varieties will greatly increase farmers' income. In order to better cooperate with the breeding and cultivation techniques of "safe overwintering" and "northward migration of winter peas", it is necessary to select winter-hardy materials from pea genetic resources for further study on development of winter-hardy pea varieties. For such kind of reasons, researches on the mechanism of cold tolerance were conducted.

According to the research experience of other cool-season legumes, such as chickpeas (Srinivasan et al. 1998; Toker 2005), studies on low-temperature stress of peas can be started from two growth stages: vegetative stage and flowering and podding stage. Five years ago, the frost damage in flowering and pod bearing stage of peas is the only case that has been studied. France launched a winter pea breeding program aiming at frost resistance. They tried to link the freeze resistance ability with the flowering period and found that there was a linear relationship between flowering period starting date and temperature (Lejeune-Hénaut et al. 1999). At the same time, Lejeune-Hénaut and others found a late-flowering gene *Hr* that was responsible for the coseparation with a frost resistance quantitative trait locus (QTL) (Lejeune-Hénaut et al. 2004). The results suggest that late-blooming peas may be more resistant to freezing. However, Eujayl et al. identified a similar antifreeze related gene in lentils and found that *Hr* gene and antifreeze related gene were 9.1 cM apart by random amplified polymorphic DNA (RAPD) molecular marker (Eujayl et al. 1999). Many other scientists study cold domestication of peas from the perspective of cold physiology. Cold acclimation is closely related to temperature, sowing time, light intensity, and pea type. It is believed that pea domestication needs at least 2 °C cold treatment for 2 days, and arabidopsis at least needs 4 °C processing for 2 days (Weiser et al. 1990). Bourion and others found that sowing time associated with yield of peas, as the yield of autumn sowing peas is higher and more stable than that of spring sowing peas. Cold acclimation temperature above 0 °C for cold resistance can obviously increase the tolerant ability to cold temperature in winter. Winter growing forage and grain types of peas have better cold acclimation effect than that of spring type; the domestication effect was significantly different under different light conditions; at low light intensity, there was little difference in domestication effect between different types of peas (Bourion et al. 2003). Voican et al. determined the injury index of freezing injury to cell membrane, which was used to distinguish different pea varieties (Voican et al. 1995). Through cold domestication, the photosynthetic rate of pea was slowed down, resulting in short plant size, reduced tissue water content and significantly increased soluble substances (Margesin et al. 2007). Bourion et al. found that there was a close relationship between the sugar content in leaves and frost resistance (Bourion et al. 2003). Exogenous glycine, glycine betaine, proline can protect the light system II of peas from injury (Busheva and Apostolova 1997), and exogenous ABA can increase the frost tolerance of peas (Welbaum et al. 1997). Dumont et al. measured the sugar content, cell membrane electrolyte leakage, osmotic pressure, and RuBisCO activity of Champagne x Terese and its subpopulation containing only *Hr* sites, compared and analyzed the detected QTL with protein quantitative loci (PQL), and found that cottonseed sugar and RuBisCO activity played a major role in cold acclimation of pea (Dumont et al. 2009).

Pea, as a cold season legume crop, has a certain degree of cold resistance, but the ordinary varieties are not strong in winter hardy ability, which requires careful screening and repeated tests to select varieties with superior winter tolerance (Weightman 2005; Stoddard et al. 2006). Pea can tolerate frost at the vegetative stage but experiences yield loss when freezing stress occurs at the reproductive stage. Cold tolerance improvement of pea varieties is important for stable yield and expansion of the winter pea planting area. Under natural low-temperature conditions during winter in Qingdao, Shandong, China, we evaluated the cold tolerance of 3,672 pea germplasm accessions in the field and categorized them as displaying high resistance (214), moderate resistance (835), or susceptibility (2,623). The highly and moderately resistant genotypes were validated in the following year. We found that genotypes from the winter production region showed higher cold tolerance than genotypes from the spring production region. The accessions identified as having high levels of cold tolerance are recommended as potential genetic resources in cold tolerance breeding of pea (Zhang et al. 2016).

Frost stress is one of the major abiotic stresses causing seedling death and yield reduction in winter pea. To improve the frost tolerance of pea, field evaluation of frost tolerance was conducted on 672 diverse pea accessions that selected from the high resistance and moderate resistance accessions we identified before, at three locations in Northern China in three growing seasons from 2013 to 2016 and marker-trait association analysis of frost tolerance were performed with 267 informative simple sequence repeat (SSR) markers in this study. Sixteen accessions were identified as the most winter-hardy for their ability to survive in all nine field experiments with a mean survival rate of 0.57, ranging from 0.41 to 0.75. Population structure analysis revealed a structured population of two subpopulations plus some admixtures in the 672 accessions. Association analysis detected seven markers that repeatedly had associations with frost tolerance in at least two different environments with two different statistical models. One of the markers is the functional marker EST1109 (expressed sequence tag 1109) on linkage group (LG) VI which was predicted to colocalize with a gene involved in the metabolism of glycoproteins in response to chilling stress and may provide a novel mechanism of frost tolerance in pea. These winter-hardy germplasms and frost tolerance associated markers will play a vital role in marker-assisted breeding for winter-hardy pea cultivar (Liu et al. 2017).

In the winter of 2017/2018, we screened and identified another 2,887 new accessions of peas at an experimental station (35°0′14″N, 117°24′51″E, 137 m altitude) in the south boundary of Shandong province. Two hundred forty-six pea resources with high resistance (survival rate >80%) were obtained, and it was proved that the above genetic resources could be normally matured before May 30. Millet varieties were sown in June and harvested normally at the beginning of October, which proved that there was no problem in the connection of seasons and time between overwintering peas and summer millet. On this basis, the rotation systems of "pea (winter and spring)-millet (summer and autumn)" and "pea (spring)-millet (summer and autumn)" were designed and are verifying in the winter of 2018/2019. Once the above innovation mode is verified successfully, on the basis of further improving and optimizing the supporting cultivation technology, a sustainable "double reduction" (reduction of fertilizer and farm chemical) and high-efficient large-scale sustainable cropping system will be demonstrated to support the healthy and sustainable development of peas and millet industry. This model is complementary to the existing "wheat (winter and spring)–soybean (summer and autumn)" rotation model and is conducive to solve the major problems of the combination of arable land use and cultivation and the sustainable development of crop industry. And our innovation for "pea (winter and spring)–millet (summer and autumn)" and "pea (spring)–millet (summer and autumn)" rotation systems allow dry pea, fresh pea, and fresh pods production, and add high value to the local farmers for peas and millet production, for increasing farmers' income and poverty reduction for great significance.

6.2.3 Constrains and Prospects of Pea Breeding

As a food source, peas are well accepted for animal nutrition and for human food with relatively little processing of the raw ingredient necessary in a wide variety of cases in human nutrition. Peas and other pulses have been shown to have an important role in preventing illnesses such as cancer, heart disease, and diabetes. Peas contain a higher level of protein than cereals and high levels of both soluble and non-soluble fiber with a low glycaemic index. In developing countries, peas are an important source of vegetable proteins and constitute the main source of protein for most populations. As a crop, the role of peas in farm rotation has become even more important as the pressure to reduce the use of chemical fertilizers and pesticides and becomes greater in many areas of the developed and developing world. Their value as a non-cereal crop grown in rotation to reduce the risk of soil-borne diseases and weeds and the non-reliance on the use of chemical nitrogen fertilizer is a major advantage. It has now been established that growing legumes, like peas, has an overall effect on reducing greenhouse gas emissions over the whole farming system, as the production of nitrous oxide is much lower in the absence of the requirement for nitrogen fertilizer (Jeuffroy et al. 2013).

Constraints in pea breeding and production due to a range of agro-limitations, like yield plateauing and susceptibility to pests and diseases, will affect the development of peas industry. Therefore, in peas breeding, several factors should be taken into consideration, including varietal performance, yield, quality, resistance to pests and diseases, pesticide availability, cold tolerance, drought tolerance, demand for human food and vegetables, demand for animal feed, climate change, cropping systems, food use and processing development, and breeding technologies. Considering the coming needs for pea industry in the world, the future breeding objectives of peas should be described as winter hardy, heat tolerance, herbicide resistance, high yield, better digest quality, suitable for small-scale machine sowing and machine harvesting, BNF (biotic nitrogen fixation) efficiency, high photosynthetic efficiency, better flavor and palatability, dual usage both for dry peas and green (fresh) peas, dual usage for forage grass and for green pods, as well as for market price. The role of biotechnology in pea breeding can also be applied through maker assistant selection in breeding programs. This relies on breakthrough in-depth studies on pea genetic resources, high-density SSR/SNP-based genetic linkage map construction and important traits

QTLs analysis. There are many research programs in progress that are aiming at overcoming the main constraints to pea production both in the developed and in the developing countries of the world. All are aiming primarily at extending the yield and increasing the quality of the harvested products through plant breeding and agronomic management and at extending the demand through innovation for food to meet the growing human population, increasing human health and reducing the dependence on non-sustainable commodities, as well as food for animal and fish production. In addition, work is ongoing to exploit peas as a source of pharmaceutical and nutraceutical products.

Peas have been an important part of the human diet for thousands of years. Despite the fact that they provide a healthy and sustainable source of protein, there has been only a slight increase in production worldwide, unlike other major crops such as cereals, rice, maize, and soya, which have had increases of production of 200–800% over the past 50 years. Future crops must be developed to provide varieties that are adaptable to a wide range of growing conditions, able to withstand periods of high or low soil moisture availability as well as cold or high temperature, must be resistant to commonly found pests and diseases and reduce the risk of variability of production from year to year. Consumption has been declining slowly in both developing and developed countries as demand for meat and dairy products grow. However, such an increase in animal and dairy products as well as a rising demand in farmed fish is putting a strain on the more traditional sources of plant protein, namely, soya. The role of peas as a soya substitute is beginning to increase and recent work in the developed nations has shown the value of peas as soya replacement in rations for pigs, poultry, fish, and ruminants (Anthony 2017).

This demand is already encouraging a growth in international trade large dry pea producers such as Canada are supplying product to China, India, and the Far East at increasing levels. This trend is continuing and a country such as China which was self-sufficient in dry peas is now a net importer as both the amount and price of peas continue to rise (FAO 2018). The future for peas as adaptable, sustainable, environmentally friendly and with high production potential, together with increasing demand, is positive.

6.3 Prioritizing Climate-Smart (CS) Agronomic Traits

6.3.1 Flowering Time

Pea is a valuable food capable of meeting the global dietary needs of proteins, vitamins, minerals, and carbohydrates. In recent years, its nutritional and health value is accepted by more and more consumers and producers (FAOSTAT http://faostat3.fao.org). Among agronomic traits, the flowering time has drawn the attention of researchers. Stipules are lateral organs of angiospermic plants formed on nodes. The stipule can perform photosynthesis and provide protection to leaf and branch-bud and inflorescence in the axil of leaf. In recent studies in *Pisum sativum*, stipules are found to contribute to 30% of plant's photosynthesis (Kumar et al. 2011a, b). Furthermore, stipules associated inflorescence in the axil of leaf and leaf are found to determine the flowering time, and presence of stipules maintains pods and maximizes the harvest index (Sharma and Kumar 2012). Moreover, the timing of flowering, in particular the degree to which it is responsive to the environment, is a key factor in the adaptation of pea to various ecogeographic locations and agricultural practices. The genes and regulators controlled the transition to flower have been identified in pea (e.g., Weller et al. 2009). Three genes were well described their effects on flowering time in pea: HIGH RESPONSE TO PHOTOPERIOD (HR, Weller et al. 2012), LATE FLOWERING (LF, Foucher et al. 2003) and STERILE NODES (SN, Liew et al. 2014). The HR and SN act in the leaf, while LF acts in the shoot apex in the flowering model. At the HR locus, the dominant HR allele is known to make long day (LD) a requirement for flowering. In spring, during short day (SD) conditions, hr allows early flowering (Weller et al. 2009). The mechanism of HR control of flowering is through HR interaction with the circadian clock gene SN (Weller et al. 2012). SN is the pea ortholog of the LUX ARRHYTHMO (LUX) transcription factor in Arabidopsis (Liew et al. 2014). Under SD, SN has been shown to influence flowering time in pea through mutations accelerating the transition to flowering (Liew et al. 2014; Murfet 1971). LF was the first pea flowering time locus to be identified at the molecular level, it produces a TFL1 homologue that regulates transcription factor activity in the shoot apex and its expression is strongly correlated with flowering time (Foucher et al. 2003). Furthermore, these three genes produce complex flowering time adaption in Swedish landrace pea and links genetic diversity of the three genes and growing season at the site of origin (GSO) and are less clear-cut (Vanhala et al. 2016). Thus, genes and regulators related to vernalization and photoperiod are affecting flowering time. Further, three flowering loci, namely, SN, DNE (Day neutral), and HR was reported controlling multi-flowering trait (Murfet 1990), which along with LF allele determines the actual floral node number in pea. Murfet (1985) suggested these four genes (SN, DNE, HR, and LF) were related with late flowering in pea. This indicated the assumption that multi-flowering genes are very tightly linked with late-flowering genes may be truly associated. Based on genetic analysis and field investigation, conversion of multi-flowers per peduncle to multi-pods per peduncle is linked with genotypes, peduncle diameter (thick), and temperature (around 20 °C) (Devi et al. 2018). Therefore, genotypic background and pea-growing locations are also known to cause variations in the flowering time.

6.3.2 Root Characters

Plant roots are crucial for water and nutrient and probiotics supply, as well as for pathogen resistance. Pea root characters, i.e., root length, number of lateral roots, and root biomass, have been reported to be controlled under polygenic factors. A large root system (larger number of roots and longer roots) in pea varieties were proved to be less susceptible to *Aphanomyces euteiches* and *Fusarium solani* (Desgroux et al. 2018; Kraft and Boge 2001). The diversity of genetic determinants of root architecture in pea natural variability and a recombinant inbred line (RIL) population was studied using comparative genome-wide-association mapping (Desgroux et al. 2018) and a QTL approach (Bourion et al. 2010). Eleven of the LA-QTL were detected and confirmed to control pea root character (Bourion et al. 2010; Desgroux et al. 2018). One significant locus, mapped to the major QTL Ae-Ps7.6, was associated with both root character and resistance to *Aphanomyces euteiches*. In pea, three SNP markers associated with pleiotropic effects on root character variation were designed either in the sequence or in the same genomic block as three cloned pea genes, i.e., *Le*, which encodes gibberellin 3b-hydroxylase and controls internode length, *PsTFL1a* (= *DET*) and *PsTFL1b* both involved in flowering regulation (Weeden and Moffet 2002; Bourion et al. 2010; Desgroux et al. 2018). This indicated the growth potential of roots could also have been affected by accelerated flowering traits. For example, *PsTFL1a* is expressed in roots regardless of the developmental stage and *PsTFL1b* is expressed in roots during both vegetative and reproductive stages, but *PsTFL1a,* known to induce a reduction in the flowering period, is expressed in the shoot apex only after the transition to flowering, and *PsTFL1b* was never found to be expressed in flowers (Foucher et al. 2003). These genetic loci associated with both root character and resistance would be useful for breeding pea types limiting disease development.

6.3.3 Nutrient-Use Efficiency

Pea is grown in many countries and provides important vegetable proteins. As a grain legume, peas are capable of fixing the vast majority of their own N through biological nitrogen fixation (BNF). Nitrogen from BNF in agriculture can make a major contribution to sustainable agriculture, such as soil fertility restoration and cereal rotation to minimize low quantities of organic and inorganic sources of soils. Over 80% of nitrogen in pea plants was contributed by BNF, and 25 kg ha^{-1} of nitrogen on average to the soil system was added by BNF (Ruisi et al. 2012). In northwestern China, pea–maize intercropping system is popularized (Chai et al. 2013). This system can enhance agronomic productivity, atmospheric N_2 fixation, and reduce carbon emission (Chai et al. 2013; Hu et al. 2017). In this cropping system, pea is more competitive than maize (Hu et al. 2016), indicating the potential for increased N remobilization to grains. Phosphorus (P), an important constituent of protein and phospholipids, is another key nutrient for increasing productivity of peas (Rani et al. 2016). Sulfur (S) is essential for synthesis of proteins, vitamins, and sulfur containing essential amino acids and is also associated with nitrogen metabolism (Kumar 2011). Kumar (2011) performed a pea field experiment at four levels of P_2O_5 (0, 20, 40, and 60 kg/ha) and three levels of sulfur (0, 20, and 40 kg/ha), respectively. As expected, the highest mature green pod yield (73.83 q/ha) was obtained from the application of 60 kg P_2O_5. Similarly, the highest mature green pod yield (66.51 q/ha) was obtained from the application of 40 kg S. This indicated that yield attributing traits and yield of pea were improved by the application of phosphorus and sulfur. Furthermore, a pot experiment suggested peas had maximum uptake of N, P, and S when the nutrients were combined application of Co, B, and Mo or sole application of Co (Singh et al. 2015a, b). Thus, nutrients uptake of P and S by pea would be more efficient when combined with use of micronutrients.

6.3.4 Water Use Efficiency

Producers may achieve high yields and high resource use efficiencies by intercropping systems (Farahani et al. 1998; Gao et al. 2009). However, intercropping annually carries considerable water shortages in semiarid areas (Zentner et al. 2002; Lyon and Peterson 2005). For example, crop/wheat intercropping system was impracticable because wheat consumes more water than other crops (Sun et al. 2006). Field pea is a drought-tolerant cash crop and has the potential for intercropping systems with crop to obtain both a high productivity and a low water use (Siddique et al. 2001). Thus, a maize/pea intercropping system is rapidly exploited under water constraint. Mao et al. (2012) proved intercropping of maize and pea could enhance water use efficiency and grain yields compared to growing them as sole crops. When the row arrangement of the intercropping system was suitable, the maize and pea intercropping could save 255 (2010) and 120 mm (2011) irrigation water, respectively, compared to total 630 mm irrigation water generally used in intercrops of maize and wheat (Hu et al. 2010). Furthermore, temperature increases reduced water-use efficiency of all crops in the rotation system. On a pea–spring wheat–potato (Pe–Sw–Po) rotation system at the Guyuan Experimental Station in a semiarid region of China during 2000–2005, with a 1.2 °C temperature increase, water-use efficiency decreased by 7.3%; and with a 2.0 °C temperature increase, decreased by 12.5% (Xiao et al. 2007). In semiarid climates, the previous crop stubble changed the microclimate near the soil surface and improved growing conditions for the subsequent crop. Averaging across the 4 yr of field experiments, pea had the highest water-use efficiency and yield from the tall stubble and lowest from the cultivated stubble treatments (Cutforth et al. 2002). Therefore, water-use efficiency of pea could benefit from appropriate management of field trials.

6.3.5 Carbon and Nitrogen Sequestration

Soil organic C and N dynamics caused by cultivation have been of increasing concern in recent years. Carbon and N sequestration are usually stored in plant stems leaves, roots, and soil (Sherrod et al. 2003). Previous studies suggested that intensive cropping has increased C storage in the soil, combined with one reduced- and no-till systems (Peterson et al. 1998). Data from an ongoing winter wheat–spring pea long-term experiment indicated spring plow could maintain soil organic carbon

(SOC), soil total nitrogen (STN) (Awale et al. 2018). Further, wheat–pea system could increase SOC, STN, and pea yields under no-till tillage system and disk tillage and chisel plow system relative to conventional fall plow, but might reduce surface soil pH and decrease wheat yields (Awale et al. 2018). Furthermore, pea had higher N content and increased potential C and N mineralization (PNM) than wheat or fallow (Wang and Sainju 2014). In contrast, spring wheat had increased SOC, particulate organic C (POC), microbial biomass C (MBC) and N and MBN), MBC, STN, and microbial biomass N (MBN) than pea or fallow (Wang and Sainju 2014). In pea plants, C compounds are mainly transported as sucrose in the phloem and delivered to the root system (Kühn and Grof 2010), whereas N is mainly taken up from the soil solution and transported in the plant xylem and transferred to plant sinks to build storage or structural proteins (Witte 2011; Salon and Munier-Jolain 2010). Therefore, incorporation of pea–nonlegume residues into the soil using conventional tillage can increase N and C sequestration in the soil.

6.3.6 *Greenhouse Gas Emission*

Global atmospheric concentrations of the major greenhouse gases (GHG) such as carbon dioxide (CO_2), methane (CH_4) and nitrous oxide (N_2O) have been rising since the beginning of the industrial era (Cristina et al. 2010). The emissions of greenhouse gases are stemmed from various human activities, farming among them (Janzen et al. 2006). Agriculture has been identified as a significant source of greenhouse gases (Cole et al. 1997). Globally, agriculture is responsible for > 13% of anthropogenic GHG emissions (WRI 2014). As such, agriculture may contribute to global and national efforts toward reducing GHG emissions.

There are large annual exchanges of CO_2 between agricultural lands and the atmosphere, and global net fluxes of CO_2 from agricultural lands are estimated to be only 0.04 Gt CO_2 yr^{-1} (Smith et al. 2007). The fact is that a large amount of GHG emissions associated with farming activities results mainly from nitrous oxide (N_2O) (Janzen et al. 2006). Nearly two-thirds of the total emissions in agriculture occurred as N_2O, a potent greenhouse gas with 300 times global warming potential (Forster et al. 2007). Reducing the emissions of GHG would reduce the potential impact on the global climate. Mitigations of these gases from agricultural ecosystems can be facilitated by more efficient management of C and N flows (Smith et al. 2007).

Policy makers, producers, and researchers urge to develop effective farming practices to reduce GHG emissions, while maximizing the potential economic returns from farming. One of the promising strategies in mitigating GHG emissions from farming is to adopt diversified cropping systems where cereal, oilseed, and pulses (i.e., legume) crops are arranged in well-defined crop sequences in crop rotation systems (Gan et al. 2011). Pea (*Pisum sativum* L., Fabaceae Lindl.) is one of the most commonly used grain legumes that may provide a new opportunity to develop a production system that fulfills both economic and environmental interests. The study by Chapagain and Riseman (2014) demonstrated that intercropping barley and pea is an efficient strategy to achieve higher land productivity, N and C yields, and higher soil C sequestration than when barley was grown as a monoculture (Chapagain and Riseman 2014). Conservation farming systems that include reduced tillage or no tillage (NT) vis a vis conventional tillage (CT) is promoted as an agricultural practice that could increase the soil organic C (SOC) pool, and thus mitigate GHG emissions (Sanderman et al. 2010). Conventional plow-based farming systems could accelerate carbon mineralization, and thus reduce SOC content, which is attributed to disruption of soil aggregates and increased oxidization (Ussiri et al. 2009). Crop residue management has a crucial impact on the C and N cycles, and thus an increased potential to enhance the sequestration of C in soils. In a modeling study, Yan et al. (2007) estimated that practicing no-tillage on 50% of arable lands and by returning 50% of the crop residue to the soils would lead to an annual soil C sequestration of 32.5 Tg C in China. Yeboah et al. (2016) find that no-tillage and conventional tillage measures with residue could increase SOC stocks in northwestern China, and the accumulation of soil organic carbon (SOC), microbial biomass carbon (MBC) and light fraction organic carbon (LFOC) in no-till with residue retention (NTS) soils were greater than conventional tillage with straw removed (T) soils in a spring wheat–field pea crop sequence (Yeboah et al. 2016).

6.3.7 *Genome Plasticity*

At present, there are not many studies on the plasticity of pea genome, but there are related studies on the endophytic fungi of legumes, the symbiotic nitrogen-fixing bacteria of legumes, and the plasticity of leguminous-rhizobial nodules (del Val et al. 2007; MacLean et al. 2007; Burghardt et al. 2017).

Clausen, Keck, and Heisey's classic work (reviewed by Nu~nez-Farfán and Schlichting 2001) recognize the importance of genetic variation in plant plasticity. Bradshaw (1965) was one of the first authors to recognize the importance of genetic variation in plasticity. Subsequently, many authors explored the conceptual importance of heterozygosity, allelic sensitivity, epistasis, and regulatory genes in the generation and evolution of plastic reactions (reviewed by Pigliucci 2001). Smith et al. (1990) provided a molecular model of plasticity by suggesting that repetitive sequences from gene families may exhibit environmentally specific expression, allowing for environmental sensitivities of key phenotypes. He emphasized that the causal relationship between environmental perception and receptors, signal transduction pathways, and downstream molecular responses is a key aspect of any active adaptive response. These discussions eventually led to debates about the existence of plasticity genes, their molecular basis, and the choice of targets that lead to adaptive plasticity (Via 1993; Via and Conner 1995). Legume roots show a remarkable plasticity to adapt their architecture to biotic and abiotic constraints, including symbiotic interactions.

In the context of legume/rhizobia nodulation, the NCRs are regulators of bacterial (symbiont) differentiation and viability (Pan and Wang 2017), some of which are required for functional symbiosis (Shabab et al. 2016). When host genes necessary for early signaling between partners are mutated, the nodule transcriptome exhibits extensive alterations relative to wild-type plants (Larrainzar et al. 2015). Similarly, when host genes necessary for late nodule development and functional nitrogen fixation are mutated, nodules arrest at different stages of development and vary extensively in terms of symbiont and host transcriptomes (Starker et al. 2006; Lang and Long 2015). Taken together, these data suggest that nodule transcriptomes differ based on developmental stage and the amount of nitrogen the rhizobia inside are converting into a usable form. The expression of defense genes in legumes is altered by rhizobial vaccination (Kouchi et al. 2004; Lohar et al. 2006) and hosts differ strikingly in the extent to which they induce defense genes in nodules (Burghardt et al. 2017). Furthermore, differential expression of host genes that alter the bacteroid environment may represent a mechanism by which the host can control differentiation and nitrogen fixation in bacteroids (Oono et al. 2011).

6.3.8 Specific Traits for Vegetable Purposes

Pea (*Pisum sativum* L.) is a cool-season crop, which is among the most consumed vegetables worldwide, with a registered global production of 15 million tons in 2010 (FAO 2013). Pea is a highly nutritious crop and grown for a variety of usages. For centuries the crop has been an important raw material for feed and food purposes in many forms including forage for animal feeding, fresh seeds for canning and freezing, dry seeds, partly for human consumption, but mostly for animal feeding, and pods as a fresh vegetable for human consumption (Grant and Cooper 1997).

There are two major types of pea cultivated in the world, namely, vegetable and grain type. The cultivation of the grain-type pea aims to the production of either dry seeds consumed by humans, also known as pulses, or animal fodder. However, the vegetable-type pea cultivated for fresh consumption either as pods or as immature seeds.

The grain-type pea is generally white and round, and is not as sweet as the vegetable-type at green stage. The mature dry seed of grain type is widely consumed as vegetable, dal, and flour in developing countries, and pea, as a protein source, has a favorable image with additional value in terms of environment, reducing fertilizer use in agriculture, and food safety (Bovine Spongiform Encephalopathy disease). It is a safe raw material with no specific problems of mycotoxin, pesticide or fungicide residues (Bourdillon 1998; Santalla Ferradás et al. 2001). Dry pea is a good source of proteins, vitamins, and minerals (Martins 2010), and it also has sufficient carbohydrates, total sugar, and starch. Apart from that the grain-type pea also has high levels of the essential amino acids, lysine, and tryptophan, which are usually low in cereal grains (Kumari et al. 2015). It also plays a significant role in human health as many nutrients in dry peas help lower the risk of heart disease, stroke, diabetes, and various cancers.

Another type of pea is vegetable pea, which is a popular green shelled vegetable of the temperate and subtropical areas of the world. The vegetable-type pea is widely consumed as a fresh vegetable in the growing season, mainly with its pods, shoots, beans. Farmers harvest fresh green pods for vegetable purpose when the pods are fully filled (Shanmugasundaram 1991). Immature seeds in green pods are sweet and green, which are mainly consumed as fresh, frozen or canned form. Pea shoots were recently presented as a ready-to-eat vegetable. The consumption of pea shoots is not as common as eating the peas. Pea shoots are recognized as a popular specialty vegetable in some parts of Asia and Africa that is gaining popularity in the United States and Europe (Miles and Sonde 2003). The consumption of green leafy vegetables is recommended due to their high content of vitamins, minerals, and antioxidant phytochemicals, as well as low content of fat and carbohydrates (Rico et al. 2007). Pea shoots are harvested in a very early maturation stage when the leaves and tendrils are tender, crispy and have an intense pea flavor (Miles and Sonde 2003). This green leafy vegetable can be eaten raw in salads or cooked with other ingredients. From a nutritional point of view, legumes are considered important sources of plant protein, carbohydrates, essential minerals, vitamins, and several other antioxidants and health-promoting compounds (Souci et al. 2001; Bouchenak and Lamri-Senhadji 2013). Pea consumed as vegetables contain more water and less proteins than those consumed as dry pulses. On the other hand, soluble carbohydrates are higher and starch content is lower in fresh vegetable pea, which makes them more palatable than dry pulses. Moreover, vegetable peas are richer sources of antioxidants and other health-promoting compounds contained mainly or only in fresh plant biomass, such as carotenoids, phenolics, chlorophyll, vitamin A, and vitamin C (Bhattacharya and Malleshi 2012). Furthermore, vegetable legumes, which contain much more water than dry pulses, are short-season crops which can be grown more than once a year being offered to the market as a fresh food with a limited shelf life.

In conclusion, the grain type mainly provides protein for humans. Vegetable pea is rich in nutrients, and the consumption of vegetable pea is mainly intended to provide more balanced nutrition full of healthy compounds rather than to serve as a primary protein source. Pea consumed as fresh vegetables are indeed short-season crops with all the benefits on crop rotations, costs of production and overall a more sustainable production since the required general inputs are much lower. Growing pea to be consumed as fresh vegetables renders a high-quality product when compared to other vegetables; the product can be much faster prepared than dry pulses fitting into the modern and more demanding consumption habits of consumers (Ntatsi et al. 2018). There are substantial differences in quality features between legumes used either as vegetables or as dry pulses. The genetic features involved in the quality of legumes, which are very complex, are linked to relevant agronomic aspects and any change might influence the overall adaptation and performance of the crop to biotic and abiotic factors. Further work will be necessary for Fabaceae to elucidate links between genetic changes and the respective effects on the crop.

6.4 Prioritizing CS Stress Tolerance Traits

Pea (*Pisum sativum* L.) is an important cool-season legume crop that is affected by various biotic and abiotic stresses during all growth stages (Ali et al. 1993; Wang et al. 2007). Abiotic stresses are referred to as noninfectious disease, which is caused by adverse environmental conditions such as cold, drought, frost, improper nutrition, and air pollution, etc. Biotic stresses are those caused by infectious pathogens or pests. These stresses are major factors limiting the yield and quality of pea.

6.4.1 Cold Tolerance

As one of the abiotic stress factors, low temperatures represent the major constraints limiting agricultural production in temperate climate (Kosová et al. 2015). Pea is a cool-season legume which is well-adapted to wide ranges of temperature ranging from 7 to 30 °C (Shereena and Salim 2006). For spring sowing pea, the increasingly low temperature at reproductive stage results in the damage of the development of flowers, pods, and seeds (Shafiq et al. 2012; Siddique et al. 2013). Autumn sowing in pea can increase the yields (Karkanis et al. 2016), extend growing season (Mikić et al. 2011), escape terminal heat, and drought stress (Vocanson and Jeuffroy 2008) and provide an alternative rotational crop. So improved winter hardiness of pea has become a major trait of interest for breeders (Zhang et al. 2016; Vann et al. 2018).

According to the purpose of the research, different treatments, and methods were used to screen and evaluate the cold tolerance. Shafiq et al. (2012) discriminate the frost tolerance among field pea genotypes under a controlled environment, viz. frost chamber using temperatures as low as 4.8 °C. The adult pea plants were submitted to winter frost and winter frost damages was evaluated with scores varying from 0 to 5 in the field experiment (Lejeune-Hénaut et al. 2008). The winter hardiness of pea was determined using the overwintering rate (number of plants after winter divided by number before winter) (Urbatzka et al. 2012). Mortality and three levels of cold tolerance scoring were used to screen pea accessions at the natural field (Zhang et al. 2016). In the controlled chamber, at the end of the recovery period, seedlings at the four-leaf stage and frost damage were evaluated through the extent of yellowing and necrotic areas on leaves with five levels (Dumont et al. 2009).

With the methods of screening for cold tolerance, some cultivars and germplasm of pea were assessed and used in many countries (McPhee and Muehlbauer 2007). Cultivars such as Dove and 5174 were released for their early maturity, high tolerance to cold and high grain yield in France (Mikić et al. 2011). Three sowing dates and six winter genotypes were examined on the winter hardiness and productivity (Urbatzka et al. 2012). Two hundred and fourteen field pea accessions with high resistance to cold were selected from 3,672 pea germplasm in China (Zhang et al. 2016). Sixteen accessions were identified as the most winter-hardy for their ability to survive in all nine field experiments in China from 672 diverse pea accessions (Liu et al. 2017).

6.4.2 Drought Tolerance

For pea production in dry areas, the most common water shortage occurs either at the vegetative stage, which mainly affects establishment and survival, or during grain-filling (terminal drought) that has adverse effects on grain yield and quality. Identification of cultivars with more efficient water use and greater drought resistance is considered as sustainable and economically viable approach (Condon et al. 2004). According to Petrović et al. (2016), pea cultivars Dukat and Parner showed lower susceptibility to drought stresses during the period of seed germination (Petrović et al. 2016). Pea cultivar HR1 was regarded as a wide-adapted genotype, whereas Desso showed the best adaptation for drought environments (Iglesias-García et al. 2017). P665 has been characterized as tolerant to water stress according to a visual scale and some drought-related traits (Iglesias-García et al. 2012a), while Messire has been described as a moderately water stress-tolerant genotype (Iglesias-García et al. 2012a, b; Castillejo et al. 2016; Iglesias-García et al. 2017). Several wild pea (*Pisum fulvum* Sibth. & Smith, Pf) accessions exhibited lower drought susceptibility indices and high productivity (Naim-Feil et al. 2017).

Long-term drought was more critical to reduce growth and yield than drought at flowering stage. In other words, long-term water shortage increased the susceptibility to drought in peas. Drought stress markedly enhanced the activities of superoxide dismutase (SOD), catalase (CAT) and peroxidase (POX) but slightly changed the activity of ascorbate peroxidase (APX) (Karataş et al. 2014). A later study showed that foliar application of glycine betaine (GB) increased the yield and soluble protein concentration under drought at vegetative stage, in addition to proline that led to the maximum increase in nonenzymatic antioxidant defense system under long-term drought (Osman 2015). GB and proline were reported to enhance the activity of SOD, APX, and catalase in leaves, while increased SOD activity in seeds under long-term drought stress. APX activity in seeds under drought decreased by GB application. The maximum positive effect was for GB under unstressed condition and drought at vegetative stage, by maximizing APX activity, in addition to enhancing the production and translocation of assimilation from source to sink. Besides, Arshad et al. (2008) and Zahir et al. (2008) about simultaneously found that inoculation with plant growth-promoting rhizobacteria (PGPR) containing 1-aminocyclopropane-1-carboxylate (ACC) deaminase could be partially or completely eliminating effects of drought stress on growth, yield, and ripening of pea. Phosphorus application and elevated CO_2 interactively enhance periodic drought tolerance in field-grown pea as a result of decreased stomatal conductance, deeper rooting, and high Pi availability for carbon assimilation in leaves in a phosphorus-deficient vertisol (Jin et al. 2014).

QTLs underlying drought tolerance was identified using 98 $F_{7:8}$ RILs developed from parental lines P665 and cv. Messire, and 10 QTLs associated with the traits explaining individually from 9 to 33% of the phenotypic variance depending on the variable assessed and altogether from 20 to 57%. In addition, a set of reproducible markers linked to these QTLs (A6, AA175, AC74, AD57, AB141, AB64, Psblox2, PsAAP2_SNP4, and DipeptIV_SNP1) were identified for selecting the individuals harboring the desired QTLs in pea breeding programs for drought adaptation (Iglesias-García et al. 2015).

Eleven and seven nodule proteins, encoded by *Pisum sativum* and *Rhizobium leguminosarum* genomes, respectively, were regulated by drought. These 18 proteins were further analyzed and three RNA-binding proteins, three proteins involved in flavonoid metabolism and two in sulfur metabolism were identified (Irar et al. 2014). COP1, an E3 ubiquitin-ligase, negative regulated of photomorphogenesis-related factors of Arabidopsis (*Arabidopsis thaliana*), including long hypocotyl 5 (HY5), phytochrome A (PHYA), phytochrome B (PHYB) and cryptochrome (CRY) to repress stomatal opening (Osterlund et al. 2000; Smirnova et al. 2012; Jia et al. 2014). Sullivan and Gray (2000) found partial duplication within the pea homolog of COP1 and gave rise to a created lip1 mutant that displayed short stems along with open and expanded shoots when grown in darkness (Seyyedi et al. 1999; Stoop-Myer et al. 1999). Using the lip1 mutant, it was very recently demonstrated that COP1 plays a fundamental role in the regulation of stomatal movements in response to dehydration and its function was conserved during evolution in pea (Moazzam-Jazi et al. 2018).

6.4.3 Salinity Tolerance

Soil salinization is a major limiting factor for sustainable agriculture. It gradually disrupts homeostasis in plant water potential and ion distribution at cellular, as well as, whole plant level, and strongly influences biological molecules, causing the arrest of growth and even death of the plant (Munns and Tester 2008). According to an estimate of FAO, salinity limits the production of nearly 34 million ha (11% of irrigated area) of the world's land (FAO 2012). Peas are relatively more salt sensitive to salinity than other legumes (Duzdemir et al. 2009). Severe salinity (solution electrical conductivity of 24.9 dS m^{-1}) limits the production and growth of pea globally even resulting in death (Steppuhn et al. 2001; Najafi et al. 2007).

Many ions present in the soil solution, particularly chloride and sodium ions resulting in soil salinity in different plant parts were revealed. It was indicated that these ions impair plant survival by disturbing several physiological and biochemical mechanisms such as photosynthetic rate (Hayat et al. 2010), transpiration rate (Cambrolle et al. 2011), stomatal conductance (Perez-Perez et al. 2009), water-use efficiency (Grewal 2010), content of sugars (Noreen and Ashraf 2009), content of osmolytes proteins and water relations. At present, germination, emergence, plant fresh/dry biomass and accumulation of leaf inorganic osmolytes (Na+, K+, Ca2+, K+/Na+, Ca2+/Na+) are significant screening criteria for salt tolerance in pea genotypes (Shahid et al. 2012).

Different approaches are tested to improve physiological processes and enhance crop yield under salt environment, and seed priming with plant growth regulators (PGR), such as ivin, kinetin, 6-BAP, fusicoccin, ethylen, salicylic acid, methyl-jasmonate, makes a seed and seedling more resistant to salt exposure for maize, barely, corn, sunflower sorgo and sweet sorghum (Lutsenko et al. 2005; Kaya et al. 2006; Patanè et al. 2009; Carvalho et al. 2011; Ribchenko and Palladina 2012; Ehab 2016). In respect of pea, growth regulator Biolan applied for seeds priming promoted water uptake by seeds and germination, restored K/Na ratio and then slight alleviated the total salt inhibitory effects (Dmytruk et al. 2016). The growth regulator 5-aminolevulinic acid (ALA) improved the photosynthetic pigments, biosynthesis, activities of antioxidant enzymes, and expression of stress-related genes in the pea germplasm Nain Ordinaire and Elatius 3 (El-Esawi et al. 2018). Husen et al. (2016) investigated that foliar application of indole acetic acid (IAA) significantly reduced the loss caused by salinity stress by enhancing capacity to withstand the salt stress in pea plants. Ahmad et al. (2009, 2010) found that under saline conditions, the production of ROS was increased, and ROS-mediated membrane damage has been regarded as a major cause of the cellular toxicity by salinity in peas (Ahmad et al. 2009, 2010). Ali et al. (2015) observed that salt-tolerant plant growth-promoting rhizobacteria (PGPR) improved salt stress tolerance of transgenic pea plants overexpressing the Na^{+}/H^{+} gene.

Some interesting landraces were identified as more salt-tolerant and could be used in breeding programs to improve the salinity tolerance in field pea crops. Cerda et al. (1982) noted that the cultivar SP-290 was classified as moderately salt-sensitive and Durana cultivar as moderately salt-tolerant. Francois and Maas (1994) rated pea crops as moderately salt-sensitive. Upon screening of pea (*Pisum sativum* L.) genotypes for salt tolerance based on early growth stage attributes and leaf inorganic osmolytes. Shahid et al. (2012) categorized the pea varieties Samarina Zard, Climax, 9800-5, 9800-10 and 2001-55 as the most highly salt-tolerant genotypes. Seven hundred eighty globally distributed *Pisum sativum* accessions originating from China and Greece were found to exhibit significant variation in response to applied NaCl, based on plant symptoms (Leonforte et al. 2013a). Lines with relatively higher tolerance as compared to commercial varieties grown in Australia were most frequently identified within landraces originating from the central, eastern and southern provinces of China. The most tolerant accession identified was an unadapted landrace "ATC1836" originating from Greece.

The genetics of plant salinity tolerance is complex and governed by multiple genes with small effects (Arraouadi et al. 2012; Leonforte et al. 2013b), and is highly affected by environmental conditions. However, with molecular markers, the genetic analysis of quantitative inheritance of salinity tolerance is possible (Koyama et al. 2001), and Leonforte et al. (2013b) have identified molecular markers for two salinity tolerance QTLs on linkage groups Ps III and Ps VII, explaining 12% and 19% of phenotypic variance for salt index score, and 12% and 17% for the symptom score, respectively, from the moderately tolerant field pea cultivar Parafield.

An increased relative abundance of 14-3-3-like proteins and alcohol dehydrogenase has been found in salt-stressed grass pea leaves (Chattopadhyay et al. 2011; Parihar et al. 2015). In addition, an increased relative abundance of adenylate kinase ADK involved in ATP biosynthesis and nucleoside diphosphate kinase 1 (NDPK1) involved in interconversion between ATP and CTP, GTP, and UTP have been observed in salt-tolerant pea (Kav et al. 2004; Parihar et al. 2015).

6.4.4 Disease Resistance

Pea disease caused by pathogens includes fungi, bacteria, viruses and nematodes (Kraft and Pfleger 2001). Based on tissue and organic distribution, the diseases were classified into foliar, stem or vascular, root, and seed diseases (Rubiales et al. 2015). Foliar diseases were caused by biotrophic and necrotrophic pathogens. To date, there are three foliar diseases, rusts, powdery mildew, and downy mildews, caused by biotrophic pathogens, which are major limiting factors in legume crops production including peas (Kraft and Pfleger 2001; Biddle and Cattlin 2007; Wang et al. 2007). Long-distance aerial dispersal is an important survival strategy for these biotrophic pathogens.

Pea rust has been reported to be caused by *Uromyces viciae-fabae*, particularly in subtropical areas (Kushwaha et al. 2006), whereas in cooler regions *U. pisi* is predominant (Barilli et al. 2009a, b). *Uromyces viciae-fabae* also causes rusts of other legumes such as faba bean and lentil (Rubiales et al. 2015). Pea powdery mildew, caused by the fungus, *Erysiphe pisi* DC, is a major disease worldwide, which can considerably limit global pea production (Smith et al. 1996). *E. pisi* can cause yield losses of up to 80% under heavy infection pressure in susceptible cultivars (Smith et al. 1996; Ghafoor and McPhee 2012). Recently, *E. baeumleri* and *E. trifolii* have also been reported to cause powdery mildew and induce similar symptoms on pea (Ondřej et al. 2005; Attanayake et al. 2010). Downy mildew of pea, caused by the pathogen *Peronospora viciae* f. sp. *pisi*, is a widespread disease, but it is most frequent and severe in cool, moist, maritime climates (Kraft and Pfleger 2001).

Ascochyta blight caused by necrotrophic fungi is one of the most important diseases affecting field peas. The disease occurs in almost all pea-growing regions of the world and can cause significant crop losses when conditions are favorable for an epidemic. Ascochyta blight is incited by different pathogens in the various legumes. In pea, the disease of Ascochyta blight is caused by a complex of fungal species that includes *Ascochyta pisi*, *Didymella pinodes*, and *Phoma medicaginis* var. *pinodella* (Tivoli et al. 2006; Bretag et al. 2006; Khan et al. 2013). *Phoma koolunga* and *P. glomerata* have also been reported to be part of this complex in pea in Australia (Davidson et al. 2009). Of the Ascochyta complex in pea, *D. pinodes* is the most prevalent and damaging (Tivoli and Banniza 2007).

Stem or vascular wilt, and root or foot rot are usually caused by soil-borne pathogens including fungi and oomycetes. Epidemic of these diseases are highly dependent on environmental conditions. Fusarium wilt is a vascular disease caused by the ascomycete *Fusarium oxysporum* f. sp. *pisi* (Kraft 1994), which is a major constraint in the production of pea in most growing areas. Vascular wilt in pea is mainly caused by *F. oxysporum* f.sp. *pisi*. Apart from Fusarium wilt, vascular wilt disease has also been induced by *Verticillium dalhiae* in pea that results in similar wilt symptoms (Isaac and Rogers 1974).

Fusarium root rot of pea caused by *Fusarium solani* f. sp. *pisi* is a troublesome problem in all pea-producing areas. Fusarium root rot is distinct from Fusarium wilt and can occur in conjunction with other root diseases of pea such as

Aphanomyces, Rhizoctinia, or Pythium root rot (Kraft and Pfleger 2001; Wang et al. 2007). Aphanomyces root rot caused by the oomycete *Aphanomyces euteiches* is also an important threat to legumes and is one of the major constraints of pea production in North America and Europe (Gaulin et al. 2007). Like most soil-borne pathogens, *A. euteiches* has a very wide host range and it is able to cause disease to many legume crops including common bean, vetch, lentil, faba bean, alfalfa, and red clover (*Trifolium pratense*) (Levenfors et al. 2003). Among the different root rot pathogens, *F. solani* and related species are a major threat to most legumes including pea.

In pea, there are two destructive foliar diseases caused by bacteria including bacterial leaf blight and brown spot, induced by *Pseudomonas syringae* pv. *pisi* and *P. syringae* pv. *syringae*, respectively (Kraft and Pfleger 2001). Interestingly, different pathovars of *P. syringae* can cause different diseases on the same host. Disease is more severe under conditions of high temperature and rainfall. *P. syringae* pv. *pisi* is a seed-borne pathogen. Infected seed, produced on infected plants, serves as a particularly important inoculum source and seed is the main pathway by which the bacterium is introduced into new production areas (Wang et al. 2007). *P. syringae* pv. *syringae* is less important on pea than *P. syringae* pv. *pisi*, which can be found more commonly in pea planted in the autumn or winter or under overhead irrigation (Biddle and Cattlin 2007).

In pea, the most important virus diseases of pea are caused by a few viruses, including Alfalfa Mosaic Virus, Pea Enation Mosaic Virus, Bean Leaf Roll Virus, Pea Streak Virus, Pea Seed-borne Mosaic Virus, Bean Yellow Mosaic Virus, and Broad Bean Wilt Virus (Kraft and Pfleger 2001; Biddle and Cattlin 2007; Wang et al. 2007). Viruses may cause different symptoms on different host species or host cultivars, whereas different viruses may cause the same or similar symptoms on the same host plants. Virus diseases cause serious losses in yield and quality of cultivated plants worldwide. Because aphids can transmit the viruses, epidemics of virus diseases always coincide with aphid epidemics. Thus, these losses caused by viruses can be reduced by monitoring aphid populations or controlling aphid epidemics (Rubiales et al. 2015).

Several parasitic nematode species infect most important legume crops including pea. Among these species, the cyst and root-knot nematodes are the most important worldwide (Rubiales et al. 2015). Pea cyst nematode (*Heterodera goettingiana*) is a major parasite of pea and is a soil-borne pest. Soil-borne population of cysts can remain viable for up to 20 years.

Root-knot nematodes (*Meloidogyne incognita*) are also important nematodes of pea, which can infect nearly all legume and vegetable crops. Damage caused by nematodes is often associated with root infection by pathogens, such as *Phythium* spp. and *F. solani*. Thus, it is important to control root rot infection in the field (Kraft and Pfleger 2001; Biddle and Cattlin 2007; Wang et al. 2007).

6.4.5 Insect Resistance

Insects are important pests of legumes throughout the world. A major class of insect pests is chewing insects of which in the case of legumes the most important tend to be storage insect pests bruchid beetles (Keneni et al. 2011). In pea, the pea seed beetle, pea weevil (*Bruchus pisorum* L.), is the most important insect pest of dry pea, which is widely distributed throughout the areas of the world where dry pea is grown. Damage is done by the larvae, which consume the interior of seed in storage period. However, all infestations originate in the field because only developing peas can be successfully attacked. Seeds damaged by pea seed beetle pea weevil, suffer from poor germination and are unacceptable to the food industry, which affect human consumption or seed markets (Kraft and Pfleger 2001; Biddle and Cattlin 2007; Wang et al. 2007).

Another important class of pests is sap-sucking insects. In pea, sap-sucking insect pests, pea aphid is a major pest in all temperate areas of world where peas or other legumes are grown. Aphids can cause serious losses to peas when populations colonize the growing points. In many cases, these aphids cause damage both by direct feeding and through virus transmission accelerating virus disease.

In addition, there are also several important insect pests of pea, including pea thrip (*Kakothrips pisivorus*), pea leafminer (*Liriomyza Huidobrensis* and *L. sativae* Blanchard), *Chromatomyia horticola* and other worms, etc.

6.5 Biological Nitrogen Fixation and Soil Resources as CS Traits

6.5.1 *Rhizobium* for Nodulation

Rhizobia are a kind of gram-negative bacteria widely distributed in soil. They can infect legumes to form root nodules, fix the molecular nitrogen in the air to form ammonia and provide nitrogen nutrition for plants. The amount of annual biological nitrogen fixation is 1.75×10^8 t in the world (Shen and Jing 2003) and is 4.37 times the world industrial nitrogen fertilizer output. Legumes in symbiosis with soil rhizobia are reported to fix 65% of nitrogen each year in agricultural production systems (Dénarié and Roche 1991; Zabran et al. 1999; Chen et al. 2002; Herridge et al. 2008). It is the most powerful biological nitrogen fixation system.

Nitrogen fixation is the process by which certain plants, including legumes, take nitrogen gas from the atmosphere; incorporate the molecules into their tissue, and subsequently into the ground, thus improving their own growth as well as soil health and overall productivity of the farming systems.

Symbiotic nitrogen fixation (SNF) is, therefore, a natural process of significant importance in world agriculture. The N_2 fixed by the legume crops represents a renewable source of nitrogen for agriculture soils.

Legume crops differ in biological nitrogen fixation (BNF) potential in the field, which is greatest for soybean followed by faba bean, pea, chickpea, lentil, and common bean (Herridge et al. 2008). As a good nitrogen-fixing crop, pea in symbiosis with rhizobia can fix about 75 kg of nitrogen per hectare, equivalent to 375 kg of ammonium sulfate or 225 kg of urea (Zheng et al. 1998).

6.5.2 Characterization for Rhizobium

The role of host plant–rhizobium symbiosis system in ecological agriculture can be expressed as follows: at normal temperature and pressure, nitrogen in the air can be fixed as ammonia, which meets all their own nitrogen requirements. In addition, the roots of host plants contain 30–35% of their total nitrogen, which can be left for later crop use. The host plant–rhizobium symbiosis can provide 30–60% of the nitrogen needed for the seasonal growth of other plants mixed with it (Chen and Chen 2016). The secretion of rhizobia has the function of dissolving phosphorus, calcium, magnesium, and other minerals (Alikhani et al. 2006). Some rhizobium bacteria can produce such progesterone as IAA, nodulation factor (LCO), and promote the growth of plants such as grasses and strawberries (Stokkermans and Peters 1994; Flores-Félix et al. 2015). Rhizobia inoculated with legumes and intercropping with other crops can enhance disease resistance of crops and reduce the use of pesticides (Yang et al. 2009).

In Brazil, little or no nitrogen fertilizer is applied on 23 million ha of soybeans, which are successfully reliant on both applied rhizobia inoculum and other nitrogen-fixing organisms (PGRB) such as *Azospirilla,* for up 80% of the crop nitrogen requirements to 300 kg/ha, with very large savings in crop input costs (Herridge et al. 2008). BNF benefits for grain yield have been demonstrated in Vietnam with rhizobial inoculation across a variety of legume crops (Herridge 2002). Rotation benefits for following wheat crops have been shown in Australia from BNF in chickpea and in faba bean crops (David and Hari 2009). These ranged from 10 to 40% increase in wheat grain yield with up to 40 kg/ha of additional nitrogen available from BNF.

The levels of indigenous rhizobia in the soils of China can be built up by initial cultivation of legumes in the absence of nitrogen fertilizer, to provide satisfactory levels of rhizobia inoculum for subsequent legume crops (Peoples et al. 1992).

6.5.3 Interaction Between Pea and Its Anchored Rhizobium

Studies have shown that the symbiotic nitrogen fixation of host plant–rhizobium system is a very complex interaction process. Vincent (1982) proposed that the process of symbiosis is influenced by three factors: rhizobium genome, plant genome, and environmental factors. Chen et al. (2004) also pointed out that the establishment of symbiosis between rhizobia and host plant was the result of the interaction between rhizobia, plants, and environment. Even though the relationship among legume varieties was very similar, there were significant differences between different varieties and different rhizobia (Gibson and Newton 1981).

The symbiotic system of rhizobia and legumes relies on a series of signals exchanged and responded at the molecular level. The nodulation gene of rhizobia, induced by special compound such as flavonoids and exotic brass secreted from the roots of legumes, is responsible for the synthesis and secretion of lipochitin oligosaccharide of nodulation factor. Conversely, nodulation factor can induce plant root hair curl and the formation of Nodules primordium root oncogen. Subsequently, rhizobia infected plants. The signals of the symbiotic system continue to exchange to maintain the normal growth of root nodules and carry out the nitrogen fixation reaction (Long 1989; Brewin et al. 1991; Christoph and Bisseling 1997).

6.5.4 Interaction Between Rhizobium and Soil

Soil is the main limiting factor that restricts the growth and symbiotic nitrogen fixation activity of host plants and rhizobia. In the rhizobia-host plant's symbiosis system, nitrogen fixation is significantly correlated with host plants. Therefore, in the presence of factors that limit the viability of host plants, such as soil salinity, soil pH, nutrient stress, soil temperature, soil moisture, indigenous rhizobia, etc. (Peoples et al. 1995; Walsh 1995), the competitive persistent strain would be unable to express all its nitrogen fixation capacity (Zahran 1999).

Soil salinity limits the internal processes of the symbiotic system. Under the presence of 170 mM NaCl, the root hairs of soybean showed curvature or deformation. Under the condition of 210 mM NaCl, nodulation formation was completely inhibited (Tu 1981). The infection rate of rhizobia was significantly reduced when soil moisture decreased from 5.5 to 3.5% (Worrall and Roughley 1976). Sellstedt et al. (1993) showed that nitrogen from nitrogen fixation reduced by about 26% under water stress by acetylene reduction method. Soil temperature affects nodules infection, nodules structure and nodules function (Roughley 1970). The optimum temperature is 28–31 °C for growing of most rhizobia. At 37 °C, many strains cannot grow (Piha and Munns 1987). The nitrogenase activity of rhizobia will be at a full stop in soil temperature of 8 °C. At soil temperature of 21 °C, the number of nodule reaches the maximum (Rice and Olsen 1988).

Soil pH in the neutral range optimizes the availability of all nutrients. In acidic pH soils, the availability of nutrients such as Ca, Mg, and P becomes limiting; on the other hand in soils with pH in the alkaline range the toxicity of sodium is likely to be stressful that affects nodulation and nitrogen fixation. Thus soil pH is an important soil characteristic that indicates the availability of plant nutrients (Dwivedi et al. 2015). Moreover, soil pH also directly influences nodulation and SNF through its effect on the numbers of naturally occurring rhizobium in noncultivated soils (Brockwell et al. 1991).

Slattery and coworkers (Slattery et al. 2004) have shown that there were differences in the effect of nodule inoculation on rhizobia among different types of soil. Compared with sandy soil, the survival rate of rhizobia in loam is higher (Li et al. 1997). Indigenous rhizobia with low nitrogen fixation efficiency and strong competitiveness widely exist in soil, which is easy to form a root nodule with low nitrogen fixation efficiency. Therefore, it can affect the percentage of inoculate rhizobia and reduce effective nitrogen fixation. In fact, soil microbial load is limited under certain soil conditions (Ding 1992). Therefore, in order to inoculate successfully, it is necessary to screen the strains with stronger competitive nodulation ability than local indigenous bacteria.

In addition, in order to reduce agricultural economic losses, large quantities of pesticides are applied to soil. Herbicides, fungicides, and other pesticides all have damaging effects and cause limitations on nitrogen fixation.

6.5.5 Optimized Operation for Rhizobium Fertilization

Mineral nutrition of the host plant can affect SNF via host plant growth and development as well as through the process of nodule development and function as these processes rest on the symbiosis between the rhizobium and the legume. The essential mineral nutrients required for legume SNF are those required for a normal establishment and functioning of the symbiosis. They are carbon (C), hydrogen (H), oxygen (O), nitrogen (N), phosphorus (P), potassium (K), sulfur (S), calcium (Ca), magnesium (Mg), iron (Fe), zinc (Zn), manganese (Mn), copper (Cu), boron (B), molybdenum (Mo), chlorine (Cl), nickel (Ni), and cobalt (Co). Each essential nutrient performs specific physiological and biochemical roles and is required in optimum concentration in the medium for the establishment and function of symbiosis between the legume host and the rhizobium. The role of mineral nutrients on SNF has been reviewed elsewhere (O'Hara et al. 1988; Zahran 1999; O'Hara 2001; Weisany et al. 2013).

A review of the published literature on the effects of starter N application on SNF by legumes indicates mixed results relative to the basal application of small amount of mineral N. However, it is widely accepted that in high fertility soils, especially those rich in organic matter, the application of starter N is not necessary; and at times it can reduce nodulation and SNF in crops such as soybean (Hungria et al. 2006) and bean (Vargas et al. 2000). On the other hand, in soils of low to very low in fertility and organic matter, the application of starter N at rate of 20–30 kg ha^{-1} has generally been reported to be beneficial to the growth and yield of several legumes (Sogut et al. 2013).

It has long been established that nitrates in the soil inhibit root infection, nodule development, and nitrogenase activity. Likewise, adequate nodulation is necessary for maximizing SNF by a legume (Sanginga et al. 1996). Moreover, poor and scanty nodulation is generally not able to satisfy the N needs of the plants, and they rely on soil N to grow and produce (Zahran 1999).

Among the major nutrients, phosphorus is essential for both nodulation and N_2 fixation. Indeed, nodules are strong sinks for phosphorus; as a consequence, symbiotic nitrogen-fixing plants require more phosphorus than those supplied with mineral fertilizers. The mode of nitrogen nutrition of legumes affects their phosphorus requirement (Cassman et al. 1981a, b). For achieving the potential of SNF by legumes, an adequate supply of phosphorus is a prerequisite because some legumes do not get established in conditions of insufficient soil P (Sahrawat et al. 2001). Mycorrhizal infection of roots of legumes stimulates both nodulation and nitrogen fixation under low phosphorus soil conditions (Redecker et al. 1997).

In addition, to optimize operation for rhizobium fertilization, the following points must be noted: Different rhizobia were selected for different legume species. The most efficient rhizobium was selected for different plant varieties. Rhizobia adapted to local soil environment were selected in different ecological zones (Chen and Chen 2016).

6.6 Genetic Resources of CS Genes

6.6.1 Primary Gene Pool

Based on the current accepted version of *Pisum* taxonomic classification (Smýkal et al. 2011), the primary gene pool consists of *Pisum sativum* subsp. *sativum* (includes var. *sativum* and var. *arvense)* and *P. sativum* subsp. *elatius* (includes var. *elatius*, var. *brevipedunculatum* and var. *pumilio*). Because of cross compatibility and abundant genetic diversity, the primary gene pool is the main resources of climate-smart genes against biotic and abiotic stresses. Resistance sources have been identified for powdery mildew (Ghafoor and McPhee 2012; Sun et al. 2016a, b), Ascochyta blight (Khan et al. 2013; Tran et al. 2014a, b), rust (Barilli et al. 2014), Aphanomyces root rot (Malvick and Percich 1999; Wu et al. 2018a, b), Fusarium root rot (Grünwald et al. 2003; Bodah et al. 2016), Fusarium wilt (McPhee et al. 1999; Bani et al. 2012), white mold (Porter et al. 2009; Porter 2012), bacterial blight (Hollaway et al. 2007; Martín-Sanz et al. 2012; Rodda et al. 2015), pea cyst nematode (Vito and Perrino,1978), some diseases caused by viruses (Kraft and Pfleger 2001; Congdon et al. 2016), broomrape (Rubiales et al. 2005, 2009), pea weevil (Pesho et al. 1977; Teshome et al. 2015a, b; Aznar-Fernández et al. 2018a), and pea aphid (Aznar-Fernández and Rubiales 2018b). Some germplasms tolerant to salt, drought, cold, frost have been screened as well (Sánchez et al. 1998; Okçu et al. 2005; Shahid et al. 2012; Humplík et al. 2015; Liu et al. 2017).

Many resistance genes and QTLs have been reported in traditional and primitive landraces of *P. sativum* against major diseases and pests. Since the first powdery mildew resistance gene *er1* was discovered in the pea landrace Huancabamba by Harland in 1948, the gene has been widely used in pea breeding in the world. The gene *er1* is recessive and can provide stable and durable powdery mildew resistance

(Tiwari et al. 1997). Humphry et al. (2011) indicate that resistance provided by *er1* is due to a loss of function of *PsMLO1*, a *MLO* gene. Based on homologous sequence analysis of *PsMLO1* and molecular mapping, so far 10 natural allelic variations at *er1* locus have been identified in the world *P. sativum* germplasm collections, and functional markers cosegregated with these *er1* alleles are available (Sun et al. 2016a, b; Ma et al. 2017a, b). Another recessive powdery mildew resistance gene, *er2*, also has been identified in pea germplasms SVP952 and JI2480 (Tiwari et al. 1997). Molecular markers linked to *er2* have been detected (Katoch et al. 2010). However, since expression of *er2* is strongly influenced by temperature and leaf age, and its effective resistance to powdery mildew was specific to particular geographic regions (Fondevilla et al. 2006), its application in pea breeding is limited.

Ascochyta blight complex caused by several fungal pathogens is a destructive disease in pea worldwide (Khan et al. 2013). Partial or incomplete resistance to the disease complex has been identified in *P. sativum* (Tivoli et al. 2006; Zhang et al. 2007). To date, a number of resistance QTLs to *Ascochyta pisi* or *Didymella pinodes* have been detected on all the seven linkage groups of pea (Timmerman-Vaughan et al. 2002; Tar'an et al. 2003; Timmerman-Vaughan et al. 2004; Prioul et al. 2004; Fondevilla et al. 2008; Carrillo et al. 2014; Timmerman-Vaughan et al. 2016; Gali and Shunmugam 2018). Some QTLs might be common for resistance to Ascochyta blight complex in the different populations. Further, candidate gene approaches have revealed the colocalization of QTLs for resistance to Ascochyta blight and resistance gene analogs, defense-related genes (Prioul-Gervais et al. 2007; Carrillo et al. 2014; Timmerman-Vaughan et al. 2016). Recently, Jha et al. (2015) genotyped diverse *P. sativum* accessions using 15 single nucleotide polymorphisms (SNPs) within candidate genes associated with reaction to Ascochyta blight and found that SNP loci *PsDof1p308* and *RGA-G3Ap103* had significant associations with Ascochyta blight scores.

Pea rust, caused by *Uromyces fabae* (Syn. *Uromyces viciae-fabae*) or by *Uromyces pisi. U. fabae,* leads to rust in tropical and subtropical regions, while *U. pisi* is epidemic in temperate regions (Barilli et al. 2009a). A few *P. sativum* germplasms were reported to have partial resistance to *U. pisi* (Barilli et al. 2009b), but the characteristic for resistance to *U. pisi* has not been determined. Many efforts have been made to identify sources of resistance in pea against *U. fabae*, and immune or partially resistant germplasms have been identified (Barilli et al. 2014). Complete or partial resistance to *U. fabae* has been reported in different studies (Katiyar and Ram 1987; Chand et al. 2006). Vijayalakshmi et al. (2005) found a single partially dominant gene *Ruf* and two or QTLs conferring resistance to *U. fabae* in the pea cross FC 1 × HUVP 1, and identified two flanking random amplified polymorphic DNA (RADP) markers linked to the gene *Ruf*. Rai et al. (2011) molecularly mapped one major (*Qruf*) and one minor (*Qruf1*) QTL for rust resistance on LGVII in RILs derived from the cross HUVP 1 × FC 1, and simple sequence repeat (SSR) markers flanking *Qruf* and *Qruf1,* respectively, were identified. Recently, Rai et al. (2016) further revealed that rust resistance was conditioned by four QTLs, two major (*Qruf* on LGVII, *Qruf2* on LG I), and two minor (*Qruf1* on LG VII and *Qruf3* on LG VI) in the same RILs, and the minor QTL *Qruf3* was contributed by the susceptible parent

HUVP 1. Singh et al. (2015a, b) and Upadhyay et al. (2017) demonstrated that SSR markers flanking *Qruf* and *Qruf1* were effective in marker-assisted selection (MAS) of pea rust resistance.

Resistance to Fusarium root rot caused by *Fusarium solani* f. sp. *pisi* is inherited as a quantitative trait regulated by more than one gene (Grünwald et al. 2003). Nine QTLs have been identified for this pathogen (Weeden and Porter 2008; Feng et al. 2011; Coyne et al. 2015), and the SSR markers closely linked to the QTLs are available for MAS to develop resistance cultivars. Fusarium wilt of pea is caused by race-specific *Fusarium oxysporum* f. sp. *pisi*, so far four races of the pathogen, race 1, 2, 5 and 6, were described in the world (Haglund and Kraft 1979; Bani et al. 2012), Genetic resistance to the four races are conferred by independent single dominant genes (Kraft and Pfleger 2001). Up to now, three Fusarium wilt resistance genes, *Fw*, *Fnw*, and *Fwf*, have been identified. *Fw* for resistance to race 1 has been placed on LG III, three user-friendly markers and one functional codominant marker linked to the *Fw* gene have been developed for MAS (Kwon et al. 2013; Jain et al. 2015). *Fnw* gene conditioning resistance to race 2 has located the gene on LG IV between SSR markers *AC22_185* and *AD171_197* using an RIL population, meanwhile two minor resisatnce loci, *Fnw 3.1* and *Fnw 3.2* also were detected on LG III in the same RIL population, which revealed that additional genetic factors is associated with resistance to race 2 (McPhee et al. 2012). The gene *Fwf* conferring resistance to race 5 has been mapped on LG II, and the locus coding the plastid isozyme of aspartate aminotransferase (*Aatp*) was found to be linked to *Fwf* at a distance of 9.1 cM (Coyne et al. 2000).

Aphanomyces root rot, caused by *Aphanomyces euteiches* Drechs., is one of the major limitations to pea production worldwide (Kraft and Pfleger 2001). Genetic resistance is the most promising means to control the disease. Great efforts have been made to identify resources resistant to Aphanomyces root rot, and a number of *P. sativum* germplasms with high levels of partial resistance to Aphanomyces root rot have been identified and used for breeding (Malvick and Percich 1999; Wicker et al. 2003). Hamon et al. (2013) identified 27 meta-QTLs associated with resistance to *A. euteiches* in four resistant sources by a comprehensive meta-QTL mapping program, and revealed seven main genomic regions with high position consistency over locations, years, isolates and populations (Pilet-Nayel et al. 2002, 2005; Hamon et al. 2011). The seven main QTLs, *Ae-Ps1.2*, *Ae-Ps2.2*, *Ae-Ps3.1*, *Ae-Ps4.1*, *Ae-Ps4.5*, *Ae-Ps5.1*, and *Ae-Ps7.6*, were located on six of the seven pea linkage groups of a consensus map, and linked SSR markers were available (Hamon et al. 2013). Further, the seven main resistance QTLs were evaluated to be effective in different pea genetic backgrounds (Lavaud et al. 2015, 2016). Desgroux et al. (2016a, b), validated six QTLs, *Ae-Ps1.2*, *Ae-Ps2.2*, *Ae-Ps3.1*, *Ae-Ps4.1*, *Ae-Ps4.5*, and *Ae-Ps7.6*, and detected novel resistance loci in a pea-Aphanomyces collection of 175 pea lines using genome-wide association analysis.

Chang et al. (2018) combined GWAS and RNA-Seq to identify white mold resistance in pea and found a transcript encoding a glutathione S-transferase associated with resistance variant for both lesion and nodal resistance.

Bacterial blight, caused by *Pseudomonas syringae* pv. *syringae* and *P. syringae* pv. *pisi*, has been reported in most of the pea-growing areas worldwide. Eight pathogenic races of *P. syringae* pv. *pisi* have been described (Taylor et al. 1989; Martín-Sanz et al. 2011). Resistance to *P. syringae* pv. *pisi* in pea is conferred by specific single dominant genes, while inheritance of resistance to *P. syringae* pv. *syringae* in pea is quantitative (Fondevilla et al. 2012). To *P. syringae* pv. *pisi*, five dominant resistance genes, *Ppi1*for races 1, *Ppi2* for races 2, *Ppi3* for races 3, *Ppi4* for races 4 and *Ppi8* for races 4 and 8, respectively, were identified (Hunter et al. 2001; Martín-Sanz et al. 2016). *Ppi1* was mapped on LG VI; *Ppi2* was located LG VII; *Ppi3* and *Ppi4* was mapped on LG II; *Ppi8* was located LG III. Eight QTLs associated with resistance to *P. syringae* pv. *syringae* have been detected in three RIL populations, and placed on LG II, III, VI, and VII (Fondevilla et al. 2012; Sudheesh et al. 2015a, b). Sudheesh et al. (2015a, b) identified a single common genomic region associated with *P. syringae* pv. *pisi* resistance on LG III in the same mapping populations as well.

Crenate broomrape (*Orobanche crenata*) is the major constraint for pea cultivation in the Mediterranean Basin and Middle East. Four QTLs associated with field broomrape resistance have been identified in wild pea (Fondevilla et al. 2010).

Four recessive resistance genes, *sbm-1*, *sbm-2*, *sbm-3* and *sbm-4*, to Pea Seed-borne Mosaic Virus (PSbMV) have been reported in pea (Provvidenti and Hampton 1991). The gene *sbm-1* is a mutant allele of eukaryotic translation initiation factor 4E (*eIF4E*) on LG VI, while *sbm-2* closely links with the isoform of *eIF4E* (*eIF(iso)4E*) on LG II (Gao et al. 2004). The genes *sbm-3* and *sbm-4* map close to *sbm-1* on LG VI. Nine alleles of *eIF4E* were identified to confer resistance to PSbMV (Konečná et al. 2014). The specific molecular markers for the gene *sbm-1* (*eIF4E* allele) were available for marker-assisted pea breeding (Smýkal et al. 2010). Resistance genes in pea to other potyviruses also have been reported to be recessive and closely linked and clustered with *sbm-2* on LG II and *sbm-1* on LG VI (Provvidenti and Muehlbauer 1990). The genes on LG II include *bcm*, *cyv1*, *mo*, and *pmv*, confer resistance to Bean Common Mosaic Virus (BCMV), Clover yellow vein virus (ClYVV), Bean Yellow Mosaic Virus (BYMV), and Pea Mosaic Virus (PMV), respectively. The genes on LG VI are *cyv2* and *wlv* which mediate resistance to ClYVV and BYMV, respectively. The *cyv2* and *wlv* correspond to the same *sbm-1* allele of *eIF4E* (Gao et al. 2004; Andrade et al. 2009). Resistance to Pea Enation Mosaic Virus (PEMV) is controlled by a single dominant gene *En* which was mapped on LG III (Marx et al. 1985). Two sequence tagged site markers *CNGC* and *tRNAMet2* linked to *En* have been developed for efficient selection of PEMV resistance in pea (Jain et al. 2013). Resistance to Bean Leaf Roll Virus (BLRV) is inherited as a single recessive gene, designated *lr* (Drijfhout 1968). Another recessive gene, *lrv* confers tolerance to BLRV in pea (Baggett and Hampton 1991).

Tolerance QTLs for abiotic stresses have been reported in pea. Leonforte et al. (2013a, b) identified several salinity tolerance QTLs and candidate genes on LG III and VII. Iglesias-García et al. (2015) identified 10 QTLs associated with drought which explained individually from 9 to 33% of the phenotypic variance depending on the variable assessed and altogether from 20 to 57%. Javid et al. (2017) detected a QTL associated with metribuzin tolerance located on LG IV accounting for 12–21%

of phenotypic variance. Liu et al. (2017) detected seven markers repeatedly associated with frost tolerance in at least two different environments by marker-trait association analysis on 672 worldwide pea collections. Genetic markers closely flanking these stress tolerance QTLs are suitable for implementation in pea breeding programs.

6.6.2 Secondary Gene Pool

The secondary gene pool of pea consists of *Pisum fulvum*, and *P. abyssinicum*. *P. fulvum*, a wild relative of pea, is an important source to improve the genetic resistance of pea against biotic and abiotic stresses. *P. fulvum* has been shown to be resistant to major pea fungal pathogens such as *Erysiphe pisi* (Fondevilla et al. 2007a, b), *Uromyces pisi* (Barilli et al. 2009b), *Mycosphaerella pinodes* (Gurung et al. 2002; Fondevilla et al. 2005; Jha et al. 2012). Recently, Jha et al. (2016) identified nine resistance QTLs for Ascochyta blight in an RIL population derived from interspecific cross between the resistant *P. fulvum* accession P651 and the susceptible *P. sativum* cultivar Alfetta. A high-density integrated DArTseq SNP-based genetic map of *P. fulvum* was constructed and three rust resistance QTLs *UpDSII*, *UpDSIV* and *UpDSIV.2* were located in the LGs II and IV, respectively, (Barilli et al. 2018). A pea rust resistance QTL explaining 63% of the total phenotypic variance was located on LG III, and sequenced tag site (STS) markers flanking this QTL were developed for MAS for rust resistance (Barilli et al. 2010). A new dominant gene for resistance to powdery mildew, *Er3*, has been identified in a *P. fulvum* line P660-4 selected from the ICARDA accession IFPI3261 originating from Idlib, Syria (Fondevilla et al. 2007a).

P. fulvum accessions have been found to be valuable sources of resistance to the pea weevil (Hardie et al. 1995; Clement et al. 2002; Aznar-Fernández et al. 2018a). Pea weevil resistance from *P. fulvum* has been introgressed into cultivated field pea through backcrossing (Aryamanesh et al. 2012). The genome regions controlling cotyledon, pod wall/seed coat and pod wall resistance to pea weevil from *P. fulvum* were identified through QTL mapping, and the QTL markers to probe *Pisum* germplasm for pea weevil resistance genes were developed (Aryamanesh et al. 2014).

Incomplete levels of resistance to crenate broomrape (*Orobanche crenata*) are available in *P. fulvum* (Pérez-de-Luque et al. 2005; Rubiales et al. 2009), which should provide diversified alleles for resistance breeding in pea.

In addition, *P. fulvum* could represent a gene pool which could be used for improving salt and drought tolerance in field pea (Miljus-Djukic et al. 2013; Naim-Feil et al. 2017).

Pisum abyssinicum has a much more restricted distribution in the highlands of Ethiopia and Yemen. Although genetic diversity within *P. abyssinicum* is extremely limited (Weeden 2018), several surveys have shown that *P. abyssinicum* accessions could be a valuable source for resistance to some major pea disease pathogens such as the nematode *Heterodera goettingiana* (Vito and Perrino 1978), Fusarium wilt race 2 (McPhee et al. 1999), *Pseudomonas syringae* pv. *pisi* causing pea bacterial blight

(Elvira-Recuenco et al. 2003; Martín-Sanz et al. 2012; Rodda et al. 2015), *Erysiphe pisi* (Fondevilla et al. 2007b) and *Orobanche cernata* (broomrape) (Rubiales et al. 2009). Resistance to pea bacterial blight in *P. abyssinicum* was race-nonspecific to all races of *P. syringae* pv. *pisi*, more likely a major recessive gene together with a number of modifiers (Elvira-Recuenco et al. 2003; Hollaway et al. 2007). Due to lack of specific resistance to race 6, the race-nonspecific resistance in *P. abyssinicum* and its recessive characteristic offer the possibility to achieve a durable resistance to pea bacterial blight.

6.6.3 Tertiary Gene Pool

Vavilovia formosa (Stev.) Fed. belongs to the tribe Fabeae and is a scientifically valuable common ancestor of the plant tribe Fabeae (Mikić 2014). Vavilovia has a close phylogenetic relationship with pea, so it is considered to be the tertiary gene pool of pea (Smýkal et al. 2011). Vavilovia is a diploid species with the same number of chromosomes as pea, $2n = 14$. Golubev (1990) indicated that it was possible for producing hybrids between Vavilovia and pea, but no further researches have been done after Golubev's work. Recently, Ochatt et al. (2016) developed a range of biotechnology approaches for in vitro propagation of Vavilovia which will be useful for accelerating taxonomical and breeding researches. So far, the agronomic importance of Vavilovia have not been evaluated, but a hypothetical gene for perenniality in Vavilovia could be valuable for its cultivated relatives, specifically pea (Mikić et al. 2010).

Lathyrus sativus L., known as the grasspea, is a close relative of pea and Vavilovia, so it might be considered to be a tertiary gene pool of pea (Smýkal et al. 2011). Although intergenric hybridization between *Pisum* and *Lathyrus* is incompatible, electrical and chemical fusion of protoplasts between pea and *L. sativus* has achieved (Durieu and Ochatt 2000). *L. sativus* has been reported to be resistant sources to several fungal pathogens of pea, such as *Mycosphaerella pinodes*, *Erysiphe pisi*, *Uromyces pisi*, *U. viciae-fabae*. Genetic analysis of an accession ATC80878 resistant to Ascochyta blight suggested the resistance may be controlled by two independently segregating genes, operating in a complementary epistatic manner (Skiba et al. 2004). Vaz Patto and Rubiales (2009) indicated a collection of Iberian-cultivated *L. sativus* germplasm had different resistant levels to powdery mildew caused by *E. pisi*, accessions with reduced disease severity despite being of a high infection type were identified. Vaz Patto and Rubiales (2009) identified a germplasm collection of *L. sativus* that were highly resistant to pea rust pathogen *U. viciae-fabae*, and had partial resistant to *U. pisi*.

6.6.4 Artificially Induced/Incorporated Traits/Genes

Induced mutations for improving characters such as disease resistance via chemical mutagens and physical irradiation have been applied in pea. Pereira et al. (2010) induced two powdery mildew resistant mutants, *er1mut1* and *er1mut2,* in pea using ethylnitrosourea mutagenesis which affected the same locus *er1* identified in natural powdery mildew resistance sources. Mutant *er1mut2* was found to carry a new mutation site in *er1* locus and the powdery mildew resistance allele was designed as *er1–10* (Santo et al. 2013; Ma et al. 2017a, b). A STS marker and a KASPar marker cosegregated with *er1–10* were developed for MAS as well (Santo et al. 2013; Ma et al. 2017a, b). A novel *er1* resistance allele, *er1–5,* was obtained by diethyl sulfate mutagenesis in pea and codominant cleaved amplified polymorphic sequence (CAPS) marker and HRM-based marker suitable for MAS were developed on the mutation site (Pavan et al. 2011, 2013). Sharma et al. (2010) used ethyl methane sulfonate (0.2 and 0.3%) or γ-rays (5–22.5 kR) mutagenized two Fusarium wilt-susceptible pea genotypes and obtained 25 mutants exhibiting complete or enhanced Fusarium wilt resistance.

Genetic resources carrying important climate-smart genes have been created through genetic engineering in pea. The ∞-amylase inhibitor in bean seed is toxic to cowpea weevil, *Callosobruchus maculatus* (Huesing et al. 1991). The gene encoding *αAI-1* has been transferred by genetic engineering into pea to protect against some bruchids. Shade et al. (1994) found that transgenic pea seeds expressing the α-amylase inhibitor of the common bean were sufficient to provide complete resistance to bruchid beetles, *C. chinensis* and *C. maculatus*. Schroeder et al. (1995) transferred ∞-amylase inhibitor *(a-AI)* gene into pea and obtained transgenic peas resistant to pea weevil, and under field conditions, the bean gene in the transgenic peas provided complete protection from pea weevil (*Bruchus pisorum*) (Morton et al. 2000).

Jones et al. (1998) generated transgenic pea lines carrying PSbMV replicase (*NIb*) sequences that exhibited resistance to PSbMV. Transformed pea lines expressing viral coat protein (CP) resistant to Pea Enation Mosaic Virus and Alfalfa Mosaic Virus were created (Chowrira et al. 1998; Timmerman-Vaughan et al. 2001). Transgenic peas expressing a proteinase inhibitor (*Na-PI*) from *Nicotiana alata* increased resistance to *Helicoverpa armigera* (Charity et al. 1999). Transgenic pea plants harboring salt stress-tolerant gene (*AtNHX1*) from *Arabidopsis thaliana* enhanced salt stress tolerance and frost tolerance (Ali et al. 2018).

6.7 Glimpses on Classical Genetics and Traditional Breeding for CS Traits

6.7.1 Classical Mapping Efforts

In the last four decades, there were many studies of classical genetic linkage map in pea. Different kind of mutants with morphological traits were collected and studied. In 2002, Ellis and Poyser summarized the previous work (Blixt 1972; Weeden et al. 1993, 1998; Rozov et al. 1999) and constructed an integrated linkage map with 105 markers and 7 groups.

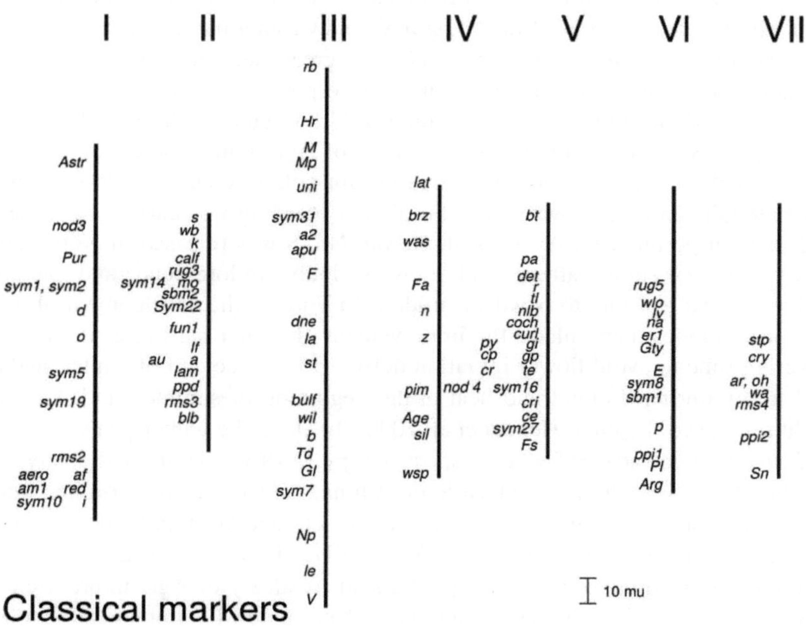

Cited from Ellis and Poyser (2002)

6.7.2 Limitations of Classical Endeavors and Utility of Molecular Mapping

The classical genetic linkage map has limited morphologic markers and the density of number is not enough to undertake gene cloning. Moreover, the morphologic difference between two parents is very limited. As a result, molecular genetic linkage map was constructed with AFLP, RAPD, SSR and SNP markers. In 2014, the genetic linkage map based on the Chinese varieties were constructed including 157 SSR markers, 11 linkage groups and covering 1518 cM in total length (Sun et al. 2014). Moreover, another 33 polymorphic SSR markers were added to the genetic linkage map of the same population (Yang et al. 2015).

6.7.3 Breeding Objectives: Positive and Negative Selection

There were many breeding objectives in the world for pea breeding. Resistance and tolerance for biotic stress and abiotic stress are very important breeding objectives, respectively. Winter pea is traditionally sown in autumn in the South and Centre China (south of north latitude 33°). In recent years, more and more northern areas plant winter pea in autumn. However, severe cold weather has affected the historical pea-producing area in recent years, leading to great yield fluctuation. Thus, cold-tolerant varieties are very important in the safe production of winter pea (Zhang et al. 2017). From 2009 to 2011, Institute of Crop Science, Chinese Academy of Agricultural Sciences and Qingdao Academy of Agricultural Sciences cooperated and finished 3677 pea accessions evaluation for cold tolerance in the field during winter in Qingdao. Moreover, the cold-tolerant breeding was carried out based on cold-tolerant germplasm. As a result, Kewan No. 8 was released in North China. This variety was semi-leafless, white flowers, double inflorescence and tolerant to cold stress and resistant to powdery mildew. In Europe, the development of winter field pea varieties, particularly the improvement of winter hardiness, the fitting of flowering time to avoid flower initiation during frost at the end of winter, and also seed filling during drought and heat at the beginning of summer, and Ascochyta blight resistance (Lejeune-Hénaut et al. 2008). In USA, the winter-pea variety were bred and near infrared reflectance spectroscopy (NIRS) was used to analyze the breeding trait (Saha et al. 2018). In South China, the weather is warm and humid. Improving resistance to multiple disease resistances and keeping high yield are the most important local breeding targets. Yunwan No. 17 and Yunwan No. 37, which were bred by Institute of Grain Crops, Yunnan Academy of Agricultural Sciences (IGC-YAAS), were both high resistant to powdery mildew (Communication to Prof. Yuhua He).

6.7.4 Achievements of Traditional Breeding

Following the classical breeding method, pea breeders have made much progress in yield, quality, stress and resistance breeding in the world. As we known, 30 years ago, landrace in rainfed areas dominated the dry pea production in China. Until now, a total of 45 varieties have been bred and released from nine public institutes. 34 of them were used as dry peas and 11 were vegetable peas. For dry pea varieties, 26 were normal leaf types and 8 were semi-leafless types. Among vegetable varieties, six were snow peas, three were snap peas, and two were tendrill-free vegetable pea

(Thomas et al. 2015). For example, Kewan No. 1 bred by Ms. Ling Li and Dr. Zong was high yield with the average number about 3182 kg per hectare. It is resistant to PSbMV, downy mildew and lodging. Yunwan No. 1 was bred by Institute of Grain Crops, Yunnan Academy of Agricultural Sciences (IGC-YAAS). This variety was high yield with the average number about 3020 kg/ha in the field and it is moderate resistant to powdery mildew. For vegetable variety, Shijiadacaiwan No. 1 was the first snow pea variety in China, which was sweet and of good quality with improved tolerance to frost and lodging (Cheng and Wang 2009; Li et al. 2017).

6.7.5 Limitations of Traditional Breeding and Rationale for Molecular Breeding

Traditional breeding approaches are vital for producing novel genetic variants, hybridization between contrasting parental lines and mutation. In traditional breeding program, various methods such as pedigree method, backcrossing, recurrent selection and mutation breeding are used. However, the traditional breeding has apparent weakness, such as long breeding cycle, low selection efficiency and difficulty in distant crossing. Meanwhile it is difficult to pyramid many useful genes into a single elite breeding line, and follow up them in subsequent breeding programs.

Molecular breeding, containing genomic-assisted breeding, marker-assisted breeding, and genetic engineering has experienced significant innovations and advances during the past three decades (Xu et al. 2012; Poland 2018). Breeding by design and genome editing have received great attention as designing a desirable plant based on marker and associated gene information becomes increasingly possible (Chen et al. 2018; Scheben and Edwards 2018).

6.8 Genetic Diversity Analysis

6.8.1 Phenotype-Based Diversity Analysis

There have been many previous studies on phenotype-based diversity analysis to evaluate pea genetic resources, providing important data for screening elite and excellent genotypes such as high yield, high quality, stress resistance, and disease resistance. In the year 2005, analysis of 148 Ethiopian *P. sativum* genotypes for 12 agronomically important traits detected significant differences in all traits studied, except for the number of seeds per pod. Based on these traits, clusters of genotypes were formed but there was no obvious relationship between agronomic clusters and geographic origin (Keneni et al. 2005). In the year 2011, 624 pea genotypes sampled from a world collection of various geographical origins were evaluated based on 20 morphological traits, and a large genetic variation among geographic origin based groups of genetic resources was revealed by comparison of mean value, CV, genetic diversity index of their 20 morphological traits (He and Zong 2011). Two gene pools, a Chinese gene pool and an oversea gene pool were detected and defined by three-dimensional PCA graph method. Genetic distance based UPGMA clustering analysis provided the same results for support (He and Zong 2011). Cold tolerance improvement of pea varieties is important for stable yield and expansion of the winter pea planting area. Under natural low-temperature conditions during winter in Qingdao, Shandong, China, recently, 3,672 pea germplasm accessions were evaluated for cold tolerance in the field and categorized as displaying high resistance (214 accessions), moderate resistance (835 accessions), or susceptibility (2623 accessions). The highly and moderately resistant genotypes were validated in the following year; the genotypes from the winter production region showed higher cold tolerance than genotypes from the spring production region; the accessions identified as having high levels of cold tolerance are recommended as potential genetic resources in cold tolerance breeding of pea (Zhang et al. 2016).

In addition, some studies combined morphological and molecular markers for correlation analysis and screening of molecular markers related to important agronomic traits and stress resistance characteristics, laying an important foundation for pea molecular assisted breeding. In 2008, 25 varieties of pea (*Pisum sativum* L.) were selected from the list of recommended varieties for cultivation in the Czech Republic, and made both a standard classification by 12 morphological descriptors and a classification by biochemical-molecular markers. The results showed that molecular identification could be used to assess distinctness and complement morphological assessment (Smýkal et al. 2008a). In year 2010, 30 genotypes and 10 commercial varieties of pea from Turkey were analyzed by UPOV (the International Union for the Protection of New Varieties of Plants) phenotypic criteria and 10 SSR primer pairs, and the UPGMA dendrogram revealed genetic relatedness of tested genotypes (Sarıkamış et al. 2010). More recently, in year 2017, genetic structure, diversity, and inter-relationships in a worldwide collection of 151 pea accessions were analyzed using 21 morphological descriptors and 20 simple sequence repeat (SSR) primers (Rana et al. 2017); genetic relationships inferred from a neighbor-joining tree separated accessions into three groups; Bayesian model-based STRUCTURE analysis detected three gene pools for the analyzed pea germplasm and showed these three gene pools coexisted in accessions belonging to different geographic regions indicating frequent transference and exchange of pea germplasm during its domestication history (Rana et al. 2017).

6.8.2 Genotype-Based Diversity Analysis Using Molecular Markers

Many molecular marker-based studies have been conducted on *Pisum* germplasm collections to investigate genetic and trait diversities. Molecular variation in the John

Innes Centre *P. sativum* core collection was assessed by comparing Ty1-copia retrotransposable element sequences in 1998 (Ellis et al. 1998), and Pearce et al. (2000) went on to further characterize the Ty1-copia retrotransposable element sequences in pea in 2000. The usefulness of the Ty1-copia sequence was proposed due to being structurally conserved but having high insertion site polymorphism. Diversity among accessions within the John Innes *Pisum* core collection (Matthews et al. 1995) was assessed by examining differences in the insertional sequence of the PDR1 element. Using sequence-specific amplification polymorphism (SSAP) within various retrotransposon element variable regions, a detailed *Pisum* phylogeny was developed (Pearce et al. 2000). In 2012, characterization of genetic diversity among 4,538 *Pisum* accessions held in seven European Genebanks has identified sources of novel genetic variation, and both reinforces and refines previous interpretations of the overall structure of genetic diversity in *Pisum*. Molecular marker analysis was based upon the presence/absence of polymorphism of retrotransposon insertions scored by a high-throughput microarray and SSAP approaches. The diversity of *Pisum* constitutes a broad continuum, with graded differentiation into sub-populations which display various degrees of distinctness. The most distinct genetic groups correspond to the named taxa. The cultivars and landraces of *Pisum sativum* can be divided into two broad types, one of which is strongly enriched for modern cultivars (Jing et al. 2012). In 2000, 45 *Pisum sativum* cultivars were analyzed by the combination of six enzyme systems (acid phosphatase, amylase, esterase, leucine aminopeptidase, shikimate dehydrogenase and phosphoglucomutase) with two electrophoretic techniques (NATIVE-PAGE, isoelectric focusing) through isozyme electrophoresis with the objective to find protein markers for exact and reproducible discrimination of individual genotypes, by use of seed and leaf tissue enabled to identify all 45 studied cultivars; critical factors which may affect utilization of isozyme electrophoresis for commercial applications in pea breeding and seed production and testing are discussed (Zdeněk and Miroslav 2000).

In 2004, protein and PCR-based molecular markers were employed to assess genetic diversity among 148 *Pisum* accessions that represented both primitive and modern cultivated forms (Baranger et al. 2004). Cluster analysis enabled discrimination of pea types corresponding to end use such as fodder, food and animal feed, and separated the wild and primitive forms. More specifically, the spring-sown feed types were clearly differentiated from the winter-sown feed types (Baranger et al. 2004). They also reported that released cultivars and breeding lines contained far less diversity than the rest of the collection that represented the wider *Pisum* genepool (Baranger et al. 2004). In 2007, a diversity study among 24 elite Indian *P. sativum* genotypes using 60 random amplified polymorphic DNA (RAPD) markers separated groups corresponding to tall and dwarf type varieties, and the similarity detected between pairs of accessions ranged from 60 to 87% (Choudhury et al. 2007). In 2014, a set of 83 pea accessions with maximum ecological and geographical diversity (including those from the centres of origin of the species) obtained from the Vavilov Research Institute (VIR) collection including representatives of three subspecies, was studied using the amplified fragment length polymorphism (AFLP) method and a number of their morphological and biological characteristics were evaluated. The affiliation of the samples to the certain subspecies was not confirmed by the obtained data, and the ecogeographic differentiation of the samples was not reflected by the data. Factor analysis allowed to identify the sample groups of European and Asian origin and the intermediate nature of most of the samples from the studied set of pea accessions (Dyachenko et al. 2014). According to an initial study, PCR-based microsatellite markers were generally found to be more informative for assessing genetic relationships within *Pisum* than restriction fragment length polymorphism (RFLP) type markers (Lu et al. 1996).

In the year 2009, 21 SSR markers were applied to compare 1,243 cultivated pea genetic accessions from China with that of 774 accessions from the rest of world and 103 wild pea genotypes. It was found that there were significant differences between Chinese and world pea gene banks; and the core germplasm of Chinese pea was constructed (Zong et al. 2009a, b). In the year 2014, 96 pea cultivars widely grown or used in breeding programs in the USA and Canada were analyzed for genetic diversity using 31 microsatellite or simple sequence repeat (SSR) and 11 novel EST-derived genic markers. Genetic diversity was assessed through unweighted neighbor-joining method, and 96 varieties were grouped into three main clusters based on the dissimilarity matrix; four subpopulations were determined through STRUCTURE analysis with no significant geographic separation of the subpopulations (Jain et al. 2014). A recent study in 2017 using 30 SSR markers for comparison of 295 Chinese genetic resources and 305 world pea genetic resources, the genetic diversity and population genetic structure of germplasm resources revealed the similar conclusion; it also found that Chinese peas genetic resources can be divided into two groups, the significant difference existed between genetic resources from spring sowing area and from winter sowing area (Wu et al. 2017). In 2015, 46 accessions, of which 43 were from Ethiopia, were subjected to genetic diversity analysis using 15 newly developed EST-SSR markers developed from publicly available ESTs; a total of 37 alleles were detected across all accessions, and high levels of genetic variation were detected in field pea accessions from Ethiopia using these markers (Teshome et al. 2015a, b).

To evaluate the genetic diversity and the scale of linkage disequilibrium (LD) decay in pea, a collection of 917 accessions were genotyped in 2017, gathering elite cultivars, landraces, and wild relatives using an array of about 13,000 SNPs (single nucleotide polymorphisms). Genetic diversity is broadly distributed across three groups corresponding to wild/landraces peas, winter types, and spring types. At a finer subdivision level, genetic groups relate to local breeding programs and type usage. LD decreases steeply as genetic distance increases. When considering subsets of the data, LD values can be higher, even if the steep decay remains. It is revealed that the genomic regions exhibiting high level of differentiation between wild/landraces, winter, and spring pea, respectively. Two regions on linkage groups 5 and 6 containing 33 SNPs exhibit stronger differentiation between winter and spring peas than would be expected under neutrality. Interestingly, QTL for resistance to cold acclimation and frost resistance have been identified in the same regions (Mathieu et al. 2017).

6.8.3 Relationship with Other Cultivated Species and Wild Relatives

In 2009, 197 *Pisum* accessions from 62 counties of five continents were employed for SSR analysis using 21 polymorphic primer pairs in this study. Except for cultivated field pea *Pisum sativum* subsp. *sativum* var. *sativum* (94 genotypes), it also included wild relative genotypes that were classified as *P. fulvum*, *P. sativum* subsp. *abyssinicum*, *P. sativum* subsp. *asiaticum*, *P. sativum* subsp. *transcaucasicum*, *P. sativum* subsp. *elatius* var. *elatius*, *P. sativum* subsp. *elatius* var. *pumilio* and *P. sativum* subsp. *sativum* var. *arvense* (103 genotypes). SSR alleles were uniformly distributed among botanical taxon units under *pisum* genus, but significant difference appeared in most pairwise comparisons for genetic diversity between taxon units based groups of genetic resources. Genetic diversity level of wild species *P. fulvum* was much lower than the cultivated species *P. sativum*. Under species *P. sativum*, *P. sativum* ssp. *sativum* var. *sativum* and *P. sativum* ssp. *asiaticum* were the highest in gentic diversity, followed by *P. sativum* ssp. *elatius* var. *elatius* and *P. sativum* ssp. *transcaucasicum*, *P. sativum* ssp. *elatius* var. *pumilio*, *P. sativum* ssp. *sativum* var. *arvense* and *P. sativum* ssp. *abyssinicum* were the lowest. Four gene pool clusters were detected under *Pisum* genus by using PCA analysis. Gene pool "*fulvum*" mainly consisted of wild species *Pisum fulvum*, gene pool "*abyssinicum*" mainly consisted of *P. sativum* ssp. *abyssinicum*, and gene pool "*arvense*" mainly consisted of *P. sativum* ssp. *sativum* var. *arvense*. While gene pool "*sativum*" were composed by five botanical taxon units, they are *P. sativum* ssp. *asiaticum*, *P. sativum* ssp. *elatius* var. *elatius*, *P. sativum* ssp. *transcaucasicum*, *P. sativum* ssp. *elatius* var. *pumilio* and *P. sativum* ssp. *sativum* var. *sativum*. "*sativum*" gene pool constructed the primary gene pool of cultivated genetic resources; "*fulvum*" gene pool, "*abyssinicum*" gene pool and "*arvense*" gene pool together constructed the secondary gene pool of cultivated genetic resources. Pairwise Nei78 genetic distance among botanical taxon based groups of pea genetic resources ranged from 7.531 to 35.956, three large cluster groups were identified based on the UPGMA dendrogram. Group I equals to "*sativum*" and "*arvense*" gene pools, Group II equals to "*abyssinicum*" gene pool, and Group III equals to "*fulvum*" gene pool. The UPGMA clustering results generally support the PCA clustering results (Zong et al. 2009).

In 2015, the 372 pea accessions, including landraces and cultivars of garden, field or fodder peas as well as wild peas, were characterized at the molecular level using newly developed SNP markers, as well as SSR markers and RBIP (Retrotransposon Based Insertion Polymorphism) markers. The three types of markers were used to describe the structure of the collection and revealed different pictures of the genetic diversity among the collection. SSR showed the fastest rate of evolution and RBIP the slowest rate of evolution, pointing to their contrasted mode of evolution (Burstin et al. 2015).

6.8.4 Relationship with Geographical Distribution

In 2008, a set of 731 pea accessions from 67 countries except China were analyzed using 21 SSR primer pairs, 109 polymorphic bands were amplified, SSR alleles were not uniformly distributed among continents, and the number of effective alleles and Shannon's information index (I) were much varied among continental based groups of genetic resources. Significant difference appeared in the pairwise comparisons for genetic diversity between continental based groups of genetic resources. Asia group had the highest level of genetic diversity (I = 1.1753), followed by Europe (I = 1.1387), USSR (I = 1.0285), America (I = 1.0196), Africa (I = 0.9254), and Oceania (I = 0.8608) groups. Two large cluster groups and four cluster subgroups were identified based on the dendrogram of pairwise Nei78 genetic distance. The clustering results of genetic resources revealed geographically broad correlation to their genetic diversity. Three types of population structure within 731 pea accessions were inferred by Structure analysis, which also broadly correlated to their geographic origins (Zong et al. 2008a). In a parallel study, 1221 pea (*Pisum sativum* L.) landraces from 19 provinces in spring and winter sowing areas in China were employed for SSR analysis. One hundred and four polymorphic bands were detected by using 21 SSR primer pairs, of which, 62.52% were effective alleles for diversity. SSR alleles were uniformly distributed in the landraces among provinces, while the number of effective alleles and Shannon's information index (I) varied much among provincial based groups of genetic resources. Significant difference appeared in the majority of pairwise comparisons for genetic diversity between provincial based groups of genetic resources. Inner Mongolia possessed the highest level of genetic diversity (I = 1.066), followed by Gansu (I = 1.041), Sichuan (I = 1.026), Yunnan (I = 1.017) and Tibet (I = 0.996); Liaoning was the lowest (I = 0.515) in genetic diversity. Three gene pools were detected in Chinese pea landraces, which were polarized during its adaptation and cultivation in thousands years. Gene Pool I mainly consisted of the pea landraces from Inner Mongolia and Shaanxi provinces, Gene Pool II mainly consisted of the pea landraces from Henan province, and Gene Pool III mainly consisted of the pea landraces from other provinces except Shaanxi, Inner Mongolia, and Henan. Pairwise Nei78 genetic distance among provincial based groups of pea landraces ranged from 5.159 to 27.586, two large cluster groups and eight cluster subgroups were identified based on the dendrogram interacting with the three gene pools. The clustering results of Chinese landraces revealed ecologically and geographically close correlation to their genetic diversity (Zong et al. 2008b).

6.9 Association Mapping Studies

Compared to QTL analysis, association mapping has several advantages. It uses a broader genetic resource and there is no need to develop RIL/F_2 populations. The resolution of association mapping is much higher than QTL mapping.

There was some work of association mapping for peas and the most of them were genome-wide association studies. Two hundred eighty-five USDA core collections and 137 DNA markers including SSR, RAPD and SCAR were used for association mapping study. Kwon and colleagues discovered three subgroups of the USDA *Pisum* core collection and 10 markers associated with seed micronutrients, 11 markers associated with disease/pest resistance, 42 markers associated with morphological/agronomic characters, 4 markers associated with phenology and 10 markers associated with production, respectively (Kwon et al. 2012; Smýkal and Konečná 2014). In 2015, 384 accessions of USDA pea collection, including 330 landraces and cultivars, 28 *P. sativum* subsp. *elatius* var. *elatius*, 16 *P. sativum* subsp. *sativum* var. *arvense*, 4 *P. sativum* subsp. *elatius* var. *pumilio*, 3 *P. abyssinicum*, 2 *P. fulvum*, and 1 *P. sativum* subsp. *transcaucasicum*, were genotyped with 256 SNP markers and phenotyped with 25 valuable traits. As a result, 71 significant marker–trait associations were validated (Cheng et al. 2015). A panel of 94 accessions, including 92 cultivars and breeding lines, one *P. fulvum* and one *P. sativum* subsp. *elatius*, were genotyped with 1,233 EST-based SNP markers and phenotyped with iron, zinc and selenium concentrations in seeds of pea. 9 SNP markers were significantly associated with iron, 2 SNP markers with zinc concentration in seeds, and no marker was associated with seed Se concentration (Diapari et al. 2015).

In the field of frost tolerance study of pea, association analysis of frost tolerance was performed with 267 SSR markers and 672 diverse pea accessions at three locations for three years. As a result, seven SSR markers were found to be associated with frost tolerance in at least two different environments with two different statistical models (Liu et al. 2017). Target gene-based association studies were very scarce. A panel of 92 diverse pea lines and 25 candidate genes in the pea starch metabolic pathway affects starch structure and percent amylose were used and associations were found for polymorphisms in seven candidate genes plus Mendel's r locus (Carpenter et al. 2017). Mapping a nucleotide sequence with a specific trait offers an opportunity for pea breeders to exploit the genetic variation present in germplasm resources and apply marker-assisted breeding in the future.

6.10 Molecular Mapping of CS Genes and QTLs

As a significant legume crop, there was a long history of molecular mapping studies in pea (Gilpin et al. 1997; Laucou et al. 1998; Ellis and Poyser 2002; Loridon et al. 2005; Deulvot et al. 2010). Different types of molecular markers were used: RFLP, RAPD, SSR, STS, and SNP. Genetic maps of pea were built from different kind of populations, such as RILs (Boutte et al. 2016; Ma et al. 2017a, b), and F_2 derived from the cross (Sun et al. 2014; Yang et al. 2015). Moreover, the "consensus" molecular marker pea maps were constructed by the pea research community (Loridon et al. 2005; Tayeh et al. 2015a, b, c). According to full length de novo assembly of RNA-seq data, the GenoPea 13.2 K SNP Array was newly developed and 12 pea RIL populations were genotyped using the GenoPea 13.2 K SNP Array. Then, individual and consensus genetic maps were built. In all, this was a vital tool for QTL mapping and marker-assistant breeding (Alves-Carvalho et al. 2015; Tayeh et al. 2015a).

Because of lack of reference genome, mapping of simply-inherited traits was very difficult in pea. However, there were still some advances in this field. Cross-species identification was a very important method for gene discovery in pea. *SGR* gene homologs were identified in pea and other plant species, mutations of which partially disable plant senescence. The biochemical characterization and map location of this gene in pea indicated that it was the same locus that determined yellow (*I*) and green (*i*) cotyledon color, as originally described by Mendel (Armstead et al. 2007; Sato et al.2007). The genetic map of pea was aligned to genomic sequences of *Medicago* using the sequences of cDNA probes known to flank the *A* locus which encodes a bHLH transcription factor (Hellens et al. 2010). *LeLe* plants are tall and *lele* plants are dwarf, and this difference was due to internode length. Then, subsequent studies found that *Le* locus encodes a GA 3-oxidase (EC 1.14.11.15) (Lester et al. 1997; Martin et al. 1997). The wrinkled phenotype to be characterized corresponded to a mutation in a gene encoding a biosynthetic enzyme (EC 2.4.1.18) (Bhattacharyya et al. 1990). *Lf* was the first locus controlling flowering in pea to be identified through a candidate gene approach as a homolog of the Arabidopsis inflorescence identity gene *TFL1* (Foucher et al. 2003). Then there was another locus with the phenotype late-flowering, photoperiod-insensitive that has been identified, and the pea *LATE BLOOMER1* (*LATE1*) gene was an ortholog of Arabidopsis *GIGANTEA* (Hecht et al. 2007).

A lot of QTL mapping studies focused on biotic stresses and abiotic stresses. However, these results have been summarized in the Section on **Prioritizing CS Stress Tolerance Traits**. For both in the F_2 population and RIL population, only one common genomic region was identified as containing seed weight QTLs (Timmerman-Vaughan et al. 1996). In 2004, the map, consisting of 204 different types of markers, was used for interval mapping of QTLs controlling seed number, pod number, 1000-seed weight, 1000-yield, and seed protein content (Irzykowska and Wolko 2004).

Seed weight QTLs are associated with nine genomic regions, and seed number QTL have been detected in association with nine genomic regions. Seed yield QTLs are associated with genomic regions on LG III, IV (two QTLs), and VII (two QTLs), Harvest index QTLs were detected in association with markers on LG I, II, III, and IIIa (Timmerman-Vaughan et al. 2005). New loci with alleles coming from the protein-rich Wt11238 line, positive for yield components, were identified in two populations under multi-environment (Krajewski et al. 2012). In 2017, using genotyping by sequencing (GBS) genotype technology, 46 seed mineral concentration QTLs, 37 seed mineral content QTLs, and 6 seed weight QTLs were discovered (Ma et al. 2017a, b). Available individual and consensus genetic maps constructed for biparental populations and quantitative trait loci positioned on these maps in pea were summarized in Table 6.6.

Table 6.6 Available individual and consensus genetic maps constructed for biparental populations and quantitative trait loci positioned on these maps

Parents	Interval (cM)	LOD	#ind.	#LGs	#markers	Types of markers	Map length (cM)	Mapped traits	References
(A) F2 Populations									
Primo × OSU442-15	7.5	–	227	11	108	RFLP, RAPD, AFLP, STS	1510	Dry seed weight, seed color, seed yield, yield components, flowering nodes, total node number	Timmerman-Vaughan et al. (1996, 2005), Gilpin et al. (1997), Tayeh et al. (2015a, b, c)
Wt10245 × Wt11238	12	–	114	2	204	Morphological, isozyme, AFLP, ISSR, STS, CAPS, RAPD	2416	Seed number, pod number, 1000-seed weight, 1000-yield, and seed protein content	Irzykowska and Wolko (2004)
G0003973 × G0005527	9.7		190	11	157	SSR	1518	–	Sun et al. (2014), Yang et al. (2015), Tayeh et al. (2015a, b, c)

(continued)

Table 6.6 (continued)

Parents	Interval (cM)	LOD	#ind.	#LGs	#markers	Types of markers	Map length (cM)	Mapped traits	References
(B) RIL Populations									
Slow × JI1794	–	–	53	14	235	RFLP, RAPD, AFLP, morphological	1289	Tolerance to Fusarium solani f. sp. Pisi, Pod dehiscence, dry seed weight	Timmerman-Vaughan et al. (1996), Tayeh et al. (2015a, b, c)
Térèse × K586	–	5	139	9	255	SSR, RAPD, RFLP, morphological	1139	Yield components, seed protein content, number, and volume of cotyledon cells, flowering time, plant height, number of basal branches, plant biomass and nitrogen content, nitrogen nutrition index, harvest index	Laucou et al. (1998), Tayeh et al. (2015a, b, c), Loridon et al. (2005)
Champagne × Terese	–	–	164	7	189	SSR, RAPD, morphological	552.2	Photoperiod response, frost tolerance	Loridon et al. (2005), Tayeh et al. (2015a, b, c)

(continued)

Table 6.6 (continued)

Parents	Interval (cM)	LOD	#ind.	#LGs	#markers	Types of markers	Map length (cM)	Mapped traits	References
Shawnee × Bohatyr	–	–	187	8	302	SSR, RAPD, isozymes, morphological	1716	Seed Mineral Content, partial resistance to Fusarium wilt Race 2	Loridon et al. (2005)
Wt11238 × Wt3557	20	–	110	11	91	Morphological, isozymic , AFLP, RAPD, STS, CAPS, ISSR	853	Yield components, seed protein content	Krajewski et al. (2012, Tayeh et al. (2015a, b, c)
Wt10245 × Wt11238	11	–	101	12	191	Morphological, isozymic , AFLP, RAPD, STS, CAPS, ISSR	1086	Stem length, internode number, yield components, seed protein content	Krajewski et al. (2012), Tayeh et al. (2015a, b, c)
VavD265 × Cameor	0.56	–	176	7	6781	SNP, SSR, RAPD, AFLP, Protein, morphological, genomic	752.6	Seed protein quality/quantity	Tayeh et al. (2015a, b, c)

(continued)

Table 6.6 (continued)

Parents	Interval (cM)	LOD	#ind.	#LGs	#markers	Types of markers	Map length (cM)	Mapped traits	References
Ballet × Cameor	0.64	–	159	7	6644	SNP, SSR, RAPD, AFLP, Protein, morphological, genomic	854.5	Flowering time, leaf appearance rate, radiation use efficiency, leaf area, leaf chlorophyll content, shoot length, seed weight, plant N nutrition indices (biomass, nitrogen content, symbiotic N fixation and nodule efficiency), seed protein quality/quantity	Tayeh et al. (2015a, b, c)
VavD265 × Ballet	0.61	–	168	7	6031	SNP, SSR, RAPD, AFLP, Protein, morphological, genomic	850.1	Seed protein quality/quantity	Tayeh et al. (2015a, b, c)
Cameor × Melrose	0.61	–	120	7	8503	SNP, SSR, RAPD, AFLP, Protein, morphological, genomic	736.3	–	Tayeh et al. (2015a, b, c)

(continued)

Table 6.6 (continued)

Parents	Interval (cM)	LOD	#ind.	#LGs	#markers	Types of markers	Map length (cM)	Mapped traits	References
Kazar × Cameor	0.91	–	84	7	7013	SNP, SSR, RAPD, AFLP, Protein, morphological, genomic	700.7	–	Tayeh et al. (2015a, b, c)
Kazar × Melrose	1.03	–	118	7	3917	SNP, SSR, RAPD, AFLP, Protein, morphological, genomic	682.5	–	Tayeh et al. (2015a, b, c)
China × Cameor	0.74	–	124	7	7641	SNP, SSR, RAPD, AFLP, Protein, morphological, genomic	833.1	Flowering time, winter frost damage, branching type, leaf area, leaf chlorophyll content, plant height, yield components, plant biomass, harvest index and seed protein content	Tayeh et al. (2015a, b, c)
Champagne × Terese	0.92	–	76	7	5984	SNP, SSR, RAPD, AFLP, Protein, morphological, genomic	888.2	Photoperiod response, frost tolerance	Tayeh et al. (2015a, b, c)

(continued)

Table 6.6 (continued)

Parents	Interval (cM)	LOD	#ind.	#LGs	#markers	Types of markers	Map length (cM)	Mapped traits	References
JI296 × DP	1.48	–	48	7	4830	SNP, SSR, RAPD, AFLP, Protein, morphological, genomic	552.2	Resistance to Mycosphaerella pinodes, plant height, flowering date	Tayeh et al. (2015a, b, c)
Baccara × PI180693	1.25	–	47	7	4245	SNP, SSR, RAPD, AFLP, Protein, morphological, genomic	705.2	Aphanomyces root rot resistance, earliness at flowering	Tayeh et al. (2015a, b, c)
Cameor × Sommette	0.91	–	144	7	5537	SNP, SSR, RAPD, AFLP, Protein, morphological, genomic	769.1	–	Tayeh et al. (2015a, b, c)
Cameor × Cerise	1	–	120	7	7206	SNP, SSR, RAPD, AFLP, Protein, morphological, genomic	523.8	–	Tayeh et al. (2015a, b, c)
Baccara × PI180693	–	–	48	7	64,263	SNP	–	–	Boutte et al. 2016

(continued)

Table 6.6 (continued)

Parents	Interval (cM)	LOD	#ind.	#LGs	#markers	Types of markers	Map length (cM)	Mapped traits	References
PI648006 × PI 357292	1.3	–	158	7	1683	SNP, SSR	1310.1	Seed mineral concentration, seed mineral content, and seed weight	Ma et al. (2017a, b)
(C) Consensus Maps Based on RIL Populations									
Terese × K586; Champagne × Terese; Shawnee × Bohatyr	–	–	490	7	579	Microsatellite	1430	–	Loridon et al. (2005), Tayeh et al. (2015a, b, c)
Champagne × Terese; VavD265 × Cameor; Ballet × Cameor; VavD265 × Ballet; Cameor × Melrose; Kazar × Cameor; Kazar × Melrose; China × Cameor; Cameor × Sommette; Cameor × Cerise; Baccara × PI180693; I296 × DP	0.24	–	1384	7	15,079	SNP, SSR, RAPD, AFLP, Protein, morphological, genomic	794.9	–	Tayeh et al. (2015a, b, c)

6.11 Marker-Assisted Breeding for CS Traits

This section mainly deals with the application of marker-assisted breeding for genetic improvement of pea. First, the germplasm characterization and distinctness, uniformity and stability (DUS) for a new variety were discussed. Second, it describes about the research progress of marker-assisted gene introgression and gene pyramiding. Third, the limitations and prospects of MAS and marker-assisted backcrossing breeding (MABCB) were then reviewed at the end.

6.11.1 Germplasm Characterization and DUS

To ensure the agricultural sustainability and food security, one main task for plant breeders is developing elite varieties with the feature of high productivity and more tolerance to biotic and abiotic stresses, which depends on the abundant genetic diversity in the germplasm maintained in gene banks (Johal et al. 2008; McCouch et al. 2013; Byrne et al. 2018). Germplasm resources are carriers of biological genetic information with actual or potential use value, which is regarded as the material basis for crop genetic improvement (Li et al. 2015; Liu et al. 2018). It is important to characterize germplasm comprehensively and accurately that mainly lies in the following aspects: (1) providing theoretical guidance for the effective collection and protection of crop germplasm; (2) helping to clarify the origin and evolution of crops; (3) laying foundation for the discovery of excellent germplasm and alleles (Tanksley and McCouch 1997; Li et al. 2015; Liu et al. 2018). Therefore, germplasm characterization can result in a better understanding and utilization of germplasm resources (Upadhyaya et al. 2006).

Pea is one of annual cool-season legumes with important economic value and ecological advantages, which was domesticated by Neolithic farmers about 10,000 years ago (Zohary and Hopf 2000; Smýkal et al. 2011, 2015). To improve the utilization of pea germplasm with the goal of production and adaptation improvement, previous studies on germplasm evaluation of pea have been reported extensively. Some studies have used morphological and agronomic traits to evaluate pea germplasm resources, providing important information for screening for superior germplasm resources such as high yield, high quality, stress resistance, and disease resistance (Ali et al. 2007; Sardana et al. 2007; Smýkal et al. 2008a, b; Sarikamis et al. 2010; Handerson et al. 2014). Other studies combined with morphological and molecular markers to screen markers associated with important agronomic and abiotic stress resistance traits have laid an important foundation for pea marker-assisted breeding (Kwon et al. 2012; Cheng et al. 2015; Diapari et al. 2015; Liu et al. 2017). There are also a large number of studies using different molecular markers to detect the genetic diversity and population genetic structure of pea germplasm from different sources. Rich genetic diversity and obvious genetic structure were found in pea germplasm and its closely related wild species, and some core collections of pea germplasm have been constructed (Jing et al. 2012; Smýkal et al. 2008a, b; Zong et al. 2009a, b; Jain et al. 2014; Burstin et al. 2015; Teshome et al. 2015a, b). All the above researches have provided an important theoretical basis for the conservation and utilization of pea germplasm resources.

Crop genetic improvement is a process in which human's directional reform crop target traits purposefully and systematically; the process is called breeding and the resulting products are called varieties (Zhang et al. 2017). To grant a plant breeders' right as well as the protection of new plant varieties, UPOV (International Union for the Protection of New Varieties of Plants) has developed a detailed identification guidelines of distinctness, uniformity and stability (DUS) for new plant variety, referred to as DUS test (Smýkal et al. 2008a; Furones-Pérez and Fernández-López 2009). Distinctness is determined by comparing the test variety with reference varieties and demonstrating clear differences between them. Uniformity refers to the intra-cultivar homogeneity and the analysis of the mutation rate of the entire community. Stability refers to no significant temporal or spatial variation after repeating test (Furones-Pérez and Fernández-López 2009).

Traditionally, DUS tests are mainly based on the evaluation of morphological and physiological characters called descriptors obtained by growing multiple varieties side by side under the same growth conditions (Kwon et al. 2005; Furones-Pérez and Fernández-López 2009). It is generally necessary to repeat observations for 2–3 years before finally making a reasonable and objective evaluation. However, traditional method of DUS test is considered time-consuming and expensive and often subjective (Kwon et al. 2005). In addition, the number of morphological traits is limited and most of them are multigenic or quantitative those are easily influenced by environmental factors (Kwon et al. 2005; Smýkal et al. 2008a). With the development of molecular markers, they are introduced as characters in distinctness assessment for the following advantages: (1) abundant markers to be used as characters; (2) easy to be observe and score; (3) avoid genotype-environment interaction; (4) free from being affected by growth stage, season, location, and agronomic practice (Kwon et al. 2002, 2005; Smýkal et al. 2008a). Among the molecular markers, SSR markers have been identified as the most widely used marker system for plant variety characterization due to the feature of highly polymorphic, reproducible, codominant, and multi-allelic types of variation (Kwon et al. 2002, 2005; Smýkal et al. 2008a). A study on variety discrimination in pea utilized morphological descriptors and biochemical-molecular markers to make classification of 25 varieties of pea, and the results showed that molecular identification could be used to assess distinctness and complement morphology-based DUS procedure (Smýkal et al. 2008a).

6.11.2 Marker-Assisted Gene Introgression and Gene Pyramiding

Over the past decades, plant breeding has played a vital role and made remarkable progress in improving crop yields and food security, which has been benefited from the development of biotechnology (Tester and Langridge 2010). Marker-assisted selection (MAS) involves using variation of molecular markers associated with the desired trait to assist selection in plant breeding, providing an important alternative to phenotypic selection for the advantages of more reliable, more convenient as well as less labor and time-consuming (Collard and Mackill 2008; Moose and Mumm 2008; Tester and Langridge 2010). Marker-assisted gene introgression is a method of introducing a favorable gene with the aid of linked marker from a donor variety to a recipient variety (i.e., elite cultivar) while maintaining the original genetic background of the recipient variety as much as possible (Visscher et al. 1996). Gene pyramiding is the process of combining several genes (i.e., multiple disease resistance genes) together into a single genotype and the efficiency can be greatly facilitated by means of MAS (Collard and Mackill 2008).

The implementation of all the above methods is inseparable from the gene discovery of important traits and the development of reliable markers. QTL mapping and association mapping are two commonly used methods in MAS (Fig. 6.1). QTL mapping depending on genetic linkage maps based on molecular markers developed with segregating population has been used to identify QTLs for many years and laid a solid foundation for MAS (Tanksley 1993; Tanksley and McCouch 1997; Morgante and Salamini 2003). On the other hand, association mapping, utilizing a broader gene pool including germplasm or natural population, identifies functional polymorphisms by examining the marker-trait associations, which has emerged as an alternative approach to traditional QTL mapping (Zhu et al. 2008; Rafalski 2010; Hamblin et al. 2011). There are several advantages of association mapping over traditional QTL mapping: (1) more diverse genetic variations for marker-trait correlations, (2) higher resolution mapping due to more recombination events, (3) available to take advantage of previous phenotyping data, and (4) less time, labor and cost for no need to develop biparental populations (Abdurakhmonov and Abdukarimov 2008; Hamblin et al. 2011).

In pea, a number of studies have been made using QTL mapping with different molecular markers for the localization of genetic loci related to important agronomic traits as well as biotic and abiotic stress resistance (Rubiales et al. 2015; Jacob et al. 2016). For agronomic traits, several studies of QTL mapping based on genetic linkage maps have found and located some important QTLs controlling different traits, such as stem length and number of nodes (Irzykowska et al. 2002), seed weight (Timmerman-Vaughan et al. 1996), seed yield and seed protein content (Tar'an et al. 2004; Timmerman-Vaughan et al. 2005); seed mineral concentrations and contents (Ma et al. 2017a, b). Due to the yield reduction caused by diseases and pests, great efforts have been made on QTL mapping of biotic resistance in pea and substantial researches have been reported on QTLs or genes responsible for disease resistance

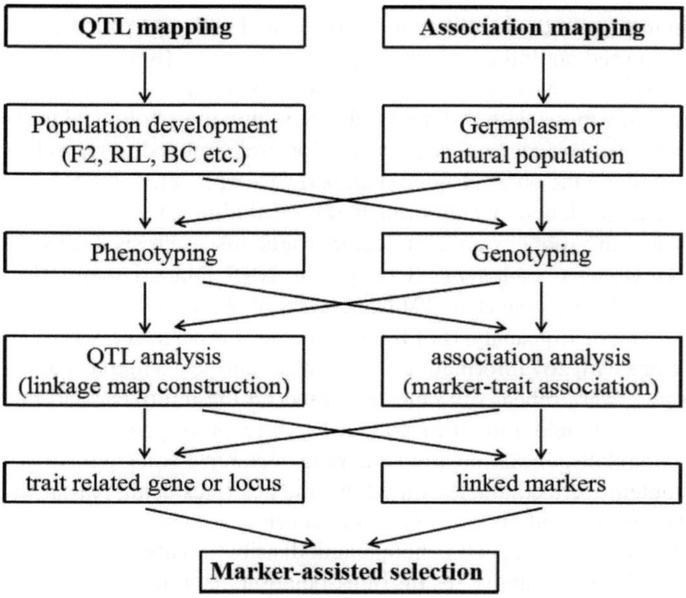

Fig. 6.1 'Pipline' of Marker-assisted selection

to Aphanomyces root rot (Pilet-Nayel et al. 2002), Ascochyta blight (Timmerman-Vaughan et al. 2002; Prioul et al. 2004; Fondevilla et al. 2008; Timmerman-Vaughan et al. 2016; Jha et al. 2017), powdery mildew (Katoch et al. 2010; Sun et al. 2016a, b), pea rust (Rai et al. 2011; Barilli et al. 2018) as well as pest resistance to pea weevil (Aryamanesh et al. 2014). In addition, some other studies identified QTLs related to salinity tolerance (Leonforte et al. 2013a, b) and freezing tolerance (Tayeh et al. 2013). A good example of MAS in pea is the development of resistance to powdery mildew in pea breeding. Since 1969, three genes (*er1*, *er2,* and *Er3*) were reported to be conferring resistance to powdery mildew, caused by *Erysiphe pisi* in pea and its wild relative *Pisum fulvum*. Moreover, various DNA markers linked to *er1*, *er2,* and *Er3* have been developed and used in breeding programs of pea (Ghafoor and McPhee 2012). Recently, several novel *er1* alleles were found in Chinese pea germplasm and cosegregating functional markers were detected and validated (Sun et al. 2016a, b). All the above researches provide powerful tools for breeding cultivars resistance to powdery mildew in pea via MAS.

With the advances of genomic technology, association mapping studies in pea combined with morphological and molecular markers to screen markers associated with important agronomic and stress resistance traits have also laid an important foundation for pea marker-assisted breeding (Kwon et al. 2012; Cheng et al. 2015; Diapari et al. 2015; Desgroux et al. 2016a, b, 2018; Carpenter et al. 2017; Jiang et al. 2017; Liu et al. 2017). In an early study, genome-wide association using various DNA markers including SSR, RAPD, and SCAR with USDA pea core collection and certain markers were identified to be associated with eight mineral nutrient concentrations in seed and other important phenotypic traits (Kwon et al. 2012). After that, several genome-wide association mapping studies using SNP markers identified significant associations with different traits including morphological traits related to flower color, seed surface, and seed coat color; traits related to reproductive development containing the onset of flowering, pod development, number of reproductive nodes; mineral nutrient concentrations in seed including calcium, magnesium, iron, zinc and selenium; root system architecture traits; disease resistance such as Fusarium wilt, *Aphanomyces euteiches* (Cheng et al. 2015; Diapari et al. 2015; Desgroux et al. 2016a, b, 2018; Jiang et al. 2017). In terms of abiotic stresses, a recent study of marker-trait association analysis of frost tolerance was performed on 672 worldwide pea collections with 267 informative SSR markers and seven markers associated with frost tolerance were repeatedly detected based on the multiyear and multilocation phenotypic data in field trails (Liu et al. 2017). Except for genome-wide association mapping, candidate gene association mapping of starch chain length distribution and amylose content were conducted on 92 diverse pea lines using partial sequences of 25 candidate genes, and polymorphisms in seven candidate genes were found to be associated with the amylopectin chain length distribution and percent amylose (Carpenter et al. 2017). The above results of association mapping will play a vital role in the marker-assisted breeding of pea.

6.11.3 Limitations and Prospects of MAS

There is a general consideration that MAS can greatly increase the efficiency and effectiveness in plant breeding compared to the traditional phenotypic selection with the following advantages: (1) time saving through molecular tests without temporal and spatial constraints; (2) more precision for free from the environmental effects in field trials; (3) more cost-effective for selection of genotypes at seedling stage; 4) more efficient by means of gene pyramiding; (4) traits with low heritability or not feasible for phenotypic selection can be selected; (5) selecting for single plant based on the genotype (Collard et al. 2005; Collard and Mackill 2008; Gupta et al. 2010). However, despite the great success of marker-assisted breeding in major cereal crops, the progress of applying MAS in legume crops breeding has been slow and limited to a few legume crops such as chickpea and common bean (Kumar et al. 2011a, b; Jacob et al. 2016). The limitations of MAS in pea mainly lies in the following aspects: (1) lack of sufficient genetic analysis and genomic data, which is a major constraint to develop reliable and user-friendly marker closely linked to traits of interest; (2) markers associated with a trait must be validated before using in different genetic backgrounds; (3) "knowledge and application gap" between the molecular biologists and plant breeder, hindering the wide use of MAS in breeding program (Collard and Mackill 2008; Kumar et al. 2011a, b; Jacob et al. 2016). However, with the development of sequencing technology and the emergence of new biotechnological tools such as GBS and GWAS, the accumulation of genetic analysis and genomic data of pea and other legume crops will greatly promote the marker development and gene discovery, which will encourage the more widespread use of MAS in pea and other legume crops (Collard and Mackill 2008; Kumar et al. 2011a, b; Jacob et al. 2016).

6.12 Genomics-Aided Breeding for CS Traits

Although the complete reference genome sequence of pea has not been published yet till now, with the advent of modern genotyping technologies and approaches, more and more genomic resources and tools including transcriptome, genotyping and mapping were developed in pea, which will pave the way for genomics-aided breeding in pea (Smýkal et al. 2012; Tayeh et al. 2015b).

6.12.1 Structural and Functional Genomic Resources Developed

The advances of next-generation sequencing and new bioinformatics methods make it possible to generate transcriptome resources for nonmodel species such as pea without a sequenced genome (Trapnell et al. 2010; Franssen et al. 2011b; Grabherr et al. 2011). As mentioned in the previous section, reliable and user-friendly markers are the primary prerequisites in the strategy of MAS. To realize such target, numerous SSR and SNP markers have been developed by means of high-throughput transcriptome sequencing in pea recently (Kaur et al. 2012; Duarte et al. 2014; Sindhu et al. 2014; Alves-Carvalho et al. 2015; Yang et al. 2015), which are useful in the genetic diversity assessment, genetic mapping, marker-trait association analysis and so on. In addition, several gene or transcript-based SNP datasets has been successfully used to design SNP arrays with marker density ranging from 384 to 15,000 based on different platforms including Illumina GoldenGate (Deulvot et al. 2010; Leonforte et al. 2013a, b; Duarte et al. 2014; Sindhu et al. 2014), Illumina Infinium (Tayeh et al. 2015a) and Sequenom MassARRAY (Cheng et al. 2015) for high-throughput genotyping, which provide powerful tools to gene discovery, high-density genetic mapping and genome-wide association studies (GWAS) (Tayeh et al. 2015b). Except for transcriptome sequencing, alternative sequencing technology such as genotyping by sequencing (GBS) and the diversity arrays technology sequencing (DArTseq) have been also utilized in pea and its wild species to identify genome-wide SNP markers and construct high-density genetic linkage maps (Ma et al. 2017a, b; Barilli et al. 2018).

High-density genetic linkage maps are important tools for functional gene localization, map-based gene cloning, comparative genomics research, assisting de novo genome assembly and marker-assisted breeding (Semagn et al. 2006). There is a long history on genetic linkage mapping in pea using different markers including isozymes (Weeden and Marx 1987), RFLP (Ellis et al. 1992), AFLP (Vos et al. 1995), RAPD (Laucou et al. 1998), SSR (Loridon et al. 2005), EST-SSR (DeCaire et al. 2012; Mishra et al. 2012) and SNP (Aubert et al. 2006; Deulvot et al. 2010; Leonforte et al. 2013a, b) with different populations. With the development of next-generation sequencing technology, high-throughput development of genomic SSR markers, EST-SSR markers, and SNP markers based on transcriptome sequencing and simplified genome sequencing laid an important foundation for the construction of high-density genetic linkage maps and functional gene mapping (Duarte et al. 2014; Sindhu et al. 2014; Sun et al. 2014; Tayeh et al. 2015a; Boutte et al. 2016; Ma et al. 2017a, b). In addition to conduct genetic mapping of individual population, a consensus genetic linkage map with a higher density and more completed genome coverage can be obtained by integrating information from multiple mapping populations (Sudheesh et al. 2015a, b; Tayeh et al. 2015a). Until now, more than 52 genetic linkage maps containing 8,503 markers at most are available in pea, all of which have provided powerful tools in gene discovery related to important traits in pea and played a vital role in molecular breeding of pea (Tayeh et al. 2015b).

6.12.2 Details of Genome Sequencing

In the development of pea genome sequencing, two recent studies have provided good examples and reference for others using transcriptome (Alves-Carvalho et al. 2015) and whole genome sequencing, respectively (Tayeh et al. 2015a). The details are as follows.

In spite of the lack of full genome sequence, gene atlas of pea has been recently produced as a reference for the pea exome (Alves-Carvalho et al. 2015). In the study, 20 cDNA libraries of the pea cultivar "Cameor" from various plant tissues at diverse developmental stages under distinct nitrogen conditions were sequenced, which generated more than one billion reads about 100 Gb data. After a specific strategy of de novo assembly and redundancy reduction, 46,099 contigs were identified with N50 length of 1,667 nt, constituting a unigene set representing a comprehensive full-length gene catalog of pea. The unigene set provides a powerful functional tool for pea orthologous gene searching, transcript expression patterns determination, uncharacterized gene identification, gene ontology pathways, comparison between tissues, which will undoubtedly promote the SNP and gene discovery and the development of transcriptome and proteome in pea.

Except for transcriptome sequencing, whole genome sequencing was also utilized to develop two important genomic resources in pea (Tayeh et al. 2015a). In combination of a gene space assembly generated by RNA-seq and 23.9X whole genome sequences of the same genotype Cameor as well as whole genome sequences of 15 other genotypes within the *Pisum* genus (sequencing depths ranging from 4.4 to 8.1), a total of 2,48,617 nonredundant SNPs for all the above genotypes compared to Cameor were identified with a range of 18,997 to 59,243 robust SNPs per accession. After a series of filtered procedures, a final set of 13,024 SNPs were selected to develop the GenoPea 13.2 K SNP array, which was then used to genotype 1,384 individuals from 12 RIL populations. A high-density and high-resolution consensus genetic linkage map was obtained containing 15,079 markers distributed in seven likage groups and covering a length of 794.9 cM with an average marker density of 0.24 cM. The GenoPea 13.2 K SNP array and the high-density genetic linkage maps of pea have provided powerful genomic tools and laid an important foundation for the genomic-aided breeding in pea.

6.12.3 Gene Annotation

Genome annotation includes structural annotation (identifying gene and their intron–exon structures) and functional gene annotation (attaching biological information such as gene ontology to gene) (Yandell and Ence 2012). Gene annotation has been performed in several transcriptome analysis studies for pea and various results were obtained due to the differences of sequencing materials, annotation strategies and reference databases. An early study of pea transcriptome resulted in 3,24,428 and 81,449 unigenes with different assembly strategies, respectively (Franssen et al. 2011a). In another study, a total of 11,737 and 22,295 unigenes were obtained by comparing all the consensus sequences of pea to *Medicago* coding sequences and the NCBI nonredundant (nr) database of GenBank, respectively (Kaur et al. 2012). After that, transcriptome sequencing of eight pea genotypes generated 68 K unigene set, of which 41 K unigenes were annotated by homology search against the model species *Medicago truncatula* (Duarte et al. 2014). Recently, comprehensive transcriptome analysis in pea with 20 cDNA libraries from various plant tissues at diverse developmental stages under distinct nitrogen conditions identified 46,099 unigenes classified into a low-copy-number unigene set (40,204) and a high-copy-number unigene set (5,704) as well as organelle set (191). In addition, Gene Ontologies (GO) functional annotation revealed a different functional constitution of the low-copy-number and the high-copy-number unigene set (Alves-Carvalho et al. 2015). With the accumulation of genomic data and the increasing availability of public databases, gene annotation has been greatly improved for functional genomic research. A comparative transcriptomic study on the seed development of vegetable and grain pea identified various number of unigenes by combining five public databases including NCBI nonredundant (nr), Swiss-prot, Pfam, Kyoto Encyclopedia of Genes and Genomes (KEGG) and Clusters of Orthologous Groups of proteins (COG) databases. In addition, the functional annotation of the unigenes was further conducted using GO assignments, COG classifications and KEGG pathway distribution (Liu et al. 2015). All the above-annotated unigenes provide valuable resources for genetic and functional genomic research in pea.

6.12.4 Impact of Genomics Research on Germplasm Characterization and Gene Discovery

Plant genomic resources have shown great value in the improvement of crops, including exploitation of genetic variation, understanding molecular genetic basis of complex traits and increasing the efficiency of crop breeding (Bevan et al. 2017). Unlike the major cereals and model legumes, the deployment of genomics tool and technologies in pea is in the initial stage but with great potential. Except for transcriptome sequencing, new technology of NGS-based GBS and DArTseq promising marker discovery as well as high-throughput genotyping without a reference genome have also been successfully utilized in pea, which provides alternative SNP detection methods with low-cost investments for pea (Ma et al. 2017a, b; Barilli et al. 2018). In addition, large-scale genome-wide molecular markers have been developed and used to establish high-throughput genotyping platforms such as SNP array in pea, making it possible for genotyping large populations more precise, rapid and cost-effective in pea (Tayeh et al. 2015a; Varshney et al. 2017). Furthermore, the availability of large-scale markers and new genotyping platform further facilitate the construction of high-density and high-resolution genetic linkage map of pea, which will undoubtedly improve the precision and efficiency of gene discovery in pea (Tayeh et al. 2015a, b). All these genomic advances in pea will play a vital role in the exploitation of genetic variation of germplasm and accelerating the gene discovery of complex traits and will further improve the genetic gain of genomics-aided breeding (Varshney et al. 2017).

6.12.5 Application of Structural and Functional Genomics in Genomics-Assisted Breeding

Genomic selection (GS) has emerged as one of effective genomics-aided breeding approaches with less time and cost as well as high accuracy in parallel with the advances of sequencing technology (Tayeh et al. 2015c). GS help breeders to select the superior lines according to the genomic-estimated breeding values (GEBV) of a testing population with only genotyping data based on a prediction model built in combination of phenotyping and genotyping data of a training population (Crossa et al. 2017; Varshney et al. 2017). Examples of GS application in pea have been described below. In one case of pea, a limited number of 331 SNP markers were used to predict phenotypes including the date of flowering (BegFlo), the number of seeds per plant (Nseed) and thousand seed weight (TSW) for 372 accessions of diverse pea collection. Different statistical methods were tested in the study taking into account of population structure of the collection and results showed that TSW can be reliably predicted (Burstin et al. 2015). In another study, high-density genotyping was conducted on a 339 pea collection utilizing the GenoPea 13.2 K SNP Array for genomic prediction of phenotypes—TSW, NSeed and BegFlo, and results showed that prediction quality was best for BegFlo followed by TSW. In addition, factors influenced the prediction accuracy has been also investigated, and large effect was found in the size and composition of the training population but limited for the statistical method and marker density (Tayeh et al. 2015c). Results of these studies have laid an important foundation for the GS application in pea.

6.13 Brief Account on Social, Political and Regulatory Issues

6.13.1 Concerns and Compliances

With the completion of whole-genome sequencing of more and more species, the application of genomic editing technology in agriculture will be more and more extensive and has already produced huge economic, social and environmental benefits. Pea genome editing breeding technology is no exception. In 2014, Nature Methods ranked gene editing as one of the ten most influential research methods in biology in the past decade. The regulation of gene editing breeding (including peas) lags behind the development of technology. Gene editing breeding for peas may be just a tool or a means, with no inherent special risks and no new risks compared to conventional breeding techniques. However, due to missing target effect, it also has some risks. Therefore, a new subject of how to evaluate and supervise safety is put forward. Meanwhile, advanced genetic engineering raises a worldwide regulatory issue by creating indistinct boundaries in genetically modified organism (GMO) regulations because without introducing new genetic material, genome editing can be used to make modifications similar to naturally occurring mutations.

6.13.2 Patent and IPR Issues

In the big data and intelligent service system of intellectual property in China under the state intellectual property office, there will be 4,152 patent documents for inquiring the keyword "pea". From the perspective of patent type, there are 3,326 invention applications, 433 invention authorization applications, 221 designs, and 129 utility models, as well as 18 Taiwan inventions and 25 others. Judging from the current status of rights, there are 1,777 cases under review, 1,606 cases without power, 736 cases with power, and 8 others. From the year of application, there were 686 pieces in 2014, 660 pieces in 2015, 787 pieces in 2016, 461 pieces in 2017, 127 pieces in 2018 and 1,398 pieces in other years. In terms of patent type, pea planting and processing technology is the main technology. For example, Tianmen Xinmanyuan Modern Agriculture Development Co., LTD. applied for a pea cultivation method, Qingdao Shoutai Agricultural Science and Technology Co., LTD. applied for the production process of fried crispy pea food, Shandong Jianyuan Biological Engineering Co., LTD. applied for an improved pea separation protein preparation process and Liuzhou Liunan Mingda Pigeon Breeding Association applied for a pigeon feed and its preparation method and other 30 related feed processing technology patents. In the system, 145 pieces of software copyright related with "pea" as the keyword, and 561 pieces of related copyright works.

6.13.3 Disclosure of Sources of GRs, Access and Benefit Sharing

As an important grain, vegetable and forage multipurpose crop, pea plays an increasingly important role in the improvement of people's living standard and the sustainable development of national economy and agriculture. But at present, the cultivated area of pea is decreasing, and its economic benefit, yield, and product quality are not high. One of the main reasons is the development, improvement and variety management of pea germplasm resources. The strict examination and scientific popularization of breeding varieties can guarantee the resources of pea varieties to better serve the development of pea production.

Local variety resources in China are divided into three gene pools: gene pool I mainly consists of spring sowing area of Inner Mongolia, Shaanxi resources; gene pool II mainly consists of autumn sowing area's northernmost resources in Henan; in addition, gene pool III is mainly composed of the province of Anhui, Guizhou, Yunnan, Hubei, Sichuan, Guangxi, Qinghai, Shanxi, Gansu, Guangdong, Hunan, and Liaoning, Shanghai, Beijing and Xinjiang's resources.

The evaluation of agronomic characters of Chinese pea germplasm resources mainly includes the following two aspects. One is the growth period, plant morphology, yield characteristics, and other indicators of nutritional quality identification including protein content, fat content, amylopectin, and amylopectin content. The post-qualification germplasm resource can be direct as a cultivation variety in a specific area, such as No. 23 in Zhangye in Gansu province and No. 4 in Taizhong county of Fujian province, which has been a local main crop variety once with its obvious stimulation effect.

6.13.4 Famers' Rights

Farmers have less say in the application of gene editing technology in peas, so it is often up to the government authorities to decide the application scope of gene editing technology, while farmers' rights are less protected. In the application of pea gene editing technology, farmers should have the right to decide whether to plant gene editing varieties or non-gene editing varieties.

6.13.5 Traditional Knowledge

Pea (*Pisum sativum* L.), is a leguminous climbing herb, 0.5–2 m high. The whole plant is green, smooth and hairless, and is creamy. Leaves consists of 4–6 leaflets, stipules heart-shaped, lower teeth with fine teeth. The leaflets are ovoid, and the flowers are solitary or severally arranged as racemes. Calyx campanulate, lobulate lanceolate. Corollas vary in color and vary from breed to breed, but are mostly white and purple. The ovary is glabrous and the style is flat. The pods are swollen and oblong. The seeds are round, turquoise, and turn yellow when dry. Flowering period lasts from June to July, while fruiting period from July to September.

The peas are native to the Mediterranean and Central Asia and are one of the world's most important cultivated crops. Seeds, tender pods and tender seedlings can be eaten. The seeds contain starch, oil and fat, which can be used for medicinal purposes, and have strong, diuretic and antidiarrheal effects. Stems and leaves can cool off the heat and make green manure, feed or fuel.

6.13.6 Treaties and Conventions

The research and application of gene editing technology in China is still in a relatively disordered state. Although the country has invested a lot of money in gene editing research and development, relevant supervision, management and laws and regulations are relatively weak, and there are no treaties or conventions specifically for pea gene editing. To fill these gaps, Chinese authorities and scientists will also need to work together to formulate rules to clarify the scope and scope of gene editing techniques.

6.13.7 Participatory Breeding

Participatory breeding of peas is a method by which researchers of pea breeding work with farmers to improve or breed varieties. Many of the researches on peas are carried out in farmers' fields, with the aim of ensuring that the research being carried out actually meets farmers' needs. Farmers who have been on the front line of production for a long time know most about their ecological environment, regional climate, production habits and requirements for products. Therefore, farmers participating in pea breeding can make non-centralized selection in farmers' fields to avoid the risk of weeding out useful strains of peas, which is a very effective method to successfully introduce crops into specific natural and socioeconomic environments.

6.14 Future Perspectives

6.14.1 Potential for Expansion of Productivity

According to the latest statistics of FAO (FAO 2018), in 2016, 95 countries in the world produced dry peas (Fig. 6.2), with a cultivated area of 6.396 million ha and a total output of 14.36 million tons. Fresh peas are produced in 82 countries (Fig. 6.3), covering an area of 2.589 million ha and a total yield of 19.052 million tons. The dry peas harvesting area in the top five countries were Canada (1.697 million ha), India (1.100 million ha), the Russian federation (1.040 million ha), China (0.834 million ha), the United States (0.402 million ha), which covered 66% of world total; The dry peas productivity of the top five countries are Canada (4.611 million tons), the Russian federation (2.199 million tons), China (1.194 million tons), India (1.02 million tons) and the United States (0.782 million tons), which covered 68% of world total. The harvesting area of fresh peas in the top five countries were China (1.523 million ha), India (0.497 million ha), the United States (57,600 ha), France (35,667 ha), Britain (35,533 ha), which covered 83% of world total; The Fresh peas productivity in the top five countries were China (12.208 million tons), India (4.814 million tons), the United States (312,000 tons), France (233,000 tons) and Egypt (194,000 tons), which covered 89% of world total. In 2016, the harvesting area of pea in the world totaled 8.985 million ha, while China shared 2.357 million ha, accounting for 26.2% of the global total; Canada shared 1.769 million ha, accounting for 19.7% of the global total. The top two producer, China and Canada, covered 45.9% harvesting area of the world total. China, Canada, India, Rasia, USA, are the major producer of dry and fresh peas in the world.

The global total harvesting area of dry and fresh peas kept relatively stable during the past 55 years from 1961 (Fig. 6.4), however the dry pea area decreased continuously in the past 55 years, while the area of fresh peas increased sharply year by year from 1990, according to FAO statistics (FAO 2018). The contribution to the global increase of fresh pea production largely relied on the expansion of fresh productivity in developing countries, especially in China (Fig. 6.5) and India.

The demand for fresh peas will be markedly increased along with the improvement of living standard of the people and awareness for the healthy food in both developing and developed countries. This resulted in low benefits from dry pea production of peas in developing countries and decreased the sowing area and total production of dry pea. At the same time, the need for vegetable pea consumption, as well as better benefits from vegetable peas, caused sharp increase of fresh pea production globally, which also benefits the cropping system by reducing cropping duration as vegetable. In the future cropping systems that including peas, the dry pea production will keep stable to low rate decreasing, and the vegetable pea production will be increasing quickly. So that, the breeding researches on genetic resources and genetic studies to support vegetable pea production, will be strengthened.

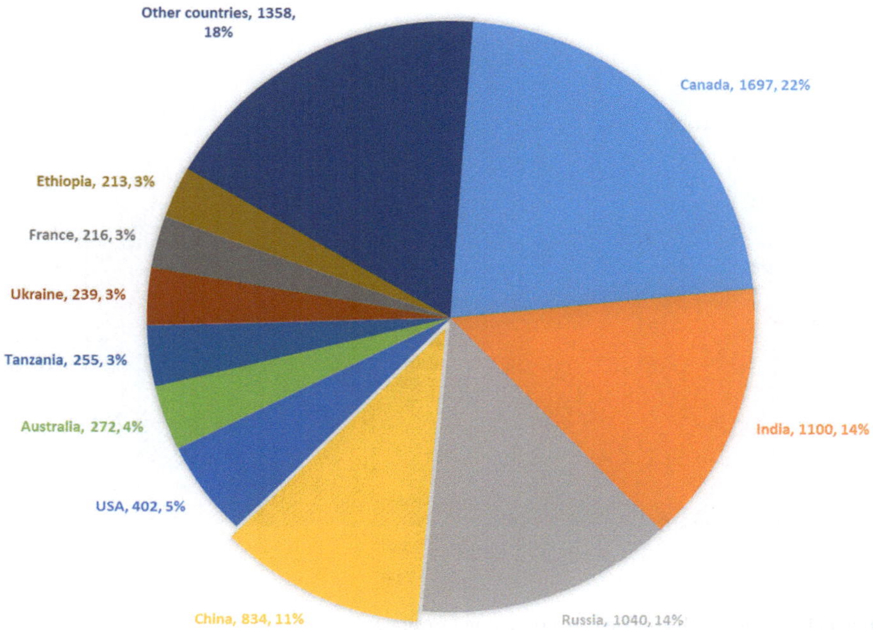

Fig. 6.2 Major producers of dry peas in kha and shares in percentage in 2016

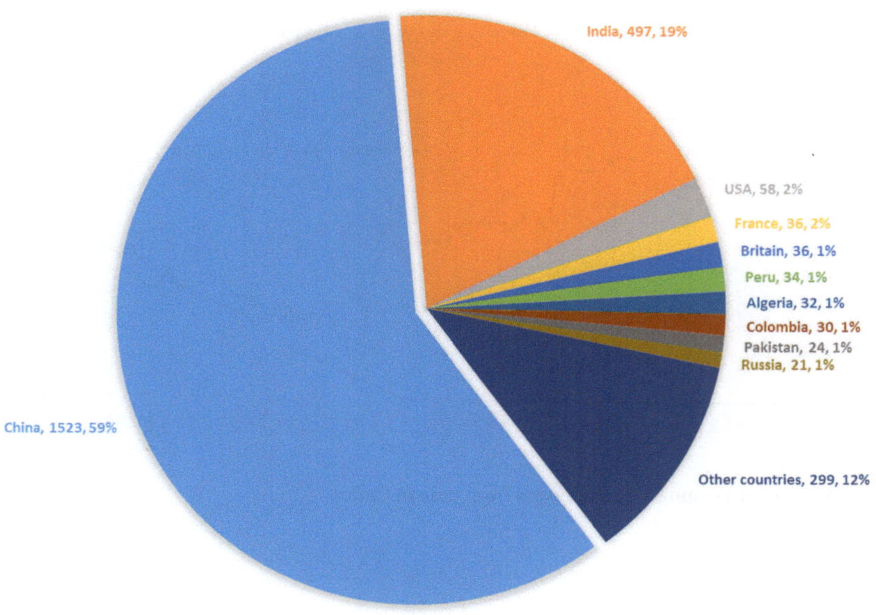

Fig. 6.3 Major producers of fresh peas in kha and shares in percentage in 2016

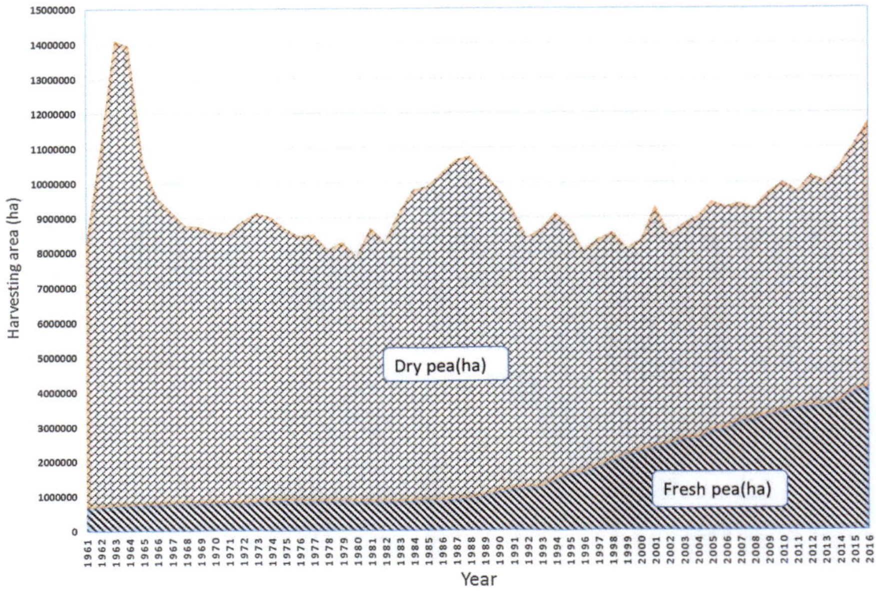

Fig. 6.4 The global harvesting area of dry and fresh peas

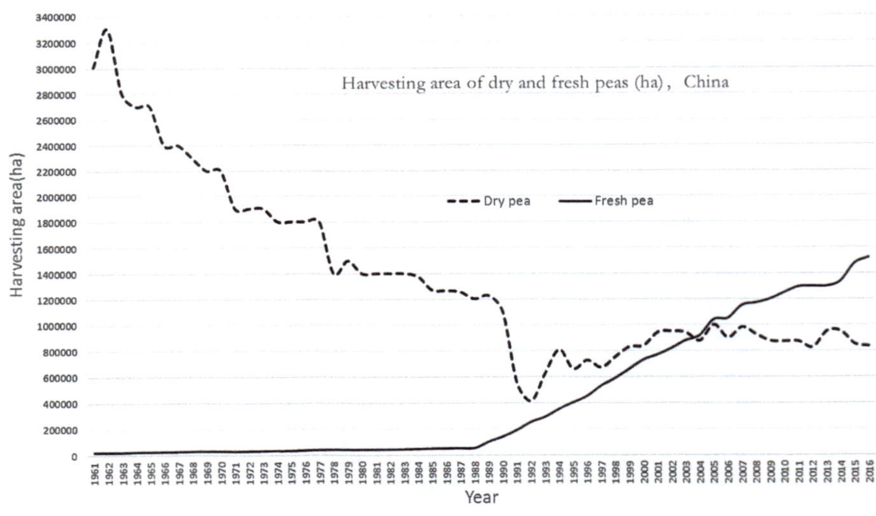

Fig. 6.5 The harvesting area of dry and fresh peas in China

6.14.2 Potential for Expansion into Nontraditional Areas

Pea has adapted to a wide range of climates and altitudes. It is commonly recognized as consisting of spring, Mediterranean, and winter types (Stoddard et al. 2006). Both winter and spring types are grown in many countries. Winter pea has higher yield potential than spring pea owing to its longer growth period, higher efficiency of radiation use in early spring, and escape from drought stress at harvest stage (Stoddard et al. 2006; Urbatzka et al. 2011). Winter pea is conventionally sown in autumn in the area south of 33° north latitude. In recent years, the northern boundary of winter pea has been moved northward in China and other countries to achieve yield increases by enlargement of the winter pea region (Zhang et al. 2016). The sowing area of vegetable peas has expanded to orchard in both winter and spring sowing season when fruit trees are in their dormancy period.

Rotation, intercropping, and mixed cropping involving peas are normal cropping systems in many developing countries. Distributed in winter sowing area like Sichuan and Yunnan provinces in China, with warm seasons, natural water resources, and a high multi-crop index, can be more potential for peas in nontraditional areas (Li et al. 2017). Average winter temperature is 9–14 °C, which is ideal for pea, one of the main winter crops. The following is a common rotation system involving peas. The first year: peas—early rice—late rice; the second year: barley (wheat)—early rice—late rice; the third year: canola—early rice—late rice.

A new cropping system of three-dimensional agriculture has been developed along the southeastern coast of China, such as in Jiangsu, Shanghai, Zhejiang, and cotton-producing areas of Anhui and Henan, which have a new comprehensive configuration cropping system of "early-late", "tall–short" and "legume–nonleguminous crops", such as peas intercropping with vegetables, maize, wheat, or cotton (Li et al. 2017). Moreover, to make full use of natural resources in spring sowing area of peas, intercropping system is very common in these areas. Peas may be intercropped with maize, potato, sunflower, wheat, and canola in many countries. In rainfed cropping systems, a creative "winter pea-summer millet" rotation system has been established in China, to emphasize N fixation and on expanding use of BNF in new farming system (Li et al. 2017). Soil quality, fertility and the quantity of arable soil have declined significantly, in part due to long-term use of chemical fertilizers affecting pH (acidic soils are hostile to the majority of legumes) and cation exchange profiles, plus pesticide-related declines in soil-renewing earthworms (Liu and Diamond 2005). It is time to change the situation of overdependence on chemical fertilizer and cereal monocropping by introduction of cool-season legume crops, like peas, in nontraditional areas in the world. As cool-season legume crop, peas will become more important for its BNF nitrogen contribution to intercropping and rotation with cereals and other crops in the future (Jensen and Hauggaard-Nielsen 2003).

References

Abdurakhmonov IY, Abdukarimov A (2008) Application of association mapping to understanding the genetic diversity of plant germplasm resources. Intl J Plant Genom 574927

Ahmad P, Jaleel CA, Sharma S (2010) Antioxidative defence system, lipid peroxidation, proline metabolizing enzymes and biochemical activity in two genotypes of *Morus alba* L. subjected to NaCl stress. Russ J Plant Physiol 57:509–517

Ahmad P, Jeleel CA, Azooz MM, Nabi G (2009) Generation of ROS and non-enzymatic antioxidants during abiotic stress in plants. Bot Res Intl 2:11–20

Ali SM, Sharma B, Ambrose MJ (1993) Current status and future strategy in breeding pea to improve resistance to biotic and abiotic stresses. Euphytica 1:115–126

Ali Z, Ullah N, Naseem S, Haq, MIU, Jacobsen HJ (2015) Soil bacteria conferred a positive relationship and improved salt stress tolerance in transgenic pea (*Pisum sativum* L.) harboring Na+/H+ antiporter. Turk J Bot 39(6):962–972

Ali Z, Qureshi AS, Ali W, Gulzar H, Nisar M, Ghafoor A (2007) Evaluation of genetic diversity present in pea (*Pisum sativum* L.) germplasm based on morphological traits, resistance to powdery mildew and molecular characteristics. Pak J Bot 39:2739–2747

Ali Z, Saeed W, Naseem S, Ahmad F, Akrem A, Yasmeen N, Jacobsen HJ (2018) Phenotypic evaluation of transgenic peas (*Pisum sativum* L.) harboring AtNHX1 demonstrates stable gene expression and conserved morphology in subsequent generations. Turk J Bot 42(2):150–158

Alikhani HA, Saleh-Rastin N, Antoun H (2006) Phosphate solubilization activity of rhizobia native to Iranian soils. Plant Soil 287:35–41

Alves-Carvalho S, Aubert G, Carrere S, Cruaud C, Brochot AL, Jacquin F, Klein A, Martin C, Boucherot K, Kreplak J, da Silva C, Moreau S, Gamas P, Wincker P, Gouzy J, Burstin J (2015) Full-length *de novo* assembly of RNA-seq data in pea (*Pisum sativum* L.) provides a gene expression atlas and gives insights into root nodulation in this species. Plant J 84:1–19

Andrade M, Abe Y, Nakahara KS, Uyeda I (2009) The cyv-2 resistance to clover yellow vein virus in pea is controlled by the eukaryotic initiation factor 4E. J Gen Plant Pathol 75:241–249

Anthony JB (2017) Peas and beans. In: Atherton J (ed) Crop production science in horticulture (series). CABI, Nosworthy Way, Wallingford, Oxfordshire OX10 8DE, UK, pp 1–14

Armstead I, Donnison I, Aubry S, Harper J, Hörtensteiner S, James C, Mani J, Moffet M, Ougham H, Roberts L, Thomas A, Weeden N, Thomas H, King I (2007) Cross-species identification of Mendel's I locus. Science 315:73

Arraouadi S, Badri M, Abdelly C, Huguet T, Aouani ME (2012) QTL mapping of physiological traits associated with salt tolerance in *Medicago truncatula* recombinant inbred lines. Genomics 9:118–125

Arshad M, Shaharoona B, Mahmood T (2008) Inoculation with *Pseudomonas* spp. containing ACC-deaminase partially eliminates the effects of drought stress on growth, yield, and ripening of pea (*Pisum sativum* L.). Pedosphere 18(5):611–620

Aryamanesh N, Byrne O, Hardie DC, Khan T, Siddique KHM, Yan G (2012) Large-scale density-based screening for pea weevil resistance in advanced backcross lines derived from cultivated field pea (*Pisum sativum*) and *Pisum fulvum*. Crop Pasture Sci 63(7):612–618

Aryamanesh N, Zeng Y, Byrne O, Hardie DC, Al-Subhi AM, Khan T, Siddique KHM, Yan G (2014) Identification of genome regions controlling cotyledon, pod wall/seed coat and pod wall resistance to pea weevil through QTL mapping. Theor Appl Genet 127(2):489–497

Attanayake KPR, Glawe N, McPhee DA, Dugan KE, Chen W (2010) *Erysiphe trifolii*—a newly recognized powdery mildew pathogen of pea. Plant Pathol 59:712–720

Aubert G, Morin J, Jacquin F, Loridon K, Quillet MC, Petit A, Rameau C, Lejeune-Hénaut I, Huguet T, Burstin J (2006) Functional mapping in pea, as an aid to the candidate gene selection and for investigating synteny with the model legume *Medicago truncatula*. Theor Appl Genet 112:1024–1041

Awale R, Machado S, Rhinhart K (2018) Soil carbon, nitrogen, pH, and crop yields in winter wheat-spring pea systems. Agron J 110(4):1523–1531

Aznar-Fernández T, Carrillo-Perdomo E, Flores F, Rubiales D (2018a) Identification and multi-environment validation of resistance to pea weevil (*Bruchus pisorum*) in *Pisum* germplasm. J Pest Sci 91(2):505–514
Aznar-Fernández T, Rubiales D (2018b) Identification and characterisation of antixenosis and antibiosis to pea aphid (*Acyrthosiphon pisum*) in *Pisum* spp. germplasm. Ann Appl Biol 172(3):268–281
Baggett JR, Hampton RO (1991) Inheritance of viral bean leaf roll tolerance in peas. J Amer Soc Hort Sci 116(4):728–731
Baranger AG, Aubert G, Arnau G, Lainé AL, Deniot G, Potier J, Weinachter C, Lejeune-Hénaut J, Lallemand J, Burstin J (2004) Genetic diversity within *Pisum sativum* using protein- and PCR-based markers. Theor Appl Genet 108:1309–1321
Barilli E, Sillero JC, Serrano A, Rubiales D (2009a) Differential response of pea (*Pisum sativum*) to rusts incited by *Uromyces viciae-fabae* and *U. pisi*. Crop Protec 28:980–986
Barilli E, Sillero JC, Fernández-Aparicio M, Rubiales D (2009b) Identification of resistance to *Uromyces pisi* (Pers.) Wint. in *Pisum* spp. germplasm. Field Crop Res 114:198–203
Barilli E, Cobos MJ, Carrillo E, Kilian A, Carling J, Rubiales D (2018). A high-density integrated DArTseq SNP-based genetic map of *Pisum fulvum* and identification of QTLs controlling rust resistance. Front Plant Sci 9:167
Bani M, Rubiales D, Rispail N (2012) A detailed evaluation method to identify sources of quantitative resistance to *Fusarium oxysporum* f. sp. *pisi* race 2 within a *Pisum* spp, germplasm collection. Plant Pathol 61(3):532–542
Barilli E, Satovic Z, Rubiales D, Torres AM (2010) Mapping of quantitative trait loci controlling partial resistance against rust incited by *Uromyces pisi* (Pers.) Wint. in a *Pisum fulvum* L. intraspecific cross. Euphytica 175:151–159
Barilli E, Sillero JC, Prats E, Rubiales D (2014) Resistance to rusts (*Uromyces pisi* and *U. viciae-fabae*) in pea. Czech J Genet Plant Breed 50(2):135–143
Bateson W (1901) Experiments in plant hybridization by Gregor Mendel. J Roy Hort Soc 26(1):1–32
Bhattacharyya MK, Smith AM, Ellis TH, Hedley C, Martin C (1990) The wrinkled-seed character of pea described by Mendel is caused by a transposon-like insertion in a gene encoding starch-branching enzyme. Cell 60:115–22
Bhattacharya S, Malleshi NG (2012) Physical, chemical and nutritional characteristics of premature-processed and matured green legumes. J Food Sci Technol 49(4):459–466
Biddle AJ, Cattlin ND (2007) Pests, diseases and disorders of peas and beans, a coloured handbook. Academic Press, Elsevier, Boston, USA
Blixt S (1972) Mutation genetics in *Pisum*. Agri Hort Genet 30:1–293
Bretag TW, Keane PJ, Price TV (2006) The epidemiology of ascochyta blight in field peas: a review. Aus J Agri Res 57:883–902
Bevan MW, Uauy C, Wulff BBH, Zhou J, Krasileva K, Clark MD (2017) Genomic innovation for crop improvement. Nature 543:346
Bodah ET, Porter LD, Chaves B, Dhingra A (2016) Evaluation of pea accessions and commercial cultivars for Fusarium root rot resistance. Euphytica 208(1):63–72
Bouchenak M, Lamri-Senhadji M (2013) Nutritional quality of legumes, and their role in cardiometabolic risk prevention: a review. J Med Food 16(3):185–198
Bourdillon A (1998) Advantages and constraints of grain legumes for the feed market. In: 3rd European conference on grain legumes, AEP, Valladolid, Spain, 14–19 Nov 1998, pp 5–6
Bourion V, Lejeune-Hénaut I, Munier-Jolain N, Salon C (2003) Cold acclimation of winter and spring peas: carbon partitioning as affected by light intensity. Eur J Agron 19:535–548
Bourion V, Rizvi SMH, Fournier S, de Larambergue H, Galmiche F, Marge P (2010) Genetic dissection of nitrogen nutrition in pea through a QTL approach of root, nodule, and shoot variability. Theor Appl Genet 121:71–86
Boutte G, Alves Carvalho S, Falque M, Peterlongo P, Lhuillier E, Bouchez O, Lavaud C, Pilet-Nayel M-L, Rivière N, Baranger A (2016) SNP discovery and genetic mapping using genotyping by sequencing of whole genome genomic DNA from a pea RIL population. BMC Genom 17:121

Bradshaw AD (1965) Evolutionary significance of phenotypic plasticity in plants. Adv Genet 13:115–155

Brewin NJ, van Kammen A, Horvanth B (1991) Development of the legume root nodule. Annu Rev Cell Biol 7:191–226

Brockwell J, Pilka A, Holliday RA (1991) Soil PH is a major determinant of the numbers of naturally-occurring *Rhizobium meliloti* in non-cultivated soils in Central New South Wales. Aust J Exp Agri 31:211–219

Burghardt LT, Guhlin J, Chun CL, Liu JQ, Sadowsky MJ, Young ND, Tiffin P (2017) Transcriptomic basis of genome by genome variation in a legume-rhizobia mutualism. Mol Ecol 26:6122–6135

Burstin J, Salloignon P, Chabert-Martinello M, Magnin-Robert J-B, Siol M, Jacquin F, Chauveau A, Pont C, Aubert G, Delaitre C, Truntzer C, Duc G (2015) Genetic diversity and trait genomic prediction in a pea diversity panel. BMC Genom 16:105

Busheva M, Apostolova E (1997) Influence of saccharides and glycine betaine on freezing of photosystem 2 enriched particles: a chlorophyll fluorescence study. Photosynthetica 34:591–594

Byrne PF, Volk GM, Gardner C, Gore MA, Simon PW, Smith S (2018) Sustaining the future of plant breeding: the critical role of the USDA-ARS national plant germplasm system. Crop Sci 58:451–468

Cambrolle J, Redondo-Gomez S, Mateos-Naranjo E, Luque T, Figueroa ME (2011) Physiological responses to salinity in the yellow-horned poppy, *Glaucium flavum*. Plant Physiol Biochem 49:186–194

Carpenter MA, Shaw M, Cooper RD, Frew TJ, Butler RC, Murray SR, Moya L, Coyne CJ, Timmerman-Vaughan GM (2017) Association mapping of starch chain length distribution and amylose content in pea (*Pisum sativum* L.) using carbohydrate metabolism candidate genes. BMC Plant Biol 17:132

Carrillo E, Satovic Z, Aubert G, Boucherot K, Rubiales D, Fondevilla S (2014) Identification of quantitative trait loci and candidate genes for specific cellular resistance responses against *Didymella pinodes* in pea. Plant Cell Rep 33:1133–1145

Carvalho RF, PiottoI FA, SchmidtI D (2011) Seed priming with hormones does not alleviate induced oxidative stress in maize seedlings subjected to salt stress. Sci Agri (Braz) 68(5):598–602

Cassman KG, Munns DN, Beck DP (1981a) Phosphorus nutrition of rhizobium japonicum: strain differences in phosphorus storage and utilization. Soil Sci Soc Amer J 45(3):517–520

Cassman KG, Whitney AS, Fox RL (1981b) Phosphorus requirements of soybean and cowpea as affected by mode of N nutrition. Agron J 73(1):17–22

Castillejo MÁ, Iglesias-García R, Wienkoop S, Rubiales D (2016) Labelfree quantitative proteomic analysis of tolerance to drought in *Pisum sativum*. Proteomics 16:2776–2787

Cerda A, Caro M, Fernandez FG (1982) Salt tolerance of two pea cultivars. Agron J 74(5):796–798

Chai Q, Qin A, Gan Y, Yu A (2013) Higher yield and lower carbon emission by intercropping maize with rape, pea, and wheat in arid irrigation areas. Agron Sustain Dev 34:535–543

Chapagain T, Riseman A (2014) Barley–pea intercropping: effects on land productivity, carbon and nitrogen transformations. Field Crops Res 166(9):18–25

Chattopadhyay A, Subba P, Pandey A, Bhushan D, Kumar R, Datta A, Chakraborty S, Chakraborty N (2011) Analysis of the grasspea proteome and identification of stress-responsive proteins upon exposure to high salinity, low temperature, and abscisic acid treatment. Phytochemistry 72:1293–1307

Chand R, Srivastava CP, Singh BD, Sarode SB (2006) Identification and characterization of slow rusting components in pea (*Pisum sativum* L.). Genet Resour Crop Evol 53:219–224

Chang HX, Sang H, Wang J, McPhee K, E, Zhuang X, Porter LD, Chilvers MI (2018) Exploring the genetics of lesion and nodal resistance in pea (*Pisum sativum* L) to *Sclerotinia sclerotiorum* using genome-wide association studies and RNA-Seq. Plant Direct 2(6): e00064

Charity JA, Anderson MA, Bittisnich DJ, Whitecross M, Higgins TJV (1999) Transgenic tobacco and peas expressing a proteinase inhibitor from *Nicotiana alata* have increased insect resistance. Mol Breed 5(4):357–365

Chen WX, Chen WF (2016) The leguminous plants-rhizobium symbiosis nitrogen fixation plays an important role in ecological agriculture. In: Chinese research society of well-off construction (eds), Worrying about China's food security and the way out (in Chinese). Tsinghua University Press, Beijing, pp 120–137

Chen MJ, Liu GF, Yu H, Wang B, Li JY (2018) Towards molecular design of rice plant architecture and grain quality (in Chinese). Chin Sci Bull 63:1276–1289

Chen W, Wang E, Chen W (2004) The relationship between the symbiotic promiscuity of rhizobia and legumes and their geographical environments. China Agri Sci 37(1):81–86

Chen D, Zeng Z, Shui X (2002) Screening of highly effective symbiotic rhizobia of alfalfa. Pratacult Sci 19(6):27–31

Cheng X, Wang S (2009) Food legumes breeds in China. China Agricultural Science and Technology Press, Beijing, pp 333–377

Cheng P, Holdsworth W, Ma Y, Coyne CJ, Mazourek M, Grusak MA, Fuchs S, McGee RJ (2015) Association mapping of agronomic and quality traits in USDA pea single-plant collection. Mol Breed 35:75

Choudhury PR, Tanveer H, Dixit GP (2007) Identification and detection of genetic relatedness among important varieties of pea (*Pisum sativum* L.) grown in India. Genetica 130(2):183–191

Chowrira GM, Cavileer TD, Gupta SK, Lurquin PF, Berger PH (1998) Coat protein-mediated resistance to pea enation mosaic virus in transgenic *Pisum sativum* L. Transgen Res 7(4):265–271

Christoph F, Bisseling T (1997) Molecular basis of symbiosis between Rhizobium and legumes. Nature 387(7):394–401

Clement SL, Hardie DC, Elberson LR (2002) Variation among accessions of *Pisum fulvum* for resistance to pea weevil. Crop Sci 42(6):2167–2173

Cole CV, Duxbury J, Freney J, Heinemeyer O, Minami K, Mosier A (1997) Global estimates of potential mitigation of greenhouse gas emissions by agriculture. Nutr Cycl Agroecosyst 49(1–3):221–228

Collard BCY, Jahufer MZZ, Brouwer JB, Pang ECK (2005) An introduction to markers, quantitative trait loci (QTL) mapping and marker-assisted selection for crop improvement: the basic concepts. Euphytica 142:169–196

Collard BCY, Mackill DJ (2008) Marker-assisted selection: an approach for precision plant breeding in the twenty-first century. Phil Trans Roy Soc B-Biol Sci 363:557–572

Condon AG, Richards RA, Rebetzke GJ, Farquhar GD (2004) Breeding for high water-use efficiency. J Exp Bot 55:2447–2460

Congdon BS, Coutts BA, Renton M, Banovic M, Jones RAC (2016) Pea seed-borne mosaic virus in field pea: widespread infection genetic diversity and resistance gene effectiveness. Plant Dis 100(12):2475–2482

Coyne CJ, Inglis DA, Whitehead SJ, McClendon MT, Muehlbauer FJ (2000) Chromosomal location of Fwf the Fusarium wilt race 5 resistance gene in *Pisum sativum*. Pisum Genet 32:20–22

Coyne CJ, Pilet-Nayel ML, McGee RJ, Porter LD, Smýkal P, Grünwald NJ (2015) Identification of QTL controlling high levels of partial resistance to *Fusarium solani* f. sp. *pisi* in pea. Plant Breed 134(4):446–453

Cristina M, Leandro P, Carlos M, Erick Z (2010) Greenhouse gas (CO_2 and N_2O) emissions from soils: a review. Chil J Agri Res 70(3)

Crossa J, Perez-Rodriguez P, Cuevas J, Montesinos-Lopez O, Jarquin D, de los Campos G, Burgueno J, Gonzalez-Camacho JM, Perez-Elizalde S, Beyene Y, Dreisigacker S, Singh R, Zhang X, Gowda M, Roorkiwal M, Rutkoski J, Varshney RK (2017) Genomic selection in plant breeding: methods, models, and perspectives. Trends Plant Sci 22:961–975

Cutforth HW, McConkey BG, Ulrich D, Miller PR, Angadi SV (2002) Yield and water use efficiency of pulses seeded directly into standing stubble in the semiarid Canadian prairie. Can J Plant Sci 82:681–686

Cutforth HW, Mcginn SM, Mcphee KE, Miller PR (2007) Adaptation of pulse crops to the changing climate of the northern great plains. Agron J 99:1684–1699

David WE, Hari K (eds) (2009) Nitrogen fixation in crop production. GRDC Publication, Australia, pp 349–385

Davidson JA, Hartley D, Priest M, Herdina MKK, Mckay A, Scott ES (2009) A new species of Phoma causes ascochyta blight symptoms on field peas (*Pisum sativum*) in South Australia. Mycologia 101:120–128

DeCaire J, Coyne CJ, Brumett S, Shultz JL (2012) Additional pea EST-SSR markers for comparative mapping in pea (*Pisum sativum* L.). Plant Breed 131:222–226

del Val C, Rivas E, Torres-Quesada O, Toro N, Jiménez-Zurdo JI (2007) Identification of differentially expressed small non-coding RNAs in the legume endosymbiont Sinorhizobium meliloti by comparative genomics. Mol Microbiol 66(5):1080–1091

Dénarié J, Roche P (1991) Rhizobium nodulation signals. In: Verma DPS (eds), Molecular signals in plant-microbe communications. CRC Press, London, pp 295–324

Deulvot C, Charrel H, Marty A, Jacquin F, Donnadieu C, Lejeune-Hénaut I, Burstin J, Aubert G (2010) Highly-multiplexed SNP genotyping for genetic mapping and germplasm diversity studies in pea. BMC Genom 11:468

Desgroux A, Baudais VN, Aubert V, Le Roy G, de Larambergue H, Miteul H, Aubert G, Boutet G, Duc G, Baranger A, Burstin J, Manzanares-Dauleux M, Pilet-Nayel M-L, Bourion V (2018) Comparative genome-wide-association mapping identifies common loci controlling root system architecture and resistance to *Aphanomyces euteiches* in pea. Front Plant Sci 8:2195

Desgroux A, L'Anthoene V, Roux-Duparque M, Riviere JP, Aubert G, Tayeh N, Moussart A, Mangin P, Vetel P, Piriou C, McGee RJ, Coyne CJ, Burstin J, Baranger A, Manzanares-Dmytruk MV, Vasyliv NV, Derkach IV, Mykiyevych IM, Romanyuk ND (2016a) Effects of seed priming with growth regulator biolan on salt tolerance of *Pisum sativum* L. Біологічні студії 10(3–4):107–118

Desgroux A, L'anthoëne V, Roux-Duparque M, Rivière, JP, Aubert G, Tayeh N, Moussart A, Mangin P, Vetel P, Piriou C, McGee RJ, Coyne CJ, Burstin J, Alain Baranger A, Manzanares-Dauleux M, Virginie Bourion V, Pilet-Nayel ML (2016b). Genome-wide association mapping of partial resistance to *Aphanomyces euteiches* in pea. BMC Genomics 17(1):1–21

Devi J, Mishra GP, Sanwal SK, Dubey RK, Singh PM, Singh B (2018) Development and characterization of penta-flowering and triple-flowering genotypes in garden pea (*Pisum sativum* L. var. *hortense*). PLoS ONE 13(7):e0201235

Diapari M, Sindhu A, Warkentin TD, Bett K, Tar'an B (2015) Population structure and marker-trait association studies of iron, zinc and selenium concentrations in seed of field pea (*Pisum sativum* L.). Mol Breed 35:30

Ding W (1992) Analysis of ecological factors influencing nodule formation of rhizobia. Chin J Ecol 11(4):50–54

Dmytruk VM, Vasyliv VN, Derkach VI, Mykiyevych MI, Romanyuk N (2016) Effects of seed priming with growth regulator biolan on salt tolerance of *Pisum sativum* L. Studia Biologica 10:107–118

Doležel J, Greilhuber J (2010) Nuclear genome size: are we getting closer? Cytometry Part A 77A(7):635–642

Dolezel J, Greilhuber J, Lucretti S, Meister A, Lysak MA, Nardi L, Obermayer R (1998) Plant genome size estimation by flow cytometry: Inter-laboratory comparison. Ann Bot 82:17–26

Drijfhout E (1968) Testing for pea leafroll virus and inheritance of resistance in peas. Euphytica 17(2):224–235

Druery CT, Bateson W (1901) Experiments in plant hybridization. J Roy Hort Soc 26:1–32

Duarte J, Riviere N, Baranger A, Aubert G, Burstin J, Cornet L, Lavaud C, Lejeune-Henaut I, Martinant JP, Pichon JP, Pilet-Nayel ML, Boutet G (2014) Transcriptome sequencing for high throughput SNP development and genetic mapping in pea. BMC Genom 15:126

Dumont E, Fontaine V, Vuylsteker C, Sellier H, Bodèle S, Voedts N, Devaux R, Frise M, Avia K, Hilber JL, Bahrman N, Hanocq E, Lejeune-Hénaut I, Delbreil B (2009) Theor Appl Genet 118:1561–1571

Durieu P, Ochatt SJ (2000) Efficient intergeneric fusion of pea (*Pisum sativum* L.) and grass pea (*Lathyrus sativus* L.) protoplasts. J Exp Bot 51(348):1237–1242

Duzdemir O, Kurunc A, Unlukara A (2009) Response of pea (*Pisum sativum*) to salinity and irrigation water regime. Bulg J Agri Sci 15(5):400–409

Dwivedi SL, Sahrawat KL, Upadhyaya HD, Mengoni A, Galardini M, Bazzicalupo M, Biondi EG, Hungria M, Kaschuk G, Blair MW, Ortiz R (2015) Chapter One-Advances in host plant and rhizobium genomics to enhance symbiotic nitrogen fixation in grain legumes. In: Sparks DL (eds), Advances in Agronomy. Academic Press, New York, pp 1–116

Dyachenko EA, Ryzhova NN, Vishnyakova MA, Kochieva EZ (2014) Molecular genetic diversity of the pea (*Pisum sativum* L.) from the Vavilov Research Institute collection detected by the AFLP analysis. Russ J Genet 50(9):916–924

Ehab AI (2016) Seed priming to alleviate salinity stress in germinating seeds. J Plant Physiol 192(15):38–46

El-Esawi M, Al-Ghamdi A, Ali H, Alayafi A, Witczak J, Ahmad M (2018) Analysis of genetic variation and enhancement of salt tolerance in French pea (*Pisum Sativum* L.). Intl J Mol Sci 19(8):2433

Ellis THN, Hofer JMI, Timmerman-Vaughan GM, Coyne CJ, Hellens RP (2011) Mendel, 150 years on. Trends Plant Sci 16(11):590–596

Ellis TH, Turner L, Hellens RP, Lee D, Harker CL, Enard C, Domoney C, Davies DR (1992) Linkage maps in pea. Genetics 130:649–663

Ellis THN, Poyser SJ (2002) An integrated and comparative view of pea genetic and cytogenetic maps. New Phytol 153:17–25

Ellis THN, Poyser SJ, Knox MR, Vershinin AV, Ambrose MJ (1998) Polymorphism of insertion sites of Ty1-copia class retrotransposons and its use for linkage and diversity analysis in pea. Mol Gen Genet 260:9–19

Elvira-Recuenco M, Bevan JR, Taylor JD (2003) Differential responses to pea bacterial blight in stems, leaves and pods under glasshouse and field conditions. Eur J Plant Pathol 109(6):555–564

Eujayl I, Erskine W, Baum M, Pehu E (1999) Inheritance and linkage analysis of frost injury in lentil. Crop Sci 39:639–642

Farahani HJ, Peterson GA, Westfall DG (1998) Dryland cropping intensification: a fundamental solution to efficient use of precipitation. Adv Agron 64:197–223

FAO (2012) International year of pulses-nutritious seeds for a sustainable future. www.fao.org/pulses-2012

FAO (2013) International year of pulses-nutritious seeds for a sustainable future. www.fao.org/pulses-2013

FAOSTAT (2018). http://www.fao.org/faostat/en/#data

Feng J, Hwang R, Chang KF, Conner L, Hwang SF, Strelkov SE, Gossen BD, McLaren DL, Xue AG (2011) Identification of microsatellite markers linked to quantitative trait loci controlling resistance to Fusarium root rot in field pea. Can J Plant Sci 91:199–204

Flavell RB, Bennett MD, Smith JB, Smith DB (1974) Genome size and the proportion of repeated nucleotide sequence DNA in plants. Biochem Genet 12:257–269

Flores-Félix JD, Silva LR, Rivera LP, Marcos-Garía M, Garcia-Fraile P, Martínez-Molina E, Mateos PF, Velázquez E, Andrade P, Rivas R (2015) Plants probiotics as a tool to produce highly functional fruits: the case of Phyllobacterium and vitamin C in strawberries. PLoS ONE 10(4):e0122281

Fondevilla S, Satovic Z, Rubiales D, Moreno MT, Torres AM (2008) Mapping of quantitative trait loci for resistance to *Mycosphaerella pinodes* in *Pisum sativum* subsp syriacum. Mol Breed 21:439–454

Forster P, Ramaswamy V, Artaxo P, Berntsen T, Betts R, Fahey DW, Haywood J, Lean J, Lowe DC, Myhre G (2007) Changes in atmospheric constituents and in radiative forcing. In: Solomon S, Qin D, Manning M, Chen Z, Marquis M, Averyt KB, Tignor M, Miller HL (eds) Climate change 2007: the physical science basis. Contribution of Working Group I to the fourth assessment report of the intergovernmental panel on climate change. Cambridge University Press, New York, pp 129–234

Foucher F, Morin J, Courtiade J, Cadioux S, Ellis N, Banfield MJ, Rameau C (2003) Determinate and late flowering are two terminal FLOWER1/CENTRORADIALIS homologs that control two distinct phases of flowering initiation and development in pea. Plant Cell 15:2742–2754

Francois LE, Maas EV (1994) Crop response and management on salt-affected soils. In: Pessarakli M (ed) Handbook of plant and crop stress. Marcel Dekker, New York, pp 149–181

Franssen SU, Shrestha RP, Bräutigam A, Bornberg-Bauer E, Weber AP (2011a) Comprehensive transcriptome analysis of the highly complex *Pisum sativum* genome using next generation sequencing. BMC Genom 12:227

Franssen SU, Shrestha RP, Brautigam A, Bornberg-Bauer E, Weber APM (2011b) Comprehensive transcriptome analysis of the highly complex *Pisum sativum* genome using next generation sequencing. BMC Genomics 12

Fondevilla S, ÅVila CM, Cubero JI, Rubiales D (2005) Response to Mycosphaerella pinodes in a germplasm collection of *Pisum* spp. Plant Breed 124(3):313–315

Fondevilla S, Carver TLW, Moreno MT, Rubiales D (2006) Macroscopic and histological characterisation of genes *er1* and *er2* for powdery mildew resistance in pea. Eur J Plant Pathol 115:309–321

Fondevilla S, Carver TLW, Moreno MT, Rubiales D (2007a) Identification and characterization of sources of resistance to *Erysiphe pisi* Syd in Pisum spp. Plant Breed 126(2):113–119

Fondevilla S, Torres AM, Moreno MT, Rubiales D (2007b) Identification of a new gene for resistance to powdery mildew in *Pisum fulvum*, a wild relative of pea. Breed Sci 57(2):181–184

Fondevilla S, Fernández-Aparicio M, Satovic Z, Emeran AA, Torres AM, Moreno MT, Rubiales D (2010). Identification of quantitative trait loci for specific mechanisms of resistance to *Orobanche crenata* Forsk. in pea (*Pisum sativum* L.). Mol Breed 25(2):259–272

Fondevilla S, Martín-Sanz A, Satovic Z, Fernández-Romero MD, Rubiales D, Caminero C (2012) Identification of quantitative trait loci involved in resistance to *Pseudomonas syringae* pv. *syringae* in pea (*Pisum sativum* L.). Euphytica 186(3):805–812

Furones-Pérez P, Fernández-López J (2009) Usefulness of 13 morphological and phenological characteristics of sweet chestnut (*Castanea sativa* Mill.) for use in the DUS test. Euphytica 167:1–21

Gali K, K, Liu Y, Sindhu A, Diapari M, Shunmugam A, S, Arganosa G, Daba K, Caron C, Lachagari RVB, Tar'an B, Warkentin TD (2018) Construction of high-density linkage maps for mapping quantitative trait loci for multiple traits in field pea (*Pisum sativum* L). BMC Blant Biol 18(1):172

Gan Y, Liang C, Wang X, Mcconkey B (2011) Lowering carbon footprint of durum wheat by diversifying cropping systems. Field Crops Res 122(3):199–206

Gao Y, Duan AW, Sun JS, Li FS, Liu ZG, Liu H, Liu ZD (2009) Crop coefficient and water-use efficiency of winter wheat/spring maize strip intercropping. Field Crops Res 111:65–73

Gao Z, Johansen E, Eyers S, Thomas CL, Ellis THN, Maule AJ (2004) The Potyvirus recessive resistance gene, sbm1, identifies a novel role for translation initiation factor eIF4E in cell-to-cell trafficking. Plant J 40:376–385

Gaulin E, Jacquet C, Bottin A, Dumas B (2007) Root rot disease of legumes caused by *Aphanomyces euteiches*. Mol Plant Pathol 8:539–548

Genetic diversity in European Pisum germplasm collections (2012). Theor Appl Genet 125:367–380

Ghafoor A, McPhee K (2012) Marker assisted selection (MAS) for developing powdery mildew resistant pea cultivars. Euphytica 186:593–607

Gibson AH, Newton WE (1981) Current perspectives in nitrogen fixation. Elsevier/North Holland Biomedical Press, Canberra, Australia and New York, USA, pp 14–30

Gilpin BJ, McCallum JA, Frew TJ, Timmerman-Vaughan GM (1997) A linkage map of the pea (*Pisum sativum* L.) genome containing cloned sequences of known function and expressed sequence tags (ESTs). Theor Appl Genet 95:1289–1299

Godfray HC, Beddington JR, Crute IR, Haddad L, Lawrence D, Muir JF, Pretty J, Robinson S, Thomas SM, Toulmin C (2010) Food security: the challenge of feeding 9 billion people. Science 327:812–818

Golubev AA (1990) Habitats, collection, cultivation and hybridization of Vavilovia formosa Fed. Bull Appl Bot Genet Plant Breed 135:67–75

Grabherr MG, Haas BJ, Yassour M, Levin JZ, Thompson DA, Amit I, Adiconis X, Fan L, Raychowdhury R, Zeng QD, Chen ZH, Mauceli E, Hacohen N, Gnirke A, Rhind N, di Palma F, Birren BW, Nusbaum C, Lindblad-Toh K, Friedman N, Regev A (2011) Full-length transcriptome assembly from RNA-Seq data without a reference genome. Nat Biotechnol 29:644–654

Grant J, Cooper P (1997) Peas (*pisum sativum* L.). Field Crops Res 53:111–130

Grewal HS (2010) Water uptake, water use efficiency, plant growth and ionic balance of wheat, barley, canola and chickpea plants on a sodic vertosol with variable subsoil NaCl salinity. Agri Water Manag 97:148–156

Grünwald NJ, Coffman VA, Kraft JM (2003) Sources of partial resistance to Fusarium root rot in the *Pisum* core collection. Plant Dis 87:1197–1200

Gupta PK, Langridge P, Mir RR (2010) Marker-assisted wheat breeding: present status and future possibilities. Mol Breed 26:145–161

Gurung AM, Pang ECK, Taylor PWJ (2002) Examination of *Pisum* and *Lathyrus* species as sources of ascochyta blight resistance for field pea (*Pisum sativum*). Australas Plant Pathol 31(1):41–45

Haglund WA, Kraft JM (1979) *Fusarium oxysporum* f. sp. *pisi* race 6: occurrence and distribution. Phytopathology 69:818–820

Hamblin MT, Buckler ES, Jannink J-L (2011) Population genetics of genomics-based crop improvement methods. Trends Genet 27:98–106

Hayat S, Hasan SA, Yusuf M, Hayat Q, Ahmad A (2010) Effect of 28-homobrassinolide on photosynthesis, fluorescence and antioxidant system in the presence or absence of salinity and temperature in *Vigna radiata*. Environ Exp Bot 69:105–112

Hamon C, Baranger A, Coyne CJ, McGee RJ, Le Goff I, L'Anthoene V, Esnault R, Riviere JP, Klein A, Mangin P, McPhee KE, Roux-Duparque M, Porter L, Miteul H, Lesne A, Morin G, Onfroy C, Moussart A, Tivoli B, Delourme R, Pilet-Nayel ML (2011) New consistent QTL in pea associated with partial resistance to *Aphanomyces euteiches* in multiple French and American environments. Theor Appl Genet 123:261–281

Hamon C, Coyne CJ, McGee RJ, Lesné A, Esnault R, Mangin P, Hervé M, Le Goff I, Deniot G, Roux-Duparque M, Morin G, McPhee KE, Delourme R, Alain Baranger A, Pilet-Nayel ML (2013) QTL meta-analysis provides a comprehensive view of loci controlling partial resistance to *Aphanomyces euteiches* in four sources of resistance in pea. BMC Plant Biol 13:45

Handerson C, Noren SK, Wricha T, Meetei NT, Khanna VK, Pattanayak A, Datt S, Choudhury PR, Kumar M (2014) Assessment of genetic diversity in pea (*Pisum sativum* L.) using morphological and molecular markers. Indian J Genet 74:205–212

Hardie DC, Baker GJ, Marshall DR (1995) Field screening of Pisum accessions to evaluate their susceptibility to the pea weevil (Coleoptera: Bruchidae). Euphytica 84(2):155–161

Hecht V, Knowles CL, Vander Schoor JK, Liew LC, Jones SE, Lambert MJ, Weller JL (2007) Pea *LATE BLOOMER1* is a *GIGANTEA* ortholog with roles in photoperiodic flowering, deetiolation, and transcriptional regulation of circadian clock gene homologs. Plant Physiol 144:648–661

Hellens RP, Moreau C, Lin-Wang K, Schwinn KE, Thomson SJ, Fiers MWEJ, Frew TJ, Murray SR, Hofer JMI, Jacobs JME, Davies KM, Allan AC, Bendahmane A, Coyne CJ, Timmerman-Vaughan GM, Ellis THN (2010) Identification of Mendel's white flower character. PLoS ONE 5:e13230

Herridge D (2002) Inoculants and nitrogen fixation of legumes in Vietnam. ACIAR Proc 109e. www.aciar.gov.au/publications. ISBN: 1 86320 335 4

Herridge DF, Peoples MB, Boddey RM (2008) Global inputs of biological nitrogen fixation in agricultural systems. Plant Soil 311:1–18

He CB, Zong XX (2011) Genetic diversity of pea (*Pisum sativum* L.) germplasm resources revealed by morphological traits. J Plant Genet Resour 12(1):42–48

Hollaway GJ, Bretag TW, Price TV (2007) The epidemiology and management of bacterial blight (*Pseudomonas syringae* pv. *pisi*) of field pea (*Pisum sativum*) in Australia: a review. Aust J Agri Res 58(11):1086–1099

Hu F, Gan Y, Chai Q, Feng F, Zhao C, Yu A (2016) Boosting system productivity through the improved coordination of interspecific competition in maize/pea strip intercropping. Field Crop Res 198:50–60

Hu F, Zhao C, Feng F, Chai Q, Mu Y, Zhang Y (2017) Improving N management through intercropping alleviates the inhibitory effect of mineral N on nodulation in pea. Plant Soil 412:235–251

Hu Z, Ma Z, Bao X, Zhang J (2010) Effect of regulated deficient irrigation (RDI) on yield and water consumption of major crops in Shiyang river area. Water Sav Irrig 7:10–13 (in Chinese with English abstract)

Huesing JE, Shade RE, Chrispeels MJ, Murdock LL (1991) α-amylase inhibitor, not phytohemagglutinin, explains resistance of common bean seeds to cowpea weevil. Plant Physiol 96(3):993–996

Humplík JF, Lazár D, Fürst T, Husičková A, Hýbl M, Spíchal L (2015) Automated integrative high-throughput phenotyping of plant shoots: a case study of the cold-tolerance of pea (*Pisum sativum* L.). Plant Methods 11(1):20

Humphry M, Reinstädler A, Ivanov S, Bisseling T, Panstruga R (2011) Durable broad-spectrum powdery mildew resistance in pea *er1* plants is conferred by natural loss-of-function mutations in PsMLO1. Mol Plant Pathol 12:866–878

Hungria M, Franchini JC, Campo RJ, Crispino CC, Moraes JZ, Sibaldelli RNR, Mendes IC, Arihara J (2006) Nitrogen nutrition of soybean in Brazil: contributions of biological N_2 fixation and N fertilizer to grain yield. Can J Plant Sci 86(4):927–939

Hunter PJ, Ellis N, Taylor JD (2001) Association of dominant loci for resistance to Pseudomonas syringae pv. pisi with linkage groups II, VI and VII of *Pisum sativum*. Theor Appl Genet 103:129–135

Husen A, Iqbal M, Aref IM (2016) IAA-induced alteration in growth and photosynthesis of pea (*Pisum sativum* L.) plants grown under salt stress *Pisum sativum*. J Environ Biol 37:421–429

Iglesias-García R, Prats E, Rubiales D (2012a) Use of physiological parameters as tools to discriminate water stress tolerance in pea. In: International conference on bioscience: biotechnology and biodiversity, Novi Sad, Serbia, 18–20 June 2012, p 107

Iglesias-García R, Prats E, Rubiales D (2012b) Identification and characterization of drought resistance sources in pea. University of Córdoba, Córdoba, Spain

Iglesias-García R, Prats E, Flores F, Amri M, Mikić A, Rubiales D (2017) Assessment of field pea (*Pisum sativum* L.) grain yield, aerial biomass and flowering date stability in Mediterranean environments. Crop Pasture Sci 68(11):915–923

Iglesias-García R, Prats E, Fondevilla S, Satovic Z, Rubiales D (2015) Quantitative trait loci associated to drought adaptation in pea (*Pisum sativum* L.). Plant Mol Biol Rep 33(6):1768–1778

Irar S, González EM, Arrese-Igor C, Marino D (2014) A proteomic approach reveals new actors of nodule response to drought in split-root grown pea plants. Physiol Plant 152(4):634–645

Isaac I, Rogers WG (1974) Verticillium wilt of pea (*Pisum sativum*). Ann Appl Biol 76:27–35

Irzykowska L, Wolko B, Swiecicki W (2002) Interval mapping of QTLs controlling some morphological traits in pea. Cell Mol Biol Lett 7:417–422

Irzykowska L, Wolko B (2004) Interval mapping of QTLs controlling yield-related traits and seed protein content in *Pisum sativum*. J Appl Genet 45:297–306

Jacob C, Carrasco B, Schwember AR (2016) Advances in breeding and biotechnology of legume crops. Plant Cell Tiss Org Cult 127:561–584

Rana JC, Rana M, Sharma V, Nag A, Chahota RK, Sharma TR (2017) Genetic diversity and structure of pea (*Pisum sativum* L.) germplasm based on morphological and SSR markers. Plant Mol Biol Rep 35(1):118–129

Jain S, Weeden NF, Porter LD, Eigenbrode SD, McPhee K (2013) Finding linked markers to En for efficient selection of pea enation mosaic virus resistance in pea. Crop Sci 53(6):2392–2399

Jain S, Kumar A, Mamidi S, McPhee K (2014) Genetic diversity and population structure among pea (Pisum sativum L.) cultivars as revealed by simple sequence repeat and novel genic markers. Mol Biotechnol 56:925–938

Jain S, Weeden NF, Kumar A, Chittem K, McPhee K (2015) Functional codominant marker for selecting the Fw gene conferring resistance to Fusarium wilt race 1 in pea. Crop Sci 55(6):2639–2646

Janzen HH, Angers DA, Boehm M, Bolinder M, Wang H (2006) A proposed approach to estimate and reduce net greenhouse gas emissions from whole farms. Can J Soil Sci 86(3):401–418

Javid M, Noy D, Sudheesh S, Forster JW, Kaur S (2017) Identification of QTLs associated with metribuzin tolerance in field pea (*Pisum sativum* L.). Euphytica 213:91

Jensen ES, Hauggaard-Nielsen H (2003) How can increased use of biological N_2 fixation in agriculture benefit the environment? Plant Soil 252:177–186

Jeuffroy MH, Baranger E, Carrouee B, de Chezelles E, Gosme M et al (2013) Nitrous oxide emissions from crop rotations including wheat, oilseed rape and dry peas. Biogeosciences 10:1787–1797

Jha AB, Gali KK, Tar'an B, Warkentin TD (2017) Fine mapping of QTLs for Ascochyta Blight resistance in pea using heterogeneous inbred families. Front Plant Sci 8

Jha AB, Warkentin TD, Gurusamy V, Tar'an B, Banniza S (2012) Identification of mycosphaerella blight resistance in wild Pisum species for use in pea breeding. Crop Sci 52(6):2462–2468

Jha AB, Tar'an B, Diapari M, Sindhu A, Shunmugam A, Bett K, Warkentin TD (2015) Allele diversity analysis to identify SNPs associated with ascochyta blight resistance in pea. Euphytica 202(2):189–197

Jha AB, Tar'an B, Stonehouse R, Warkentin TD (2016) Identification of QTLs associated with improved resistance to ascochyta blight in an interspecific pea recombinant inbred line population. Crop Sci 56(6):2926–2939

Jia K-P, Luo Q, He S-B, Lu X-D, Yang H-Q (2014) Strigolactone-regulated hypocotyl elongation is dependent on cryptochrome and phytochrome signaling pathways in Arabidopsis. Mol Plant 7:528–540

Jiang YF, Diapari M, Bueckert RA, Tar'an B, Warkentin TD (2017) Population structure and association mapping of traits related to reproductive development in field pea. Euphytica 213:20

Jing R, Ambrose MA, Knox MR, Smykal P, Hybl M, Ramos A, Caminero C, Burstin J, Duc G, van Soest LJM, Swiecicki WK, Pereira MG, Vishnyakova M, Davenport GF, Flavell AJ, Ellis THN (2012) Genetic diversity in European *Pisum* germplasm collections. Theor Appl Genet 125:367–380

Jin J, Lauricella D, Armstrong R, Sale P, Tang C (2014) Phosphorus application and elevated CO2 enhance drought tolerance in field pea grown in a phosphorus-deficient vertisol. Ann Bot 116(6):975–985

Johal GS, Balint-Kurti P, Well CF (2008) Mining and harnessing natural variation: a little MAGIC. Crop Sci 48:2066–2073

Jones AL, Johansen IE, Bean SJ, Bach I, Maule AJ (1998) Specificity of resistance to pea seed-borne mosaic potyvirus in transgenic peas expressing the viral replicase (*Nlb*) gene. J Gen Virol 79(12):3129–3137

Kankanen J (2015) Reducing use of fossil energy by biological N fixation. Legume Perspect 8:24–25

Karataş İ, Öztürk L, Demir Y, Ünlükara A, Kurunç A, Düzdemir O (2014) Alterations in antioxidant enzyme activities and proline content in pea leaves under long-term drought stress. Toxicol Indust Health 30(8):693–700

Karkanis A, Ntatsi G, Kontopoulou CK, Pristeri A, Bilalis D, Savvas D (2016) Field pea in European cropping systems: adaptability, biological nitrogen fixation and cultivation practices. Not Bot Hort Agrobot 44(2):325–336

Katiyar RP, Ram RS (1987) Genetics of rust resistance in pea. Indian J Genet 47:46–48

Katoch V, Sharma S, Pathania S, Banayal DK, Sharma SK, Rathour R (2010) Molecular mapping of pea powdery mildew resistance gene *er2* to pea linkage group III. Mol Breed 25:229–237

Kaur S, Pembleton LW, Cogan NO, Savin KW, Leonforte T, Paull J, Materne M, Forster JW (2012) Transcriptome sequencing of field pea and faba bean for discovery and validation of SSR genetic markers. BMC Genom 13:104

Kav NNV, Srivastava S, Goonewardende L, Blade SF (2004) Proteome-level changes in the roots of *Pisum sativum* in response to salinity. Ann Appl Biol 145:217–230

Kaya MD, Okçu G, Atak M, Çıkılı Y, Kolsarıcı Ö (2006) Seed treatments to overcome salt and drought stress during germination in sunflower (*Helianthus annuus* L.). Eur J Agron 24(4):291–295

Keneni G, Bekele E, Getu E, Imtiaz M, Damte BM, Dagne K (2011) Breeding food legumes for resistance to storage insect pests: potential and limitations. Sustainability 3:1399–1415

Keneni G, Jarso M, Wolabu T, Dino G (2005) Extent and pattern of genetic diversity for morpho-agronomic traits in ethiopian highland pulse landraces: i. field pea (*Pisum sativum* L.). Genet Resour Crop Ev 52:539–549

Khan TN, Timmerman-Vaughan GM, Rubiales D, Warkentin TD, Siddique KHM, Erskine W, Barbetti MJ (2013) Didymella pinodes and its management in field pea: challenges and opportunities. Field Crop Res 148:61–77

Kosová K, Vítámvás P, Urban MO, Klíma M, Roy A, Prášil IT (2015) Biological networks underlying abiotic stress tolerance in temperate crops—a proteomic perspective. Intl J Mol Sci 16:20913–20942

Kouchi H, Shimomura K, Hata S, Hirota A, Wu GJ, Kumagai H, Tajima S, Suganuma N, Suzuki A, Aoki T, Hayashi M, Yokoyama T, Ohyama T, Asamizu E, Kuwata C, Shibata D, Tabata S (2004) Large-scale analysis of gene expression profiles during early stages of root nodule formation in a model legume, *Lotus japonicus*. DNA Res 11(4):263–274

Koyama ML, Levesley A, Koebner RM, Flowers TJ, Yeo AR (2001) Quantitative trait loci for component physiological traits determining salt tolerance in rice. Plant Physiol 125:406–422. https://doi.org/10.1104/pp.125.1.406

Kraft JM (1994) Fusarium wilt of peas (a review). Agronomie 14:561–567

Kraft JM, Pfleger FL (2001) Compendium of pea diseases and pests. American Phytopathological Society, St. Paul, MN, USA

Kraft JM, Boge W (2001) Root characteristics in pea in relation to compaction and *Fusarium* root rot. Plant Dis 85:936–940

Kühn C, Grof CP (2010) Sucrose transporters of higher plants. Curr Opin Plant Biol 13:287–297

Kumar S, Sharma V, Chaudhary S, Kumari R, Kumari N, Mishra P (2011a) Interaction between *COCHLEATA* and *UNIFOLIATA* genes enables normal flower morphogenesis in the garden pea, *Pisum sativum*. J Genet 90:309–314

Kumar J (2011a) Effect of phosphorus and sulphur application on performance of vegetable ea (*Pisum sativum* L.) CV. Pant matar-2. Legume Res 34 (4):292–295

Kumar J, Choudhary AK, Solanki RK, Pratap A (2011b) Towards marker-assisted selection in pulses: a review. Plant Breed 130:297–313

Kumari J, Dikshit HK, Singh B, Singh D (2015) Combining ability and character association of agronomic and biochemical traits in pea (*pisum sativum* L.). Sci Hort 181:26–33

Kushwaha C, Chand R, Srivastava C (2006) Role of aeciospores in outbreaks of pea (*Pisum sativum*) rust (*Uromyces fabae*). Eur J Plant Pathol 115:323–330

Konečná E, Šafářová D, Navrátil M, Hanáček P, Coyne C, Flavell A, Vishnyakova M, Ambrose M, Redden R, Smýkal P (2014) Geographical gradient of the eIF4E alleles conferring resistance to potyviruses in pea (Pisum) germplasm. PLoS ONE 9(3):e90394

Krajewski P, Bocianowski J, Gawłowska M, Kaczmarek Z, Pniewski T, Święcicki W, Wolko B (2012) QTL for yield components and protein content: a multienvironment study of two pea (*Pisum sativum* L.) populations. Euphytica 183:323–336

Kwon S-J, Brown AF, Hu J, McGee R, Watt C, Kisha T, Timmerman-Vaughan G, Grusak M, McPhee KE, Coyne CJ (2012) Genetic diversity, population structure and genome-wide marker-trait association analysis emphasizing seed nutrients of the USDA pea (*Pisum sativum* L.) core collection. Genes Genom 34:305–320

Kwon YS, Kim KM, Kim DH, Eun MY, Sohn JK (2002) Marker-assisted introgression of quantitative trait loci associated with plant regeneration ability in anther culture of rice (*Oryza sativa* L.). Mol Cells 14:24–28

Kwon YS, Lee JM, Yi GB, Yi SI, Kim KM, Soh EH, Bae KM, Park EK, Song IH, Kim BD (2005) Use of SSR markers to complement tests of distinctiveness, uniformity, and stability (DUS) of pepper (*Capsicum annuum* L.) varieties. Mol Cells 19:428–435

Kwon SJ, Smýkal P, Hu J, Wang M, Kim SJ, McGee RJ, McPhee K, Coyne CJ (2013) User-friendly markers linked to Fusarium wilt race 1 resistance Fw gene for marker-assisted selection in pea. Plant Breed 132(6):642–648

Lang C, Long SR (2015) Transcriptomic analysis of *Sinorhizobium meliloti* and *Medicago truncatula* symbiosis using nitrogen fixation-deficient nodules. Mol Plant Microbe Interact 28(8):856–868

Larrainzar E, Riely BK, Kim SC, Carrasquilla-Garcia N, Yu HJ, Hwang HJ, Oh M, Kim GB, Surendrarao AK, Chasman D, Siahpirani AF, Penmetsa RV, Lee GS, Kim N, Roy S, Mun JH, Cook DR (2015) Deep sequencing of the *Medicago truncatula* root transcriptome reveals a massive and early interaction between nodulation factor and ethylene signals. Plant Physiol 169(1):233–265

Laucou V, Haurogne K, Ellis N, Rameau C (1998) Genetic mapping in pea. 1. RAPD-based genetic linkage map of *Pisum sativum*. Theor Appl Genet 97:905–915

Lavaud C, Lesne A, Piriou C, Le Roy G, Boutet G, Moussart A, Poncet C, Delourme R, Baranger A, Pilet-Nayel ML (2015) Validation of QTL for resistance to *Aphanomyces euteiches* in different pea genetic backgrounds using near-isogenic lines. Theor Appl Genet 128(11):2273–2288

Lavaud C, Baviere M, Le Roy G, Hervé MR, Moussart A, Delourme R, Pilet-Nayel ML (2016) Single and multiple resistance QTL delay symptom appearance and slow down root colonization by *Aphanomyces euteiches* in pea near isogenic lines. BMC Plant Biol 16(1):166

Lejeune-Hénaut I, Hanocq E, Be´thencourt L, Fontaine V, Delbreil B, Morin J, Petit A, Devaux R, Boilleau M, Stempniak JJ, Thomas M, Laine´ AL, Foucher F, Baranger A, Burstin J, Rameau C, Giauffret C (2008) The flowering locus Hr colocalizes with a major QTL affecting winter frost tolerance in *Pisum sativum* L. Theor Appl Genet 116(8):1105–1116

Lejeune-Hénaut I, Bourion V, Eteve G, Cunot E, Delhaye K, Desmyter C (1999) Floral initiation in field grown forage peas is delayed to a greater extent by short photoperiods, than in other types of European varieties. Euphytica 109:201–211

Lejeune-Hénaut I, Morin J, Fontaine V, Etévé G, Devaux R, Thomas M, Boilleau M, Stempniak JJ, Petit A, Rameau C, Baranger A (2004) Towards genes to breed for freezing resistance in pea. In: AEP (ed) AEP Conference 2004, Dijon, AEP, Paris, France, p 127

Leonforte A, Forster JW, Redden, RJ, Nicolas ME, Salisbury PA (2013a) Sources of high tolerance to salinity in pea (*Pisum sativum* L.). Euphytica 189(2):203–216

Leonforte A, Sudheesh S, Cogan NO, Salisbury PA, Nicolas, ME, Materne M, Kaur S (2013b) SNP marker discovery, linkage map construction and identification of QTLs for enhanced salinity tolerance in field pea (*Pisum sativum* L.). BMC Plant Biol 13(1):161

Lester DR, Ross JJ, Davies PJ, Reid JB (1997) Mendel's stem length gene (*Le*) encodes a gibberellin 3b-hydroxylase. Plant Cell 9:1435–1443

Levenfors JP, Wikstrom M, Persson L, Gerhardson B (2003) Pathogenicity of *Aphanomyces* spp. from different leguminous crops in Sweden. Eur J Plant Pathol 109:535–543

Li L, Yang T, Liu R, Redden B, Maalouf F, Zong XX (2017) Food legume production in China. Crop J 5(2):115–126

Li X, Dou X, Liu Q, Wang Y (1997) Study on the quantity distribution and colonization ability of soybean rhizobium in Heilongjiang province. Heilongjiang Agri Sci 5:24–27

Li Y, Li Y, Yang Q, Zhang J, Zhang J, Qiu L, Wang T (2015) Genomics-based crop germplasm research: Advances and perspectives. Sci Agri Sin 48:3333–3353

Liew LC, Hecht V, Sussmilch FC, Weller JL (2014) The pea photoperiod response gene *STERILE NODES* is an ortholog of *LUX ARRHYTHMO*. Plant Physiol 165:648–657

Lipper L, Thornton P, Campbell BM, Baedeker T, Braimoh A, Bwalya M, Caron P, Cattaneo A, Garrity D, Henry K, Hottle R, Jackson L, Jarvis A, Kossam F, Mann W, McCarthy N, Meybeck A, Neufeldt H, Remington T, Pham Thi S, Sessa R, Shula R, Tibu A, Torquebiau EF (2014) Climate-smart agriculture for food security. Nat Clim Change 4:1068–1072

Liu J, Diamond J (2005) China's environment in a globalizing world. Nature 435:1179–1186

Liu N, Zhang G, Xu S, Mao W, Hu Q, Gong Y (2015) Comparative transcriptomic analyses of vegetable and grain pea (*Pisum sativum* L.) seed development. Front Plant Sci 6:1039

Liu R, Fang L, Yang T, Zhang X, Hu J, Zhang H, Han W, Hua Z, Hao J, Zong X (2017) Marker-trait association analysis of frost tolerance of 672 worldwide pea (*Pisum sativum* L.) collections. Sci Rep 7(1):5919

Liu X, Li L, Li Y, Fang W (2018) Crop germplasm resources: Adv Trends J Agri 8:1–6

Lobell DB, Schlenker W, Costa-Roberts J (2011) Climate trends and global crop production since 1980. Science 333:616–620

Lohar DP, Sharopova N, Endre G, Peñuela S, Samac D, Town C, Silverstein KA, VandenBosch KA (2006) Transcript analysis of early nodulation events in Medicago truncatula. Plant Physiol 40(1):221–234

Long SR (1989) Rhizobium-legume nodulation: life together in the underground. Cell 56(2):203–217

Loridon K, McPhee K, Morin J, Dubreuil P, Pilet-Nayel ML, Aubert G, Rameau C, Baranger A, Coyne C, Lejeune-Hénaut I, Burstin J (2005) Microsatellite marker polymorphism and mapping in pea (*Pisum sativum* L.). Theor Appl Genet 111:1022–1031

Lu J, Knox MR, Ambrose MJ, Brown JKM, Ellis THN (1996) Comparative analysis of genetic diversity in pea assessed by RFLP and PCR-based methods. Theor Appl Genet 93:1103–1111

Lutsenko EK, Marushko YA, Kononenko TG (2005) Influence of fusicoccin on early growth of sorgo under high NaCl concentrations. Plant Physiol 52(3):378–383

Lyon DJ, Peterson GA (2005) Continuous dryland cropping in the Great Plains: What are the limits? Agron J 97:347–348

Ma Y, Coyne CJ, Grusak MA, Mazourek M, Cheng P, Main D, McGee RJ (2017) Genome-wide SNP identification, linkage map construction and QTL mapping for seed mineral concentrations and contents in pea (*Pisum sativum* L.). BMC Plant Biol 17:17

Ma Y, Clarice J, Coyne CJ, Main D, Pavan S, Sun S, Zhu Z, Zong X, José Leitão J, McGee RJ (2017) Development and validation of breeder-friendly KASPar markers for er1 a powdery mildew resistance gene in pea (*Pisum sativum* L.). Mol Breed 37:151

Macas J, Neumann P, Navrátilová A (2007) Repetitive DNA in the pea (*Pisum sativum* L.) genome: comprehensive characterization using 454 sequencing and comparison to soybean and *Medicago truncatula*. BMC Genomics 8, 427

Macwilliam S, Wismer M, Kulshreshtha S (2014) Life cycle and economic assessment of Western Canadian pulse systems: The inclusion of pulses in crop rotations. Agri Syst 123(2):43–53

Malvick DK, Percich JA (1999) Identification of *Pisum sativum* germ plasm with resistance to root rot caused by multiple strains of *Aphanomyces euteiches*. Plant Dis 83(1):51–54

Mao LL, Zhang LZ, Li WQ, van der Werf W, Sun JH, Spiertz H, Li L (2012) Yield advantage and water saving in maize/pea intercrop. Field Crops Res 138:11–20

Margesin R, Neuner G, Storey KB (2007) Cold-loving microbes, plants, and animals-fundamental and applied aspects. Naturwissenschaften 94:77–99

Martins I (2010) Tabela da Composição de Alimentos. Instituto Nacional de Saúde Ricardo Jorge, Lisboa, Portugal

Martín-Sanz A, Palomo JL, de la Vega MP, Caminero C (2011) Identification of pathovars and races of *Pseudomonas syringae*, the main causal agent of bacterial disease in pea in North-Central Spain, and the search for disease resistance. Eur J Plant Pathol 129(1):57–69

Martín-Sanz A, Pérez de la Vega M, Caminero C (2012) Resistance to Pseudomonas syringae in a collection of pea germplasm under field and controlled conditions. Plant Pathol 61(2):375–387

Martín-Sanz A, Aparicio T, Santana JC, García P, Winter P, Caminero C, de la Vega MP (2016) Mapping genes for resistance to bacterial blight (*Pseudomonas syringae* pv. *pisi*) in pea and identification of genes involved in resistance by DeepsuperSAGE transcriptome profiling. Euphytica 210(3):375–392

Marx GA, Weeden NF, Provvidenti R (1985) Linkage relationships among markers in chromosome 3 and En, a gene conferring virus resistance. Pisum Newsl 17:57–60

Matthews P, Ambrose MJ. Development and use of a "core" collection for the John Innes *Pisum* Collection. In: Proceedings of 2nd European conference on grain legumes, Copenhagen, Denmark, 1995, pp 194–195

Mathieu S, Jacquin F, Chabert-Martinello M, Smýkal P, Paslier MCL, Aubert G, Burstin J (2017) Patterns of genetic structure and linkage disequilibrium in a large collection of pea germplasm. Genes Genomes Genet 7:2461–2471

Martin DN, Proebsting WM, Hedden P (1997) Mendel's dwarfing gene: cDNAs from the Le alleles and function of the expressed proteins. Proc Natl Acad Sci USA 94:8907–8911

MacLean AM, Finan TM, Sadowsky MJ (2007) Genomes of the symbiotic nitrogen-fixing bacteria of legumes. Plant Physiol 144:615–622

McPhee KE, Muehlbauer FJ (2007) Registration of 'specter' winter feed pea. J Plant Regist 1:118–119

McPhee KE, Tullu A, Kraft JM, Muehlbauer FJ (1999) Resistance to Fusarium wilt race 2 in the *Pisum* core collection. J Amer Soc Hort Sci 124(1):28–31

McPhee KE, Inglis DA, Gundersen B, Coyne CJ (2012) Mapping QTL for Fusarium wilt race 2 partial resistance in pea (*Pisum sativum*). Plant Breed 131(2):300–306

McCouch S, Baute GJ, Bradeen J, Bramel P, Bretting PK, Buckler E, Burke JM, Charest D, Cloutier S, Cole G, Dempewolf H, Dingkuhn M, Feuillet C, Gepts P, Grattapaglia D, Guarino L, Jackson S, Knapp S, Langridge P, Lawton-Rauh A, Lijua Q, Lusty C, Michael T, Myles S, Naito K, Nelson RL, Pontarollo R, Richards CM, Rieseberg L, Ross-Ibarra J, Rounsley S, Hamilton RS, Schurr U, Stein N, Tomooka N, van der Knaap E, van Tassel D, Toll J, Valls J, Varshney RK, Ward J, Waugh R, Wenzl P, Zamir D (2013) Feeding the future. Nature 499:23–24

Mendel JG (1866) Versuche uber Pflanzenhybriden Verhandlungen des naturforschenden Vereines. In: Brunn, Bd. IV fur das Jahr, 1865 Abhandlungen 3–47 English translation, see: Druery CT, Bateson W (1901) Experiments in plant hybridization. J Roy Hort Soc 26:1–32

Mikić A, Mihailović V, Ćupina B, Đorđević V, Milić D, Duc G, Stoddard FL, Lejeune-Hénaut I, Marget P, Hanocq E (2011) Achievements in breeding autumn-sown annual legumes for temperate regions with emphasis on the continental Balkans. Euphytica 180:57–67

Mikić A, Smýkal P, Kenicer G, Sarukhanyan N, Akopian J, Gabrielyan I, Vanyan A, Sinjushin A, Demidenko N, Ć upina B, Mihailović V, Vishnyakova M, Ambrose M (2010) Achievements in research on vavilovia (*Vavilovia formosa* (Stev.) Fed.), a legume crop wild relative. Ratar Povrt 47:387–394

Mikić A, Smýkal P, Kenicer G, Vishnyakova M, Sarukhanyan N, Akopian JA, Vanyan A, Gabrielyan I, Smýkalová I, Sherbakova E, Zorić L, Atlagić J, Zeremski-Škorić T, Ćupina B, Krstić Đ, Jajić I, Antanasović S, Đordević V, Mihailović V, Ivanov A, Ochatt S, Toker C, Zlatković B, Ambrose M (2014) Beauty will save the world, but will the world save beauty? The case of the highly endangered *Vavilovia formosa* (Stev.). Fed Planta 240(5):1139–1146

Miles CA, Sonde M (2003) Pea Shoots. Pacific Northwest Extension Publications, PNW 567:1–8

Miljus-Djukic J, Stanisavljevic N, Radovic S, Jovanovic Z, Mikic A, Maksimovic V (2013) Differential response of three contrasting pea (*Pisum arvense*, *P. sativum* and *P. fulvum*) species to salt stress: assessment of variation in antioxidative defence and miRNA expression. Aust J Crop Sci 7(13):2145–2153

Mishra RK, Gangadhar BH, Nookaraju A, Kumar S, Park SW (2012) Development of EST-derived SSR markers in pea (*Pisum sativum*) and their potential utility for genetic mapping and transferability. Plant Breed 131:118–124

Moazzam-Jazi M, Ghasemi S, Seyedi SM, Niknam V (2018) COP1 plays a prominent role in drought stress tolerance in Arabidopsis and Pea. Plant Physiol Biochem 130:678–691

Moose SP, Mumm RH (2008) Molecular plant breeding as the foundation for 21st century crop improvement. Plant Physiol 147:969–977

Morgante M, Salamini F (2003) From plant genomics to breeding practice. Curr Opin Biotechnol 14:214–219

Morton RL, Schroeder HE, Bateman KS, Chrispeels MJ, Armstrong E, Higgins TJ (2000) Bean α-amylase inhibitor 1 in transgenic peas (*Pisum sativum*) provides complete protection from pea weevil (*Bruchus pisorum*) under field conditions. Proc Natl Acad Sci USA 97(8):3820–3825

Murfet IC (1971) Flowering in *Pisum*: a three-gene system. Heredity 27:93–110

Murfet IC (1985) *Pisum sativum*. In: Halevy AH (ed) Handbook of flowering, vol IV. CRC Press. Boca Raton, Florida, pp 97–126

Murfet IC (1990) Flowering genes in pea and their use in breeding. Pisum Newsl 22:78–86

Murray MG, Peters DL, Thompson WF (1981) Ancient repeated sequences in the pea and mung bean genomes and implications for genome evolution. J Mol Evol 17:31–42

Munns R, Tester M (2008) Mechanisms of salinity tolerance. Annu Rev Plant Biol 59:651–681

Naim-Feil E, Toren M, Aubert G, Rubinstein M, Rosen A, Eshed R, Sherman A, Ophir R, Saranga Y, Abbo S (2017) Drought response and genetic diversity in *Pisum fulvum*, a wild relative of domesticated pea. Crop Sci 57(3):1145–1159

Najafi F, Khavari-Nejad RA, Rastgar-jazii F, Sticklen M (2007) Growth and some physiological attributes of pea (*Pisum sativum* L.) as affected by salinity. Pak J Biol Sci 10:2752–2755

Noreen Z, Ashraf M (2009) Assessment of variation in antioxidative defense system in salt- treated pea (*Pisum sativum*) cultivars and its putative use as salinity tolerance markers. J Plant Physiol 166:1764–1774

Novák P, Neumann P, Macas J (2010) Graph-based clustering and characterization of repetitive sequences in next-generation sequencing data. BMC Bioinform 11:378

Ntatsi G, Gutiérrez-Cortines ME, Karapanos I, Barros A, Weiss J, Balliu A (2018) The quality of leguminous vegetables as influenced by preharvest factors. Sci Hort 232:191–205

Ochatt S, Conreux C, Smýkalová I, Smýkal P, Mikić A (2016) Developing biotechnology tools for 'beautiful'vavilovia (*Vavilovia formosa*), a legume crop wild relative with taxonomic and agronomic potential. Plant Cell Tiss Org Cult 127(3):637–648

O'Hara GW, Boonkerd N, Dilworth MJ (1988) Mineral constraints to nitrogen fixation. Plant Soil 108:93–110

O'Hara GW (2001) Nutritional constraints on root nodule bacteria affecting symbiotic nitrogen fixation: a review. Aust J Exp Agri 41:417–433

Okçu G, Kaya MD, Atak M (2005) Effects of salt and drought stresses on germination and seedling growth of pea (*Pisum sativum* L.). Turk J Agri For 29(4):237–242

Ondřej M, Dostálová R, Odstrčilová L (2005) Response of *Pisum sativum* germplasm resistant to *Erysiphe pisi* to inoculation with *Erysiphe baeumleri*, a new pathogen of peas. Plant Protect Sci 41:95–103

Oono R, Anderson CG, Denison RF (2011) Failure to fix nitrogen by non-reproductive symbiotic rhizobia triggers host sanctions that reduce fitness of their reproductive clonemates. Proc Biol Sci 278:2698–2703

Osman HS (2015) Enhancing antioxidant–yield relationship of pea plant under drought at different growth stages by exogenously applied glycine betaine and proline. Ann Agri Sci 60(2):389–402

Osterlund MT, Hardtke CS, Wei N, Dong XW (2000) Targeted destabilization of HY5 during light-regulated development of Arabidopsis. Nature 405:462

Pan H, Wang D (2017) Nodule cysteine-rich peptides maintain a working balance during nitrogen-fixing symbiosis. Nat Plants 3(5):17048

Parihar P, Singh S, Singh R, Singh VP, Prasad SM (2015) Effect of salinity stress on plants and its tolerance strategies: a review. Environ Sci Pollut R 22(6):4056–4075

Patanè C, Cavallaro V, Cosentino SL (2009) Germination and radicle growth in unprimed and primed seeds of sweet sorghum as affected by reduced water potential in NaCl at different temperatures. Ind Crop Prod 30(1):1–8

Pathak AR, Pithia MS, Javia RM, Mehta DR (2017) Challenges and options for meeting the needs of pulses-A review. Agri Rev 38:103–111

Pavan S, Schiavulli A, Appiano M, Marcotrigiano AR, Cillo F, Visser RGF, Yuling Bai Y, Lotti C, Luigi Ricciardi L (2011) Pea powdery mildew er1 resistance is associated to loss-of-function mutations at a MLO homologous locus. Theor Appl Genet 123:1425–1431

Pavan S, Schiavulli A, Appiano M, Miacola C, Visser RGF, Bai YL, Lotti C, Ricciardi L (2013) Identification of a complete set of functional markers for the selection of er1 powdery mildew resistance in *Pisum sativum* L. Mol Breed 31:247–253

Pearce SR, Knox M, Ellis THN, Flavell AJ, Kumar A (2000) Pea Ty1-copia group retrotransposons: transpositional activity and use as markers to study genetic diversity in *Pisum*. Mol Gen Genet 263:898–907

Peoples MB, Bell MJ, Bushby HVA (1992) Effect of rotation and inoculation with Bradyrhizobium on nitrogen fixation and yield of peanut (*Arachis hypogaea* L., cv. Virginia Bunch). Aust J Agri Res 43(3):595–607

Peoples MB, Ladha JK, Herridge DF (1995) Enhancing legume N_2 fixation through plant and soil management. Plant Soil 174:83–101

Pereira G, Marques C, Ribeiro R, Formiga S, Dâmaso M, Sousa T, Farinhó M, Leitão JM (2010) Identification of DNA markers linked to an induced mutated gene conferring resistance to powdery mildew in pea (*Pisum sativum* L.). Euphytica 171:327–335

Pérez-de-Luque A, Jorrín J, Cubero JI, Rubiales D (2005) Orobanche crenata resistance and avoidance in pea (*Pisum* spp.) operate at different developmental stages of the parasite. Weed Res 45(5):379–387

Perez-Perez JG, Robles JM, Tovar JC, Botia P (2009) Response to drought and salt stress of lemon 'Fino 49' under field conditions: water relations, osmotic adjustment and gas exchange. Sci Hort 122:83–90

Pesho GR, Muehlbauer FJ, Harberts WH (1977) Resistance of pea introductions to the pea weevil. J Econ Entomol 70(1):30–33

Peterson GA, Halvorson AD, Havlin JL, Jones OR, Lyon DJ, Tanaka DL (1998) Reduced tillage and increasing cropping intensity in the Great Plains conserves soil C. Soil Till Res 47:207–218

Petrović G, Jovičić D, Nikolić Z, Tamindžić G, Ignjatov M, Milošević D, Milošević B (2016) Comparative study of drought and salt stress effects on germination and seedling growth of pea. Genetika 48(1):373–38

Piha MI, Munns DN (1987) Sensitivity of the common bean (*Phaseolus vulgaris* L.) symbiosis to high soil temperature. Plant Soil 98:183–194

Pilet-Nayel ML, Muehlbauer FJ, McGee RJ, Kraft JM, Baranger A, Coyne CJ (2002) Quantitative trait loci for partial resistance to Aphanomyces root rot in pea. Theor Appl Genet 106:28–39

Pilet-Nayel ML, Muehlbauer FJ, McGee RJ, Kraft JM, Baranger A, Coyne CJ (2005) Consistent quantitative trait loci in pea for partial resistance to *Aphanomyces euteiches* isolates from the United States and France. Phytopathology 95:1287–1293

Poland J (2018) Breeding-assisted genomics. Curr Opin Plant Biol 24:119–24

Porter LD, Hoheisel G, Coffman VA (2009) Resistance of peas to *Sclerotinia sclerotiorum* in the *Pisum* core collection. Plant Pathol 58(1):52–60

Porter L (2012) Selection of pea genotypes with partial resistance to *Sclerotinia sclerotiorum* across a wide range of temperatures and periods of high relative humidity. Euphytica 186(3):671–678

Praça-Fontes MM, Carvalho CR, Clarindo WR (2014) Karyotype revised of *Pisum sativum* using chromosomal DNA amount. Plant Syst Evol 300:1621–1626

Prioul S, Frankewitz A, Deniot G, Morin G, Baranger A (2004) Mapping of quantitative trait loci for partial resistance to *Mycosphaerella pinodes* in pea (*Pisum sativum* L.), at the seedling and adult plant stages. Theor Appl Genet 108:1322–1334

Prioul-Gervais S, Deniot G, Receveur EM, Frankewitz A, Fourmann M, Rameau C, Pilet-Nayel ML, Baranger A (2007) Candidate genes for quantitative resistance to Mycosphaerella pinodes in pea (*Pisum sativum* L.). Theor Appl Genet 114:971–984

Provvidenti R, Muehlbauer FJ (1990) Evidence of a cluster of linked genes for resistance to pea seedborne mosaic virus and clover yellow vein virus on chromosome 6. Pisum Newsl 22:43–45

Provvidenti R, Hampton RO (1991) Chromosomal distribution of genes for resistance to seven potyviruses in *Pisum sativum*. Pisum Genet 23:26–28

Rafalski JA (2010) Association genetics in crop improvement. Curr Opin Plant Biol 13:174–180

Rai R, Singh AK, Singh BD, Joshi AK, Chand R, Srivastava CP (2011) Molecular mapping for resistance to pea rust caused by *Uromyces fabae* (Pers.) de-Bary. Theor Appl Genet 123:803–813

Rai R, Singh AK, Chand R, Srivastava CP, Joshi AK, Singh BD (2016) Genomic regions controlling components of resistance for pea rust caused by *Uromyces fabae* (Pers.) de-Bary. J Plant Biochem Biotechnol 25(2):133–141

Rani S, Kumar P, Kumar A, Sewhag M (2016) Effect of biofertilizers on nodulation, nutrient uptake, yield and energy use efficiency of field pea (*Pisum sativum* L.). J Agrometeorol 18(2):330–332

Redecker D, von Berswordt-Wallrabe P, Beck DP, Werner D (1997) Influence of inoculation with arbuscular mycorrhizal fungi on stable isotopes of nitrogen in *Phaseolus vulgaris*. Biol Fertil Soils 24:344–346

Reid J, Ross J (2011) Mendel's genes: toward a full molecular characterization. Genetics 189(1):3–10

Ribchenko ZI, Palladina TA (2012) The action of adaptogenic preparations on vacuolar proto n pumps activity in corn root cells under salt stress conditions. Ukr Biochem J 84(3):88–93

Rice WA, Olsen PE (1988) Root-temperature effects on competition for nodule occupancy between two Rhizobium meliloti strains. Biol Fertil Soils 6:137–140

Rico D, Martín-Diana AB, Barat JM, Barry-Ryan C (2007) Extending and measuring the quality of fresh-cut fruit and vegetables: a review. Trends Food Sci Technol 18(7):0–386

Rodda MS, Kant P, Lindbeck KD, Gnanasambandam A, Hollaway GJ (2015) A high-throughput glasshouse based screening method to evaluate bacterial blight resistance in field pea (*Pisum sativum*). Australas Plant Pathol 44(5):515–526

Roughley RJ (1970) The influence of root temperature, Rhizobium strain and host selection on the structure and nitrogen-fixing efficiency of the root nodules of Trifolium subterraneum. Ann Bot 34(3):631–646

Rozov SM, Kosterin Borisov AY, Tsyganov V (1999) The history of the pea gene map: last revolutions and the new symbiotic genes. Pisum Genet 31:55–56

Rubiales D, Fondevilla S, Chen W, Gentzbittel L, Higgins TJV, Castillejo MA, Singh KB, Rispail N (2015) Achievements and challenges in legume breeding for pest and disease resistance. Crit Rev Plant Sci 34:195–236

Rubiales D, Moreno MT, Sillero JC (2005) Search for resistance to crenate broomrape (*Orobanche crenata* Forsk.) in pea germplasm. Genet Resour Crop Evol 52(7):853–861

Rubiales D, Fernández-Aparicio M, Pérez-de-Luque A, Castillejo MA, Prats E, Sillero JC, Rispaila N, Fondevilla S (2009) Breeding approaches for crenate broomrape (*Orobanche crenata* Forsk.) management in pea (*Pisum sativum* L.). Pest Manage Sci 65(5):553–559

Ruisi P, Giambalvo D, Di Miceli G, Frenda AS, Saia S, Amato G (2012) Tillage effects on yield and nitrogen fixation of legumes in mediterranean conditions. Agron J 104:1459–1466

Saha U, Vann RA, Chris Reberg-Horton S, Castillo MS, Mirsky SB, McGee RJ, Sonon L (2018) Near-infrared spectroscopic models for analysis of winter pea (*Pisum sativum* L.) quality constituents. J Sci Food Agri 98(11):4253–4267

Sahrawat KL, Abekoe MK, Diatta S (2001) Application of inorganic phosphorus fertilizer. Soil Science Society of America and American Society of Agronomy, Madison, WI53711, USA, Soil Science Society of America Special Publication no. 58

Salon C, Munier-Jolain N (2010) Carbon and nitrogen fluxes within the plant. In: Munier-Jolain N, Biarnès V, Chaillet I, Lecoeur J, Jeuffroy MH (eds) Physiology of the Pea Crop. CRC Press, Boca Raton, FL, pp 88–95

Sanderman J, Farquharson R, Baldock J (2010) Soil carbon sequestration potential: a review for Australian agriculture. CSIRO

Santalla Ferradás M, Amurrio JM, Ron Pedreira AMD (2001) Food and feed potential breeding value of green, dry and vegetable pea germplasm. Can J Plant Sci 81(4):601–610

Sánchez FJ, Manzanares M, de Andres EF, Tenorio JL, Ayerbe L (1998) Turgor maintenance, osmotic adjustment and soluble sugar and proline accumulation in 49 pea cultivars in response to water stress. Field Crop Res 59(3):225–235

Sanginga N, Wirkom LE, Okogun A, Akobundu IO, Carsky RJ, Tian G (1996) Nodulation and estimation of symbiotic nitrogen fixation by herbaceous and shrub legumes in Guinea savanna in Nigeria. Biol Fertil Soils 23:441–448

Santo T, Rashkova M, Alabaça C, Leitão J (2013) The ENU-induced powdery mildew resistant mutant pea (*Pisum sativum* L.) lines S (er1mut1) and F (er1mut2) harbour early stop codons in the PsMLO1 gene. Mol Breed 32(3):723–727

Sardana S, Mahajan RK, Gautam NK, Ram B (2007) Genetic variability in pea (*Pisum sativum* L.) germplasm for utilization. SABRAO J Breed Genet 39:31–41

Sarıkamış G, Yanmaz R, Ermiş S, Bakir M, Yüksel C (2010) Genetic characterization of pea (*Pisum sativum*) germplasm from Turkey using morphological and SSR markers. Genet Mol Res 9(1):591–600

Sarikamis G, Yanmaz R, Ermis S, Bakir M, Yuksel C (2010) Genetic characterization of pea (*Pisum sativum*) germplasm from Turkey using morphological and SSR markers. Genet Mol Res 9:591–600

Sato Y, Morita R, Nishimura M, Yamaguchi H, Kusaba M (2007) Mendel's green cotyledon gene encodes a positive regulator of the chlorophyll-degrading pathway. Proc Natl Acad Sci USA 104:14169–14174

Scheben A, Edwards D (2018) Towards a more predictable plant breeding pipeline with CRISPR/Cas-induced allelic series to optimize quantitative and qualitative traits. Curr Opin Plant Biol 45:218–225

Schroeder HE, Gollasch S, Moore A, Tabe LM, Craig S, Hardie DC, Chrispeels MJ, Spencer D, Higgins TJ (1995) Bean α-amylase inhibitor confers resistance to the pea weevil (*Bruchus pisorum*) in transgenic peas (*Pisum sativum* L.). Plant Physiol 107(4):1233–1239

Sellstedt A, Stahl L. Mattsson U, Josson K, Hogborg P (1993) Can the ^{15}N dilution technique be used to study N_2 fixation in tropical tree symbioses as affected by water deficit? J Exp Bot 44:1749–1755

Semagn K, rnstad B, Ndjiondjop M (2006) Principles, requirements and prospects of genetic mapping in plants. Afr J Biotechnol 5:2569–2587

Seyyedi M, Timko MP, Sundqvist C (1999) Protochlorophyllide, NADPH-protochlorophyllide oxidoreductase, and chlorophyll formation in the lip1mutant of pea. Physiol Plant 106:344–354

Shabab M, Arnold MF, Penterman J, Wommack AJ, Bocker HT, Price PA, Griffitts JS, Nolan EM, Walker GC (2016) Disulfide cross-linking influences symbiotic activities of nodule peptide NCR247. Proc Natl Acad Sci USA 113(36):10157–10162

Shade RE, Schroeder HE, Pueyo JJ, Tabe LM, Murdock LL, Higgins TJ, Chrispeels MJ (1994) Transgenic pea seeds expressing the α-amylase inhibitor of the common bean are resistant to bruchid beetles. Nat Biotechnol 12(8):793–796

Shahid MA, Pervez MA, Balal RM, Abbas T, Ayyub CM, Mattson NS, Riaz A, Iqbal Z (2012) Screening of pea (*Pisum sativum* L.) genotypes for salt tolerance based on early growth stage attributes and leaf inorganic osmolytes. Aust J Crop Sci 6(9):1324

Shafiq S, Mather DE, Ahmad M, Paull J (2012) Variation in tolerance to radiant frost at reproductive stages in field pea germplasm. Euphytica 186:831–845. https://doi.org/10.1007/s10681-012-0625-0

Shanmugasundaram (1991) Mungbean varietal improvement. Asian Vegetable Research and Development Center, Shanhua, Taiwan, pp 30–77

Sharma V, Kumar S (2012) Roles of stipules include determination of flowering time and pod harvest index in garden pea grain Legume *Pisum sativum*. Natl Acad Sci Lett 35(5):449–456

Sharma A, Rathour R, Plaha P, Katoch V, Khalsa G, S, Patial V, Singh Y, Pathania NK (2010) Induction of Fusarium wilt (*Fusarium oxysporum* f. sp. *pisi*) resistance in garden pea using induced mutagenesis and in vitro selection techniques. Euphytica 173(3):345–356

Shen S, Jing Y (2003) Research and prospect of biological nitrogen fixation in China. Chin Sci Bull 48(6):532–540

Shereena J, Salim N (2006) Chilling tolerance in *Pisum sativum* L. seeds: an ecological adaptation. Asian J Plant Sci 5:1047–1050

Sherrod LA, Peterson GA, Westfall DG, Ahuja LR (2003) Cropping intensity enhances soil organic carbon and nitrogen in a no-till agroecosystem. Soil Sci Soc Amer J 67:1533–1543

Siddique KHM, Regan KL, Tennant D, Thomson BD (2001) Water use and water use efficiency of cool season grain legumes in low rainfall Mediterranean-type environments. Eur J Agron 15:267–280

Siddique KHM, Erskine W, Hobson K, Knights EJ, Leonforte A, Khan TN, Paull JG, Redden R, Materne M (2013) Cool-season grain legume improvement in Australia—Use of genetic resources. Crop Pasture Sci 64:347–360

Sindhu A, Ramsay L, Sanderson L-A, Stonehouse R, Li R, Condie J, Shunmugam ASK, Liu Y, Jha AB, Diapari M, Burstin J, Aubert G, Tar'an B, Bett KE, Warkentin TD, Sharpe AG (2014) Gene-based SNP discovery and genetic mapping in pea. Theor Appl Genet 127:2225–2241

Singh AK, Rai R, Singh BD, Chand R, Srivastava CP (2015) Validation of SSR markers associated with rust (Uromyces fabae) resistance in pea (*Pisum sativum* L.). Physiol Mol Biol Plants 21(2):243–247

Singh DK, Singh AK, Singh SK, Singh M, Srivastava O (2015) Effect of balanced nutrition on yield and nutrient uptake of pea (*Pisum stivum* L.) under Indo-gangetic plains of India. The Bioscan 10(3):1245–1249

Skiba B, Ford R, Pang ECK (2004) Genetics of resistance to *Mycosphaerella pinodes* in *Lathyrus sativus*. Aust J Agri Res 55(9):953–960

Slattery JF, Pearce DJ, Slattery WJ (2004) Effects of resident Rhizobial communities and soil type on the effective nodulation of pulse legumes. Soil Biol Biochem 36(8):1339–1346

Smirnova O, Stepanenko I, Shumny V (2012) Mechanism of action and activity regulation of COP1, a constitutive repressor of photomorphogenesis. Russ J Plant Physiol 59:155–166

Smith RI, Harvey FJ, Cannell MGR (1990) Pattern of tea shoot growth. Exp Agr 26(2):197–208

Smith P, Martino D, Cai Z, Gwary D, Janzen H, Kumar P, McCarl B, Ogle S, O'Mara F, Rice C, Scholes B, Sirotenko O (2007) Agriculture. In: Metz B et al. (eds) Climate change 2007: mitigation. Contribution of Working Group III to the Fourth Assessment Report of the Intergovernmental Panel on Climate Change. Cambridge University Press, Cambridge, United Kingdom and New York, NY, USA, pp 498–540

Smith PH, Foster EM, Boyd LA, Brown JKM (1996) The early development of *Erysiphe pisi* on *Pisum sativum* L. Plant Pathol 45:302–309

Smýkal P, Aubert G, Burstin J, Coyne CJ, Ellis NTH, Flavell AJ, Ford R, Hýbl M, Macas J, Neumann P, McPhee KE, Redden RJ, Rubiales D, Weller JL, Warkentin TD (2012) Pea (*Pisum sativum* L.) in the genomic era. Agronomy 2:74–115

Smýkal P, Kenicer G, Flavell A, J, Corander J, Kosterin O, Redden RJ, Ford R, Coyne CJ, Maxted N, Ambrose MJ, Ellis NTH (2011) Phylogeny phylogeography and genetic diversity of the *Pisum* genus. Plant Genet Resour 9(1):4–18

Smýkal P, Coyne CJ, Ambrose MJ, Maxted N, Schaefer H, Blair MW, Berger J, Greene SL, Nelson MN, Besharat N, Vymyslický T, Toker C, Saxena RK, Roorkiwal M, Pandey MK, Hu J, Li YH, Wang LX, Guo Y, Qiu LJ, Redden RJ, Varshney RK (2015) Legume crops phylogeny and genetic diversity for science and breeding. Crit Rev Plant Sci 34:43–104

Smýkal P, Horáèek J, Dostálová R, Hýbl M (2008a) Variety discrimination in pea (*Pisum sativum* L.) by molecular, biochemical and morphological markers. J Appl Genet 49(2):155–166

Smýkal P, Hybl M, Corander J, Jarkovsky J, Flavell AJ, Griga M (2008b) Genetic diversity and population structure of pea (*Pisum sativum* L.) varieties derived from combined retrotransposon, microsatellite and morphological marker analysis. Theor Appl Genet 117:413–424

Smýkal P, Konečná E (2014) Advances in pea genomics. In: Gupta S, Nadarajan N, Gupta DS (eds) Legumes in the Omic Era. Springer, New York, NY, pp 301–338

Smýkal P, Šafářová D, Navrátil M, Dostalová R (2010) Marker assisted pea breeding: eIF4E allele specific markers to pea seed-borne mosaic virus (PSbMV) resistance. Mol Breed 26(3):425–438

Sogut T, Oztturk F, Temiz G, Toncer O, Zaimoglu B (2013) The effects of rhizobial inoculation and nitrogen fertilizer application on nodulation, yield and yield components of groundnut (*Arachis hypogaea* L.). Glob J Adv Pure Appl Sci 1:156–165

Srinivasan A, Johansen C, Saxena NP (1998) Cold tolerance during early reproductive growth of chickpea (*Cicer arietinum* L.): characterisation of stress and genetic variation in pod set. Field Crops Res 57:181–193

Souci SW, Fachman W, Kraut M (2001) Food composition and nutrition tables. MEDPHARM Scientific Publishers, Stuttgart, Germany, p 1091

Starker CG, Parra-Colmenares AL, Smith L, Mitra RM, Long SR (2006) Nitrogen fixation mutants of Medicago truncatula fail to support plant and bacterial symbiotic gene expression. Plant Physiol 140(2):671–80

Steppuhn H, Volkmar KM, Miller PR (2001) Comparing canola, field pea, dry bean, and durum wheat crops grown in saline media. Crop Sci 41(6):1827–1833

Stoddard FL, Balko C, Erskine W, Khan HR, Link W, Sarker A (2006) Screening techniques and sources of resistance to abiotic stresses in cool-season food legumes. Euphytica 147:167–186

Stokkermans TJ, Peters NK (1994) Bradyrhizobium elkanii lipo-oligosaccharide signals induce complete nodule structures on *Glycine soja* Siebold et Zucc. Planta 193(3):413–433

Stoop-Myer C, Torii KU, McNellis TW, Coleman JE, Deng XW (1999) The N-terminal fragment of Arabidopsis photomorphogenic repressor COP1 maintains partial function and acts in a concentration-dependent manner. Plant J 20:713–717

Sudheesh S, Lombardi M, Leonforte A, Cogan NOI, Materne M, Forster JW, Kaur S (2015a) Consensus genetic map construction for field pea (*Pisum sativum* L.), trait dissection of biotic and abiotic stress tolerance and development of a diagnostic marker for the *er1* Powdery mildew resistance gene. Plant Mol Biol Rep 33:1391–1403

Sudheesh S, Rodda M, Kennedy P, Verma P, Leonforte A, Cogan NOI, Materne M, Forster JW, Kaur S (2015b) Construction of an integrated linkage map and trait dissection for bacterial blight resistance in field pea (*Pisum sativum* L.). Mol Breed 35:185

Sullivan JA, Gray JC (2000) The pea light-independent photomorphogenesis1 mutant results from partial duplication of COP1 generating an internal promoter and producing two distinct transcripts. Plant Cell 12:1927–1937

Sun HY, Liu CM, Zhang XY, Shen YJ, Zhang YQ (2006) Effects of irrigation on water balance, yield and WUE of winter wheat in the North China Plain. Agri Water Manag 85:211–218

Sun X, Yang T, Hao J, Zhang X, Ford R, Jiang J, Wang F, Guan J, Zong X (2014) SSR genetic linkage map construction of pea (*Pisum sativum* L.) based on Chinese native varieties. Crop J 2:170–174

Sun S, Deng D, Wang Z, Duan C, Wu X, Wang X, Zong X, Zhu Z (2016a) A novel er1 allele and the development and validation of its functional marker for breeding pea (*Pisum sativum* L.) resistance to powdery mildew. Theor Appl Genet 129(5):909–919

Sun S, Fu H, Wang Z, Duan C, Zong X, Zhu Z (2016b) Discovery of a novel er1 allele conferring powdery mildew resistance in Chinese pea (*Pisum sativum* L.) landraces. PLoS ONE 11(1):e0147624

Tanksley SD (1993) Mapping polygenes. Annu Rev Genet 27:205–233

Tanksley SD, McCouch SR (1997) Seed banks and molecular maps: unlocking genetic potential from the wild. Science 277:1063–1066

Tar'an B, Warkentin T, Somers DJ, Miranda D, Vandenberg A, Blade S, Bing D (2004) Identification of quantitative trait loci for grain yield, seed protein concentration and maturity in field pea (*Pisum sativum* L.). Euphytica 136:297–306

Tar'an B, Warkentin T, Somers DJ, Miranda D, Vandenberg A, Blade S, Woods S, Bing D, Xue A, DeKoeyer D, Penner G (2003) Quantitative trait loci for lodging resistance plant height and partial resistance to mycosphaerella blight in field pea (*Pisum sativum* L.). Theor Appl Genet 107(8):1482–1491

Taylor JD, Bevan JR, Crute IR, Reader SL (1989) Genetic relationship between races of *Pseudomonas syringae* pv. *pisi* and cultivars of *Pisum sativum*. Plant Pathol 38:364–375

Tayeh N, Bahrman N, Devaux R, Bluteau A, Prosperi JM, Delbreil B, Lejeune-Hénaut I (2013) A high-density genetic map of the *Medicago truncatula* major freezing tolerance QTL on chromosome 6 reveals colinearity with a QTL related to freezing damage on *Pisum sativum* linkage group VI. Mol Breed 32:279–289

Tayeh N, Aubert G, Pilet-Nayel ML, Lejeune-Hénaut I, Warkentin TD, Burstin J (2015a) Genomic tools in pea breeding programs: status and perspectives. Front Plant Sci 6:1037

Tayeh N, Aluome C, Falque M, Jacquin F, Klein A, Chauveau A, Berard A, Houtin H, Rond C, Kreplak J, Boucherot K, Martin C, Baranger A, Pilet-Nayel M-L, Warkentin TD, Brunel D, Marget P, Le Paslier M-C, Aubert G, Burstin J (2015b) Development of two major resources for pea genomics: the GenoPea 13.2 K SNP array and a high-density, high-resolution consensus genetic map. Plant J 84:1257–1273

Tayeh N, Klein A, Le Paslier M-C, Jacquin F, Houtin H, Rond C, Chabert-Martinello M, Magnin-Robert J-B, Marget P, Aubert G, Burstin J (2015c) Genomic prediction in pea: effect of marker density and training population size and composition on prediction accuracy. Front Plant Sci 6:941–941

Teshome A, Bryngelsson T, Dagne K, Geleta M (2015a) Assessment of genetic diversity in Ethiopian field pea (*Pisum sativum* L.) accessions with newly developed EST-SSR markers. BMC Genetics 16:102

Teshome A, Mendesil E, Geleta M, Andargie D, Anderson P, Rämert B, Seyoum E, Hillbur Y, Dagne K, Bryngelsson T (2015b) Screening the primary gene pool of field pea (*Pisum sativum* L. subsp. *sativum*) in Ethiopia for resistance against pea weevil (*Bruchus pisorum* L). Genet Resour Crop Evol 62(4):525–538

Tester M, Langridge P (2010) Breeding technologies to increase crop production in a changing world. Science 327:818–822

Thomas DW, Smýkal P, Coyne CJ, Weeden N, Domoney C, Bing DJ, Leonforte A, Zong XX, Dixit GP, Boros L, McPhee KE, McGee RJ, Burstin J, Ellis THN (2015) Chapter 2. Pea. In: De Ron AM (ed) Grain legumes: handbook for plant breeding. Springer, New York, Heidelberg, Dordrecht, London, pp 37–84

Tivoli B, Banniza S (2007) Comparison of the epidemiology of ascochyta blights on grain legumes. Eur J Plant Pathol 119:59–76

Tivoli B, Baranger A, Avila CM, Banniza S, Barbetti M, Chen W, Davidson J, Lindeck K, Kharrat M, Rubiales D, Sadiki M, Sillero JC, Sweetingham M, Muehlbauer FJ (2006) Screening techniques and sources of resistance to foliar diseases caused by major neurotrophic fungi in grain legumes. Euphytica 147(1–2):223–253

Tilman D, Balzer C, Hill J, Befort BL (2011) Global food demand and the sustainable intensification of agriculture. Proc Natl Acad Sci USA 108:20260–20264

Timmerman-Vaughan GM, Frew TJ, Russell AC, Khan T, Butler R, Gilpin M, Murray S, Falloon K (2002) QTL mapping of partial resistance to field epidemics of ascochyta blight of pea. Crop Sci 42:2100–2111

Timmerman-Vaughan GM, Mills A, Whitfield C, Frew T, Butler R, Murray S, Lakeman M, McCallum J, Russell A, Wilson D (2005) Linkage mapping of QTL for seed yield, yield components, and developmental traits in pea. Crop Sci 45:1336–1344

Timmerman-Vaughan GM, Moya L, Frew TJ, Murray SR, Crowhurst R (2016) Ascochyta blight disease of pea (*Pisum sativum* L.): defence-related candidate genes associated with QTL regions and identification of epistatic QTL. Theor Appl Genet 129:879–896

TimmermanVaughan GM, McCallum JA, Frew TJ, Weeden NF, Russell AC (1996) Linkage mapping of quantitative trait loci controlling seed weight in pea (*Pisum sativum* L). Theor Appl Genet 93:431–439

Timmerman-Vaughan GM, Pither-Joyce MD, Cooper PA, Russell AC, Goulden DS, Butler R, Grant JE (2001) Partial resistance of transgenic peas to alfalfa mosaic virus under greenhouse and field conditions. Crop Sci 41(3):846–853

Timmerman-Vaughan GM, Frew TJ, Butler R, Murray S, Gilpin M, Falloon K, Johnston P, Lakeman MB, Russell A, Khan T (2004) Validation of quantitative trait loci for Ascochyta blight resistance in pea (*Pisum sativum* L.) using populations from two crosses. Theor Appl Genet 109(8):1620–1631

Tiwari KR, Penner GA, Warkentin TD (1997) Inheritance of powdery mildew resistance in pea. Can J Plant Sci 77:307–310

Toker C (2005) Preliminary screening and selection for cold tolerance in annual wild Cicer species. Genet Resour Crop Evol 52:1–5

Tran HS, You MP, Lanoiselet V, Khan TN, Barbetti MJ (2014a) First report of Phoma glomerata associated with the ascochyta blight complex on field pea (*Pisum sativum*) in Australia. Plant Dis 98(3):437

Tran HS, You MP, Khan TN, Pritchard I, Barbetti MJ (2014b) Resistance in field pea (Pisum sativum) to the black spot disease complex in Western Australia. Eur J Plant Pathol 140(3):597–605

Trapnell C, Williams BA, Pertea G, Mortazavi A, Kwan G, van Baren MJ, Salzberg SL, Wold BJ, Pachter L (2010) Transcript assembly and quantification by RNA-Seq reveals unannotated transcripts and isoform switching during cell differentiation. Nat Biotechnol 28:511–517

Tu JC (1981) Effect of salinity on Rhizobium-root hair interaction, nodulation and growth of soybean. Can J Plant Sci 61(2):231–239

Upadhyaya HD, Gowda CLL, Buhariwalla HK, Crouch JH (2006) Efficient use of crop germplasm resources: identifying useful germplasm for crop improvement through core and mini-core collections and molecular marker approaches. Plant Genet Resour Characteriz Utiliz 4:25–35

Upadhyay V, Kushwaha KPS, Pandey P (2017) Molecular screening of pea germplasm for rust disease resistance using SSR Marker. J Pure Appl Microbiol 11(1):343–348

Urbatzka P, Graß R, Haase T, Schüler C, Trautz D, Heß J (2011) Grain yield and quality characteristics of different genotypes of winter pea in comparison to spring pea for organic farming in pure and mixed stands. Org Agri 1(2011):187–202

Urbatzka P, Graß R, Haase T, Schüler C, Heß J (2012) Influence of different sowing dates of winter pea genotypes on winter hardiness and productivity as either winter catch crop or seed legume. Eur J Agron 40:112–119

Ussiri DAN, Lal R, Jarecki MK (2009) Nitrous oxide and methane emissions from long-term tillage under a continuous corn cropping system in ohio. Soil Till Res 104(2):247–255

Vanhala T, Normann KR, Lundström M, Weller JL, Leino MW, Hagenblad J (2016) Flowering time adaption in Swedish landrace pea (*Pisum sativum* L.). BMC Genetics 17:117

Vann RA, Reberg-Horton SC, Castillo MS, Mirsky SB, McGee RJ (2018) Winter pea cultivar/breeding line screening for grain crop potential in the Southeastern United States. Agron J 110:1217–1225

Vargas MAT, Mendes IC, Hungria M (2000) Response of field-grown bean (*Phaseolus vulgaris* L) to Rhizobium inoculation and nitrogen fertilization in two Cerrados soils. Biol Fertil Soils 32(3):228–233

Varshney RK, Roorkiwal M, Sorrells ME (eds) Genomic selection for crop improvement: new molecular breeding strategies for crop improvement. Springer International, Cham, Swirzerland, pp 131–147

Vaz Patto MC, Rubiales D (2009) Identification and characterization of partial resistance to rust in a germplasm collection of *Lathyrus sativus* L. Plant Breed 128(5):495–500

Via S (1993) Adaptive phenotypic plasticity: target or by-product of selection in a variable environment? Am Nat 142(2):352–365

Via S, Conner J (1995) Evolution in heterogeneous environments: genetic variability within and across different grains in *Tribolium castaneum*. Heredity 74(1):80–90

Vijayalakshmi S, Yadav K, Kushwaha C, Sarode SB, Srivastava CP, Chand R, Singh BD (2005) Identification of RAPD markers linked to the rust (*Uromyces fabae*) resistance gene in pea (*Pisum sativum*). Euphytica 144:265–274

Vincent JM (1982) Nitrogen fixation in legumes. Academic Press, New York, pp 1–284

Visscher PM, Haley CS, Thompson R (1996) Marker-assisted introgression in backcross breeding programs. Genetics 144:1923–1932

Vito MD, Perrino P (1978) Reaction of *Pisum* spp. to the attacks of Heterodera goettingiana. Nematol Mediterr 6(1):113–118

Vocanson A, Jeuffroy MH (2008) Agronomic performance of different pea cultivars under various sowing periods and contrasting soil structures. Agron J 100(3):748–759

Voican V, Scurtu I, Voican AV, Racaru V (1995) Features of seed germination and growth of young pea plants in conditions of moisture and thermal stress. II. Effect of frost on the plants. An Inst Cercetari pentru Legumicult Floricult Vidra 13:387–396 (English summary)

Vos P, Hogers R, Bleeker M, Reijans M, van de Lee T, Hornes M, Frijters A, Pot J, Peleman J, Kuiper M (1995) AFLP: a new technique for DNA fingerprinting. Nucl Acids Res 23:4407–4414

Walsh KB (1995) Physiology of the legume nodule and its response to stress. Soil Biol Biochem 27(4):637–655

Wang J, Sainju UM (2014) Soil carbon and nitrogen fractions and crop yields affected by residue placement and crop types. PLoS ONE 9(8):e105039

Wang X, Zhu Z, Duan C, Zong X (2007) Identification and control technology of disease and pest on faba bean and pea. China Agricultural Science and Technology Press, Beijing, pp 54–101

Weeden NF, Marx GA (1987) Further genetic analysis and linkage relationships of isozyme loci in the pea: Confirmation of the diploid nature of the genome. J Hered 78:153–159

Weeden NF, Porter L (2008) The genetic basis of Fusarium root rot tolerance in the 'Afghanistan' pea. Pisum Genet 39:35–36

Weeden NF (2018) Domestication of pea (*Pisum sativum* L.): the case of the Abyssinicum pea, Front Plant Sci 9:515

Weeden NF, Ellis THN, Timmerman-Vaughan GM, Swiêcicki WK, Rozov SM, Berdnikov VA (1998) A consensus linkage map for *Pisum staivum*. Pisum Genet 30:1–4

Weeden NF, Moffet M (2002) Identification of genes affecting root mass and root/shoot ratio in a JI1794 x "Slow" RILL population. Pisum Genet 34:28–31

Weeden NF, Swiêcicki WK, Ambrose M, Timmerman GM (1993) Linkage groups of pea. Pisum Genet 25:4

Weightman RM (2005) The rise and fall of grain legumes and their predicted yields in UK agriculture. In: Wiseman J, Sylvester-Bradley R (eds) Yields of farmed species: constraints and opportunities in the 21st century. Nottingham University Press, Nottingham, Chap 12, pp 261–288

Weisany W, Raei Y, Allahverdipoor KH (2013) Role of some mineral nutrients in biological nitrogen fixation. Bull Environ Pharmacol Life Sci 2(4):77–84

Weiser CJ, Walner SJ, Waddell JW (1990) Cell wall and extensin mRNA changes during cold acclimation of pea seedlings. Plant Physiol 93:1021–1026

Welbaum GE, Bian D, Hill DR, Grayson RL, Gunatilaka MK (1997) Freezing tolerance, protein composition and abscisic acid localization and content of pea epicotyl, shoot, and root tissue in response to temperature and water stress. J Exp Bot 48:643–654

Weller JL, Hecht V, Liew LC, Sussmilch FC, Wenden B, Knowles CL, Vander Schoor JK (2009) Update on the genetic control of flowering in garden pea. J Exp Bot 60:2493–24999

Weller JL, Liew LC, Hecht VF, Rajandran V, Laurie RE, Ridge S, Wenden B, Vander Schoor JK, Jaminon O, Blassiau C et al (2012) A conserved molecular basis for photoperiod adaptation in two temperate legumes. Proc Natl Acad Sci USA 109:21158–21163

Wicker E, Moussart A, Duparque M, Rouxel F (2003) Further contributions to the development of a differential set of pea cultivars (*Pisum sativum*) to investigate the virulence of isolates of Aphanomyces euteiches. Eur J Plant Pathol 109(1):47–60

Witte CP (2011) Urea metabolism in plants. Plant Sci 180:431–438

Worrall VS, Roughley RJ (1976) The effect of moisture stress on infection of Trifolium subterraneum L. by Rhizobium trifolii Dang. J Exp Bot 27(6):1233–1241

Wu L, Chang KF, Conner RL, Strelkov S, Fredua-Agyeman R, Hwang SF, Feindel D (2018a) *Aphanomyces euteiches*: a threat to Canadian field pea production. Engineering 4:542–551

Wu L, Chang KF, Hwang SF, Conner R, Fredua-Agyeman R, Feindel D, Strelkov SE (2018) Evaluation of host resistance and fungicide application as tools for the management of root rot of field pea caused by *Aphanomyces euteiches*. Crop J. https://doi.org/10.1016/j.cj.2018.07.005

Wu XB, Li NN, Hao JJ, Hu JG, Zhang XY, Blair MW (2017) Genetic diversity of Chinese and global pea (*Pisum sativum* L.) collections. Crop Sci 57:1–11

Xiao GJ, Zhang Q, Yao, YB Yang, SM, Wang, RY, Xiong, YC, Sun ZJ (2007) Effects of temperature increase on water use and crop yields in a pea-spring wheat-potato rotation. Agri Water Manage 91:86–91

Xu YB, Li ZK, Thomson MJ (2012) Molecular breeding in plants: moving into the mainstream. Mol Breed 29(4):831–832

Yandell M, Ence D (2012) A beginner's guide to eukaryotic genome annotation. Nat Rev Genet 13:329

Yang H, Yang J, Yin K (2009) Study on effects of barley *Vicia faba* intercropping on control the *Liriomyza sativae*. China Seed Ind S1:66–67

Yang T, Fang L, Zhang X, Hu J, Bao S, Hao J, Li L, He Y, Jiang J, Wang F, Tian S, Zong X (2015) High-throughput development of SSR markers from pea (*Pisum sativum* L.) based on next generation sequencing of a purified Chinese commercial variety. PLoS ONE 10:e0139775

Yan H, Cao M, Liu J, Tao B (2007) Potential and sustainability for carbon sequestration with improved soil management in agricultural soils of china. Agri Ecosyst Environ 121(4):325–335

Yeboah S, Zhang R, Cai L, Song M, Li LL, Xie J (2016) Greenhouse gas emissions in a spring wheat–field pea sequence under different tillage practices in semi-arid northwest china. Nutr Cycl Agroecosyst 106(1):77–91

Zabran HH (1999) Rhizobium-legume symbiosis and nitrogen fixation under severe condition and in arid climate. Microbiol Mol Biol Rev 63:968–989

Zahran HH (1999) Rhizobium-legume symbiosis and nitrogen fixation under severe conditions and in arid climate. Microbiol Mol Biol Rev 63(4):968–989

Zahir ZA, Munir A, Asghar HN, Shaharoona B, Arshad M (2008) Effectiveness of rhizobacteria containing ACC deaminase for growth promotion of peas (*Pisum sativum*) under drought conditions. J Microbiol Biotechnol 18(5):958–963

Zdeněk P, Miroslav G (2000) Utilization of isozyme polymorphism for cultivar identification of 45 commercial peas (*Pisum sativum* L.). Euphytica 113(3):249–256

Zentner RP, Wall DD, Smith DG, Young DL, Miller PR, Campbel CA, Lafond GP, Brandt SA, Johnston AM (2002) Economics of crop diversification opportunities for the northern Great Plains. Agron J 94:216–230

Zhang R, Hwang SF, Gossen BD, Chang KF, Turnbull GD (2007) A quantitative analysis of resistance to Mycosphaerella blight in field pea. Crop Sci 47(1):162–167

Zhang XY, Wan SW, Hao JJ, Hu JG, Yang T, Zong XX (2016) Large-scale evaluation of pea (*Pisum sativum* L.) germplasm for cold tolerance in the field during winter in Qingdao. Crop J 4:377–383

Zhang X, Ma L, Zheng J (2017) Characteristics of genes selected by domestication and intensive breeding in crop plants. Acta Agron Sin 43:157–170

Zheng ZJ, Zong XX, Liu FY (1998) Questions and answers on the cultivation techniques of legume (in Chinese). China Agricultural Press, Beijing, pp 34–67

Zhu CS, Gore M, Buckler ES, Yu JM (2008) Status and prospects of association mapping in plants. Plant Genome 1:5–20

Zohary D, Hopf M (1973) Domestication of pulses in the Old World. Science 182:887–894

Zohary D, Hopf M (2000) Domestication of plants in the old world, 3rd edn. Oxford University Press, Oxfod, UK

Zong XX (ed) (2002) High-yielding cultivation and food processing of food legumes, 1st edn. China Agricultural Science and Technology Press, Beijing, China, pp 56–144

Zong XX, Guan JP, Wang SM, Liu QC, Redden RJ, Ford R (2008a) Genetic diversity and core collection of alien *Pisum sativum* L. germplasm. Acta Agron Sin 34(9):1518–1528

Zong XX, Guan JP, Wang SM, and Liu QC (2008b) Genetic diversity among Chinese Pea (*Pisum sativum* L.) landraces revealed by SSR markers. Acta Agron Sin 34(8):1330–1338

Zong XX, Ford R, Redden RJ, Jian-ping G, Wang SM (2009a) Identification and analysis of genetic diversity structure within *Pisum* genus based on microsatellite markers. Sci Agri Sin 42(1):36–46
Zong XX, Redden RJ, Liu QC, Wang SM, Guan JP, Liu J, Xu YH, Liu XJ, Gu J, Yan L, Ades P, Ford R (2009b) Analysis of a diverse global *Pisum* sp collection and comparison to a Chinese local *P. sativum* collection with microsatellite markers. Theor Appl Genet 118(2):193–204

Chapter 7
Genomics-Aided Breeding for Climate-Smart Traits in Faba Bean

Ahmed Sallam and Sami Ul-Allah

Abstract Faba bean (*Vicia faba* L.) is an important pulse crop, which provides useful source of protein for human and animal consumption. The faba bean cultivation area around world has been slightly decreased due to the lack of cultivars adaptable to various biotic and abiotic stresses effects of which tend to gradually increase as consequences of climate change. Breeding for improved faba bean with biotic and abiotic stress tolerance will maintain and increase the cultivation area of faba bean by producing new cultivars having high tolerance to these stresses combined with high yield. Climate-smart traits (CSTs) can be used to evaluate faba bean genotypes for stress tolerance and to select the true promising genotypes for target traits. Moreover, the advances in genetic research in faba bean should be exploited in accelerating breeding programs to genetically improve CSTs. Unfortunately, the progress of molecular breeding research is slow due to the complexity of faba bean genome and few studies, compared to those reported in other important crops (e.g., wheat, maize, etc.), have been conducted to detect quantitative trait loci (QTLs) controlling CSTs. This chapter sheds light on the recent breeding research for CSTs in faba bean. The most important QTLs controlling CSTs detected by QTL mapping and genome-wide association study (GWAS) methods and promising validated QTL have been discussed. Moreover, an overview in faba bean genome sequencing and gene annotation for candidate genes controlling CSTs has been presented.

Keywords Faba bean · Abiotic stress · Climate-smart traits · QTLs · Genomewide association study · Gene annotation

---

A. Sallam (✉)
Faculty of Agriculture, Department of Genetics, Assiut University, Assiut 71526, Egypt
e-mail: amsallam@aun.edu.eg

S. Ul-Allah
Department of Plant Breeding and Genetics, College of Agriculture, BZU Bahadur Sub-Campus, Layyah, Pakistan

C. Kole (ed.), *Genomic Designing of Climate-Smart Pulse Crops*,
https://doi.org/10.1007/978-3-319-96932-9_7

7.1 Introduction

Faba bean is one of the most important legume crops and it ranks fourth after dry beans, dry peas, and chickpeas (Toker et al. 2007). The main producers of faba bean are China, Ethiopia, Australia, France, Egypt, Morocco, Sudan, and the UK. According to FAO (2017), the global acreage of faba bean decreased from 3.7 to 2.1 million ha during the last 34 years from 1980 to 2014. Faba bean is mainly used for food due to its high protein content in the Middle East and North Africa. On the other hand, it is used for feeding animals in Europe (Link et al. 2010; Sallam 2014). In the agricultural sector, faba bean plays a vital role in maintaining soil fertility by its ability to fix nitrogen in the soil. So, it significantly contributes to the sustainability of agriculture systems (Karkanis et al. 2018).

Like other crops, various biotic and abiotic stresses affect faba bean production and productivity. The major biotic factors affecting faba bean yield are aphids, rust, chocolate spot, aphids, leaf miners, and ascochyta blight. The major abiotic factors, on the other hand, are frost damage (for winter faba bean) during both flowering and podding stages, and drought stress or waterlogging at maturity (Redden et al. 2014). The severity of effects of these stresses is expected to increase due to climate change. Breeding research can improve the target traits in faba bean; however, it takes a lot of time and efforts to achieve the goals. Advances in molecular genetics and genomics, especially DNA sequencing methods, can pave the way for improving the target traits effectively in combination with breeding research. In the breeding program of faba bean, improving grain yield, climate-smart traits (CSTs) including tolerance or resistance to various abiotic and biotic stresses, adaption to wide range of environments, plant growth, and quality of seeds are the main objectives of faba bean breeding for dealing with the serious problem of climate change. Developing and studying new climate-smart traits will help research to produce faba bean cultivars having a capability to survive if they are exposed to biotic and abiotic stresses.

Faba bean is a diploid plant with $2n = 2x = 12$ chromosomes and has one of the largest genomes among crops (~13,000 Mb), more or less similar to the hexaploid genome of wheat (Ammar et al. 2017). This could complicate the development of physical and genetic maps, as well as map-based cloning (Ellwood et al. 2008). Recently, many researchers have brought about a massive progress in detecting some important quantitative trait loci (QTLs) associated with CSTs due to the advances in DNA technology. However, more efforts should be paid in the genomics and genetics research of faba bean to improve and understand the inheritance of the important traits.

7.2 Advances in Breeding for Climate-Smart Traits (CSTs) in Faba Bean

The key point for breeding of CSTs is to have a good phenotyping, which is considered a very important method to genetically improve CSTs. Genetic data produced by DNA markers should be linked with carefully measured phenotypes to maximize the benefits of using marker-assisted selection. Phenotypes should be repeated over replications or/and locations or/and years and good heritability estimates should be obtained before using CSTs in breeding and genetics programs. In the agricultural research, biotic and abiotic stresses are main problems that limit yield and quality of crops. Because yield is controlled by a very complex network of genes, therefore, improvement of yield-associated traits might be more fruitful but for that purpose complete knowledge of correction among yield-associated traits is very important. Producing cultivars having more tolerance to biotic and abiotic stresses should be combined with breeding for high yield. Famers need to have high yielding cultivars in addition to resistance to various biotic and abiotic stresses.

CSTs associated with biotic stresses such as *Orobanche crenata*, ascochyta blight, leaf rust, etc. are assessed using visual scoring (VS) of the symptoms of the diseases. The resistance to ascochyta blight and chocolate spot (*Botrytis fabae*) can be scored using disease severity based on a visual evaluation of the percentage of symptoms in leaf area and infection type scale (from 1 to 5) (Díaz-Ruiz et al. 2009; Atienza et al. 2016; Ocaña-Moral et al. 2017; Beyene et al. 2018). The resistance to *Orobanche crenata* is assessed using simple regression method (Román et al. 2002; Díaz-Ruiz et al. 2010; Ocaña-Moral et al. 2017). Disease development (VS from 1 to 9) and pustule size (VS from 1 to 5) were suggested to evaluate the resistance to leaf rust (Herath et al. 2001).

The CSTs associated with abiotic stresses (such as drought, heat, frost, etc.) are more than those reported in biotic stresses. The list of most CSTs recently reported in abiotic stress studies is illustrated in Fig. 7.1. Unlike phenotyping of biotic stress resistance, each abiotic stress tolerance can be assessed using many traits which are used to separate and discriminate between susceptible and tolerant genotypes. For frost tolerance, previous studies focused on scoring loss of leaf turgidity, dry and fresh matter, and cell membrane stability (Herzog 1987, 1988; Arbaoui and Link 2007). Recently, new CSTs associated with frost tolerance were reported. For example, regrowth after frost is a very important trait that reflects the ability of plants to produce new shoots after exposing faba bean plants to frost stress (Sallam et al. 2015). Moreover, a high genetic variation was found among the tolerant genotypes for this trait which makes selection to the most tolerant genotypes useful. Days to recovery after frost ("disposition to survive") was developed by Roth and Link (2010) and used in the study of Sallam et al. (2015). Root traits including root frost susceptibility (RFS), root fresh matter, and root dry matter were scored after frost in a set of 200 genotypes by Sallam and Moursi (2016).

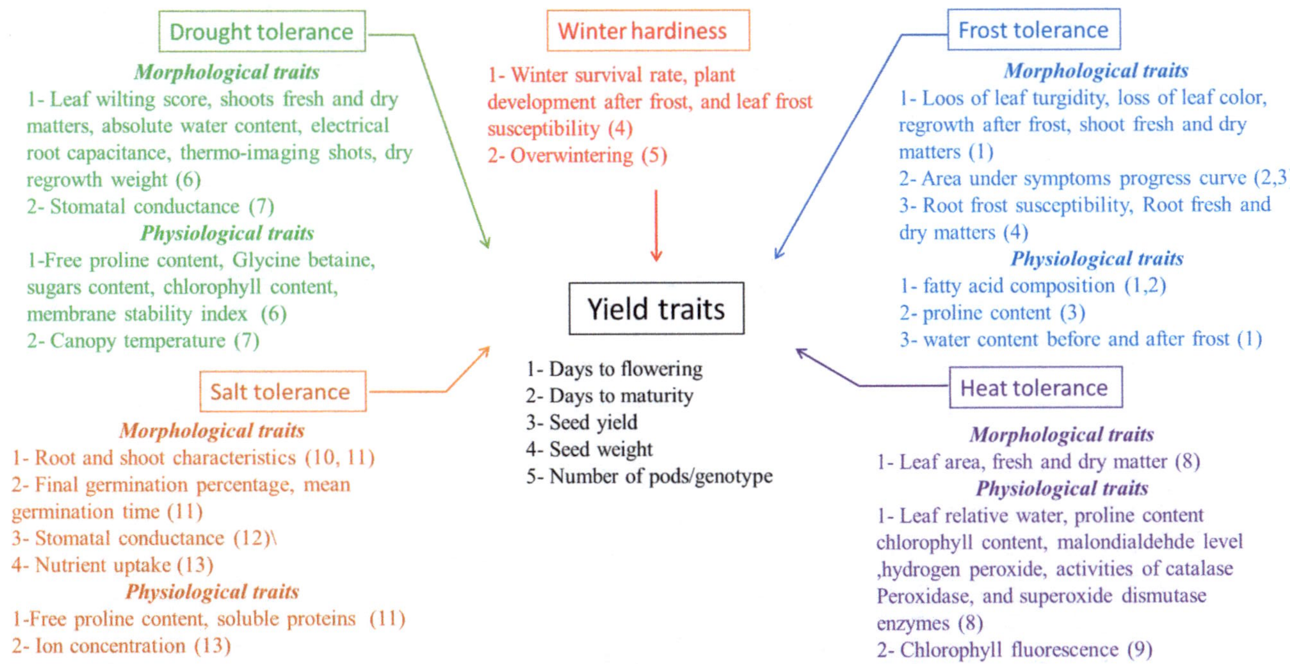

Fig. 7.1 List of the most important climate-smart traits scored on faba bean plants. (1) Sallam and Martsch (2015), (2) Sallam et al. (2016a), (3) Arbaoui et al. (2008b), (4) Sallam and Moursi (2016), (5) Arbaoui et al. (2008a), (6) Ali (2015), (7) Khazaei et al. (2014), (8) Siddiqui et al. (2015), (9) Zhou et al. (2018), (10) Abdelhamid et al. (2010), (11) Gaballah and Gomaa (2004), (12) Katerji et al. (2005), (13) El Fouly et al. (2001)

CSTs associated with heat tolerance such as leaf area, leaf fresh weight, leaf dry weight, leaf water content, pollen damage, and floral development stages were reported (Siddiqui et al. 2015; Bishop et al. 2016). Moreover, response of flowering time to high temperature in faba bean was studied by Catt et al. (2017). Drought tolerance in faba bean was assessed also using the traditional CSTs such as leaf wilting, dry matter, fresh matter, and relative water content. Ali (2015) cut the plant of faba bean after drought treatment and re-irrigated the plants to test their ability to regrowth after drought period. Moreover, he used thermo-imaging shots to evaluate leaf temperature depression during drought stress. Root and shoots characteristics, final germination percentage, stomatal frequency, stomatal conductance, and nutrient uptake are CSTs associated with salt and drought stress tolerance (Gaballah and Gomaa 2004; Katerji et al. 2005; Abdelhamid et al. 2010; Khazaei et al. 2014).

Most of the aforementioned CSTs were scored under greenhouses and controlled conditions. Yield traits could also be good indicators as CSTs associated with abiotic stress tolerance. For example, reduction in yield or seed yield due to stress can be calculated from values under favorable and stress conditions. CSTs in both field and controlled conditions should be considered to improve biotic and abiotic stress tolerance in faba bean. Such information gained from both types of experiments is very useful to develop new cultivars having a combination of high tolerance to biotic or/and abiotic stresses and high yield. For example, days to regrowth after frost ("disposition to survive") were tested under controlled condition (artificial frost) in a set of 200 highly diverse genotypes at seeding stage, and it was significantly correlated with winter survival rate which was scored on the same genotypes under field conditions after natural frost ($r = 0.53^{**}$) (Sallam et al. 2016b). They reported that such correlations could be useful for improving frost tolerance in faba bean.

More importantly, understanding physiological changes in response to abiotic and biotic stress tolerance are useful in producing cultivars having more tolerance to these stresses. For example, leaf fatty acid composition and its correlation with frost tolerance were investigated by Arbaoui and Link (2007), Sallam (2014), and Sallam et al. (2015). These studies reported that the accumulation of unsaturated fatty acid in faba bean leave is associated with increased frost tolerance in faba bean. Proline accumulation increases the tolerance to drought, heat, and salt (El Fouly et al. 2001; Gaballah and Gomaa 2004; Ali 2015; Siddiqui et al. 2015). Therefore, it is highly recommended to consider the analysis of physiological traits in combination with scoring CSTs.

Selection of the promising genotypes is the most important step in the breeding programs to improve biotic and abiotic stress tolerance. Most of the earlier studies focused on screening faba bean genotypes for one or two traits to distinguish the tolerant genotypes from susceptible ones to specific stress. However, studying a single trait could be useless to select the promising tolerant genotypes for biotic or abiotic stress. Most of the CSTs associated with biotic or abiotic stress tolerance are complex and controlled by polygenes. Therefore, it is important to screen the same set of genotypes to many CSTs associated with a specific biotic or abiotic stress. This will undoubtedly help in selecting the real tolerant genotypes to be included in breeding programs. For instance, five CSTs (scored on shoots and roots) associ-

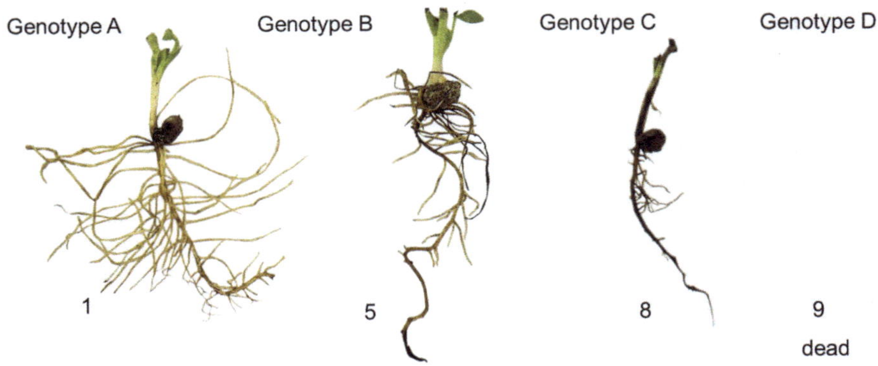

Fig. 7.2 Phenotypic variation in root frost susceptibility (RFS) as described by Sallam and Moursi (2016). RFS was visually scored with a scale extended from 1 (no frost symptoms) to 9 (dead roots)

ated with frost tolerance, fatty acid compositions (seven fatty acids), and changes in relative water content (before and after frost) were scored in a set of 200 highly homozygous faba bean lines (Sallam et al. 2015). The authors selected the most 20 tolerant genotypes for each CST (five). A group of 12 genotypes were found to be superior in more than one trait. Interestingly, two genotypes were among the best 20 frost-tolerant genotypes desirable with regard to in all CSTs. In addition to the previous point, regrowth (REG) after frost was among the five traits. The tolerant genotypes selected based on REG had genetic differences in root frost sensitivity (Sallam and Moursi 2016). They reported significant differences in root frost sensitivity among the surviving genotypes after frost (Fig. 7.2). Although genotypes A and B did survive after frost test and they were considered as tolerant genotypes, based on regrowth after frost, they did differ in root frost susceptibility (RFS) traits. Genotype A was highly promising to continue regrowing as its roots seemed very healthy as compared to genotype B. Therefore, it was so important to consider more CSTs to select precisely the most promising frost-tolerant genotypes.

The CSTs scored to address the genetic variation in winter hardiness were overwintering, winter survival rate, plant development after frost, and leaf frost susceptibility (Arbaoui et al. 2008a; Sallam et al. 2016b). Based on CSTs associated with frost-tolerant, winter hardiness, and seed yield (after frost), Sallam et al. (2016b) and Sallam and Moursi (2016) were able to select the most promising genotypes in winter faba bean in a combination with high yield.

With regard to drought tolerance, 58 faba bean genotypes were screened for seven CSTs and two physiological traits by Ali (2015) who found two promising drought-tolerant genotypes (S062 and S252) which can be used to improve drought tolerance in winter faba bean. A selection index developed by Baker (1986) provides also an effective tool to improve a group of CSTs in breeding program. It includes the most promising CSTs which have strong phenotypic and genotypic correlations and high heritability estimates. A set of five CSTs associated with frost tolerance were grouped in on selection index and named as frost tolerance index (FTI) (Sallam et al. 2015), and it was also used in selecting the most frost-tolerant genotypes with high values of FTI. Selection of index including more than one traits was also used in selection of the most drought-tolerant genotypes in wheat (Sallam et al. 2018).

Information from different types of CSTs associated with stress tolerance, physiological traits, and yield traits (that can be used as a good indicator to the effect of stress) should be combined to improve faba bean production and productivity under various stresses (Fig. 7.1).

7.3 Recent Advances in QTL Mapping and GWAS for Improving Climate-Smart Traits

Faba bean breeding programs deal with improvement in seed yield, tolerance/resistance to abiotic and biotic stresses, and acclimatization to the target environment, appropriate plant growth habit, phenology, seed quality, and crop management especially in climate change scenario. It takes several years to accomplish abovementioned objectives through a conventional breeding program. However, genomic tools and plant molecular breeding techniques could accelerate the faba bean breeding process to achieve the required objectives (Gnanasambandam et al. 2012). Therefore, it is very important to understand the genetics and genomic of the faba bean.

DNA molecular markers allow the detection of quantitative traits by mapping QTLs. QTL mapping requires a segregating population which is derived from the cross between two different parents in terms of the traits of interest. The QTL can be detected if the target gene is contrasting. Therefore, the choice of the parental lines is very important as the trait of interest should have a segregation pattern in the progeny. The QTLs for CSTs related to biotic and abiotic stress tolerance were detected in many earlier studies using different types of molecular markers. The list of important QTLs associated with CSTs is presented in Table 7.1. QTL mapping depends on the number and the type of DNA markers that are used to construct the genetic maps. In early studies, random amplified polymorphic DNA (RAPD) and restriction fragment length polymorphism (RFLP) markers were used to construct useful genetic linkage maps of faba beans using early segregating generations, i.e., F_2 populations (Torres et al. 1993). Afterward, simple sequence repeats (SSRs) were used to construct the linkage maps (Požárková et al. 2002). These markers were used to map gene loci related to complex traits such as pod length, flowering time, seeds/pod and ovules/pod (Cruz-Izquierdo et al. 2012), biotic stresses such as *Ascochyta* blight, crenate broomrape (*Orobanche crenata* Forsk) and rust resistance (Patto et al. 1999; Román et al. 2002, 2003; Avila et al. 2004; Díaz-Ruiz et al. 2009; Sillero et al. 2010). With regard to abiotic stresses, expressed sequence-tagged sites (ESTs) have been identified for drought avoidance in *Medicago truncatula*, chickpea, and pea (Jayashree et al. 2005; Buhariwalla et al. 2005). The highest number of DNA markers used in mapping CSTs of faba bean was 687 single-nucleotide polymorphism (SNP)

Table 7.1 List of the recent and most important QTL mapping and GWAS studies on biotic and abiotic stress tolerance in faba bean

Trait	Number of QTL	Population	Method	R^2	References
Frost tolerance	16	Biparental	QTL mapping	3.36–11.49%	Arbaoui et al. (2008b)
	71	Diverse	GWAS	9.7–35.9%	Sallam and Martsch (2015)
	17	Biparental	QTL mapping	11.8–29.41%	Sallam et al. (2016a)
	25	Diverse	GWAS	3.74–11.89%	Sallam et al. (2016a)
Winter hardiness and yield traits	25	Diverse	GWAS	3.97–9.27%	Sallam et al. (2016b)
Drought tolerance and its related traits	15	Biparental	QTL mapping	5.7–9.3%	Khazaei et al. (2014)
	13	Diverse	GWAS	6.60–14.66%	Ali et al. (2016)
Heat tolerance and photoperiod (flowering time)	8	Biparental	QTL mapping	8.6–24.1%	Catt et al. (2017)
Orobanche crenata	7	Biparental	QTL mapping	22–33%	Gutiérrez et al. (2013)
	3	Biparental	QTL mapping	11.5–35%	Román et al. (2002)
	4	Biparental	QTL mapping	17–34%	Ocaña-Moral et al. (2017)
	3	Biparental	QTL mapping	11–43%	Díaz-Ruiz et al. (2010)
Ascochyta blight	2	Biparental	QTL mapping	21.0–25.2%	Román et al. (2003)
	6	Biparental	QTL mapping	6.3–36.1% for C099-01 8.8–44.7% for LO98-01	Avila et al. (2004)
	10	Biparental	QTL mapping	9.8–15.9%	Atienza et al. (2016)
	9	Biparental	QTL mapping	10.6–21.4%	Ocaña-Moral et al. (2017)
Rust	5	Biparental	QTL mapping	–	Avila et al. (2003)

markers (Webb et al. 2015). This number is considered very low compared to the number of markers that were mapped for other crops such as wheat, barley, etc. The small number of markers used for genetic maps construction in faba bean is due to the complexity of its genome size which hindered the progress of genetic improvement for CSTs. Although many genetic maps and QTL mapping works were performed to detect genomic regions controlling CSTs, the faba bean consensus map (FBCM) developed by Webb et al. (2015) is one of the most important genetic maps for identifying QTLs and candidate genes for CSTs. The FBCM was developed using kompetitive allele-specific PCR (KASP) method which is an uniplex SNP genotyping platform. Recently, SNP markers are extensively used instead of other DNA molecular markers (e.g. SSR, AFLP, RFLP, etc.) in crops especially those that have been fully sequenced. Compared to other DNA markers, KASP genotyping provides many features and advantages (Semagn et al. 2013). Another important feature of the FBCM was that the SNPs that used its genetic map were derived from *Medicago truncatula*, a legume model that has been fully sequenced. Many genetic maps in faba bean were constructed for QTL mapping using SNPs mapped in the FBCM to detect genomic regions associated with drought and frost tolerance (Khazaei et al. 2014; Ali et al. 2016; Sallam et al. 2016a, b). The list of QTLs associated with important CSTs controlling abiotic and/or biotic stress tolerance and mapped, in different studies, in FBCM is illustrated in Fig. 7.3 and listed in Table 7.3 . All SNPs controlling CSTs in FBCM were found to be distributed on all the six chromosomes. Many SNPs were associated with more than one CST that could be related to the same or different stress tolerances. Chromosome 2 is the most important chromosome carrying important QTLs controlling stomatal traits, frost tolerance and winter hardiness, and yield traits (scored under frost stress) (Khazaei et al. 2014).

GWAS was extensively used in the last 10 years to identify alleles associated with target traits in crops. The GWAS has many advantages over QTL mapping including higher resolution in localizing QTLs controlling traits of interest and more precisely identifying more superior alleles. Unlike QTL mapping, GWAS uses diverse and important genotypes in which the target genes should be well segregating (Tian et al. 2011). Most importantly, GWAS can detect usual polymorphisms in a gene that are accountable to the difference between two individuals regarding a phenotypic trait (Palaisa et al. 2003). GWAS has expanded because of the significant advances in DNA sequencing technologies which allow identifying a large number of molecular markers such as SNPs. For example, recent studies in maize used GWAS to dissect the quantitative genetic nature of leaf blight resistance and other traits using ~1.6 million SNPs (Kump et al. 2011). However, the complexity of faba bean genome hindered producing such a large number of SNPs. In faba bean, the highest number of markers used in GWAS was 1,322 polymorphic markers consisting of 175 SNPs and 1,147 AFLPs (amplified fragment length polymorphisms) to identify QTLs for reproductive features and vicine–convicine (Puspitasari 2017). The same set of markers was used in GWAS for CSTs (Ali et al. 2016).

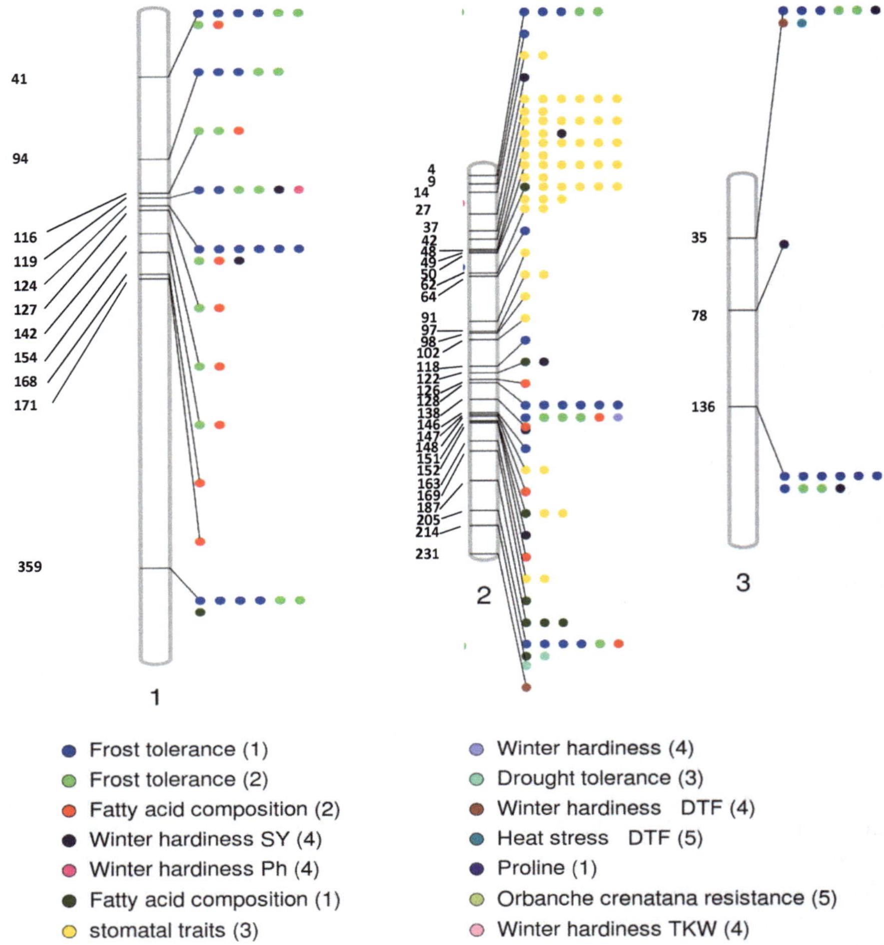

Fig. 7.3 The most important QTL controlling CSTs associated with biotic and abiotic stress tolerance on faba bean consensus map developed by Webb et al. (2015). (1) Sallam and Martsch (2015), (2) Sallam et al. (2016a), (3) Khazaei et al. (2014), (4) Sallam et al. (2016b), (5) Ocaña-Moral et al. (2017). Distances are in centimorgan. The list of SNP names is presented in Table 7.3

7.3.1 QTLs Controlling CSTs in Faba Bean

7.3.1.1 Frost Tolerance and Winter Hardiness

Frost stress is a serious problem for winter faba bean. In central and north Europe, faba bean is mainly planted as a spring crop due to insufficient winter hardiness of the germplasm in use (Arbaoui et al. 2008b; Sallam 2014). Growing winter beans have many features compared to the spring types, including higher yield, excellent use

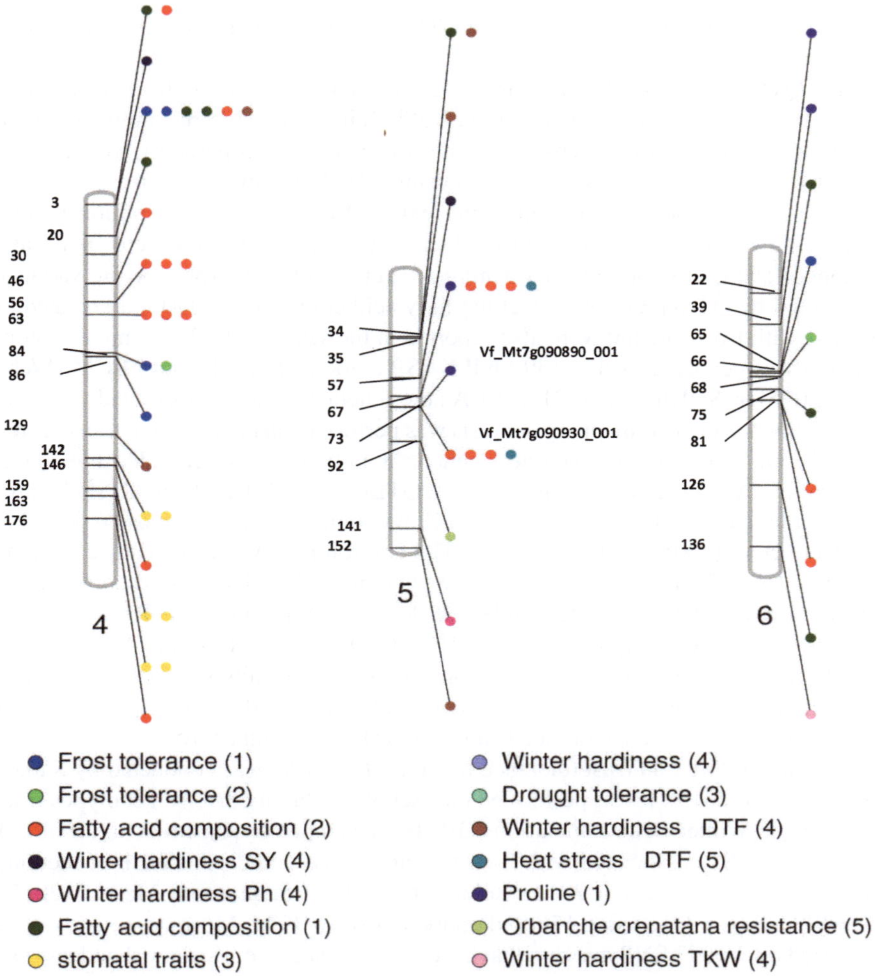

Fig. 7.3 (continued)

of soil moisture, higher tolerance to drought, and some pests (Maqbool et al. 2009). However, the insufficient winter hardiness of faba bean genotypes exposes faba bean to kill due to low freezing temperature in winter (Arbaoui and Link 2007). Therefore, faba bean in cool-temperate regions is often sown as a spring crop. Hence, improving frost tolerance in winter faba bean is crucially needed. The freezing temperature causes a nucleation of the intracellular fluid and affects the plasma membrane (Maqbool et al. 2009). The damage can be extended to cause dehydration of cells due to the freezing of extracellular spaces (Andrews 1996). Moreover, susceptibility to pathogen (e.g., bacterial blight) is possible to increase as a consequence of frost stress. Assessing CSTs can be a useful tool for improving frost tolerance in faba bean. Few studies focused on detecting QTLs for CSTs that are associated with frost tolerance (Table 7.1).

For QTL mapping studies, loss of leaf turgidity and loss of leaf color were assessed as area under symptom progress curve (AUSPC) in a set of 101 recombinant inbred lines (RILs) derived from a cross between two frost-tolerant parents (Arbaoui et al. 2008b). A genetic map was constructed using 131 RAPD markers. Five QTLs controlling AUSPC measured on genotypes exposed to freezing temperature without hardening (three QTLs) and with hardening (two QTLs) phases were found and mapped. The QTLs found for hardening AUSPC had low R^2 (phenotypic variance explained by QTLs). QTLs controlling fatty acid composition (FAC) associated (7 QTLs) with frost-tolerant were also reported in the same study. The same RIL population was genotyped using 189 SNP KASP markers derived from FBCM (Webb et al. 2015) by Sallam et al. (2016a, b). A new genetic map was constructed with a set of 117 SNPs and QTL mapping analysis was performed using the same traits scored in Arbaoui et al. (2008b). The new genetic map-based SNPs revealed five QTLs controlling AUSPC after hardening and 12 QTLs for FAC. The R^2 of AUSPC (hardening) ranged from 13.15 to 18.79% which was much higher than those reported in Arbaoui et al. (2008b). Out of the 12 QTLs controlling FAC, 11 had an R^2 ranging from 12.07 to 29.41%. Although the lower number of SNPs mapped compared to Arbaoui et al. (2008a, b), the number of detected QTLs and their R^2 were much higher in the study of Sallam et al. (2016a, b). This result indicated the power of SNPs markers in detecting QTLs responsible for target traits with a high mapping resolution. Unfortunately, only these two studies reported some important QTLs controlling frost tolerance in faba bean using QTL mapping approach.

The first GWAS on frost tolerance in winter faba bean was conducted by Sallam and Martsch (2015). It was performed on a set of 189 highly diverse genotypes from Goettingen Winter Bean Population (GWBP) developed by Gasim et al. (2004). A total of 11 CSTs associated with frost tolerance was scored on GWBP at seedling stage (Table 7.1) (Sallam 2014; Sallam et al. 2015; Sallam and Martsch 2015). The GWAS was done using 156 SNP markers from FBCM. More importantly, they reported a set of 13 SNP markers that have a significant association with at least two CSTs. Out of the 13 SNPs, two SNP markers *Vf_Mt3g086600* and *Vf_Mt1g105040* had a significant association with seven CSTs associated with frost tolerance. The association mapping revealed 74 QTLs controlling CSTs associated with frost tolerance (Sallam and Martsch 2015).

Frost tolerance is a basic component of winter hardiness (Rizza et al. 1994). Winter hardiness in faba bean needs a lot of improvement. Three main characters are considered very important to address the winter hardiness of a genotype including (1) frost tolerance degree, (2) resistance to biotic stresses, and (3) resistance to other abiotic stresses (Arbaoui and Link 2007). Unfortunately, there is no QTL mapping study in faba bean for mapping QTLs controlling winter hardiness. Genomewide association study (GWAS) was also performed on GWBP which was tested at the Experimental Field Station of Goettingen University under natural frost stress (Sallam et al. 2016b). Three CSTs associated with winter hardiness were scored, viz., leaf frost susceptibly, survival rate, and plant development after frost. Only one QTL controlling winter survival was detected in this study. Interestingly, the SNP marker (*Vf_Mt3 g086600*) linked to this QTL was reported before as it controlled seven CSTs associated with frost tolerance (Sallam and Martsch 2015). Strong correlation was established between frost tolerance at seedling stage and winter hardiness in GWBP (Sallam et al. 2016b). Notably, all CSTs associated with frost tolerance reported so far had high board-sense heritability estimates. Moreover, the CSTs associated, and scored on GWBP, with frost tolerance and winter hardiness had strong and significant correlation with grain yield (g per 2-m^2). This result was very promising to improve CSTs related to frost tolerance and winter hardiness through breeding programs.

Climate-smart traits associated with frost tolerance and winter hardiness were used to identify and select the most tolerant genotypes having a combination of frost tolerance, winter hardiness, and yield. Based on regrowth after frost stress in a frost growth chamber (three nights of freezing temperatures of −16, −18, and −20 °C), the 10 most frost-tolerant and 10 frost susceptible genotypes from GWBP were selected and tested for their winter hardiness and yield attributes (Sallam et al. 2017). On average, the tolerant group had higher winter survival rate and less leaf frost susceptibility compared to the susceptible group. A high genetic distance (GD) was found among genotypes in the tolerant group ($GD > 0.50$), indicating that these genotypes may have different frost genes controlling CSTs associated with frost tolerance.

7.3.1.2 Drought Tolerance

In the world, 40% of lands have a dry environment (United Nations 2011). Faba bean is relatively more sensitive to drought among legume crops (Dai 2013). In most production areas, faba bean depends on rainfall and soil moisture for its growth in the field. Therefore, producing drought-tolerant genotypes is an urgent mission to improve faba bean yield and production in drylands (Stoddard et al. 2006). Drought stress can be assessed in field and under controlled conditions. However, evaluating drought tolerance in field is more costly due to more space, labor, and management requirements compared to drought experiments under controlled conditions. Very little information on genes and genomic regions controlling CSTs associated with drought tolerance is available in faba bean.

QTLs controlling stomatal morphology and function, and seed weight under normal conditions were identified in faba bean (Khazaei et al. 2014). Moreover, they detected QTLs controlling canopy temperature under drought stress. The QTL mapping was performed in a mapping population of 211 F_5 recombinant inbred lines (RILs) using 188 SNPs from FBCM. The study revealed 15 minor QTLs for six CSTs that may have an association with drought tolerance. Interestingly, a QTL cluster including eight QTLs was found in LG04, indicating an important genomic region that may include important genes for CSTs.

Drought tolerance in faba bean was studied also in GWBP and GWAS analyses performed to detect alleles associated with CSTs (Ali et al. 2016). The GWAS was performed using 1,322 SNP and AFLP markers. A total of 13 QTLs controlling glycine betaine, soluble sugar content, leaf chlorophyll content, and proline content under drought stress were detected. Two AFLP markers controlling soluble sugar content had high R^2 values of 14.66 and 14.49%.

7.3.1.3 Heat Tolerance

With climate change, increase in temperature is the most alarming threat to agriculture which in one way may somewhat increase the growth of crops due to CO_2 elevation, but on the other hand it can harshly affect heat-sensitive crops like faba bean. Bishop et al. (2016) concluded from his 2-year experiment that with climate change heat stress will have negative impact on yield of faba bean due to sensitivity of floral parts to heat stress which results in less seed formation. Heat stress affects plant growth and development by influencing different physiological and biochemical processes. Tolerant genotypes tend to maintain physiological and biochemical processes (Siddiqui et al. 2015) and thus maintain growth and yield under heat stress conditions.

Breeders used various techniques to identify heat tolerance genes in germplasm or related species. Researchers investigated morphological, physiological, and enzymatic and genetic response of the crop plants to assess heat stress tolerance. Alterations in photosynthetic systems have been considered a reliable indication of heat stress tolerance/susceptibility in faba beans (McDonald and Paulsen 1997). Siddiqui et al. (2015) reported genotypes with better osmotic adjustments and ability to accumulate more proline with increased activities of antioxidant enzymes and protect themselves from free radicals and thus maintain growth and development. Similarly, Siddiqui et al. (2016) reported that application of magnesium in faba beans improves enzymatic activities, improves accumulation of organic solutes like proline and glycine betaine, and decreases DNA damage and H_2O_2 concentrations which maintain plant growth and yield even in stress conditions.

Molecular response of plant to heat stress is mostly assessed by the gene expression in the form of heat shock proteins (HSPs). These proteins are located in different cellular parts like endoplasmic reticulum, mitochondria, chloroplast, cytoplasm, etc. (Singh and Grover 2010). These heat shock proteins are responsible for regulating different physiological and biochemical processes during heat stress and ensure protection of cell homeostasis (Kotak et al. 2007; Haslbeck and Vierling 2015; Wang et al. 2015). Regulation of plant processes through production of heat stress proteins, in stress conditions, has also been reported in faba bean (Kumar et al. 2015, 2016). Kumar et al. (2016) reported accumulation of *ClpB/Hsp100* protein which induces heat stress tolerance. Kumar et al. (2015) identified and characterized a small heat shock protein *17.9-CII* gene from faba bean which was associated with pollen viability under heat stress conditions. Siddiqui et al. (2015) characterized ten genotypes of faba bean and reported physiological and biochemical differences among them in response to heat stress, and they identified a genotype C5 resistant to heat stress due to having more relative water contents and accumulation of proline contents with improved enzymatic activities.

In the climate change scenario, with reference to heat stress, identification of genes responsible for the abovementioned physiological, biochemical, and molecular traits is crucial. Genotypes having these heat stress adoptive traits will be successful in heated climate. It is need of the time to breed the faba bean genotypes with modern genomic tools for adaptation in the climate change scenario. Serval QTLs and molecular markers have been identified associated with heat tolerance traits in faba bean. Marker-assisted breeding along with validation of genes associated with markers is the best way to introduce genes for heat stress tolerance in modern cultivars. Markers associated with heat stress tolerance genes have been identified in various crops (Kidokoro et al. 2015; Zhu 2016; Ohama et al. 2017). But still heat-responsive genes specific to faba bean have not been reported in literature. The best strategy to identify stress-responsive genes may be screening of faba bean germplasm and its wild relatives in heat stress environments. Following development of a population from the cross of susceptible and tolerant genotypes, QTLs may be identified to develop markers associated with heat stress tolerance. Once markers have been detected, these can be used to identify heat-tolerant genotypes to introduce in hot climatic regions.

In addition, heat stress tolerance is a quantitative trait and involves many genes; therefore, it should be manipulated using combination of genetic, physiological, and morphological traits, instead of selecting for one gene. In future, different genes responsible for somatic and reproductive physiology should be combined to develop heat-tolerant genotypes adaptive to broad climatic conditions in the climate change scenario.

For QTL controlling heat tolerance in faba bean, a biparental population (Icarus × Ascot) was used to evaluate flowering time under high temperatures and different photoperiods (Catt et al. 2017). Under high temperature (HT), they reported five QTLs for days to flowering with R^2 ranging from 8.6 to 16.5% and four QTLs for node of the first flower with R^2 ranging from 11.1 to 24.1%. Unfortunately, there is not any GWAS for heat tolerance in faba bean.

7.3.1.4 Salt Stress Tolerance

Salinity stress is an important abiotic stress and a threat in the climate change scenario. Due to waterlogging or uprise of water table, salt accumulates in root zone of the crop and hinders in nutrient uptake. Salinity stress limits crop growth, water uptake, uptake of mineral nutrients and cause oxidative stress, and hamper smooth process of biochemical reactions (Carillo et al. 2011). All these factors limit crop from germination till maturity and result in lower yields.

Salt stress affects faba bean in many ways. Delgado et al. (1994) reported that salt stress reduces nitrogen fixation of faba beans by inhibiting acetylene reduction activity, leghemoglobin contents, and respirations of bacteroides which lead to reduced growth and production. Tavakkoli et al. (2010) reported that accumulation of NaCl in plant tissue reduced growth of crop but effects of Na^+ and Cl^- are not same. Furthermore, they reported that Cl^- cause chlorophyll degradation, which results in inefficient photosynthesis and the quantum yield, whereas Na^+ interferes with K^+ ions and Ca^{++} ions and results in disturbed stomatal regulation that leads to inefficient transpiration and photosynthesis and thus reduced yield. Salinity stress imbalances the plant water relations which result in lower osmotic potential and chlorophyll content, photosynthetic efficiency, stomatal conductance, and transpiration rate. But genotypes differ in their response to the salt stress (Abdul Qados 2011; Taïbi et al. 2016). The concentration of Na^+ Cl^-, soluble carbohydrates, total phenolic, proline and free amino acids, and activities of antioxidant enzymes increased in plant leaves in response to salinity stress (Dawood et al. 2014).

These biochemical and physical changes are due to molecular response of plant to the salinity. Tolerant plants can detoxify the toxic-free radicals by different procedures. The ability of a plant to tolerate external harsh environment like soil salinity can be assessed by multiple biochemical means. Ismail et al. (2016) reported superoxide dismutase (SOD) activity as a key antioxidant enzyme for adaptation of a plant in saline condition. In salt-tolerant plants, salt stress induces higher SOD production which in return produce H_2O_2 to protect themselves from adverse effects of salinity (Bose et al. 2014). In salt-tolerant plants, proline is accumulated in plant parts and increases the activity of antioxidant enzymes which regulate photosynthesis, plant growth, and homeostasis (Ben Ahmed et al. 2010). Gene regulations mechanisms have also been identified for salt stress tolerance. Johnson et al. (2002) reported downregulation of *bZIP* gene expression in salt-tolerant wheat genotypes. Similarly, Nakashima et al. (2007) reported upregulation of *NAC* gene in salt-tolerant genotypes of wheat and rice. *OsNAC5* and *ZFP179* transcription factors have been reported to play a role in accumulation of osmoprotectants like proline, sugar, etc. which play a role in adaptation of crop plant in saline soils (Song et al. 2011). Quantitative proteomics is an emerging technique to assess the adaptation of plants to abiotic stress. More than two thousand proteins have been identified which are upregulated or downregulated with salinity stress (Zhang et al. 2011). Understanding of all above-mentioned mechanisms is necessary to breed faba bean genotypes for salinity stress tolerance in the climate change scenario.

Various efforts have been made to identify and breed faba bean genotypes for salinity stress tolerance. It is well known that salt tolerance in faba bean is related to the ability of the genotypes to accumulate osmotic solutes, and has antioxidant enzymatic activities to keep higher K^+/N^+ ratio (Tavakkoli et al. 2010; Hanafy et al. 2013). Del Pilar Cordovilla et al. (1995) reported slow growth rate of salt-tolerant genotypes and suggested nitrogen fixation as an unreliable criterion for selection of faba bean under salinity stress because enzymes responsible for nitrogen fixation were found sensitive to salinity stress. Hanafy et al. (2013) suggested transformation of faba bean genotypes with *PR10a* gene from potato as transgenic faba bean from this gene showed enhanced tolerance to salt stress and osmotic stress.

Like other abiotic stresses, salt tolerance is also a multigenic trait and involves many complex processes from biochemical and physiological processes to symbiotic relationship of nitrogen-fixing bacteria. But still this is a neglected crop in terms of use of modern genomic tools for improvement in salinity tolerance. Neither QTL mapping nor GWAS was reported earlier for salt tolerance in faba bean. Cultivated and wild germplasm must be screened out to identify tolerant genotypes to use in breeding programs. Molecular markers and transgenic techniques should be exploited to develop salt stress-tolerant faba bean genotypes for the changing climates.

7.3.1.5 Disease Resistance

Biotic stress represents a significant constraint to faba bean production. The main biotic stresses in faba bean are foliar disease (such as ascochyta blight, chocolate spot, rust, gall diseases), insects, viruses, pests, and parasitic weeds. Very little information on genes controlling CSTs associated with biotic stress tolerance is available in faba bean.

Broomrape (*Orobanche crenata* Forsk), a root parasite, has highly devastating effects on faba bean crop in the Mediterranean region. Resistance/tolerance against broomrape is not an easy trait to assess due to polygenetic nature and environmental influence (Rubiales et al. 2006). This has made selection for the resistance tough and has mired the breeding process (Gutiérrez et al. 2013). An F_2 population of 196 genotypes derived from the cross between a susceptible and a resistant parent was used to construct a genetic map (121 markers) and detect QTLs associated with *O. crenata* resistance using microsatellites, RAPDs, seed protein genes, and isozymes (Román et al. 2002). They mapped three QTLs for broomrape resistance with R^2 ranging from 11.2 to 35%. Useful simple sequence repeat (SSR) markers for studies focusing on resistance to *Orobanche crenata* were developed by Zeid et al. (2009) based on 10 tolerant and 10 susceptible genotypes. The most promising SSR loci were originally developed from Giza 402, a resistant cultivar. The SSR loci described in Zeid et al. (2009) could be applicable to QTL mapping studies that focus on resistance to *O. crenata*. These SSR loci were used along with random amplified polymorphic DNA (RAPD) primers in constructing a genetic map including 171 markers (SSRs and RAPDs) by Gutiérrez et al. (2013). They reported seven QTLs controlling *O. crenata* resistance in a population of 119 $F_{7\text{-}8}$ recombinant inbred lines derived from the cross 29H × Vf136. The seven QTLs had an R^2 ranging from 22 to 33% which could be used for marker-assisted breeding to improve broomrape resistance in faba bean. Three genomic regions associated with resistance to broomrape were identified in an F_2 population derived from a cross between VF6 and VF136. The R^2 ranged from 0.11 to 0.43% (Díaz-Ruiz et al. 2010). Recently, four QTLs controlling broomrape resistance were detected by Ocaña-Moral et al. (2017) in a recombinant inbred line population of 119 $F_{7:9}$ with R^2 extending from 17 to 34%.

Ascochyta blight is a serious fungal disease in faba bean that is caused by *Ascochyta fabae* (*A. fabae*) Speg. It can cause a huge reduction in yield with a range extending from 35 to 90% in susceptible genotypes when the environmental conditions are suitable for disease spread (Hanounik 1980). A set of 196 F_2 individuals were examined for *A. fabae* resistance by Román et al. (2003). They detected two QTLs controlling resistance to *A. fabae*. The two QTLs jointly contributed an R^2 of 46%. Six QTLs controlling *A. fabae* resistance were detected using a linkage map constructed from an F_2 population derived from a cross between 29H (resistant) and VF136 (susceptible) (Avila et al. 2004). The population was tested for the resistance to two pathogenically distinct *Ascochyta* races (CO99-01 and LO98-01). The R^2 for resistance to CO99-01 race ranged from 6.6 to 36.1%, while it ranged from 8.8 to 44.7 for the resistance to LO98-01 race. A recombinant inbred line population (119 $F_{7:8}$) from the same cross (29H × VF136) was used to identify QTLs associated with *A. faba* resistance. They reported 10 QTLs controlling *A. faba* resistance with R^2 extending from 9.8 to 14%. Recently, Ocaña-Moral et al. (2017) used an advanced generation (119 $F_{7:9}$) from the same cross (29H × VF136). The QTL mapping was performed using 92 SNPs which were combined with a previous data set of Gutiérrez et al. (2013) and Atienza et al. (2016). Eight QTLs were found to be associated with *A. faba* resistance with R^2 ranging from 10.6 to 21.4%.

Rust in faba bean is incited by *Uromyces viciae-fabae* (Pers.) J. Schröt which can be prevalent in all areas where bean are grown reducing faba bean yield significantly. Avila et al. (2003) identified three QTLs controlling rust resistance. Unfortunately, there is no other study on the CSTs associated with rust resistance in faba bean. There is no genomewide association study for CSTs associated with disease resistance in faba bean.

7.3.2 *QTL Validation for CSTs in Faba Bean*

Marker-assisted selection (MAS) is one of the useful tools for developing new verities in shorter time compared to classical plant breeding methods. It includes a combination of traditional genetics and molecular biology. Validating QTLs associated with CSTs reported in previous studies is an essential step in MAS to genetically improve target traits. The first reported QTLs are normally considered as putative QTLs. These QTLs need to be validated before using them in MAS. In faba bean, very few studies have been conducted as validation studies for previously reported QTLs that are associated with CSTs. The list of validated QTLs for CST is presented in Table 7.2.

There are different ways by which QTL scan be validated. First, a QTL can be validated if the same QTL can still be detected in the same genetic background when grown in other locations or/and years (Landi et al. 2005). For example, for biotic stress tolerance, some QTLs controlling resistance to *Ascochyta fabae* has been identified in faba bean for the same populations (Díaz-Ruiz et al. 2009). They detected QTLs for resistance to *A. fabae* in F_2 and validated them in F_3 and F_6 in the

Table 7.2 List of the most validated QTL for CST associated with biotic and abiotic stress tolerances

Trait	Putative markers associated with QTL	Type of tested population	Population of validation
Frost tolerance	*F15-476, I10-661, and E20-1556*	BPP (Arbaoui et al. 2008b)	Diverse, frost tolerance (Sallam and Martsch 2016)
	VF_Mt5g026780, VF_Mt3g086600, VF_Mt4g127690, VF_Mt4g125100, and VF_Mt2g027240	BPP (Sallam et al. 2016a)	Diverse 1. Frost tolerance (Sallam et al. 2016a) 2. Winter hardiness)
Flowering time	*Vf-Mt1g056180* *Vf-Mt7 g084010*	Diverse, winter hardiness (Sallam et al. 2016a)	BPP, heat tolerance (Catt et al. 2017).
Broomrape (*Orobanche crenata* Forsk.)	*Oc2, Oc3, Oc4, Oc5*	BPP, F_2 population (Vf6 × Vf136) (Román et al. 2002)	BPP, F_6 RILs (Vf6 × Vf136) (Díaz-Ruiz et al. 2010)
	Oc7	BPP, $F_{7\text{-}8}$ RILS population (29H × Vf136) (Gutiérrez et al. 2013; Ocaña-Moral et al. 2017)	Same population in the three seasons at Córdoba, Spain. (Gutiérrez et al. 2013; Ocaña-Moral et al. 2017)
Ascochyta blight	*Af1 and Af2*	BPP, two different F_2 population – 29H × Vf136. (Avila et al. 2004) – Vf6 × Vf136. (Román et al. 2003)	BPP, F_6 RILs (Vf6 × Vf136). (Díaz-Ruiz et al. 2009)

population derived from the same cross. QTLs controlling resistance to *Orobanche crenata* were valued across environments and generations (Díaz-Ruiz et al. 2010). These studies validated the QTLs using QTL mapping methods.

Second, validation is done in a population with a different genetic background (e.g., backcross, multiparent advanced generation intercross, recombinant inbred lines). QTL for a trait associated with biotic or abiotic stress tolerance could also be validated in a different genetic background with the same stress tolerance (Sallam et al. 2016a, 2017). Sallam and Martsch (2016) validated QTLs associated with increased frost tolerance that was previously published by Arbaoui et al. (2008b). These QTLs were first detected by RAPD markers in a biparental population (BPP). In the validation study, the same markers were tested for their association to frost tolerance in GWBP (a diverse population). As a result, a set of three QTLs were found to be associated with frost tolerance and fatty acid composition after hardening in the diverse population. Recently, some studies used the features of QTL mapping and GWAS to identify and validate QTLs controlling CSTs. Sallam et al. (2016a) reported the first study in faba bean of using QTL mapping and GWAS on two different genetic background populations (biparental and diverse population) to identify and validate QTLs controlling frost tolerance in faba bean. They genotyped the two populations with the same KASP markers that were mapped previously in the FBCM. Hence, significant markers associated with frost tolerance can be easily found and validated. They found five significant KASP markers associated with frost tolerance and fatty acid composition in both populations. These five KSAP markers were found also to be significantly associated with winter hardiness and yield traits in a study of Sallam et al. (2016a). Out of these five SNP markers, one marker (*VF_Mt3g086600*) was found to be associated with increased frost tolerance (10 frost tolerance traits), winter hardiness, and seed yield (after frost stress).

Third, some DNA markers were found to be associated with a CST trait under different conditions. For example, two SNP markers (*Vf-Mt1g056180* and *Vf-Mt7 g084010*) were associated with days to flowering under winter hardiness conditions (Sallam et al. 2016b). The same SNPs were reported as they associated with flowering time under high temperature (Catt et al. 2017).

Validating QTLs in different genetic backgrounds such as in biparental and diverse populations is more efficient than using the same population in different generations and/or locations and/or years. The diverse populations include genotypes which most likely are genetically dissimilar compared to biparental population. This offers more support for the true association between the marker and trait. Hence, these validated QTLs in narrow and broad genetic backgrounds could be used for further populations. Many QTLs detected in a biparental population are specific for that population, and they may not be detectable in other populations. Therefore, using different genetic backgrounds in detecting QTLs by two different statistical methods gives more power for QTL detection and it is very useful for a long-use in MAS to improve target traits.

7.4 Details of Faba Bean Genome Sequencing

Faba bean is diploid and has six somatic chromosomes pairs ($2n = 2x = 12$). As discussed earlier, genetic maps in faba bean included a few DNA markers compared to other crops. The advances in next-generation sequencing have improved a little bit the genetic maps by including more DNA markers. However, the complexity of faba bean genome acts as an obstacle for constructing a high-density genetic map as it was achieved in wheat and barley, etc. Expressed sequence tags (EST) (Yang et al. 2012) and genome survey sequences (GSS) (Gong et al. 2010) are large-scale sequence resources which could be useful in maker development, especially SSR and SNP markers, with a significant improvement of genetic map resolution. Expanding the sequence information in faba bean will accelerate the transition to genomic selection for CSTs (Meuwissen et al. 2001). Basically, a reference whole-genome sequence is used in conjunction with sequencing of selected genotypes in training populations (Braich et al. 2017). The large genome of faba bean is a big challenge for having a whole-genome sequence assembly because it contains a lot of repetitive DNA sequences (Moreton et al. 2016). Transcript sequencing could be an alternative approach through the use of RNA-seq technology as it is one of the second-generation DNA sequencing methods which have been recently used in crops. The transcript sequencing results shed light on the gene regulation, the isolation of gene, the magnitude of gene expression, and the annotation of gene (Moreton et al. 2016). Importantly, it allows understanding the comparative genomics that are very useful in transferring genes of target traits (Garg and Jain 2013).

Ray and Georges (2010) prepared an expressed sequence tag (EST) library in faba bean from a developing embryo of two garden varieties, i.e., Windsor Broad and Exhibition Long. This library has been used for identifying SSR markers and sequencing different tissues of these two cultivars (Windsor Broad and Exhibition Long). The sequencing resulted in identifying 18,000 genes (Kaur et al. 2012). Moreover, many QTLs controlling ascochyta blight resistance were detected from this library. The EST libraries allow synteny study between species with large genome (e.g., faba bean) to more established genomes (e.g., lentil, *M. truncatula*, etc.) (Ray et al. 2015). EST libraries were developed from tissue of root and shoots of three faba bean cultivars having differences in vicine contents (VC) to locate DNA markers and genes associated with low VC (Ray et al. 2015). Generated variants among the three cultivars were compared, and candidate markers for low VC were identified. Furthermore, many genes controlling phytate pathways, the proanthocyanidin pathway, and the raffinose family oligosaccharides (RFOs) synthetic pathway were analyzed. The sequence data and variant identification can be used for further faba bean genomic studies to understand the biochemical pathways (Ray et al. 2015).

A good reference to transcriptome assembly will undoubtedly improve the application of MAS and genomic selection for CSTs in breeding programs. This offers a good advantage especially if the crop does not have a whole-genome assembly such as faba bean (Ray et al. 2015). Recently, comprehensive transcriptome assemblies were accomplished from two different faba bean cultivars (Doza and Farah) using RNA-Seq technology (Braich et al. 2017). The two genotypes present variation in growth habit, disease resistance, and adaptation characteristics which are good examples to detect possible genes controlling these traits. The number of unigenes generated from the transcriptome analysis was compared to those in chickpea, lentil, and *M. truncatula*. As expected, the highest match was found between faba bean and *M. truncatula*. Interestingly, these transcriptome assemblies had a high proportion of transcripts from Webb et al. (2015) with a 95.5% and about 98% of contigs and 78% singletons from Kaur et al. (2012), indicating the usefulness of using these assemblies for further genetic studies on faba bean crop. The transcriptomes derived from Doza and Farah can be used for generating a lot of SNPs for not only between these two cultivars but also for other related genotypes. Bearing that in mind the highest number of markers used for GWAS in faba bean was 1,322 (AFLPs and SNPs). The high number of SNPs, disturbed on all the six chromosome pairs, will be suitable for genetic diversity studies and GWAS which are very important to identify alleles controlling target traits (e.g., CSTs). More importantly, the transcriptomes generated by RNA-seq technology can be used for genotyping-by-sequencing (GBS) as reported (He et al. 2014). Using the unigenes detected in Braich et al. (2017), an alignment to a high-quality reference assembly can be utilized to generate a lot of markers that can be used to genetically improve target traits in faba bean.

7.5 Gene Annotation for CSTs in Faba Bean

Gene annotation is an important step in molecular breeding to identify the possible locations of genes and coding genomic regions controlling target traits (e.g., CSTs). Three important information can be obtained from gene annotation of target marker including (1) identifying genomic regions that do not code for proteins, (2) predicting genes in the genome, and (3) understanding biological information on the function of annotated gene (Stein 2001). Gene annotation can be easily done in many crops for which genomes are sequenced (e.g., wheat, barley, arabidopsis, etc.). However, the situation for faba bean is still difficult and complicated due to its large genome size. Looking for a model plant could be an alternative way by which gene annotation in faba bean can be facilitated. Recently, some studies suggested that research on model legumes such as *Medicago truncatula* could lead to new avenues for enhancing faba bean breeding efforts (Duc 2004; Ellwood et al. 2008; Braich et al. 2017).

Studying a model species will help us to better understand many important biological processes such as plant development (Rispail et al. 2010; Cruz-Izquierdo et al. 2012), response of plants to abiotic and biotic stresses (Jones and Dangl 2006; Swindell et al. 2007), and the physiological adaptations of the plants to threatening stresses. In legumes, *Medicago truncatula* is a model plant as it has a small genome (*M. truncatula* around 500 Mb; Gnanasambandam et al. 2012) and therefore, it is useful for the researchers. This small genome size is very suitable to conduct genetic and genomic research than large genomes species such as *Vicia faba* (around 13,000 Mb, Ellwood et al. 2008). Candidate genes, especially those that contribute to stress tolerance and quality traits, identified by gene annotation may be useful in developing tolerant/resistant transgenic lines (Rispail et al. 2010). Therefore, *M. truncatula* seed represents a suitable model for identification of CSTs associated genes. Burstin et al. (2007) reported that faba bean has homologous loci for biotic and abiotic stresses and identification of genes for the abiotic stress mechanism may also select biotic stress tolerance (Rispail et al. 2010). Mapping of important QTLs associated with increased frost tolerance traits by Avia et al. (2013) in *M. truncatula* may lead to identification of genes resistance to frost stress. A high-level match was observed between faba bean sequence annotations and sequences from chickpea and *M. truncatula* (Braich et al. 2017).

The FBCM developed by Webb et al. (2015) consisted of 845 SNPs that can be annotated from *M. truncatula*. The map has six linkage groups which are presumed to correspond to the six *V. faba* chromosomes. This map is the first genetic map for which all SNP markers were derived from *M. truncatula*. The name of each SNP marker includes the chromosome number in *M. truncatula* and the position of that SNP on the respective chromosome. For example, *Vf_Mt5g026780* marker; Vf refers to *vicia faba*, Mt refers to *M. truncatula*, 5 refers to the number of chromosome in which this SNP was located in *M. truncatula*, and 26780 the position of that SNP on the chromosome. The steps of gene annotation of this SNPs are described as follow:

1. Go to LegumeIP website:
   https://plantgrn.noble.org/LegumeIP/
2. Click on "Gene", then select "Search Gene by keyword".
3. Change the SNP name. For example, Vf_Mt5g026780 should be changed to Medtr5g026780.
4. Submit.

The FBCM was successfully used to identify gene annotation for important QTLs controlling CSTs associated with frost tolerance and winter hardiness (Sallam et al. 2016a, b) and drought tolerance (Khazaei et al. 2014) in faba bean. Out of 20 SNP markers identified using GWAS for yield traits under winter hardiness conditions, 15 annotated genes were identified for seed yield, thousand-seed weight, plant height, and flowering time (Sallam et al. 2016b). Out of five important QTLs associated with frost tolerance, four candidate genes for these QTLs were identified and their biological functions were linked to frost stress. Interestingly, one important SNP marker *VF_Mt3g086600* was found to be associated with increased frost tolerance (11 CSTs), winter hardiness, and seed yield (Sallam et al. 2016a, c). The gene annotation of that marker was encoded to a hypothetical protein. This marker could be very important in improving frost tolerance and winter hardiness through MAS. This information is useful and it should take the attention of faba bean geneticists to work further on that gene to better understand its biological function in frost tolerance and winter hardiness (Sallam et al. 2016a). Khazaei et al. (2014) found 12 QTLs controlling CSTs associated with drought tolerance. They annotated eight genes controlling CSTs in faba bean.

7.6 Conclusion

Climate change has started to affect the agricultural crops by the increasing degree of effects from biotic and abiotic stresses singly or in combination. Faba bean is one of the important crops that large human populations in the poor and developing countries depend upon as a valuable source of food. Climate-smart trait (CST), recommended by Prof. C. Kole, is a new term for a measurable trait that has a direct association with biotic or/and abiotic stress tolerance. The recent interest of researchers is to improve CSTs through breeding and genetics programs. In faba bean, CSTs were reported for biotic and abiotic stress tolerance. However, little breeding research has been done for improving CSTs in faba bean. In order to genetically improve CSTs in faba bean, genetic maps with high-density markers are urgently needed to target important QTLs associated with CSTs. Generally, QTL studies are very few compared to the other important crops (e.g., wheat, barley, maize, etc.). For example, there are huge number of unidentified QTL for CSTs controlling heat and salt tolerance in faba bean. Although some QTLs were reported for frost and drought, they are too few to dissect the genetics of these stresses. For biotic stress tolerance, there is no QTL study for some diseases such as chocolate spot and Fusarium root rot except only one study on rust disease (with RAPD which is not a reliable marker type as compared to codominant gene-specific markers such as SNP and SSR). The advancement in DNA sequencing has improved the understanding of molecular genetics in faba bean to some extent. Using the faba Bean Consensus Map (FBCB) developed by Webb et al. (2015) could be the best option to continue QTL studies by using SNP markers. The feature of GWAS should be exploited in faba bean research to detect alleles for important CSTs. The KASP markers in FBCM can be used to genotype diverse populations to conduct GWAS. Faba bean researchers should pay more attention to the research gaps especially those related to QTL studies for abiotic and biotic stresses.

Appendix

Table 7.3 List of important SNP markers that controlling different biotic and abiotic stress tolerance

SNP	Chromosome	Position (cM)	phenotype (reference)
Vf_Mt2g027240_001	1	41	Frost tolerance (1)
Vf_Mt2g027240_001	1	41	Frost tolerance (1)
Vf_Mt2g027240_001	1	41	Frost tolerance (1)
Vf_Mt2g027240_001	1	41	Frost tolerance (1)
Vf_Mt2g027240_001	1	41	Frost tolerance (2)
Vf_Mt2g027240_001	1	41	Frost tolerance (2)
Vf_Mt2g027240_001	1	41	Frost tolerance (2)
Vf_Mt2g027240_001	1	41	Fatty acid composition (2)
Vf_Mt5g015280_001	1	94	Frost tolerance (1)
Vf_Mt5g015280_001	1	94	Frost tolerance (1)
Vf_Mt5g015280_001	1	94	Frost tolerance (1)
Vf_Mt5g015280_001	1	94	Frost tolerance (2)
Vf_Mt5g015280_001	1	94	Frost tolerance (2)
19a15_3	1	116	Frost tolerance (2)
19a15_3	1	116	Frost tolerance (2)
19a15_3	1	116	Fatty acid composition (2)
Vf_Mt5g026780_001	1	119	Frost tolerance (1)
Vf_Mt5g026780_001	1	119	Frost tolerance (1)
Vf_Mt5g026780_001	1	119	Frost tolerance (2)

(continued)

Table 7.3 (continued)

SNP	Chromosome	Position (cM)	phenotype (reference)
Vf_Mt5g026780_001	1	119	Frost tolerance (2)
Vf_Mt5g026780_001	1	119	Winter hardiness SY (4)
Vf_Mt5g026780_001	1	119	Winter hardiness Ph (4)
Vf_Mt5g046030_001	1	124	Frost tolerance (1)
Vf_Mt5g046030_001	1	124	Frost tolerance (1)
Vf_Mt5g046030_001	1	124	Frost tolerance (1)
Vf_Mt5g046030_001	1	124	Frost tolerance (1)
Vf_Mt5g046030_001	1	124	Frost tolerance (1)
Vf_Mt5g046030_001	1	124	Frost tolerance (1)
Vf_Mt5g046030_001	1	124	Frost tolerance (2)
Vf_Mt5g046030_001	1	124	Fatty acid composition (2)
Vf_Mt5g046030_001	1	124	Winter hardiness SY (4)
Vf_Mt5g044980_001	1	127	Frost tolerance (2)
Vf_Mt5g044980_001	1	127	Fatty acid composition (2)
Vf_Mt5g037120_001	1	142	Frost tolerance (2)
Vf_Mt5g037120_001	1	142	Fatty acid composition (2)
Vf_Mt5g033880_001	1	154	Frost tolerance (2)
Vf_Mt5g033880_001	1	154	Fatty acid composition (2)
Vf_Mt5g005120_001	1	168	Fatty acid composition (2)
Vf_Mt5g098060_001	1	171	Fatty acid composition (2)
Vf_Mt2g086880_001	1	359	Frost tolerance (1)
Vf_Mt2g086880_001	1	359	Frost tolerance (1)
Vf_Mt2g086880_001	1	359	Frost tolerance (1)
Vf_Mt2g086880_001	1	359	Frost tolerance (1)
Vf_Mt2g086880_001	1	359	Frost tolerance (2)
Vf_Mt2g086880_001	1	359	Frost tolerance (2)
Vf_Mt2g086880_001	1	359	Fatty acid composition (1)
Vf_Mt4g007030_001	2	4	Frost tolerance (1)
Vf_Mt4g007030_001	2	4	Frost tolerance (1)
Vf_Mt4g007030_001	2	4	Frost tolerance (1)
Vf_Mt4g007030_001	2	4	Frost tolerance (2)
Vf_Mt4g007030_001	2	4	Frost tolerance (2)
Vf_Mt4g014710_001	2	9	Frost tolerance (1)
Vf_Mt4g014430_001	2	14	stomatal traits (3)
Vf_Mt4g014430_001	2	14	stomatal traits (3)
Vf_Mt4g025120_001*	2	27	Winter hardiness SY (4)

(continued)

Table 7.3 (continued)

SNP	Chromosome	Position (cM)	phenotype (reference)
Vf_Mt4g035200_001	2	37	stomatal traits (3)
Vf_Mt4g035200_001	2	37	stomatal traits (3)
Vf_Mt4g035200_001	2	37	stomatal traits (3)
Vf_Mt4g035200_001	2	37	stomatal traits (3)
Vf_Mt4g035200_001	2	37	stomatal traits (3)
Vf_Mt4g035200_001	2	37	stomatal traits (3)
Vf_Mt4g035200_001	2	37	stomatal traits (3)
Vf_Mt4g035200_001	2	37	stomatal traits (3)
Vf_Mt3g117120_001	2	42	stomatal traits (3)
Vf_Mt3g117120_001	2	42	stomatal traits (3)
Vf_Mt3g117120_001	2	42	stomatal traits (3)
Vf_Mt3g117120_001	2	42	stomatal traits (3)
Vf_Mt3g117120_001	2	42	stomatal traits (3)
Vf_Mt3g117120_001	2	42	stomatal traits (3)
Vf_Mt3g117120_001	2	42	stomatal traits (3)
Vf_Mt3g117120_001	2	42	stomatal traits (3)
Vf_Mt3g117120_001	2	42	Winter hardiness SY (4)
Vf_Mt3g116080_001	2	48	stomatal traits (3)
Vf_Mt3g116080_001	2	48	stomatal traits (3)
Vf_Mt3g116080_001	2	48	stomatal traits (3)
Vf_Mt3g116080_001	2	48	stomatal traits (3)
Vf_Mt3g116080_001	2	48	stomatal traits (3)
Vf_Mt3g116080_001	2	48	stomatal traits (3)
Vf_Mt3g116080_001	2	48	stomatal traits (3)
Vf_Mt3g116080_001	2	48	stomatal traits (3)
Vf_Mt3g115870_001	2	49	stomatal traits (3)
Vf_Mt3g115870_001	2	49	stomatal traits (3)
Vf_Mt3g115870_001	2	49	stomatal traits (3)
Vf_Mt3g115870_001	2	49	stomatal traits (3)
Vf_Mt3g115870_001	2	49	stomatal traits (3)
Vf_Mt3g115870_001	2	49	stomatal traits (3)
Vf_Mt3g115870_001	2	49	stomatal traits (3)
Vf_Mt3g115870_001	2	49	stomatal traits (3)
Vf_Mt3g114780_001	2	50	Fatty acid composition (1)
Vf_Mt3g114780_001	2	50	stomatal traits (3)

(continued)

Table 7.3 (continued)

SNP	Chromosome	Position (cM)	phenotype (reference)
Vf_Mt3g114780_001	2	50	stomatal traits (3)
Vf_Mt3g114780_001	2	50	stomatal traits (3)
Vf_Mt3g114780_001	2	50	stomatal traits (3)
Vf_Mt3g114780_001	2	50	stomatal traits (3)
Vf_Mt3g114780_001	2	50	stomatal traits (3)
Vf_Mt3g114780_001	2	50	stomatal traits (3)
Vf_Mt3g114780_001	2	50	stomatal traits (3)
Vf_Mt5g024090_001	2	62	stomatal traits (3)
Vf_Mt5g024090_001	2	62	stomatal traits (3)
Vf_Mt3g110600_001	2	64	Frost tolerance (1)
Vf_Mt3g102180_001	2	91	stomatal traits (3)
Vf_Mt3g100500_001	2	97	stomatal traits (3)
Vf_Mt3g100500_001	2	97	stomatal traits (3)
Vf_Mt3g099130_001	2	98	stomatal traits (3)
Vf_Mt3g098530_001	2	102	stomatal traits (3)
Vf_Mt3g087760_001	2	118	Frost tolerance (1)
Vf_Mt3g090670_001	2	122	Fatty acid composition (1)
Vf_Mt3g090670_001	2	122	Winter hardiness SY (4)
Vf_Mt3g087150_001	2	126	Fatty acid composition (2)
Vf_Mt3g086600_001	2	128	Frost tolerance (1)
Vf_Mt3g086600_001	2	128	Frost tolerance (1)
Vf_Mt3g086600_001	2	128	Frost tolerance (1)
Vf_Mt3g086600_001	2	128	Frost tolerance (1)
Vf_Mt3g086600_001	2	128	Frost tolerance (1)
Vf_Mt3g086600_001	2	128	Frost tolerance (1)
Vf_Mt3g086600_001	2	128	Frost tolerance (1)
Vf_Mt3g086600_001	2	128	Frost tolerance (2)
Vf_Mt3g086600_001	2	128	Frost tolerance (2)
Vf_Mt3g086600_001	2	128	Frost tolerance (2)
Vf_Mt3g086600_001	2	128	Fatty acid composition (2)
Vf_Mt3g086600_001	2	128	Winter hardiness (4)
Vf_Mt3g086600_001	2	128	Winter hardiness SY (4)
Vf_Mt3g084090_001	2	138	Fatty acid composition (2)
Vf_Mt2g014220_001	2	146	Frost tolerance (1)
Vf_Mt3g062540_001	2	147	stomatal traits (3)
Vf_Mt3g062540_001	2	147	stomatal traits (3)

(continued)

Table 7.3 (continued)

SNP	Chromosome	Position (cM)	phenotype (reference)
Vf_Mt3g077670_001	2	148	Fatty acid composition (2)
Vf_Mt3g076650_001	2	151	Fatty acid composition (1)
Vf_Mt3g076650_001	2	151	stomatal traits (3)
Vf_Mt3g076650_001	2	151	stomatal traits (3)
Vf_Mt3g076660_001	2	151	Winter hardiness SY (4)
Vf_Mt3g076590_001	2	152	Fatty acid composition (2)
Vf_Mt3g072080_001	2	163	stomatal traits (3)
Vf_Mt3g072080_001	2	163	stomatal traits (3)
RBPC_0SNP	2	169	Fatty acid composition (1)
Vf_Mt3g061590_001	2	187	Fatty acid composition (1)
Vf_Mt3g061590_001	2	187	Fatty acid composition (1)
Vf_Mt3g061590_001	2	187	Fatty acid composition (1)
Vf_Mt5g075540_001	2	205	Frost tolerance (1)
Vf_Mt5g075540_001	2	205	Frost tolerance (1)
Vf_Mt5g075540_001	2	205	Frost tolerance (1)
Vf_Mt5g075540_001	2	205	Frost tolerance (1)
Vf_Mt5g075540_001	2	205	Frost tolerance (2)
Vf_Mt5g075540_001	2	205	Fatty acid composition (2)
Vf_Mt5g075540_001	2	205	Fatty acid composition (1)
Vf_Mt5g075540_001	2	205	Drought tolerance (3)
Vf_Mt3g026020_001	2	214	Drought tolerance (3)
Vf_Mt3g010290_001	2	231	Winter hardiness DTF (4)
Vf_Mt1g056180_001	3	35	Frost tolerance (1)
Vf_Mt1g056180_001	3	35	Frost tolerance (1)
Vf_Mt1g056180_001	3	35	Frost tolerance (1)
Vf_Mt1g056180_001	3	35	Frost tolerance (2)
Vf_Mt1g056180_001	3	35	Frost tolerance (2)
Vf_Mt1g056180_001	3	35	Winter hardiness SY (4)
Vf_Mt1g056180_001	3	35	Winter hardiness DTF (4)
Vf_Mt1g056180_001	3	35	Heat stress DTF (5)
GLPSNP	3	78	Winter hardiness SY (4)
Vf_Mt1g105040_001	3	136	Frost tolerance (1)
Vf_Mt1g105040_001	3	136	Frost tolerance (1)
Vf_Mt1g105040_001	3	136	Frost tolerance (1)
Vf_Mt1g105040_001	3	136	Frost tolerance (1)

(continued)

Table 7.3 (continued)

SNP	Chromosome	Position (cM)	phenotype (reference)
Vf_Mt1g105040_001	3	136	Frost tolerance (1)
Vf_Mt1g105040_001	3	136	Frost tolerance (1)
Vf_Mt1g105040_001	3	136	Frost tolerance (1)
Vf_Mt1g105040_001	3	136	Frost tolerance (2)
Vf_Mt1g105040_001	3	136	Frost tolerance (2)
Vf_Mt1g105040_001	3	136	Winter hardiness SY (4)
Vf_Mt3g117800_001	4	3	Fatty acid composition (1)
Vf_Mt3g117800_001	4	3	Fatty acid composition (2)
Vf_Mt3g118320_001	4	3	Winter hardiness SY (4)
Vf_Mt4g127690_001	4	20	Frost tolerance (1)
Vf_Mt4g127690_001	4	20	Frost tolerance (1)
Vf_Mt4g127690_001	4	20	Fatty acid composition (1)
Vf_Mt4g127690_001	4	20	Fatty acid composition (1)
Vf_Mt4g127690_001	4	20	Fatty acid composition (2)
Vf_Mt4g127690_001	4	20	Winter hardiness DTF (4)
Vf_Mt4g125100_001	4	30	Fatty acid composition (1)
Vf_Mt4g118420_001	4	46	Fatty acid composition (2)
Vf_Mt4g114900_001	4	56	Fatty acid composition (2)
Vf_Mt4g114900_001	4	56	Fatty acid composition (2)
Vf_Mt4g114900_001	4	56	Fatty acid composition (2)
Vf_Mt4g113270_001	4	63	Fatty acid composition (2)
Vf_Mt4g113270_001	4	63	Fatty acid composition (2)
Vf_Mt4g113270_001	4	63	Fatty acid composition (2)
Vf_Mt4g101130_001	4	84	Frost tolerance (1)
Vf_Mt4g101130_001	4	84	Frost tolerance (2)
Vf_Mt4g100760_001	4	86	Frost tolerance (1)
GLIP253SNP	4	129	Winter hardiness DTF (4)
Vf_Mt8g020800_001	4	142	stomatal traits (3)
Vf_Mt8g020800_001	4	142	stomatal traits (3)
Vf_Mt8g022290_001	4	146	Fatty acid composition (2)
CNGC4	4	159	stomatal traits (3)
CNGC4	4	159	stomatal traits (3)
Vf_Mt7g038120_001	4	163	stomatal traits (3)
Vf_Mt7g038120_001	4	163	stomatal traits (3)
Vf_Mt8g039690_001	4	176	Fatty acid composition (2)
Vf_Mt7g051360_001	5	34	Fatty acid composition (1)
Vf_Mt7g051360_001	5	34	Winter hardiness DTF (4)

(continued)

Table 7.3 (continued)

SNP	Chromosome	Position (cM)	phenotype (reference)
Vf_Mt7g080890_001	5	35	Winter hardiness DTF (4)
Vf_Mt7g078800_001	5	57	Winter hardiness SY (4)
Vf_Mt7g084010_001	5	67	Proline (1)
Vf_Mt7g084010_001	5	67	Fatty acid composition (2)
Vf_Mt7g084010_001	5	67	Fatty acid composition (2)
Vf_Mt7g084010_001	5	67	Fatty acid composition (2)
Vf_Mt7g084010_001	5	67	Heat stress DTF (5)
Vf_Mt7g090890_001	5	73	Proline (1)
Vf_Mt7g090930_001	5	73	Fatty acid composition (2)
Vf_Mt7g090930_001	5	73	Fatty acid composition (2)
Vf_Mt7g090930_001	5	73	Fatty acid composition (2)
Vf_Mt7g090930_001	5	73	Heat stress DTF (5)
Vf_Mt7g098440_001	5	92	Orbanche crenatana resistance (5)
Vf_Mt7g112640_001	5	141	Winter hardiness Ph (4)
Vf_Mt7g118320_001	5	152	Winter hardiness DTF (4)
Vf_Mt8g101390_001	6	22	Proline (1)
Vf_Mt8g100120_001	6	39	Proline (1)
Vf_Mt8g086470_001	6	65	Fatty acid composition (1)
GLIP265SNP	6	66	Frost tolerance (1)
GLIP081SNP	6	68	Frost tolerance (2)
Vf_Mt4g085900_001	6	75	Fatty acid composition (1)
Vf_Mt4g087540_001	6	81	Fatty acid composition (2)
Vf_Mt4g088010_001	6	81	Fatty acid composition (2)
HYPTE3SNP	6	128	Fatty acid composition (1)
Vf_Mt4g053880_001	6	163	Winter hardiness TKW (4)

The SNPs were mapped in FBCM (Webb et al. 2015)

References

Abdelhamid MT, Shokr MB, Bekheta MA (2010) Effects of induced salinity on four *Vicia faba* cultivars differing in their broomrape tolerance. In: Fourteenth international water technology conference, Cario, Egypt, 21–23 March 2010

Abdul Qados AMS (2011) Effect of salt stress on plant growth and metabolism of bean plant *Vicia faba* (L.). J Saudi Soc Agric Sci 10:7–15. https://doi.org/10.1016/J.JSSAS.2010.06.002

Ali MB (2015) Physiological response of German winter faba bean (*Vicia faba* L.) to drought. J Crop Improv 29:319–332. https://doi.org/10.1080/15427528.2015.1022918

Ali MBM, Welna GC, Sallam A, Martsch R, Balko C, Gebser B, Sass O, Link W (2016) Association analyses to genetically improve drought and freezing tolerance of faba bean (*Vicia faba* L.). Crop Sci 56. https://doi.org/10.2135/cropsci2015.08.0503

Ammar MH, Khan AM, Migdadi HM, Abdelkhalek SM, Alghamdi SS (2017) Faba bean drought responsive gene identification and validation. Saudi J Biol Sci 24:80–89. https://doi.org/10.1016/J.SJBS.2016.05.011

Andrews CJ (1996) How do plants survive ice? Ann Bot 78:529–536. https://doi.org/10.1006/anbo.1996.0157

Arbaoui M, Balko C, Link W (2008a) Study of faba bean (*Vicia faba* L.) winter-hardiness and development of screening methods. Field Crop Res 106:60–67. https://doi.org/10.1016/j.fcr.2007.10.015

Arbaoui M, Link W (2007) Effect of hardening on frost tolerance and fatty acid composition of leaves and stems of a set of faba bean (*Vicia faba* L.) genotypes. Euphytica 162:211–219. https://doi.org/10.1007/s10681-007-9521-4

Arbaoui M, Link W, Satovic Z, Torres AM (2008b) Quantitative trait loci of frost tolerance and physiologically related trait in faba bean (*Vicia faba* L.). Euphytica 164:93–104. https://doi.org/10.1007/s10681-008-9654-0

Atienza SG, Palomino C, Gutirrez N, Alfaro CM, Rubiales D, Torres AM, Avila CM (2016) QTLs for ascochyta blight resistance in faba bean (*Vicia faba* L.): validation in field and controlled conditions. Crop Pasture Sci 67:216–224. https://doi.org/10.1071/CP15227

Avia K, Pilet-Nayel M-L, Bahrman N, Baranger A, Delbreil B, Fontaine V, Hamon C, Hanocq E, Niarquin M, Sellier H, Vuylsteker C, Prosperi J-M, Lejeune-Hénaut I (2013) Genetic variability and QTL mapping of freezing tolerance and related traits in *Medicago truncatula*. Theor Appl Genet 126:2353–2366. https://doi.org/10.1007/s00122-013-2140-7

Avila CM, Satovic Z, Sillero JC, Rubiales D, Moreno MT, Torres AM (2004) Isolate and organ-specific QTLs for ascochyta blight resistance in faba bean (*Vicia faba* L.). Theor Appl Genet 108:1071–1078. https://doi.org/10.1007/s00122-003-1514-7

Avila CM, Sillero JC, Rubiales D, Moreno MT, Torres AM (2003) Identification of RAPD markers linked to the Uvf-1 gene conferring hypersensitive resistance against rust (*Uromyces viciae-fabae*) in *Vicia faba* L. Theor Appl Genet 107:353–358. https://doi.org/10.1007/s00122-003-1254-8

Baker RJ (1986) Selection indices in plant breeding. CRC Press, Boca Raton, FL

Ben Ahmed C, Ben Rouina B, Sensoy S, Boukhriss M, Ben Abdullah F (2010) Exogenous proline effects on photosynthetic performance and antioxidant defense system of young olive tree. J Agri Food Chem 58:4216–4222. https://doi.org/10.1021/jf9041479

Beyene AT, Derera J, Sibiya J (2018) Genetic variability of faba bean genotypes for chocolate spot (*Botrytis fabae*) resistance and yield. Euphytica 214:132. https://doi.org/10.1007/s10681-018-2210-7

Bishop J, Potts SG, Jones HE (2016) Susceptibility of faba bean (*Vicia faba* L.) to heat stress during floral development and anthesis. J Agron Crop Sci 202:508–517. https://doi.org/10.1111/jac.12172

Bose J, Rodrigo-Moreno A, Shabala S (2014) ROS homeostasis in halophytes in the context of salinity stress tolerance. J Exp Bot 65:1241–1257. https://doi.org/10.1093/jxb/ert430

Braich S, Sudheesh S, Forster J, Kaur S, Braich S, Sudheesh S, Forster JW, Kaur S (2017) Characterisation of faba bean (*Vicia faba* L.) transcriptome using RNA-Seq: sequencing, de novo assembly, annotation, and expression analysis. Agronomy 7:53. https://doi.org/10.3390/agronomy7030053

Buhariwalla HK, Jayashree B, Eshwar K, Crouch JH (2005) Development of ESTs from chickpea roots and their use in diversity analysis of the Cicer genus. BMC Plant Biol 5:16

Burstin J, Marget P, Huart M, Moessner A, Mangin B, Duchene C, Desprez B, Munier-Jolain N, Duc G (2007) Developmental genes have pleiotropic effects on plant morphology and source capacity, eventually impacting on seed protein content and productivity in pea. Plant Physiol 144:768–781. https://doi.org/10.1104/pp.107.096966

Carillo P, Grazia M, Pontecorvo G, Fuggi A, Woodrow P (2011) Salinity stress and salt tolerance. In: Shanker A, Venkateswarlu B (eds) Abiotic stress in plants—mechanisms and adaptations. InTechOpen, Rijeka, Croatia

Catt SC, Braich S, Kaur S, Paull JG (2017) QTL detection for flowering time in faba bean and the responses to ambient temperature and photoperiod. Euphytica 213:125. https://doi.org/10.1007/s10681-017-1910-8

Cruz-Izquierdo S, Avila CM, Satovic Z, Palomino C, Gutierrez N, Ellwood SR, Phan HTT, Cubero JI, Torres AM (2012) Comparative genomics to bridge Vicia faba with model and closely-related legume species: stability of QTLs for flowering and yield-related traits. Theor Appl Genet 125:1767–1782. https://doi.org/10.1007/s00122-012-1952-1

Dai A (2013) Increasing drought under global warming in observations and models. Nat Clim Change 3:52–58. https://doi.org/10.1038/nclimate1633

Dawood MG, Taie HAA, Nassar RMA, Abdelhamid MT, Schmidhalter U (2014) The changes induced in the physiological, biochemical and anatomical characteristics of Vicia faba by the exogenous application of proline under seawater stress. S Afr J Bot 93:54–63. https://doi.org/10.1016/J.SAJB.2014.03.002

del Pilar Cordovilla M, Ligero F, Lluch C (1995) Influence of host genotypes on growth, symbiotic performance and nitrogen assimilation in faba bean (*Vicia faba* L.) under salt stress. Plant Soil 172:289–297. https://doi.org/10.1007/BF00011331

Delgado MJ, Ligero F, Lluch C (1994) Effects of salt stress on growth and nitrogen fixation by pea, faba-bean, common bean and soybean plants. Soil Biol Biochem 26:371–376. https://doi.org/10.1016/0038-0717(94)90286-0

Díaz-Ruiz R, Satovic Z, Ávila CM, Alfaro CM, Gutierrez M V., Torres AM, Román B (2009) Confirmation of QTLs controlling *Ascochyta fabae* resistance in different generations of faba bean (*Vicia faba* L.). Crop Pasture Sci 60:353. https://doi.org/10.1071/cp08190

Díaz-Ruiz R, Torres AM, Satovic Z, Gutierrez MV, Cubero JI, Román B (2010) Validation of QTLs for Orobanche crenata resistance in faba bean (*Vicia faba* L.) across environments and generations. Theor Appl Genet 120:909–919. https://doi.org/10.1007/s00122-009-1220-1

Duc G (2004) Seed composition of Medicago truncatula (line J5), compared to other seed legumes. In: 5th European conference on grain legumes/2nd ICLGG conference. Dijon

El Fouly MM, Zeinab MM, Zeinab AS (2001) Micronutrient sprays as a tool to increase tolerance of faba bean and wheat plants to salinity. Plant nutrition. Springer, Dordrecht, Netherlands, pp 422–423

Ellwood SR, Phan HTT, Jordan M, Hane J, Torres AM, Avila CM, Cruz-Izquierdo S, Oliver RP (2008) Construction of a comparative genetic map in faba bean (*Vicia faba* L.); conservation of genome structure with *Lens culinaris*. BMC Genomics 9:380. https://doi.org/10.1186/1471-2164-9-380

FAO (2017) FAOSTAT database. Food and Agriculture Organization of the United Nations, Available

Gaballah MS, Gomaa AM (2004) Performance of faba bean varieties grown under salinity stress and biofertilized with yeast. J Appl Sci 4:93–99. https://doi.org/10.3923/jas.2004.93.99

Garg R, Jain M (2013) RNA-Seq for transcriptome analysis in non-model plants. In: Rose RJ (ed) Legume genomics. Humana Press, Totowa, NJ, pp 43–58

Gasim S, Abel S, Link W (2004) Extent, variation and breeding impact of natural cross-fertilization in German winter faba beans using hilum colour as marker. Euphytica 136:193–200

Gnanasambandam A, Paull J, Torres A, Kaur S, Leonforte T, Li H, Zong X, Yang T, Materne M (2012) Impact of molecular technologies on faba bean (*Vicia faba* L.) breeding strategies. Agronomy 2:132–166. https://doi.org/10.3390/agronomy2030132

Gong Y-M, Xu S-C, Mao W-H, Hu Q-Z, Zhang G-W, Ding J, Li Z-Y (2010) Generation and characterization of 11 novel est derived microsatellites from *Vicia faba* (Fabaceae). Am J Bot 97:e69–e71. https://doi.org/10.3732/ajb.1000166

Gutiérrez N, Palomino C, Satovic Z, Ruiz-Rodríguez MD, Vitale S, Gutiérrez MV, Rubiales D, Kharrat M, Amri M, Emeran AA, Cubero JI, Atienza SG, Torres AM, Avila CM (2013) QTLs for *Orobanche* spp. resistance in faba bean: identification and validation across different environments. Mol Breed 32:909–922. https://doi.org/10.1007/s11032-013-9920-2

Hanafy MS, El-Banna A, Schumacher HM, Jacobsen H-J, Hassan FS (2013) Enhanced tolerance to drought and salt stresses in transgenic faba bean (*Vicia faba* L.) plants by heterologous expression of the PR10a gene from potato. Plant Cell Rep 32:663–674. https://doi.org/10.1007/s00299-013-1401-x

Hanounik S (1980) Effect of chemical treatments and host genotypes on disease severity/yield relationships of Ascochyta blight in faba beans (*Vicia faba*). FABIS Newsl 2:50

Haslbeck M, Vierling E (2015) A first line of stress defense: Small heat shock proteins and their function in protein homeostasis. J Mol Biol 427:1537–1548. https://doi.org/10.1016/j.jmb.2015.02.002

He J, Zhao X, Laroche A, Lu Z-X, Liu H, Li Z (2014) Genotyping-by-sequencing (GBS), an ultimate marker-assisted selection (MAS) tool to accelerate plant breeding. Front Plant Sci 5:484. https://doi.org/10.3389/fpls.2014.00484

Herath IHMHB, Stoddard FL, Marshall DR (2001) Evaluating faba beans for rust resistance using detached leaves. Euphytica 117:47–57

Herzog H (1988) Winter hardiness in faba beans: varietal differences and interrelations among selection criteria. Plant Breed 101:269–276. https://doi.org/10.1111/j.1439-0523.1988.tb00299.x

Herzog H (1987) Freezing resistance and development of faba beans as affected by ambient temperature, soil moisture and variety. J Agron Crop Sci 159:90–100. https://doi.org/10.1111/j.1439-037X.1987.tb00617.x

Ismail H, Dragišic Maksimovic J, Maksimovic V, Shabala L, Živanovic BD, Tian Y, Jacobsen S-E, Shabala S (2016) Rutin, a flavonoid with antioxidant activity, improves plant salinity tolerance by regulating K^+ retention and Na^+ exclusion from leaf mesophyll in quinoa and broad beans. Funct Plant Biol 43:75. https://doi.org/10.1071/FP15312

Jayashree B, Punna R, Prasad P, Bantte K, Hash CT, Chandra S, Hoisington DA, Varshney RK (2006) A database of simple sequence repeats from cereal and legume expressed sequence tags mined in silico: survey and evaluation. In Silico Biol 6:0054

Johnson RR, Wagner RL, Verhey SD, Walker-Simmons MK (2002) The abscisic acid-responsive kinase PKABA1 interacts with a seed-specific abscisic acid response element-binding factor, TaABF, and phosphorylates TaABF peptide sequences. Plant Physiol 130:837–846. https://doi.org/10.1104/pp.001354

Jones JD, Dangl JL (2006) The plant immune system. nature 444(7117):323

Karkanis A, Ntatsi G, Lepse L, Fernández JA, Vågen IM, Rewald B, Alsiņa I, Kronberga A, Balliu A, Olle M, Bodner G, Dubova L, Rosa E, Savvas D (2018) Faba bean cultivation—revealing novel managing practices for more sustainable and competitive European cropping systems. Front Plant Sci 9:1115. https://doi.org/10.3389/fpls.2018.01115

Katerji N, van Hoorn JW, Hamdy A, Mastrorilli M, Oweis T (2005) Salt tolerance analysis of chickpea, faba bean and durum wheat varieties. Agri Water Manage 72:177–194. https://doi.org/10.1016/j.agwat.2004.09.015

Kaur S, Pembleton LW, Cogan NO, Savin KW, Leonforte T, Paull J, Materne M, Forster JW (2012) Transcriptome sequencing of field pea and faba bean for discovery and validation of SSR genetic markers. BMC Genomics 13:104. https://doi.org/10.1186/1471-2164-13-104

Khazaei H, O'Sullivan DM, Sillanpää MJ, Stoddard FL (2014) Use of synteny to identify candidate genes underlying QTL controlling stomatal traits in faba bean (*Vicia faba* L.). Theor Appl Genet 127:2371–2385. https://doi.org/10.1007/s00122-014-2383-y

Kidokoro S, Watanabe K, Ohori T, Moriwaki T, Maruyama K, Mizoi J, Myint Phyu Sin Htwe N, Fujita Y, Sekita S, Shinozaki K, Yamaguchi-Shinozaki K (2015) Soybean DREB1/CBF-type transcription factors function in heat and drought as well as cold stress-responsive gene expression. Plant J 81:505–518. https://doi.org/10.1111/tpj.12746

Kotak S, Larkindale J, Lee U, von Koskull-Döring P, Vierling E, Scharf K-D (2007) Complexity of the heat stress response in plants. Curr Opin Plant Biol 10:310–316. https://doi.org/10.1016/j.pbi.2007.04.011

Kumar R, Lavania D, Singh AK, Negi M, Siddiqui MH, Al-Whaibi MH, Grover A (2015) Identification and characterization of a small heat shock protein 17.9-CII gene from faba bean (*Vicia faba* L.). Acta Physiol Plant 37:190. https://doi.org/10.1007/s11738-015-1943-3

Kumar R, Singh AK, Lavania D, Siddiqui MH, Al-Whaibi MH, Grover A (2016) Expression analysis of *ClpB/Hsp100* gene in faba bean (*Vicia faba* L.) plants in response to heat stress. Saudi J Biol Sci 23:243–247. https://doi.org/10.1016/j.sjbs.2015.03.006

Kump KL, Bradbury PJ, Wisser RJ, Buckler ES, Belcher AR, Oropeza-Rosas MA, Zwonitzer JC, Kresovich S, McMullen MD, Ware D, Balint-Kurti PJ, Holland JB (2011) Genome-wide association study of quantitative resistance to southern leaf blight in the maize nested association mapping population. Nat Genet 43:163–168. https://doi.org/10.1038/ng.747

Landi P, Sanguineti MC, Salvi S, Giuliani S, Bellotti M, Maccaferri M, Conti S, Tuberosa R (2005) Validation and characterization of a major QTL affecting leaf ABA concentration in maize. Mol Breed 15:291–303. https://doi.org/10.1007/s11032-004-7604-7

Link W, Balko C, Stoddard FL (2010) Winter hardiness in faba bean: physiology and breeding. Field Crop Res 115:287–296. https://doi.org/10.1016/j.fcr.2008.08.004

Maqbool A, Shafiq S, Lake L (2009) Radiant frost tolerance in pulse crops—a review. Euphytica 172:1–12. https://doi.org/10.1007/s10681-009-0031-4

McDonald GK, Paulsen GM (1997) High temperature effects on photosynthesis and water relations of grain legumes. Plant Soil 196:47–58. https://doi.org/10.1023/A:1004249200050

Meuwissen TH, Hayes BJ, Goddard ME (2001) Prediction of total genetic value using genome-wide dense marker maps. Genetics 157:1819–1829

Moreton J, Izquierdo A, Emes RD (2016) Assembly, assessment, and availability of de novo generated eukaryotic transcriptomes. Front Genet 6:361. https://doi.org/10.3389/fgene.2015.00361

Nakashima K, Tran L-SP, Van Nguyen D, Fujita M, Maruyama K, Todaka D, Ito Y, Hayashi N, Shinozaki K, Yamaguchi-Shinozaki K (2007) Functional analysis of a NAC-type transcription factor OsNAC6 involved in abiotic and biotic stress-responsive gene expression in rice. Plant J 51:617–630. https://doi.org/10.1111/j.1365-313X.2007.03168.x

Nations U (2011) Global drylands: a UN system-wide response. Prepared by the Environment Management Group. http://www.zaragoza.es/contenidos/medioambiente/onu//issue07/1107-eng.pdf

Ocaña-Moral S, Gutiérrez N, Torres AM, Madrid E (2017) Saturation mapping of regions determining resistance to Ascochyta blight and broomrape in faba bean using transcriptome-based SNP genotyping. Theor Appl Genet 130:2271–2282. https://doi.org/10.1007/s00122-017-2958-5

Ohama N, Sato H, Shinozaki K, Yamaguchi-Shinozaki K (2017) Transcriptional regulatory network of plant heat stress response. Trends Plant Sci 22:53–65. https://doi.org/10.1016/j.tplants.2016.08.015

Palaisa KA, Morgante M, Williams M, Rafalski A (2003) Contrasting effects of selection on sequence diversity and linkage disequilibrium at two phytoene synthase loci. Plant Cell 15:1795–1806. https://doi.org/10.1105/TPC.012526

Patto MCV, Torres AM, Koblizkova A, Macas J, Cubero JI (1999) Development of a genetic composite map of *Vicia faba* using F_2 populations derived from trisomic plants. Theor Appl Genet 98:736–743. https://doi.org/10.1007/s001220051129

Požárková D, Koblížková A, Román B, Torres AM, Lucretti S, Lysák M, Doležel J, Macas J (2002) Development and characterization of microsatellite markers from chromosome 1-Specific DNA libraries of *Vicia Faba*. Biol Plant 45:337–345. https://doi.org/10.1023/A:1016253214182

Puspitasari W (2017) Association analyses to genetically study reproduction and seed quality features of faba bean (*Vicia faba* L.). Doctoral dissertation, Niedersächsische Staats- und Universitätsbibliothek, Göttingen, Germany

Ray H, Bock C, Georges F (2015) Faba bean: transcriptome analysis from etiolated seedling and developing seed coat of key cultivars for synthesis of proanthocyanidins, phytate, raffinose family oligosaccharides, vicine, and convicine. Plant Genome 8:0. https://doi.org/10.3835/plantgenome2014.07.0028

Ray H, Georges F (2010) A genomic approach to nutritional, pharmacological and genetic issues of faba bean (*Vicia faba*): prospects for genetic modifications. GM Crops 1:42–49. https://doi.org/10.4161/gmcr.1.2.11891

Redden R, Paull J, Zong X, Sass O, Yang T, Ling L (2014) Faba Bean. In: Singh M, Singh I, Dutta M (eds) Broadening the genetic base of grain legumes. Springer, New Delhi, pp 75–90

Rispail N, Kaló P, Kiss GB, Ellis THN, Gallardo K, Thompson RD, Prats E, Larrainzar E, Ladrera R, González EM, Arrese-Igor C, Ferguson BJ, Gresshoff PM, Rubiales D (2010) Model legumes contribute to faba bean breeding. Field Crop Res 115:253–269. https://doi.org/10.1016/j.fcr.2009.03.014

Rizza F, Crosatti C, Stanca AM, Cattivelli L (1994) Studies for assessing the influence of hardening on cold tolerance of barley genotypes. Euphytica 75:131–138. https://doi.org/10.1007/BF00024540

Román B, Satovic Z, Avila CM, Moreno MT, Rubiales D, Torres AM (2003) Locating genes associated with Ascochyta fabae resistance in *Vicia faba*. Aust J Agri Res 54:85. https://doi.org/10.1071/AR02034

Román B, Torres AM, Rubiales D, Cubero JI, Satovic Z (2002) Mapping of quantitative trait loci controlling broomrape (*Orobanche crenata* Forsk.) resistance in faba bean (*Vicia faba* L.). Genome 45:1057–1063

Roth F, Link W (2010) Selection on freezing-tolerance of faba bean (*Vicia faba* L.): improvement of methods and results. In: 60. Jahrestagung der Vereinigung der Pflanzenzüchter und Saatgutkaufleute Österreichs, 24–26 Nov 2009, Raumberg-Gumpenstein, pp 31–37

Rubiales D, Pérez-de-Luque A, Fernández-Aparico M, Sillero JC, Román B, Kharrat M, Khalil S, Joel DM, Riches C (2006) Screening techniques and sources of resistance against parasitic weeds in grain legumes. Euphytica 147:187–199. https://doi.org/10.1007/s10681-006-7399-1

Sallam A (2014) Detailed genetic approach to improve frost tolerance of German winter faba beans. Doctoral dissertation, Niedersächsische Staats- und Universitätsbibliothek, Göttingen, Germany

Sallam A, Arbaoui M, El-Esawi M, Abshire N, Martsch R (2016a) Identification and verification of QTL associated with frost tolerance using linkage mapping and GWAS in winter faba bean. Front Plant Sci 7:1098. https://doi.org/10.3389/fpls.2016.01098

Sallam A, Dhanapal AP, Liu S (2016b) Association mapping of winter hardiness and yield traits in faba bean (*Vicia faba* L.). Crop Pasture Sci 67:55. https://doi.org/10.1071/cp15200

Sallam A, Ghanbari M, Martsch R (2017) Genetic analysis of winter hardiness and effect of sowing date on yield traits in winter faba bean. Sci Hort 224:296–301. https://doi.org/10.1016/j.scienta.2017.04.015

Sallam A, Martsch R (2015) Association mapping for frost tolerance using multi-parent advanced generation inter-cross (MAGIC) population in faba bean (*Vicia faba* L.). Genetica 143:501–514. https://doi.org/10.1007/s10709-015-9848-z

Sallam A, Martsch R (2016) Validation of RAPD markers associated with frost tolerance in winter faba bean (*Vicia faba L.*). Turk J Bot 40. https://doi.org/10.3906/bot-1508-21

Sallam A, Martsch R, Moursi YS (2015) Genetic variation in morpho-physiological traits associated with frost tolerance in faba bean (*Vicia faba* L.). Euphytica 205:395–408. https://doi.org/10.1007/s10681-015-1395-2

Sallam A, Mourad AMI, Hussain W, Stephen Baenziger P (2018) Genetic variation in drought tolerance at seedling stage and grain yield in low rainfall environments in wheat (*Triticum aestivum* L.). Euphytica 214:169. https://doi.org/10.1007/s10681-018-2245-9

Sallam A, Moursi Y (2016) Selection index for seedling root traits to improve frost tolerance in winter faba bean. Ann Agri Crop Sci 1:10–14

Semagn K, Babu R, Hearne S, Olsen M (2013) Single nucleotide polymorphism genotyping using competitive allele specific PCR (KASP): overview of the technology and its application in crop improvement. Mol Breed 33:1–14. https://doi.org/10.1007/s11032-013-9917-x

Siddiqui MH, Al-Khaishany MY, Al-Qutami MA, Al-Whaibi MH, Grover A, Ali HM, Al-Wahibi MS (2015) Morphological and physiological characterization of different genotypes of faba bean under heat stress. Saudi J Biol Sci 22:656–663. https://doi.org/10.1016/J.SJBS.2015.06.002

Siddiqui MH, Alamri SA, Al-Khaishany MYY, Al-Qutami MA, Ali HM, Al-Whaibi MH, Al-Wahibi MS, Alharby HF (2016) Mitigation of adverse effects of heat stress on Vicia faba by exogenous application of magnesium. Saudi J Biol Sci. https://doi.org/10.1016/j.sjbs.2016.09.022

Sillero JC, Villegas-Fernández AM, Thomas J, Rojas-Molina MM, Emeran AA, Fernández-Aparicio M, Rubiales D (2010) Faba bean breeding for disease resistance. Field Crops Res 115(3):297–307

Song S-Y, Chen Y, Chen J, Dai X-Y, Zhang W-H (2011) Physiological mechanisms underlying OsNAC5-dependent tolerance of rice plants to abiotic stress. Planta 234:331–345. https://doi.org/10.1007/s00425-011-1403-2

Stein L (2001) Genome annotation: from sequence to biology. Nat Rev Genet 2:493–503. https://doi.org/10.1038/35080529

Stoddard FL, Balko C, Erskine W, Khan HR, Link W, Sarker A (2006) Screening techniques and sources of resistance to abiotic stresses in cool-season food legumes. Euphytica 147:167–186. https://doi.org/10.1007/s10681-006-4723-8

Swindell WR, Huebner M, Weber AP (2007) Plastic and adaptive gene expression patterns associated with temperature stress in Arabidopsis thaliana. Heredity 99(2):143

Taïbi K, Taïbi F, Ait Abderrahim L, Ennajah A, Belkhodja M, Mulet JM (2016) Effect of salt stress on growth, chlorophyll content, lipid peroxidation and antioxidant defence systems in *Phaseolus vulgaris* L. S Afr J Bot 105:306–312. https://doi.org/10.1016/J.SAJB.2016.03.011

Tavakkoli E, Rengasamy P, McDonald GK (2010) High concentrations of Na^+ and Cl^- ions in soil solution have simultaneous detrimental effects on growth of faba bean under salinity stress. J Exp Bot 61:4449–4459. https://doi.org/10.1093/jxb/erq251

Tian F, Bradbury PJ, Brown PJ, Hung H, Sun Q, Flint-Garcia S, Rocheford TR, McMullen MD, Holland JB, Buckler ES (2011) Genome-wide association study of leaf architecture in the maize nested association mapping population. Nat Genet 43:159–162. https://doi.org/10.1038/ng.746

Toker C, Canci H, Yildirim T (2007) Evaluation of perennial wild Cicer species for drought resistance. Genet Resour Crop Evol 54:1781–1786. https://doi.org/10.1007/s10722-006-9197-y

Torres AMM, Weeden NFF, Martín A (1993) Linkage among isozyme, RFLP and RAPD markers in Vicia faba. Theor Appl Genet 85:937–945. https://doi.org/10.1007/BF00215032

Wang H, Tang X, Wang H, Shao H-B (2015) Proline accumulation and metabolism-related genes expression profiles in Kosteletzkya virginica seedlings under salt stress. Front Plant Sci 6:792. https://doi.org/10.3389/fpls.2015.00792

Webb A, Cottage A, Wood T, Khamassi K, Hobbs D, Gostkiewicz K, White M, Khazaei H, Ali M, Street D, Duc G, Stoddard FL, Maalouf F, Ogbonnaya FC, Link W, Thomas J, O'Sullivan DM (2015) A SNP-based consensus genetic map for synteny-based trait targeting in faba bean (*Vicia faba* L.). Plant Biotechnol J. https://doi.org/10.1111/pbi.12371

Yang T, Bao S, Ford R, Jia T, Guan J, He Y, Sun X, Jiang J, Hao J, Zhang X, Zong X (2012) High-throughput novel microsatellite marker of faba bean via next generation sequencing. BMC Genomics 13:602. https://doi.org/10.1186/1471-2164-13-602

Zeid M, Mitchell S, Link W, Carter M, Nawar A, Fulton T, Kresovich S (2009) Simple sequence repeats (SSRs) in faba bean: new loci from Orobanche resistant cultivar 'Giza 402'. Plant Breed 128:149–155. https://doi.org/10.1111/j.1439-0523.2008.01584.x

Zhang Y-J, Yang J-S, Guo S-J, Meng J-J, Zhang Y-L, Wan S-B, He Q-W, Li X-G (2011) Over-expression of the Arabidopsis CBF1 gene improves resistance of tomato leaves to low temperature under low irradiance. Plant Biol 13:362–367. https://doi.org/10.1111/j.1438-8677.2010.00365.x

Zhou R, Hyldgaard B, Yu X, Rosenqvist E, Ugarte RM, Yu S, Wu Z, Ottosen CO, Zhao T (2018) Phenotyping of faba beans (*Vicia faba* L.) under cold and heat stresses using chlorophyll fluorescence. Euphytica 214(4):68

Zhu J-K (2016) Abiotic stress signaling and responses in plants. Cell 167:313–324. https://doi.org/10.1016/j.cell.2016.08.029

Chapter 8
Bambara Groundnut (*Vigna Subterranea* (L) Verdc)—A Climate Smart Crop for Food and Nutrition Security

Sean Mayes, Wai Kuan Ho, Hui Hui Chai, Bo Song, Yue Chang and Festo Massawe

Abstract Bambara groundnut (*Vigna subterranea* (L) Verdc.) is an underutilized legume native to sub-Saharan Africa, where it is grown at low levels by many farmers as a component of household food and nutritional security. It is generally regarded as drought tolerant and fills the same agroecological niche as peanut (*Arachis hypogaea* L). Molecular research in this crop really began only in the early 2000s but has gathered pace and the recent publication of the first genome draft as part of the AOCC drive to sequence 101 African crop species marks an important milestone towards the application of genome-enabled breeding. This crop has potential to contribute to the climate-smart agriculture of the future. The current article traces the progress made in recent years and highlights the challenges that still remain.

Keywords Bambara groundnut · *Vigna subterranea* (L) Verdc · Drought tolerance · Genetic analysis · Crop breeding · Nitrogen fixation, yield modeling

8.1 Introduction

Bambara groundnut [*Vigna subterranea* (L) Verdc] is a minor tropical African legume which is cultivated at low levels throughout sub-Saharan Africa (Dalziel 1937; Doku

S. Mayes (✉)
Plant and Crop Sciences, Biosciences University of Nottingham, Sutton Bonington Campus, Loughborough, Leicestershire LE12 5RD, UK
e-mail: sean.mayes@nottingham.ac.uk; sean.mayes@cffresearch.org

W. K. Ho · H. H. Chai · F. Massawe
Biosciences, University of Nottingham Malaysia, Crops for the Future, Jalan Broga, 43500 Semenyih, Selangor, Malaysia

S. Mayes · W. K. Ho · H. H. Chai · F. Massawe
Crops for the Future Research Centre, Jalan Broga, 43500 Semenyih, Selangor, Malaysia

B. Song · Y. Chang
BGI-Shenzhen, Beishan Industrial Zone, Yantian District, Shenzhen 518083, China

China National GeneBank, BGI-Shenzhen, Jinsha Road, Shenzhen 518120, China

C. Kole (ed.), *Genomic Designing of Climate-Smart Pulse Crops*,
https://doi.org/10.1007/978-3-319-96932-9_8

and Karikari 1971; Duke 1981; Azam-Ali et al. 2001; Effa and Uko 2017; Feldman et al. 2019). It is a close relative of cowpea (*V. unguiculata*), but morphologically it is similar to peanut (groundnut; *Arachis hypogaea* L.) and has been partly displaced by the introduction of this South American cash crop species (Azam-Ali et al. 2001). Bambara groundnut is well adapted for growing in low-input agriculture on marginal land and on poor, free draining, soils. It is traditionally used as part of household food and nutritional security, while providing some additional income (Berchie et al. 2010; Abu and Buah 2011; Hillocks et al. 2012; Adzawla et al. 2016a, b; Olayide et al. 2018; Feldman et al. 2019).

8.1.1 Economic Importance of Bambara Groundnut

The crop has been adapted to a wide range of environmental conditions from semi-arid regions with unimodal rains, as found in Namibia, through to humid tropical conditions in the island of Java, Indonesia. As such, it has potential to be a far more widely employed crop for future climates, as predicted under climate change. However, photoperiod requirement for pod-filling can be a critical limitation for some origins (Linnemann 1993; Linnemann and Craufurd 1994; Linnemann et al. 1995; Brink 1997, 1998, 1999; Brink et al. 2000; Jørgensen et al. 2009; Berchie et al. 2013; Kendabie et al. 2016).

As a legume, it is able to fix atmospheric nitrogen (Dakora 1998; Sprent et al. 2010; Dakora et al. 2015) with a range of rhizobia (e.g., fixation in central Peninsula Malaysia, presumably using cowpea (*Vigna spp*) rhizobial strains; Musa et al. 2016).

Typical grain composition is quite similar to chickpea, with between 18 and 24% protein and 6% oil, with the remainder largely starches (Minka and Bruneteau 2000; Amarteifio et al. 2010; Atoyebi et al. 2018; Azman Halimi et al. 2019). Amino acid composition is reported to be favorable compared to other legumes and like other legumes, it can act as a nutritional complement to a diet based on cereals (Brough et al. 1992; Yetunde et al. 2009; Okpuzor et al. 2010; Uvere et al. 2010; Baptista et al. 2017). Consumer preferences vary from country to country, with Bambara groundnut being used in stews, in tempeh, as a flour supplement, as a milk for infants or eaten fresh after boiling in the pods (Mubaiwa et al. 2018). In Indonesia, it is processed into a value-added snack, 'Kacang Bogor' ('Bogor nut') similar to dry roasted peanuts (Endah Sri Redjeki, Pers Comm.). As with all foods, there is the potential for allergenicity to be present and some cross reactivity has been shown to human sera samples (Astutia et al. 2016). In some countries, there are also strong cultural components to preference (Hillocks et al. 2012; Feldman et al. 2019).

8.1.2 Effects of Global Warming and Climate Change

As a nitrogen fixing legume component of low-input agriculture, Bambara groundnut and similar crops can be seen as being 'climate friendly' and are often resilient to environmental stresses, having been grown under low-input agricultural systems for long periods of time (Tadele 2018). Bambara groundnut grows on poor and acidic soils, e.g., on low pH 4.2 and highly mineralized soils after three rounds of oil palm in Malaysia (Musa et al. 2016) and fills the niche for companion and rotation cropping with cereals.

Crop modeling work (Azam-Ali 1998; http://www.fao.org/land-water/databases-and-software/aquacrop/; Karunaratne et al. 2010, 2011, 2015) has explored the potential water productivity and yield for Bambara groundnut under climate change scenarios and it has been predicted that reductions in rainfall over the next 80 years in western South Africa will lead to increased water productivity and yield (Mabhaudhi and Modi 2013; Mabhaudhi et al. 2013, 2018). In Malaysia initial modeling suggests that it could be grown far more extensively and yields may even increase under the current climate change scenarios for certain regions (www.cropbase.org). In both cases, the modeling is using landrace parameters, rather than advanced lines or cultivars, which do not yet exist.

The major characteristic that suits Bambara groundnut for future climates is its ability (in some landraces) to withstand quite long periods of intermittent drought, with such conditions predicted to become more common in the future. It is likely that all three mechanisms of drought resistance (escape, avoidance, tolerance) are employed in the different agroecologies where adapted germplasm is found (Collinson et al. 1997; Mwale et al. 2007; Berchie et al. 2012; Al-Shareef et al. 2014; Chibarabada et al. 2015; Muhammad et al. 2015; Berchie et al. 2016; Chai et al. 2016; Mabhaudhi et al. 2018).

8.1.3 Limitations of Traditional Breeding

In essence, Bambara groundnut is still almost exclusively grown as landraces—mixes of inbred, but heterogeneous, lines (as determined by marker analysis; see below) which have been adapted to the conditions of the local environment in which they have been repeatedly grown. It is expected that the overall genetic composition will potentially change with every growth cycle as lines within the landrace are differentially selected by annual weather conditions. There is no question that farmers have been selecting these landraces by eye for hundreds, if not thousands, of years, but there has been little 'controlled cross' breeding in this species until recent decades (Massawe et al. 2003c; Suwanprasert et al. 2006). The flowers, which are very small, do produce nectar and it has been observed that ants often visit the flowers (Doku and Karikari 1971). However, the genetic evidence is that this rarely leads to detectable cross-pollination events in this cleistogamous species (see below).

A number of groups in Africa and Southeast Asia have been actively involved in line selection, leading to the registration of a limited number of 'varieties' such as 'Mana' and 'Kazuma' in Zimbabwe in 2004. One plant of Kazuma forms the basis for the recently produced draft genome sequence (Chang et al. 2018). However, even these 'released' varieties can still retain significant heterogeneity within the seed bulks (Ho et al. 2016). Additionally, 'Nalbam 3', 'Nalbam 4', 'Nalbam 6' and 'Myao' were registered in Tanzania in 2014 and 'Songkhla' in Thailand around 2010. More recent efforts at line selection are likely to lead to registered varieties in a number of countries and regions. Successful controlled cross-pollination has been reported from UK, Thailand and Malaysia (Massawe et al. 2003c; Suwanprasert et al. 2006; https://www.youtube.com/watch?reload=9&v=usPlhsGhN3M; Kendabie et al. 2016) and there is a 'community-research institute' project for producing crosses for selection by farmers underway at the Bogor Agricultural University (BAU), Indonesia (Plant Breeders Without Borders; http://pbwob.org/). Such efforts will advance breeding in this species from what is currently a very low base.

One consideration before advocating crossing, pedigree breeding and selection is the nature of landraces (Zeven 1998) with the genetic heterogeneity potentially providing a degree of environmental buffering (Massawe et al. 2005). For some forms of low-input agriculture, landraces may be more resilient than pure lines and the possibility of constructing 'artificial landraces' needs further exploration Massawe et al. 2005).

8.2 Prioritizing Climate-Smart (CS) Traits

8.2.1 Flowering Time and Pod-Filling

Flowering times (and maturity dates) vary between landraces and genotypes and are key components of crop phenology; important to fit this minor crop into the annual cycle with major crops. Some landraces/genotypes are clearly early (e.g. S19-3 from Namibia; around 110 days in Nottingham tropical glasshouses or late maturity, e.g. Gresik, from Indonesia; around 150 days in Nottingham tropical glasshouses; see comment above).

In many other crops, the duration of flowering and seed set are important determinants of carbon sink construction and, thus, harvest index and yield (Slafer et al. 2006; Song et al. 2010; Botwright et al. 2015). There is no obvious stem storage tissue in Bambara groundnut for late filling of carbon sinks, as exists in the form of stem fructan reserves in some wheats, and it is currently assumed that pod-filling occurs from carbon dioxide fixed during the same phenological stage. In Bambara groundnut, the situation is also complicated by landraces/genotypes being variable for the level of determinacy present. Progress in other legumes has identified likely candidate genes for testing/diagnostics in Bambara groundnut (Repinski et al. 2012; Kwak et al. 2012; Mir et al. 2014). S19-3 is a strongly determinant genotype, while others, such as Ankpa4 can be indeterminate, particularly under long photoperiods and can lead to synchronous or asynchronous maturity and pod production, respectively (P. Kendabie, Pers Comm.)

Photoperiod, however, in Bambara groundnut is a far more important trait than in other legumes. Flowering date can be effected by photoperiod (Linnemann 1993), but the effects are limited. However, pod-filling can be very strongly effected, with some genotypes, such as Ankpa4, not filling any pods under 16 h (Linnemann 1993; Linnemann and Craufurd 1994; Kendabie et al. 2015; Kendabie, unpublished data), with fertilized zygotes eventually degenerating. This has led to the characterization of different genotypes into classes, from quantitative long days through to qualitative short-day requirements. It has also been suggested to be one of the reasons that yields can be erratic in this species, with rainfall patterns shifting in recent centuries so that planting is no longer optimal to give the required photoperiod at pod-filling. Clearly, given the predictions for major changes in patterns of rain fall due to climate change, this trait needs to be fully understood to ensure that the potential of Bambara groundnut as a climate-smart crop can be realized.

Indeed, a good understanding and manipulation of photoperiod requirements for flowering in soybean has been one of the key steps for the current extensive use of this crop around the world (Raper and Thomas 1978; Summerfield et al. 1991, 1993, 1998)—it is likely to be an even more important trait for Bambara groundnut.

8.2.2 *Root Characters*

As noted, Bambara groundnut is found in many countries of sub-Saharan Africa and has adapted to a wide range of growing conditions. Limited work has been done on this crop for root morphology, largely due to the difficulty of direct root assessment, although raised soil beds (Nautiyal et al. 2017) and root columns (Mwale et al. 2007) have been used in initial attempts. In the future, understanding how the above ground and below ground components interact is a major challenge in all crops and new technologies (such as X-ray Micro-CT and NMR; e.g., https://www.nottingham.ac.uk/hiddenhalf/crop/crop-roots.aspx) are likely to begin elucidating root behavior without the need to remove the soil or disrupt structure. Alongside 'shovelomics' (Trachse et al. 2011) and other quick or early stage phenotyping methods, joint analysis above and below ground are likely to be a major focus in the coming years. Studies on root front velocity have shown that landraces adapted to limited soil moisture reserves can produce long tap-roots (such as S19-3; Mwale et al. 2007). Moreover, in this species, morphology overall appears to be very dependent on internode length, with the presence of a strong tap-root dependent on soil conditions and how bunched the above ground morphology is. Indeed, the same QTL location has been seen in both a wild/domesticated cross and also in a domesticated/domesticated crosses, suggesting extensive allelic variation for a single gene or a cluster of genes related to this trait (Basu et al. 2007a, b Mayes unpublished data; Fig. 8.3).

8.2.3 *Heat Tolerance*

There have been very few studies published on heat tolerance in Bambara groundnut partly because it is difficult to uncouple drought and heat stress experimentally (Sesay et al. 2013; Soni et al. 2015), although several experimental methods have been used in other legumes (Srinivasan et al. 1996; Talwar et al. 1999; 2002). Transpirational loss of water is an important mechanism to reduce leaf temperature and drought can often lead to reduced transpiration and an increase in leaf temperature. This is being investigated in a number of crop species as a surrogate for root depth/spread. A number of Bambara groundnut accessions come from growing countries where the crop would regularly endure high temperatures during the growing season. However, the effects of heat stress are dependent (as are those of drought) on the phenological stage at which the stress acts and a better definition of growth stages in Bambara groundnut is needed (Dhanaraj 2018). For many crops, the most sensitive stage is pollen development. For a determinant crop, where there is an irreversible switch underway from flowering to sink filling, heat-based pollen sterility can lead to significant yield losses, due to lack of pollination. For some crops, it may be possible to reinitiate flowering ('reflush') or, for indeterminant crops, flowering overlaps significantly with sink filling and seed set, so that there can be a clear recovery from the stress and continued flowering and fruit-set, even if yield is reduced. Initial experiments involving a controlled heat shock (36 °C over 3 days in controlled environment rooms) to seven genotypes of Bambara groundnut suggests that while there are differences in the genotype response, most of the genotypes are able to recover fertility, reflush and set new pods, albeit with a delay in maturity (Dhanaraj 2018). The ability to reflush (whether or not reversing the phenological stage) may be an important component of the resilience seen in Bambara groundnut and the ability to produce some yield under many circumstances. Optimizing this for the typical growing environment (if one exists, given climate volatility) could be an important step towards security of production (Mayes et al. 2011). However, as mentioned, very limited literature currently exists in this crop on the sole effects of heat.

In an attempt to produce quantitative and comparative data on the potential value of drought tolerant legumes, the Kirkhouse Trust has developed the 'Stress Tolerant Orphan Legumes project (STOL; https://www.kirkhousetrust.org/stolprojects) which aims to identify the optimal niches of a range of potential legume crops in Africa.

8.2.4 Cold Tolerance

Cold tolerance is a problem in a few countries, generally away from the equator. In arid environments, while day temperatures can be high, night temperatures can be low (and certainly below the permissive temperature for growth). That there is genetic variation for response to low temperatures was demonstrated by an experiment where the temperature was lowered from 28 °C down to 18 °C in a series of steps, while leaf RNA was extracted for a cross-species microarray experiment (Bonthala et al. 2016). This identified a number of 'modules' of gene expression during cold stress of two genotypes of Bambara groundnut, while also identifying differences between genotypes.

8.2.5 Drought Tolerance

Drought resistance is the major climate change trait associated with Bambara groundnut, as mentioned above and there has been extensive work to begin to understand the basis for this trait, both under controlled conditions and in the field (Collinson et al. 1997; Mwale et al. 2007; Jørgensen et al. 2010; Berchie et al. 2012; Sesay et al. 2013; Al-Shareef et al. 2014; Chibarabada et al. 2015; Muhammad et al. 2015; Berchie et al. 2016; Chai et al. 2016; Nautiyal et al. 2017; Mabhaudhi et al. 2018). However, different lines and landraces may have been adapted to drought at different phenological stages and it seems likely that all three resistance mechanisms (escape, avoidance, tolerance) are present within the germplasm available, with directed breeding potentially allowing the selection of more hardy lines for some of the most stressful conditions, particularly in/below the African Sahel region. The trait is covered in more detail in Mayes et al. (in revision for Planta).

8.2.6 Flooding and Submergence Tolerance

Generally speaking, flooding and submergence tolerance tend to be the greatest weakness in many legumes and Bambara groundnut is no exception. In experiments in Indonesia, there was some observational evidence for a differential response to flooding, with DodR potentially being better able to withstand the stress (E. Sri Redjeki, Pers Comm), but generally Bambara groundnut must be planted on free draining soils with very limited periods of waterlogging.

As a better understanding of how the plant perceives stress is gained, it may be possible that options will arise to manipulate this system to help stabilize the plants under waterlogging for longer periods of time. *Sub*-1 exploited in rice (Ismail et al. 2013) is an example where greater stress resilience to intermittent waterlogging can be achieved and more generally, genes with the N-end rule pathway may provide opportunities to manipulate plant responses to short term stress, improving adaptation to submergence (Graciet and Wellmer 2010; Gibbs et al. 2012; Mendiondo et al. 2016; Vicente et al. 2018).

8.2.7 Salinity Tolerance

At this point in time, relatively little is known about the variability within this species to salt stress and salinity in general. In many ways, salinity mimics drought stress, so it might be expected that a drought tolerant species would also display some components of salinity tolerance, but as far as we are aware there has been no formal work to date. The current development of an Association Panel between CFF and IITA should provide a better resource for such studies (see below).

8.2.8 *Disease Resistance*

Bambara groundnut is often cited as being 'relatively disease and pest free', although there are clearly issues with *Cercospora canescens* in Southeast Asia and with Fusarium wilt in Zimbabwe (Obagwu 2003; Wakhungu et al. 2017) especially. Recent molecular analysis of diseased tissues in Malaysia indicated the presence of *Fusarium equiseti* and *Lasiodiplodia theobromae* samples from the Balau Estate, although further investigation is needed to determine whether these are the pathogenic causes of symptoms or merely opportunistic colonizers.

Mkandawire (2007) summarized pest and disease issues which had been observed in sub-Saharan Africa and these include *Cerscospora canescens* (leaf spot), powdery mildew (*Erysiphe* sp.), *Phyllosticta voandzeia* leaf spot, wilt (*Fusarium* sp.), leaf blotch (*Phomopsis* sp. and *Sclerorotium rolfsii*) and reports of viral (aphid borne diseases) such as chlorotic and green rosette viruses. Along with more common generalist viruses also exist potential problems with nematodes, particularly *Meloidogynae javanica* (Mkandawire 2007).

If Bambara groundnut is to become more widely grown or geographically spread, then an active program of disease identification and resistance breeding is needed. However, most studies of disease to date suggest that there is genetic variation available to tackle these targets within species.

8.2.9 *Insect Resistance*

As with many bean crops, bruchids are a problem and can often lead to damage in the field and significant damage in storage (Srinives et al. 2007) if conditions for emergence are right. Bambara groundnut is reported to support the life cycle of the two main species; *Callosobruchus maculatus* and *C. chinensis*. Differences in resistance to *C. maculatus* (F.) between Nigerian accessions of Bambara groundnut (Ajayi and Lale 2001; Echezona et al. 2013) have been reported, with the possibility that testa coat color could be a component of the response. Kosini et al. (2017) also report extensive variation for bruchid resistance in material from Cameroons. Together with reports of bruchid resistance in other species (e.g., cowpea (*Vigna unguiculata*) and *V. radiata*; Miesho et al. 2018; Chotechung et al. 2016) as well as wild relatives (Srinives et al. 2007) it should be possible to breed for better tolerance to this major pest within Bambara groundnut.

8.2.10 *Resource Use Efficiency*

Resource use efficiency has been investigated at a number of levels, but with a major focus on drought tolerance and specifically in terms of water productivity (Mabhaudhi and Modi 2013; Mabhaudhi et al. 2013, 2018). Research into drought tolerance has been discussed above. One surrogate for integrated stomatal transpiration has been tested in a controlled cross experiment. CID (Farquhar et al. 1989; Condon et al. 2004) measures the relative abundance of the ^{12}C and C_{13} molecules of carbon dioxide. ^{13}C is naturally discriminated against by photosynthetic enzymes so when the ratio of $^{13}C/^{12}C$ rises, it suggests that the stomata have been closed. Chai et al. (2016, 2017) reported two QTLs at different locations for CID in drought conditions, but not irrigated conditions. However, CID is not a simple surrogate for water use efficiency as different water use strategies may be needed in different environments, which would lead to different desirable CID values. CID has been used in wheat to breed the variety 'Drysdale' in Australia and recent modeling using genetically similar wheat cultivars in Australia (Dysdale and Hertzog) suggest selection for high transpiration efficiency could also be an advantage under climate change predictions for Australia (Christy et al. 2018).

8.3 Genetic Resources of Climate-Smart Genes

8.3.1 Primary Gene Pool

Bambara groundnut (*Vigna subterranea* (L) Verdc.) has been classified into two forms; spp. *subterranea* and *spontanea*. *V. subterranea* spp. *spontanea* is considered to be the ancestral type (Pasquet and Fotso 1997; Pasquet et al. 1999; Pasquet 2004) although Basu and others have shown through crossing an extreme spreading *spontanea* to a domesticated *subterranea* that the number of controlling gene changes between the morphological types is probably limited to 2–3 (Basu et al. 2007a, b) *V. subterranea* spp. *spontanea* has been identified in collections in Northern Nigeria, through to the Cameroons, around the Jos Plateau (Dalziel 1937; Hepper 1963; Temegne et al. 2018). However, the limited genetic differences (initially investigated by Pasquet using isozymes; Pasquet et al. 1999) and the existence of a clear genetic differentiation between West African and South&Eastern African accessions (Olukolu et al. 2012; Somta et al. 2011; Molosiwa et al. 2015) perhaps opens up the possibility of two centers of domestication (Aliyu et al. 2016) or even that some *spontanea* types are revertants from domesticated types. Overall, given that the spreading habit has been shown to be a dominant trait (Basu et al. 2007a), it is difficult to distinguish between the revertant theory and the idea that *V. subterranea* spp. *spontanea* is the undomesticated form.

The major collection of the ex situ germplasm is held in the IITA (www.iita.org) by the Genetic Resources Center, with 1890 accessions from 28 sub-Saharan African countries. These accessions are derived from a wide range of environmental growing conditions, with local adaptation. This material could form the basis of a breeding response to climate change over the next 30–40 years.

8.3.2 Secondary Gene Pool

The closest relative of Bambara groundnut is cowpea, which is also an important legume in sub-Saharan Africa. Begemann reported successful crossing of Bambara groundnut with cowpea, but this has yet to be repeated by other groups (Begemann 1988). The recent release of a good quality draft genome sequence for cowpea will directly benefit work in Bambara groundnut and a series of comparative mapping efforts within *Vigna* have helped to clarify the relationships between the species within the genus (Ho et al. 2017; Chang et al. 2018), potentially allowing cross-species localization of candidate genes.

8.3.3 Closely Related Vigna

Vigna represent an important class of legumes and have been particularly used in Asia, with mung bean, azuki bean and others part of many local diets. However, there is no evidence to suggest sexual compatibility between the different *Vigna* species, although cell fusion and genetic transformation techniques might both be applied. In addition, elucidation of trait controlling genes in other *Vigna* species may allow such genes to be introduced into Bambara groundnut through gene editing approaches in the future. At this point of time, there is relatively little information available on attempts at genetic modification in Bambara groundnut, with the report by Karmeswaree Govinthan Naiken-O Lochlainn (http://eprints.nottingham.ac.uk/14279/1/555429.pdf) being the only attempts at extensive tissue culture work that we are aware of.

8.3.4 Artificially Induced/Incorporated Traits/Genes

A number of initial attempts at developing mutagenized populations have been undertaken in India (Bharatkumar et al. 2015; Chandra et al. 2017) and Africa (O. Molosiwa, pers comm) and further efforts are underway in Namibia.

To date, the number of lines stabilized has been too few to have a good chance of obtaining desired mutants—although note that the Indian group report a line with 27% protein—and a more systematic community effort is needed to establish these resources.

8.4 Glimpses on Classical Genetics and Traditional Breeding for CS Traits

8.4.1 *Classical Mapping Efforts*

As already mentioned, pedigree and controlled cross breeding is only now beginning to be established, although quite extensive diversity analysis has used morphological and trait descriptors, together with some initial isozyme work focused on the relationship between *subterranea* and *spontanea* (see below).

8.4.2 *Limitations of Classical Endeavors and Utility of Molecular Mapping*

A wide range of molecular markers have been applied to Bambara groundnut. RAPD, AFLP and their variants have been used to investigate population structure (Massawe et al. 2003a, b; Fatimah et al. 2018) and to contribute to initial genetic mapping work (Massawe et al. 2003b). Codominant markers, such as SSR markers have been characterized cross-species as well as developed within species (Basu et al. 2007c; Somta et al. 2011; Molosiwa et al. 2015), initially through hybridization-based repeat capture techniques. With the advent of NGS larger SSR markers sets have been designed and tested (e.g., from 454 Titanium reagent on hybridization captured amplicons). More recently, the availability of de novo transcriptome data for Bambara groundnut has made very large numbers of SSRs possible (Chapman 2015; Molosiwa et al. 2015). The recent announcement and publication of the AOCC sponsored, Illumina sequencing-based, genome draft has produced a significant resource (Chang et al. 2018). When the first 100 resequenced lines data are released, it should be possible to choose SSRs in silico which distinguish specific lines.

In addition to the higher throughput achieved in SSR development, NGS has also permitted the development of GBS approaches, with over 500 diversity lines characterized by DArT Seq (S. Mayes Pers. comm.) as part of an ITPGRFA project led by CFF, working with IITA (Nigeria), CSIR (Ghana) and BAU (Indonesia).

Over 1,000 lines from a range of controlled crosses have also been genotyped with DArT Seq for map construction and trait dissection. Once the 100 re-sequenced lines data is released, then the development of a dedicated SNP chip will be possible. Initial testing of four Bambara groundnut genomic samples on the dedicated cowpea chip (iSelect Cowpea Consortium Array; 51,128 SNPs) (Timothy Close pers. comm.) suggested good technical replication (i.e., sequence similarity), but very limited polymorphism, with polymorphism having presumably arisen since divergence of the two species, as might be expected. A dedicated Bambara groundnut chip would allow the current density of markers (by GBS; approximately 5000 good quality markers) to be increased at least 10-, if not 100-fold.

8.4.3 *Breeding Objectives*

Some breeding objectives in this species are common with most crop species, such as disease resistance, improved yield, optimized nitrogen fixation and high harvest index, alongside stress tolerance under low-input agricultural systems. However, with the recent development of controlled crossing techniques it now becomes possible to take a more structured approach to genetic improvement, using the available germplasm and an understanding of its genetic structure (e.g., Massawe et al. 2005; Aliyu et al. 2015, 2016).

8.5 Diversity Analysis

8.5.1 *Phenotype-Based Diversity Analysis*

In the absence of molecular markers, phenotype has been the basis for characterization of germplasm collections. In Bambara groundnut there is a range of traits which can be assessed, with perhaps the most adopted one being seed coat color, which is highly polymorphic in this species. Phenology and morphology have also been widely used to differentiate landrace material into groups and quantitative selection traits or seed compositional traits are also of value (Gonné et al. 2013; Karikari 2004; Ouedraogo et al. 2008; Ntundu et al. 2006; Nofita et al. 2015; Atoyebi et al. 2017; Unigwe et al. 2016; Ofori 1996; Ofori et al. 2001; Shegro et al. 2013; Mohammed et al. 2016).

8.5.2 *Genotype-Based Diversity Analyses*

While phenotypic trait analysis is perhaps the most relevant in terms of selection, the number of traits available can be limited (particularly simply inherited traits) and many are a reflection of the results of farmers' selection. The advent of molecular markers, initially isozymes (Odeigah and Osanyinpeju 1998; Pasquet et al. 1999) and more recently using DNA based techniques, such as RAPD (Massawe et al. 2003a; Fatimah et al. 2018) and AFLP (Massawe et al. 2003b) has allowed a more trait-independent assessment of genetic diversity. The development of SSR-based markers (Basu et al. 2007c; Somta et al. 2011; Aliyu and Massawe 2013; Chapman 2015; Molosiwa et al. 2015; Odongo et al. 2015), facilitated by next-generation sequencing of repeat enriched libraries or transcriptomes has permitted the development of codominant and highly polymorphic markers in this species, as well as the transfer of a number of markers from related species. More recently, NGS has allowed the application of GBS approaches and allowed the identification of large numbers of SNP markers for diversity analysis, with DArT Arrays and then DArT Seq being employed extensively for genetic diversity and genetic mapping exercises (Olukolu et al. 2012; Ho et al. 2017), with datasets able to be combined from different DArT Seq runs. As the genome sequence (Chang et al. 2018) becomes available and more developed, it should be possible to assign markers to specific loci within the assembly, allowing a better understanding of marker distribution for GBS. The approach of using the sequence tag attached to each marker (up to 64 bp) has already allowed extensive cross-species comparison between linkage order in Bambara groundnut and physical order in related legumes which already have good genome sequences (Ho et al. 2017) and the initial development of genotyped lines for an association panel of around 400 genotyped lines, through an ITPGRFA Round 3, Window 3 (Pr26) project. The diversity data for this has already been generated and an initial analysis completed (WK Ho, unpublished data).

The work of Molosiwa et al. (2015) and more recent work have allowed a comparative analysis between a range of the same genotypic lines using SSR and DArT Seq markers (WK Ho, unpublished data). Perhaps, as would be expected, the SSR marker being multiallelic and also evolving at a far higher rate than single point mutations shows a greater depth of diversity than SNPs. However, in both cases the molecular divergence between geographical groups is clear and was also evident from the early use of the DArT Array technique (Stadler et al. 2007; Stadler 2008; Mayes et al. 2009). In addition to genotype-based analysis, initial work by Santos (2018) suggests that there is a significant linguistic association between different landraces, which could be the consequence of migration patterns from West Africa.

8.5.3 Relationship with Other Cultivated Species and Wild Relatives

Ho et al. (2017) presents an analysis using the sequence tag of DArT Seq based genetic maps to align the genetic order in Bambara groundnut to the physical order in other related legumes species. Overall, this confirms the taxonomical relationships expected, but does also identify which related legume is likely to have the best chromosome model for each of the 11 chromosomes of Bambara groundnut.

8.5.4 Relationship with Geographical Distribution

As mentioned above, there is a clear geographical differentiation between Regions, although whether this represents a transfer of domesticated material from West Africa where there is almost certainly a center of domestication or the existence of a secondary center of domestication is in Southern/Eastern Africa unclear. There is also some evidence for a linguist association, but larger dataset needs to be analyzed (Santos 2018).

8.5.5 Extent of Genetic Diversity

Despite Bambara groundnut being a highly cleistogamous species [(average observed heterozygosity (H_o) across over 500 genotype lines was 1.8%)], there appears to be a strong geographical signal in the datasets, suggesting a long-term adaptation and change within specific regions, alongside potential founder effects (see Figs. 8.1 and 8.2). Moreover, the very wide range of environmental conditions to which Bambara groundnut has adapted (from sub-Sahelian and Namib Desert conditions through to West Java in Indonesia) suggests significant trait variability. For pests and diseases, initial screens have also reported the potential to address a wide range of problems through screening of existing germplasm. The fact that the vast majority of the growing material is still in the form of landraces is also encouraging in terms of developing elite lines through breeding and even artificial landraces. There seem to be few problems with developing crosses between *V. subterranea spp. subterranea* and *V. subterranea spp. spontanea*, so there is the ability to reach back into the supposed ancestral material for new genes. However, it is still unclear whether *V. subterranea spp. spontanea* represents a distinct species or in some cases may represent a revertant type and while Figs. 8.1 and 8.2 suggests a single center of origin, followed by founder effects in South and East Africa and then again in Southeast Asia, this still needs to be formally tested.

8.6 Association Mapping Studies

Until recently, two factors have delayed the development of this approach. The first of these is the lack of distinct genotype lines which are (largely) homozygous. The initial development of SSR markers (Basu et al. 2007c) allowed an accurate estimate of heterozygosity of individuals within a landrace. This led to the conclusion that a landrace consists of a number of highly inbred lines (>98%), but with heterogeneity between lines of the landrace. This enabled the 'single plant descent' approach, with seed from a single plant being used to establish a highly inbred line (although not completely inbred line) representing the landrace. The second requirement was the development of high-density marker coverage of the genome. While 100 SSR markers have been developed in species (Molosiwa et al. 2015), alongside DArT Array markers (Stadler 2008) and also transferred SSRs from related species (Somta et al. 2011), it was with the advent of DArT Seq GBS based on NGS that sufficient markers became available to begin testing this approach. With funding from ITPGRFA, CFF and IITA are currently developing the first association panel, with genotyping by DArT Seq complete and the first rounds of phenotyping and multiplying underway.

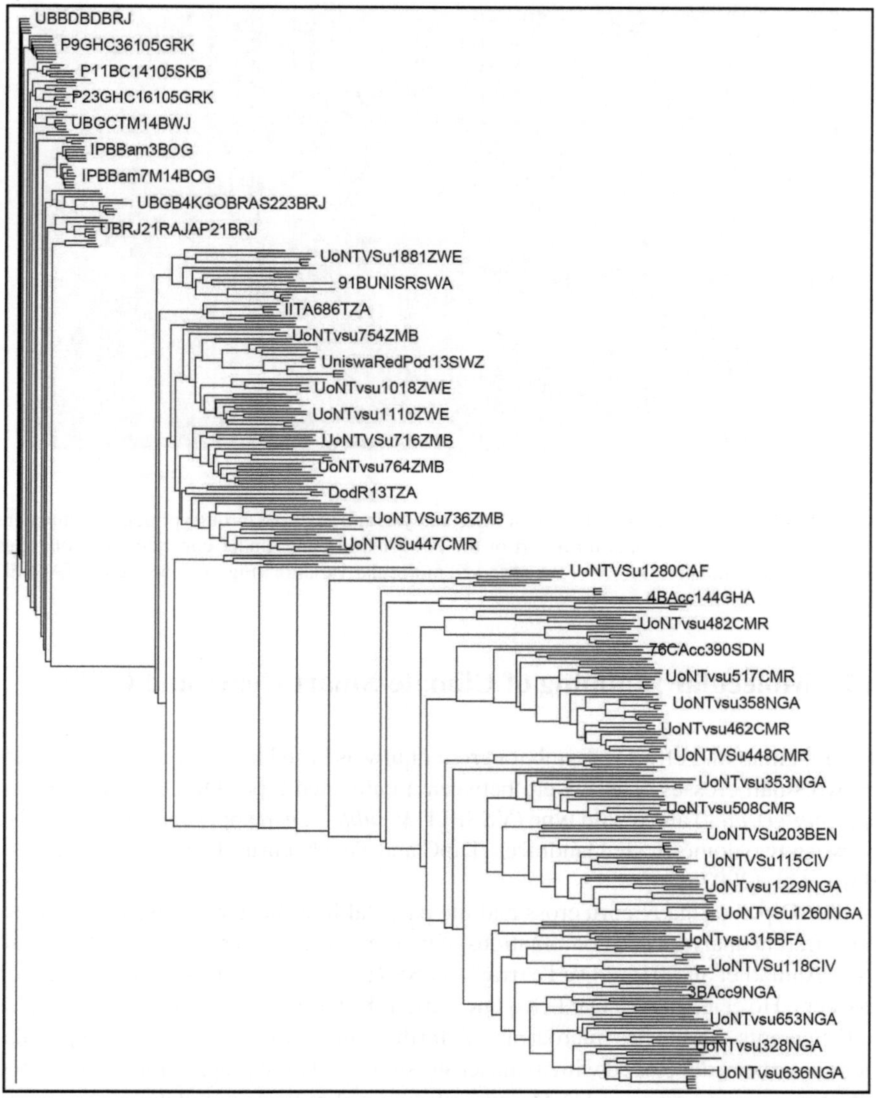

Fig. 8.1 4398 SNP loci derived from DArT Seq analysis used to construct a Neighbor Joining dendrogram from 368 individuals forming part of the Association Panel under construction. The last three digits of the selected sample codes represent country of origin. Named accession represent examples across the tested germplasm. Figure generated in TASSEL 5.2.51

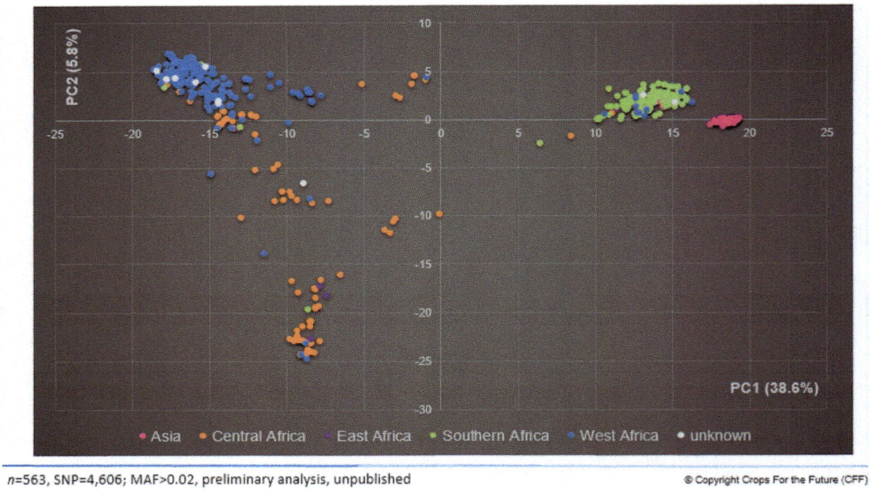

Fig. 8.2 4398 SNP loci derived from DArT Seq analysis and analyzed through principle component analysis. The 368 individuals form part of the Association Panel under construction. Both axes account for a cumulative 44.4% of the observed molecular variation. Figure generated in TASSEL 5.2.51

8.7 Molecular Mapping of Climate-Smart Genes and QTLs

The first mapping effort in Bambara groundnut was based largely on AFLP markers in two small crosses. These were between a cultivated type (DipC; *V. subterranea* spp. *subterranea*) and a wild type (VSSP11; *V. subterranea* spp. *spontanea*) and also between two domesticated landraces (DipC and Tiga Nicuru). Both were reported in Basu et al. (2007a, b).

The DipC x Tiga Nicuru cross and the parental lines have been used extensively for different analyses and combinations of marker types (Chai et al. 2016—trait analysis; Ahmad et al. 2016—DarT Array and SSR; Chai et al. 2017—gene expression markers; Ho et al. 2017—DarT Seq integration; Khan et al. 2017—drought tolerance in the parents—Xspecies microarray). A further range of F_1 and segregating populations have been developed with manuscripts submitted or in preparation. A concerted effort is now underway at UNM and CFF to develop these populations through to inbred lines and once completed, these will be made available to the community.

The standard approach adopted now (Ho et al. 2017; Fig. 8.3) is to reanalyze previous DArT Seq datasets to identify common markers between different maps and allow a common framework to be generated with each additional map, including integration to the cowpea physical map. This will facilitate integration of trait information and also the comparison of trait loci across germplasm. When the genome sequence is complete as pseudomolecules, it should be possible to evaluate genetic order between different germplasm sources and the extent of large-scale rearrangements, which will assist in interpreting G x G effects in crosses.

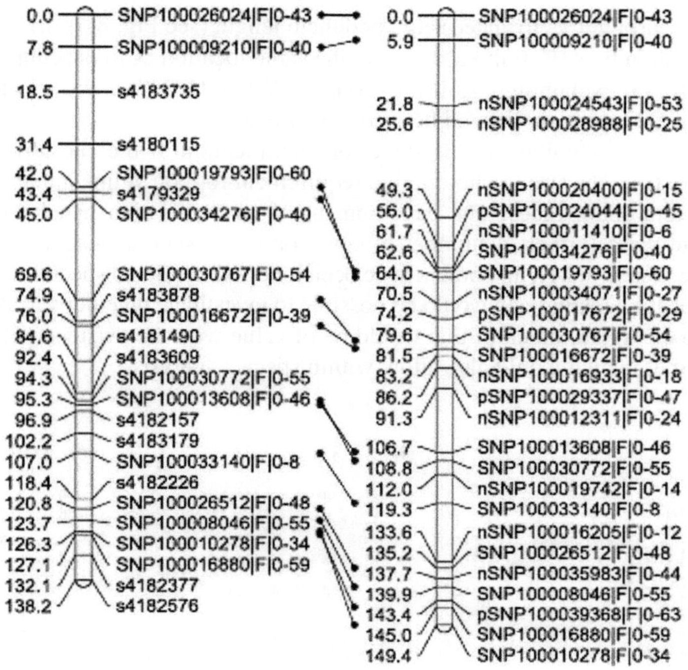

Fig. 8.3 Integration through common marker tags from DArT Seq of two genetic maps in Bambara groundnut; IITA686 x Ankpa4 ($n = 263$) and IITA x LunT ($n = 184$). Both populations were mapped at the F_2 level

8.7.1 *Mapping of Simply Inherited Climate-Smart Traits*

The main (relatively) simply inherited traits in Bambara groundnut relate to testa color (mapping in progress; P Kendabie, unpublished data) and photoperiod requirement (from the strong Ankpa-4 allele; Kendabie et al. 2016). Other traits may be identified and mapped in the future (e.g., red pod and petiole color in genotypes such as DodR), but most traits of importance appear to be quantitative in the domesticated lines.

8.7.2 *Mapping of Quantitative Trait Loci (QTLs) Underlying Climate-Smart Traits*

Chai et al. (2017) carried out a comparative QTL analysis on a small F_5 population of DipC x Tiga Nicuru ($n = 73$) with either a drought treatment or continued irrigation. Some of the QTLs identified appear intrinsic to the cross (occurring in both treatments, drought vs irrigated) e.g., peduncle length (see Fig. 8.4); Irrigated LOD = 8.93 explaining 41% trait variation, at the same location as in drought treatment; LOD = 10.83, explaining 54% trait variation. While CID (see above) identified a number of QTLs which differed between treatments.

The best opportunities at this stage for identification of the causative genes in species are from the strong photoperiod requirement for pod-filling allele in Ankpa4 and for testa color genes. The development of an association panel will help to investigate these further. A number of common traits, such as testa color, bruchid resistance, determinacy, and others have been reported in related species and a cross-species locational approach would be possible to investigate whether variation exists at these loci. Again, much of this would be of value with the existence of the association panel, which would also allow within species analysis.

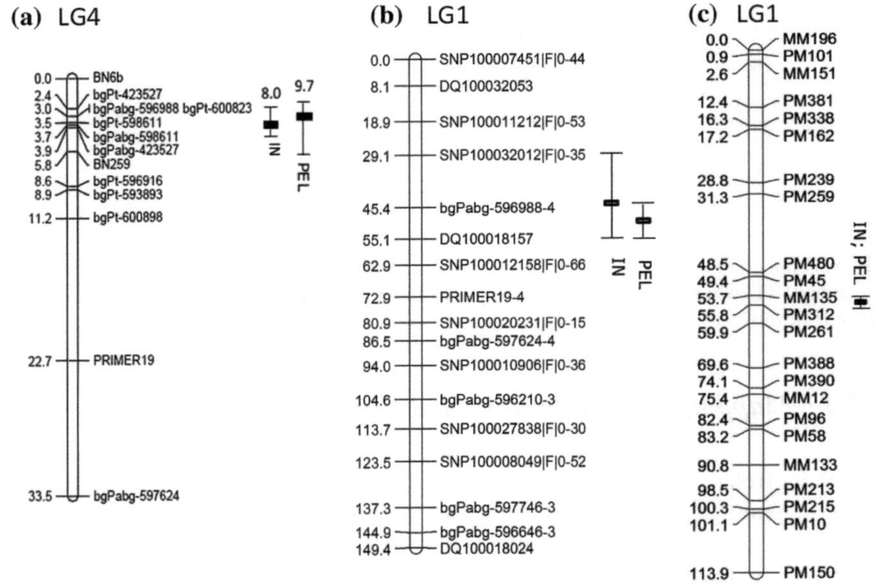

Fig. 8.4 Common mapping of the internode length (IN) and peduncle length (PEL) traits in three mapping populations (before integration and linkage group harmonization in Ho et al. (2017). **a** Molecular map of a linkage group from DipC x Tiga Nicuru (n = 73) at F_2; DArT Array markers and SSRs **b** from DipC x Tiga Nicuru at F_3; DArT Array, SSR and DArT Seq markers **c** from DipC x Tiga Nicuru at F_5; GEMs based on leaf RNA hybridized to an Affymetrix soybean microarray (Basu et al. 2007a, b; Ahmad et al. 2016; Chai et al. 2016, 2017; Ho et al. 2017)

8.8 Genomics-Aided Breeding for Climate-Smart Traits

8.8.1 Structural and Functional Genomic Resources

At the moment, there are around 100 SSR markers available, a fully developed (and openly available) Bambara groundnut DArT Seq system established as a service provider, a limited amount of XSpecies microarray data and a small amount of RNA-seq data from a number of sources (Chapman 2015; https://doi.org/10.5061/dryad.k9h76; Bonthala et al. 2016; Bonthala 2018 (Geo database acc: GSE75982); Chang et al. 2018; (https://db.cngb.org/cnsa; accession number CNP0000096); CFF—Khan et al. paper in preparation).

Transcriptomic analysis has taken two major routes so far; Cross-species hybridization to Affymetrix chips and in species RNAseq.

Bonthala et al. (2016) used an experimental analysis of the effects of progressive cold on the growth of two genotypes of Bambara groundnut (S19-3 and Uniswa Red) derived from different environments. Previous work (S. Noah. Pers comm) had demonstrated that UniSwa Red ceased root growth at higher temperatures than S19-3 and RNA was extracted at three temperatures to evaluate cold effects on transcription using a cross-species hybridization to the soybean chip. Co-expression analysis was used to identify key modules for the response. Chai et al. (2016, 2017) used leaf RNA from a small controlled cross to generate cross-species information from the Affymetrix GeneChip, allowing evaluation of the population response to imposed drought compared to continued irrigation. This permitted both QTL analysis and map construction using GEMs.

Khan et al. (2017) evaluated the transcription response to drought in two contrasting parental genotypes using a cross-species approach and identified a number of potential candidate genes indicative of a differential response in the two lines.

Initial transcriptomes have been provided for hyacinth bean (*Lablab purpureus* (L.) Sweet), grasspea (*Lathyrus sativus* L.), winged bean (*Psophocarpus tetragonolobus* (L.) DC.), and Bambara groundnut (*Vigna subterranea* (L.) Verdc.) by Chapman (2015) with an analysis of commonalities.

8.8.2 Details of Genome Sequencing

The genome draft sequence for Bambara groundnut (Kazuma, single plant) was produced as part of the AOCC effort to sequence 101 underutilized nutritious African crop genomes (www.africanorphancrops.org) and the details for Bambara groundnut can be found alongside those for *Lablab purpureus*, *Faidherbia albida*, *Sclerocarya birrea*, and *Moringa oleifera* in Chang et al. (2018) and at https://db.cngb.org/cnsa; accession number CNP0000096. http://gigadb.org/dataset/101055.

https://www.ncbi.nlm.nih.gov/bioproject?LinkName=biosample_bioproject&from_uid=9356032; Accession: PRJNA474418.

The basic approach was to carryout deep Illumina PE sequencing to high depth for each species and to produce a further 100 genotype sequences at approximately ×10 depth. So far, 67 additional genomes have been sequenced by the World Agroforestry Centre (ICRAF) (www.icraf.org; P. Hendre, pers. comm.). The genotypes came from a community call for lines to sequence through the Bambara Network (www.bambaragroundnut.org). As well as DNA sequencing, a limited number of tissues were RNA sequenced and used along with available data to help with the genome annotation. Detailed methods are given in Chang et al. (2018).

Overall, the results suggest good coverage and assembly of coding regions, with a BUSCO of 92.1, an N50 of 19,154 and 640,666 bp for contigs and scaffolds, respectively. From this, 31,701 genes were predicted with average gene size of 3,287 bp and coding size of 1,163 bp, with five exons. Of the predicted coding regions, 70.95% were identified by SwissPro matches, 69.83% by KEGG, 34.1% by COG and combined with other methods, this left a total of 633 putative genes unidentified.

Repeat classes identified within Bambara groundnut (bearing in mind that this is a short read dataset) were LTRs (105,828,735, representing 19.77% of the genome), with a major contribution by other DNA repeat types (38,294,871, representing 7.15% of the genome). LINES, satellites and simple repeats all represented minor components of the genome (<0.25%) and SINES were not detected. Other repeat types accounted for a further 11.94% of the genome, so that 38.35% of the genome is composed of recognized repeat sequences. This is reasonably consistent with the relatively small genome size of this legume, with an estimate 1C value of 880 MB (data.kew.org/cvalues/).

While this is an excellent starting point, the use of Illumina sequencing only is likely to mean that the assembly focuses on the genic regions. Some initial work to use the existing DArT Seq genetic map data to assign scaffolds to linkage groups has been done (Y. Chang and W. K. Ho, unpublished) but further work is needed with long read NGS to convert this assembly into pseudomolecules. This will hopefully happen in 2019. With the release and completion of the additional low-level (× 10) sequencing of genotypes, it should become possible to generate a sequence based association panel and construct a SNP-based genotyping array, to allow very high density exploration of LD patterns and trait inheritance, alongside a thorough exploration of the germplasm available and to implement genome-enabled breeding work.

8.9 An Account on Social, Political and Regulatory Issues

Bambara groundnut falls under the *Vigna* ssp entry in Annex 1 of the ITPGRFA (www.fao.org/plant-treaty), so germplasm is available under the conditions of the Treaty and by signing of an SMTA. As such this is a relatively simple crop to access germplasm.

Clearly, working closely with the end users (whether they be small farmers or small businesses) to focus breeding programs on desirable traits and ensuring that any products are likely to be accepted is critical. This requires participatory approaches and also a good understanding of why people still grow Bambara groundnut (where they do).

8.10 Future Perspectives

Bambara groundnut is currently quite widely spread and grown at low levels throughout sub-Saharan Africa. Modeling suggests that this can become more widely spread and that climate change may lead to further use of it. As a legume, it is an important part of low-input farming contributing N to companion crops and also providing a relatively protein rich alternative to meat. However, there are clearly a wide range of challenges which need to be addressed through breeding, although it is encouraging that the necessary genetic variation is present in the germplasm for many traits. The other issues will relate to scaling-up production and the development of further value-added products, supply chains and markets which will underpin the expansion of this crop (Hillocks et al. 2012; Feldman et al. 2019). Mechanization will be a key requirement for large-scale production, for planting, harvesting and processing and large-scale innovation can hopefully be modified for household or village scale mechanization, as Bambara groundnut is seen as being a labor-intensive crop.

References

Abu H, Buah S (2011) Characterization of Bambara groundnut landraces and their evaluation by farmers in the upper West Region of Ghana. J Dev Sustain Agri 6:64–74

Adzawla W, Donkoh SA, Nyarko G, O'Reilly PJ, Olayide O, Mayes S, Feldman AB, Azman Halimi R (2016a) Adoption of Bambara groundnut production and its effects on farmers' welfare in North Ghana. Afr J Agri Res 11(7):583–594

Adzawla W, Donkoh SA, Nyarko G, O'Reilly P, Mayes S (2016b) Use patterns and perceptions about the attributes of Bambara groundnut (*Vigna subterranea* (L.) Verdc.) in Northern Ghana. Ghana J Sci Technol Dev 4(2):56–71

Ahmad NS, Sri Redjeki E, Ho WK, Aliyu S, Mayes K, Massawe F, Kilian A, Mayes S (2016) Construction of a genetic linkage map and QTL analysis in Bambara groundnut. Genome 59(7):459–472

Ajayi FA, Lale NES (2001) Susceptibility of unprotected seeds and seeds of local Bambara groundnut cultivars protected with insecticidal essential oils to infestation by *Callosobruchus maculatus* (F.) (Coleoptera: Bruchidae). J Stored Prod Res 37:47–62

Al Shareef I, Sparkes D, Azam-Ali S (2014) Temperature and drought stress effects on growth and development of Bambara groundnut (*Vigna subterranea* L.). Exp Agri 50(1):72–89

Aliyu S, Massawe F (2013) Microsatellites based marker molecular analysis of Ghanaian Bambara groundnut (*Vigna subterranea* (L.) Verdc.) landraces alongside morphological characterization. Genet Resour Crop Evol 60:777–787

Aliyu S, Massawe F, Mayes S (2015) Beyond landraces: developing improved germplasm resources for underutilized species—a case for Bambara groundnut. Biotechnol Genet Eng Rev 30(2):127–141

Aliyu S, Massawe F, Mayes S (2016) Genetic diversity and population structure of Bambara groundnut (*Vigna subterranea* (L.) Verdc.): synopsis of the past two decades of analysis and implications for crop improvement programmes. Genet Resour Crop Evol 63:925–943

Amarteifio JO, Tibe O Njogu R (2010) The nutrient composition of Bambara groundnut [*Vigna subterreanea* (L.) Verdc.] Landraces Cultivated in Southern Africa. Agri Trop Subtrop 43:1–5

Astutia RM, Sri Palupib N, Zakariab FR (2016) Allergic reactivity of Bambara groundnut (*Vigna subterranea*) proteins. Food Agri Immunol 27(4):535–546

Atoyebi JO, Oyatomi O, Osilesi O, Adebawo O, Abberton M (2017) Morphological characterisation of selected African Accessions of Bambara groundnut (*Vigna subterranea* (L.) Verdc.). Intl J Plant Res 7(2):29–35

Atoyebi JO, Osilesi O, Abberton M, Adebawo O, Oyatomi O (2018) Quantification of selected anti—nutrients and bioactive compounds in African Bambara groundnut (*Vigna subterranea* (L.) Verdc.). Am J Food Nutr 6(3):88–95

Azam-Ali SN (1998) Bambara groundnut yield/ecology. Evaluating the potential for Bambara groundnut as a food crop in semi-arid Africa. An approach for assessing the yield potential and ecological requirements of an underutilized crop. Summary report of European Commission supported STD-3 Projects. Published by CTA

Azam-Ali SN, Sesay A, Karikari SK, Massawe F, Aguilar-Manjarrez J, Bannayan M, Hampson KJ (2001) Assessing the potential of an underutilized crop—a case study using Bambara groundnut. Exp Agri 37:433–472

Azman Halimi R, Mayes S, Barkla B, King G (2019) The potential of the underutilized pulse Bambara groundnut (*Vigna subterranea* (L.) Verdc.) for nutritional food security. J Food Compos Analys 77:47–59

Baptista A, Pinho O, Pinto E, Casal S, Mota C, Ferreira IMPLVO (2017) Characterization of protein and fat composition of seeds from common beans (*Phaseolus vulgaris* L.), cowpea (*Vigna unguiculata* L. Walp) and Bambara groundnuts (*Vigna subterranea* L. Verdc) from Mozambique. Food Meas 11:442

Basu SM, Mayes S, Davey M, Roberts JA, Azam-Ali SN, Mithen R, Pasquet RS (2007a) Inheritance of 'domestication' traits in Bambara groundnut (*Vigna subterranea* (L.) Verdc.). Euphytica 157:59–68

Basu S, Roberts JA, Azam-Ali SN, Mayes S (2007b) Bambara groundnut. In: Kole C (ed) Genome mapping and molecular breeding in plants, vol Pulses. Sugar and Tuber Crops. Springer, New York, pp 159–173

Basu S, Roberts SN, Azam-ali SN, Mayes S (2007c) Development of microsatellite markers for bambara groundnut (Vigna subterranea L. Verdc.) - an underutilised African legume crop species. Mol Ecol Notes 7(6):1326–1328

Begemann F (1988) Ecogeographic differentiation of Bambara groundnut (*Vigna subterranea*) in the collection of the International Institute of Tropical Agriculture (IITA). Wissenschaftlicher Fachverlag Dr Fleck, Niederkleen, Germany

Berchie JN, Adu-Dapaah HK, Dankyi AA, Plahar WA, NelsonQuarety F, Haleegoah J, Asafu-Agyei JN, Addo JK (2010) Practices and constraints in Bambara groundnut production, marketing and consumption in the brong ahafo and upper-East regions of Ghana. J Agron 9(3):111–118

Berchie JN, Opoku M, Adu-Dapaah H, Agyemang A, Sarkodie-Addo J, Asare E, Addo J, Akuffo H (2012) Evaluation of five Bambara groundnut (*Vigna subterranea* (L) Verdc.) landraces to heat and drought stress at Tono-Navrongo, Upper East Region of Ghana. Afr J Agri Res 7(2):250–256

Berchie JN, Amelie G, McClymont S, Raizada M, Adu-Dapaah H, Sarkodie-Addo J (2013) Performance of 13 Bambara groundnut (*Vigna subterranea* (L.) Verdc.) landraces under 12 H and 14 H photoperiod. J Agron 12(1):20–28

Berchie JN, Dapaah HA, Agyemang A, Sarkodie Addo J, Addo JK, Addy S, Blankson E (2016) Performance of five Bambara groundnut (*Vigna subterannea* (L.) Verdc.) landraces in the Transition Agroecology of Ghana under different sowing dates. Agri Food Sci J Ghana 9:718–729

Bharatkumar C, Nandini R, Bhanuprakash K, Dhanapal GN, Shashidhar H, Savithramma D (2015) Genetic Enhancement of protein and methionine content in Bambara groundnut (*Vigna subterranea* (L.) Verdc.) through mutation breeding. Intl J Res Agri For 2(11):8–13

Bonthala VS (2018) Translating nucleic acid binding protein function from model species to minor crops using transfer learning. PhD thesis, University of Nottingham

Bonthala VS, Mayes K, Moreton J, Blythe M, Wright V, May ST, Massawe F, Mayes S and Twycross J (2016) Identification of gene modules associated with low temperature response in Bambara groundnut by network-based analysis. PloS One 11(2):e0148771

Botwright Acuna T, Riffkin P, Merry A, Christy B, Richards R, Zhang H, Berger J, O'Leary G, Partington D (2015) Can the duration of the spike construction phase increase the yield of wheat? In: Proceedings of the 17th ASA conference, 20–24 Sept 2015, Hobart, Australia. www.agronomy2015.com.au

Brink M (1997) Rates of progress towards flowering and podding in Bambara groundnut (*Vigna subterranea*) as a function of temperature and photoperiod. Ann Bot 80(4):505–513

Brink M (1998) Matching crops and environments: quantifying photothermal influences on reproductive development in Bambara groundnut (*Vigna subterranea* (L.) Verdc. PhD thesis, Wageningen University, Netherlands

Brink M (1999) Development, growth and dry matter partitioning in Bambara groundnut *Vigna subterranea*) as influenced by photoperiod and shading. J Agri Sci (Camb.) 133:159–166

Brink M, Sibuga KP, Tarimo AJP, Ramolemana GM (2000) Quantifying photothermal influences on reproductive development in Bambara groundnut (*Vigna subterranea*): models and their validation. Field Crops Res 66(1):1–14

Brough SH, Azam-Ali SN, Taylor AJ (1992) The potential of Bambara groundnut (*Vigna subterranea* (L.) Verdc.) in vegetable milk production and basic protein functionality system. Food Chem 47:277–283

Chai HH, Massawe F, Mayes S (2016) Effects of mild drought stress on the morpho-physiological characteristics of a Bambara groundnut segregating population. Euphytica 208(2):225–236; Chandra K, Nandini R, Pranesh, Kumar CB, Gobu R (2017) Improving nutritional security of India through a potential underutilised legume—Bambara groundnut (*Vigna subterranea* (L) Verdc) Environ Ecol 35(2):606–610

Chai HH, Ho WK, Graham N, May S, Massawe F, Mayes S (2017) A cross-species gene expression marker-based genetic map and QTL analysis in bambara groundnut. Genes 8:84

Chang Y, Liu H, Liu M, Liao X, Sahu SK, Fu Y, Song B, Cheng S, Kariba R, Muthemba S, Hendre PS, Mayes S, Ho WK, Kendabie K, Wang S, Li L, Muchugi A, Jamnadass R, Lu R, Peng S, Van Deynze A, Simons A, Yana-Shapiro H, Xu X, Yang H, Wang J, Liu X (2018) The draft genomes of five agriculturally important African orphan crops. GigaScience. https://doi.org/10.1093/gigascience/giy152

Chapman M (2015) Transcriptome sequencing and marker development for four underutilized legumes. Appl Plant Sci 3(2):1400111

Chibarabada TP, Modi AT, Mabhaudhi T (2015) Water use characteristics of a Bambara groundnut (*Vigna subterranea* L. Verdc) landrace during seedling establishment. Water SA 41(4):472–482

Chotechung S, Somta P, Chen J, Yimram T, Chen X, Srinives P (2016) A gene encoding a polygalacturonase-inhibiting protein (PGIP) is a candidate gene for bruchid (Coleoptera: *bruchidae*) resistance in mungbean (*Vigna radiata*). Theor Appl Genet 129(9):1673–1683

Christy B, Tausz-Posch S, Tausz M, Richards R, Rebetzke G, Condon A, McLean T, Fitzgerald G, Bourgault M, O'Leary G (2018) Benefits of increasing transpiration efficiency in wheat under elevated CO2 for rainfed regions Glob Change Biol 24(5):1965–1977; Collinson ST, Clawson EJ, Azam-Ali SN, Black CR (1997) Effects of soil moisture deficits on the water relations of Bambara groundnut (*Vigna subterranea* L. Verdc.). J Exp Bot 48:877–884

Condon AG, Richards R, Rebetzke G, Farquhar G (2004) Breeding for high water-use efficiency. J Exp Bot 55:2447–2460; Dakora FD (1998) Nodule function in symbiotic Bambara groundnut (*Vigna subterranea* L.) and kersting's bean (*Macrotyloma geocarpum* l.) is tolerant of nitrate in the root medium. Ann Bot 82:687–690

Dakora FD, Belane AK, Mohale KC, Makhubedu TI, Makhura P, Pule-Meulenberg F, Mapope N, Mogkelhe SN, Gyogluu C, Phatlane GP, Muhaba S, Mokobane F, Oteng-Frimpong R (2015) Food grain legumes: their contribution to soil fertility, food security, and human nutrition/health in Africa. In: Frans J de Bruijn (ed) Biological nitrogen fixation. Wiley-Blackwell. https://doi.org/10.1002/9781119053095

Dalziel JM (1937) *Voandzeia* Thou. In: The useful plants of west tropical Africa. Crown Agents for the Colonies, London, Agents, London, pp 269–271

Dhanaraj B (2018) Effect of short duration high temperature stress on bambara groundnut (Vigna subterranea (L.) Verdc.) plant reproduction. PhD thesis, University of Nottingham

Doku EV, Karikari SK (1971) Operational selection in wild Bambara groundnut. Ghana J Sci 11:45–56

Duke J (1981) *Voandzeia subterranea* (L). Thouars. In: Handbook of legumes of world economic importance. Plenum Press, New York, pp 307–310

Echezona BC, Amuji CF, Eze S (2013) Evaluation of some accessions of Bambara groundnut (*Vigna subterranea* L. Verdc.) for resistance to bruchid infestation, based on grain source and seed coat colour. J. Plant Protec Res 53(3):210–213

Effa EB, Uko AK (2017) Food security potentials of Bambara groundnut (*Vigna subterranea* (L.) Verdc.). Intl J Dev Sustain 6 (12):1919–1930

Farquhar GD, Ehleringer JR, Hubick KT (1989) Carbon isotope discrimination and photosynthesis. Annu Rev Plant Physiol Plant Mol Biol 40:503–537

Fatimah S, Ariffin, Ardiarini NR, Kuswanto (2018) Genetic diversity of madurese Bambara groundnut (*Vigna subterranea* L. Verdc.) lines based on morphological and RAPD markers. SABRAO J Breed Genet 50(2):101–114

Feldman A, Ho WK, Massawe F, Mayes S (2019) Bambara groundnut is a climate-resilient crop: how could a drought-tolerant and nutritious legume improve community resilience in the face of climate change? In: Sarkar A, Sensarma S, vanLoon G (eds) Sustainable solutions for food security. Springer, Cham, Switzerland, pp 151–167

Gibb DJ, Lee SC, Isa MN, Gramuglia S, Fukao S, Bassel GW, Correia CS, Corbineau F, Theodoulou FL, Bailey-Serres J, Holdsworth MJ (2012) Homeostatic response to hypoxia is regulated by the N-end rule pathway in plant. Nature 479(7373):415–418

Gonné S, Félix-Alain W, Benoît KB (2013) Assessment of twenty Bambara groundnut (*Vigna subterranea* (L.) Verdcourt) landraces using quantitative morphological traits. Intl J Plant Res 3(3):39–45. https://doi.org/10.5923/j.plant.20130303.04

Graciet E, Wellmer F (2010) The plant N-end rule pathway: Structure and functions. Trends Plant Sci 15:447–453

Hepper FN (1963) Plants of the 1957–58 West African Expedition: II. The Bambara groundnut (*Voandzeia subterranea*) and Kersting's groundnut (*Kerstingiella geocarpa*) Wild in West Africa. Kew Bull 16:395. https://doi.org/10.2307/4114681

Hillocks RJ, Bennett C, Mponda OM (2012) Bambara nut: a review of utilisation, market potential and crop improvement. Afr Crop Sci J 20(1):1–16

Ho WK, Muchugi A, Muthemba S, Kariba R, Mavenkeni BO, Hendre P, Song B, Deynze AV, Massawe F, Mayes S (2016) Use of microsatellite markers for the assessment of Bambara groundnut breeding system and varietal purity before genome sequencing. Genome. 59(6):427–431

Ho WK, Chai HH, Kendabie P, Ahmad NS, Jani J, Massawe F, Kilian A, Mayes S (2017) Integrating genetic maps in Bambara groundnut [*Vigna subterranea* (L) Verdc.] and their syntenic relationships among closely related legumes. BMC Genomics 18(1):192

Ismail AM, Singh US, Singh S, Darb MH, Mackill DJ (2013) The contribution of submergence-tolerant (Sub1) rice varieties to food security in flood-prone rainfed lowland areas in Asia. Field Crops Res 152:83–93

Jørgensen ST, Aubanton M, Harmonic C, Dieryck C, Jacobsen S, Simonsen H, Ntundu W, Stadler F, Basu S, Christiansen J (2009) Identification of photoperiod neutral lines of Bambara groundnut (*Vigna subterranea*) from Tanzania. In: IOP Conference Series: earth and environmental science, vol 6, 2023

Jørgensen ST, Liu F, Ouédraogo M, Ntundu WH, Sarrazin J, Christiansen JL (2010) Drought responses of two Bambara groundnut (*Vigna subterranea* L. Verdc.) landraces collected from a dry and a humid area of Africa. J Agron Crop Sci 196:412–422

Karikari SK (2004) Variability between local and exotic Bambara groundnut landraces in Botswana. Afr Crop Sci J 8:145–152

Karunaratne AS, Azam-Ali SN, Al-Shareefa I, Sesay A, Jørgensen ST, Crout NMJ (2010) Modelling the canopy development of Bambara groundnut. Agri For Meterol 150(7–8):1007–1015

Karunaratne AS, Azam-Ali SN, Izzi G, Steduto P (2011) Calibration and validation of fao-aquacrop model for irrigated and water deficient Bambara groundnut. Exp Agric 47(3):509–527

Karunaratne AS, Walker S, Ruane AC (2015) Modelling Bambara groundnut yield in Southern Africa: towards a climate-resilient future. Clim Change 65:193–203

Kendabie P, Massawe F, Mayes S (2015) Developing genetic mapping resources from landrace-derived genotypes that differ for photoperiod sensitivity in Bambara groundnut (*Vigna subterranea* L.). Asp Appl Biol 124:1–8

Khan F, Chai HH, Ajmera I, Hodgman C, Mayes S, Lu C (2017) A transcriptomic comparison of two bambara groundnut landraces under dehydration stress. Genes 8(4):533–546

Kosini D, Saidou C, Nukenine EN (2017) Physico-chemical properties and resistance of ten Bambara groundnut (*Vigna subterranea)* varieties to attack by *Callosobruchus maculatus* (Fabricius) (Coleoptera: Chrysomelidae) in the Sudano-Sahelian and Sudano-Guinean Zones of Cameroon. J Exp Agri Intl 15(1):1–14

Kwak M, Toro O, Debouck DG (2012) Gepts P (2012) Multiple origins of the determinate growth habit in domesticated common bean (*Phaseolus vulgaris*). Ann Bot 110:1573–1580

Linnemann AR (1993) Phenological development of Bambara groundnut (*Vigna subterranea* (L) Verdc.) at constant exposure to photoperiod of 10 h and 16 h. Ann Bot 71:445–452

Linnemann AR, Craufurd PQ (1994) Effects of temperature and photoperiod on phenological development in three genotypes of Bambara groundnut (*Vigna subterranea*). Ann Bot 74:675–681

Linnemann AR, Westphal E, Wessel M (1995) Photoperiod regulation of development and growth in Bambara groundnut (*Vigna subterranea*). Field Crop Res 40:39–47

Mabhaudhi T, Modi AT (2013) Growth, phenological and yield responses of a Bambara groundnut (*Vigna subterranea* (L.) Verdc.) landrace to imposed water stress under field conditions. S Afr J Plant Soil 30(2):69–79. Mabhaudhi T, Modi AT, Beletse YG (2013) Growth, phenological and yield responses of a Bambara groundnut (*Vigna subterranea* L. Verdc) landrace to imposed water stress: II. Rain shelter conditions. Water SA 39:191–198

Mabhaudhi T, Chibarabada TP, Chimonyo VGP, Modi AT (2018) Modelling climate change impact: a case of Bambara groundnut (*Vigna subterranea*). Phys Chem Earth 105:25–31

Massawe F, Roberts J, Azam-Ali S, Davey MR (2003a) Genetic diversity in Bambara groundnut (*Vigna subterranea* (L.) Verdc) landraces assessed by random amplified polymorphic DNA (RAPD) markers. Genet Resour Crop Evol 50:737–741

Massawe F, Dickinson M, Roberts J, Azam-Ali S (2003b) Genetic diversity in Bambara groundnut (*Vigna subterranea* (L.) Verdc) landraces revealed by AFLP markers. Genome 45(6):1175–1180

Massawe F, Schenkel W, Basu S, Temba EM (2003c) Artificial hybridisation in bambara groundnut (*Vigna subterranea* (L.) Verdc.). In: Proceedings of the international Bambara groundnut symposium, vol 1, Botswana, pp 193–209

Massawe FJ, Mwale SS, Azam-Ali SN Roberts JA (2005) Breeding in Bambara groundnut (*Vigna subterranea* (L.) Verdc.): strategic considerations. Afr J Biotechnol 4:463–471

Mayes S, Basu S, Murchie E, Roberts JA, Azam-Ali SN, Stadler F, Mohler V,Wenzel G, Massawe F, Kilian A, Bonin A, Beena A, Sheshshayee MS (2009) BAMLINK—a cross disciplinary programme to enhance the role of Bambara groundnut (*Vigna subterranea* L. Verdc.) for food security in Africa and India. Acta Hort 806:137–150

Mayes S, Massawe FJ, Alderson PG, Roberts JA, Azam-Ali SN, Hermann M (2011) The potential for underutilized crops to improve security of food production. J Exp Bot 63(3):1075–1079

Mendiondo GM, Gibbs DJ, Szurman-Zubrzycka M, Korn A, Marquez J, Szarejko I, Maluszynski M, King J, Axcell B, Smart K, Corbineau F, Holdsworth MJ (2016) Enhanced waterlogging tolerance in barley by manipulation of expression of the N-end rule pathway E3 ligase PROTEOLYSIS 6. Plant Biotechnol J 14:40–50

Miesho WB, Gebremedhin HM, Msiska UM, Mohammed KE, Malinga GM, Sadik K, Odong TL, Rubaihayo P, Kyamanywa S (2018) New sources of cowpea genotype resistance to cowpea bruchid *Callosobruchus maculatus* (F.) in Uganda. Intl J Agron Agri Res 12(4):39–52

Minka SR, Bruneteau M (2000) Partial chemical composition of Bambara pea [*Vigna subterranea* (L.) Verdc.]. Food Chem 68:273–276

Mir RR, Kudapa H, Srikanth S, Saxena RK, Sharma A, Azam S, Saxena K, Penmetsa RV, Varshney RK (2014) Candidate gene analysis for determinacy in pigeonpea (*Cajanus* spp.). Theor Appl Genet 127:2663–2678

Mkandawire CH (2007) Review of Bambara groundnut production in sub-Saharan Africa. Agri J 2(4):464–470

Mohammed MS, Shimelis HA, Laing MD (2016) Preliminary investigation on some agronomic and morphological variations of within and between Bambara groundnut landraces. J Agri Sci Technol 18:1909–1920

Molosiwa OO, Aliyu S, Stadler F, Mayes K, Massawe F, Kilian A, Mayes S (2015) SSR marker development, genetic diversity and population structure analysis of Bambara groundnut [*Vigna subterranea* (L.) Verdc.] landraces. Genet Resour Crop Evol 62:1225–1243

Mubaiwa J, Fogliano V, Chidewe C, Bakker EJ, Linnemann AR (2018) Utilization of bambara groundnut (Vigna subterranea (L.) Verdc.) for sustainable food and nutrition security in semi-arid regions of Zimbabwe. PLoS One 13(10):e0204817

Muhammad YY, Mayes S, Massawe F (2015) Effects of short term water deficit stress on physiological characteristics of Bambara groundnut (*Vigna subterranea* (L) Verdc.). S Afr J Plant Soil 33(1):1–8

Musa M, Massawe F, Mayes S, Alshareef I and Singh A (2016) Nitrogen fixation and N-balance studies on Bambara groundnut (*Vigna subterranea* L. Verdc) landraces grown on tropical acidic soils of Malaysia. Commun Soil Sci Plant Analys 47(4):533–542

Mwale SS, Azam-Ali SN, Massawe FJ (2007). Growth and development of Bambara groundnut (*Vigna subterranea*) in response to soil moisture 2. Resource capture and conversion. Eur J Agron 26(4):354–362

Nautiyal PC, Kulkarni G, Singh AL, Basu MS (2017) Evaluation of water-deficit stress tolerance in Bambara groundnut landraces for cultivation in sub-tropical environments in India. Indian J Plant Physiol 22(2):190–196

Nofita ANF, Sri Lestari P, Kuswanto (2015) Agronomical characters on 18 selected genotypes of Bambara groundnut (*Vigna subterranea* (L.) Verdcourt). J Produksi Tanaman 3(2):157–163

Ntundu WH, Shillah SA, Marandu WYF, Christiansen JL (2006) Morphological diversity of Bambara groundnut [*Vigna subterranea* (L.) Verdc.] landraces in Tanzania. Genet Resour Crop Evol 53:367–378

Obagwu J (2003) Evaluation of Bambara groundnut (*Vigna subterranea* (L.) Verdc) lines for reaction to *Cercospora* leaf spot. J Sustain Agri 22(1):93–100

Odeigah PGC, Osanyinpeju AO (1998) Evaluating the genetic biodiversity of Bambara groundnut accessions from Nigeria using SDS-polyacrylamide gel electrophoresis. Genet Resour Crop Evol 45:451–458

Odongo FO, Oyoo FE, Wasike V, Owouche JO, Karanje K, Korir, P (2015) Genetic diversity of Bambara groundnut (*V. subterranea* (L) Verdc) landraces in Kenya using microsatellite markers. Afr J Biotechnol 14(4):283–291

Ofori I (1996) Correlation and path-coefficient analysis of components of seed yield in Bambara groundnut (*Vigna subterranea*). Euphytica 91:103–107

Ofori K, Kumaga FK, Bimi LK (2001) Variation in seed size, seed protein and tannin content of Bambara groundnut. Trop Sci 41:41–44

Okpuzor J, Ogbunugafor HA, Okafor U, Sofidiya M (2010) Identification of protein types in bambara nut seeds: perspectives for dietary protein supply in developing countries EXCLI J 9:17–28. http://www.excli.de/vol9/okpuzor02_2010/okpuzor_bambara_010210_proof.pdf

Olayide OE, Donkoh SA, Ansah IGK, Adzawla W, O'Reilly PJ, Mayes S, Feldman A, Halimi RA, Nyarko G, Ilori CO, Alabi T (2018) Assessing socioeconomic factors influencing production and commercialisation of Bambara groundnut as an indigenous climate resilient crop in Nigeria In: Leal Filho W (ed) Handbook of climate change resilience. Springer Nature. https://doi.org/10.1007/978-3-319-71025-9_158-1

Olukolu BA, Mayes S, Stadler F, Ng NQ, Fawole I, Dominique D, Azam-Ali SN, Abbott AG, Kole C (2012) Genetic diversity in Bambara groundnut (*Vigna subterranea* (L.) Verdc.) as revealed by phenotypic descriptors and DArT marker analysis. Genet Resour Crop Evol 59:347–358

Ouedraogo M, Oeudraogo TJ, Tignere BJ, Balama D, Dabiere BC, Konate G (2008) Characterization and evaluation of accessions of Bambara groundnut (*Vigna subterranea* (L.) Verdcourt) from Burkina Faso. Sci Nat 5:191–197

Pasquet RS (2004) Bambara groundnut and cowpea genepool organization and domestication. In: Proceedings of the International Bambara Groundnut Symposium, 5–12 Sept 2003, Gaborone, Botswana, pp 265–274

Pasquet RS, Fotso M (1997) The ORSTOM Bambara groundnut collection. In: Heller J, Begemann F, Mushonga J (eds) Bambara Groundnut, *Vigna subterranea* (L.) Verdc. IPGRI, Rome, Italy, pp 119–123

Pasquet RS, Schwedes S, Gepts P (1999) Isozyme diversity in Bambara groundnut. Crop Sci 39:1228–1236

Raper DC Jr, Thomas JF (1978) Photoperiodic alteration of dry matter partitioning and seed yield in soybean. Crop Sci 18:654–656

Repinski SL, Kwak M, Gepts P (2012) The common bean growth habit gene *PvTFL1y* is a functional homolog of Arabidopsis *TFL1*. Theor Appl Genet 124:1539

Santos R (2018) PhD thesis submitted in partial fulfilment of the requirements of a Doctor of Philosophy Degree, University of Nottingham, UK

Sesay A, Mpuisang T, Morake TS, Al-Shareef I, Chepete HJ, Moseki B (2013) Preliminary assessment of Bambara groundnut (*Vigna subterranea* L.) landraces for temperature and water stress tolerance under field conditions in Botswana. S Afr J Plant Soil 27(4):312–321

Shegro WS, van Rensburg J, Adebola PO (2013) Assessment of genetic variability in Bambara groundnut (*Vigna subterrenea* L. Verdc.) using morphological quantitative traits. Acad J Agri Res 1(3):45–51

Slafer GA, González FG, Kantolic AG, Whitechurch EM, Abeledo LG, Miralles DJ, Savin R. Basra AS (2006) Grain number determination in major grain crops. In: Basra AS (ed) Handbook of seed science and technology. Haworth Press, New York, pp 95–123

Somta P, Chankaew S, Rungnoi O, Srinives P (2011) Genetic diversity of the Bambara groundnut (*Vigna subterranea* (L.) Verdc.) as assessed by SSR markers. Genome 54(11):898–910

Song YH, Ito S, Imaizumi T (2010) Similarities in the circadian clock and photoperiodism in plants. Curr Opin Plant Biol 13(5):594–603

Soni ML, Yadava ND, Talwar HS, Nathawat NS, Rathore VS, Gupta K (2015) Variability in heat tolerance in Bambara groundnut (*Vigna subterranea* (L.) Verdc.) Indian J Plant Physiol 20(1):92–96

Sprent JI, Odee DW, Dakora FD (2010) African legumes: a vital but under-utilized resource. J Exp Bot 61(5):1257–1265

Srinivasan A, Takeda H, Senboku T (1996) Heat tolerance in food legumes as evaluated by cell membrane thermostability and chlorophyll fluorescent technique. Euphytica 88:35–43

Srinives P, Somta P, Somta C (2007) Genetics and breeding of resistance to bruchids (*Callosobruchus spp.*) in Vigna Crops: a review. NU Sci J 4(1):1–17

Stadler F (2008) Analysis of differential gene expression under water-deficit stress and genetic diversity in Bambara groundnut (*Vigna subterranea* (L.) Verdc.) using novel high-throughput technologies. PhD thesis, Technische Universität München, Wissenschaftszentrum Weihenstephan, für Ernährung, Landnutzung und Umwelt Lehrstuhl für Pflanzenzüchtung, Germany

Stadler F, Mohler V, Kilian A, Bonin A, Thümmler F, Mayes S, Ros B, Wenzel G (2007) Applying biotechnology to underutilised crops: initial results from applying Diversity Arrays Technology

(DArT) and massively parallel signature sequencing to Bambara groundnut. New approaches to plant breeding of orphan crops in Africa, 19–21 Sept, Bern, Switzerland. Abstracts 66

Summerfield RJ, Roberts EH, Ellis RH, Lawn RJ (1991) Towards the reliable prediction of time to flowering in 6 annual crops 1. The development of simple-models for fluctuating field environments. Exp Agri 27(1):11–31

Summerfield RJ, Lawn RJ, Qi A, Ellis RH, Roberts EH, Chay PM, Brouwer JB, Rose JL, Shanmugasundaram S Yeates SJ, Sandover S (1993). Towards the reliable prediction of time to flowering in 6 annual crops 2. Soybean (*Glycine-max*). Exp Agri 29(3):253–289

Summerfield RJ, Asumadu H, Ellis RH, Qi A (1998) Characterization of the photoperiodic response of post-flowering development in maturity isolines of soyabean [*Glycine max* (L.) Merrill] 'Clark'. Ann Bot 82(6):765–771

Suwanprasert J, Toojinda T, Srinives P, Chanprame S (2006) Hybridization technique for Bambara groundnut. Breed Sci 56:125–129

Tadele Z (2018) African orphan crops under abiotic stresses: challenges and opportunities. Scientifica 1451894. https://doi.org/10.1155/2018/1451894

Talwar HS, Takeda H, Yashima S, Senboku T (1999) Growth and photosynthetic responses of groundnut genotypes to high temperature. Crop Sci 39:460–466

Talwar HS, Chandra Sekhar A, Nageswara Roa RC (2002) Genotypic variability in membrane thermostability in groundnut. Indian J Plant Physiol 7(2):97–102

Temegne NC, Gouertoumbo WF, Wakem GA, Nkou FTD, Youmbi E, Ntsomboh-Ntsefong G (2018) Origin and ecology of Bambara groundnut (*Vigna subterranea* (L.) Verdc. J Ecol Nat Resour 2(4):000140

Trachse S, Kaeppler SM, Brown KM, Lynch JP (2011) Shovelomics: high throughput phenotyping of maize (*Zea mays* L.) root architecture in the field. Plant Soil 341(1–2):75–87

Unigwe AE, Gerrano AS, Adebola P, Pillay M (2016) Morphological variation in selected accessions of Bambara groundnut (*Vigna subterranea* L. Verdc) in South Africa. J Agri Sci 8(11):69–80

Uvere P, Onyekwere EU, Ngoddy PO (2010) Production of maize–Bambara groundnut complementary foods fortified pre-fermentation with processed foods rich in calcium, iron, zinc and provitamin A. J Sci Food Agri 90:566–573

Vicente J, Mendiondo G, Pauwels J, Pastor V, Izquierdo Y, Naumann C, Movahedi M, Rooney D, Gibbs DJ, Smart K, Bachmair A, Gray JE, Dissmeyer N, Castresana C, Ray RV, Gevaert K, Holdsworth MJ (2018) Distinct branches of the N-end rule pathway modulate the plant immune response. New Phytol. https://doi.org/10.1111/nph.15387

Wakhungu CN, Tabu IM, Otaye OD, Wasike WV (2017) Distribution of Fusarium wilt of Bambara nut (*Vigna subterranea* (L.) Verdc.) in farmers' fields' of Busia County in Western Kenya and its management using farmyard manure. Arch Phytopathol Plant Protec 50(7–8):398–414

Yetunde EA, Iyam MA, Lawal O, Udofia U, Ani IF (2009) Utilization of Bambara groundnut flour blends in bread production. J Food Technol 4:111–114

Zeven AC (1998) Landraces: a review of definitions and classifications. Euphytica 104(2):127–139

Chapter 9
Grass Pea: Remodeling an Ancient Insurance Crop for Climate Resilience

Abhimanyu Sarkar, Peter M. F. Emmrich, Ashutosh Sarker, Xuxiao Zong, Cathie Martin and Trevor L. Wang

Abstract Grass pea (*Lathyrus sativus*) is a hardy legume grown for food, feed, and fodder. It is an ancient crop which has been cultivated for more than 8000 years because of its tolerance of drought, flooding, salinity, and poor soils, its ability to fix nitrogen, and its seeds with high levels of protein. These traits make it an outstanding crop for ensuring nutritional security (particularly for protein) for resource-poor farmers, especially in the face of impending changes in climate. However, the presence of β-N-oxalyl-L-α,β-diaminopropionic acid (β-ODAP or ODAP), a neurotoxin present in the seeds and vegetative tissues of grass pea, has limited its breeding and modern-day cultivation. β-ODAP causes lathyrism, a paralysis of the lower limbs that occurs in epidemics in undernourished communities. This has resulted in grass pea being an "orphan crop" whose potential has not been fully realized due to lack of markets and research funding. The recent emphasis on climate smart crops has refocused attention on this very promising crop. Genomic resources and low-ODAP

A. Sarkar · P. M. F. Emmrich · C. Martin (✉) · T. L. Wang
John Innes Centre, Norwich Research Park, Norwich NR4 7UH, UK
e-mail: cathie.martin@jic.ac.uk

A. Sarkar
e-mail: abhimanyu.sarkar@jic.ac.uk

P. M. F. Emmrich
e-mail: peter.emmrich@jic.ac.uk

T. L. Wang
e-mail: trevor.wang@jic.ac.uk

P. M. F. Emmrich
Biosciences Eastern and Central Africa - International Livestock Research Institute Hub, P.O. 30709, Nairobi, Kenya

A. Sarker
South Asia & China Program & Principal Food Legume Breeder, International Center for Agricultural Research in the Dry Areas (ICARDA), NASC Complex, DPS Marg, New Delhi 110012, India
e-mail: a.sarker@cgiar.org

X. Zong
Institute of Crop Sciences, Chinese Academy of Agricultural Sciences, Beijing 100081, China
e-mail: zongxuxiao@caas.cn

C. Kole (ed.), *Genomic Designing of Climate-Smart Pulse Crops*,
https://doi.org/10.1007/978-3-319-96932-9_9

lines are being developed, and it is hoped that these will soon allow grass pea to reach its full potential as a resilient protein crop for food and nutritional security through sustainable agriculture in the face of climate change.

Keywords *Lathyrus* · Protein · Climate smart · Food security · Nutrition · Genetic maps · β-ODAP

9.1 Challenges, Priorities, and Prospects of Recent Plant Breeding

9.1.1 Food, Nutrition, Energy and Environmental Security (FNEES)

Food security is defined by the Food and Agriculture Organization (FAO) of the United Nations (UN) and the World Health Organisation (WHO) (https://www.wfp.org/node/359289) as the state of having reliable access to a sufficient quantity of affordable, nutritious food. This is a global aspiration as defined in the Sustainable Development Goals (SDG) of the United Nations (https://www.un.org/sustainabledevelopment/sustainable-development-goals/). In many developing countries, meat, dairy, and fish are expensive and out of the reach of the poor. These populations therefore depend on plant foods to cover their protein needs. Protein and energy deficiencies, in both quantity and quality, are often the cause of widespread malnutrition. In the developing world, around 30% of the populations are currently undernourished according to the United States Department of Agriculture's (USDA) 2014–2024 International Food Security Assessment (http://farmpolicy.com/wp-content/uploads/2014/06/gfa25_final-0626.pdf). In addition, Low and Medium Income Countries (LMICs) regularly suffer from drought. Ethiopia, for example, is currently suffering from its worst drought in decades, with more than 10 million people needing food aid. Many of the regions cultivating legumes as sources of plant proteins rely on rainfed crop growth with limited access to resources or fertilizers, so the requirement for resilient legume crops in such areas is paramount.

Pulses are the ideal crops to address food, nutrition, energy, and environment (FNEE) security (Kole 2017). Pulses are a hugely important component of sustainable food production systems aimed at increasing food and nutritional security. Protein obtained from pulses is significantly less expensive and is markedly more sustainable than animal protein, requiring substantially lower inputs, especially of water. Pulses are crucial components of multiple cropping systems, namely intercropping, crop rotation and agroforestry. These cropping systems have substantially greater species diversity than monocrop systems, which translates into not only a more efficient use of resources, such as light, water, and nutrients but also into higher outputs as yields are increased, alongside lower risk of overall crop failure. In multiple cropping systems, nutrient recycling and soil formation are improved through the ability of pulse crops to fix nitrogen and release phosphorous.

In recognition of the importance of legumes worldwide for food, nutrition, energy and environment security, the United Nations declared 2016 the International Year of Pulses, to promote grain legume production, consumption, and value chain development and to improve global food security, health and agricultural sustainability.

Pulse crops are of particular importance in India because of the high proportion of vegetarians, variously estimated as between 20 and 42% of the population. There are as many as 500 million people in India solely dependent on nonanimal protein sources. India is the world's largest producer of pulses but still needs to import considerable amounts of these crops every year because productivity levels do not match those achieved in other parts of the world. For example, India's pulse production was around 17.5 million tons in 2015, representing a shortfall of 20% compared to a requirement of at least 22 million tons. The area of pulse production in India is around 22–23 million hectares with little change since 1990–1991 (Singh et al. 2015).

One strategy to address food insecurity (and particularly protein insecurity) in LMICs is by providing resources to improve locally accepted pulse crops, such as grass pea (*Lathyrus sativus* L.). Grass pea is a robust crop that can be used as a primary source of protein for animal and human consumption. Grass pea has been cultivated for thousands of years because of its remarkable ability to tolerate drought, excess precipitation and flooding (Campbell et al. 1994; Malek 1998; Girma and Korbu 2012) and it also requires lower inputs than other crops for its cultivation. It has a penetrating root system and can be cultivated in very poor to heavy clay soil (Campbell et al. 1994). As a legume, grass pea has the ability to fix atmospheric nitrogen (Campbell et al. 1994; Dixit et al. 2016) and these qualities make it an attractive crop for adverse agricultural conditions (Ahlawat et al. 1981; Campbell et al. 1994; Joshi 1998). It can be cultivated both in poor soils and in environments prone to flooding or drought. For these reasons, it is often used as an insurance crop since it continues to yield when all other crops fail.

No plant can survive entirely without water, but physiological and morphological adaptations allow grass pea to:

(1) avoid drought by means of early flowering and a fast life cycle and by restricting vegetative growth under water stress to free up resources to complete grain filling, and to
(2) avoid and tolerate dehydration by accessing water deeper in the soil and osmotically adjusting to maintain turgor. Its penetrating root system allows the plant to be cultivated on poor and heavy soils with minimal management. Grass pea has a high nutritional value due to its high protein content of around 30% of dry seed mass and an amino acid profile high in lysine (Fikre et al. 2008). This makes it an ideal complement to drought tolerant cereals, e.g., sorghum, millets, which are low in lysine but relatively high in sulfur-containing amino acids (methionine and cysteine), which are low in grass pea and other legumes.

Grass pea currently is disadvantaged, however, by being a poorly researched orphan crop with very few genetic resources. This neglect may be because it contains a toxin, β-*N*-oxalyl-L-α,β-diaminopropionic acid (β-ODAP or ODAP) associated with an irreversible, neurodegenerative disorder, neurolathyrism, in human and domestic animals (Patto and Rubiales 2014; Dixit et al. 2016). The reason behind lathyrism is due to the presence of antinutritive factor β-ODAP, which has neurotoxic properties (Rao et al. 1964; Xu et al. 2017; Getahun et al. 2003). Grass pea can be safely consumed as part of a normal diet, but if consumed excessively (>40% of caloric intake for several months) during times of malnourishment, it may cause neurolathyrism, marked by permanent paralysis of the legs following the death of motor neurones. The production of the toxin is highly dependent on environmental conditions and increases under a number of stress conditions, including drought. This raises the risk of neurolathyrism, as such conditions are likely to coincide with malnutrition and with grass pea forming a relatively larger part of people's diets.

9.1.2 Effects of Global Warming and Climate Change

Many of the regions of the world cultivating pulse crops rely on rainfed growth, and farmers are typically already resource-poor with limited access to irrigation infrastructure, making their crops highly vulnerable to heat and drought stress that can cause up to 50% yield losses (Singh et al. 2015). Cool season legumes are especially vulnerable to terminal drought due to shortening springs and drought conditions that are occurring earlier in the season. This situation is likely only to worsen in the future because of climate change and is one of the most important factors limiting the production of cool season legumes. The requirement for more resilient, climate smart pulse crops is therefore paramount.

Grass pea meets these requirements well because of its tolerance of both drought and flooding, thought to result principally from its large and penetrating root structure, allowing it to use deeper lying water in times of drought compared to close relatives like pea (*Pisum sativum*) or other pulses like chickpea (*Cicer arietinum*), and to explore deeper, more aerated soils in times of waterlogging (Fig. 9.1). The resilience of grass pea and its replenishment of nitrogen in soils as a legume crop, means that it is also well suited for development as a forage crop using varieties suited to interplanting (for example with rice or with finger millet) for improved agricultural sustainability, especially in LMICs in the face of climate change.

9.1.3 Limitations of Traditional Breeding

Traditional breeding relies heavily on germplasm diversity whether natural or induced. Currently, the genetic and genomic resources for the improvement of grass pea are poor with non-standardized methods of germplasm curation, dispersed collec-

Fig. 9.1 A field of grass pea growing in Oromia region of Ethiopia illustrating the resilience of grass pea. The field was originally sown with chickpea, but none germinated after 3 weeks without rain. The field was resown with grass pea and the image was captured after a further 7 weeks without rain. The inset shows one of very few rogue chickpea plants that eventually germinated

tions, and even one collection at the Conservatoire Botanique National des Pyrénées et de Midi Pyrénées (CNB PMP) in the Pyrenees in danger of being lost as a result of lack of financial resources (see Sect. 9.3.1). This hinders significantly the use of such resources for breeding and grass pea improvement.

The most important single trait to be addressed for grass pea is the content of the toxin, β-ODAP. Its removal is vital to any improvement in the crop, and a number of varieties with β-ODAP contents of seed of <0.1% have been released, such as Wasi, developed by Dr. Ali Abd El-Moneim at the International Center for Agriculture Research in the Dry Areas (ICARDA), and Mahateora, promoted by India's National Food Security Mission and OCP Foundation projects (https://www.icarda.org/crop/grasspeas). For further rapid progress in grass pea breeding, conservation and maintenance of germplasm collections that ensure its return to active research are needed urgently.

The identification of new markers that can be used in breeding programs utilizing globally diverse germplasm, must also be a priority. In addition, facilitated access to the variation available in the ICARDA collection of ~4000 grass pea accessions, through an ecoTILLING platform (see Sect. 9.9.1) or sequencing of the core collection of accessions, would greatly augment the ability to develop this under-resourced orphan crop for both food and fodder applications.

New lower β-ODAP traits, with levels that remain low in stressful environments together with climate smart traits that can improve the resilience of grass pea productivity and nutritional quality, are important targets for future breeding programs.

9.2 Prioritizing Climate Smart Traits

Arguably, grass pea is already climate smart. Its resilience to adverse conditions means that it is likely to survive and produce a crop even when the temperatures rise as predicted and climatic conditions become more extreme. One would also expect that it could be brought into more widespread use under such conditions and become a staple legume in, for example, Northern Europe. Even if it were not a crop already, it is also a useful model for obtaining climate smart genes because of its resilience. The barrier to such acceptance is, of course, its toxicity and hence we would include this as a crop-specific climate smart target for grass pea (see Sect. 9.2.14). Examination of its inherent climate smart traits for transfer into other less climate smart crops will become much more feasible once the appropriate genomic tools are fully in place (see Sects. 9.8, 9.9, 9.10 and 9.11). Many of these features of grass pea were outlined by Campbell (1997).

9.2.1 Flowering Time

Flowering time is a crucial trait in domestication and a major determinant of time to maturity. The optimal flowering time for a crop depends on the geographical and agronomic environment, including factors like sowing time, day length, rainfall patterns, and crop rotations, so differential tuning may be necessary for different regions. There is an impetus to grow short duration varieties that mature quickly and are photoperiod insensitive to fit into a shortened growing season due to climate change. Much flowering time research on grass pea's close relative garden pea has been carried out over the last 50 years by researchers at the University of Tasmania (Weller and Ortega 2015). They have identified many of the genes important for modifying the trait in legumes. Altering the expression of these genes in combination may help to develop varieties suitable to different environments. Target genes include *Late Flowering (LF), Early Flowering 1 (ELF1), High Response to Photoperiod (HR),* and *Short Vegetative Phase (SVP).*

9.2.2 Root Characters

Grass pea has a "hardy and penetrating root system" (Campbell et al. 1994), so it has been hypothesized that this is the basis for its drought and flood tolerance (Campbell

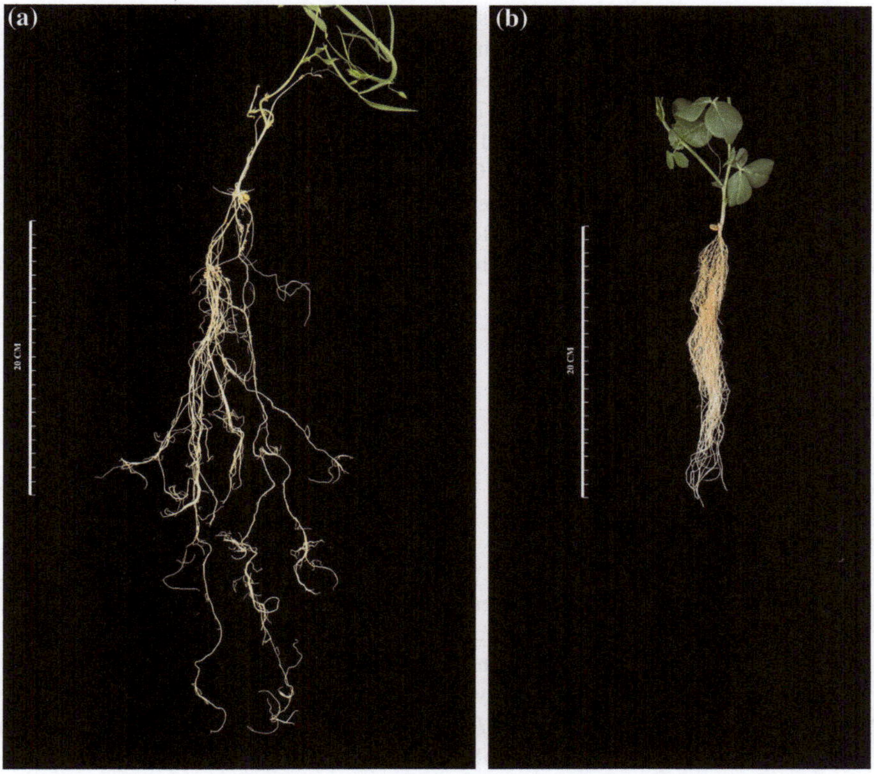

Fig. 9.2 Comparison of roots of **a** grass pea and **b** pea of plants of the same age. Scale bars show 20 cm

1997) (Fig. 9.2). Deeper tap roots generally translate into better adaptability to water scarcity while the growth of lateral roots near the surface helps the plant to better mobilize phosphorus, a key limiting nutrient for legume productivity. A comparison of six food legume species (Wahiduzzaman et al. 1996) indicated that drought tolerance in legumes appears to be associated with larger, often deeper, root systems. Understanding the characters that underline this property of grass pea would make it a useful model system for other crops and potential donor of genes for drought tolerance.

9.2.3 Heat Tolerance

Although tolerant to heat, grass pea is a cool season legume and therefore is not immune to the effect of high temperature. High temperature (>30 °C) during flowering reduces pollen viability, increases flower drop and reduces seed set/pod filling, thereby limiting grain yield. An indirect effect of heat is terminal drought resulting from field moisture loss. The effects of rising temperatures have been explored recently for legumes (Sita et al. 2017). Selection for heat tolerant varieties is one option, but the trait is often associated with multiple factors and so tolerance may be difficult to achieve (Sita et al. 2017). One target gene (*OsSIZ*, a homolog of *AtSIZ1*, encoding the *E3* ligase, *SUMO*) has been identified in rice. When *OsSIZ* was overexpressed in bent grass or cotton, it led to increased growth under drought and thermal stresses (Mishra et al. 2017). The same article cites several other target genes that may be of use for heat tolerance, such as the heat shock protein of cotton, *GhHSP26*.

9.2.4 Cold Tolerance

Cold tolerance assumes importance if there is frost and extreme cold during the growing season because grass pea is grown as a cool season legume in the Indian subcontinent, whereas it is grown as a spring-sown summer crop in Europe. Cold tolerance permits the range of grass pea to be moved northwards and for it to be adopted as a winter cover crop. There is a gene, *SCOF-1* encoding *a* C_2H_2 zinc finger protein in soybean that is inducible by cold but not by other abiotic stresses, and so may be a useful target. When *SCOF-1* was overexpressed in sweet potato responses to low-temperature stress were efficiently modulated (Kim et al. 2011). Yang et al. (2010) have shown that *CRLK-1*, a calcium/calmodulin-regulated member of the receptor-like kinase family, confers cold tolerance in *Arabidopsis*. In an extensive search for rice cold tolerance quantitative trait loci (QTLs), Xiao et al. (2018) found a single-nucleotide polymorphism (SNP), $SNP2^G$, in *LOC-Os10g34840* was responsible for conferring cold tolerance at the seedling stage in rice. Equivalents of all these genes could be potential targets for improving the range of grass pea cultivation.

9.2.5 Drought Tolerance

Grass pea is substantially drought tolerant. Genes have been identified in model systems that could be investigated to determine whether they underpin this capacity. If not, they could be used to enhance grass pea's capabilities. Examples could be knocking out *SPL8* (see below), overexpression of *SUMO* (see above) and using *HARDY* (Karaba et al. 2007), all of which affect tolerance to several abiotic stresses.

9.2.6 Flooding and Submergence Tolerance

Grass pea is also flooding tolerant and this characteristic could be understood and modified by examining the expression of genes identified in other species (Bailey-Serres and Voesenek 2008) for example from rice *Sub1*A and other low-oxygen response *ERF* transcription factors (Singh et al. 2017).

9.2.7 Salinity Tolerance

Potential targets for salinity tolerance are covered above. Other crops, such as sweet sorghum, that are highly tolerant to salinity may also yield gene targets in the future (Ding et al. 2018a, b).

9.2.8 Disease Resistance

Resistance to diseases has been examined extensively in legumes including grass pea (Rubiales et al. 2015). Some genes have been identified, e.g., *PsMLO1* for susceptibility to powdery mildew. Powdery mildew is an agriculturally relevant disease affecting many crops including grass pea. Recessive loss-of-function mutants of pea at the e*r1* locus have been identified as *MLO1* alleles and confer durable broad-spectrum resistance in pea and other crops (Humphry et al. 2011). *MLO1* mutations in grass pea could confer resistance to this disease.

9.2.9 Insect Resistance

Rubiales et al. (2015) have reviewed this subject in legumes. Trichomes have been identified as conferring some resistance to insects and their density can be enhanced by overexpressing *MIXTA*-like genes (*MYB* transcription factors) and may increase resistance to insect pests (Dubos et al. 2010; Plett et al. 2010).

9.2.10 Nutrient Use/Acquisition Efficiency

The nutrient use response of legumes to potential changes in climate has been reviewed (García-Hernández et al. 2010). Phosphorus is a key limiting nutrient for legume productivity and several potential target genes have been identified that might improve phosphorus use efficiency, especially the *SPX* family of which there are multiple genes in legumes (Liu et al. 2018) thus making them a climate smart target for grass pea. The expression of these genes increases under low phosphate conditions and in soybean they interact with *MYB*-like (*GARP*) transcription factors as part of the phosphorus signaling network (Zhang et al. 2016). Other potential targets involved in phosphorus signaling would be gene equivalent to rice *PHO2* (encoding a phosphatase) and *PHR2* (regulator of the phosphate starvation response).

9.2.11 Water Use Efficiency

This trait is linked to drought, flooding, submergence, and salinity tolerance, whereby some genes (Karaba et al. 2007) influence all four traits and thus become obvious climate smart targets for manipulation. An additional recently described target (Glowacka et al. 2018) for use in grass pea could be the *Photosystem II Subunit S* (*PsbS*) gene. The gene from *N. benthamiana,* when overexpressed in tobacco plants, affected a chloroplast-derived signal for stomatal opening in response to light leading to a 25% reduction in water loss per CO_2 assimilated under field conditions. If grass pea shows normal expression levels of this gene in field drought conditions and thus uses a different mechanism to survive drought, *PsbS* could be used to enhance its water use efficiency.

9.2.12 Carbon Sequestration and Greenhouse Gas Emission

Legumes play important roles in carbon sequestration in soils and in limiting greenhouse gas emission (Abberton et al. 2010). Their cultivation limits the need for artificial fertilizer production and use, and hence the generation of large amounts of carbon dioxide during the manufacturing process. Legumes also improve the soil composition and structure and augment the activities of microbial communities for the subsequent crops (Kumar et al. 2018). The climate smart target here is not so much in the modification of grass pea, but in increasing its utilization, especially in rotations, or transferring its natural ability to fix atmospheric nitrogen to other plants such as cereals thus reducing carbon loss from soils.

9.2.13 Genome Plasticity

Crop legumes including grass pea have large genomes with potentially mobile elements and many repetitive sequences that have contributed to genome rearrangements and new phenotypes. The estimated size of the genome of grass pea is 8.12 Gbp, of which a large proportion is repetitive elements including retrotransposons. Induction of movement of retro elements by genome shock, such as occurs during interspecific

Fig. 9.3 *Gracias á la almorta*: Starving figures collecting grain (grass pea) during the siege of Madrid; from an unbound album of first edition impressions by Francisco de Goya. While the image acknowledges the nutritional value of grass pea, the figure at the front is prostrate, probably as a result of neurolathyrism (https://en.wikipedia.org/wiki/Lathyrism)

hybridization could give rise to rapid genome remodeling. Such events allow organisms to adapt to large-scale changes in the environment and provide new variation for breeding (McClintock 1984).

9.2.14 Other Crop-Specific Traits

The orphan crop status of grass pea means the improvement in its agronomic characteristics lags behind most established crops. It is in much need of improvements geared mainly towards the structure of the individual plants as they relate to the canopy and towards seed quality traits. Fortunately, for most traits, there are ample examples to target from other legumes and especially from its close relative garden pea (for recent reviews see Tayeh et al. 2015; Patto et al. 2015).

As mentioned above, the first target for grass pea would be to decrease or eliminate the toxin. β-ODAP is a compound produced by grass pea that can cause a paralytic condition called neurolathyrism (Fig. 9.3) if consumed excessively over a long period of malnourishment. Research at the John Innes Centre has identified enzymes encoded by genes *LsAAE3* and *LsBOS,* involved in the biosynthesis of this toxin. Another has been identified by other researchers (*LsCAS*; Xu et al. 2017). Disruption of the genes encoding these enzymes could reduce or eliminate β-ODAP biosynthesis in the plant, thus improving the safety of food prepared from this crop.

Methionine is an essential amino acid found in low concentrations in legume seeds that are otherwise high in protein. Enhancing sulfur-rich amino acid content would improve the nutritional quality of grass pea and has been a target in legumes for many years (Casey and Davies 1993). In addition, methionine is also known to be a protective factor preventing neurolathyrism (Getahun et al. 2003). Thus, enhancing methionine content would be a complementary route to reducing the risk of this disease, in addition to reducing β-ODAP content. Unfortunately, there is no direct route to this character and it relies currently on modifying the ratio of storage proteins using transgenesis (Patto et al. 2015).

Modifying other storage product components of the seeds could also be beneficial. Obvious targets are trypsin inhibitors since they are major antinutritional factors in legumes, reducing the digestibility and bioavailability of proteins. Reduction of the expression of the genes (*TI1*, *TI2*) encoding trypsin inhibitors could enhance the nutritional value of grass pea, especially as a feed and for food processing methods involving minimal cooking. Mutants for the two trypsin inhibitor genes have been identified in garden pea that exhibit lower trypsin inhibitor activity (Clemente et al. 2015). Similar mutations could be sought in the homologous genes in grass pea.

Although not a high priority, but nevertheless desirable, altering the starch composition in grass pea may be helpful nutritionally (Patto et al. 2015) and provide increased consumer acceptability. Starch biosynthesis in pea has been studied extensively (Wang et al. 1998) and numerous mutations have been identified. *STARCH BRANCHING ENZYME 1* (*SBE1*; Bhattacharyya et al. 1993) has been identified as the enzyme responsible for the wrinkled phenotype tracked by Gregor Mendel in his seminal experiments on pea (Mendel 1865). This enzyme is involved in starch synthesis in legume seeds and mutations in the encoding gene give rise to high seed sugar content as well as wrinkled seed coats. High sugar content is a customer-preferred trait in grass pea, while the wrinkled seed coats represent an easily trackable phenotype for the identification of improved varieties Other genes are also known to modify the starch content, digestibility, and composition (Bogracheva et al. 1999) and could also be targeted in grass pea.

For seed traits, increasing seed size would be advantageous for human and animal consumption. Seed size is an important yield-related trait that is also relevant to customer preference. *BIG SEEDS 1* (*BS1*; Ge et al. 2016) is a negative controller of plant cell proliferation and knockouts and knockdowns of this gene are associated with increased plant organ size and yield.

Plant structure is often modified in crop plants. In legumes, leaf architecture is regulated by a series of genes (Gourlay et al. 2000), but the most useful in the development of the pea crop was *afila* (Goldenberg 1965), the so-called semi-leafless character. An *afila* mutant showing leaflets converted to tendrils was incorporated into the pea breeding program at the John Innes Centre in the 1980s (Snoad 1981), its advantage being that the plants remain more erect because of the support each gives to the other through the enhanced number of tendrils. It also enhances light capture by the lower leaves and stipules on the plant. Grass pea similarly collapses as the grain load develops and this increases the risk of disease and loss of yield. Increasing yield can also be achieved via other features of plant architecture. Stem branching determines the number of tillers produced by a plant and is a major determinant of biomass production. Several branching genes have been identified in legumes including the *ramosus* genes in pea (Ligerot et al. 2017) that are involved in auxin-strigolactone interactions and *SQUAMOSA PROMOTER BINDING PROTEIN-LIKE 8* (*SPL8*; Gou et al. 2018) in alfalfa. Knocking out *SPL8*, a suppressor of branching, greatly increases branching and biomass in alfalfa, while also enhancing resprouting after harvest and tolerance to drought and salinity. There is evidence for the conservation of the control of branching, so it is likely a similar impact could be achieved in grass pea to enhance its value as a forage crop. Modification to plant architecture, therefore, would be advantageous in grass pea.

In addition, an indirect response to plant architecture would be to modify (reduced or enhanced) shade-avoidance response via phytochrome genes (*PHYA, PHYB, PHYC, PHYD, PHYE*). While shade-avoidance responses are an important trait for plants to outcompete others in a natural environment, it has been hypothesized that a reduced shade response could enhance yield in a monoculture setting. Conversely, an enhanced shade response may increase biomass production, making the mutated plants more suitable for fodder production (Carriedo et al. 2016).

9.3 Genetic Resources of Climate Smart Genes

9.3.1 Primary and Secondary Gene Pools

As mentioned, grass pea has inherent climate smart traits that have not been well exploited. The diversity in *Lathyrus* is large—there are estimated to be up to 187 species in the genus (Allkin et al. 1986), although 160 is more often quoted (Patto and Rubiales 2014)—but little use has been made of the collections to leverage climate smart genes in breeding programs and few interspecific hybridizations have been carried out to determine the gene pool composition accurately (see Sects. 9.3.2 and 9.5.3 for more details).

The ICARDA holds the main collection of *Lathyrus* germplasm, with nearly 4200 accessions at its research centre in Morocco. There are also large collections at CBN PMP (about 4000 accessions) and the National Bureau of Plant Genetic Resources (NBPGR), India (about 2600 accessions), although these collections are not readily accessible. The large collection from the CBN PMP originates from the Université de Pau, France, but has ceased active distribution and is not being regularly regenerated (personal communication, Jocelyne Cambecèdes, CBN PMP, France) which may place some grass pea diversity at risk. There are also other smaller collections around the globe (Patto and Rubiales 2014; see below) (Table 9.1).

The Genesys (www.genesys-pgr.org) database supported by the Crop Diversity Endowment Fund (www.croptrust.org/our-mission/crop-diversity-endowment-fund/) contains 6580 accessions under the name of grass pea. Many accessions (around half) of these are not *L. sativus*, but cover *ca.* 80 other species, *L. aphaca* (yellow pea or yellow vetchling) being the next largest group with more than 500 accessions. About half the *L. sativus* collection is available for distribution. The major *L. sativus* collection is held at ICARDA (2715 accessions) with USDA's Agricultural Research Service (ARS) centre at Pullman, Washington, USA next (337). ICARDA holds the largest collection of species accessions (around 1900) with Kew, London, UK holding around 1100. Accessions have been obtained from the Balkans, Central

Table 9.1 Grass pea accessions in the Genesys collection by country and holding institute (Data accessed through GENESYS Global Portal on Plant Genetic Resources, https://www.genesys-pgr.org, 2018-11-25)

Country	Holding institute	No. of accessions
Syria	ICARDA	4184
UK	Royal Botanic Gardens Kew	1115
USA	ARS-GRIN Pullman, ARS Ft Collins, Boyce Thompson	949
	Arboretum, Arizona, ARS National Arboretum, Washington D.C.	
China	National Genebank, Beijing	704
Ethiopia	ILRI	155
Greece	Agricultural Research Center of Macedonia & Thrace	47
Italy	Universitá degli studi Perugia, CRA-Centro di Ricerca per	42
Spain	l'Orticoltura,	34
	Centro Nacional de Recursos Fitogenéticos,	
	Universidad Politécnica de Madrid, Centro de Conservación de la	
	Biodiversidad Agrícola de Tenerife	
Portugal	Universidade da Madeira,	22
Azerbaijan	Genetic Resources Institute	14
Russia	N.I. Vavilov Research Institute of Plant Industry	8
Slovenia	Agricultural Institute of Slovenia	4
Taiwan,	World Vegetable Center	4
China	Plant Breeding and Acclimatization Institute	2
Poland		

Asia, Europe, Near and Middle East, North and South America, Indian subcontinent (although not India), Mediterranean regions, UK. India and China are not well represented in the collection although they are areas where grass pea is cultivated. More than 2000 accessions in the collection are wild species as opposed to cultivars, landraces or breeding material and many (1664) are not classified. Table 9.1 shows the accession origins by country.

There have been numerous studies of *Lathyrus* diversity based on limited (e.g., Yang et al. 2014), often country-specific collections (e.g., Lioi et al. 2011). The latest study used a greater range and employed simple sequence repeats (Wang et al. 2015) but relied less on the main collections. The analysis used 17 species relatives, but only 15 of the total 266 wild and cultivated *L. sativus* accessions were from the ICARDA collection. Of the remaining, 20 were from China, 98 from the remainder of Asia and 33 from Africa and were sourced from the Institute of Crop Science, Beijing, China and the Vavilov Research Institute in St Petersburg, Russia. This study separated the accessions into three distinct groups—the species, the Africa/European and the Asian. It was also able to separate annual (including *L. sativus*) and perennial species. This emphasizes the need to open up global grass pea diversity and make all collections available for breeding purposes to improve the crop. If this does not happen, some unique and potentially useful sources of climate smart traits may be lost.

9.3.2 Tertiary Gene Pool

There is a history of attempting to introduce genes by interspecific hybridization between *Lathyrus* spp. This occurred most notably in the ornamental *L. odoratus* to generate novel flower coloration, but there is potential for some useful trait transfers into *L. sativus*, such as disease resistance (Gurung and Pang 2013) if crosses can be successful. There have been relatively few examples of attempting to rescue embryos from interspecies crosses. One example where numerous attempts were made to cross *L. sativus* with a range of *Lathyrus* spp. was the study undertaken by Addis and Narayan (2000). Crosses with *L. pseudo-cicera* were the most successful in generating viable F_1 seeds and embryo rescues with *L. cicera* created some calli with buds. There does not appear to be any more recent studies which are not surprising given the transformation and gene editing approaches now dominant in research. These are likely to lead to more wide adoption of useful traits from related genera, as mentioned in the previous section.

9.3.3 Artificially Induced Mutations

Both chemical mutagens and gamma irradiation have been used to generate additional variation in grass pea. Most mutagenesis programs (Talukdar 2009b; Rybiński 2003; Rybinski et al. 2006) have focused on morphological or seed traits and growth habit (e.g., branching pattern). Some useful modified traits such as dwarfism (Talukdar 2009a) and others have been found. In their review of attempts on crop improvement focussed on research in India, Dixit and colleagues (2016) reported on many of them, but there have been no reported attempts to include the mutants from these programs into the grass pea crop. Early attempts to isolate low-ODAP lines—a major target of breeding programs—in grass pea (Nerkar 1972) were unsuccessful, however, small populations were used and not targeted directly at the toxin, i.e., by screening for ODAP. In a recent study, Emmrich (2017) developed a high throughput assay for ODAP that allowed him to screen tens of thousands of plants directly for their ODAP content. He found a range of mutants, some of which had lower β-ODAP contents in their seeds than lines already in agricultural use.

In the genomics era, however, once they are identified, it should be relatively easy to isolate induced mutants for climate smart traits using mutagenized populations followed by conventional TILLING or TILLING-by-sequencing (see Sect. 9.9.1).

9.4 Classical Genetics and Traditional Breeding

9.4.1 *Basic Botanical, Cytological Information, Genetics of Key Traits, and Traditional Breeding*

Grass pea is a diploid (2n = 2x = 14 chromosomes) that shows some degree of variation in karyotype (Barpete et al. 2012). There are sharp variations in chromosome size, centromere location, size and location of secondary constrictions, despite the identical haploid number of seven chromosomes (Battistin and Fernandez 1994). Aneuploid and polyploid plants have also been reported in *Lathyrus* species that showed the same basic chromosome number (Talukdar and Biswas 2008).

The genus *Lathyrus* consists of 187 species (Allkin et al. 1986). The floral biology is such that it favors self-pollination. The flowers are bright blue, reddish purple, red pink or white, axillary, and solitary. They bear diadelphous (9 + 1) stamens, the 9 fused nearly 1.5 cm long including their tube with the 10th, a 9-mm long vexillary stamen, free and winged at base (Campbell 1997). Seeds are 4–7 mm in diameter, angled and wedge shaped, colored white, gray-brown, gray-white, yellow-white, yellow-brown, pink, red, purple, and black as well as spotted and mottled forms. Seed germination is hypogeal, the epicotyl being purplish-green. Plant height varies between 15 and 172 cm and growth habit ranges from prostrate to erect.

An initial study on genetic control of ODAP content revealed a simple Mendelian inheritance (Nerkar 1972). However, an in-depth study indicated that ODAP content is quantitatively inherited with pronounced influence of environment (Tiwari and Campbell 1996). They also reported a digenic inheritance of flower color (a 13:3) with inhibiting gene interactions. They postulated the symbol *LB* for blue flower color (dominant) and *LW* for white flower color. Most of the reported work on genetic enhancement of grasspea has been on the reduction of ODAP content in seed besides its yield improvement. Conventional breeding approach has been employed to develop low-ODAP/ODAP-free grasspea varieties for various edapho-climatic conditions. Besides, tissue culture, mutation breeding approaches contributed to development of low-ODAP varieties with desirable seed yield (see Sect. 9.4.4).

9.4.2 *Limitations of Classical Endeavors and Utility of Molecular Mapping*

Genomic data about grass pea is essential for planning of breeding activities as well as for acceptance of new hereditary variation in available germplasm (Hao et al. 2017). There are few reports available on the development of grass pea genomic resources; this may be due to its large genome size (8.12 Gbp) and the limited characterization of available germplasm (Bennett and Leitch 2012; Sarkar et al. 2019; manuscript under preparation). However, until recently, a grass pea reference genome sequence was not available (Hao et al. 2017). There are a few studies available on molecular markers, Inter Simple Sequence Repeats (ISSR), Sequence-tagged Sites (STS), Random Amplified Polymorphic DNA (RAPD), Amplified Fragment Length Polymorphism (AFLP), and Restriction Fragment Length Polymorphism (RFLP) that could be used in molecular breeding and phylogenetic studies (Almeida et al. 2014a, b; Hao et al. 2017). Recently, 200–300 Expressed Sequence Tag Simple Sequence Repeats (EST-SSR) primer pairs were identified by Sun et al. (2012) and Almeida et al. (2014a, b). However, novel traits with closely linked molecular markers are scarce and this limits breeding efficiency.

9.4.3 Breeding Objectives: Positive and Negative Selection

Grass pea is an under-researched but nutritionally important crop for humans and animals. The major breeding efforts are devoted to developing high seed and biomass yields with low-ODAP (<0.1%) content and the ability to withstand harsh environments. However, soil nutrient status and environmental factors like drought limit expression of traits at the field level and contribute to negative selection. For example, the same genotype may show variable ODAP content in drought compared to available soil moisture during stages of flowering to pod setting; saline soil causes low-ODAP content in seeds. Although grass pea is a self-pollinated crop, outcrossing by bees (Rahman et al. 1995; Chowdhury and Slinkard 1997) contributes to genetic contamination, thus affecting breeding objectives. Major efforts in breeding have been underway in Nepal, Bangladesh, India, Pakistan, Ethiopia, Turkey, and Russia over the last few decades (Kumar et al. 2011; Sarker et al. 2017).

9.4.4 Classical Breeding Achievements

Major grass pea breeding programs are being carried out in India, Bangladesh, Nepal, Ethiopia, and the ICARDA following intra- and interspecific crosses and tissue culture with the major objectives mentioned above. From these endeavors, 0.04–0.1% ODAP content varieties have been released globally for commercial cultivation (Dixit et al. 2016; Sarker et al. 2017). Seed yields up to 2.9 tons/ha and residues of up 5.2 tons/ha together with resistance to downy and powdery mildew and tolerance to drought have been reported. Globally released varieties include: BARI Khesari-1, BARI Khesari-2, BARI Khesari-3, BARI Khesari-4, BARI Khesari-5 and BINA Khesari-1 in Bangladesh; Ratan, Prateek, Mahateora, and Nirmal in India; Wasie in Ethiopia; CLIMA pink, 19A, 20B in Nepal; Ceora and Chalus in Australia; Ali Bar in Kazakhstan; LS 8246 and AC-Greenfix in Canada; Derek and Krab 9 in Poland; Strandja in Bulgaria, and Gurbuz-1 in Turkey. Many of these varieties have been developed using ICARDA-supplied genetic materials. ICARDA continues to share improved germplasm with national programs based on their requirements.

9.4.5 *Limitations of Traditional Breeding and Rationale for Molecular Breeding*

Many quantitative traits, and more specifically ODAP content, are influenced by environmental and edaphic factors. Moreover, a significant amount of cross-pollination occurs due to bees, and crossing needs to be done under controlled conditions that require extra costs and effort. Thus, precise phenotyping has become critical for key traits. In this context, identification of closely linked markers for major genes and QTLs with novel traits and their use in marker-aided selection will enhance breeding efficiency.

Recently, grass pea has received more attention for cultivation in problematic soils and in new niches like rice-fallow lands in South Asia. Moreover, low-income consumers prefer grass pea due to its low price compared to other legumes available in the market. Thus, grass pea is essential for human nutritional security and its residues and green fodder are important for animal feed.

9.5 Brief on Diversity Analysis

9.5.1 *Phenotype-Based Diversity*

Globally, grass pea is predominantly grown in India, Nepal, Bangladesh, Pakistan, China, and Ethiopia and to a lesser extent in the Middle East, Ghana, Canada, Australia, Sudan, Niger, Ivory, Coast, Mauritania, France, and Spain. Enormous diversity exists among landraces with respect to morphological and agronomic traits: flower color (blue, pink, red, white), flower initiation (34–62 days), 50% flowering (63–103 days), duration of flowering (15–26 day), days to maturity (97–156 days), seed size (7.21–17.6 gm/100-seeds), shape (wrinkled, round), seed coat color (ash, gray, cream), seed yield (863–3100 kg/ha), and ODAP content (0.04–2.6%). Considerable variation in economic traits has encouraged breeders to improve the crop through genetic selection.

9.5.2 *Genotype-Based Diversity Analysis Employing Molecular Markers*

To identify variation in germplasm phenotypically and/or genotypically, molecular markers play an important role. Although several molecular markers (ISSR, STS, RAPD, AFLP and RFLP, EST-SSR) have been identified by various laboratories (Almeida et al. 2014a, b; Hao et al. 2017), their systematic application is yet to be achieved. As an under-researched crop, grass pea has received very little attention.

9.5.3 Relationship with Other Cultivated Species and Wild Relatives

Harlan and de Wet's (1971) gene pool concept using crossability information for *L. sativus* was applied to elucidate its gene pools. The cultivated and wild races of *L. sativus* are included in the primary gene pool (GP1). Townsend and Guest (1974) suggested that the primary gene pool is not adequately differentiated in terms of morphological traits, as there are no clear-cut discontinuities between the cultivated and wild forms of grass pea. Although Smartt (1984) concluded that the white flowered and white-seeded grass pea varieties were the most highly selected, others have suggested the blue flowered, small speckled seeded forms are primitive. Therefore, white flower and white-seeded varieties are in GP1A and blue flowered, small seeded forms are in GP1B. If *L. sativus* wild and cultivated races form the primary gene pool, then the secondary gene pool (GP2) includes up to ten other species that can cross, but with some difficulty, with grass pea. Heywood et al. (2007) defined GP2 to include *L. chrysanthus, L. gorgoni, L. marmoratus* and *L. pseudo-cicera*, with which *L. sativus* can cross to produce ovules, and possibly more remotely *L. amphicarpos, L. blepharicarpus, L. chloranthus, L. cicera, L. hierosolymitanus* and *L. hirsutus*, with which *L. sativus* can cross and form pods. The tertiary gene pool includes species that can cross with the original crop species only with the use of specialized techniques such as embryo rescue and culture (see Sect. 9.3.2) or the use of bridging species and the remaining species of the genus can be considered in the tertiary gene pool (GP3), although many have yet to be examined.

9.5.4 Future Aspects for Research

In any future development of grass pea as a safe crop for human consumption, zero or a very low level (<0.1%) ODAP content will be essential. A comprehensive characterization of all available genetic resources available in international and national gene banks needs to be undertaken. The genetic control of economic traits needs to be elucidated and a dense linkage map along with genomic information for *Lathyrus* species needs to be developed. Research on inter-relationships between different *Lathyrus* species using genetic, cytogenetic techniques, interspecific hybridization, and molecular approaches needs to be undertaken. On the applied and adaptive research side, low-toxin and high yielding varieties with resistance to various biotic and abiotic stresses need to be developed and delivered to farmers.

9.6 Association Mapping

In establishing genetic maps, specific populations are usually created between parents differing for a particular trait. Such populations are frequently used for QTL analysis (see Sect. 9.7). In this instance, only two alleles will be segregating for a character. Association mapping (or linkage disequilibrium), however, relies on a collection of individual plants often from germplasm collections where several alleles at a number of loci can be associated with a trait. The lack of integrated genetic maps with genotyping data generated from the grass pea collections (see Sect. 9.3) means linkage disequilibrium patterns across the genome cannot be investigated. This is a prerequisite for association mapping as has been applied to other legumes. For example, this has been implemented in *Phaseolus vulgaris* for bacterial blight resistance (Shi et al. 2011) and plant architecture lodging and productivity (Resende et al. 2018), in *Vigna unguiculata*, where association mapping was combined with conventional biparental QTL analysis, for the stay-green phenomenon (Muchero et al. 2013) and in *Medicago sativa* for forage traits (Biazzi et al. 2017).

One would anticipate that once grass pea genomic tools catch up with other legumes and, as proposed by Patto and Rubiales (2014), high-throughput germplasm screens become feasible, the potential of association mapping will be harnessed to accelerate breeding of grass pea (Almeida et al. 2015). One recent attempt, looking at phenolic compounds in grass pea, has been reported briefly by Patto et al. (2018) recently using 100 accessions and Diversity Arrays Technology Sequencing (DArT-seq) based SNP markers. They measured a range of 46–102 mg gallic acid equivalents across their genotypes and found one significant marker-trait association for total phenolic content. Unfortunately, since there is no genome sequence available for grass pea, they could not associate the markers with a chromosome position.

9.7 Brief Account of Molecular Mapping of Climate Smart Genes and QTLs

Given the exceptionally large size (1C = 8215 Mb; http://data.kew.org/cvalues/) of the grass pea genome and its restricted cultivation on marginal lands coupled with a small size of the associated research community, grass pea is characterized by a dearth of molecular tools and technologies. Mapping efforts were nonexistent until 1999 when the world first published genetic linkage map in *Lathyrus* was constructed from 100 F_2 individuals (derived from a single F_1 plant), assigned 71 RAPDs, three isozymes and one morphological marker to 14 linkage groups spanning 898 centi-Morgans (cM) (Chowdhury and Slinkard 1999; Fig. 9.4). In 2004, another reported *Lathyrus* genetic linkage map was established from a population of 92 backcrossed individuals based on 47 RAPD, 7 EST-SSR, 13 STS/Cleaved Amplified Polymorphic Sequence (CAPS) markers, with genetic distance of 803.1 cM (Skiba et al. 2004). In 2018, the world first genetic linkage map of a widely used wild grass pea relative "red pea" (*L. cicera*) was constructed based on 189 SNP, 113 EST-derived Simple Sequence Repeats (E-SSR), and 5 Intron Targeted Amplified Polymorphism (ITAP) markers in a 103 F_5 Recombinant Inbred Lines (RIL) population derived by single seed descent from a cross between the two previously described *L. cicera* genotypes, BGE008277 and BGE023542. The resultant map covered a total of 724.2 cM, with an average density of one marker every 2.4 cM, organized in 9 linkage groups, seven longer than 40 cM and 2 shorter groups (Santos et al. 2018).

RAPD markers were first developed and used for genetic linkage map construction of grass pea in 1999 (Chowdhury and Slinkard 1999) followed by work on construction of another grass pea linkage map (work using combinations of RAPD, EST-SSR, STS/CAPS markers; Skiba et al. 2004). More recently, SNP and E-SSR markers were developed and used for the establishment of the genetic linkage map of *L. cicera* (Santos et al. 2018). RAPD markers were also used for comparative genetic diversity studies of grass pea in 2001 (Chtourou-Ghorbel et al. 2001), for analysis of genetic diversity among selected grass pea genotypes in 2007 (Barik et al. 2007) and for genetic variation among different accessions of grass pea in 2012 (Nosrati et al. 2012). Then, RFLP markers were applied for genetic diversity studies in 2001 (Chtourou-Ghorbel et al. 2001), and AFLP markers were used for genetic variation characterizing a grass pea collection in 2007 (Tavoletti and Iommarini 2007). Moreover, CAPS and derived-CAPS (dCAPS) were also designed by sequencing the monomorphic simple sequence repeats (SSR) fragments and examined for segregation RIL population in *L. cicera* (Almeida et al. 2014a, b).

The first nine polymorphic EST-SSR markers for *L. sativus* were selected from the 24 *Medicago truncatula* specific markers after comparative analysis in three legume species, and four of them were then employed to investigate the population structure and gene flow across 240 grass pea individuals belonging to seven diverse regions of Ethiopia in 2005 (Gutierrez et al. 2005). In another attempt in 2011, the transferability of seven EST-SSRs developed from the ESTs of *M. truncatula* were also tested across 19 accessions belonging to 11 different genera including one accession of *L. sativus* (EC539028) (Chandra 2011). The first set of simple sequence repeat (SSR) marker comprising 20 SSRs in *Lathyrus* was developed using in silico survey of European Molecular Biology Laboratory (EMBL) database in year 2011 (Lioi et al. 2011). SSR and EST-SSR markers for grass pea have been developed and published and only 17 of them were size polymorphic (Ponnaiah et al. 2011). Another researcher group developed a set of SSR markers derived from ESTs and 19 EST-SSR primer pairs were designed, the EST-SSR markers were mined in silico from a *L. sativus* Expressed Sequence Tag (EST) database (Shiferaw et al. 2012).

Following the work of Shiferaw et al. (2012), EST sequences within the public domain databases were screened, and 44 novel polymorphic and 117 monomorphic EST-SSR markers for *L. sativus* were identified and characterized for size polymorphism on 24 genotypes originating from around the world (Sun et al. 2012). In 2014, using the 454 FLX Titanium pyrosequencing technique, 651,827 simple sequence repeat (SSR) loci were identified and 50,144 nonredundant primer pairs were successfully designed, of which 288 were randomly selected for validation among 23 *L. sativus* and one *L. cicera* accessions of diverse provenance. Seventy-four were

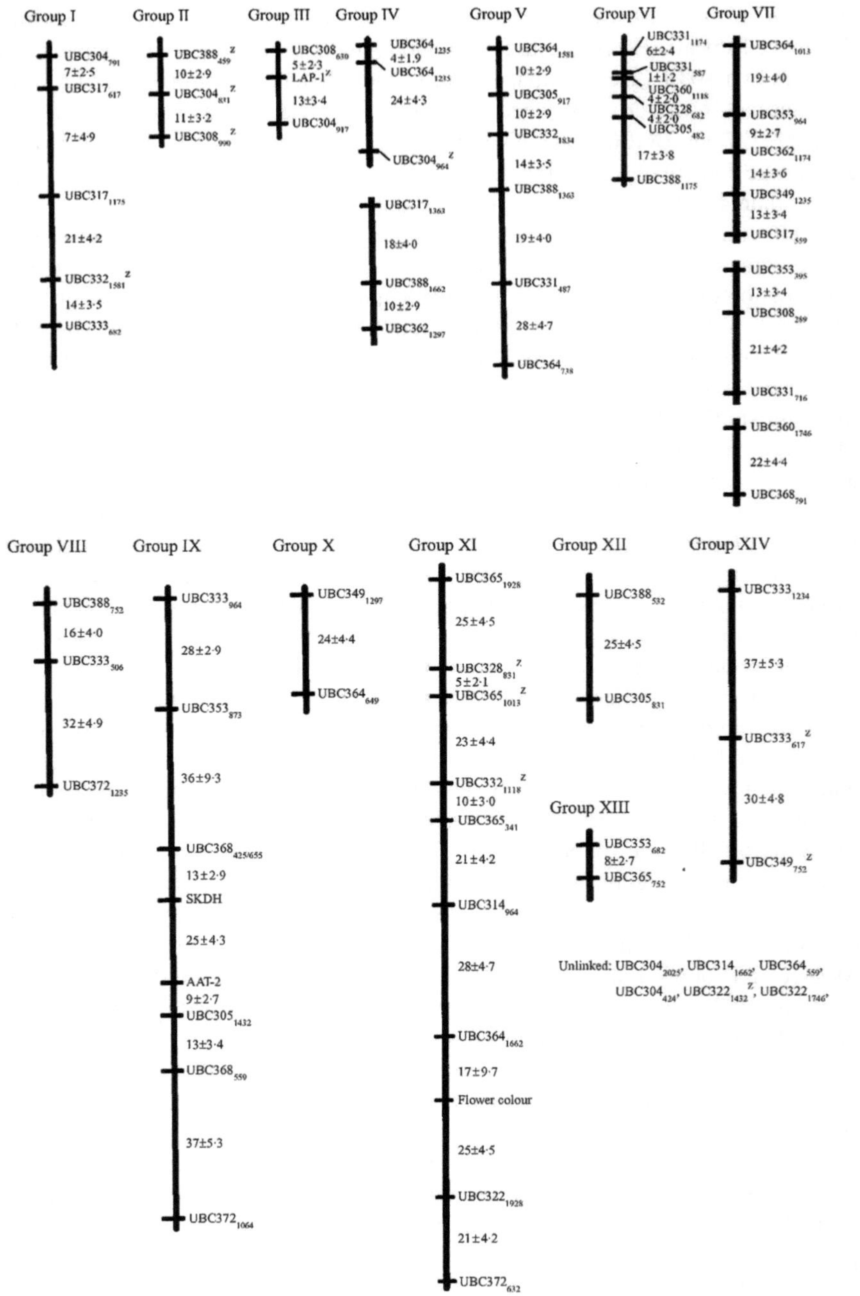

Fig. 9.4 The world's first genetic linkage map of grass pea (*Lathyrus sativus*) (Chowdhury and Slinkard 1999) based on F_2 population assigned 69 markers (one morphological, three isozyme and 65 RAPD) to 14 linkage groups comprising 898 cM with average distance of 17 ± 2 cM between two adjacent markers

found to be polymorphic, 70 monomorphic, and 144 with no polymerase chain reaction (PCR) product (Yang et al. 2014). More recently, (Hao et al. 2017) 570 million quality-filtered and trimmed cDNA sequence reads with total length of over 82 billion bp were obtained using an Illumina NextSeq^{TM} 500 platform. Approximately two million contigs and 142,053 transcripts were assembled from RNA-Seq data, which resulted in 27,431 unigenes with an average length of 1250 bp and maximum length of 48,515 bp. Among these unigenes, 3204 EST-SSR primers were designed, 284 of which were randomly chosen for validation. Of these validated unigenes, 87 (30.6%) EST-SSR primers produced polymorphic amplicons among 43 grass pea accessions selected from different geographical locations. Meanwhile, 146,406 SNPs were screened and 50 SNP loci were randomly chosen for the kompetitive allele-specific PCR (KASP) validation. Over 80% (42) SNP loci were successfully transformed to KASP markers.

During the past few years, experimental populations derived from two genetically diverse parents have been created in *Lathyrus* including F_2 (Chowdhury and Slinkard 1999), backcrossed (Skiba et al. 2004), and RIL (Santos et al. 2018) populations established in different research groups around the world (Bohra et al. 2014). Linkage analysis, segregation, and recombination frequency analysis in F_2-like individuals has led to the identification of some linked isozyme loci (Gutierrez et al. 2001).

For the world first published genetic linkage map in *Lathyrus* (Chowdhury and Slinkard 1999), MAPMAKER (Lander et al. 1987) was used to detect linkage and construct linkage maps. In case of construction of a linkage map based on a *L. sativus* backcross population and preliminary investigation of QTLs associated with resistance to ascochyta blight (Skiba et al. 2004), MapManager QTX (Manly et al. 2001) was applied. For the first genetic linkage map of *Lathyrus cicera* based on RNA sequencing-derived markers, JoinMap 4.0 software (Van Ooijen 2006) was used for linkage analysis and segregation distortion tests (Santos et al. 2018).

The estimated numbers of linkage groups differ from the chromosome number of grass pea (7) 14 (Chowdhury and Slinkard 1999), 9 (Skiba et al. 2004), and 9 (Santos et al. 2018).

QTLs for stem resistance to ascochyta blight in *L. sativus* at the seedling stage has been detected (Skiba et al. 2004) while the *L. cicera* transcriptome has been analyzed in response to rust (*Uromyces pisi*) infection (Santos et al. 2018). No reports on mapping the inheritance of the most important trait—β-ODAP or ODAP content—have yet been published.

Another preliminary genetic linkage map with a total of 64 markers mapped on a backcross population, included 47 RAPD, seven sequence-tagged microsatellite sites and 13 STS/CAPS markers (Fig. 9.5). QTLs associated with ascochyta blight resistance were detected using single-point analysis and simple and composite interval mapping. The backcross population was evaluated for stem rust resistance in temperature-controlled growth room trials. One significant QTL, *QTL1*, was located on linkage group 1 and explained 12% of the phenotypic variation in the backcross population. A second suggestive QTL, *QTL2*, was detected on linkage group 2 and accounted for 9% of the trait variation (Skiba et al. 2004).

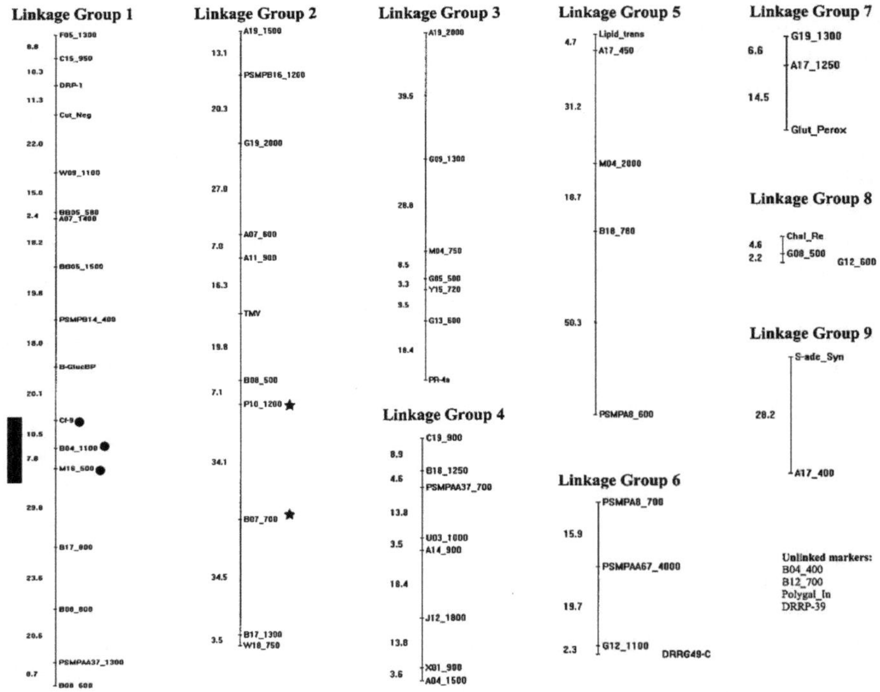

Fig. 9.5 A genetic linkage map of grass pea (*Lathyrus sativus*) (Skiba et al. 2004). This version still has more linkage groups than there are chromosomes

Based on a F_5 RIL population with 103 individuals, the world first genetic linkage map of red pea (*Lathyrus cicera*) successfully anchored 307 polymorphic loci, namely, 189 SNPs, 113 E-SSRs and 5 ITAPs (Fig. 9.6). Markers with distorted segregation ratios are marked with asterisks for their significance levels (* = 0.05, ** = 0.01 and *** = 0.005).

Other functional genomics resources such as expressed sequence tags (ESTs) amount to only a small number (178) available at the National Center for Biotechnology Information (NCBI) GeneBank for cultivated species of grass pea (*L. sativus*). There are no ESTs for wild relatives of grass pea at NCBI (*L. cicera, L. ochrus* and *L. tingitanus)*. However, comparatively higher numbers of ESTs (8702) have been reported for the grass pea relative *L. odoratus* (http://www.ncbi.nlm.nih.gov/genbank/dbest/dbest_summary/).

Single-point analysis, simple interval mapping and composite interval mapping have been performed using MapManager QTX (Bohra et al. 2014) for QTL mapping of stem resistance to ascochyta blight at the seedling stage of grass pea (Skiba et al. 2004). To detect significant interactions between any QTLs detected, a general linear model (GLM) analysis has been conducted using Minitab (Minitab, State College, Penn., USA), release 11.2 (Skiba et al. 2004).

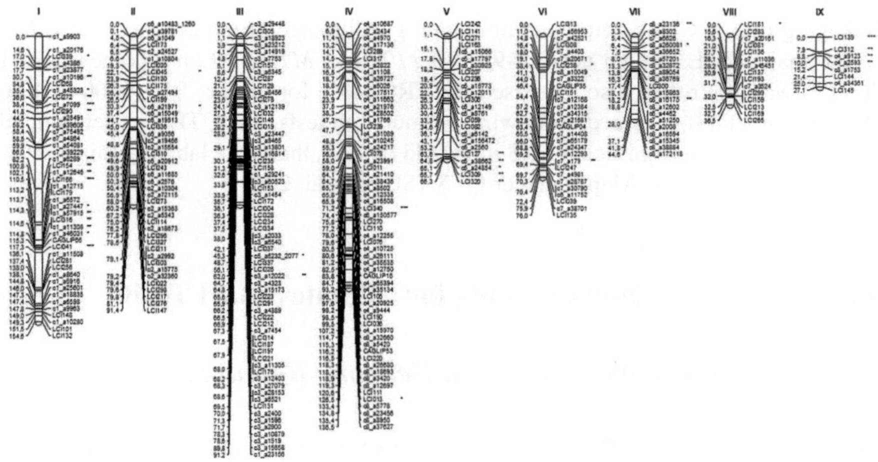

Fig. 9.6 A genetic linkage map of red pea (*Lathyrus cicera*) (Santos et al. 2018)

The only reference to climate smart QTLs of grass pea was on ascochyta blight resistance. Three markers (*Cf-9, B04_1100, M16_500*) were identified as being associated with ascochyta blight resistance, at an likelihood ratio statistics (LRS) threshold of 9.22 by single-point analysis, calculated from the log of the odds ratio (LOD). These markers were located in a single, continuous region on linkage group 1 (Fig. 9.4). Two additional markers, positioned together on linkage group 2, *P10_1200* and *B07_700*, fell short of the LRS threshold; however, the P-values were <0.05, suggesting that these two markers may be significantly associated with a QTL (Fig. 9.5).

Simple interval mapping detected one region significantly associated with resistance on linkage group 1 (*QTL1*), composed of the same three markers identified in the single-point analysis (Fig. 9.4), with the highest LRS value detected 3 cM away from marker *M16_500* (LRS = 11.4) (Skiba et al. 2004). *QTL1* accounted for 11% of the phenotypic variance observed in the backcross population, and a second potential QTL was detected on linkage group 2 (*QTL2*), approximately 10 cM from marker *P10_1200*, which accounted for up to 8% of the phenotypic variation (Skiba et al. 2004), although this region fell below the LRS threshold of significance using interval mapping (LRS = 8.3, LOD = 1.8, Fig. 9.5). Composite interval mapping of linkage group 1, where the effects of marker *P10_1200* (*QTL2*) were controlled, increased the LRS values for markers *Cf-9*, *B04_1100*, and *M16_500* (Skiba et al. 2004). The permutation test set a significant LRS value of 11.6, which these markers exceeded; this supported the possibility that a QTL may lie within this region on linkage group 1. Composite interval mapping of linkage group 2, where the effect of the major QTL located near *Cf-9*, *B04_1100* and *M16_500* on linkage group 1 (*QTL1*) was controlled also increased the LRS value for marker P10_1200 to 10.6 (Skiba et al. 2004). However, following permutation tests, this QTL was determined to be just below the significant LRS threshold of 12.1, therefore labeling this only as a 'suggestive' QTL by MapManager QTX (Skiba et al. 2004).

9.8 Genomics-Aided Breeding for Climate Smart Traits

9.8.1 Structural and Functional Genomic Resources

The presence of β-ODAP led to the banning of grass pea cultivation and sale in India in the early 1960s. This in turn adversely impacted the funding for grass pea research, resulting in few genomic resources being developed before the turn of the century. The first detailed genetic maps of grass pea (Chowdhury and Slinkard 1999) have been mentioned. In the last few years, more genomic resources have been created. About 50 SSR markers were identified by a computational search of the EST databases at EMBL and NCBI (Ponnaiah et al. 2011; Sun et al. 2012). About seven polymorphic genomic sequence based SSR markers have been created from an enriched genomic library (Lioi and Galasso 2013) by affinity capture of restriction fragments to biotinylated microsatellite oligonucleotides, whereas a larger set of 74 validated SSR markers was developed by Yang et al. (2014).

Transcriptome sequencing of an Indian grass pea line, (Rewa 2; Chapman 2015) identified about 1106 potential SSR markers and also resulted in the development of a set of 12 validated, PCR-based conserved orthologous set (COS) markers (Fulton et al. 2002) for the legume family. A large RNA sequencing (RNAseq) dataset from African and European grass pea accessions has recently been used to develop over 87 validated polymorphic EST SSR markers and 42 KASP markers that were successfully tested on a global collection of 43 different accessions (Hao et al. 2017).

9.8.2 Details of Genome Sequencing

Recently, the sequencing of the genome of a European grass pea accession (LS007) has been completed (Sarkar et al. 2019; manuscript under preparation). Genomic DNA from etiolated seedlings was initially used to make amplification free libraries and sequenced using the Illumina HiSeq platform. The sequences were assembled using Discovar. Following this, Long Mate pair (LMP) libraries with 2, 5, 8, and 14 kbp average insert sizes were made and sequenced using the Illumina HiSeq platform. The draft genome assembly is about 8.12 Gbp in size, with an N50 of about 59.7 kbp (Table 9.2).

Table 9.2 Grass pea draft genome assembly statistics

Mean scaffold length	12,120
Median scaffold length	1856
Minimum scaffold length	1000
Maximum scaffold length	1,110,364
N50 (length)	4,059,609,658
N50 (value)	59,728
L50	31,600
Total length	**8,119,118,376**
Total sequences	669,893

Table 9.3 Grass pea gene model summary statistics

	All	High confidence	Low confidence
Gene count	87,222	33,819	53,403
Total transcripts	90,253	35,500	54,753
Total exons	307,706	172,558	135,148

9.8.3 Gene Annotation

RNA-Seq data from multiple samples/tissues of three different grass pea genotypes (the high ODAP European line LS007, the high ODAP Indian line LSWT11, and the low ODAP Indian variety Mahateora) were utilized for the annotation (Sarkar et al. 2019; manuscript under preparation). Protein-coding genes were predicted using AUGUSTUS (Stanke et al. 2006) by means of a Generalized Hidden Markov Model (GHMM) that takes both intrinsic and extrinsic information into account. Protein sequences from nine plant species (*Cicer arietinum, Cucumis sativus, Fragaria vesca, Glycine max, Malus domestica, Medicago truncatula, Prunus persica, Phaseolus vulgaris, Trifolium pratense*) were aligned to the draft genome. Predicted genes were checked against the RNAseq data for expression estimation and the predicted proteins were annotated. A total of 33,819 high confidence genes were identified in the draft genome (Table 9.3).

9.8.4 Impact on Germplasm Characterization and Gene Discovery

EST-based SSR markers have been used to understand the genetic diversity present in Ethiopian grass pea germplasm (Shiferaw et al. 2012). In another study, 30 SSR loci were employed to assess the genetic diversity and population structure of 283 lines from wild and domesticated *Lathyrus* spp. populations from Africa, Europe,

Asia and ICARDA (Wang et al. 2014). The recent identification and validation of a set of 12 PCR-based conserved orthologous set (COS) markers (Fulton et al. 2002) for the legume family (Chapman 2015), as well as the availability of the SSR, RNA seq, and draft genome data should allow for the rapid and facile characterization of germplasm diversity and enhance gene discovery efforts in grass pea. The John Innes Centre's efforts towards the elucidation of the β-ODAP biosynthesis pathway, as well as other agronomically important traits has been particularly aided by the availability of the RNA-seq and draft genome data.

9.8.5 Application of Structural and Functional Genomics in Genomics-Assisted Breeding

The deepSuperSAGE (Matsumura et al. 2012) analysis of the transcriptomes of *Ascochyta lathyri* infected and uninfected grass pea resulted in the identification of 14,387 UniTags, of which 738 were significantly differentially expressed in the infected and uninfected plants (Almeida et al. 2015). There was a clear upregulation of the ethylene pathway defense genes, with other genes potentially involved also identified as targets for breeding *Ascochyta* blight resistance.

9.9 Recent Concepts and Strategies Developed

9.9.1 TILLING

TILLING or **T**argeting **I**nduced **L**ocal **L**esions **in G**enomes is a strategy to identify mutations in genes of interest in a targeted manner from a population that has been subjected to mutagenesis, usually by using the chemical ethyl methane sulphonate (EMS) that generates mainly point mutations and sometimes physical mutagens such as gamma rays or fast neutrons that generate deletions of varying sizes and require slightly different technologies to identify mutants. The original concept was developed by McCallum et al. (2000) using *Arabidopsis* and refined by Colbert et al. (2001) for high throughput using DNA sequencers. Since then it has been adopted for use on many species, both plants, and animals, and on several different platforms culminating in in silico TILLING (Wang et al. 2012) (Fig. 9.7). A variant of the approach, called EcoTILLING (Comai et al. 2004), allows assessment of diversity in germplasm collections.

TILLING is particularly applicable to organisms that lack genomic resources, are not amenable to efficient transformation or where it is not desirable to mutate by transgenesis. Grass pea falls into all these categories and hence the John Innes Centre has had created by BenchBio Pvt. Ltd. (www.benchbio.com) EMS-mutagenized populations in two grass pea varieties. One has been screened using

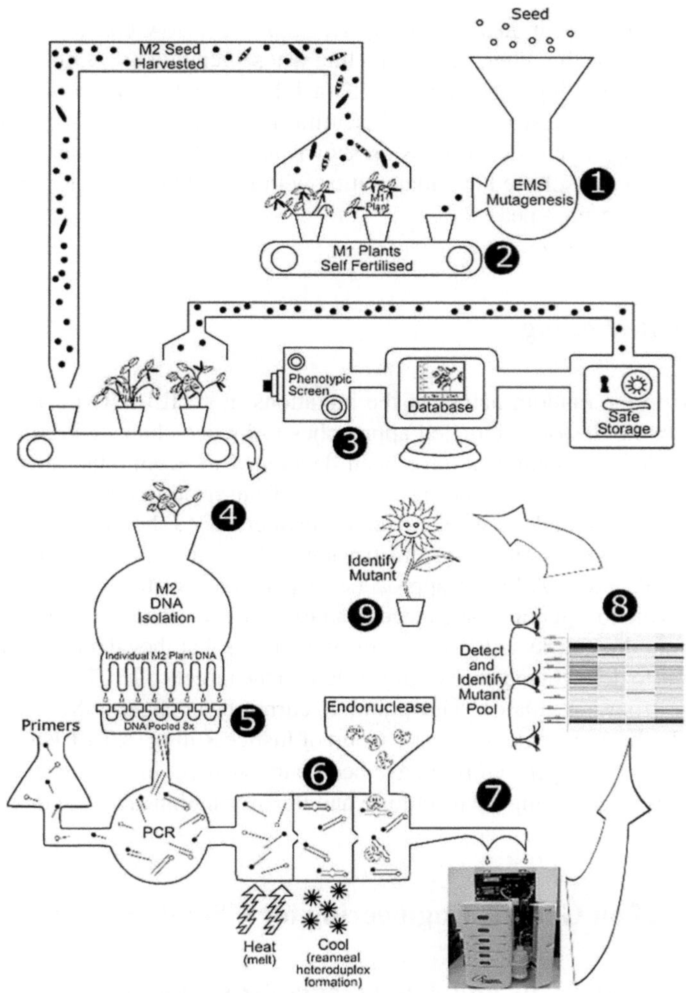

Fig. 9.7 *RevGen*UK's TILLING process including an optional forward genetic screen. Pure seed is treated with EMS (1) to induce mutations. The seeds are germinated and grown to create M1 plants (2) that are selfed and seed harvested as individuals to obtain M2 seed. The M2 seed is catalogued, resown as families (12 seed usually) and the plants screened for segregating phenotypes (3) and seed harvested and stored appropriately from one fertile plant of each family. This plant is used to collect leaves for DNA extraction (4) and the DNA stored in 96-well microtitre plates. Siblings can be screened for a variety of phenotypes by eye or by high-throughput (bio)chemical screens. Seeds from sibling plants bearing useful phenotypes can also be also stored (3). Individual DNAs are pooled (5) in microtitre plates and the pools (usually 8-fold) used for PCR amplification using gene specific primers. Products are melted, cooled and the annealed products cut (6) with CEL1 enzyme that preferentially cleaves at sites of heteroduplex mismatches between annealed wild-type and mutant DNA. The products are purified and separated by capillary electrophoresis on an Agilent Fragment Analyzer™ (7) that generates a false gel-like image to help detect the mismatched products (8). Pools containing mutant DNA are identified by eye and PCR products from individuals sequenced to identify the mutant plant (9). Image updated from Wang et al. (2012)

a new high-throughput method (Emmrich 2017) for low-ODAP mutants, (see Sect. 9.2), whereas the other, based on a line whose genome has been sequenced, is currently being developed by *RevGen*UK (a John Innes Centre service) for TILLING (www.jic.ac.uk/research-impact/scientific-facilities/genomic-services/reverse-genetics/), using its current process as shown in Fig. 9.7. A further collection of germplasm is being selected in collaboration with ICARDA to provide an ecoTILLING platform for grass pea.

9.9.2 Gene Editing

In contrast to the random nature of the mutations in a TILLING population, gene editing technologies offer targeted approaches to induce changes in the gene(s) of interest. Various technologies have been developed to accomplish this, including Meganucleases, Zinc finger nucleases (ZFN), Transcription activator-like effector nucleases (TALEN) and the clustered regularly interspaced short palindromic repeats (CRISPR)/Cas9 gene editing system (reviewed by Gaj et al. 2013; Puchta 2016; Songstad et al. 2017). All these approaches depend on a system of directing a DNA editing enzyme to a genomic sequence in a targeted, sequence-specific manner. One of the current drawbacks to these systems in plants is that they depend on achieving transgenesis to deliver the gene editing system to the target cells. This also constrains their usage from a regulatory viewpoint, as currently in Europe these systems are considered to be transgenic (European Court of Justice Ruling ECLI:EU:C:2018:583, 2018). No reports of gene editing have been published yet in grass pea, chiefly due to difficulties of achieving tissue culture and transformation easily in this species.

9.10 Brief on Genetic Engineering for Climate Smart Traits

9.10.1 Target Traits and Alien Genes—Physiological and Reproductive Traits—Biotic and Abiotic Stress Resistance

Grass pea, even more than other grain legumes, suffers a large yield gap between what is theoretically achievable compared to what resource constrained farmers in developing countries usually achieve. Decades of under-resourcing grass pea research have resulted in few agronomically improved cultivars. In the context of grass pea agriculture in the Indian subcontinent, where a major proportion of the global production of grass pea takes place, farmers face many challenges. Grass pea is usually grown as a cool season legume there (Dixit et al. 2016; Kumar et al. 2011). Yield constraints include phosphorus deficiency, terminal drought and heat, as well as diseases (downy and powdery mildew, root rot, etc.) and insect pests (aphids, bruchids, pod borer). Tolerance to these biotic and abiotic stresses could contribute substantially to yield stability. In these regions, the impetus is to develop short duration, heat, and photo-insensitive varieties that can resist diseases and escape terminal drought. Development of high biomass varieties would help in adapting this crop as a useful forage crop, especially for small ruminants such as sheep and goats. All these need to be achieved against a background of very low (preferably zero) β-ODAP content for human and animal health/safety.

9.10.2 Review on Achievements of Transgenics

The earliest report in literature on the successful genetic transformation of grass pea (Barna and Mehta 1995) relied on a biolistic method of gene transfer of a construct expressing the beta-glucuronidase (GUS) marker, followed by the induction of somatic embryogenesis. This was preceded by the development of methods for the regeneration of grass pea in tissue culture from various explants (Roy et al. 1992, 1993). A detailed study on the factors affecting *Agrobacterium* mediated transgenesis of grass pea was published in 2005 (Barik et al. 2005), with a protocol for the production of transgenic grass pea using standard *Agrobacterium* mediated transformation techniques. Finally, the expression of an oxalate-degrading enzyme, oxalate decarboxylase (*FvOXDC*) of *Flammulina velutipes* in transgenic grass pea (Kumar et al. 2016) led to the reduction in seed ODAP levels and also improved tolerance of the fungal pathogen *Sclerotinia sclerotiorum.* While transgenesis in grass pea, like other legumes, is technically challenging, improvement of transformation rates through investing further in research will likely result in the development of transgenic grass pea carrying constructs for the transgenes/gene silencing or gene editing of important traits (such as zero ODAP). Recently, methods have been developed that have improved transformation rates for other legumes such as chickpea (Tan et al. 2017).

9.10.3 Metabolic Engineering Pathways and Gene Discovery

One of the major constraints in grass pea production and consumption is the presence of the neurotoxin β-ODAP in all aerial tissues of the plant, including in the seed. While the biochemical pathway for its synthesis is fairly well understood (Kuo et al. 1998; Lambein et al. 1999; Yan et al. 2006), the genes responsible for its biosynthesis have not been well characterized. A metabolomic approach to study the ODAP biosynthesis pathway revealed genes that are up or downregulated and provided insights into mechanisms of β-ODAP accumulation and degradation in grass pea (Xu et al. 2018). Using a bioinformatics approach, a cysteine synthase that is putatively involved in ODAP biosynthesis in grass pea has been recently identified (Chakraborty et al. 2018).

9.10.4 Prospects of Cisgenics

Cisgenics are genetically modified organisms (GMO) that contain genes and regulatory sequences from the same species as the GMO. It is possible that such lines might be subject to lower regulatory hurdles for their release and use, as these GMOs do not contain any "foreign" genetic sequences. However, given the current regulatory ruling by the European Court of Justice (European Court of Justice Ruling ECLI:EU:C:2018:583, 2018) classifying gene edited organisms as transgenic, any technology that uses *Agrobacterium*-mediated transformation is likely to face the same regulatory hurdles as transgenic genetically modified organisms.

9.11 Brief Account of the Role of Bioinformatics as a Tool

The large size of the grass pea genome (approximately 8.12 Gbp) has limited research into developing genomic resources. However, several research groups have used transcriptome sequencing to reduce the complexity of the necessary assembly. A grass pea root transcriptome was included in the 1000 Plant Genome project (1KP BLAST Search Portal. [http://www.bioinfodata.org/app/Blast4OneKP/home]), which despite its name, only focused on transcriptome sequencing. Transcriptomic analyses have also been used to generate sets of SSR and KASP markers (Yang et al. 2014; Hao et al. 2017; Chapman 2015) study the response of grass pea to the pea rust pathogen *Uromyces pisi* and the leaf blight pathogen *Ascochyta lathyri* (Almeida et al. 2015) and to identify candidate genes involved in the biosynthesis of β-ODAP (Emmrich 2017). In another transcriptomic analysis currently conducted by the John Innes Centre, Earlham Institute and National Institute for Agricultural Botany, the responses of grass pea, pea, and faba bean to drought stress are being compared to identify the genetic and physiological mechanisms mediating drought tolerance in grass pea. This may reveal new ideas about how to breed for increased drought tolerance in other pulse crops.

9.12 Brief Account on Social, Political and Regulatory Issues

Concerns about the risk of neurolathyrism have historically led to some jurisdictions introducing laws against the cultivation and/or sale of grass pea. The first known ban was implemented in 1671 by Georg the Duke of Wurttemberg in Germany (Cohn and Streifler 1983). In the twentieth century, bans were introduced in several states of China and India, after outbreaks of neurolathyrism had followed periods of drought and famine (Yang and Zhang 2005). Other jurisdictions, such as Nepal, used public education campaigns to warn consumers of the risk of neurolathyrism and promoted the use of alternative legume crops (Pandey et al. 2000; Gharti et al. 2014). A notable exception to this is Ethiopia, where no ban or large-scale discouragement policies were introduced and grass pea cultivation has increased since 2000 (Haimanot et al. 2005; Girma et al. 2011). While these policies generally led to a reduction of grass pea cultivation and sale, cultivation for subsistence use has persisted due to the use of grass pea as an insurance crop. In addition, the low market price of grass pea has led to the use of grass pea flour as an adulterant for more valuable chickpea flour (besan) in India (Deshpande and Campbell 1992). However, the introduction of low-ODAP varieties and the greatly reduced incidence of neurolathyrism thanks to better nutrition has resulted in restrictions on grass pea gradually being lifted. In 2016, the Indian Council of Medical Research (ICMR) cleared three low-ODAP varieties of grass pea, effectively lifting the ban on the sale of grass pea grain (Anand 2016). While the reintroduction and promotion of grass pea remain politically controversial, cultivation is now again on the rise in India. The production of pulse crops in India has increased massively in recent years following the adoption of minimum support prices for mungbean, pigeon pea, black gram, chickpea, and lentil. Inclusion of low-ODAP grass pea varieties under this policy would likely result in a production increase for this crop as well.

9.12.1 Patent and Intellectual Property Rights Issues

Patent issues have not been a significant limitation for grass pea breeding and development as the vast majority of formal grass pea breeding has been publicly funded and done by government research agencies, such as the Bangladesh Agricultural Research Institute (BARI) in Bangladesh, Ethiopian Institute of Agricultural Research (EIAR) in Ethiopia, Indian Institute of Pulses Research (IIPR) in India and others, or through the international Consultative Group for International Agricultural Research (CGIAR) system, primarily the ICARDA. The varieties resulting from these breeding efforts are available in public genebanks. The only variety released by a commercial breeder is AC Greenfix, a green manure developed by Dakota Frontier Seeds and released in the USA (Rao and Northup 2011). In regions where grass pea is cultivated as a subsistence food crop, the seed market has traditionally been insufficient to support commercial seed production. To deliver improved varieties into the field, it may thus be necessary to engage farmers in the production of certified seed, as well as subsidizing seed sales. A collaboration between ICARDA and the Indian Council of Agricultural Research (ICAR) produced 1583 tons of improved seed between 2010 and 2015, and similar efforts are being conducted by National Agricultural Research Systems (NARS) in Bangladesh, India, and Ethiopia. Such initiatives need to be scaled up to produce sufficient improved seed for all farmers.

Across all crops, an average of 80% of seed planted in sub-Saharan Africa originates from informal seed systems, i.e., farmers saving seed from harvest to sowing or receiving noncertified farm-produced seed through swaps, gifts, or informal purchases (Byerlee et al. 2007). To date, the expectation that formal seed systems, which already exist for some crops, especially maize and cotton would soon supplant informal seed systems in sub-Saharan Africa has not come to pass (Lohr et al. 2015; Louwaars and De Boef 2012). Pulse crops in particular are still primarily traded through informal seed systems, though this may change if improved varieties provide farmers with clear economic benefits, as indicated by a survey conducted in Uganda, Burundi, Rwanda, and the Democratic Republic of Congo (David and Sperling 1999). This may be difficult to achieve for grass pea, as this crop is primarily cultivated as a food security crop by subsistence farmers and grain markets (both domestic and international) are limited. The distribution and promotion of improved grass pea seed at scale raises the risk of eroding existing grass pea diversity in farmer's fields. Hence, purposeful collection of grass pea landraces, especially from remote field sites, and conservation of these materials in local and international gene banks is necessary before this useful genetic diversity is lost.

9.12.2 Disclosure of Sources of Germplasm Resources, Access, and Benefit Sharing

The Nagoya protocol on Access and Benefit Sharing, effective as of 2014, was intended to enhance the sharing of genetic resources across national borders. However, national policies in some grass pea cultivating countries, most notably India and Ethiopia, who are both signatories of the protocol, have hindered international research collaborations to improve grass pea. In part as a result, the largest collections of grass pea are currently held outside of countries of major cultivation. This includes the ICARDA collection, which was moved from the Aleppo headquarters to bases in Morocco and Lebanon due to the Syrian civil war. To rebuild these collections, grass pea seeds were part of the first ever withdrawal of seeds from the Svalbard seed vault in 2015. Despite this major disruption, all accessions were saved, and the collection has now recommenced distribution of seeds.

As mentioned (see Sect. 9.3.1), a large collection containing over 4000 accessions of *Lathyrus* assembled by the Université de Pau has ceased active distribution and is now held at the CBN PMP.

Smaller collections are held by NARS institutions in multiple countries cultivating grass pea including, Ethiopia, Bangladesh, India, Australia, USA, Canada, Russia, and Pakistan.

Grass pea is nutritionally valuable, especially with regard to high levels of crude protein, lysine, and homoarginine in the seeds, leading to some authors advocating its use as a functional food (Rao 2011; Singh and Rao 2013). The neurotoxin β- ODAP, found in grass pea, is also present in similar concentrations in ginseng species. In this context, the compound has been investigated for its medicinal properties, particularly its haemostatic effects due to vasoconstriction (Okuda et al. 1990). The Chinese company Yunnan Baiyao has patented the use of this compound for the manufacturing of band-aids to reduce blood flow (Wang et al. 2014). Separately, the company Kunming Shenguo Pharmaceutical has applied for a patent to use β-ODAP for thrombocytopenia (Ding et al. 2018a, b; Lan et al. 2016). Other bioactive properties of β- ODAP are currently under investigation for the treatment of hypoxia (Eslavath et al. 2016) and Alzheimer's disease (Singh and Rao 2013).

9.12.3 Traditional Knowledge

Grass pea has historically been cultivated primarily by smallholders, often for local consumption. The well-known tolerance of grass pea to extreme weather events and poor soils has led to the cultivation of the crop in marginal areas and with low farming inputs.

An unusual cropping practice widely used in Bangladesh and India, known as relay, *paira* or *utera* cropping, involves the broadcasting of grass pea into a standing rice paddy, 2 weeks before the rice harvest. This allows the grass pea seeds to germinate and grow up rapidly, outcompeting weeds among the rice straw. Utilizing residual moisture in the paddy, the grass pea crop then grows to maturity during the dry and cool rabi season from October to March. This overlap of the times of the crops in the field allows the use of later maturing, higher yielding grass pea varieties (Dixit et al. 2016) Under these conditions, grass pea performs better than other legume crops (chickpea, lentil or pea) and linseed, which are also used as *utera* crops.

The potential dangers of neurolathyrism have been known since antiquity, tracing back to descriptions by Hippocrates, ancient Indian authors and Ibn Sīnā (Dastur and Iyer 1959), globally these are well known in all areas of significant grass pea cultivation. Many traditional methods of food preparation, including steeping the seeds in water before boiling and fermentation with the fungal species *Rhizopus oligosporus* and *Aspergillus oryzae* are effective at partially detoxifying grass pea seeds, but do not result in complete detoxification (Kuo et al. 1995; Padmajaprasad et al. 1997; Ramachandran et al. 2005; Yigzaw et al. 2004). Detoxification occurs through both nonenzymatic isomerization of β- ODAP into nontoxic α- ODAP and leaching of β-ODAP into the water used for boiling, which is then discarded. While consumers in Ethiopia and India deliberately select food preparation techniques that reduce toxicity (Butler et al. 1999; Girma et al. 2011), some false beliefs (for example that the steam from a pot of boiling grass peas or from discarded cooking water is toxic and must not be breathed in) persist. In addition, cooking methods for preparing grass pea as a snack food (e.g., roasting dry seeds or salting immature pods) are inefficient in removing β-ODAP.

9.12.4 Treaties and Conventions

Under the FAO's International Treaty for Plant Genetic Resources for Food and Agriculture (IT-PGRFA), a number of plant species and genera are listed as Annex 1 priority crops. This includes 47 genera of food crops and 29 genera of forage crops. Grass pea is the only species that appears in both lists, emphasizing its important role as a multifunctional crop (Visser 2013). Grass pea is also covered under the Nagoya protocol to facilitate the sharing of genetic resources.

9.12.5 Participatory Breeding

The practices of cultivating grass pea in marginal areas, as part of crop rotations and for multiple uses (food, feed and green manure) reinforce the necessity of involving farmers in the breeding process to ensure that newly released varieties meet farmers' needs under real agricultural conditions (Dixit et al. 2016; Hillocks and Maruthi 2012) Participatory breeding can also serve as a stepping stone to involving farmers in the production of certified seed and help integrate new varieties into informal seed systems (Lohr et al. 2015; Louwaars and De Boef 2012; Louwaars et al. 2013). This approach is already being taken by researchers and breeders in West Bengal, India (personal communication, Raghunath Sadhukhan), and by ICARDA in the development and extension of new grass pea varieties. To ensure that the seed produced by farmers remains genetically pure and low-ODAP, regular sampling and analysis should be undertaken by the seed certifying authorities.

9.13 Future Perspectives

9.13.1 Potential for Expansion of Productivity

Grass pea offers great potential for resilient agriculture in regions prone to drought and/or flooding, including both countries that already cultivate grass pea and those that do not. In both cases, the application of safe low/zero-ODAP varieties will be instrumental to achieve widespread acceptance of grass pea as a food crop. Grass pea acreage have been rising slowly in Ethiopia and Bangladesh, while yields have been rising more quickly, reaching an average of 1.8 t/ha in Ethiopia in 2016 according to national agricultural statistics (Bangladesh Bureau of Statistics 2016). However, production has been declining in Nepal and India, where grass pea is being increasingly replaced by lentil and other pulse crops. A major factor in this trend is the introduction of minimum support prices for pulses, which greatly reduce farmer risks.

In the countries that already cultivate grass pea as a significant crop, such as Ethiopia, India, and Bangladesh (Malek and Gazipur 1999; Fikre et al. 2011), grass pea is primarily cultivated by smallholder farmers who save seed from harvest to sow or who buy seed on the informal market (Campbell 1997). This poses the risk of improved grass pea varieties hybridizing with potentially deleterious landraces through open pollination. As the grass pea flower is cleistogamous, the great majority of seed will be produced by self-fertilization and it is unlikely that spontaneous outcrossing will result in bulk grain with levels of toxin that could pose a risk to human health. If improved seed with higher yield or other significant advantages was present in the market, it would likely suffice to incentivize farmers to buy improved seed every few cropping cycles to negate this risk.

One way to reduce the risk of high-ODAP grass pea landraces being mistaken for improved grass pea varieties (due to spontaneous outcrossing, mix-up or intentional mislabelling) would be to introgress easily distinguishable phenotypic traits into improved varieties. For example, in the development of the low-ODAP variety Mahateora, a pink-flowered parent (JRL-2) was used to give the improved variety an obvious and recessive trait for easy identification in the field (Kumar et al. 2011). Other morphological traits including pod anthocyanin production, seed color and seed shape could also be used for this purpose. However, stakeholder participation in the selection of these traits is crucial to ensure the traits used are both acceptable to farmers and consumers as well as identifiable when compared to genotypes already cultivated in a given region.

9.13.2 Potential for Expansion into Nontraditional Areas

As climate change increases the frequency and severity of extreme weather events over the course of this century (Cai et al. 2014; Dai 2013), the need for resilient crops for food and feed is going to become more pressing. At the same time, rising human and livestock populations in sub-Saharan Africa are expected to greatly increase demand for pulses in the region, with one study projecting a 50% rise in consumption between 2009 and 2030 (Clancey 2009). The same study calls for the improvement and increased use of new, locally adapted pulse varieties to limit the reliance on food imports and improve local food security. Grass pea could thus play an important role in maintaining food security in areas beyond its current range of cultivation. The unusual ability to withstand both drought and flooding stress makes grass pea suitable for rainfed agriculture in regions with highly unpredictable weather patterns. The use of safe low/zero-ODAP varieties is crucial for the introduction of grass pea to new areas of cultivation as consumers there may not be familiar with the potential risks and techniques for their mitigation, such as detoxifying cooking methods. The multiple uses of grass pea also allow multiple avenues for the introduction into new areas. For example, it may prove easier to advocate grass pea as a green manure or cover crop or as a source of animal feeds. Conversely, communities that have never consumed grass pea or have cultivated it historically but do not have a living memory of neurolathyrism may be readier to adapt new safe varieties, because the advantages of grass pea cultivation could be applied without having to overcome deeply held doubts over the safety of the crop.

References

Abberton M, Conant R, Batello C (eds) (2010) Grassland carbon sequestration: management, policy and economics. Integrated Crop Management. In: Proceedings of the workshop on the role of grassland carbon sequestration in the mitigation of climate change, FAO, Rome, Apr 2009

Addis G, Narayan RKJ (2000) Interspecific hybridisation of *Lathyrus sativus* (guaya) with wild *Lathyrus* species and embryo rescue. Afr Crop Sci J 8:129–136

Ahlawat IPS, Singh A, Saraf CS (1981) Effects of winter legumes on the nitrogen economy and productivity of succeeding cereals. Exp Agri 17:57–62

Allkin R, Goyder DJ, Bisby FA, White RJ (1986) Names and synonyms of species and subspecies in the Vicieae. Vicieae Database Proj 7:1–75

Almeida NF, Leitao ST, Krezdorn N, Rotter B, Winter P, Rubiales D et al (2014a) Allelic diversity in the transcriptomes of contrasting rust-infected genotypes of *Lathyrus sativus*, a lasting resource for smart breeding. BMC Plant Biol 14:376. https://doi.org/10.1186/s12870-014-0376-2

Almeida NF, Leitão ST, Caminero C, Torres AM, Rubiales D, Patto MCV (2014b) Transferability of molecular markers from major legumes to *Lathyrus* spp. for their application in mapping and diversity studies. Mol Biol Rep 41:269–283

Almeida NF, Rubiales D, Patto MCV (2015) Grass pea. In: De Ron AM (ed) Grain legumes, handbook of plant breeding. Springer, Heidelberg, Germany, pp 251–265

Anand U (2016) ICMR panel clears 'unsafe' khesari dal banned in 1961. The Indian Express. http://indianexpress.com/article/india/india-news-india/icmr-panel-clears-unsafe-khesari-dal-banned-in-61/

Bailey-Serres J, Voesenek LACJ (2008) Flooding stress: acclimations and genetic diversity. Annu Rev Plant Biol 59:313–339

Bangladesh Bureau of Statistics (2016) Yearbook of Agricultural Statistics-2015 Dhaka, Bangladesh p 27

Barik DP, Mohapatra U, Chand PK (2005) Transgenic grasspea (*Lathyrus sativus* L.): factors influencing *Agrobacterium*-mediated transformation and regeneration. Plant Cell Rep 24:523. https://doi.org/10.1007/s00299-005-0957-5

Barik D, Acharya L, Mukherjee A, Chand P (2007) Analysis of genetic diversity among selected grass pea (*Lathyrus sativus* L.) genotypes using RAPD markers. Z Naturforsch 62:869–874

Barna K, Mehta SL (1995) Genetic transformation and somatic embryogenesis in *Lathyrus sativus*. J Plant Biochem Biotechnol 4:67. https://doi.org/10.1007/BF03262955

Barpete S, Parmar D, Sharma NC, Kumar S (2012). Karyotype analysis in grass pea (*Lathyrus sativus* L.). J Food Legum 25:14–17

Battistin A, Fernandez A (1994) Karyotypes of four species of South America natives and one cultivated species of *Lathyrus L.* Caryologia 47:325–330

Bennett MD, Leitch IJ (2012) Angiosperm DNA C-values database, release 6.0, Dec 2012 edn. London, UK

Bhattacharyya M, Martin C, Smith A (1993) The importance of starch biosynthesis in the wrinkled seed shape character of peas studied by Mendel. Plant Mol Biol 22:525–531

Biazzi E, Nazzicari N, Pecetti L, Brummer EC, Palmonari A et al (2017) Genome-wide association mapping and genomic selection for alfalfa (*Medicago sativa*) forage quality traits. PLoS ONE 12(1):e0169234. https://doi.org/10.1371/journal.pone.0169234

Bogracheva TY, Cairns P, Noel TR, Hulleman S, Wang TL et al (1999) The effect of mutant genes at the *r*, *rb*, *rug3*, *rug4*, *rug5* and *lam* loci on the granular structure and physico–chemical properties of pea seed starch. Carbohyd Polym 39:303–314

Bohra A, Jha UC, Kavi Kishor PB, Pandey S, Singh NP (2014) Genomics and molecular breeding in lesser explored pulse crops: current trends and future opportunities. Biotechnol Adv 32:1410–1428

Butler A, Tesfay Z, D'Andrea C, Lyons D (1999) The ethnobotany of *Lathyrus sativus* L. in the highlands of Ethiopia. In: Van der Veen M (ed) The exploitation of plant resources in ancient Africa. Springer, Heidelberg, Germany, pp 123–136

Byerlee D, De Janvry A, Sadoulet E, Townsend R, Klytchnikova I (2007). World Development Report 2008: agriculture for development. World Bank, Washington, DC, USA
Cai W, Borlace S, Lengaigne M, van Rensch P, Collins M et al (2014) Increasing frequency of extreme El Niño events due to greenhouse warming. Nat Clim Chang 4:111. https://doi.org/10.1038/nclimate2100
Campbell CG (1997) Grass pea. *Lathyrus sativus* L. Promoting the conservation and use of underutilized and neglected crops, vol 18. Institute of Plant Genetics and Crop Plant Research, Gatersleben/International Plant Genetic Resources Institute, Rome, Italy
Campbell CG et al (1994) Current status and future strategy in breeding grasspea (*Lathyrus sativus*). In: Muehlbauer FJ, Kaiser WJ (eds) Expanding the production and use of cool season food legumes. Current plant science and biotechnology in agriculture, vol 19. Springer, Dordrecht. https://doi.org/10.1007/978-94-011-0798-3_3
Carriedo LG, Maloof JN, Brady SM (2016) Molecular control of crop shade avoidance. Curr Opin Plant Biol 30:151–158. https://doi.org/10.1016/j.pbi.2016.03.005
Casey R, Davies DR (eds) (1993) Peas: genetics, molecular biology and biotechnology. CABI, Wallingford, Oxford, UK
Chakraborty S, Mitra J, Samanta MK, Sikdar N, Bhattacharyya J et al (2018) Tissue specific expression and in-silico characterization of a putative cysteine synthase gene from *Lathyrus sativus* L. Gene Express Patt 27:128–134. https://doi.org/10.1016/j.gep.2017.12.001
Chandra A (2011) Use of EST database markers from *Medicago truncatula* in the transferability to other forage legumes. J Environ Biol 32:347–354
Chapman MA (2015) Transcriptome sequencing and marker development for four underutilized legumes. Appl Plant Sci 3(2):1400111. https://doi.org/10.3732/apps.1400111
Chowdhury MA, Slinkard AE (1997) Natural outcrossing in grass pea. J Hered 88:154–156
Chowdhury MA, Slinkard AE (1999) Linkage of random amplified polymorphic DNA, isozyme and morphological markers in grasspea (*Lathyrus sativus*). J Agri Sci 33:389–395
Chtourou-Ghorbel N, Lauga B, Combes D, Marrakchi M (2001) Comparative genetic diversity studies in the genus *Lathyrus* using RFLP and RAPD markers. *Lathyrus* Lathyrism Newsl 2:62–68
Clancey B (2009) World pulse outlook. Stat Publishing, Vancouver, Canada
Clemente A, Arques MC, Dalmais M, Le Signor C, Chinoy C et al (2015) Eliminating anti-nutritional plant food proteins: the case of seed protease inhibitors in pea. PLoS ONE 10:e0134634. https://doi.org/10.1371/journal.pone.0134634
Cohn DF, Streifler M (1983) Intoxication by the Chickling Pea (*Lathyrus sativus*): nervous system and skeletal findings. In: Chambers CM, Chambers PL, Gitter S (eds) Toxicology in the use, misuse, and abuse of food, drugs, and chemicals. Archives of toxicology (Supplement), vol 6. Springer, Berlin, Heidelberg, Germany, pp 190–193
Colbert T, Till BJ, Tompa R, Reynolds S, Steine MN et al (2001) High-throughput screening for induced point mutations. Plant Physiol 126:480–484
Comai L, Young K, Till BJ, Reynolds SH, Greene EA et al (2004) Efficient discovery of DNA polymorphisms in natural populations by Ecotilling. Plant J 37:778–786
Dai A (2013) Increasing drought under global warming in observations and models. Nat Clim Change 3:52–58
Dastur DK, Iyer CGS (1959) Lathyrism versus odoratism. Nutr Rev 17:33–36. https://doi.org/10.1111/j.1753-48871959.tb06376.x
David S, Sperling L (1999) Improving technology delivery mechanisms: lessons from bean seed systems research in eastern and central Africa. Agri Hum Values 16:381–388. https://doi.org/10.1023/a:1007603902380
Deshpande SS, Campbell CG (1992) Genotype variation in BOAA, condensed tannins, phenolics and enzyme inhibitors of grass pea (*Lathyrus sativus*). Can J Plant Sci 72(4):1037–1047
Ding T, Yang Z, Wei X, Yuan F, Yin S et al (2018a) Evaluation of salt-tolerant germplasm and screening of the salt-tolerance traits of sweet sorghum in the germination stage. Funct Plant Biol 45:1073–1081. https://doi.org/10.1071/FP18009

Ding S, Wang M, Fang S, Xu H, Fan H et al (2018b) D-dencichine regulates thrombopoiesis by promoting megakaryocyte adhesion, migration and proplatelet formation. Front Pharmacol 9:297. https://doi.org/10.3389/fphar.2018.00297

Dixit GP, Parihar AK, Bohra A, Singh NP (2016) Achievements and prospects of grass pea (*Lathyrus sativus* L.) improvement for sustainable food production. Crop J 4:407–416. https://doi.org/10.1016/j.cj.2016.06.008

Dubos C, Stracke R, Grotewold E, Weisshaar B, Martin C et al (2010) *MYB* transcription factors in *Arabidopsis*. Trends Plant Sci 15:573–581. https://doi.org/10.1016/j.tplants.2010.06.005

Emmrich PMF (2017) Genetic improvement of grass pea (*Lathyrus sativus*) for low β-L-ODAP Content. PhD Thesis, University of East Anglia, Norwich. https://ueaeprints.uea.ac.uk/63944/

Eslavath RK, Sharma D, Bin Omar NAM, Chikati R, Teli MK et al (2016) β-N-oxalyl-l-α, β-diaminopropionic acid induces HRE expression by inhibiting HIF-prolyl hydroxylase-2 in normoxic conditions. Eur J Pharmacol 791:405–411. https://doi.org/10.1016/j.ejphar.2016.07.007

European Court of Justice Ruling, ECLI:EU:C:2018:583 (2018) http://curia.europa.eu/juris/documents.jsf?num=C-528/16

Fikre A, Korbu L, Kuo YH, Lambein F (2008) The contents of the neuro-excitatory amino acid β-ODAP (β-N-oxalyl-l-α, β-diaminopropionic acid), and other free and protein amino acids in the seeds of different genotypes of grass pea (*Lathyrus sativus* L.). Food Chem 110:422–427

Fikre A, Negwo T, Kuo YH, Lambein F, Ahmed S (2011) Climatic, edaphic and altitudinal factors affecting yield and toxicity of *Lathyrus sativus* grown at five locations in Ethiopia. Food Chem Toxicol 49:623–630. https://doi.org/10.1016/j.fct.2010.06.055

Fulton TM, Van der Hoeven R, Eannetta NT, Tanksley SD (2002) Identification, analysis, and utilization of conserved ortholog set markers for comparative genomics in higher plants. Plant Cell 14(7):1457–1467. https://doi.org/10.1105/tpc.010479

Gaj T, Gersbach CA, Barbas CF (2013) ZFN, TALEN, and CRISPR/Cas-based methods for genome engineering. Trends Biotechnol 31(7):397–405

García-Hernández JL, Orona-Castillo I, Preciado-Rangel P, Flores-Hernández A, Murillo-Amador B, Troyo-Diéguez E (2010) Nutrients use efficiency in legume crops to climatic changes. In: Yadav S, Redden R (eds) Climate change and management of cool season grain legume crops. Springer, Dordrecht, Netherlands, pp 193–206

Ge L, Yu J, Wang H, Luth D, Bai G et al (2016) Increasing seed size and quality by manipulating *BIG SEEDS1* in legume species. Proc Natl Acad Sci USA 113:12414–12419. https://doi.org/10.1073/pnas.1611763113

Getahun H, Lambein F, Vanhoorne M, Van der Stuyft P (2003) Food-aid cereals to reduce neurolathyrism related to grass-pea preparations during famine. Lancet 362:1808–1810

Gharti DB, Darai R, Subedi S, Sarker A, Kumar S (2014) Grain legumes in Nepal: present scenario and future prospects. World J Agri Res 2:216–222. https://doi.org/10.12691/wjar-2-5-3

Girma D, Korbu L (2012) Genetic improvement of grass pea (*Lathyrus sativus*) in Ethiopia: an unfulfilled promise. Plant Breed 131:231–236

Girma A, Tefera B, Dadi L (2011) Grass pea and neurolathyrism: farmers' perception on its consumption and protective measure in North Shewa, Ethiopia. Food Chem Toxicol 49:668–672. https://doi.org/10.1016/j.fct.2010.08.040

Glowacka K, Kromdijk J, Kucera K, Xie J, Cavanagh AP et al (2018) Photosystem II Subunit S overexpression increases the efficiency of water use in a field-grown crop. Nat Commun 9:868. https://doi.org/10.1038/s41467-018-03231-x

Goldenberg JB (1965) *afila*, a new mutation in pea (*Pisum sativum* L.). Biol Genet 1:27–31

Gou J, Debnath S, Sun L, Flanagan A, Tang Y et al (2018) From model to crop: functional characterization of *SPL8* in *M. truncatula* led to genetic improvement of biomass yield and abiotic stress tolerance in alfalfa. Plant Biotechnol J 16:951–962

Gourlay CW, Hofer JMI, Ellis THN (2000) Pea compound leaf architecture Is regulated by interactions among the genes *UNIFOLIATA, COCHLEATA, AFILA*, and *TENDRIL-LESS*. Plant Cell 12:1279–1294. https://doi.org/10.1105/tpc.12.8.1279

Gurung AM, Pang CK (2013) *Lathyrus*. In: Kole C (ed) Wild Crop Relatives: Genomic and Breeding Resources: Legume Crops and Forages. Springer Verlag, Heidelberg, Germany, pp 117–126

Gutierrez JF, Vaquero F, Vences FJ (2001) Genetic mapping of isozyme loci in *Lathyrus sativus* L. Lathyrus Lathyrism Newsl 2:74–78

Gutierrez MV, Patto MCV, Huguet T, Cubero JI, Moreno MT et al (2005) Cross-species amplification of *Medicago truncatula* microsatellites across three major pulse crops. Theor Appl Genet 110:1210–1217

Haimanot RT, Feleke A, Lambein F (2005) Is lathyrism still endemic in northern Ethiopia?—The case of Legambo Woreda (district) in the South Wollo Zone, Amhara National Regional State. Ethiop J Heal Dev 19:230–236

Hao X, Yang T, Liu R, Hu J, Yao Y et al (2017) An RNA sequencing transcriptome analysis of grass pea (*Lathyrus sativus* L.) and development of SSR and KASP markers. Front Plant Sci 8:1873. https://doi.org/10.3389/fpls.2017.01873

Harlan JR, de Wet JMJ (1971) Towards a rational classification of cultivated plants. Taxon 20:509–517

Heywood V, Casas A, Ford-Lloyd B, Kell S, Maxted N (2007) Conservation and sustainable use of crop wild relatives. Agri Ecosyst Environ 121:245–255

Hillocks RJ, Maruthi MN (2012) Grass pea (*Lathyrus sativus*): Is there a case for further crop improvement? Euphytica 186:647–654

Humphry M, Reinstädler A, Ivanov S, Bisseling T, Panstruga R (2011) Durable broad-spectrum powdery mildew resistance in pea *er1* plants is conferred by natural loss-of-function mutations in *PsMLO1*. Mol Plant Pathol 12:866–867. https://doi.org/10.1111/j.1364-3703.2011.00718.x

Joshi M (1998) Status of grass pea (*Lathyrus sativus* L.) genetic resources in Nepal. In: Mathur PN, Rao VR, Arora RK (eds) Lathyrus genetic resources network: proceedings of a IPGRI-ICARDA-ICAR Regional Working Group Meeting, National Bureau of Plant Genetic Resources, New Delhi, India, pp 22–29, 8–10 Dec 1997

Karaba A, Dixit S, Greco R, Aharoni A, Trijatmiko KR (2007) Improvement of water use efficiency in rice by expression of *HARDY*, an *Arabidopsis* drought and salt tolerance gene. Proc Natl Acad Sci USA 104(39):15270–15275. https://doi.org/10.1073/pnas.0707294104

Kim YH, Kim MD, Park SC, Yang KS, Jeong JC et al (2011) *SCOF-1*-expressing transgenic sweet potato plants show enhanced tolerance to low-temperature stress. Plant Physiol Biochem 49:1436–1441. https://doi.org/10.1016/j.plaphy.2011.09.002

Kole C (2017) Combating climate change for FNEE security. In: International conference on the status of plant & animal genome research, San Diego, CA, 14–18 Jan 2017

Kumar S, Bejiga G, Ahmed S, Nakkoul H, Sarker A (2011) Genetic improvement of grass pea for low neurotoxin (ODAP) content. Food Chem Toxicol 49:589–600

Kumar V, Chattopadhyay A, Ghosh S, Irfan M, Chakraborty N et al (2016) Improving nutritional quality and fungal tolerance in soya bean and grass pea by expressing an oxalate decarboxylase. Plant Biotechnol J 14:1394–1405. https://doi.org/10.1111/pbi.12503

Kumar S, Meena SW, Lal R, Yadav GS, Mitran T et al (2018) Role of legumes in soil carbon sequestration. In: Meena R, Das A, Yadav G, Lal R (eds) Legumes for soil health and sustainable management. Springer, Singapore, pp 109–138

Kuo YH, Bau HM, Quemener B, Khan JK, Lambein F (1995) Solid-state fermentation of *Lathyrus sativus* seeds using *Aspergillus oryzae* and *Rhizopus oligosporus* sp T-3 to eliminate the neurotoxin β-ODAP without loss of nutritional value. J Sci Food Agri 69:8189. https://doi.org/10.1002/jsfa.2740690113

Kuo YH, Ikegami F, Lambein F (1998) Metabolic routes of β-(isoxazolin-5-on-2-yl)-l-alanine (bia), the precursor of the neurotoxin odap (β-n-oxalyl-l-α, β-diaminopropionic acid), in different legume seedlings. Phytochem 49(1):43–48. https://doi.org/10.1016/s0031-9422(97)01001-7

Lambein F, Chowdhury B, Kuo YH (1999) Biochemistry of the *Lathyrus* toxins. Lathyrus Genet Resour Netw 8:60

Lan G, Lan F, Sun X (2016) Use of dencichine in preparation of drug for treating thrombocytopenia. https://www.google.com/patents/US20160089351

Lander ES, Green P, Abrahamson J, Barlow A, Daly M et al (1987) MAPMAKER: an interactive computer package for constructing primary linkage maps of experimental populations. Genomics 1(2):174–181

Ligerot Y, de Saint Germain A, Waldie T, Troadec C, Citerne S et al (2017) The pea branching *RMS2* gene encodes the *PsAFB4/5* auxin receptor and is involved in an auxin-strigolactone regulation loop. PLoS Genet 13(12):e1007089. https://doi.org/10.1371/journal.pgen.1007089

Lioi L, Galasso I (2013) Development of genomic simple sequence repeat markers from an enriched genomic library of grass pea (*Lathyrus sativus* L.). Plant Breed 132:649–653. https://doi.org/10.1111/pbr.12093

Lioi L, Sparvoli F, Sonnante G, Laghetti G, Lupo F et al (2011) Characterization of Italian grass pea (*Lathyrus sativus* L.) germplasm using agronomic traits, biochemical and molecular markers. Genet Resour Crop Evol 58:425–437

Liu N, Shang W, Li C, Jia L, Wang X et al (2018) Evolution of the *SPX* gene family in plants and its role in the response mechanism to phosphorus stress. Open Biol 8:170231. http://dx.doi.org/10.1098/rsob.170231

Lohr K, Camacho A, Vernooy R (2015) Seed systems–an overview. Farmers' seed systems: the challenge of linking formal and informal seed systems. Deutsche Gesellschaft für Internationale Zusammenarbeit (GIZ) GmbH, Bonn, Germany, pp 3–7

Louwaars NP, De Boef WS (2012) Integrated seed sector development in Africa: a conceptual framework for creating coherence between practices, programs, and policies. J Crop Improv 26:39–59

Louwaars NP, de Boef W, Edeme J (2013) Integrated seed sector development in Africa: a basis for seed policy and law. J Crop Improv 27:186–214

Malek MA (1998) Genetic resources of grass pea (*Lathyrus sativus* L.) in Bangladesh. In: Mathur PN, Rao VR, Arora RK (eds) *Lathyrus* Genetic Resources Network: proceedings of a IPGRI-ICARDA-ICAR Regional Working Group Meeting, National Bureau of Plant Genetic Resources, New Delhi, pp 1–6, 8–10 Dec 1997

Malek MA, Gazipur B (1999) Genetic resources of grass pea (*Lathyrus sativus* L.) in Bangladesh. Lathyrus Genet Resour Netw 8:1

Manly KF, Cudmore RH Jr, Meer JM (2001) Map Manager QTX, cross-platform software for genetic mapping. Mam Genome 12:930–932

Matsumura H, Urasaki N, Yoshida K, Krüger DH, Kahl G, Terauchi R (2012) Supersage: powerful serial analysis of gene expression. In: Jin H, Gassmann W (eds) RNA Abundance analysis. methods in molecular biology (methods and protocols), vol 883. Humana Press, Totowa, NJ. https://doi.org/10.1007/978-1-61779-839-9_1

McCallum CM, Comai L, Greene EA, Henikoff S (2000) Targeted screening for induced mutations. Nat Biotechnol 18:455–457

McClintock B (1984) The significance of responses of the genome to challenge. Science 226:792–801

Mendel G (1865) Versuche über P:flanzen-Hybriden. Verh. Naturforsch. Ver. Brünn. 4:3–47

Mishra N, Sun L, Zhu X, Smith J, Srivastava AP et al (2017) Overexpression of the rice *SUMO E3* ligase gene *OsSIZ1* in cotton enhances drought and heat tolerance, and substantially improves fiber yields in the field under reduced irrigation and rainfed conditions. Plant Cell Physiol 58(4):735–746. https://doi.org/10.1093/pcp/pcx032

Muchero W, Roberts PA, Diop NN, Drabo I, Cisse N, Close TJ et al (2013) Genetic architecture of delayed senescence, biomass, and grain yield under drought stress in cowpea. PLoS ONE 8(7):e70041. https://doi.org/10.1371/journal.pone.0070041

Nerkar Y (1972) Induced variation and response to selection for low neurotoxin content in *Lathyrus sativus*. Indian J Genet 32(2):175–180

Nosrati H, Hosseinpour-Feizi MA, Nikniazi M, Razban-Haghighi A (2012) Genetic variation among different accessions of *Lathyrus sativus* (Fabaceae) revealed by RAPDs. Bot Serb 36(1):41–47

Okuda H, Lee SD, Matsuura Y, Zheng Y, Sekiya K, Takaku T et al (1990) Biological activities of non-saponin compounds isolated from Korean red ginseng. J Ginseng Res 14:157–161

Padmajaprasad V, Kaladhar M, Bhat RV (1997) Thermal isomerisation of β-N-oxalyl-L-α, β-diaminopropionic acid, the neurotoxin in *Lathyrus sativus*, during cooking. Food Chem 59:77–80

Pandey SP, Yadav CR, Sah K, Pande S, Joshi PK (2000) Legumes in Nepal. In: Johansen C, Duxburv JM, Virmani SM, Gowda CLL, Pande S, Joshi PK (eds) Legumes in rice and wheat cropping systems of the Indo-Gangetic Plain—constraints and opportunities. International Crops Research institute for the Semi-Arid Tropics; Cornell University, Patancheru, Andhra Pradesh, India; Ithaca, NY, USA, pp 71–97

Patto MCV, Rubiales D (2014) Lathyrus diversity: available resources with relevance to crop improvement—L. sativus and L. cicero as case studies. Ann Bot 113:895–908. https://doi.org/10.1093/aob/mcu024

Patto MCV, Amarowicz R, Aryee AN, Boye JI, Chung HJ et al (2015) Achievements and challenges in improving the nutritional quality of food legumes. Crit Rev Plant Sci 34(1–3):105–143. https://doi.org/10.1080/07352689.2014.897907

Patto MCV, Mecha E, Pereira AB, Leitão ST, Alves ML, Bronze MR (2018) Deciphering grain legumes quality riddle: the genomics of bioactive compounds. In: Brazauskas G, Statkevičiūtė G, Jonavičienė K (eds) Breeding grasses and protein crops in the era of genomics. Springer, Cham. https://doi.org/10.1007/978-3-319-89578-9_21

Plett JM, Wilkins O, Campbell MM, Ralph SG, Regan S (2010) Endogenous overexpression of *Populus* MYB186 increases trichome density, improves insect pest resistance, and impacts plant growth. Plant J 64:419–432. https://doi.org/10.1111/j.1365-313X.2010.04343.x

Ponnaiah M, Shiferaw E, Pe ME, Porceddu E (2011) Development and application of EST-SSRs for diversity analysis in Ethiopian grass pea. Plant Genet Resour 9:276–280

Puchta H (2016) Using CRISPR/Cas in three dimensions: towards synthetic plant genomes, transcriptomes and epigenomes. Plant J 87:5–15

Rahman MM, Kumar J, Rahaman MA, Afzal MA (1995) Natural outcrossing in *Lathyrus sativus* L. Indian J Genet 55(2):204–207

Ramachandran S, Bairagi A, Ray AK (2005) Improvement of nutritive value of grass pea (*Lathyrus sativus*) seed meal in the formulated diets for rohu, *Labeo rohita* (Hamilton) fingerlings after fermentation with a fish gut bacterium. Bioresour Technol 96:1465–1472. https://doi.org/10.1016/j.biortech.2004.12.002

Rao SLN (2011) A look at the brighter facets of beta-N-oxalyl-l-alpha, beta-diaminopropionic acid, homoarginine and the grass pea. Food Chem Toxicol 49:620–622. https://doi.org/10.1016/j.fct.2010.06.054

Rao SC, Northup BK (2011) Growth and nutritive value of grass pea in Oklahoma. Agron J 103:1692–1696. https://doi.org/10.2134/agronj2011.0178

Rao SLN, Adiga PR, Sarma P (1964) The isolation and characterization of β-Noxalyl-L-α, β-diaminopropionic acid, a neurotoxin from the seeds of *Lathyrus sativus*. Biochem 3:432–436

Resende RT, de Resende MDV, Azevedo CF, Fonseca E, Silva F, Melo LC et al (2018) Genome-wide association and regional heritability mapping of plant architecture, lodging and productivity in *Phaseolus vulgaris*. Genes Genomes Genet 8(8):2841–2854. https://doi.org/10.1534/g3.118.200493

Roy PK, Barat GK, Mehta SL (1992) In vitro plant regeneration from callus derived from root explants of *Lathyrus sativus*. Plant Cell Tiss Organ Cult 29:135. https://doi.org/10.1007/BF00033618

Roy PK, All K, Gupta A, Barat GK, Mehta SL (1993) β-N-Oxalyl-L-α, β-diaminopropionic acid in somaclones derived from internode explants of *Lathyrus sativus*. J Plant Biochem Biotechnol 2:9. https://doi.org/10.1007/BF03262914

Rubiales D, Fondevilla S, Chen W, Gentzbittel L, Higgins TJ et al (2015) Achievements and challenges in legume breeding for pest and disease resistance. Crit Rev Plant Sci 34:195–236

Rybiński W (2003) Mutagenesis as a tool for improvement of traits in grasspea *(Lathyrus sativus* L.). Lathyrus Lathyrism Newsl 27–31

Rybinski W, Blaszczak W, Fornal J (2006) Seed microstructure and genetic variation of characters in selected grass-pea mutants (*Lathyrus sativus* L.). Intl Agrophys 20:317

Santos C, Almeida NF, Alves ML, Horres R, Krezdorn N et al (2018) First genetic linkage map of *Lathyrus cicera* based on RNA sequencing-derived markers: Key tool for genetic mapping of disease resistance. Hort Res 5:45
Sarker A, Fikre A, El-Moneim AMA, Nakkoul H, Singh M (2017) Reducing anti-nutritional factor and enhancing yield with advancing time of planting and zinc application in grass pea in Ethiopia. J Sci Food Agri 98(1):27–32. https://doi.org/10.1002/jsfa.8433
Shi C, Navabi A, Yu K (2011) Association mapping of common bacterial blight resistance QTL in Ontario bean breeding populations. BMC Plant Biol 11:52. https://doi.org/10.1186/1471-2229-11-52
Shiferaw E, Pè ME, Porceddu E, Ponnaiah M (2012) Exploring the genetic diversity of ethiopian grass pea (*Lathyrus sativus* L.) using EST-SSR markers. Mol Breed 30:789–797
Singh SS, Rao SLN (2013) Lessons from neurolathyrism: a disease of the past & the future of Lathyrus sativus (Khesari dal). Indian J Med Res 138:32–37
Singh AK, Singh SS, Prakash V, Kumar S, Dwivedi SK (2015) Pulses production in India: present status, bottleneck and way forward. J Agri Search 2:75–83
Singh A, Septiningsih EM, Balyan HS, Singh NK, Rai V (2017) Genetics, physiological mechanisms and breeding of flood-tolerant rice (*Oryza sativa* L.). Plant Cell Physiol 58:185–197. https://doi.org/10.1093/pcp/pcw206
Sita K, Sehgal A, Hanumantha Rao B, Nair RM, Vara Prasad PV, Kumar S et al (2017) Food legumes and rising temperatures: effects, adaptive functional mechanisms specific to reproductive growth stage and strategies to improve heat tolerance. Front Plant Sci 8:1658. https://doi.org/10.3389/fpls.2017.01658
Skiba B, Ford R, Pang EC (2004) Construction of a linkage map based on a *Lathyrus sativus* backcross population and preliminary investigation of QTLs associated with resistance to ascochyta blight. Theor Appl Genet 109:1726–1735
Smartt J (1984) Evolution of grain legumes. I. Mediterranean pulses. Exp Agric 20(4):275–296. https://doi.org/10.1017/S0014479700017968
Snoad B (1981) Plant form, growth rate and relative growth rate compared in conventional, semi-leafless and leafless peas. Sci Hort 14:9–18. https://doi.org/10.1016/0304-4
Songstad DD, Petolino JF, Voytas DF, Reichert NA (2017) Genome editing of plants. Crit Rev Plant Sci 36(1):1–23. https://doi.org/10.1080/07352689.2017.1281663
Stanke M, Keller O, Gunduz I, Hayes A, Waack S, Morgenstern B (2006) AUGUSTUS: ab initio prediction of alternative transcripts. Nucl Acids Res 34 supp_2(1):W435–W439. https://doi.org/10.1093/nar/gkl200
Sun XL, Yang T, Guan JP, Ma Y, Jiang JY, Cao R et al (2012) Development of 161 novel EST-SSR markers from *Lathyrus sativus* (Fabaceae). Amer J Bot 99(10):e379–e390. https://doi.org/10.3732/ajb.1100346
Talukdar D (2009a) Dwarf mutations in grass pea (*Lathyrus sativus* L.): origin, morphology, inheritance and linkage studies. J Genet 88:165–175
Talukdar D (2009b) Recent progress on genetic analysis of novel mutants and aneuploid research in grass pea (*Lathyrus sativus* L.). Afr J Agri Res 14:1549–1559
Talukdar D, Biswas AK (2008) Seven different primary trisomics in grass pea (*Lathyrus sativus* L.). II. Pattern of transmission. Cytologia 73:129–136
Tan GZ, Das Bhowmik SS, Hoang TM, Karbaschi MR, Johnson AA et al (2017) Finger on the pulse: pumping iron into chickpea. Front Plant Sci 8:1755
Tavoletti S, Iommarini L (2007) Molecular marker analysis of genetic variation characterizing a grass pea (*Lathyrus sativus*) collection from central Italy. Plant Breed 126:607–611
Tayeh N, Aubert G, Pilet-Nayel ML, Lejeune-Hénaut I, Warkentin TD, Burstin J (2015) Genomic tools in pea breeding programs: status and perspectives. Front Plant Sci 6:1037. https://doi.org/10.3389/fpls.2015.01037
Tiwari KR, Campbell CG (1996) Inheritance of seed weight in grasspea (*Lathyrus sativus* L.). Faba Bean Inf Serv 38–39:30–33

Townsend CC, Guest E (1974) Flora of Iraq: *Leguminales*, vol 3. Ministry of Agriculture and Agrarian Reforms, Baghdad, Iraq

Van Ooijen JW (2006) JoinMap 4, software for the calculation of genetic linkage maps in experimental populations. Kyazma B. V., Wageningen, Netherlands

Visser B (2013) The moving scope of annex 1. In: Halewood M, Lopez I, Noriega SL (eds) Crop genetic resources as a global commons. Routledge, Oxon, UK, pp 265–282. https://www.bioversityinternational.org/fileadmin/user_upload/online_library/publications/pdfs/Crop_genetic_Resources_global_commons/14.crops_list_under_multilater_system.pdf. Accessed 12 Sep 2018

Wahiduzzaman MIA, Yamauchi A, Kono Y (1996) Root system structure of six food legume species: inter- and intraspecific variations. Jpn J Crop Sci 65(1):131–140. https://doi.org/10.1626/jcs.65.131

Wang TL, Bogracheva TY, Hedley CL (1998) Starch: as simple as A, B, C? J Exp Bot 49:481–502

Wang TL, Uauy C, Robson F, Till B (2012) TILLING *in extremis*. Plant Biotechnol J 10:761–772. https://doi.org/10.1111/j.1467-7652.2012.00708.x

Wang Z, Yang JY, Chen Q (2014) Development of dencichine plus Yunnan Baiyao band-aid. In: Advanced Materials Research, vol 881–883, pp 364–367. https://doi.org/10.4028/www.scientific.net/AMR.881-883.364

Wang F, Yang T, Burlyaeva M, Li L, Jiang J, Fang L et al (2015) Genetic diversity of grasspea and its relative species revealed by SSR markers. PLoS ONE 10(3):e0118542. https://doi.org/10.1371/journal.pone.0118542

Weller JL, Ortega R (2015) Genetic control of flowering time in legumes. Front Plant Sci 6:207. https://doi.org/10.3389/fpls.2015.00207

Xiao N, Gao Y, Qian H, Gao Q, Wu Y, Zhang et al (2018) Identification of genes related to cold tolerance and a functional allele that confers cold tolerance. Plant Physiol 177:1108–1123. https://doi.org/10.1104/pp.18.00209

Xu Q, Liu F, Chen P, Jez JM, Krishnan HB (2017) β-N-Oxalyl-l-α, β-diaminopropionic Acid (β-ODAP) Content in Lathyrus sativus: the integration of nitrogen and sulfur metabolism through β-cyanoalanine synthase. Intl J Mol Sci 18:526. https://doi.org/10.3390/ijms1803052

Xu Q, Liu F, Qu R, Gillman JD, Bi C et al (2018) Transcriptomic profiling of *Lathyrus sativus* L. metabolism of β-ODAP, a neuroexcitatory amino acid associated with neurodegenerative lower limb paralysis. Plant Mol Biol Rep 1–12. https://doi.org/10.1007/s11105-018-1123-x

Yan ZY, Spencer PS, Li ZX, Liang YM, Wang YF et al (2006) *Lathyrus sativus* (grass pea) and its neurotoxin ODAP. Phytochem 67(2):107–121. https://doi.org/10.1016/j.phytochem.2005.10.022

Yang HM, Zhang XY (2005) Considerations on the reintroduction of grass pea in China. Lathyrus Lathyrism Newsl 4:22–26

Yang T, Chaudhuri S, Yang L, Du L, Poovaiah BW (2010) A calcium/calmodulin-regulated member of the receptor-like kinase family confers cold tolerance in plants. J Biol Chem 285:7119–7126

Yang T, Jiang J, Burlyaeva M, Hu J, Coyne CJ, Kumar S et al (2014) Large-scale microsatellite development in grasspea (*Lathyrus sativus* L.), an orphan legume of the arid areas. BMC Plant Biol 14:65. https://doi.org/10.1186/1471-2229-14-65. PMID: 24635905

Yigzaw Y, Gorton L, Solomon T, Akalu G (2004) Fermentation of seeds of tef (*Eragrostis teff*), grass-pea (*Lathyrus sativus*), and their mixtures: aspects of nutrition and food safety. J Agri Food Chem 52:1163–1169. https://doi.org/10.1021/jf034742y

Zhang J, Zhou X, Xu Y, Yao M, Xie F et al (2016) Soybean *SPX1* is an important component of the response to phosphate deficiency for phosphorus homeostasis. Plant Sci 248:82–91. https://doi.org/10.1016/j.plantsci.2016.04.010

MIX
Papier aus verantwortungsvollen Quellen
Paper from responsible sources
FSC® C105338

If you have any concerns about our products,
you can contact us on
ProductSafety@springernature.com

In case Publisher is established outside the EU,
the EU authorized representative is:
Springer Nature Customer Service Center GmbH
Europaplatz 3, 69115 Heidelberg, Germany

Printed by Libri Plureos GmbH
in Hamburg, Germany